THE OXFORD HANDBOOK OF SOFT CONDENSED MATTER

The Oxford Handbook of Soft Condensed Matter

Edited by

Eugene M. Terentjev

Cavendish Laboratory
Department of Physics, University of Cambridge
Cambridge CB3 0HE, UK

David A. Weitz

School of Engineering and Applied Sciences
Department of Physics, Harvard University
Cambridge, MA 02138, USA

UNIVERSITY PRESS

Great Clarendon Street, Oxford, OX2 6DP,
United Kingdom

Oxford University Press is a department of the University of Oxford.
It furthers the University's objective of excellence in research, scholarship,
and education by publishing worldwide. Oxford is a registered trade mark of
Oxford University Press in the UK and in certain other countries

© Eugene M. Terentjev and David A. Weitz 2015

The moral rights of the authors have been asserted

Impression: 1

All rights reserved. No part of this publication may be reproduced, stored in
a retrieval system, or transmitted, in any form or by any means, without the
prior permission in writing of Oxford University Press, or as expressly permitted
by law, by licence or under terms agreed with the appropriate reprographics
rights organization. Enquiries concerning reproduction outside the scope of the
above should be sent to the Rights Department, Oxford University Press, at the
address above

You must not circulate this work in any other form
and you must impose this same condition on any acquirer

Published in the United States of America by Oxford University Press
198 Madison Avenue, New York, NY 10016, United States of America

British Library Cataloguing in Publication Data

Data available

Library of Congress Control Number: 2014959186

ISBN 978–0–19–966792–5

Printed in Great Britain by
Clays Ltd, St Ives plc

Links to third party websites are provided by Oxford in good faith and
for information only. Oxford disclaims any responsibility for the materials
contained in any third party website referenced in this work.

Preface

Soft condensed matter represents a very large class of materials whose properties have in common the feature that they are easily deformed. Susceptibility to deformation is characterized by the elastic constants of the material, and, thus, soft materials have elastic constants that are much smaller than typical 'hard' materials that cannot be so easily deformed. In fact, the elastic constants of soft materials are typically 6 to 10 orders of magnitude less than those of hard materials, such as metals, ceramics, or glass. Since an elastic constant is essentially a measure of an energy density, and since typical energies can vary between roughly thermal energy, $k_B T$, and several hundreds of $k_B T$, such a large decrease in elastic modulus of soft materials must arise from an effectively larger length scale that determines the properties of the material. Thus, the study of soft condensed matter entails developing an understanding of the structure, dynamics, and properties of the phenomena that occur on length scales that are larger than atomic or molecular.

Soft condensed matter is a relatively new topic of study. Condensed matter has typically referred to more traditional liquids, which have a long history of study, or to solids, which have focused more commonly on hard materials, driven in large measure by the importance of structural materials or the metals and semiconductors which drove the rapid evolution of microelectronics. It is only relatively recently, over the past 20 to 30 years, that soft condensed matter has matured into its own identifiable field. The term itself has been coined by Pierre-Gilles de Gennes in his Nobel Prize lecture in 1991. However, the field is, in fact, an amalgam of many sub-fields, and many of these have themselves much longer histories of study. For example, colloids are an integral component of soft condensed matter, yet they were studied scientifically by Faraday, and some of his samples still exist in the Royal Society in London. Similarly, the study of polymers dates back at least to the beginning of the last century and the study of liquid crystals has a similarly long history. Instead, it is the recognition of the similarity in much of the underlying behavior of these materials that led to the evolution of an identifiable field of study that has become soft condensed matter.

Another reason for the slower evolution of the field of soft condensed matter is the fact that many different disciplines contribute to the field. Thus, soft condensed matter is currently studied in physics, chemistry, materials science, chemical engineering, and mechanical engineering departments. Of course, this highly interdisciplinary nature of the field is also one of its great strengths and indeed, one of the great appeals of the field. The breadth of disciplines that contribute to the field also helps drive the great breadth of applications that soft condensed matter has. This is essential to the longevity of any scientific discipline: it must ultimately also contribute to advances in technology, which in turn help drive the further rounds of basic science. That is certainly the case with soft condensed matter, with, for example, the wide use of liquid crystals in display technology, the importance of polymers in so many different materials, and the wide utility of colloids to control the flow of very viscous materials when they are processed or consumed. In addition, there is a more recent evolution of the field of soft condensed matter towards biology. Indeed virtually all living organisms are 'soft', and thus the understanding of the physical properties in biology depends critically on the underlying science of soft condensed matter.

This handbook serves as an overview of many of the topics that make up the field. Because of the great breadth of these topics, it is impossible to include them all. Nevertheless, many of the key subjects of soft condensed matter are represented. Together, they form both an introduction and an overview of the field. Each topic, and its representing chapter, could have been a full size book - in fact, there are a number of such books on many of the topics covered in the handbook. Our aim here was to give a current snapshot of the field, identify the key principles at play and the most prominent (and promising) ways of its further development, certainly give essential references for anyone to follow on the subject, but not necessarily go into a lot of detail in each topic.

Each chapter has, typically, 100-150 references (with exceptions in both directions) that are intended to be the most relevant literature for the topic (rather than the historically first, or a comprehensive list). Some of these referenecs would inevitably overlap between different chapters, since the subjects naturally overlap and there is a common thread across most of the chapters. There are several monographs that have formed the foundation of the whole field of soft matter, tested by time, and they transcend most of the topics presented in this Handbook. The incomplete list of these important books is given at the end of this Preface and a newcomer to the field may want to start from these.

The handbook opens with Chapter 1 on the fundamentals of colloid science, written in a colorful and insightful manner by Wilson Poon. The chapter mainly focuses on the interplay of energy and entropy factors in forming colloidal aggregates, when a suspension changes from a liquid at low colloid concentration to various equilibrium, and metastable (including kinetically arrested) structures at higher concentrations. The review makes inroads into emerging areas in this field, such as active driven matter and the selectivity of binding in protein-based colloids.

Chapter 2, written by Dominique Langevin, is a comprehensive introduction into the principles and action of surfactants. Surfactants, also called amphiphilic molecules, are an essential part of many soft condensed matter systems where their basic role is to stabilize interfaces between different component materials. The chapter focuses on the structure and mechanical characteristics of interfaces, individually and macroscopically (in emulsions and foams), and on the structures formed by surfactant aggregation into micelles and microemulsions.

After covering the basic definitions and fundamental properties, in Chapter 3, Oleg Lavrentovich has presented modern directions of research in liquid crystals, which in itself is a topic with over 100 years of history. The chapter reviews topological defects, the hybrid systems combining liquid crystals with polymers or colloids, and also the new ways to create the mesomorphic systems (another name for liquid crystals, reflecting the partial nature of order, bridging between simple fluids and crystalline solids).

Denis Weaire and Stefan Hutzler have written Chapter 4 focusing specifically on foams, that is, heterogeneous systems maintaining their mechanical stability by balancing tension on crossing interfaces. Two distinct aspects of this topic: the mathematics of the problem of optimizing the mesh of interfaces (culminating in the 'surface evolver' algorithm to determine equilibrium structure) and the practical aspects of foam mechanics and applications, are equally covered in this chapter.

The subject of granular matter, presented in Chapter 5 by Raphael Blumenfeld, Sam Edwards, and Stephen Walley, is an emerging area of science. Although in practice, the humankind has been handling granular systems since the dawn of civilization (think of grain silos or sandpiles), the understanding of fundamental physics that controls the structure and mechanics of granular matter, determined by the large number of rigid contact and excluded-volume constraints (but not the thermal motion) is only beginning to emerge. The chapter formulates these principles and shows how a mechanical stress and the statistical sums could be constructed.

Polymer physics is a very broad field and different aspects of it are represented in this handbook. Chapter 6, written by Ronald Larson and Zuowei Wang, focuses on how the unusual and often dramatic mechanical properties of concentrated polymer systems are determined by the physics of entanglements. This area has witnessed a rapid development in the last decades, and the review serves well to outline the current state of the art.

In Chapter 7, Lee Trask, Nacu Hernandez and Eric Cochran review the complex structures formed by block-copolymers, and the methods of structure analysis. Strictly, block-copolymer is an amphiphilic molecule analogous to a surfactant and so many of the aggregated structures have the same or similar symmetry. However, there are also many distinct aspects of polymer physics at play in micro-phase separated block-copolymers and the chapter gives a detailed summary of their structure and dynamics.

Rubber elasticity is a major topic, also with a long history. Chapter 8, by James Mark and Burak Erman, is written in a style of reference summary, rather than a didactic review. The chapter describes the variety of practical ways to form and characterize a rubber-elastic network, the evolution of theories of rubber elasticity, and presents the specific properties of swollen polymer gels where the possibility of solvent exchange leads to some dramatic transformations in the system. Finally, Mark and Erman briefly overview new emerging classes of rubber-elastic materials (such as liquid crystalline elastomers) where the internal microstructure added to the random network leads to some unique mechanical properties.

In Chapter 9, Andrey Dobrynin reviews the physics of polyelectrolytes, that is, charged polymer chains suspended in a solvent with counter-ions. This topic includes both fundamentally important science and immediate practical relevance in the areas related to biological systems and processes (where most biopolymers are indeed polyelectrolytes). The chapter systematically reviews the properties of individual charged chains in solution, in more dense environments and in crosslinked gels - in each case outlining the key variables and the main physical phenomena.

Chapter 10, by Masao Doi, succinctly focuses on the solvent dynamics in polymer gels, in equilibrium and under mechanical stress. This problem has importance to many fields (not surprisingly, touched upon in the two preceding chapters), and only recently has the physics of it been understood.

The three final chapters of this handbook focus on more specifically biological topics that are derived from several areas of "basic" soft condensed matter science. Chapter 11, written by Astrid van der Horst and Gijs Wuite, reviews the hierarchical structure and characteristics of an extracellular matrix: a heterogeneous composite system in which several key biopolymers aggregate in filaments and form a crosslinked network that has very specific physical properties required for its biological function.

Similarly, Chapter 12, by Matthieu Piel and Raphael Voituriez, focuses on the hierarchical structure and resulting physical properties of the cell cytoskeleton, where one finds filaments formed by the aggregation of key proteins (mainly actin, with an additional role of microtubules and intermediate filaments). The shape, mechanics, as well as the locomotion of cells are controlled by the dynamical response of the cytoskeleton network.

And finally, in Chapter 13, Aidan Brown and Pietro Cicuta review in a more detail the properties of interfaces and membranes. Chapters 2 (Surfactants) and 4 (Foams) have already explored some basic properties of interfaces separating two immiscible phases. Here the authors summarize the ways to describe and experimentally characterize isolated membranes, characterized by tension, elasticity and viscous damping, and as well as closed vesicles - and cells where the membrane separates the cytoskeleton from the extracellular matrix.

References

[1] B. Alberts, A. Johnson, J. Lewis, *et al.*, *Molecular Biology of the Cell*, 5th edition (Garland Publishing, New York, 1994).

[2] M. Rubinstein and R. H. Colby, *Polymer Physics*, (Oxford Univer-

sity Press, Oxford, 2003).

[3] P. G. de Gennes and J. Prost, *The Physics of Liquid Crystals*, (Clarendon Press, Oxford, 1993).

[4] P. G. de Gennes, *Scaling Concepts in Polymer Physics*, (Cornell University Press, New York, 1979).

[5] M. Doi and S. F. Edwards, *The Theory of Polymer Dynamics*, (Clarendon Press, Oxford, 1986).

[6] J. P. Hansen and I. R. MacDonald, *Theory of Simple Liquids*, 3rd edition, (Academic Press, London, 2006).

[7] R. J. Hunter, *Foundations of Colloid Science*, (Oxford University Press, Oxford, 2000).

[8] J. N. Israelachvili, *Intermolecular and Surface Forces*, 3rd edition, (Academic Press, London, 2011).

[9] P. Nelson, *Biological Physics*, (Freeman, Berlin, 2003).

[10] S. A. Safran, *Statistical Thermodynamics Of Surfaces, Interfaces, and Membranes*, (Addison-Wesley, Reading, MA, 1994).

Contents

1 Colloidal Suspensions — 1
Wilson C. K. Poon

   1.1 Is matter granular? The 'colloids as atoms' paradigm — 2
   1.2 Do hard spheres crystallise? The physics of entropy — 9
   1.3 How to make a liquid: Interactions under control — 13
   1.4 What is a glass? Challenging arrest — 18
   1.5 Einstein's most useful contribution: Driven matter — 28
   1.6 'A mysterious colloid': Proteins and patchiness — 37
   1.7 Conclusion: Looking into the future — 39

2 Surfactants — 51
Dominique Langevin

   2.1 Surface tension — 52
   2.2 Surface rheology — 56
   2.3 Bulk aggregates — 62
   2.4 Microemulsions — 68
   2.5 Non-equilibrium systems — 74
   2.6 Emulsions and foams — 81

3 Liquid Crystals — 95
Oleg D. Lavrentovich

   3.1 Thermotropic and lyotropic systems — 96
   3.2 Order parameter — 106
   3.3 Elasticity — 111
   3.4 Surface anchoring — 116
   3.5 Topological defects — 118
   3.6 Polymer-liquid crystal hybrid materials — 124
   3.7 Liquid crystal-colloid hybrid systems — 125
   3.8 Dynamics — 129
   3.9 Applications of liquid crystals — 134

4 Foams — 147
Denis Weaire and Stefan Hutzler

   4.1 Statics and dynamics — 149
   4.2 The rules of equilibrium — 150
   4.3 Structure of foam — 153
   4.4 Key properties — 154
   4.5 Analogous systems — 161

5 Granular Systems — 167
Raphael Blumenfeld, Sam F. Edwards and Stephen M. Walley

- 5.1 Introduction to granular matter — 168
- 5.2 Structural description — 173
- 5.3 Stress transmission — 179
- 5.4 Statistical mechanics — 204
- 5.5 Future directions — 222

6 Dynamics of Entangled Polymers — 233
Ronald G. Larson and Zuowei Wang

- 6.1 Foundations of entangled polymer dynamics — 234
- 6.2 Models of polymer dynamics in the linear regime — 245
- 6.3 Models of polymer dynamics in the nonlinear regime — 257

7 Block Copolymers — 271
Lee M. Trask, Nacú Hernandez and Eric W. Cochran

- 7.1 Phase behavior — 272
- 7.2 Block copolymer dynamics — 288
- 7.3 Small-angle X-ray and neutron scattering — 297

8 Elastomers and Rubberlike Elasticity — 333
James E. Mark and Burak Erman

- 8.1 Preparation and structure of networks — 334
- 8.2 Elasticity experiments and theories — 344
- 8.3 Stress-strain-temperature relationships — 359
- 8.4 Swelling and gel collapse — 362
- 8.5 Elastomers with internal microstructure — 365

9 Polyelectrolyte Solutions and Gels — 383
Andrey V. Dobrynin

- 9.1 Dilute salt-free polyelectrolyte solutions — 384
- 9.2 A polyelectrolyte chain in salt solutions — 399
- 9.3 Semidilute polyelectrolyte solutions — 411
- 9.4 Phase separation in polyelectrolyte solutions — 429
- 9.5 Polyelectrolyte gels — 433

10 Fluid Transport in Gels — 451
Masao Doi

- 10.1 Equilibrium state of non-ionic gels — 452
- 10.2 Dynamics of non-ionic gels — 458
- 10.3 Dynamics of ionic gels — 462

11 Extracellular Matrix — 475
Astrid van der Horst and Gijs J. L. Wuite

- 11.1 The building blocks: Protein synthesis and structure — 476
- 11.2 Hierarchical supramolecular organization — 480
- 11.3 Linking structure and function — 484

12 Cell Cytoskeleton — **497**
Matthieu Piel and Raphael Voituriez

12.1 Cytoskeleton components — 499
12.2 Coarse-grained models of the cytoskeleton — 503
12.3 Cell migration — 513
12.4 Cell polarity — 519

13 Interfaces and Membranes — **535**
Aidan T. Brown and Pietro Cicuta

13.1 Fluid interfaces and films — 539
13.2 Membrane mechanics — 548
13.3 Dynamics — 560
13.4 Multicomponent membranes — 567
13.5 Membranes in cell biology — 571

Index — **589**

Colloidal Suspensions

Wilson C. K. Poon

Scottish Universities Physics Alliance (SUPA)
School of Physics and Astronomy, The University of Edinburgh
King's Buildings, Edinburgh EH9 3JZ, UK

Introduction

> '**colloid** *Chem.* Applied by Graham, 1861, to describe a peculiar state of aggregation in which substances exist; opposed to *crystalloid*. Substances in the colloid state are characterized by little or no tendency to diffuse through animal membranes or vegetable parchment, do not readily crystallize, are inert in their chemical relations, but are highly changeable. So called because gelatin may be taken as the type of the class.'
> *Oxford English Dictionary* (2010)

In modern scientific usage, a colloid is a dispersion of particles or droplets in a liquid that, despite the density difference between the dispersed phase and the dispersing medium, is stable against sedimentation or creaming. In the majority of real-life suspensions, the dispersed particles or droplets have sizes in the range of ~ 10nm to $\sim 1\mu$m. Historically, it was the work of Albert Einstein [1] and Jean Baptiste Perrin [2] that revealed the physical origin of this so-called colloidal length scale, which in fact defines the science of soft condensed matter – the study of liquids containing structures on the scale of ~ 10nm to 1μm.

While the work of Einstein and Perrin laid the foundations of modern colloid physics, they themselves were motivated by something quite different. They wanted to show that matter is granular, i.e. to demonstrate the reality of atoms and molecules. This piece of history provides the inspiration for the approach taken in this chapter – we will review colloid physics from the perspective of how the subject has provided broader insights into fundamental scientific questions of relevance well beyond the study of colloids, or even soft matter physics in general. Under this perspective, colloids are *model systems*, especially when experiments are carried out using very well-characterised particles with well-defined size and shape and other properties, or 'model colloids'.

This chapter is organised as follows. We survey colloid physics by way of examining six 'big questions' in science that the study of suspensions has contributed towards answering: the granularity of matter, the crystallisation of hard spheres, the conditions for the existence of the liquid state, the nature of kinetic arrest, the non-linear physics of driven matter, and the nature of proteins. We take a deliberately historical approach to situate modern advances in their contexts; moreover, many issues addressed by pioneers of the subject remain pertinent today. We end the chapter by introducing a number of emergent areas.

Our chapter can do no more than give a flavour of the field and explain briefly a number of key concepts and results of wide applicability. Longer accounts exist in reviews and books. Amongst these, Peter Pusey's Les Houches lectures [3] remain the definitive account of the basics of modern colloid physics. Given our pedagogical goal, the size of the total literature and the availability of the above-mentioned sources, we will be highly selective in our citations. In particular, when a large body of work from one group of researchers is available, we often only cite the latest and/or the most detailed and well-referenced paper.

Throughout, we focus on fundamental science. There are, of course, more practical reasons for being interested in colloids. Suspensions of all kinds are ubiquitous in applications. Indeed, very many industrial products are either colloids in their final state or would have passed through the colloidal stage in their processing history. Colloidal products are everywhere to be seen: toothpaste, salad dressing, and paint are obvious examples; some pesticide formulations are dispersions of micro-crystals of the active compounds in water. Molten chocolate is a suspension of sugar crystals and cocoa solids in oil, while ceramics are manufactured by sintering moulded concentrated dispersions of particles of refractory materials (TiO_2, Al_2O_3, etc.) known as the 'green body'. Slurries occur in many industrial processes either as intermediates or as waste products. Various nanotechnologies increasingly depend on the synthesis and processing of suspensions of particles with sizes in the lower end of the colloidal length scale. Finally, many biological systems can be viewed as colloidal suspensions, e.g. blood is a suspension of cells and globular proteins. We will not dwell on these many applied aspects.

1.1 Is matter granular? The 'colloids as atoms' paradigm

'It would not become physical science to see in its self created, changeable, economical tools, molecules and atoms, realities behind phenomena ... The atom must remain a tool for representing phenomena.'
Ernst Mach, *The Economical Nature of Physics* (1882).

The statistical mechanical approach for explaining the bulk properties of matter developed by James Clerk Maxwell, Rudolf Clausius, Ludwig Boltzmann, and Josiah Willard Gibbs relied on averaging over the behaviour of very large numbers of atoms. These authors all assumed that the atoms they appealed to in their calculational schemes, which no one had seen, were real constituents of matter; but others like Ernst Mach (see quotation above) held that atoms were merely convenient fictional devices for performing calculations. What eventually convinced the scientific community of the reality of atoms was not techniques for 'seeing' individual atoms, which did not become available until the late twentieth century, but the explanation of a phenomenon first brought to scientific attention by the Scottish botanist Robert Brown.

In 1827 Brown reported his observation of the random, incessant motion of particles taken from inside the pollen grains of the plant *Clarckia pulchella* suspended in water [4], eventually called Brownian motion. Some of these particles were 'from nearly $\frac{1}{4000}$th to about $\frac{1}{3000}$th of an inch in length', or $\sim 6-8\mu$m, 'and of a figure between cylindrical and oblong'; others were 'of much smaller size, apparently spherical' [5]. Brown carefully ruled out various possible causes (e.g. convective currents), but was unable to give a positive explanation.

In the early years of the 20th century Albert Einstein derived the first testable theory of Brownian motion [1]. Beginning in 1910, Jean Perrin's careful, exhaustive experiments on gum colloids confirmed Einstein's theory [2]. In Perrin's words, watching and measuring the motion of colloidal particles had 'made the world of molecules finally tangible'. Colloids had, effectively, proven the granularity of matter, settling one of the fiercest controversies of nineteenth-century science, and demonstrating in a direct, incontestable way the reality of Boltzmann's statistical mechanics (for a popular account see [4]). In this section, we introduce a number of key ideas and survey some recent developments in colloid physics by way of reviewing Einstein and Perrin's results.

What Einstein and Perrin established

The mean-squared displacement (MSD) of a diffusing particle in $\mathcal{N}$ dimensions, $\langle \Delta r^2(t) \rangle$, is given by[1]

$$\langle \Delta r^2 \rangle = 2\mathcal{N} D t, \tag{1.1}$$

where D is the diffusion coefficient. Einstein predicted that D for a single sphere of radius a in an infinite fluid of viscosity η at temperature T is given by the Stokes-Einstein relation:

$$D = \frac{k_B T}{6\pi \eta a}, \tag{1.2}$$

where k_B is Boltzmann's constant.[2]

Einstein also predicted that the number density of particles, n, as a function of height, z, in a suspension that has reached sedimentation equilibrium follows an exponential distribution:

$$n(z) = n(0) e^{-z/z_0}, \tag{1.3a}$$
$$\text{where } z_0 = \frac{k_B T}{\Delta m g} \tag{1.3b}$$

defines the sedimentation height, with Δm being the buoyant mass of a single particle, given by the particle's volume times its density difference with the liquid, and g is the acceleration due to gravity.[3]

Equations (1.3) can be derived from flux balance. There is a downward flux of particle due to gravity, $J_g = n v_s$, where v_s is the sedimentation

---

[1] A straightforward argument suggests why the MSD scales as t. Denote the displacement after N steps of a discrete walk by $\vec{R}_N$. Then $\vec{R}_N = \vec{R}_{N-1} + \vec{\ell}$, where $\vec{\ell}$ is the displacement in a single step. Taking the dot product of each side with itself and averaging, we find $\langle R_N^2 \rangle = \langle R_{N-1}^2 \rangle + 2\langle \vec{R}_{N-1} \cdot \vec{\ell} \rangle + \langle \ell^2 \rangle$. If the walk is random, then $\langle \vec{R}_{N-1} \cdot \vec{\ell} \rangle = 0$. Mathematical induction then implies $\langle R_N^2 \rangle = N \langle \ell^2 \rangle$, i.e. the MSD is linear in N, so that passage to the continuous limit should give MSD $\propto$ time.

[2] Strictly, a in equation (1.2) is the *hydrodynamic* radius: the radius that determines the Stokes drag. Dynamic light scattering measures D by determining the quantity given in equation (1.30), which, for a dilute suspension, reduces to $f(q,\tau) = \langle e^{i\vec{q}\cdot\Delta\vec{r}} \rangle = e^{-Dq^2\tau}$. To be historically accurate, equation (1.2) is the Stokes-Einstein-Sutherland relation. William Sutherland, a Scot who emigrated to Australia at a young age, published it just before Einstein did [6].

[3] The time it takes to reach sedimentation equilibrium is $\gtrsim$ the time it takes one particle to fall through the height of the sample.

speed. As the sediment builds up, there is a concentration gradient driving a diffusive flux in the opposite direction, given by Fick's Law, viz. $J_d = -D\partial n/\partial z$. In the steady state, $J_g = J_d$, giving

$$\frac{\partial n}{\partial z} = -\left(\frac{v_s}{D}\right) n , \qquad (1.4)$$

which solves trivially to equation (1.3a) with

$$z_0 = \frac{D}{v_s} . \qquad (1.5)$$

Using $v_s = \Delta mg/6\pi\eta a$ and the Stokes-Einstein relation, equation (1.2), we arrive at equation (1.3b).

Interestingly, Einstein did not take this route to arrive at the exponential distribution of particles with height. Instead, he appealed directly to the fact that the colloids are in thermal equilibrium with the molecules in the liquid, so that there is equipartition of energy and the colloids and the liquid molecules have the same thermodynamic temperature. Then his result follows from the Boltzmann distribution: a particle has probability e^{-U/k_BT} to be at a potential energy of $U = \Delta mgz$ above the ground state of being at the bottom of the container.

Perrin verified both of Einstein's predictions experimentally: by tracking the motion of individual diffusing particles and by counting the number of particles as a function of height in a suspension that had reached sedimentation equilibrium. In each case he determined k_B, and, via the known value of the ideal gas constant $R = 8.314$J/mol, estimated Avogadro's number $N_A = R/k_B$. The convergence of the values of N_A obtained from these and other, unrelated, methods powerfully evidenced the existence of atoms [2].

Note that z_0 serves as an upper bound for the colloidal length scale: a colloid is a particle (radius a) for which $a \lesssim z_0$. Using typical values for the densities of everyday solids and liquids, we then arrive at the estimate that the colloidal length scale encompasses $10\text{nm} \lesssim a \lesssim 1\mu\text{m}$.[4]

That k_BT is present in both of Einstein's results, equations (1.2) and (1.3), serves to remind us that thermally-driven fluctuations dominate in the colloidal domain. Since elastic moduli have the dimensions of energy per unit volume (1 Pa = 1 J/m^3), we may estimate a typical modulus in colloidal materials as $\sim k_BT/a^3$, which, for a in the colloidal length scale, gives mPa to kPa as the range of elasticity. Colloids are therefore mechanically weak; in comparison, a typical modulus of metals is $\lesssim 10^2$ GPa ($\sim$ eV/Å^3, i.e. Fermi energy per atomic volume).

Equation (1.3) also describes the height distribution of the number of molecules in an isothermal ideal gas placed in the earth's gravitational field, where it is known as the barometric distribution. (For oxygen at 300K, z_0 is just over 8km, comparable to the summit of Everest.) The distribution can be derived by considering mechanical equilibrium, balancing the difference in pressure (P) (hence 'barometric') between the top and bottom surface of a slab of gas of thickness δz, $-(\partial P(z)/\partial z)\delta z$, with the weight per unit area of molecules in that slice, $[n(z)\delta z]mg$,

---

[4] Normalising z_0 by the particle radius gives the gravitational Péclet number, which, apart from numerical constants, is Pe$_g \sim a/z_0 \sim \Delta\rho g a^4/k_BT$. This estimates the ratio of the time taken for a particle to diffuse its own radius to the time taken to sediment over the same length scale. For a colloidal particle, Pe$_g < 1$.

where n here is the number density of molecules. Using the ideal gas equation of state (EOS), $P = nk_B T$, we obtain $n(z) = n(0)e^{-mgz/k_B T}$.

In the corresponding 'mechanical' derivation of the barometric distribution for colloids, the gas pressure, P, is replaced by the osmotic pressure of the suspension Π. This is the pressure that must be applied to a suspension in contact with a bath of pure solvent across a semi-permeable membrane impassable to the particles in order to stop the flow of solvent from the bath to the suspension. It has been known since van't Hoff that for a dilute suspension, $\Pi = nk_B T$, which is the colloidal analogue of the ideal gas EOS. Thus, a dilute suspension of colloidal particles behaves thermodynamically like a collection of 'big atoms': the particles have the same equation of state and show the same sedimentation equilibrium as a classical ideal atomic gas. This 'colloids as atoms' analogy [3] is a deep and far reaching one.[5] Thermalised by the liquid bath, colloidal particles can explore configuration space and (given time) reach thermal equilibrium. They therefore model many generic aspects of the behaviour of atomic and molecular materials, e.g. crystallisation and glassy arrest. This paradigm has inspired many important developments in colloid physics since its 1980s renaissance.

[5] I justify this analogy formally at the end of this section. Essentially, one makes certain approximations and integrates out the solvent degrees of freedom; thereafter, the suspended particles behave like a thermodynamic system in their own right.

Rather than invoking mechanical equilibrium and the ideal gas EOS to derive the barometric distribution, one may work backwards from the latter and invoke mechanical equilibrium to derive the ideal gas EOS. This procedure has proven fruitful in studying concentrated colloidal suspensions, whose equilibrium sedimentation profile, $\phi(z)$, is no longer exponential due to inter-particle interactions. The requirement of mechanical equilibrium then gives the following expression for Π:

$$\Pi(z) = \Delta \rho g \int_z^h \phi(z)\, dz \,, \qquad (1.6)$$

which, after inverting $\phi(z)$ to yield $z(\phi)$, gives the full EOS, $\Pi(\phi)$. Note, however, that measuring $\phi(z)$ in a concentrated suspension is experimentally non-trivial.

Perrin's techniques

We now examine two aspects of Perrin's experimental techniques that remain relevant today. First, Perrin's work illustrates the importance of well-characterised colloids. His starting material, particles of gamboge (a gum resin extract), had very high polydispersity (defined as the standard deviation of the particle size distribution normalised by the mean). Perrin needed quasi-monodisperse particles, especially for sedimentation experiments, because $z_0 \sim a^{-3}$, equation (1.3b). Perrin used repeated centrifugation to reduce the polydispersity. Preparation of well-characterised suspensions is a top priority in colloid physics today.

Next, Perrin used direct imaging both to track particles to determine their diffusion coefficient, and to count them at different heights to determine their sedimentation equilibrium. Direct imaging is once again becoming very popular in colloid physics, mainly because the confocal

[6] Note that simulating real-life polydispersity, i.e. a quasi-continuous particle size distribution, is difficult because of the relatively small number of particles that can be simulated compared to the number of particles in typical bulk samples.

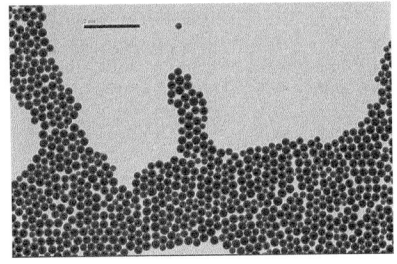

Fig. 1.1 Transmission electron micrograph of PMMA particles, with a 2μm scale bar. Image courtesy of Siobhan Liddle.

microscope can deliver good 'optical sectioning' in thick samples.

Model colloids

Preparation of well-characterised particles was crucial for Perrin, and remains of key importance in modern colloid physics. One of the reasons is that the use of colloids with uniquely-defined size, shape and interaction dramatically simplifies the comparison with theory and simulations.[6] A number of very well-characterised model systems of spherical particles exist — polystyrene (PS), silica, and poly(methymethacrylate) (PMMA) being the most common. Given the ubiquitous presence of van der Waals attraction between all particles suspended in non-refractive-index matched solvents [7], stabilisation of the particles against aggregation is necessary, commonly achieved by having either ionisable moieties or short polymer chains present on the surface. The former gives rise to charged particles repelling each other via so-called electrical double layers, while the latter leads to steric repulsion when neighbouring 'hairs' overlap. The book by Israelachvili [7] provides further details on colloidal stability and interparticle forces.

A widely-used model system consists of PMMA spheres sterically stabilised by oligomers of poly-12-hydroxy-stearic acid (PHSA), Figure 1.1. These behave as nearly-perfect hard spheres in a range of slightly non-index-matched hydrocarbons (such as decalin). The only relevant thermodynamic variable in a hard-particle system is the volume fraction ϕ, see equation (1.25); for N spherical colloids (radius a) in volume V,

$$\phi = \frac{4}{3}\pi a^3 \left(\frac{N}{V}\right). \tag{1.7}$$

The experimental determination of ϕ is not straightforward. A recent review [8] concludes that systematic uncertainties in the absolute value of ϕ of $\gtrsim 3\%$ are hard to avoid due to the steric stabilising layer and polydispersity (typically $\gtrsim 5\%$), though much higher *relative* accuracies in a series of samples are routinely achievable.

In the PMMA system, solvents such as cycloheptyl bromide added to achieve density matching between particles and dispersion medium lead to charging of the particles. For small enough particles, this charge can be screened out to recover nearly-hard-sphere behaviour by dissolving salts such as tetrabutyl ammonium bromide (TBAB) in the dispersing medium. However, since the total charge increases with the particle radius and the solubility of TBAB and similar salts in hydrocarbons is limited, it is difficult to screen out the charges adequately to recover nearly-hard-sphere behaviour on particles large enough for confocal imaging [9].

The quest for hard sphere colloids is driven by the desire for a model system for studying excluded volume effects in its barest form. Excluded volume is conceptually the simplest imaginable inter-particle interaction (see also note 11 on page 13), and spheres give mathematical tractability. Hard spheres therefore function as a reference system. Starting from this reference point, many innovations have come from either 'tuning' the inter-particle interaction away from pure hardness (see Section 1.3) or exploring non-spherical shapes (see Section 1.6).

Imaging

Hard condensed matter physics, the study of atomic and molecular liquids and solids, largely relies on scattering and spectroscopic techniques to reveal structure and dynamics respectively. Due to the length and time scales set by atoms (nm and ps, respectively), various forms of direct imaging are likely to remain specialist methods.

Scattering is also important for soft matter physics [10]. But the upper end of the colloidal length scale is also well within the resolution of optical microscopes ($\gtrsim 300$nm). Colloids are also intrinsically slow. The elementary timescale is the Brownian time, τ_B, which can be defined as the time it takes a free particle to diffuse its own radius, i.e. $6D\tau_B = a^2$. Using equation (1.2) we obtain

$$\tau_B = \frac{\pi \eta a^3}{k_B T}, \qquad (1.8)$$

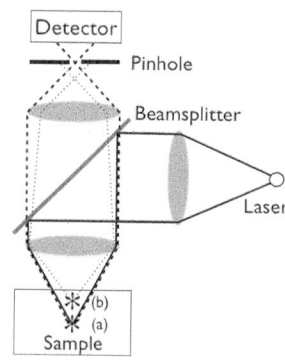

Fig. 1.2 Schematic of the principle of fluorescent confocal microscopy. Laser light, full lines, is focused at position (a) in the sample, and causes material at that position to fluoresce. The fluorescent light from this focal spot, dashed lines, is focussed at a pinhole in front of the detector and reaches the detector. Fluorescence is also excited at out-of-focus positions, such as (b). Fluorescent light from position (b), dotted line, is largely prevented from reaching the detector by the pinhole. The laser focal spot is scanned in the sample to build up a full image.

which works out to be $\lesssim$ 1s for an $a \sim 1\mu$m particle in water at room temperature ($\eta \approx 10^{-3}$Pa.s). The size and (related) slowness mean that optical microscopy coupled with even rather slow recording methods can yield useful images of $a \gtrsim 1\mu$m particles in real space and real time.

It is striking that Perrin used direct imaging in his two pioneering experiments to study both static (or average) structure (sedimentation equilibrium) and dynamics (diffusion). After a hiatus of nearly a century, direct imaging has returned to become a mainstream method in colloid physics. Perrin worked on dilute samples, so that even with the lack of refractive-index matching, he was able to image his samples and count or track particles using film as the recording medium. Today, confocal microscopy permits the imaging of concentrated samples of fluorescent particles that are somewhat turbid (although index matching greatly improves the depth to which imaging can usefully be carried out). The method relies on a pinhole to reject out-of-focus information, and gives good 'optical sectioning', Figure 1.2. This ability coupled with increasingly sophisticated algorithms for particle tracking now enable the study of concentrated samples (up to $\phi \approx 0.64$) at single-particle resolution in (almost) real time. Reviews of confocal microscopy of colloids exist, for both quiescent [11] and flowing [12] systems.

Justifying 'colloids as atoms'

Here we offer a statistical mechanical justification of the 'colloids as atoms' analogy. This more formal section may be omitted at a first reading.

Consider N colloids (at $\{\vec{r}^N\}$) and N_0 solvent molecules (at $\{\vec{r}^{N_0}\}$) in a fixed volume V at temperature T interacting via $U(\vec{r}^N, \vec{r}^{N_0})$. The

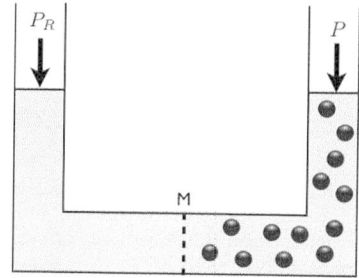

Fig. 1.3 A semi-permeable membrane, M, separates a colloidal suspension on its right from a reservoir of pure solvent (at pressure P_R) on its left.

canonical partition function is:

$$Z(N,V,T,N_0) = \frac{1}{N!\Lambda^{3N} N_0!\Lambda_0^{3N_0}} \int e^{-U(\vec{r}^N, \vec{r}^{N_0})/k_B T} d\vec{r}^N d\vec{r}^{N_0}, \quad (1.9)$$

where Λ and Λ_0 are de Broglie wavelengths. At equilibrium, the free energy, $F(N,V,T,N_0) = -k_B T \ln Z$ is minimised. Transform to a *semi-grand canonical* ensemble with a variable number of solvent molecules, N_0; the state variables are now (N,V,T,μ_0), where μ_0 is the (fixed) chemical potential of the solvent.[7] An experimental realisation is shown in Figure 1.3. The semi-grand partition function is:

$$\Xi(N,V,T,\mu_0) = \sum_{N_0=0}^{\infty} [\exp(\mu_0/k_B T)]^{N_0} Z(N,V,T,\mu_0). \quad (1.10)$$

[7] In a full grand canonical ensemble description, the colloids are also at a constant chemical potential, μ, so that the system's variables are (μ, V, T, μ_0).

The equilibrium behaviour of the system is obtained by minimising the semi-grand potential $\Omega = -k_B T \ln \Xi$. We rewrite equation (1.10) as

$$\Xi(N,V,T,\mu_0) = \frac{1}{N!\Lambda^{3N}} \int e^{-W(\vec{r}^N;\mu_0)/k_B T} d\vec{r}^N \quad (1.11)$$

by defining the 'potential of mean force'[8]:

$$W(\vec{r}^N;\mu_0)/k_B T = -\ln \sum_{N_0 \geq 0} \frac{z_0^{N_0}}{N_0!\Lambda_0^{3N_0}} \int e^{-U(\vec{r}^N, \vec{r}^{N_0})/k_B T} d\vec{r}^{N_0}, \quad (1.12)$$

[8] So that the average force on (typical) particle j is given by [13] $\langle \vec{F}_j \rangle = -\partial W/\partial \vec{r}_j$.

where $z_0 = e^{\mu_0/k_B T}/\Lambda_0^3$ is the solvent activity. Now decompose U into colloid-colloid (cc), colloid-solvent (cs) and solvent-solvent (ss) parts:

$$U(\vec{r}^N, \vec{r}^{N_0}) = U_{cc} + U_{cs} + U_{ss}. \quad (1.13)$$

Colloids are significantly larger than the solvent molecules, so we neglect depletion-type effects (cf. Figure 1.7, with the 'polymer coils' now being solvent molecules). In the limit of point-like solvent molecules,

$$U_{cs} = NA\gamma, \quad (1.14)$$

where A = surface area of a single particle and γ is an energy. Now

$$e^{-W/k_B T} \approx e^{-U_{cc}(\vec{r}^N)/k_B T} \cdot e^{-NA\gamma/k_B T} \cdot \Xi' \quad (1.15)$$

$$\text{where} \quad \Xi' = \sum_{N_0 \geq 0} \frac{z_0^{N_0}}{N_0!\Lambda_0^{3N_0}} \int e^{-U_{ss}(\vec{r}^N, \vec{r}^{N_0})/k_B T} d r^{N_0}. \quad (1.16)$$

[9] For finite-sized solvent molecules, the volume over which the integral is non-zero depends on the coordinates of the colloids, $\{\vec{r}^N\}$.

Again neglecting depletion effects, the integral over $\{r^{N_0}\}$ in Ξ' simply ranges over $(1-\phi)V = V - Nv$, where v is the volume of a single particle.[9] In this case, Ξ' is the grand partition function of the solvent in a volume $(1-\phi)V$, so that, defining ω_s to be the grand potential of the solvent per unit volume, we have[10]

[10] Note that ω_s is a constant for any given solvent chemical potential μ_0.

$$\Xi' = \exp[-\omega_s(1-\phi)V/k_B T]. \quad (1.17)$$

Putting it all together, we find that:

$$\Xi = \left\{ \frac{1}{N!\Lambda^{3N}} \int e^{-U_{cc}(\vec{r}^N)/k_BT} d\vec{r}^N \right\} \cdot e^{-NA\gamma/k_BT} \cdot e^{-\omega_s(1-\phi)V/k_BT}, \tag{1.18}$$

where $\{\ldots\}$ is the canonical partition function Z_c of N colloids at $\{\vec{r}^N\}$ in volume V at temperature T interacting via $U_{cc}(\{\vec{r}^N\})$. The semi-grand potential of the system $\Omega = -k_BT \ln \Xi$ is therefore given by

$$\Omega = F_c(N,V,T) + NA\gamma + (V - Nv)\omega_s, \tag{1.19}$$

where $F_c(N,V,T) = -k_BT \ln Z_c$ is the Helmholtz free energy of a fixed number of N colloids in volume V at temperature T. Under these canonical conditions, the last two terms on the right of equation (1.19) are constant. Thus, the behaviour of the system in Figure 1.3 is obtained by minimising the *canonical* partition function of the colloids alone: the solvent degrees of freedom have been successfully integrated out.

Since $\omega_s V = \Omega_s(V) = -P_R V$ [13], the suspension pressure is:

$$P = -\left(\frac{\partial \Omega}{\partial V}\right)_{N,T,\mu_0} = -\left(\frac{\partial F_c}{\partial V}\right)_{N,T} + P_R, \tag{1.20}$$

$$\text{where} \quad \Pi = -(\partial F_c/\partial V)_{N,T}, \tag{1.21}$$

is the osmotic pressure of the colloid: the suspension needs to be under a pressure Π higher than that of the pure solvent in order for the solvent molecules in the suspension to have the same chemical potential as (i.e 'in osmotic equilibrium with') the solvent molecules in the reservoir. In absence of this extra pressure, solvent will flow from the bath into the suspension across the semi-permeable membrane until sufficient extra pressure is built up for chemical potential equality.

The 'colloids as atom' pictures emerges from equation (1.21), which shows that the osmotic pressure Π plays the role of pressure in the thermodynamics of colloids, so that, for example, the coexistence of two phases I and II requires $\mu_I = \mu_{II}$ and $\Pi_I = \Pi_{II}$.

1.2 Do hard spheres crystallise? The physics of entropy

> 'I would like to close this discussion, for I am quite sure that the transition goes a little bit against intuition; that is why so many people have difficulty with it, and surely I am one of those. But this transition – it still might be true, you know – and I don't think one can decide by general arguments..'
> G. E. Uhlenbeck, Round table discussion at the *Symposium on the many-body problem*, Stevens Institute of Technology, January 28-29 (1957)

An ideal gas, a collection of N non-interacting point particles in volume V at temperature T, is one of the few exactly solvable models in statistical mechanics. Its equation of state (EOS), giving its pressure P, is $PV = Nk_BT$. Any interaction between the particles leads to

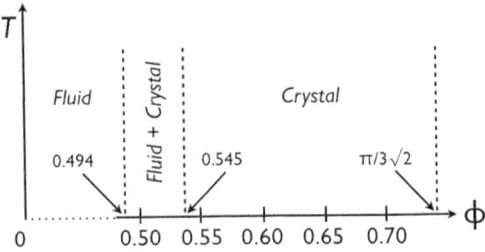

Fig. 1.4 The hard-sphere phase diagram on the temperature-volume fraction plane, (T, ϕ). Since all phase boundaries are parallel to the T axis, it is usually not shown.

more complex EOS, which can be expanded in a power series (a virial expansion) of the number density $n = N/V$, viz.

$$P = nk_BT \left(1 + \sum_{j=2}^{\infty} B_j(T) n^{j-1}\right), \qquad (1.22)$$

where the j-th virial coefficient $B_j(T)$ takes account of j-body interactions. In particular, B_2 is related to the pair interaction $U(r)$ by

$$B_2(T) = \frac{1}{2}\int \left(1 - e^{-U(r)/k_BT}\right) d^3r. \qquad (1.23)$$

Conceptually, the simplest interaction is excluded volume: point particles give way to ones that take up space and cannot overlap. Thus, for hard spheres (HS) of radius a, $B_2^{\text{HS}} = 4 \times (\frac{4}{3}\pi a^3)$. Amazingly, a semi-empirical relation, the Carnahan-Starling formula, captures the EOS of HS fluids up to $\phi \approx 0.5$ accurately [13]:

$$\frac{PV}{Nk_BT} = \frac{1 + \phi + \phi^2 - \phi^3}{(1-\phi)^3}. \qquad (1.24)$$

Note that comparing a power series expansion of this expression with equation (1.22) recovers the above-given expression for B_2^{HS}.

The thermodynamic behaviour of HS (or indeed of a collection of hard particles of any shape) is exclusively due to entropy, S, since the Helmholtz free energy of ν moles of hard particles is given by

$$F = U - TS = \frac{3}{2}\nu RT - TS = T\left(\frac{3}{2}\nu R - S\right). \qquad (1.25)$$

Thus, minimisation of F is equivalent to the maximisation of S, with temperature merely scaling the free energy as a multiplicative factor.

Hard spheres have played a vital role in the development of statistical mechanics, both classical and quantum mechanical. Thus, the properties of many liquids near the triple point and crystallisation are dominated by excluded volume. Neither claim is obvious. In this section we focus on crystallisation (see Section 1.3 for liquids, especially note 11 on page 13).

As late as 1957, the question of whether a collection of HS would undergo an ordering transition as ϕ increased was hotly debated at a gathering of some of the top theoretical physicists of the time, including G. E. Uhlenbeck, T. D. Lee, C. N. Yang, and B. J. Alder [14]. The quotation from Uhlenbeck at the beginning of this section sums up the mood of a round-table discussion he chaired. Later, William Hoover and Francis Ree [15] confirmed using Monte Carlo simulations that there was a crystallisation transition in the HS fluid at the freezing concentration of $\phi_f = 0.494$; the crystals that coexisted with this fluid were found to be at the melting concentration of $\phi_m = 0.545$, Figure 1.4.

Note that the value of ϕ_m may be related to the so-called Lindemann melting criterion, namely, that an atomic crystal should melt when the amplitude of the thermal vibrations, $\langle x^2 \rangle^{1/2}$, reaches a certain (fixed) fraction of the lattice constant a_0, i.e. when $\langle x^2 \rangle^{1/2} = c_L a_0$, where $c_L \sim 0.1\text{-}0.15$. Equipartition of energy tells us that

$$M\omega_D^2 \langle x^2 \rangle = k_B T, \qquad (1.26)$$

where M is the atomic mass and ω_D is the Debye frequency, which we use to estimate a typical (angular) frequency of lattice vibrations, and is related to the Debye temperature θ_D by $k_B T_D = \hbar \omega_D$. If Lindemann is right, then the melting temperature T_m should be proportional to $M\theta_D^2 a^2$. With very few outlying exceptions, this is approximately true for a large number of elements. For HS, the close-packed crystal expands (linearly) by a fraction of $\left\{ \left[(\pi/3\sqrt{2})/\phi_m \right]^{1/3} - 1 \right\} \approx 0.11$ at melting, where $\pi/3\sqrt{2} \approx 0.74$ is the volume fraction at crystalline close packing. This is indeed within the range of $c_L \approx 0.1\text{-}0.15$, which suggests that melting in atomic materials may also be dominated by excluded volume effects (i.e. entropy).

The key experiment

Nearly two decades after Hoover and Ree's simulations, Peter Pusey and William (Bill) van Megen [16] experimentally verified their predictions. Samples of sterically-stabilised PMMA particles with increasing ϕ were homogenised by tumbling and left to reach equilibrium. In a ϕ range in reasonable agreement with computer simulations, coexistence of fluid and crystal states was observed. Since the particles used were in the range of visible wavelengths, crystallisation could be seen visually by the 'opalescent' Bragg diffraction of white light from polycrystalline regions. While face-centred cubic packing is marginally more stable than hexagonal close packing for HS [17], in practice the crystals are more or less randomly stacked [18].

Uhlenbeck is indeed right that the spontaneous ordering of HS into crystals appears counterintuitive, since the physics is driven by entropy maximisation. The resolution of this puzzle lies in the fact that as ϕ increases, there comes a point when arranging the particles on a lattice in fact gives them more local freedom of movement ('free-volume entropy'), to the extent that the loss of configuration entropy associated with ordering is more than compensated for. That such a point should

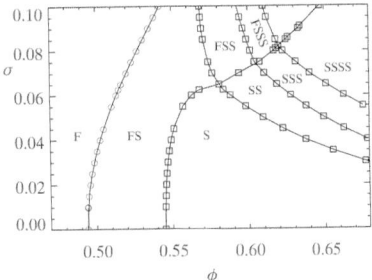

Fig. 1.5 The calculated phase diagram of hard spheres as a function of polydispersity, σ. 'F' and 'S' stand for fluid and (crystalline) solid. Thus, 'FSS' stands for the coexistence of a fluid phase with two solid phases having different lattice parameters. Redrawn from Wilding and Sollich [19].

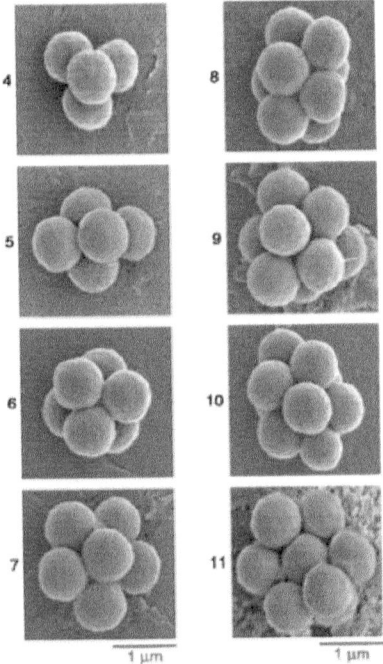

Fig. 1.6 Self-assembled particle clusters with 4 to 11 particles. From Manoharan et al. [22]. Reprinted with permission from AAAS.

exist is certainly suggested by the observation that random close packing (RCP) occurs at $\phi \approx 0.64$, while a close-packed crystal has $\phi \approx 0.74$, so that rearranging an amorphous RCP structure at $\phi = 0.64$ (which is jammed) into a crystal evidently gives rise to free-volume entropy.

Comparison between Pusey and van Megen's experiments and simulations of HS is not in fact straightforward, due to uncertainties in ϕ determination [8], residual softness in the inter-particle interaction [9], and polydispersity. Detailed experiments and simulations since then have shown that sterically-stabilised PMMA particles in decalin do indeed behave as hard spheres in many circumstances, provided a correct 'effective radius' is used. There has been also been much progress in understanding theoretically the effect of polydispersity. Results from the latest calculations of phase boundaries of HS as a function of polydispersity [19] are plotted in Figure 1.5: even small polydispersities can give rise to significant changes. Apart from altering the position and width of the fluid-solid coexistence gap, polydispersity also stabilises multiple solid phases. In practice, such solid-solid coexistence has never yet been observed in experiments and may in fact be kinetically unreachable.

Despite these complications, colloidal experiments have played a definitive role in settling the issue that Uhlenbeck and his colleagues could not agree on in 1957. Entropically-driven ordering is now well established, and colloids have become the standard experimental test-bed for such phenomena. An immediate generalisation from Pusey and van Megan [16] is the study of binary HS mixtures at various size ratios [20], which give rise to various crystals structures such as AB_2 and AB_{13} already known from metallic alloys. On the other hand, the kinetics of crystallisation can be followed in real time, e.g. by static light scattering, and has therefore be studied intensively in the last decade using model colloids. Nevertheless, significant uncertainties remain, e.g. discrepancies between measured and simulated nucleation rates [21].

Shape and entropy

Interestingly, the prediction of entropy-driven ordering in anisotropic particle systems pre-dated the round-table discussion on HS [14]. In 1949, Lars Onsager [23] explained why at high concentrations, a collection of rods might prefer to switch from isotropic packing to nematic ordering: a more or less aligned population of rods gives more room for local orientational fluctuations. The generic similarity to the explanation given above for HS crystallisation is immediately recognisable. A review of rod-shape colloids has recently been published [24].

Particles of more complex shapes also undergo entropically-driven ordering, and the literature is growing fast. We simply mention that the entropic ordering of various hard polyhedra has recently been explored *in silico* [25], and that synthetic strategies for many kinds of shape-anisotropic colloids, such as those shown in Figure 1.6, have been surveyed by Sacanna and Pine [26].

1.3 How to make a liquid: Interactions under control

> 'Under ordinary terrestrial conditions, matter appears in three states of aggregation: solids, liquids and gases. Because of the close connection between liquids and life itself, the liquid state is perhaps the most important of these. But it also provides physics with the most severe challenge of the three. ...Why should there be a state with density nearly equal to that of solids, but with such relatively low viscosity that the shape can vary readily?'
> Herbert J. Bernstein and Victor F. Weisskopf [27]

The liquid state has a fragile existence. In the P-T phase diagram of a simple substance, it occupies a niche between the triple and critical points, while the solid and gas (vapour) phases occupy essentially semi-infinite regions of this parameter space. According to Victor Weisskopf, an early pioneer of quantum electrodynamics and Director General of CERN in the 1960s, the existence of the liquid state is a fundamental challenge to theoretical physics.

There is no liquid state on the HS phase diagram, Figure 1.4. It has been known since van der Waals that inter-particle attraction is necessary for the liquid phase to exist. For spherical particles of diameter $\sigma = 2a$ interacting via a Lennard-Jones (LJ) potential, viz.

$$U_{\mathrm{LJ}}(r) = 4\epsilon \left[\left(\frac{r}{\sigma}\right)^{12} - \left(\frac{r}{\sigma}\right)^{6} \right], \quad (1.27)$$

a stable liquid phase exists below a critical temperature T_c set by the depth of $U_{\mathrm{LJ}}(r)$: $k_B T_c = 1.326\epsilon$. The stability range of the LJ liquid can be gauged by $T_c/T_{tr} = 2.0$, where $T_{tr} = 0.649\epsilon/k_B$ is the triple temperature. The LJ model provides a reasonable description of liquid rare gases such as argon (for which $T_c/T_{tr} = 1.80$).[11] Work on colloid-polymer mixtures has shown that the presence of an attraction is necessary but not sufficient for the existence of the liquid state; a sufficient condition is that the *range* of the attraction must be long enough.

Attraction through repulsion

An inter-particle attraction can be induced in a suspension of HS colloids by adding a non-adsorbing polymer, which is an example of 'tuning' the inter-particle interaction. Polymer molecules are excluded from a region between the surfaces of two close-by particles, Figure 1.7, giving a net osmotic pressure pushing them together.[12]

If the liquid is a so-called theta solvent for the polymer, then by definition there is negligible coil-coil interaction (i.e. $B_2 = 0$ for the polymers) [31], and we may reasonably model the polymer molecules as geometric points as far as their mutual interaction is concerned. However, polymer molecules are repelled from particle surfaces because a polymer coil approaching a surface closer than something like its radius of gyration, r_g, will stretch and lose configurational entropy. Thus, in

[11] The long-range nature of the LJ potential and the parent van der Waals interaction that it models explain why hard spheres provide a good starting point for perturbative treatments of the properties of simple liquids near the triple point [28]. Here, each neighbour is surrounded by order 10 other atoms. The sum of their LJ potentials at the central atom is essentially flat. To the lowest order, this then contributes only a constant to the free energy, whose minimisation is therefore once more equivalent to the maximisation of entropy; cf. equation (1.25).

[12] Much of the following material has been reviewed before [29], and is expounded in detail in a recent text [30].

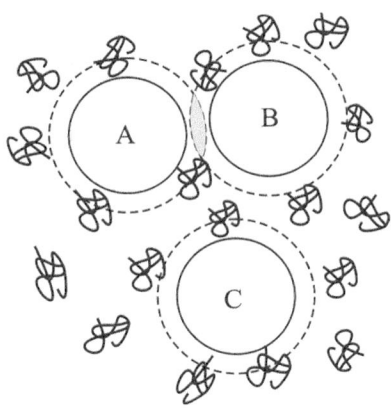

Fig. 1.7 Schematic illustration of the depletion interaction. The centre of a polymer molecule (coil) is excluded from coming closer than a certain distance, approximately its own radius of gyration, to the surface of a colloid (full circle) because of the high entropic cost of configurational distortion. Each colloid is therefore surrounded by a depletion zone (dotted circle) within which there are essentially no polymer centres of mass. If a colloid (such as C) is far away from other particles, the osmotic pressure of the polymer on the particle is isotropic. If, however, the surfaces of two colloids (such as A and B) are closer than twice the size of a depletion zone, then there is no polymer in the lens-shaped (shaded) region, and a net (osmotic) force presses the particles together – the depletion attraction. Taken from Poon [29] with permission.

the Asakura-Oosawa (AO) model of the depletion effect [30], each point-like polymer molecule may not approach the surface of a particle (radius a) by a distance less than (say) r_g. The resulting effective inter-particle attraction can be written down analytically. Its range is given by $2r_g$, while its depth is proportional to the activity of the polymer, a_p, which is the (number) density of coils in a pure polymer solution in osmotic equilibrium with the colloid-polymer mixture, $n_p^{(R)}$. If the dimensionless range is small, viz. $\xi = r_g/a \ll 1$, then a_p is related to the colloid volume fraction ϕ by $a_p \approx n_p/(1-\phi)$, where n_p is the number density of polymer coils in the colloid-polymer mixture. In the same limit, a power series expansion gives the following convenient approximation:

$$U_{\rm AO}(r) = -\eta_p^{(R)} k_B T \cdot \frac{3}{2}\left(\frac{1+\xi}{\xi^3}\right)\left(\frac{r}{2a} - 1 - \xi\right)^2, \quad (1.28)$$

where $\eta_p^{(R)} = \frac{4}{3}\pi r_g^3 a_p$ is the volume fraction of coils in the reservoir.

Note that since the polymer is non-adsorbing, all interactions in such a mixture are repulsive. Depletion is therefore a mechanism for generating 'attraction through repulsion'. This inter-particle potential can be 'tuned' separately in its range (by changing r_g) and its depth (by changing a_p). It is instructive to contrast this with the LJ potential, equation (1.27). We may estimate the range of its repulsive and attractive parts by σ and $r_{\rm inf} - \sigma$ respectively, where $r_{\rm inf} = (26/7)^{1/6}\sigma \approx 1.24\sigma$ is the potential's point of inflection, so that the LJ potential has a *fixed* dimensionless range of $\xi_{\rm LJ} \approx 0.24$ at all depths ϵ.

Unmaking the liquid state

The AO model can be approximated in the test tube by a mixture of hard colloids and non-adsorbing polymers in a near-theta solvent. Experiments using sterically-stabilised PMMA particles and linear polystyrene (PS) dispersed in decalin [32] demonstrate that when the range of the depletion attraction is large enough, the phase diagram is topologically identical to that of a simple atomic substance, showing gas (vapour), liquid, and (crystalline) solid phases, Figure 1.8(a). As the range drops, the size of the gas-liquid coexistence region shrinks, Figure 1.8(b). When the range drops below some critical value, $\xi_c \approx 0.25$ in experiments using PMMA colloids and PS in cis-decalin, the gas-liquid coexistence region vanishes. The equilibrium phase diagram now only shows fluid, solid, and fluid-solid coexistence regions, Figure 1.8(c). The same sequence is predicted by theory and computer simulations for the AO potential and other kinds of inter-particle attraction as the attractive range is progressively decreased, although the precise value of ξ_c is dependent on the the shape of the attractive potential and the approximations used [29, 30]. Intriguingly, the gas-liquid coexistence curve does not disappear all together; instead, it becomes metastable and is 'buried' inside the equilibrium F+S coexistence region (dotted in Figure 1.8c).

Thus, the existence of a thermodynamically stable liquid state is dependent of an inter-particle attraction of long enough range. This requirement can be traced back to the entropic origin of melting, Figure 1.9. According to the Lindemann criterion, a close packed crystal has to expand (volumetrically) by $\sim (1+\delta)^3 - 1$ with $\delta \sim 0.1$, or about a third, before it loses rigidity and melts into a fluid state. At this point, if neighbours are still within range of each others' attractive potential, a dense liquid phase is possible. Whether it actually occurs then depends on whether the thermal energy is enough to enable particles to overcome their mutual attraction (i.e. whether one is above or below the critical point). If, however, the range of the inter-particle attraction is shorter than a certain critical value, expansion of the crystal to the point of loss of rigidity would bring neighbours out of range of each other's attraction and the crystal catastrophically falls apart to a low-density gas (vapour). Quantitatively, if the inter-particle attraction has a range of $\gtrsim 2\delta$, then in the crystal at melting, each particle will always stay within range of its neighbours as it executes thermal motion in the cage formed by these neighbours. The Lindemann criterion then predicts that the crossover should occur at $\xi \sim 2\delta \sim 0.2$, as indeed is the case.

The depletion attraction in many practical situations is rather short ranged: for particles with $a \gtrsim 0.5\mu$m and polymers with molecular weight $M_w \lesssim 10^5$ and therefore $r_g \lesssim 10 - 50$nm, we have $\xi \lesssim 0.1$. (A common exception is when the polymer is a bacterial exopolysaccharide such as xanthan; the stiffness of the sugar backbone gives r_g in the range of $\gtrsim 10^2$nm.) In the next section we will see that other tuneable colloidal interactions also tend to be short ranged, with $\xi \sim 0.01$-0.1. Nevertheless, gas-liquid phase separation in real colloids is not uncommon, since a small degree of polydispersity (i.e. $\gtrsim 10\%$) or non-sphericity in the particles will preclude crystallisation, so that the dotted curve in Figure 1.8(c) becomes the only phase boundary even in the case of small ξ.

It is therefore useful to note that there is a universal gas-liquid coexistence boundary for particles interacting via a short-range attraction if we plot the phase boundary in the (ϕ, B_2) plane [33], i.e. the strength of the attraction is measured by the second virial coefficient, equation

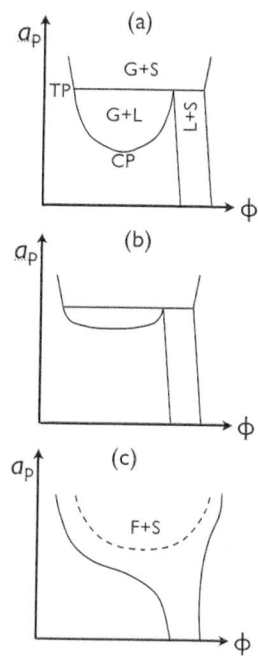

Fig. 1.8 Schematic of the effect of changing the size of the polymer, or equivalently the dimensionless range of the depletion attraction ξ, on the equilibrium phase diagram of a hard spheres + non-adsorbing polymers mixture. The axes give the colloid volume fraction ϕ (horizontal) and the polymer activity a_p (vertical). G = gas (vapour), L = liquid, S = (crystalline) solid, F = fluid (when there is only a single fluid state), '+' denotes coexistence; CP = critical point, TP = triple point (= G+L+S). (a) $\xi > \xi_c$; this phase diagram is directly comparable to that of the temperature-density phase diagram of a simple atomic substance such as argon, but plotted with an inverse temperature axis. (b) $\xi \to \xi_c$, with regions and special points the same as in (a). (c) $\xi < \xi_c$; the dashed curve here indicates the metastable gas-liquid coexistence boundary that now lies entirely 'buried' in the equilibrium F+S coexistence region.

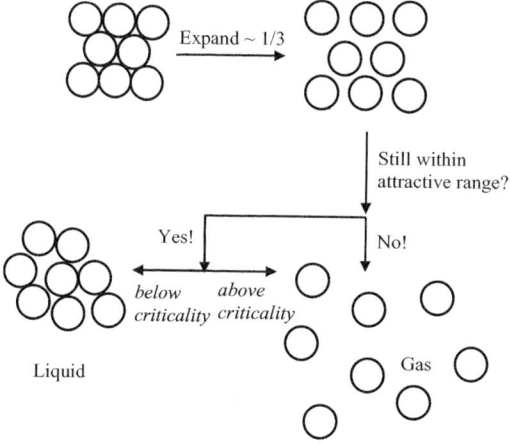

Fig. 1.9 Schematic explanation of the importance of range. See text for details. Taken from Poon [29] with permission.

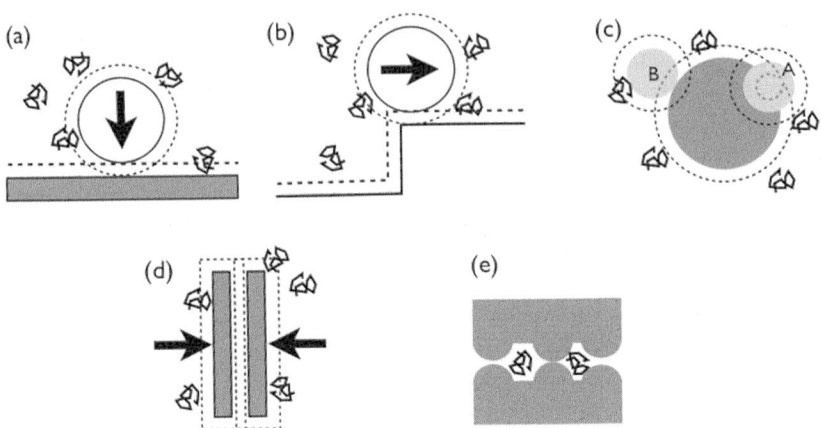

Fig. 1.10 Depletion in diverse contexts. Depletion zones from which the centres of polymers are excluded are delineated by dashed lines. Where there are overlapping depletion zones, entropic forces result (bold arrows). (a) Depletion attraction between a particle and a flat surface. (b) When a particle moves beyond a ledge, the volume of overlapping depletion zones decrease, resulting in a force pushing the particle away from the ledge. (c) Lock (large sphere) and key (small) binding: the depletion attraction is strongest when the key fits into the lock (situation A) than in any other position (such as B) (see [36] for the subtleties of what 'fits' here means precisely). (d) Depletion attraction between two rods. (d) If surface roughness is larger than the radius of the polymer coils, the depletion mechanism is no longer effective.

(1.23). Thus, knowing the phase boundary for one form of inter-particle attraction enables the boundary for other systems to be found. Perhaps the best known analytic form is due to Baxter: an infinitely deep but infinitely short range potential constructed such that B_2 is finite and well-defined. Its gas-liquid phase boundary has been accurately determined by simulations [34].

In the atomic world, van der Waals (vdW) attraction dominates. This originates in transient dipole moments due to fluctuating electron clouds. Thus the attraction is of the dipole-dipole type, and falls with distance according to $(r^{-3})^2 \sim r^{-6}$, hence the form of the attractive term in U_{LJ}, equation (1.27). We have already mentioned that the precise range of any particular form of attractive potential needed to stabilise the liquid state depends on its form. By mapping power law attractions of the form $U(r) = -\epsilon(\sigma/r)^n$ onto the so-called Yukawa potential, it can easily be verified from simulations of the latter [35] that $n < 9$ is needed to make liquids. Since the vdW attraction scales with $n = 6$, liquids are essentially inevitable in our atomic world.

Note that the results we have surveyed so far assume that the colloids are monodisperse. We have already seen how polydispersity affects the phase diagram of the pure HS system, Figure 1.5. Polydispersity has a significant effect in colloid-polymer mixtures because the depletion attraction is dependent on the size of the particles, leading to substantial fractionation – different phases displaying different polydispersities. On the fundamental level, this introduces a whole new level of complexity into the phase behaviour, giving rise to a 'zoo' of possible equilibrium phase diagrams [37]. Practically, depletion can be can be used to fractionate polydisperse colloids, although the success of this approach is dependent on the details of the size distribution [38].

Before leaving depletion, we mention examples of its operation in other contexts, Figure 1.10. Depletion can adhere particles to surfaces [39], repel particles from micro-ledges [40], bind 'lock and key' particles [36], and order rods [41]. In the latter case, the stronger depletion attraction in the side-by-side configuration of neighbouring particles leads to strong enhancement of the formation of liquid crystalline mesophases. Surface roughness can be used to tune the depletion interaction: a roughness larger than the polymer r_g turns the interaction off [42].

Other strategies for tuning colloidal interactions

One of the most interesting features of colloid physics for the experimentalist is the possibility of 'tuning' the inter-particle interaction in a way that is not available in the world of atoms and molecules. We have just introduced the use of non-adsorbing polymers to give an attraction separately tuneable in range and depth. In this section we introduce a number of other common ways of tuning colloidal interactions.

We have already pointed out that some solvents used to density match PMMA colloids confer ionisable surface groups on the colloids. Dissociation of these group gives rise to charges on the surface and oppositely-charged counter-ions in solution. The counter-ions, together with any ions from added salt, give rise to a diffused 'electrical double layer' next to the particle's surface of characteristic width [7]

$$\kappa^{-1} = \left(\frac{\epsilon_0 \epsilon_r k_B T}{\sum_{j=1}^{N} n_j Q_j^2 e^2} \right)^{1/2}. \qquad (1.29)$$

Here ϵ_0 and ϵ_r are the permittivity of free space and the dielectric constant of the suspending liquid, respectively, and the number density of ions of charge (in electronic unit e) Q_j is n_j. The ions screen the surface charges,[13] so that outside of the Debye length κ^{-1}, a particle looks essentially neutral. In limit of very small κ^{-1}, the short-range vdW attraction is exposed and the particles are unstable towards aggregation. On the other hand, when κ^{-1} is greater than the range of the inter-particle vdW attraction ($\lesssim$ nm), the colloids interact repulsively with a degree of softness that is tuneable by varying the amount of salt.

Other ways exist to tune the softness. Sterically-stabilised PMMA particles are nearly hard-sphere like because the oligomeric PHSA hairs on the surface are relatively short (10-20 nm). Longer steric stabilising 'hairs' on other particles give increasing softness.

Destabilising the steric stabilisation layer on the surface of colloids can give rise to tuneable inter-particle attraction. For stabilisation to be effective, the polymer layer has to be in a theta or good solvent. Changing the temperature or adding a second liquid may lower the solvency, eventually leading to collapse of the layers and stickiness between particles. Note that two mechanisms may operate in this context, depending on the precise nature of the polymer 'hairs' and the solvent. For example, silica particles stabilised by octadectyl chains show quite different

[13] Note the square dependence on Q_j in equation (1.29), so that polyvalent ions are very effective in screening surface charges.

behaviour in benzene and hexadecane solvents when the temperature is lowered [43]. It seems that in benzene, the octadectyl 'hairs' simply become 'sticky' at low temperatures, but the hexadecane and the 'hairs' undergo a coordinated freezing transition in the gaps between particles, giving rise to much stronger and more rigid 'bonds'.

Attraction can also be induced by the presence of a minority liquid phase that displays different wetting to the particle surfaces than the majority solvent via capillary effects between particles. Thus, it was recently found that $\gtrsim 0.2\%$ of water in a $\phi = 0.111$ dispersion of hydrophobically-modified calcium carbonate particles ($a = 0.8\mu$m) in the organic solvent diisononyl phthalate (DINP) turned the suspension from a flowing liquid into a gel able to support its own weight [44]. Trace amounts of a minority solvent can therefore be used to tune interactions and therefore rheology to dramatic effect.

Applying an electric field to a dielectric particle or a magnetic field to a paramagnetic or superparamegnetic particle induces a dipole moment of the corresponding kind. The resulting dipole-dipole interaction between particles is tuneable by the strength of the external field. These effects are exploited in electrorheological and magnetorheological fluids [45], whose mechanical properties can be rapidly switched using external fields. Such particles also give one of the simplest and most easily accessible systems with tuneable anisotropic interaction [46, 47].

A final example of tuning interactions reminds us that colloidal research can throw light on rather fundamental physics. It has been known for half a century that confined fluctuating fields give rise to long-range forces between the confining surfaces. The original prediction of Casimir pertained to zero-point fluctuations in the electromagnetic field. Recent direct measurements [48] have demonstrated that particles experience a Casimir force when immersed in a binary liquid mixture close to its critical point due to the confinement (between nearby particles) of large-scale, pre-critical concentration fluctuations. This colloidal Casimir force can be sensitively tuned, e.g. by changing the surface properties of the particles (thus affecting their interaction with the two components in the binary mixture) and by chaining the temperature (thus altering the proximity to the critical point and therefore the properties of the confined concentration fluctuations).

1.4 What is a glass? Challenging arrest

'The deepest and most interesting unsolved problem in solid-state theory is probably the theory of the nature of glass and the glass transition.'
Philip W. Anderson [49].

The nature of kinetic arrest in general, and of the glass transition in particular, is one of the 'grand challenges' in contemporary condensed matter and statistical physics. Model colloids have emerged as one of the most fruitful systems for 'clean' experiments that are susceptible to detailed comparison with theories. The first clue that colloids may be important for glass research came when Pusey and van Megen [16] found that suspensions of PMMA particles at $\phi \gtrsim 0.58$ did not reach the equilibrium state predicted by simulation, Figure 1.4, namely, full crystallisation. Instead, these samples remained essentially amorphous indefinitely (except at the suspension-air interface). They subsequently published dynamic light scattering (DLS) results to show that these samples were kinetically arrested. The data could be fitted in some detail with the predictions of mode coupling theory (MCT). But perhaps a more spectacular success of MCT is its prediction that when a short-range attraction is added, the system should display *two* distinct glassy states. These were subsequently found by experiments and simulations. This development also generated fresh interest in the nature of colloidal gelation at lower volume fractions (say, $\phi \lesssim 0.3$); the consensus now is that in many situations, the phenomenon is due to the delicate interplay between kinetic arrest and gas-liquid phase separation. We now introduce these developments in turn. [14]

[14] Note that comprehensive recent reviews of colloidal glasses [50, 51] and gels [52] are available.

The hard-sphere glass transition

The initial clue to the existence of a glass-like transition in HS colloids was visual. Pusey and van Megen [16] observed that a sample at $\phi \approx 0.59$ was filled with large, irregular crystals (compared to very small, homogeneously nucleated crystallites at lower ϕ), while samples at higher ϕ remained homogeneous and amorphous in the bulk (although they showed crystallisation at the suspension-air interface). Subsequently these authors performed DLS experiments to measure the normalised dynamic structure factor of their suspensions:

$$f(q,\tau) = \frac{\left\langle \sum_{j,k} e^{-i\vec{q}\cdot[\vec{r}_j(t)-\vec{r}_k(t+\tau)]} \right\rangle}{S(q)}. \quad (1.30)$$

This function characterises the progressive loss of correlation between the motion of pairs of particles: $\vec{r}_j(t)$ is the position of particle j at time t and $\vec{r}_k(t+\tau)$ is the position of particle k at a delay time τ later. These correlations are measured at a wavelength of $2\pi/q$, where q is the magnitude of the scattering vector. The angle brackets $\langle \ldots \rangle$ denote ensemble averaging, while the denominator is the system's static (or average) structure factor, $S(q)$, defined as the numerator with $\tau = 0$.

Trivially $f(q,0) = 1$. If the system is ergodic, i.e. its particles explore configurational space freely during the experiment, then $f(q,\tau)$ decays completely to zero during this time. On the other hand, for a non-ergodic system, i.e. one in which long-range particle motions are arrested over the measurement time window, $f(q,\tau)$ decays to a finite value in the same window. The latter is often represented as $f(q,\infty) > 0$, but the '∞' should only be taken as shorthand for 'for the longest time measured', at least as far as experiments are concerned. It was found that $f(q,\infty) > 0$ occurred for PMMA colloids at $\phi_g \gtrsim 0.58$.

Subsequent measurements of the single-particle mean-squared displacement (MSD) from tracer DLS [53], Figure 1.11, give a more in-

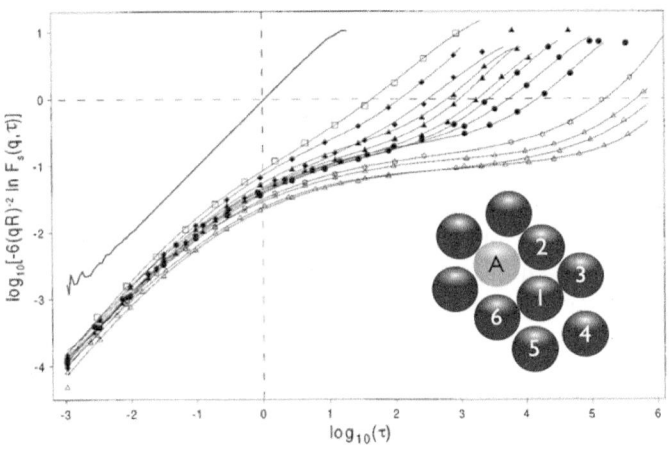

Fig. 1.11 The particle mean-squared displacement (MSD) in a PMMA suspension normalised to the squared particle radius plotted against time in units of the Brownian time, defined so that the MSD of a free particle at unit time is unity. Volume fractions from left to right: $\phi \to 0$ (solid curve), $\phi =$ 0.466 (□), 0.502 (♦), 0.519 (♦), 0.534 (▲), 0.538 (▲), 0.543 (▲), 0.548 (▲), 0.553 (●), 0.558 (●), 0.566 (open star), 0.573 (△), 0.578 (△), 0.583 (△). Taken from van Megen et al. [53]. Copyright (1998) by The American Physical Society; used with permission.

[15] The dynamic heterogeneity was reviewed in detail in a recent book by Berthier et al. [54].

tuitive understanding. At all ϕ, the short-time MSD is diffusive, but the short-time self diffusion coefficient, $D_S^{(s)}$, is below that of a free particle, D_0. At intermediate times, a plateau progressively develops, because each particle interacts directly with its 'cage' of nearest neighbours, Figure 1.11(inset). The inset shows a two-dimensional schematic of caging in concentrated HS fluids: a central particle A is caged by its six nearest neighbours; but each of these neighbours is, in turn, caged, e.g. particle 1 by particles A and 2-6. At $\phi \lesssim 0.58$, this caging is not permanent. The plateau gives way to a diffusive motion again, characterised by a long-time self diffusion coefficient $D_L^{(s)}$ that decreases precipitously as ϕ increases. At $\phi = 0.583$, the particles never recover diffusive motion, they permanently cage each other on the experimental time scale. This is the HS glass transition, which is therefore due to 'crowding'.

Apart from displaying finite $f(q, \infty)$, HS colloids at $\phi \gtrsim 0.58$ show other features that are characteristic of glassiness, such as ageing and dynamical heterogeneity. 'Ageing' refers to the dependence of sample properties, such as the decay of $f(q, \tau)$, on the time elapsed since the sample was left undisturbed (the 'waiting time'), while 'dynamic heterogeneity' points to the very marked non-uniformity in spatial distribution of dynamical behaviour,[15] such as the positions of (say, the 5%) fastest 'rattling' particles. Hard-sphere glasses display both of these characteristics [55, 56].

Nonetheless, a number of observations, both experimental and computational, appear to cast doubt on the occurrence of a glass transition in HS. In particular, HS in simulations (e.g. Zaccarelli et al. [57]) and PMMA colloids in micro-gravity [58] crystallise at $\phi \gtrsim 0.58$; even Pusey and van Megen [16] saw crystals in their non-density-matched terrestrial experiments at $\phi \gtrsim 0.59$, albeit with very much altered crystal forms compared to $\phi \lesssim 0.58$. A widely-held belief is that a system that crystallises spontaneously in the bulk (i.e. without being seeded with crystallites) cannot be a glass: since crystallisation requires long-range motion and $D_L^{(s)} = 0$ in a glass, a glass cannot crystallise. Within this framework, monodisperse HS do not vitrify, and the glass-transition-like features in experiments at $\phi \approx 0.58$ are due to the polydispersity of real colloids.

Recent computer simulations may have gone some way towards resolving this conundrum, discovering a mode of crystallisation at $\phi \gtrsim 0.58$ that does *not* require long-range transport [57]. Indeed, in this new mode, particles need to move by less than a single particle diameter for crystallisation to occur. On the other hand, before they crystallise, amorphous states at $\phi \gtrsim 0.58$ show all the signs of glassiness, including, importantly, ageing. So it appears that there is no mutual exclusivity between glassiness and crystallisation *per se* and that HS with or without polydispersity may indeed have a glass transition despite the observation of crystals appearing in glassy samples. Interestingly, in these simulations, while no effect on the particles' dynamics was observed up to a polydispersity of 8.5%, no crystallisation was seen beyond a polydispersity of 7%, consistent with the experimental observation that there is a 'terminal polydispersity' of the same approximate magnitude beyond which no colloidal crystals could be seen.

We close this section by mentioning that glassy arrest in systems of non-spherical hard particles has been studied, e.g. see the review of glassy states in rod-like suspensions by Solomon and Spicer [24].

Mode coupling theory

The mode coupling theory (MCT) of the glass transition has been outstandingly successful in predicting aspects of the measured properties of HS colloidal glasses.[16] The typical approach to deriving its results relies on the Mori-Zwanzig 'projection operator' formalism for dealing with the time evolution of dynamical variables. This literature is in general highly mathematical. One exception is an early paper by Geszti [59], in which he gives a phenomenological picture for vitrification that MCT formalises rigorously. The original discussion was in the language of glass transitions in atomic and molecular materials driven by temperature. Here we summarise it using the language of concentration-driven vitrification in HS colloids.

Geszti's key insight is to emphasise the close relationship between the glass transition and stress relaxation.[17] When a stress σ is applied to a viscoelastic liquid (such as a high-ϕ suspension), its instantaneous response is to develop a shear strain γ; the two are related by an 'infinite frequency' shear modulus $G_\infty = \sigma/\gamma$. In time, however, the system will flow at some shear rate $\dot\gamma$, with the viscosity given by $\eta = \sigma/\dot\gamma$. Following Maxwell, we write:

$$\eta = G_\infty \tau, \quad (1.31)$$

where τ is the relaxation time for the stress. In a concentrated colloid, the stress relaxes in two stages. First the 'rattling' of particles in their nearest-neighbour cages, with a (fast) relaxation time τ_{rat}, and then a much slower relaxation due to large-scale cage-level rearrangement of particles. We assume that the latter scales with the long-time diffusion

[16] Some other theoretical approaches are introduced below in Section 1.5 in the context of colloid rheology.

[17] Readers unfamiliar with rheology may wish to read Section 1.5 first before coming back to the material here.

coefficient as D_L^{-1}. Thus,

$$\eta = G_\infty \tau_{\text{rat}} + G_\infty c(\phi) D_L^{-1} = \eta_\infty(\phi) + b(\phi) D_L^{-1}, \quad (1.32)$$

where $c(\phi)$ and $b(\phi)$ are concentration-dependent coefficients to be determined from microscopic theory.

Now we make the crucial assumption that a generalised Stokes-Einstein relation between viscosity and diffusion still holds, i.e. $D = s(\phi) k_B T / \eta$, where $s(\phi)$ is another concentration-dependent coefficient, cf. equation (1.2). Substituting this into Equation (1.32) and gathering up all unknown coefficients into a single parameter $B(\phi)$, we find

$$\eta(\phi) = \eta_\infty(\phi) + B(\phi)\eta(\phi), \quad (1.33\text{a})$$

$$\text{or } \eta(\phi) = \frac{\eta_\infty(\phi)}{1 - B(\phi)}. \quad (1.33\text{b})$$

Equation (1.33a) shows that the viscosity [$\eta(\phi)$ in the left-hand side] is determined partly by itself (in the second term on the right). This mechanism of 'viscosity feedback' develops in the following manner: stress relaxes primarily by diffusion, diffusion is inversely related to the viscosity (generalised Stokes-Einstein), and viscosity is proportional to the stress relaxation time (Maxwell). The result, equation (1.33b), is that a viscosity divergence can be expected at some point, which can be identified as the glass transition.

Mode-coupling theory provides this argument rigorously. We refer the reader to the book-length treatment of Wolfgang Götze [60] for all the details and comprehensive references to the original literature, and a pedagogical review [61] for a concise introduction. Here we give a schematic summary.[18]

[18] Note that the rest of this section is the most demanding part of this chapter, and can be skipped on first reading without loss of continuity.

Mode-coupling theory for colloidal glasses starts from an exact equation of motion for the normalised dynamic structure factor, $f(q,t)$:

$$\frac{\partial f(q,t)}{\partial t} + \frac{D_0 q^2}{S(q)} \left\{ f(q,t) + \int_0^t dt' m_q(t-t') \frac{\partial f(q,t')}{\partial t'} \right\} = 0, \quad (1.34)$$

where D_0 is the free-particle diffusion coefficient. The memory kernel $m_q(t-t')$ specifies the history-dependent dynamics of the system due to interactions and is a function of the time-dependent Fourier components of the particle density at two times, $n_{\vec{q}}(t) = \sum_j e^{i\vec{q}\cdot\vec{r}_j(t)}$ and $n_{\vec{q}'}(t')$, operated on by various time-evolution operators. Without memory effects, equation (1.34) predicts, correctly, that $f(q,t)$ decays exponentially.

Equation (1.34) with the exact m_q is too complex for actual calculations. Standard MCT makes two key approximations. First, it projects the exact form of m_q onto a basis set of pairs of density fluctuations, $\delta n_{\vec{q}_1} \delta n_{\vec{q}_2}$, which turns m_q into a function of a 'four-point correlator'. Second, it decouples this latter function (four-point correlator) into a product of two two-point correlators, schematically:

$$\langle \delta n \delta n \delta n \delta n \rangle \to \langle \delta n \delta n \rangle \cdot \langle \delta n \delta n \rangle.$$

Each of the two-point correlators, properly normalised, becomes $f(q,t)$, and the resulting approximate m_q takes the form

$$m(\vec{q},t) = \sum_{\vec{k}} V_{\vec{q},\vec{k}} f(\vec{k},t) f(\vec{k}-\vec{q},t), \qquad (1.35)$$

with the 'vertex function' V being dependent solely on the system's structure factor at $\vec{k}$, $\vec{q}$, and $\vec{k}-\vec{q}$. Using the structure factor as input, the approximate MCT equation of motion for $f(q,t)$ can be solved with the initial condition $f(q,0) = 1$.

For HS, this approach predicts that $f(q,\infty)$ abruptly changes from 0 to a finite value as ϕ increases from 0 to $\phi_g^{(\mathrm{MCT})} \approx 0.52$. Note that MCT generically over-predicts the tendency of systems to verify; experimentally, $\phi_g^{(\mathrm{HS})} \approx 0.58$. The physical origins of the HS glass transition according to MCT becomes clear when one makes a crude approximation, using a delta-function at the position of the peak of the HS $S(q)$, $q_0 \approx 2\pi/a_0$ where a_0 is the average nearest-neighbour distance, as the structure input to the MCT equations. This leads to a qualitatively similar glass transition, i.e. it is structure on the length scale of a_0, or 'cages', that is responsible for dynamical arrest. Mode-coupling theory also predicts the precise functional form of $f(q,\tau)$ near ϕ_g and how quantities such as $D_L^{(s)}$ should vanish as $|\phi - \phi_g|^\gamma$ when $\phi \to \phi_g$. These predictions have been well confirmed for the HS system against both experimental data and computer simulations.

Opinion amongst theorists on the status of MCT is varied. Some do not like the fact that the approximations it makes are somewhat uncontrolled. Nevertheless, it is the only theory to date that makes concrete and detailed predictions that experimentalists can test. Thus, as one reviewer has put it (paraphrasing Winston Churchill[19]), 'Mode coupling theory is the worst theory of colloidal glasses – apart from all the others that have been tried from time to time' [62]. The way detailed MCT calculations can interact with colloidal experiments is well illustrated by our next topic.

[19] Winston Churchill said: "Democracy is the worst form of Government except all those other forms that have been tried from time to time", House of Commons, 11 Nov 1947; Oxford Dictionary of Quotations, 3rd Ed., OUP (1979).

The effect of attraction

Mode coupling theory predicts a complex scenario for kinetic arrest when short-range attraction is introduced into the HS system. Calculations [63, 64] predict two qualitatively distinct kinds of glasses in hard spheres with short-range attraction, separated by a re-entrant glass transition line. These predictions remained something of a mathematical curiosity until experiments using PMMA colloids and non-adsorbing linear PS with $\xi \approx 0.08$ [65] substantially confirmed them, Figure 1.12. In this figure, schematic drawings of particles (the shaded annulus showing the range of the attractive potential δ) illustrate (bottom to top) caging in repulsion-driven glasses (L = cage length), clustering of cages due to attraction, and the trapping of particles by nearest-neighbour potential wells in attraction-driven glasses. Consider, for example, samples with $\phi \approx 0.6$. The sample with no polymer ($c_p = 0$, filled

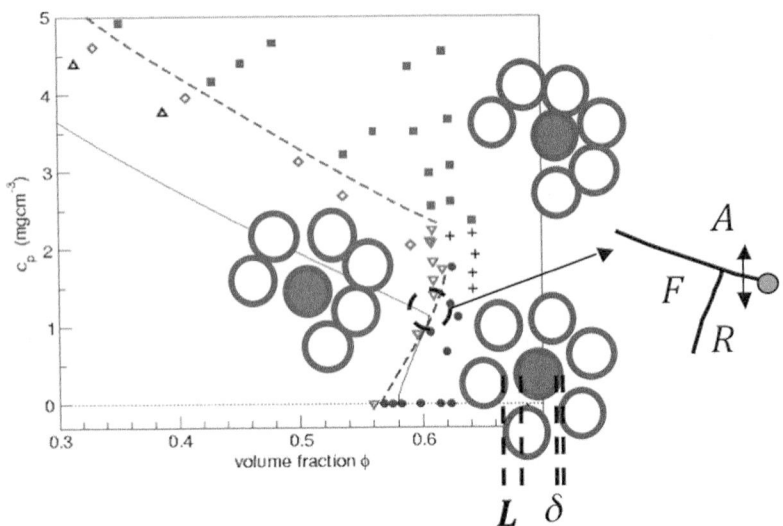

Fig. 1.12 Equilibrium and non-equilibrium behaviour of a PMMA colloid (volume fraction ϕ) + PS polymer (concentration c_p) mixture with $\xi = 0.08$. Open symbols: samples that reached thermal equilibrium in fluid ($\triangle$), fluid-crystal coexistence ($\diamond$), and fully crystallised ($\triangledown$). Other samples did not reach thermal equilibrium and failed to crystallise: some showed characteristics of repulsion-driven glasses ($\bullet$), some showed those of attraction-driven glasses ($\blacksquare$), and some showed both (+). Dashed curves: guides to the eye showing where crystallisation ceased. Continuous line: the re-entrant glass transition line predicted by MCT. Taken from Poon [66] with permission from Cambridge University Press.

circle) is a hard-sphere glass. Adding sufficient polymer (concentration $c_p = 1\text{mg/cm}^3$) induces enough attraction to give an ergodic fluid that eventually crystallises (inverted triangle). At $c_p \approx 2.5\text{mg/cm}^3$ or above, however, crystallisation again fails to occur (filled squares). Instead, these samples remain in amorphous states. Dynamic light scattering shows that the particles in these states are far more localised in space than in the hard-sphere glasses: they represent a qualitatively new kind of glass.

MCT and DLS suggest a heuristic picture for these observations, summarised in Figure 1.12. In the 'repulsion-dominated' hard-sphere glass, particles are caged by their neighbours. The length scale L in Figure 1.12 characterises the 'rattle room' each particle has in its cage (analogous to the Lindemann length scale in melting, see equation (1.26) and the associated discussion). A weak short-range attraction clusters the particles in the cage and opens up holes, ultimately melting the glass and restoring ergodicity. Increasing the attraction further produces an 'attraction-dominated' glass – particles are now localised within a short length scale δ, which is the width of the narrow interparticle attraction (set by the size r_g of the polymer in solution that provides the depletion attraction in the experiments reported here, but is more general). Simulations confirm this picture (see Götze [60] for references.)

The re-entrant glass transition line predicted by MCT is shown in Figure 1.12. The agreement with data is striking, given that all parameters were taken from experiments, with the only adjustment a ϕ scaling (to correct for the MCT ϕ_g for HS). The details of the 'cusp' are shown in an enlarged schematic in this figure: to the right of the re-entrant region separating repulsive glass (R), fluid (F), and attractive glass (A), MCT predicts a sharp glass-glass transition (double arrow), where various physical properties should change abruptly. The big dot shows the 'singularity' beyond which one can move from a repulsive to an attractive glass continuously. The distinction between the two kinds of glasses vanishes at a point known as the 'singularity'. Qualitatively, this is where the two length scales of the problem become equal; in the notation introduced in Figure 1.12, $L \to \delta$.[20]

Three caveats should be noted. First, in the scenario sketched above, particles in the attractive glass are permanently 'caged' by interparticle 'bonds'. Long-time computer simulations [67] beyond measurements times likely achievable in any DLS or confocal tracking experiment (but well within target shelf lives of industrial products), suggest that caging by bonds is not permanent, so that in the long run, particles in an attractive glass can still explore its nearest-neighbour geometric cage. This means that a sample above the 'attractive glass' line in Figure 1.12 still needs to be at high enough ϕ for a permanent geometric cage to exist for it to be truly arrested. Secondly, recent work on the crystallisation of HS glasses [57] reopens the issue of the fate of all glassy samples shown in Figure 1.12. It would also be interesting to know whether attractive glasses also crystallise, and if so, by what mechanism. Finally, polydispersity can have significant effect, especially since it leads to concomitant variation in ξ.

[20]MCT also predicts that the re-entrant glass transition disappears when δ becomes large enough; crudely, when the cage size at the HS glass transition becomes comparable to the range of the attractive well.

Gelation

It has long been known that colloids with strong short-range interparticle attraction can form solids – gels – at very low volume fractions. Two traditional explanations are easy to visualise. [21] First, gelation can be explained by appealing to the idea of percolation. Defining two particles as 'bonded' whenever they are within the range of each other's attractive potential, $U(r)$, there is a region in $(\phi, U/k_B T)$ space, to the right of line 'P' in Figure 1.13, in which the largest cluster of particles in the system will always percolate, i.e. one could traverse from one side of a macroscopic sample to another side staying entirely within mutually-bonded particles. (Note that for clarity we have not included the fluid-solid coexistence region for monodisperse particles.) If the bonds are labile, then percolation is hard to detect on the level of macroscopic properties; but if the bonds are rigid enough, then a percolated cluster will not rearrange: the system has gelled.

In another approach, one imagines single particles aggregating whenever they touch, forming dimers. Dimers can bond further with dimers, etc. If the growing clusters cannot rearrange, i.e. the bonds are permanent *and* rigid, then these clusters will be ramified (fractal) objects. Eventually, space will be filled by clusters with some characteristic size. This mechanism of gelation, known as diffusion-limited cluster-cluster aggregation (DLCA), has many variants. For example, in reaction-limited cluster-cluster aggregation (RLCA), particles and clusters encountering each other would only bond after a number of 'trials'; RLCA produces more compact clusters. Such cluster-cluster aggregation mechanisms operate in principle at arbitrarily low ϕ, because the growing frac-

[21]The literature is extensive. Some earlier work was reviewed by Poon and Haw [68].

Fig. 1.13 Schematic representation of the various boundaries in state space that affect gelation; not to scale. The axes denote the colloid volume fraction ϕ and the dimensionless contact potential U_0/k_BT. P = percolation line; AG = attractive glass line; RG = repulsive glass line; GL = gas-liquid phase boundary; Sp = gas-liquid spinodal; ab = gas-liquid tie line at the point where AG meets GL; shaded DLCA = region where the DLCA picture most clearly applies. The fate of three systems is discussed in the text: ● will phase separate into coexisting gas (vapour) and liquid; ■ will gel; ▲ will nucleate gel 'blobs'.

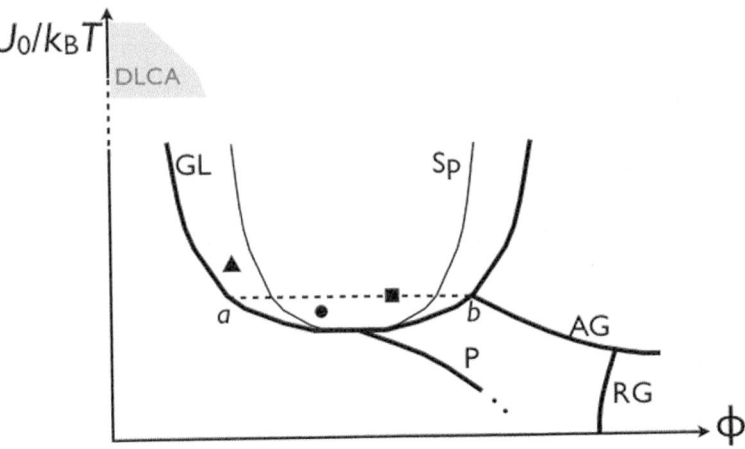

[22] Such interaction has long been known in the case of the gelation of polymer solutions: cf. Figure 1.13 with Figure 11 in the review by Keller [69].

[23] Note that in real systems, the region of complete phase separation may be vanishingly small, e.g. because the gas-liquid phase boundary is very flat.

tal clusters have ever-decreasing average density, so that space-filling will eventually occur (in an infinite system!). Indeed, the picture of more or less independent clusters repeatedly 'bumping into' each other is most easily formalised at infinitesimal ϕ (see shaded region in Figure 1.13).

After the prediction and confirmation of the existence of attractive colloidal glasses, the idea naturally suggested itself that the attractive glass line, Figure 1.12, extended to low ϕ, may be the gelation boundary. This has proved to be half correct. There is an emerging consensus that in a large class of systems, gelation occurs because of the interaction between dynamical arrest (controlled by the low-ϕ extension of the attractive glass line or something similar) and gas-liquid phase separation, because the AG and GL boundaries in Figure 1.13 inevitably intersect for spherical particles interacting with isotropic short-range attraction.[22]

Consider first a system in the GL coexistence region below the tie line ab, such as (●) in Figure 1.13. Such a sample will phase separate into coexisting gas (vapour) and liquid phases with composition on the phase boundary given by the end point of the relevant tie line.[23] If the system is monodisperse enough, this state is only metastable, and the system will eventually reach fluid-crystal coexistence (see Figure 1.8c; this is an example of the Ostwald 'rule of stages').

As U_0/k_BT increases (e.g. by adding polymer in a colloid-polymer mixture to increase the depth of the depletion attraction potential), a threshold is reached at the tie line ab. A system such as (■) in Figure 1.13 will phase separate to give a liquid phase whose composition lies on the attractive arrest line. Since this system occurs within the spinodal region, its phase separation will give rise to a bicontinuous texture, the liquid phase of which is arrested. Macroscopically, we expect such a metastable state to be a solid. Any system within the spinodal region and above the ab line in Figure 1.13 will similarly give rise to such 'arrested spinodal gels'. How this picture emerged from two decades of experiments, theory, and simulations has been reviewed in detail recently [52], where relevant references can be found.

In experiments, these gels are often gravitationally unstable and would collapse under their own weight (or 'cream' in the case of emulsion drops). A variety of collapse behaviour has been observed [70], including samples that show a delay time. Bizarrely, the collapse behaviour depends not only on the physical and chemical properties of the suspension but also on the container shape. The physical mechanisms leading to collapse and the collapse process itself are still matters under active investigation. In particular, there is as yet no fully-developed theoretical model for predicting the 'latency time' in samples that show delayed sedimentation/creaming [71]. Availability of such a model would represent a significant advance: the shelf life of a range of products may be related to this phenomenon.

The interaction of kinetic arrest and gas-liquid phase separation can give rise to other kinds of behaviour and the full parameter space represented in Figure 1.13 has not yet been explored. What has been seen, for example, is that a system such as (▲) will give rise to individual arrested clusters that have the appearance of 'beads' in the case of protein solutions (see Section 1.6 for a discussion of proteins as colloids) [72]. It seems that in this region of parameter space, the system attempts to nucleate liquid drops, but the composition of these drops lie beyond the line of kinetic arrest.

It is also interesting to enquire how the description of gelation in terms of percolation and cluster-cluster aggregation fits into this schema. The region of parameter space in which a DLCA description most naturally applies, viz. $\phi \to 0$ and $U_0/k_BT \to \infty$, is indicated in Figure 1.13. In a classic study of one such system (charged colloids at $\phi \approx 3 \cdot 10^{-4}$ destabilised by salt) [73], a brightening and collapsing small-angle 'ring' of light scattering was observed that eventually became frozen at a finite angle, indicating the presence of a characteristic length scale many particles in size. In a DLCA description, this would be taken to be the size of the clusters that finally touch and jam the system. On the other hand, such collapsing rings of scattering is characteristic of systems phase separating via the spinodal decomposition route. Reconciling these two pictures remains a task for the future.

The percolation line, P in Figure 1.13, lies wholly outside the attractive arrest line, and in many systems apparently intersects the gas-liquid boundary near the critical point. In a number of adhesive colloids this is the gelation boundary (e.g. [43]). It is likely that in such cases, the solvent and the steric-stabilising layers on the colloids co-freeze in the gaps between nearby particles, forming very rigid bonds, so that the percolating cluster is arrested and gels the system. In other systems where bonds are more labile, the percolation line has not yet been shown to have a significant effect on bulk properties.

1.5 Einstein's most useful contribution: Driven matter

'[The 1905 paper resulting from his PhD thesis] has more widespread practical applications than any other paper Einstein ever wrote ... [The] reasons for the popularity of Einstein's thesis ... are indeed not hard to find: the thesis, dealing with bulk rheological properties of particle suspensions, contains results which have an extraordinarily wide range of applications.'
Abraham Pais [74].

The widespread application of the colloidal state of matter in industry is very directly related to its behaviour under external mechanical perturbation, especially by shearing. Indeed, there are few, if any, colloids in industry that are not subjected to considerable shear either during processing and/or in use. The study of the deformation and flow of materials is traditionally within the domain of rheology. However, condensed matter and statistical physicists are also interested in rheology because it is concerned with matter driven out of equilibrium. In this perspective, the rheology of soft matter in general and of colloids in particular offer significant opportunities for advances in basic understanding, because the softness of these materials means that moderate external perturbations can drive them much further from equilibrium than is usually possible with atomic or molecular materials. Rheological experiments with soft matter can therefore explore regimes that are otherwise only accessible to simulations.

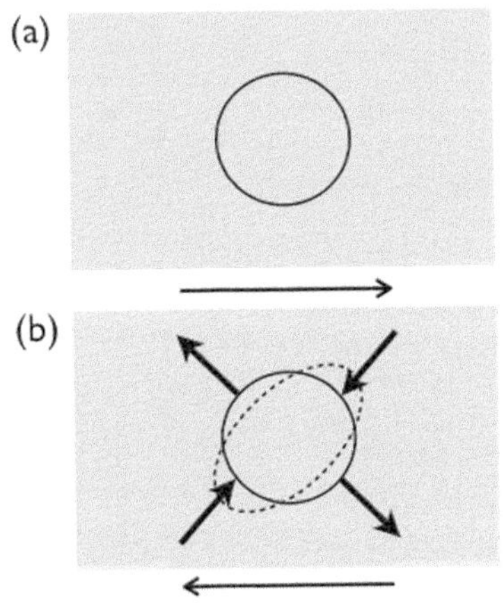

It is little appreciated that the foundations of colloid rheology were laid by Albert Einstein, as part of his quest for proofs of the granularity of matter. In his PhD thesis, Einstein solved the low-Reynolds number hydrodynamic problem of the viscosity, η, of a dilute suspension of spheres at volume fraction ϕ:

$$\eta = \eta_s \left(1 + \frac{5}{2}\phi\right), \quad (1.36)$$

where η_s is the viscosity of the dispersing solvent.[24] Figure 1.14 gives a physical picture of why it is that solid particles enhance the viscosity of a suspension and that the lowest-order enhancement is linear in ϕ, while Guyon et al. [75] give a heuristic derivation.

Einstein used his result to interpret the viscosity of sugar solutions, and found that (after correcting an arithmetic error!) it worked. Since the formula was derived for colloidal spheres whose existence no one doubted, this demonstrated that sugar molecules existed because they could be treated as very small colloids – a nice inversion of the 'colloids as atoms' paradigm that has inspired much of modern colloid physics.

In the rest of this section, we first explain how Einstein's result for a suspension in the dilute limit has been extended to more concentrated colloidal fluids. Then we explore general considerations in the deformation and flow (or rheology) of solid states, before introducing the rheology of colloidal states with solid-like rigidity under quiescent conditions,

Fig. 1.14 Schematic to explain how particles enhance the viscosity in a suspension. A spherical volume element in a quiescent liquid (a) becomes an ellipsoid (dotted line) in (b) when the liquid is subjected to shear. If the same volume element in (a) is now a rigid particle, its shape remains (to a very good approximation) unchanged, solid-lined circle in (b). This particle now exerts stresses on the liquid, represented by arrows in (b). These extra stresses over and against those developed in shearing the liquid alone at the same shear rate are manifested on the bulk level as an increase in the viscosity. At very low ϕ, each particle acts in this way independently, so that their effects are additive, hence equation (1.36) is linear in ϕ.

[24] Note that this result is one of the few in colloid physics that is *not* dependent on the particles being monodisperse!

i.e. colloidal crystals, glasses, and gels. A book-length exposition of suspension rheology is available [76].

Concentration effects in colloidal fluids

Equation (1.36) gives the first term in a power series of $\eta(\phi)$. The next term, $+5.9\phi^2$, was first calculated seven decades later by Batchelor [77]: this describes the increase in viscosity due to pair-wise interaction, so that interactions other than hard repulsion result in different coefficients for the ϕ^2 term [78]. Moreover, as soon as we move into the regime of interacting particles, polydispersity matters: the Batchelor coefficient of 5.9 applies only for monodisperse HS. The viscosity also becomes dependent on the shear rate, $\dot\gamma$, as the concentration increases. The low-shear-limit viscosity, η_0, can be defined as $\lim_{\dot\gamma \to 0} d\sigma/d\dot\gamma$, where $\sigma(\dot\gamma)$ is the 'flow curve', giving the shear stress as a function of $\dot\gamma$. Many suspensions (including HS) at $\phi \lesssim 0.5$ shear thin, i.e. η decreases with $\dot\gamma$, until a well-defined high-shear limit, η_∞ is reached.

The qualitative descriptors of 'low' and 'high' shear can be quantified by the dimensionless Péclet number, defined as the shear rate non-dimensionalised by the Brownian time, equation (1.8):

$$\mathrm{Pe} = \dot\gamma \tau_B \sim \frac{\dot\gamma \eta a^3}{k_B T}. \tag{1.37}$$

At low Pe, Brownian motion is able to homogenise the structural distortions introduced by shear, while at high Pe, shear-induced deformations are not relaxed by thermal fluctuations. For many suspensions, shear thinning sets in at Pe ~ 0.1, and η_∞ is reached by Pe ~ 1. This is the case, e.g. for the $\phi = 0.517$ HS suspension shown in Figure 1.15. On this log-log plot, linearity with unit slope denotes Newtonian behaviour (dashed lines). We see that there are low- and high-shear Newtonian regimes at Pe $\lesssim 0.1$ and Pe $\gtrsim 1$, respectively.

The low-shear viscosity of a monodisperse HS suspension as a function of concentration, $\eta_0(\phi)$, in the entire range of stability of the single-phase fluid, namely, $0 \leq \phi \leq \phi_f = 0.494$, provides an important baseline for understanding the rheology of more complex suspensions. We have already given the first two terms. Considering how important this function is, it is surprising how poorly known it was until comparatively recently. As late as two decades ago, the range of values for $\eta_0(\phi_f)/\eta_s$ in the literature spanned ≈ 20 to 400, with the value of $\gtrsim 20$ given by Papir and Krieger [79] being widely cited because they worked with particles in non-aqueous solvents, so that the presumption of HS behaviour was deemed convincing. Interestingly, these authors reported that their suspensions showed iridescence, i.e. formed colloidal crystals, at $\phi \approx 0.3$; thus their particles could not have been HS-like [16]. Other reasons for uncertainties in the determination of $\eta(\phi)$ for HS include polydispersity and inaccuracies in determining ϕ [80]. The best determinations of this function to date [81, 82] return a value of $\eta_0(\phi_f)/\eta_s \approx 50$. Note that a widely used functional form due to Krieger and Dougherty, viz.

$\eta/\eta_s = (1 - \phi/\phi_m)^{-[\eta]\phi_m}$, where $[\eta] = 2.5$ to agree with equation (1.36), does *not* give a meaningful fit to the best HS data sets, although its use for this purpose remains widespread.

Rheology of solids: general considerations

Rigidity (finite elastic moduli at zero frequency) is associated with symmetry breaking [83]. In a liquid, translation of the whole system by an infinitesimal amount, $\vec{u}$, results in another ground state of the system that is accessible from the original state by thermal fluctuations. At the freezing transition, a crystal emerges in which atoms are situated at lattice positions $\vec{L} = h\vec{a} + k\vec{b} + \ell\vec{c}$ [where (h, k, ℓ) are integers] in a lattice with basis $(\vec{a}, \vec{b}, \vec{c})$. Now, translation by $\vec{u} \neq \vec{L}$ results in a different ground state, one that is *not* accessible from the original state by thermal fluctuations. Translational symmetry has been broken. Now, different parts of a crystal can be in different ground states, i.e. there can be a position-dependent displacement $\vec{u}(\vec{r})$, storing an elastic energy that scales as $\kappa|\nabla \vec{u}(\vec{r})|^2$, where κ is a suitable elastic modulus.

Real crystal lattices are not perfect but contain defects. Stressing a real crystal beyond some limit, the yield stress σ_Y, brings about (dissipative) inelastic flows of various kinds. The role of line defects, known as dislocations, in these deformation processes has been well studied in solid state physics [84].

Glasses are, for all practical purposes, solids. At first sight at least, such mechanical rigidity is mysterious. There is no breaking of translational symmetry at the glass transition: a glass is amorphous and does not have a lattice. Whence therefore the rigidity? The answer lies in the fact that long-range motion is arrested in a glass. Thus, all but the smallest displacements $\vec{u}$ will give rise to states that are inaccessible from the original state by thermal fluctuations; hence the possibility of $\nabla \vec{u}(\vec{r})$ and stored elastic energy.

Recalling that the inelastic deformation of crystals beyond the yield stress occurs via defects, one might expect that corresponding phenomena should be absent in glasses: without a lattice, defects such as dislocations simply cannot be defined. Yet, many vitrified systems, most notably metallic glasses, do show different kinds of inelastic flows beyond a yield stress [85]. The mechanisms for such flows in glassy materials are still not completely understood. One idea is that they occur via 'shear transformation zones' (STZs).[25] An STZ is a localised region (atomic or molecular clusters) that has undergone irreversible rearrangement due to applied shear. The idea, which was developed from that of various 'flow defects' introduced earlier by Turnbull, Cohen, Spaepan, Argon, and others, has now been formalised in a theoretical structure capable of making predictions of the mechanical behaviour of metallic glasses, including the ubiquitous occurrence of shear bands (inhomogeneous shearing localised in conspicuous bands in electron micrographs). For a recent review, see Falk and Maloney [86].

---

[25] In the context of this chapter, which emphasises the use of colloids as model systems, it is interesting to note that one of the originators of the STZ idea [85] imaged bubble rafts to study deformation and flow in metallic glasses.

Yielding and flow of colloidal solids

The application of the 'colloids as atoms' paradigm to the study of solid rheology is one of the main growth areas in colloid physics today. Here we introduce the experimental rheology of colloidal crystals, hard-sphere glasses, and attractive colloidal solids as well as a number of theoretical approaches for understanding the data.

Colloidal crystals

There is great potential in studying the deformation, yielding, and flow of colloidal crystals to model corresponding phenomena in atomic and molecular crystals. Proof-of-principle experiments already exist. We mention just one example. Schall *et al.* [87] have demonstrated how to use a combination of laser diffraction microscopy (LDM) and confocal imaging to study dislocations formed by silica colloids sedimenting onto a template that has a lattice parameter different from that adopted by bulk colloidal crystals. The resulting 'misfit' dislocations induce lattice strains that show up as dark lines in LDM, whereas details on the single-particle level are revealed using confocal imaging. Perhaps surprisingly, a continuum approach was able to describe the statics of dislocation behaviour down to the length scale of a few lattice vectors. This and other recent experiments suggest that a host of new discoveries in this area can be anticipated in the next decade.

Hard-sphere glasses

Figure 1.15 shows the measured flow curves, $\sigma(\dot{\gamma})$, of concentrated HS-like PMMA suspensions in the range $0.256 < \phi < 0.63$ [88]. At the lower end of this range, the suspensions flow as Newtonian fluids (slope = 1 on this double-logarithmic plot), while by $\phi \approx 0.4$ the suspensions are shear-thinning fluids, displaying low- and high-$\dot{\gamma}$ regions in their

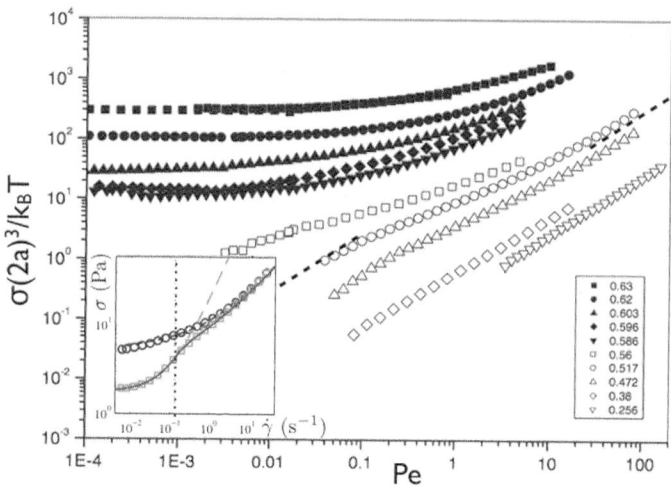

Fig. 1.15 Flow curves for concentrated HS suspensions. Taken from Petekidis *et al.* [88] with permission. Inset: the importance of boundary conditions, showing data taken with roughened cone and plate (∘) and with a roughened cone and smooth plate (□); full slip occurs in the region to the left of the vertical dotted line. The dashed and full lines represent fits to theory, cf. Ballesta *et al.* [89], from where the inset is reproduced with permission.

flow curves. Finally, a low-shear plateau in the flow curve emerges at $\phi \approx 58\%$, which coincides with where $f(q, \infty)$ first becomes finite in DLS measurements. Colloidal glasses possess rigidity against shear. The flow curves at $\phi \gtrsim 0.58$ can be fitted to a Herschel-Bulkley (HB) form for a yield-stress fluid:

$$\sigma = \sigma_Y + A\dot{\gamma}^n , \qquad (1.38)$$

where σ_Y is the yield stress and $n \approx 0.5$.

The flow curves in the main part of Figure 1.15 are measured under no-slip boundary conditions at all surfaces in a cone-plate geometry. The effect of slip is shown in the inset, where one of the data sets is directly comparable to those in the main figure (∘), while a more complex flow curve (□) was obtained with slip boundary condition at the plate [89]. Direct imaging showed that slip occurred between the first layer of particles and a smooth wall; the wall stress, σ_w, is related to the slip velocity, v_s, by

$$\sigma_w = \sigma_s + \beta v_s, \qquad (1.39)$$

where σ_s is some threshold slip stress and β is a ϕ-dependent parameter. Thus, the low-$\dot{\gamma}$ plateau in the data set with slip in the inset to Figure 1.15 is a measure of σ_s, and not of the yield stress of the bulk colloidal glass.

The flow curves in Figure 1.15 were all determined using a cone-plate geometry. While the overall bulk flow curves take on a simple form, equation (1.38), the flow inside the cone-plate geometry is complex [90]. Below a threshold shear rate, the flow profile becomes inhomogeneous, with most the shear becoming increasingly localised near the top and/or bottom surface(s) as $\dot{\gamma}$ decreases. The observed 'banded' flow profiles and the ϕ-dependent threshold shear rates for shear banding are consistent with a model invoking small concentration gradients in the system and a rather generic mechanism for coupling between shear and concentration [91].

In what we may call the 'central dogma' of conventional suspension rheology, the flow in various complex geometries can be calculated once the bulk flow curve (or, equivalently, the constitutive relation) has been determined by rheometry. For the HB fluid, equation (1.38), plug flow is predicted in a pipe. Since the stress in this geometry increases from the centre to the wall, the fluid yields beyond a boundary surface at which $\sigma = \sigma_Y$. Inside this boundary, the unyielded fluid moves as a solid 'plug'. The size of the plug decreases with applied pressure until the whole cross section is yielded to give Poiseuille flow. Interestingly, such a scenario is *not* observed for the flow of concentrated PMMA colloids in microcapillaries [92] even though their bulk flow curves indeed have the HB form. Instead, the thickness of the yielded layer appears independent of the flow rate, as also observed for the chute flow of dry grains in two dimensions. Moreover, no theory exists to explain the oscillatory instabilities observed in such micro-channel flows of concentrated HS suspensions [93]. In other words, the bulk flow curve, equation (1.38), does *not* tell the full story in the rheological behaviour of even such a

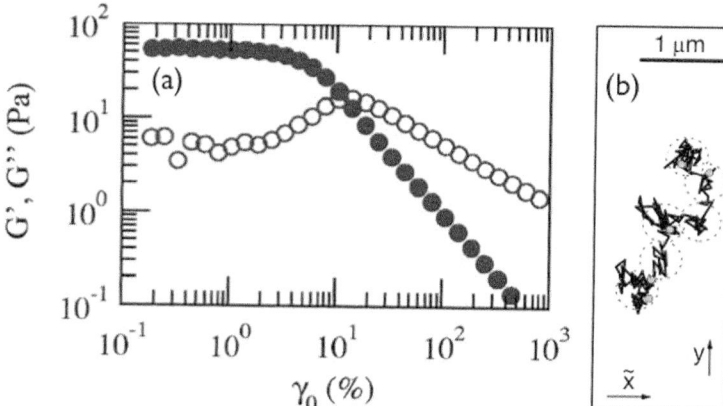

Fig. 1.16 (a) Yielding of HS-like PMMA colloids ($a = 130$nm, $\phi = 0.60$) observed in oscillatory strain experiments at $\omega = 1$rad/s. The real and imaginary parts of the complex shear moduli, G' (•) and G'' (o), are plotted as a function of strain amplitude γ_0. Taken from Pham et al. [94] with permission. (b) Shear-induced cage breaking observed by tracking a particle for 800s in a $\phi \approx 0.62$ PMMA colloidal glass ($a = 850$nm) subjected to simple shear at Pe ≈ 0.03 along the x direction (i.e. $\vec{v}$ is along x), with y being the vorticity ($\nabla \times \vec{v}$) direction. Dotted circles mark 'rattling' in successive cages. Taken from Besseling et al. [95]. Copyright (2007) by The American Physical Society; used with permission.

'simple' system as quasi-monodisperse HS — much complexity has been uncovered when the flow is imaged at the single-particle level [12]. We expect that such high-resolution imaging of colloidal flows will continue to deliver new results not trivially predictable from the flow curves of other concentrated systems.

Determining the flow curve by applying (say) a steady-state shear rate and measuring the shear stress is only one of a number of standard rheometric measurements that can be performed. In another standard measurement, a sample's response to an applied sinusoidal perturbation is determined. Consider for concreteness the application of a sinusoidal strain, $\gamma(t) = \gamma_0 e^{i\omega t}$. An elastic solid will respond with an in-phase stress, $\sigma(t) = \sigma_0 e^{i\omega t} = G\gamma(t)$. On the other hand, a Newtonian (viscous) liquid will respond with an in-phase strain $rate$, $\dot{\gamma}(t) = i\omega\gamma(t)$, so that the (measured) stress $\sigma(t) = \eta\dot{\gamma} = i(\omega\eta)\gamma(t)$ will now be $\pi/2$ out of phase with the applied strain. For a general viscoelastic material, the measured stress is at a phase angle between 0 and $\pi/2$ relative to the applied strain. In this case, the response can be captured using a complex shear modulus, $G^* = G' + iG''$, such that:

$$\sigma = G^*\gamma = G'\gamma + G''\omega^{-1}\dot{\gamma}. \tag{1.40}$$

The storage modulus G' measures the elastic response of the system, while the viscous or loss modulus G'' characterises the viscous response. $G''\omega^{-1}$ has the units of viscosity, cf. equation (1.31). The real and imaginary parts of any response function such as G^* are related (because of general arguments based on causality) by the so-called Kramers-Kronig relations [84], so that knowledge of one of them over a sufficiently large frequency window will enable the other to be calculated.

In a typical 'strain sweep' oscillatory experiment, a sinusoidal strain is applied at a fixed frequency ($\omega/2\pi$), and G' and G'' are measured as a function of the applied strain amplitude. Provided we stay in the linear regime, the moduli are independent of the strain amplitude, γ_0. At low ϕ, we find that $G'' > G'$, and the suspension is liquid like. But G' increases faster with ϕ than G'', and the point at which the two moduli cross can be identified as the glass transition [96]. In the linear regime above ϕ_g, $G' > G''$ and the system is solid-like. As γ_0 increases beyond a critical value, G' starts to decrease, until eventually it drops below G'', Figure 1.16(a) [94]. This point at which the system fluidizes as far as oscillatory rheology is concerned can be identified as the yield point. At $\phi = 0.58$ in a PMMA suspension, where this system becomes a glass, the strain amplitude at yielding is $\gtrsim 0.10$. This should be compared to the MSD, $\approx 0.3a$, measured for this system at the glass transition $\phi \approx 0.58$, Figure 1.11: a glass forms when the 'cage' is about 15% of the interparticle spacing. This length scale be used to understand the yield strain of a HS glass in much the same way the Lindemann length is used to understand the melting of crystals: at a strain amplitude of $\gamma_0 \approx 0.3a/2a = 0.15$, we expect the cages to break and therefore the glass to yield. Since the cage size decreases with ϕ (reaching 0 at $\phi \approx 0.64$), a measured yield strain of just over 10% at $\phi = 0.6$ is not unreasonable. Hard-sphere glasses yield under shear by cage breaking, Figure 1.16(b), perhaps utilising processes akin to the 'shear transformation zones' first proposed to explain the behaviour of metallic glasses [97].

Attractive colloidal solids

The presence of short-range attraction significantly complicates the rheology. While HS glasses yield in a single step, by the breaking of cages, Figure 1.16(a), there is evidence that attractive colloidal glasses yield via a more drawn-out process [94]. This process starts with the breaking of nearest-neighbour bonds. Since these 'bonds' are short-range, the beginning of the yielding process occurs at significantly lower strain amplitudes, given roughly by the range of the inter-particle potential. At this strain amplitude, G' starts to decrease with strain amplitude, but does not cross G'' until an amplitude of many tens of percent. The single-particle-level mechanisms of this drawn-out process of yielding remains to be elucidated.

The non-linear rheology of lower-ϕ attractive colloidal gels has also been investigated [98]. There are similarities and differences with high-ϕ attractive glasses. The initial breaking of bonds is still present and appears to lead to the fragmentation into clusters of particles, which then deform and fragment further at higher strain amplitudes. Single-particle-level investigations are yet to be carried out.

Theoretical approaches

The rheology of concentrated colloidal suspensions poses significant challenges to first-principles theory. Compared to the level of understanding now available for the rheology of polymer melts, the rheology of concentrated suspensions lags considerably behind. In the case of polymer

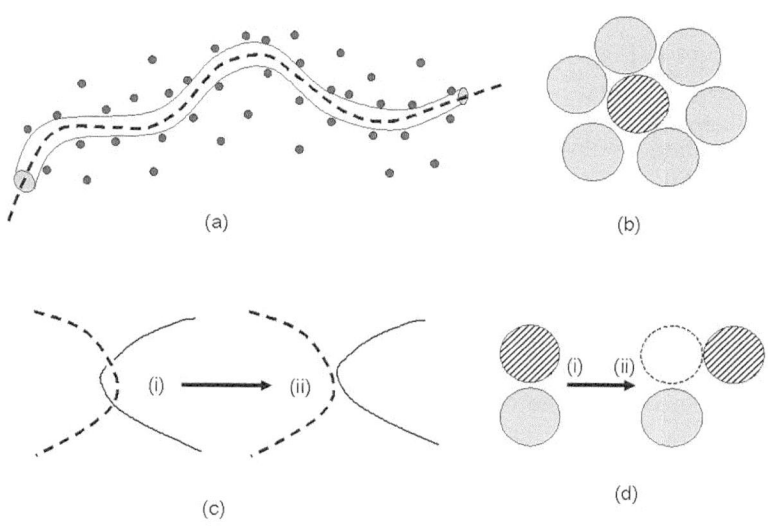

Fig. 1.17 Comparing polymer melt rheology and colloid rheology. (a) In a polymer melt, a typical chain (dashed curve) is constrained by $\sim 10^3$ other chains, here represented by small circles. This gives rise to the mean-field concept of a tube in which the chain has to move. (b) In a concentrated colloidal suspension, a typical particle (hatched) is surrounded (in 3D) by ~ 10 neighbours. This number is too small for mean-field averaging to be meaningful. (c) Large deformations in polymer melts, such as the process (i) → (ii), involves breaking covalent bonds, and so do not ordinarily occur. (d) There are no covalent constraints on order unity deformations, such as (i) → (ii), in a colloidal suspension. Taken from Poon [99], copyright Springer-Verlag Berlin Heidelberg (2010), used with permission from the publisher.

melts, it is now possible to predict with some confidence flow fields in complex geometries starting from a knowledge of molecular properties. Such a level of fundamental understanding is not yet available for concentrated colloids. The reasons for this are as follows, see Figure 1.17.

In a polymer melt, each chain moves under the topological constraints imposed by many other chains. The number of these constraints, typically of order 10^3, is sufficiently large that a mean-field picture, the so-called 'tube model' [31], can be successfully applied. Moreover, the topological entanglement between chains means that the breaking of covalent bonds is needed to impose large deformations, so that strains often remain small ($\ll 1$). In contrast, the maximum number of neighbours in a (monodisperse) suspension of spheres is of order 10^1, so that 'mean-field' averaging of nearest-neighbour 'cages' will not work, and local processes showing large spatio-temporal heterogeneities are expected to be important. Moreover, no topological constraints prevent the occurrence of strains of order unity or higher, so that very large deformations are routinely encountered. These two characteristics alone render suspension rheology much more difficult. In addition, concentrated suspensions are often 'stuck' in non-ergodic states (Section 1.4), while in a monodisperse system or in a mixture with carefully chosen size ratios, highly-ordered (crystalline) states can occur. In either case, any flow necessarily entails non-linearity (yielding, etc.). Finally, a suspension is necessarily a multiphase system, so that the relative flow of particles and solvent can, and often does, become important – a complication that does not arise in polymer melts.

Nevertheless, significant progress has been made. In terms of producing quantitative predictions, MCT has been successfully extended to deal with the rheology of concentrated colloids [100]. One of the key insights that has enabled this extension of the scope of the theory to rheological properties is that density fluctuations are advected by shear. A fluctuation mode propagating with wave vector $q_0 = 2\pi/\lambda_x$ at time zero, Figure 1.18, is advected to a mode with wave vector $q(t) = 2\pi/\lambda = q_0\sqrt{1 + (\dot\gamma t)^2} > q_0$ at time t. A mode of density fluctuation at larger wave vector is relaxed by smaller movements of the particles than would be the case in the absence of advection. This extension of MCT has given first-principles predictions of rheological properties of concentrated HS suspensions at and around the glass transition concentration ϕ_g. In particular, it predicts that a dynamic yield stress ($\sigma \to \sigma_Y > 0$ as $\dot\gamma \to 0^+$) emerges discontinuously as ϕ increases beyond ϕ_g, which is consistent with what has been observed in experiments, Figure 1.15. A schematic version of the theory that is more amenable to analytical calculations has since been implemented and can be used to predict a broader range of rheological behaviour, including non-steady shear [101]. Whether either of these versions of extended MCT can deal successfully with the rheology of attraction-induced arrested states remains to be explored.

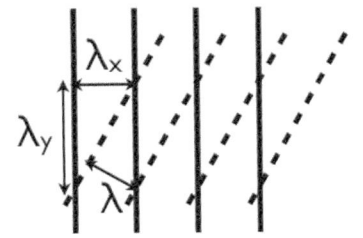

Fig. 1.18 A mode of density fluctuation travelling in the x direction with wavelength λ_x at time zero is advected to a mode with wavelength components (λ_x, λ_y) in the x, y directions at time t, with $\lambda_x/\lambda_y = \dot\gamma t$. The new wavelength can be obtained by simple geometry: $\lambda^{-1} = \sqrt{\lambda_x^{-2} + \lambda_y^{-2}}$. Redrawn after Fuchs and Cates [100].

A different theoretical approach, soft glassy rheology (SGR), is less microscopic than MCT. Instead, it focusses on generic coarse-grained structural 'elements' that can be described by an elastic strain variable. To quote from one of the founders of the theory, these elements 'should not be thought of as sharply defined physical entities, but rather as somewhat diffuse {blobs} of material' [102]. The theory starts from a collection of these 'elements' trapped in and hopping between different energy states; such a collection is known to be glassy if the density of states of these traps has an exponential tail. Deformation and flow are incorporated by allowing elements to store elastic energy locally and, when this becomes high enough, to yield and flow. The interaction between different elements is described by a variable x that plays the role of an effective temperature. Compared to MCT, SGR does *not* predict the discontinuous appearance of a yield stress, and is likely much more suitable for explaining the yielding of 'messy' systems in which arrest is 'smeared out' by polydispersity or other features. On the other hand, SGR gives a natural framework for describing the effect of ageing on rheology [103].

A third approach occupies the 'theory space' between MCT and SGR. This non-linear Langevin equation (NLE) formalism starts from 'naïve MCT' to derive the entropic potential well, $F_0(r)$, that traps each HS particle by its nearest 'cage' of neighbours. The glass transition is then a localisation transition: as ϕ increases, the entropic barrier to be surmounted for escape to occur increases rapidly, until particles are trapped for exponentially long times. The entropic potential, $F(r)$, for a particle under an externally-applied shear stress is given by $F(r) = F_0(r) - fr$, where f is the force on the particle due to the applied shear stress. This has the effect of lowering the barrier height, so that a high enough f leads to yielding [104]. The NLE approach predicts correctly a non-trivial scaling observed between the shear-induced structural relaxation time in a HS colloidal glass and the shear rate [95].

1.6 'A mysterious colloid': Proteins and patchiness

> 'The proposal of the subject for this discussion is itself a remarkable thing and a symbol of the spirit of this meeting. A few years ago the proposal would have looked preposterous. Proteins were known as a mysterious sort of colloids, the molecules of which eluded our search. What is it then that has happened in these years? Why is the most distinguished scientific society of this country inviting a discussion on the protein molecule?'
> Theodor Svedberg, Opening Address at *A Discussion on the Protein Molecule*, Royal Society, 17 November 1938 [105].

The belief prevalent until the 1930s that proteins were 'a mysterious sort of colloids' should be put in the context of the origins of the term 'colloid' (see the definition from the *Oxford English Dictionary* given at the beginning of Section 1). It was thought until less than a century ago that proteins were ill-defined amorphous aggregates. Interestingly, it was the same kind of approach that first 'tamed' colloids for physics, the study of sedimentation equilibrium, that helped convince the scientific world that proteins were in fact well-defined molecules, albeit with hitherto undreamt of molecular weights.

Theodor Svedberg invented the ultracentrifuge, and used it to study the sedimentation equilibrium of protein solutions. He showed that any particular purified protein preparation consisted of entities with well-defined molecular weight – these are *bona fide* molecules, albeit rather big ones.[26] From then on, there arose a tradition of understanding proteins as colloids (although not known in these terms): John Gamble ("Jack") Kirkwood, for example, contributed a chapter to a volume, which may still profitably be consulted by those who want to understand proteins as colloids today [107].[27] This approach, which neglects atomic-level details, became somewhat eclipsed by the molecular biology revolution ushered in by the discovery of the DNA double helix in 1953, but is now enjoying something of a renaissance, because what colloid physics may be able to give to protein science and what protein science can give to colloid physics.

[26] A highly readable history of the scientific study of proteins is available [106].

[27] The Kirkwood-Buff solution theory underlies the coarse-graining approach to colloids explained in the last part of Section 1 of this chapter.

Proteins as colloids

From the point of view of soft matter physics, a globular protein molecule displays characteristics of both polymers and colloids. Chemically, it is a more or less random copolymer of the twenty of so naturally-occurring amino acids (or 'residues'). But its physical dimensions (e.g. hydrodynamic radius measured from sedimentation in an ultracentrifuge, r_S) are much smaller than what is typically expected of such macromolecules, Figure 1.19. For a linear random-coil polymer in theta solvents, we expect its radius, r, to scale with molecular weight, M, according to

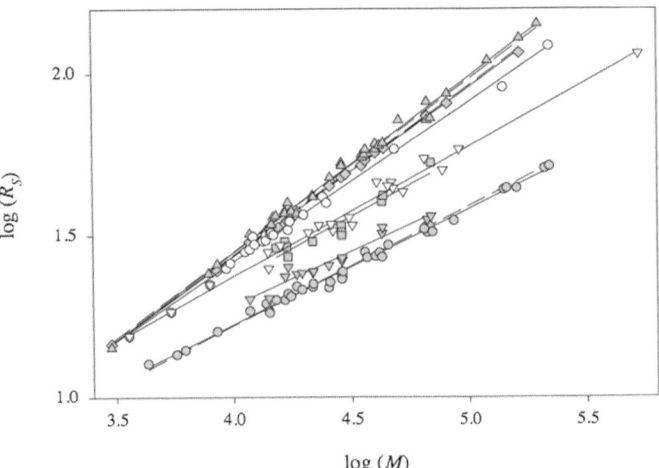

Fig. 1.19 A log-log plot of the hydrodynamic radii of various classes of proteins (R_S measured in Å) versus their molecular weights (M measured in daltons), including globular proteins (●), proteins unfolded by the addition of guanadinium chloride (▲), and natively disordered proteins (○); the fitted scaling laws were reported to have powers of 0.357±0.005, 0.543±0.007, and 0.493±0.008 respectively. Taken from Uversky [110], copyright FEBS and used with permission from John Wiley & Sons, Inc. See original for the meaning of the other symbols.

$r \sim M^\alpha$ with $\alpha = \frac{1}{2}$. This is consistent with data for so-called natively-disordered proteins ($\alpha = 0.49$) and globular proteins that have been denatured chemically ($\alpha = 0.54$). Globular proteins in their native (folded) state give $\alpha = 0.36$, close to the $\alpha = \frac{1}{3}$ expected for compact solid objects. A native globular protein is folded tightly to pack a substantial fraction of its hydrophobic residues to the inside of the molecule.

Thus, for some purposes, a globular protein can be treated in a coarse-grained perspective as a solid nano-colloid.[28] In particular, a kind of law of extended corresponding states applies for the solution phase diagram of different globular proteins. Additives such as salt or non-adsorbing polymer can be used to induce attraction between protein molecules (by screening the electrostatic repulsion or inducing depletion attraction). If the attraction induced is measured by the second virial coefficient, the crystallisation boundary for different crystallisable globular proteins collapse onto a near-universal master curve [108]. Moreover, this curve maps well onto the corresponding fluid-crystal coexistence boundary of a mixture of spherical colloids and non-adsorbing polymers in the same representation [109]. The mapping in fact extends to the gas-liquid phase boundary 'buried' in the fluid-crystal coexistence region, cf. Figure 1.8(c), which determines the onset of non-equilibrium effects such as gelation, see Figure 1.13.

Protein-inspired colloids

While there indeed is a universal crystallisation boundary for many globular proteins that do crystallise and this boundary maps onto that of a synthetic colloid-polymer mixture, it is well known that many proteins are reluctant to crystallise and need to be delicately 'coaxed' to give good crystals, or indeed any crystals at all. It has been pointed out that this may be the result of evolution. Perhaps proteins have evolved precisely *not* to turn into crystals in a cellular environment that, at first

[28] Interestingly, Sutherland [6] published Equation (1.2) before Einstein in a paper on the diffusion of the globular protein albumin. See footnote 2.

sight, should be perfect for crystallisation: crowded (and therefore high concentrations are possible), high ionic strength (so that charges are screened), and many other macromolecules are present (so that depletion is a real possibility). So a sufficiently negative B_2 is a necessary but not sufficient condition for the crystallisation of globular proteins.

Many lines of evidence suggest that the missing ingredient is the relative orientation of the molecules, because proteins are *patchy* particles [111]. While a globular protein folds to bury many of its hydrophobic residues in a compact inner core, a significant number of such residues are still exposed to an aqueous environment on the surface. These can be considered 'sticky patches'. When mutually accessible (e.g. not hidden deep inside a concave region), such patches between neighbouring molecules potentially become energy-lowering contacts in the crystal, so that optimising such contacts by specific point mutations of surface residues improves crystallisability [112].

The realisation that proteins are patchy colloids has in part inspired theoretical and simulation studies that predict novel equilibrium phases and modes of kinetic arrest in systems of such particles [111, 113], including a gel to glass transition [114]. These results have in turn stimulated attempts to prepare synthetic patchy colloids. The synthesis of Janus particles, 'two-faced' colloids each with a front and back, is now almost routine. The various strategies for such synthesis have recently been reviewed [115]. Janus particles placed in suitable solvents that can act as 'fuel' can exhibit self-propulsion, and therefore behave as 'active colloids' (see Section 1.7 for further details). Tetrahedral patches can perhaps be obtained by relying on the arrangement of defects on shells of nematic liquid crystals [116].

Many proteins function as biological catalysts, or enzymes. The simplest explanation of enzymatic action pictures the enzymes and their substrates as 'locks' and 'keys' fitting into each other by virtue of matching in shape.[29] This has inspired the creation of lock-and-key colloids [36], where the selectivity of the 'lock' for the shape of the 'key' is enhanced by taking advantage of the dependence of the depletion attraction on geometry, Figure 1.10(c).

[29] As has been repeatedly pointed out, the chief defect of this picture is that it is static, neglecting the essential role played by dynamics in enzymatic action. For a clear example of the role of dynamics in the phenomenon of allostery – the cooperative action of multiple sites on the same enzyme – see Hawkins and McLeish [117].

1.7 Conclusion: Looking into the future

'If you can look into the seeds of time,
And say which grain will grow and which will not,
Speak then to me, who neither beg nor fear
Your favours nor your hate.'
   William Shakespeare, *Macbeth* 1.3.158-161

The rigorous study of colloids by physicists started with Einstein and Perrin's work on sedimentation equilibrium and diffusion. Their work helped establish the granularity of matter. While physicists had not lost interest in colloids since that time, the modern resurgence of interest can perhaps be dated to the confirmation of the hard-sphere phase diagram predicted by simulations, Figure 1.4 from the experiments of Pusey and van Megen [16], which demonstrated that well-characterised colloidal dispersions could function as 'wet models' for studying basic questions in condensed matter and statistical physics. Our aim in this chapter has been to introduce the reader to colloid physics under this perspective. We have concentrated on hard-sphere-like suspensions, because the hard-sphere system is an important reference for understanding many-body phenomena, where excluded-volume effects are ubiquitous.

Pusey's and van Megen's experiments were made possible by the development of sterically-stabilised PMMA particles that, when dispersed in hydrocarbon solvents, often behave as nearly perfect hard spheres. They were used to settle a long-standing question in many-body physics: whether hard spheres crystallise. The very experiments that settled this question also revealed new physics, viz. the possible existence of a hard-sphere glass transition. Since then, the study of kinetic arrest in hard-sphere-like colloids has proven to be a fruitful area of interaction with theory, specifically, the MCT approach to the glass transition.

This brief recapitulation of some of the ground we have covered provides us with a typology to think about how innovations in the subject may arise. First, and most fundamentally, development of new experimental systems, especially those susceptible to detailed characterisation, almost always stimulates significant progress. These systems may make available new particle shapes, or explore hitherto inaccessible interactions given by either new particle properties and/or new ways in which the solvent can mediate interactions. Secondly, significant developments may arise when new ways are identified to use colloids to study fundamental questions in statistical physics. With this typology in mind, we close the chapter by surveying a number of areas in which significant new developments may be expected in the next decade.

New systems

In terms of developing new systems, we have already reviewed a number of emerging innovations in particle synthesis throughout this chapter: colloidal polyhedra and patchy particles are two outstanding examples. Here we point out instead that the vast majority of suspensions that have been studied in detail consists of dispersions of particles in simple liquids. These liquids are homogeneous on the colloidal length scale and flow as Newtonian fluids. Apart from providing a background with defined viscosity, density, dielectric constant, and refractive index, the dispersing medium 'does not do anything interesting'. Recently there is growing interest in studying dispersions of particles in which the solvent is itself complex, either because it displays complex rheology or because it can undergo phase transitions on its own.

Consider first particles in non-Newtonian dispersing media. One situation in which this may occur is a colloid-polymer mixture of the kind discussed in Section 1.3. When the polymer:colloid size ratio ξ is small ($\ll 1$), the polymer concentrations at which phase separation and kinetic arrest first occur are significantly below the overlap concentration

c_p^*. On the other hand, when $\xi \to 1$ and beyond, these phenomena occur at $c_p/c_p^* \sim 1$, the so-called 'semi-dilute' regime where coils begin to entangle [31]. In these cases, if the polymer has high enough molecular weight, the solution is expected to be viscoelastic. Another possibility is a dispersion of particles in a solution of surfactant worm-like micelles. Simulations [118] suggest that the stresses generated by shearing a single-phase mixture of colloids with semi-dilute polymers and a worm-like micellar solution will induce particle aggregation. In experiments [119], dispersions of particles in worm-like micelles at $\phi < 0.01$ can order into two-dimensional crystals with close packing along the flow direction, Figure 1.20(a).

Moving to the dispersion of particles in phase separating liquids, one example is the dispersion of colloids into a partially miscible binary liquid mixture. Suppose the particle surfaces are 'tuned' chemically so that they are equally wetting to both components of the mixture. Under conditions at which the two liquids are well mixed, the dispersion behave like an 'ordinary colloid'. A change of conditions (typically a sudden jump in the temperature) is used to bring the mixture into the two-phase region of the phase diagram where it phase separates with the mechanism of spinodal decomposition, so that a bicontinuous pattern develops, with a complex interface separating two phases rich in one or the other component. It is energetically favourable for a particle to migrate to this interface: the presence of a particle of radius a lowers the interfacial energy by $\Delta E \sim \pi a^2 \Gamma$, where Γ is the interfacial tension. Using $\Gamma \gtrsim 10^{-2} \mathrm{J/m^2}$ and a radius of $a \sim 10 \mathrm{nm}$ as the lower bound of the colloidal length scale, we obtain $\Delta E \gtrsim 10^3 k_B T$. Thus, any colloidal particle arriving at the interface would essentially stay there forever.[30] Computer simulations [122] and then experiments [123] show that eventually, at high enough particle concentration, the shrinking interface is jammed with particles and the spinodal texture stops coarsening. A bicontinuous interfacially jammed emulsion gel (bijel) results, where the typical length scale at arrest is linearly proportional to the initial concentration of particles.

A third example of particles in complex solvents is the dispersal of colloids into liquid crystals (LCs). The particles typically anchor the LC molecules either in parallel or perpendicular to their surfaces, so that if the LC is in one of its mesophases, say, the nematic, in which the molecules in the bulk want to be aligned more or less to a single direction (the director), a particle can only be accommodated by the generation of defects (disclinations) [124]. The energetic cost can be lowered in a multi-particle dispersion if the particles are close enough together to 'share' the elastic distortion fields associated with their attendant defects, or even to share the defects. Both mechanisms can give rise to chains of particles at low ϕ [125, 126]. At higher concentrations, nematic 'foams' [127] or arrested states with percolated entangled disclinations [120] can be prepared, Figure 1.20(b).

It is interesting to compare the two classes of colloids in phase separating liquids we have reviewed. A striking difference is that a phase

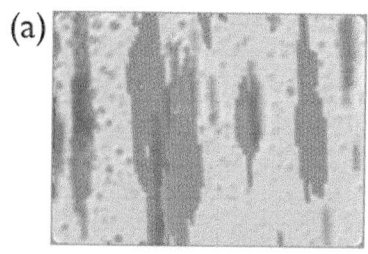

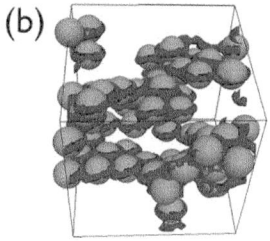

Fig. 1.20 Examples of new systems: (a) Self-assembled 2D colloidal crystals from shearing particles dispersed in a viscoelastic solution of worm-like micelles; reprinted with permission from Pasquino et al. [119]; copyright (2010) American Chemical Society. (b) Computer simulations of particles dispersed in a nematic liquid crystal ($\phi = 0.16$). The particles are entangled by a system of defects that percolate. From Wood et al. [120]. Reprinted with permission from AAAS.

[30] This is the mechanism for particle-stabilised, or Pickering, emulsions, which are enjoying a renaissance of interest at present [121]. The surface tensions encountered in bijel synthesis are lower than those in Pickering emulsions, since the experiments operate near the spinodal line.

separating binary liquid is described by a conserved order parameter, while the nematic order parameter is non-conserved. On the other hand, kinetics play an important role in both cases. Structural formation depends sensitively on the interplay between particle dynamics and the kinetics of phase separation. There are therefore multiple possibilities for generating novelties and for fine tuning the physical properties of such systems.

Modelling new phenomena

Another generic direction for new developments is the use of colloids to model new areas of fundamental many-body phenomena. We briefly sketch out three topics under this heading. First, because colloids inhabit the 'middle world' [4] between atoms and bulk materials, they are uniquely placed for mediating the atomic world to macroscopic measurements. Perrin and Einstein exploited this to prove the existence of the atomic world in the first place. Now, there is increasing interest in using colloids to probe 'small-world thermodynamics'. This operates in a regime where the certainties resulting from overwhelming numbers ($N_A = 6 \cdot 10^{23}$) in macroscopic systems can no longer be relied upon. A recent example in which this 'limit of small numbers' can be seen very clearly is the creation by Blickle and Bechinger [128] of a micrometer-sized Stirling engine in which a single colloidal particle confined to the harmonic potential of a laser trap acts as the 'working gas'. In this situation, the work done by the particle in each cycle of the heat engine (where the high temperature heat bath was provided by local laser heating) is of the same order of magnitude as the fluctuations. There is little doubt that such experiments will reveal a host of novel phenomena and pose significant challenges to theory in the next decade.

A second development in using colloids to model new statistical physics can be placed in the context of a long-standing trend in moving from equilibrium to phenomena that are progressively further away from equilibrium: from the process of returning to equilibrium (phase transition kinetics) through dynamical arrest (glasses and gels) to systems driven out of equilibrium by external forces (such as shear). Indeed, this journey from the relative *terra firma* of equilibrium properties into the increasingly uncertain territory of non-equilibrium has been the *leitmotif* of this chapter, as evidenced by the key diagrams, moving from Figures 1.4 and 1.8 through Figures 1.12 and 1.13 to Figure 1.16. The body of work represented by this sequence of figures, whether explicitly cited here or not, have provided some of the most significant data for stimulating and validating theoretical developments in non-equilibrium statistical physics. In one of the latest developments along this line, attention is now being directed to colloids that are *intrinsically non-equilibrium*, because the particles themselves are non-equilibrium entities.

In these 'active colloids' [129], each particle transduces free energy from its environment to engage in self-propelled motility and/or growth. The most obvious example is a natural one: a suspension of swimming bacteria [130] (the overwhelmingly majority of which are in the colloidal length scale) that are also growing and undergoing cell division. The synthetic analogue of self-motile particles have also become available [131, 132]. A single self-propelled particle whose orientation fluctuates due to Brownian motion performs a random walk that can be described by an effective diffusion coefficient [133]. A dilute suspension of such particles displays barometric sedimentation equilibrium in a gravitational field [134]. A suspension of active particles can be driven to aggregate and 'phase separate' by non-adsorbing polymers [135]. The activity significantly shifts the phase boundary to higher polymer concentrations (or, equivalently, stronger depletion attraction), while pre-transition finite clusters of particles show uni-directional rotation. There is no doubt that the physics of such active particles will be a fascinating field of study in its own right, but studying such suspensions will also generate many results of interest for fundamental statistical mechanics as well as materials science.

A final area under the heading of 'colloidal models' relies on the possibility of 'tuning' suspensions so that they bridge two ends of a spectrum of fascinating physics. In a colloidal suspension, the physics is dominated by Brownian motion. The Brownian relaxation time, equation (1.8), scales as the $a^3\eta$. Brownian motion therefore plays a rapidly diminishing role as a increases or when particles are dispersed in very high viscosity medium. In particular, under shear, the Péclet number also scales as $a^3\eta$, equation (1.37), and Brownian relaxation becomes irrelevant at Pe $\gg$ 1. At this non-Brownian limit, it is fruitful to draw an analogy with the behaviour of (dry) granular materials. It has been shown, for example, that a unified description is possible for the role of dilatancy in the flow of both dry grains and non-Brownian colloids [136]. Due to the relative ease in which concentrated suspensions can be imaged at the single-particle level, the possible analogy between granular materials and colloids means that using the latter to model the former should be a very fruitful avenue for exploration. Recent work on the role of 'bridges' in colloidal and granular packings [137] is one example.

Niels Bohr once famously said that 'prediction is always very difficult, especially about the future'. In the final section of this chapter, we have suggested a number of directions in which significant new developments may take place in colloid physics in the next ten years or so. Irrespective of whether (to borrow Shakespeare's almost colloidal metaphor) these 'grains will grow' or not, it is certain that the subject shows no sign of slowing down 100 years after Perrin and Einstein gave it its first experimental and theoretical tools.

Acknowledgements

We thank the ESPCI, Paris for hospitality during which most of this chapter was written, Mike Cates for many helpful discussions on theoretical aspects, Eric Weeks, Patrick Warren, Willem Kegel, and Mark Haw for helpful comments, the EPSRC for funding (EP/D071070/1), and many authors and publishers for permission to reproduce figures.

References

[1] Einstein, A. (1926). *Albert Einstein: Investigations on the Theory of the Brownian Movement.* Dover, New York.
[2] Perrin, J. (1923). *Les atomes.* Constable, London.
[3] Pusey, P. N. (1991). In *Liquids, freezing and the glass transition* (ed. J. P. Hansen, D. Levesque, and J. Zinn-Justin), pp. 765–942. Elsevier, Amsterdam.
[4] Haw, M. D. (2006). *Middle World.* Macmillan, London.
[5] Brown, R. (1866). In *The Miscellaneous Botanical Works of Robert Brown, Vol. 1* (ed. J. J. Bennett), pp. 463–486. Ray Society, London.
[6] Sutherland, W. (1905). *Phil. Mag.*, **9**, 781–785.
[7] Israelachvili, J. N. (2011). *Intermolecular and Surface Forces, 3rd edition.* Academic Press, London.
[8] Poon, W. C. K., Weeks, E. R., and Royall, C. P. (2012). *Soft Matter*, **8**, 21–30.
[9] Royall, C. P., Poon, W. C. K., and Weeks, E. R. (2013). *Soft Matter*, **9**, 17–27.
[10] Lidner, P. and Zemb, Th. (ed.) (2002). *Neutron, X-rays and Light: Scattering Methods Applied to Soft Condensed Matter.* North Holland, Amsterdam.
[11] Prasad, V., Semwogerere, D., and Weeks, E. R. (2007). *J. Phys. - Cond. Matter*, **19**, 113102.
[12] Besseling, R., Isa, L., Weeks, E. R. *et al.* (2009). *Adv. Coll. Int. Sci.*, **146**, 1–17.
[13] Hansen, J. P. and MacDonald, I. R. (2006). *Theory of Simple Liquids, 3rd edition.* Academic Press, London.
[14] Percus, J. K. (ed.) (1963). *The Many-Body Problem.* Interscience Publishers, New York.
[15] Hoover, W. G. and Ree, F. H. (1968). *J. Chem. Phys.*, **49**, 3609–3617.
[16] Pusey, P. N. and van Megen, W. (1985). *Nature*, **320**, 340–342.
[17] Bruce, A. D., Jackson, A. N., Ackland, G. J. *et al.* (2000). *Phys. Rev. E*, **61**, 906–919.
[18] Pusey, P. N., van Megen, W., Bartlett, P. *et al.* (1989). *Phys. Rev. Lett.*, **63**, 2753–2756.
[19] Wilding, N. B. and Sollich, P. (2011). *Soft Matter*, **7**, 4472–4484.
[20] Schofield, A. B., Pusey, P. N., and Radcliffe, P. (2005). *Phys. Rev. E*, **72**, 031407.
[21] Fillion, L., Ni, R., Frenkel, D. *et al.* (2011). *J. Chem. Phys.*, **134**, 134901.

[22] Manoharan, V. N., Elsesser, M. T., and Pine, D. J. (2003). *Science*, **301**, 483–487.
[23] Onsager, L. (1949). *Ann. N. Y. Acad. Sci.*, **51**, 627–659.
[24] Solomon, M. J. and Spicer, P. T. (2010). *Soft Matter*, **6**, 1391–1400.
[25] Damasceno, P. F., Engel, M., and Glotzer, S. C. (2012). *ACS Nano*, **6**, 609–614.
[26] Sacanna, S. and Pine, D. J. (2011). *Curr. Opin. Coll. Int. Sci.*, **16**, 96–105.
[27] Bernstein, H. J. and Weisskopf, V. F. (1987). *Am. J. Phys.*, **55**, 974–983.
[28] Widom, B. (1967). *Science*, **157**, 375–382.
[29] Poon, W. C. K. (2002). *J. Phys. Cond. Matter*, **14**, R859–R880.
[30] Lekkerkerker, H. N. W. and Tuinier, R. (2011). *Colloids and the Depletion Interaction*. Springer, Berlin.
[31] Rubinstein, M. and Colby, R. H. (2003). *Polymer Physics*. Oxford University Press, Oxford.
[32] Ilett, S. M., Orrock, A., Poon, W. C. K. et al. (1995). *Phys. Rev. E*, **51**, 1344–1352.
[33] Noro, M. G. and Frenkel, D. (2000). *J. Chem. Phys.*, **113**, 2941–2944.
[34] Miller, M. A. and Frenkel, D. (2004). *J. Chem. Phys.*, **121**, 535–545.
[35] Dijkstra, M. (2002). *Phys. Rev. E*, **66**, 021402.
[36] Sacanna, S., Irvine, W. T. M., Chaikin, P. M. et al. (2010). *Nature*, **464**, 575–578.
[37] Fasolo, M. and Sollich, P. (2005). *J. Chem. Phys.*, **22**, 074904.
[38] Evans, R. M. L., Fairhurst, D. J., and Poon, W. C. K. (1998). *Phys. Rev. Lett.*, **81**, 1326–1329.
[39] Dinsmore, A. D., Warren, P. B., Poon, W. C. K. et al. (1997). *Europhys. Lett.*, **40**, 337–342.
[40] Dinsmore, A. D., Yodh, A. G., and Pine, D. J. (1996). *Nature*, **383**, 239–242.
[41] Buitenhuis, J., Donselaar, L. N., Buining, P. A. et al. (1995). *J. Coll. Int. Sci.*, **175**, 46–56.
[42] Zhao, K. and Mason, T. G. (2007). *Phys. Rev. Lett.*, **99**, 268301.
[43] Grant, M. C. and Russel, W. B. (1993). *Phys. Rev. E*, **47**, 2606–2614.
[44] Koos, E. and Willenbacher, N. (2011). *Science*, **331**, 897–900.
[45] Rankin, P. J., Ginder, J. M., and Klingenberg, D. J. (1998). *Curr. Opin. Coll. Int. Sci.*, **3**, 373–381.
[46] Ebert, F., Dillmann, P., Maret, G. et al. (2009). *Rev. Sci. Inst.*, **80**, 083902.
[47] Yethiraj, A. and van Blaaderen, A. (2003). *Nature*, **421**, 513–517.
[48] Hertlein, C., Helden, L., Gambassi, A. et al. (2007). *Nature*, **451**, 172–175.
[49] Anderson, P. W. (1995). *Science*, **267**, 1615.
[50] Hunter, G. L. and Weeks, E. R. (2012). *Rep. Prog. Phys.*, **75**,

066501.
- [51] Sciortino, F. and Tartaglia, P. (2005). *Adv. Phys.*, **54**, 471–524.
- [52] Zaccarelli, E. (2007). *J. Phys. Cond. Matter*, **19**, 323101.
- [53] van Megen, W., Mortensen, T. C., Williams, S. R. *et al.* (1998). *Phys. Rev. E*, **58**, 6073–6085.
- [54] Berthier, L., Biroli, G., Bouchaud, J.-P. *et al.* (ed.) (2011). *Dynamical Heterogeneities in Glasses, Colloids and Granular Media*. Oxford University Press, Oxford.
- [55] Martinez, V., Bryant, G., and van Megen, W. (2010). *J. Chem. Phys.*, **133**, 114906.
- [56] Weeks, E. R., Crocker, J. C., Levitt, A. C. *et al.* (2000). *Science*, **287**, 727–631.
- [57] Zaccarelli, E., Valeriani, C., Sanz, E. *et al.* (2009). *Phys. Rev. Lett.*, **103**, 135704.
- [58] Zhu, J. X., Li, M., Rogers, R. *et al.* (1997). *Nature*, **387**, 883–885.
- [59] Geszti, T. (1983). *J. Phys. C.: Solid State Phys.*, **16**, 5805–5814.
- [60] Götze, W. (2009). *Complex Dynamics of Glass-Forming Liquids: a Mode-Coupling Theory*. Oxford University Press, Oxford.
- [61] Reichman, D. R. and Charbonneau, P. (2005). *J. Stat. Mech.: Theor. Exp.*, P05013.
- [62] Cates, M. E. (2003). *Ann. Henri Poincaré*, **4 Suppl. 2**, 647–661.
- [63] Bergenholtz, J. and Fuchs, M. (1999). *Phys. Rev. E*, **59**, 5706–5715.
- [64] Fabbian, L., Götze, W., Sciortino, F. *et al.* (1999). *Phys. Rev. E*, **59**, R1347–R1350.
- [65] Pham, K. N., Puertas, A. M., Bergenholtz, J. *et al.* (2002). *Science*, **296**, 104–106.
- [66] Poon, W. C. K. (2004). *MRS Bulletin*, **February**, 96–99.
- [67] Zaccarelli, E. and Poon, W. C. K. (2009). *Proc. Natl. Acad. Sci. USA*, **106**, 15203–15208.
- [68] Poon, W. C. K. and Haw, M. D. (1997). *Adv. Coll. Int. Sci.*, **73**, 71–126.
- [69] Keller, A. (1995). *Faraday Discuss.*, **101**, 1–49.
- [70] Starrs, L., Poon, W. C. K., Hibberd, D. J. *et al.* (2002). *J. Phys. Cond. Matter*, **14**, 2485–2505.
- [71] Buscall, R., Choudhury, T. H., Faers, M. A. *et al.* (2009). *Soft Mat.*, **5**, 1345–1349.
- [72] Sedgwick, H., Kroy, K., Salonen, A. *et al.* (2005). *Eur. Phys. J. E*, **16**, 77–80.
- [73] Carpineti, M. and Giglio, M. (1992). *Phys. Rev. Lett.*, **68**, 3327–3330.
- [74] Pais, A. (1982). *'Subtle is the Lord ...' The Science and the Life of Albert Einstein*. Oxford University Press, Oxford.
- [75] Guyon, É., Hulin, J. P., Petit, L. *et al.* (2001). *Physical Hydrodynamics*. Oxford University Press, Oxford.
- [76] Mewis, J. and Wagner, N. J. (2012). *Colloidal Suspension Rheology*. Cambridge University Press, Cambridge.
- [77] Batchelor, G. K. (1977). *J. Fluid Mech.*, **83**, 97–117.

[78] Dhont, J. K. G. (1996). *An Introduction to Dynamics of Colloids.* Elsevier, Amsterdam.

[79] Papir, Y. S. and Krieger, I. M. (1970). *J. Coll. Interface Sci.*, **34**, 126–130.

[80] Poon, W. C. K., Meeker, S. P, and Pusey, P. N. (1996). *J. Non-Newtonian Fluid Mech.*, **67**, 179–189.

[81] Meeker, S. P., Poon, W. C. K., and Pusey, P. N. (1997). *Phys. Rev. E*, **55**, 5718–5722.

[82] Phan, S.-E., Russel, W. B., Cheng, Z. et al. (1996). *Phys. Rev. E*, **54**, 6633–6645.

[83] Anderson, P. W. (1997). *Basic Notions of Condensed Matter Physics.* Westview Press, Boulder, CO.

[84] Kittel, C. (2005). *Introduction to Solid State Physics, 8th edition.* Wiley, New York.

[85] Argon, A. S. (1962). *J. Phys. Chem. Solids*, **43**, 945–961.

[86] Falk, M. L. and Maloney, C. E. (2010). *Eur. Phys. J. B*, **75**, 405–413.

[87] Schall, P., Cohen, I., Weitz, D. A. et al. (2004). *Science*, **305**, 1944–1948.

[88] Petekidis, G., Vlassopoulos, D., and Pusey, P. N. (2004). *J. Phys. - Cond. Matter*, **16**, S3955–S3963.

[89] Ballesta, P., Petekidis, G., Isa, L. et al. (2012). *J. Rheol.*, **56**, 1005–1037.

[90] Besseling, R., Isa, L., Ballesta, P. et al. (2010). *Phys. Rev. Lett.*, **105**, 268301.

[91] Schmitt, V., Marques, C. M., and Lequeux, F. (1995). *Phys. Rev. E*, **52**, 4009–4015.

[92] Isa, L., Besseling, R., and Poon, W. C. K. (2007). *Phys. Rev. Lett.*, **98**, 198305.

[93] Isa, L., Besseling, R., Morozov, A. N. et al. (2009). *Phys. Rev. Lett.*, **102**, 058302.

[94] Pham, K N., Petekidis, G., Vlassopoulos, D. et al. (2008). *J. Rheol.*, **52**, 649–676.

[95] Besseling, R., Weeks, Eric R., Schofield, A. B. et al. (2007). *Phys. Rev. Lett.*, **99**, 028301.

[96] Mason, T. G. and Weitz, D. A. (1995). *Phys. Rev. Lett.*, **75**, 2770–2773.

[97] Schall, P., Weitz, D. A., and Spaepen, F. (2007). *Science*, **318**, 1895–1899.

[98] Laurati, M., Egelhaaf, S. U., and Petekidis, G. (2011). *J. Rheol.*, **55**, 673–706.

[99] Isa, L., Besseling, R., Schofield, A. B. et al. (2010). *Adv. Polymer Sci.*, **236**, 163202.

[100] Fuchs, M. and Cates, M. E. (2009). *J. Rheol.*, **53**, 957–1000.

[101] Brader, J. M., Voigtman, T., Fuchs, M. et al. (2009). *Proc. Natl. Acad. Sci. USA*, **106**, 15186–15191.

[102] Sollich, Peter (1998). *Phys. Rev. E*, **58**, 738–759.

[103] Fielding, S. M., Sollich, P., and Cates, M. E. (2000). *J. Rheol.*, **44**,

323–369.

[104] Kobelev, V. and Schweizer, K. S. (2005). *Phys. Rev. E*, **71**, 021401.

[105] Svedberg, T. (1938). *Proc. Royal Soc. Lond. A*, **170**, 40–49.

[106] Tanford, C. and Reynolds, J. (2003). *Nature's Robots: A History of Proteins*. Oxford University Press, Oxford.

[107] Cohn, E. J. and Edsall, J. T. (1943). *Proteins, Amino Acids and Peptides as Ions and Dipolar Ions*. Rheinhold Publishing, New York.

[108] George, A. and Wilson, W. W. (1994). *Acta Cryst. D*, **50**, 361–365.

[109] Poon, W. C. K. (1997). *Phys. Rev. E*, **55**, 3762–3764.

[110] Uversky, V. N. (2002). *Eur. J. Biochem.*, **269**, 2–12.

[111] Doye, J. P. K., Louis, A. A., Lin, I.-C. *et al.* (2007). *Phys. Chem. Chem. Phys.*, **9**, 2197–2205.

[112] Longenecker, K. L., Garrard, S. M., Sheffield, P. J. *et al.* (2001). *Acta Cryst. D*, **57**, 679–688.

[113] Zhang, Z. and Glotzer, S. C. (2004). *Nano Lett.*, **4**, 1407–1413.

[114] Zaccarelli, E., Saika-Voivod, I., Buldyrev, S. V. *et al.* (2006). *J. Chem. Phys.*, **124**, 124908.

[115] Lattuadaa, M. and Hatton, T. A. (2011). *Nano Today*, **6**, 286–308.

[116] Fernández-Nieves, A., Vitelli, V., Utada, A. S. *et al.* (2007). *Phys. Rev. Lett.*, **99**, 157801.

[117] Hawkins, Rhoda J. and McLeish, T. C. B. (2004). *Phys. Rev. Lett.*, **93**, 098104.

[118] Santos de Oliveira, I. S., van den Noort, A., Padding, J. T. *et al.* (2011). *J. Chem. Phys.*, **135**, 104902.

[119] Pasquino, R., Snijkers, F., Grizzuti, N. *et al.* (2010). *Langmuir*, **26**, 3016–3019.

[120] Wood, T. A., Lintuvuori, J. S., Schofield, A. B. *et al.* (2011). *Science*, **334**, 79–83.

[121] Leal-Calderon, F. and Schmitt, V. (2008). *Curr. Opin. Coll. Int. Sci.*, **13**, 217–227.

[122] Stratford, K., Adhikari, R., Pagonabarraga, I. *et al.* (2005). *Science*, **309**, 2198–2201.

[123] Herzig, E. M., White, K. A., Schofield, A. B. *et al.* (2007). *Nature Mater.*, **6**, 966–971.

[124] Stark, H. (2001). *Phys. Rep.*, **351**, 387–474.

[125] Loudet, J.-C., Barois, P., and Poulin, P. (2000). *Nature*, **407**, 611–613.

[126] Ravnik, M. and Žumer, S. (2009). *Soft Mat.*, **5**, 269–274.

[127] Anderson, V. J., Terentjev, E. M., Meeker, S. P. *et al.* (2001). *Eur. Phys. J. E*, **4**, 11–20.

[128] Blickle, Valentin and Bechinger, Clemens (2012). *Nature Phys.*, **8**, 143–146.

[129] Poon, W. C. K. (2013). In *Physics of Complex Colloids* (ed. C. Bechinger, F. Sciortino, and P. Ziherl), pp. 317–386. Società Italiana di Fisica, Bologna.

[130] Berg, H. C. (2003). *E. coli in Motion*. Springer, Berlin.

[131] Ebbens, S. J. and Howse, J. R. (2010). *Soft Mat.*, **6**, 726–738.
[132] Hong, Y., Velegol, D., Chaturvedi, N. *et al.* (2010). *Phys. Chem. Chem. Phys.*, **12**, 1423–1435.
[133] Howse, J. R., Jones, R. A. L., Ryan, A. J. *et al.* (2007). *Phys. Rev. Lett.*, **99**, 048102.
[134] Palacci, J., Cottin-Bizonne, C., Ybert, C. *et al.* (2010). *Phys. Rev. Lett.*, **105**, 088304.
[135] Schwarz-Linek, J., Valeriani, C., Cacciuto, A. *et al.* (2012). *Proc. Natl. Acad. Sci. USA*, **109**, 4052–4057.
[136] Boyer, F., Guazzelli, É., and Pouliquen, O. (2011). *Phys. Rev. Lett.*, **107**, 188301.
[137] Jenkins, M. C., Haw, M. D., Barker, G. C. *et al.* (2011). *Phys. Rev. Lett.*, **107**, 038302.

Surfactants

Dominique Langevin

CNRS Laboratoire de Physique des Solides UMR 8502
Université Paris-Sud, 91405 Orsay, France

Surfactants are amphiphilic molecules, made of two parts, one hydrophilic, i.e. soluble in water, the other hydrophobic, i.e. insoluble in water. As a consequence, these molecules have a tendency to adsorb at interfaces between water and apolar media, solids, liquids, or air (Figure 2.1). By doing so, they change the interfacial energy, hence their name of *surface active* substances [1,2]. This concept can be generalized to other types of interfaces, for instance between hydrocarbons and fluorocarbons, with *surfactants* made of linked hydrocarbon and fluorocarbon chains. Similarly, copolymers A-B can play the role of surfactants for two polymers A and B. Polymers of different types are generally non miscible, but blends can be stabilized with the help of copolymers A-B [3].

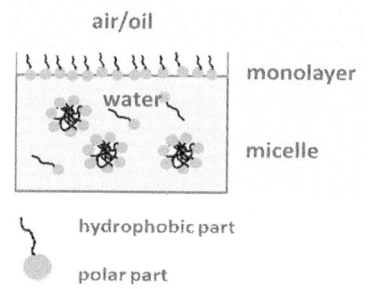

Fig. 2.1 Scheme of the behavior of surfactant in bulk water and at the interface with air or oil.

Polymers, as small surfactant molecules, can also adsorb at surfaces. Because adsorption energy is large in this case (about $k_B T$ per adsorbed monomer), adsorption is irreversible. Furthermore, adsorption is very slow, and the surface layers are frequently out of equilibrium. The same features are found with proteins, which in addition frequently change molecular conformation at the interfaces and may reticulate irreversibly into surface gels. Small particles can also behave as surfactants, when their contact angle with the interface is close to 90°: the particle volume immersed in water is comparable to the volume out of water. In this case, the adsorption energy is extremely large (about $k_B T$ times the surface area expressed in nm^2) and adsorption is also irreversible.

There are many different types of surfactants. They are generally classified according to the electrical charge that they bear when they dissociate in water: anionic surfactants for instance, dissociate into a surfactant anion and a small cation. A widely studied surfactant of this type is sodium dodecyl sulfate (SDS), which releases sodium ions when dissolved in water. Cationic surfactants dissociate into a surfactant cation and a small anion (counterion). A well-known example is dodecyl trimethyl ammonium bromide (DTAB). Other surfactants can bear either positive or negative charges in water according to pH, and are called zwitterionic surfactants. Well-known examples are the phospholipids that contain both phosphate and amine groups. Surfactants that do not bear ionizable groups are called non-ionics. A widely studied series is that of the alkyl polyethylene oxides $C_n E_m$. Table 2.1 on page 55 illustrates several examples. In general, the hydrophobic part of the molecule is made of one or two (seldom more) long hydrocarbon chains

52 *Surfactants*

and the hydrophilic part is smaller (for instance a sulfate ion in SDS). Some surfactants are different, with a hydrophobic part small compared to the hydrophobic part; this is the case of modified polydimethylsiloxanes, which are used to promote fast spreading on surfaces [4].

In this chapter, we will focus on the most general case of small surfactant molecules with long hydrocarbon chains. We will mainly address the case of adsorption at fluid surfaces, such as air–water and oil–water interfaces. Adsorption at solid surfaces is driven by similar energy considerations but in practice, significant complications may arise from surface inhomogeneity and roughness.

2.1 Surface tension

At equilibrium

When surfactant is added to water, the surface tension decreases because a surfactant monolayer spontaneously forms at the surface (Figure 2.2). The hydrophobic part is exposed to air, while the hydrophilic part remains in contact with water. This monolayer is in thermodynamic equilibrium with surfactant monomers in the bulk solution.

Thermodynamics allows the prediction of a certain number of surface properties, starting with surface tension. Let us use the Gibbs approach to describe surface tension. In systems of two bulk phases A and B, separated by an interface, the free energy density f depends on the mass density ρ, which itself depends on the distance z to the interface, with the limit conditions: $f(z \to -\infty) = f_A$ and $f(z \to +\infty) = f_B$ (Figure 2.3). The total free energy, counted per unit area, can be obtained after integration over z and be written as:

$$F = \int f[\rho(z)] dz = F_A + F_B + \Delta F,$$

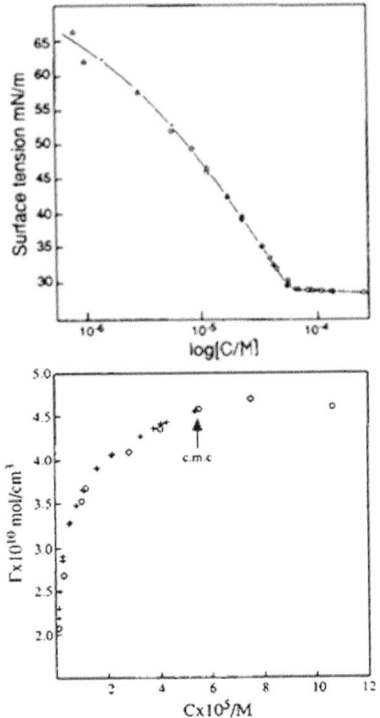

Fig. 2.2 Aqueous solutions of $C_{12}E_3$. Top: Surface tension vs bulk concentration. Bottom: Surface concentration, neutron reflectivity data (circles), and calculations using the Gibbs equation (crosses); adapted from [5].

where F_A is the free energy of an hypothetical phase A with constant density ρ_A up to the interface level, and similarly F_B is the free energy of an hypothetical phase B with constant density ρ_B. The quantity ΔF is therefore an *excess free energy* and by definition equal to the surface tension σ times the surface area A (in conditions of constant temperature and volume).

Let us now consider the case of a dilute surfactant solution. The surface tension change is usually called surface pressure Π. Since Π is the contribution of the monolayer to the energy per unit area, it represents the two-dimensional pressure in the layer (in 3D, the pressure is the derivative of the energy with respect to volume):

$$\sigma = \sigma_0 - \Pi , \qquad (2.1)$$

[1]Otherwise, there will be a depletion layer, as in salt solutions, where the ions are repelled by the surface.

σ_0 being the tension of the bare surface. Π is positive, because the surface energy must decrease during monolayer adsorption.[1] If we now express the excess free energy in terms of thermodynamic parameters,

we find:
$$\Delta F = \sigma A - T\Delta\xi + \sum \mu_i \Delta N_i,$$
where $\Delta\xi$ is the excess entropy, μ_i is the chemical potential of molecules of species i and ΔN_i the corresponding excess in number of molecules. Thermodynamics allows us to write:
$$\frac{\partial \sigma}{\partial \mu_i} = -\frac{\Delta N_i}{A} = -\Gamma_i,$$
where Γ_i is the surface excess density of molecules i. For dilute solutions: $\mu = \mu^0 + k_B T \ln c$, where c is the surfactant bulk concentration and μ^0 its standard chemical potential. It follows:
$$\Gamma = -\frac{1}{k_B T} \frac{\partial \sigma}{\partial \ln c}. \tag{2.2}$$

This equation is usually called the *Gibbs equation* and allows calculating the surface concentration when the surface tension variation with bulk concentration is known. Note that in the case of ionic surfactants, the chemical potential of the counterions has to be considered and a correction factor is needed in equation (2.1) [1].

Because the surface tension is relatively easy to measure, the Gibbs equation is widely used. Figure 2.2 shows the values of Γ calculated with equation (2.1) together with those directly measured using neutron reflectivity [5]. One sees that the agreement is excellent up to the concentration above which the surface tension levels off. This concentration is called critical micellar concentration (CMC) and corresponds to the onset of formation of micelles, which are surfactant bulk aggregates. The CMC is a solubility limit for the monomers and above it, micelles are in equilibrium with monomers at a concentration remaining close to the CMC as the total bulk concentration increases. Neutron reflectivity measurements show that the surface concentration saturates above CMC (Figure 2.2). The Gibbs equation can no longer be used because the surfactant molecules are in equilibrium with micelles and the bulk chemical potential is no longer related to the total surfactant concentration.

Micelles are usually spherical, with the hydrophobic chains hidden in the interior and the hydrophilic heads in contact with water (Figure 2.1) [2]. In some cases, they can be elongated, and sometimes are very long: in these cases they are called *giant* or *worm-like* micelles, and they give rise to very viscous solutions at very low concentrations, as polymer solutions. In other cases, micelles cannot form, and lamellar phases made of surfactant bilayers are obtained instead. In this case, dilute solutions contain vesicles, which are closed bilayer fragments, and the solutions are turbid. The surface tension nevertheless saturates above a concentration called in this case critical aggregation concentration (CAC), signalling the appearance of vesicles. Solutions may also become turbid above the concentration of surface tension saturation if slow phase separation occurs. This behavior is observed below the so-called *Krafft temperature*.

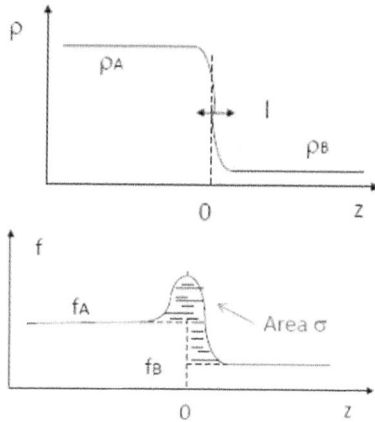

Fig. 2.3 Top: Schematic density variation ρ with z, the interface being located at $z = 0$; ρ varies sharply around $z = 0$, over a distance typically equal to a molecular size l. Bottom: Schematic free energy density variation with z, the dashed area corresponding to the integral giving the value of the surface tension.

For CTAB, the surfactant analogue to DTAB with a C_{16} chain, the Krafft temperature T_k is about 25°C: micelles are formed above T_k but precipitation of CTAB crystals is seen below T_k. Surface tension alone is unable to distinguish between this variety of behavior, which will be discussed in more details in Section 2.3.

In some cases, surface tension increases when the concentration of solute increases. According to the Gibbs equation, this means that the surface concentration Γ is negative, and that instead of a surface excess concentration, one has a deficit. This occurs for instance when the solute is a small ion and is depleted from the surface. Many simple salts behave in this way: ions are surrounded by hydration water and cannot approach too close from the air–water surface. Note that some small ions (protons in particular) behave as surfactant ions and decrease the surface tension [6].

When the chain length of the surfactant is long or if it possesses two or more chains (e.g. lipids, Table 2.1), its solubility in water will be very low. It can however be solubilized in an organic solvent and spread at the air–water interface. After solvent evaporation a so-called *insoluble monolayer* is obtained. According to the value of the surface pressure, insoluble monolayers may present different 2D phases: gas, liquid, liquid crystalline, solid [7]. Frequently, different phases coexist and do not separate at a horizontal surface: domains of one phase are dispersed in the other. In the following, we will mainly focus on the *soluble monolayers*.

Dynamic surface tension

When a fresh surface is created, the surface tension is initially equal to the bare surface tension σ_0 and decreases with time toward the equilibrium surface tension due to the progressive adsorption of surfactant. Measuring the time variation of the surface tension gives therefore information on the adsorption kinetics. This behavior is frequently called in an improper way *dynamic surface tension*: indeed surface tension is the energy per unit area and is not a dynamic property.

The time scale of the adsorption phenomena can be evaluated simply introducing the adsorption length $l_{\text{ads}} = \Gamma/c$, which is the thickness of a hypothetical layer containing an amount of surfactant equal to the adsorbed amount. The adsorption time can be traced to the time necessary for the surfactant to travel the distance l_{ads} by diffusion. If D is the diffusion coefficient, then the adsorption time t_{ads} is of order l_{ads}^2/D. Typically for small surfactants $D \sim 5 \times 10^{-10} \text{ m}^2/\text{s}$, and $\Gamma \sim 1 \text{ mg/m}^2$. For concentrations of the order of 1 g/l one gets $l_{\text{ads}} \sim 1\,\mu\text{m}$ and $t_{\text{ads}} \sim 1 \text{ ms}$: the adsorption is quite fast, even at such a small surfactant concentration. However, t_{ads} varies as the square of the bulk concentration c and increases rapidly when c is lowered: for $c = 1 \text{ mg/l}$, t_{ads} is close to one hour. The exact time variation of the surface tension has been calculated by Ward and Tordai in the case of diffusion limited adsorption [8].

Dynamic surface tension can be studied by a variety of devices, also used for standard surface tension measurements: Wilhelmy plates ($t_{\text{ads}} >$

Table 2.1 Examples of typical surfactants.

Structure	Name
	Sodium dodecylsulfate (SDS)
	Sodium diethyl-hexyl sulfosuccinate (AOT)
	Dodecyltrimethylammonium bromide (DTAB)
	Hexaoxyethylene dodecyl ether $C_{12}E_6$
	Dodecyl maltoside $C_{12}G_2$
	Triton X-100 ($n = 9\text{–}10$)
	Phosphatidylcholine
	Pluronics

1 min), shape of bubbles or dops ($t_{\text{ads}} > 1\,\text{s}$), maximum bubble pressure devices ($t_{\text{ads}} > 10\,\text{ms}$), and instabilities of liquid jets ($t_{\text{ads}} > 1\,\text{ms}$) [9]. Recently, a new method using microdroplets has been proposed [10].

When the surfactant is ionic, adsorption is slowed down. Indeed, when adsorption proceeds, the surface becomes charged and repulsive for the surfactant ions. Figure 2.4 shows an example of the dynamic surface tension of SDS at an oil–water interface, as well as the similar curve by obtained adding salt (0.1 M NaCl) in order to screen the electrostatic interactions. The curve with added salt follows the Ward and Tordai theory, whereas without salt, the surface tension decreases much more slowly, and exhibits a plateau in the intermediate time region [11].

The example above which can be modeled theoretically corresponds to one type of adsorption barrier [12]. There are other types of barriers (steric, etc.) which cannot be modeled easily. As a result, the dynamic surface tension varies from one surfactant to another, and some authors use empirical formulas to describe the process [13]. The type of aggregates present in the solution may play important roles. For instance, in the case of lipid solutions, the lipid molecule solubility is extremely small, and the corresponding t_{ads} is long. However, when the solution contains lipid vesicles, the vesicles can break near the interface and release the lipid molecules. The adsorption time is much longer for vesicles containing lipids in a L_β phase (gel-like bilayers) than in a L_α phase (liquid-like bilayers) [14]. It is clear that more work is needed to better understand the general behavior.

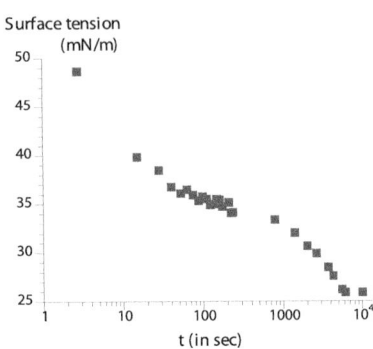

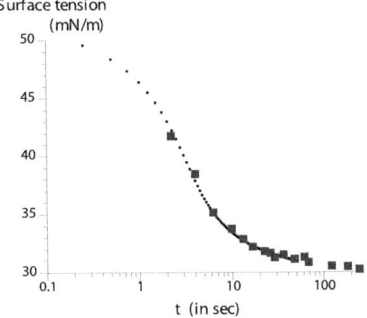

Fig. 2.4 Interfacial tension between dodecane and aqueous SDS solutions vs time. Top: $c = 0.1\,\text{g/l}$. Bottom: $c = 0.014\,\text{g/l}$, 0.1M added NaCl, the dotted line is a fit with Ward & Tordai theory [8]. Note that despite the fact that the surfactant concentration is higher in the salt-free solution, the adsorption time is slower. Adapted from [11].

2.2 Surface rheology

When surfactant monolayers are present at a liquid–gas or a liquid–liquid interface, surface tension alone cannot fully describe their response to external perturbations. In addition, several surface viscoelastic parameters need to be introduced [15]. Monolayer viscoelasticity, also called monolayer rheology, is very difficult to investigate in a fully controlled manner. This is why few data are available in the literature and why large discrepancies (up to several orders of magnitude) are found between data for the same monolayers. In some cases, the discrepancies are due to contamination by impurities (difficult to avoid in these systems, unless careful cleaning procedures are used). In other cases, they are due either to improper corrections for the motion of the surrounding bulk fluids, to intrinsic frequency dependences or to non-linearities (data corresponding to different frequencies or deformation amplitudes).

Elasticity of monolayers

The deformation of a bulk material requires elastic energy, which, in first approximation, is a quadratic function of the strain tensor elements $\mu_{ij} = (1/2)(\partial u_i/\partial x_j + \partial u_j/\partial x_i)$, a vector u being the surface displacement. By analogy, the elastic free energy per unit area needed to deform a surface covered by a monolayer (isotropic in the surface plane) can be written as:

$$F_{\text{elast}} = \frac{1}{2}(E+S)(u_{xx}+u_{yy})^2 + \\ 2S\left(u_{xy}^2 - u_{xx}u_{yy}\right) - 2\Pi\left(u_{xz}^2 + u_{yz}^2\right), \quad (2.3)$$

z being the direction of the surface normal. Any deformation can be written as the sum of pure shear and pure compression terms [16].[2] F_{elast} can be rewritten as

[2] For a motion in the surface plane $(u_z = 0)$: $u_{ij} = (1/2)(u_{xx}+u_{yy})\delta_{ij} + [u_{ij} - (1/2)(u_{xx}+u_{yy})\delta_{ij}]$.

$$F_{\text{elast in plane}} = \frac{1}{2}E(u_{xx}+u_{yy})^2 + S\sum_{ij}\left[u_{ij} - \frac{1}{2}(u_{xx}+u_{yy})\delta_{ij}\right]^2,$$

where it is seen that the compression modulus is E and the shear modulus is S. When u depends only on x, one finds from equation (2.3) that the effective compression modulus is $E+S$; this is because uniaxial compression always involves shear. The shear modulus in nonzero only in solid or gel-like monolayers [7]. This is not a frequent case for small surfactant molecules, and in practice $S = 0$ most of the time. The above distinction between uniaxial and pure compression moduli is then not necessary: both moduli are the same. [3]

[3] The compression modulus is sometimes also called *dilational* or *dilatational* modulus.

By analogy with the three-dimensional case, the compression modulus can be written as

$$E = -A\frac{\partial \Pi}{\partial A}. \quad (2.4)$$

For an insoluble monolayer $\Gamma \sim A^{-1}$ and the compression modulus is:

$$E_0 = -\Gamma\frac{\partial \sigma}{\partial \Gamma} = \Gamma\frac{\partial \Pi}{\partial \Gamma}. \quad (2.5)$$

When the monolayer is soluble, i.e. when it can exchange matter between surface and bulk, equation (2.5) can only be used when the rate of compression is faster than the rate of dissolution. When this is not the case, models have been proposed to relate E and E_0. When desorption and adsorption proceed freely, without energy barriers, and when the compression stress is sinusoidal, proportional to $e^{i\omega t}$, ω being the frequency, it was shown that the complex modulus E can be split into real and imaginary parts [17, 18]:

$$E_r = E_0\frac{1+\Omega}{1+2\Omega+2\Omega^2}, \quad E_i = E_0\frac{\Omega}{1+2\Omega+2\Omega^2}, \quad (2.6)$$

where Ω is a reduced frequency:

$$\Omega = \sqrt{\frac{D}{2\omega}}\frac{dc}{d\Gamma}$$

When $\omega \to \infty$, $\Omega \to 0$, $E_r \to E_0$, and $E_i \to 0$: the monolayer behaves as if it were insoluble as expected. When $\omega \to 0$, $\Omega \to \infty$ and both E_r and E_i go to zero: this is because if the rate of compression is slow compared to the rate of dissolution, there is no resistance to compression; the equilibrium between bulk and surface concentration has time to establish during the compression. When ω takes intermediate values, $E_r < E_0$ and $E_i \neq 0$. The meaning of E_i will be discussed thereafter, in relation to surface viscosity.

Generalizations to more complex situations, i.e. the presence of adsorption or desorption barriers, exchange with micelles, etc. were worked out [18]. These models apply in principle only to nonionic surfactant molecules. For ionic surfactants, they can be refined to take into account the influence of surface potential in the diffusion process [19].

In some particular systems, the surface tension can be very small (interfaces between microemulsions and excess oil or water for instance). In this case, higher-order terms should be added to equation (2.3). These are curvature elastic terms, very important to describe microstructures formed by surfactant mono or bilayers in bulk [20]. If the surfactant layer is replaced by a mathematical surface (the neutral surface) of principal curvatures C_1 and C_2, the curvature elastic energy per unit area in the bulk can be written as:

$$F_{\text{curv}} = \frac{1}{2}K\left(C_1 + C_2 - 2C_0\right)^2 + \bar{K} C_1 C_2 , \qquad (2.7)$$

where C_0 is the spontaneous curvature of the layer, by convention positive if the curvature is against water; flat aggregates correspond to $C_0 = 0$. The elastic constants K and $\bar{K}$ are related, respectively, to deviations to mean curvature and to Gaussian curvature. While the meaning of K is rather straightforward, the role of $\bar{K}$ is more difficult to assess. If $\bar{K}$ is negative, the second term of equation (2.7) is smaller if C_1 and C_2 have the same sign, i.e. favoring closed aggregates. If $\bar{K}$ is positive, structures with saddles (regions where C_1 and C_2 have opposite signs) are favored: these are the *sponge* phases that will be discussed in Section 2.3. It must be stressed that the description in terms of equation (2.7) is only valid for surfactant layer thicknesses much smaller than the inverse curvatures and it is not suitable for spherical micelles for instance.

In the case of monolayers adsorbed at macroscopic flat interfaces, $\bar{K}$ plays no role,[4] but K leads to an additional surface energy term. This correction can generally be neglected, except in systems with very low surface tensions and for surface deformations of small enough wavelength (less than $\sqrt{K/\sigma}$) [21].

At the difference of compression and shear moduli, many molecular models were proposed to calculate the bending moduli. The models indicate that the chains contribution to the elasticity dominates over the polar part contributions [20]. Simple scaling arguments show that $K \sim n^3/a^5$, where n is the number of carbon atoms in the chain and a is the area per molecule, in good agreement with mean-field calculations. It has also been shown that K is strongly decreased when mixtures of surfactants of different chain lengths are used, as a result of increased chain disorder. Similar effects were experimentally evidenced for the compression modulus [15].

When K is small enough and when large aggregates are formed, these aggregates are distorted due to thermal motion. The competition be-

---

[4] Saddle formation is not possible, the surface topology is invariant.

tween thermal and bending energy can be described by the notion of persistence length ξ_k. For cylinders, for which $C_0 \neq 0$:

$$\xi_k = l \frac{K}{k_B T}, \qquad (2.8)$$

where l is a molecular length ($\sim$ surfactant layer thickness); ξ_k is the length over which the axis of the cylinders is constant in direction. For lamellae, $C_0 = 0$ and [22]

$$\xi_k = l \exp \frac{2\pi K}{k_B T}. \qquad (2.9)$$

ξ_k is the length over which the normal to the lamellae is constant in direction. The above expression has been established without taking into account the scale dependence of K, due to the same thermal fluctuations. K is in fact not constant and decreases logarithmically with scale but rigorous expressions differ only by small corrections.

Monolayers viscosity

In classical linear viscoelasticity theory, when the deformations are sinusoidal in time with a frequency ω, the linear response of the system i.e. the relation stress-deformation can be described by a complex moduli: $\tilde{G} = G + i\omega\eta$ where G is the elastic modulus and η is the viscosity [16]. If we apply this description to surfaces for deformations of the type $e^{i\omega t}$, E and S can be replaced by $E + i\omega\eta_E$ and $S + i\omega\eta_S$. There are therefore two surface viscosities, η_E and η_S, which are, respectively, the compression and shear surface viscosities. Contrary to the shear modulus S that is zero except for solid and gel-like monolayers, the surface shear viscosity is non-zero for fluid monolayers; the distinction between pure and uniaxial compression viscosities is then always relevant. The imaginary part of E found in equation (2.6) for soluble monolayers can now be interpreted as a contribution of the adsorption-dissolution process to the compression viscosity: $E_i = \omega \eta_E$.

The above surface viscoelasticity model can be derived more rigorously by using the formalism of surface excess properties, as discussed in Section 2.1 for surface tension [23]. Let us note that isotropic bulk systems are characterized also by two types of viscoelastic parameters: compression and shear. The compression viscoelasticity of bulk fluids does not play a role at frequencies smaller than sound frequencies, i.e. in most practical cases. In monolayers, the sound (compression wave) frequencies are much smaller, i.e. these media are more compressible, and the compression viscoelasticity plays important roles.

Soluble surfactant monolayers usually have a zero shear modulus and their shear viscosity is very small and difficult to measure[5]: about or less than 10^{-3} mN s/m. The compression viscosities of soluble surfactant layers are frequently much larger (by several orders of magnitude) than shear viscosities, due to the exchange process between surface and bulk that introduces a contribution in the compression viscosity, absent in the shear viscosity – equation (2.6).[6]

[5]The CGS unit is the "surface poise", and is equivalent to 1 g/s or to 1mN s/m; similarly, surface tension and surface moduli CGS units are dyn/cm, equivalent to mN/m.

[6]It must be noted that even the small shear viscosities quoted above are equivalent to locally very viscous media. Indeed, the order of magnitude of the local viscosity inside the monolayer can be estimated dividing η_S by the thickness h of the monolayer. For $h \sim 1$ nm and $\eta_S \sim 10^{-3}$ mN s/m, one gets a local bulk viscosity which is 10^6 times that of water.

Condensed insoluble monolayers have larger compression elasticities and viscosities and can exhibit a non-zero shear modulus. Still larger surface shear viscosities are achieved with protein monolayers, when the unfolding of the protein at the surface leads to a very rigid structure. Values up to several mN s/m have been reported [24].

Measuring monolayer shear viscoelasticity

There are several types of experimental devices for the measurement of the various viscoelastic parameters. Surface shear properties have been the most widely studied viscoelastic properties. Popular devices among many others are channel viscosimeters for insoluble monolayers, and oscillating disk devices for both soluble and insoluble monolayers [25].

A schematic illustration of the channel viscosimeter is shown in Figure 2.5 (top). A surface pressure difference is applied between two monolayer compartments, forcing the monolayer to flow across the channel between the two compartments. The viscosity is deduced from the rate of change of the surface pressure. The instrument is the two dimensional (2D) analog of the well-known capillary viscometer. The flow can be also visualized with particles (talc, for instance). In an elegant experiment on fatty acid monolayers, the domains of a 2D condensed phase coexisting with a more expanded phase were used to visualize the flow via fluorescence microscopy (Figure 2.5 - middle) [26]. The velocity profile depends on the Boussinesq number $B = \eta_S/(\eta l)$, where l is the characteristic length scale of the velocity gradient. In the experiment of Figure 2.5, the flow profile was found to be semi-elliptical (small B) whereas parabolic (Poiseuille) profiles are expected for surface-dominated friction (large B) (Figure 2.5 - bottom). This illustrates the necessity to take the friction by the subphase into account to avoid erroneous surface viscosities determinations: in the particular case described above, the sub-phase contribution dominates, and the surface shear viscosity cannot be measured.

When the monolayer is in a two-phase region, the study of the time evolution of the shape of the domains, when slightly distorted, can also be used to measure the surface shear viscosity if B is large. In addition, the line tension of the domains can be determined [27].

Optical tweezers and videomicroscopy allow monitoring the displacement of micron-sized particles and open nowadays a new field in bulk rheology, known as *microrheology*. These techniques bring new developments in surface rheology [28]. It should be noted however, that the use of particles at surfaces introduces additional capillary forces that depend of the contact angle between the particle and the surface and need to be well characterized. Because of the small size of the particles, the Boussinesq number is large and the resolution on the surface viscosity determination is improved.

The frequency dependence of the surface viscosity and the determination of the shear modulus are difficult with the above devices. The use of oscillating disks or bicones is better for this purpose. These devices

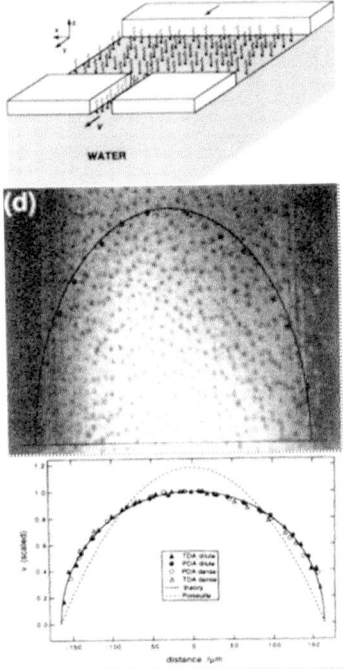

Fig. 2.5 Measurement of surface shear viscosity. Top: Scheme of the channel viscosimeter. Middle: Visualization of the velocity profile using domains of monolayer condensed phase. Bottom: Experimental velocity profile and calculations from models [26].

are now in wide-spread use as commercial instruments are available. As with bulk rheometers, Newtonian and non-Newtonian behavior can be probed as well as nonlinear behavior such as yield stress, plasticity and fracture [29]. Other instruments appeared recently, now commercially available, which make use of magnetic needles floating at the surface; instead of applying mechanical torques, the float is submitted to magnetic torques [30].

Measuring compression monolayer viscoelasticity

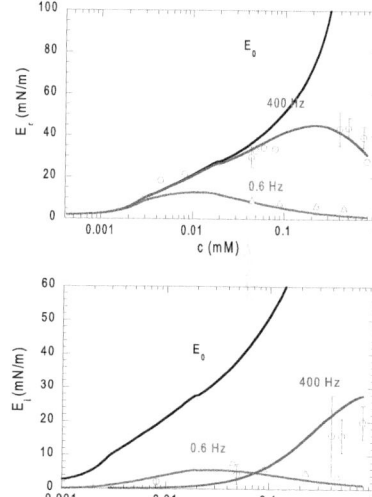

Compression properties have been investigated with moving barrier devices, including excitation of surface waves, either capillary or compression waves [18, 31]. Thermally excited capillary waves have also been studied with surface light-scattering techniques [15]. Other methods such as oscillation of bubbles or drops and overflowing cylinders are also used [9], in which the amplitude of the surface deformations is in general much larger than in surface waves methods.

Figure 2.6 shows viscoelastic parameters obtained for a soluble nonionic surfactant using excited longitudinal and capillary waves [32]. This data illustrates the greater influence of diffusion between surface and bulk at lower frequencies. The lines are calculated from the diffusion model based on equations (2.6). The characteristic diffusion time decreases when surfactant concentration increases, so the high-frequency insoluble monolayer regime is seen at small surfactant concentration, whereas at high concentration the monolayer follows the compression instantaneously and there is no more resistance to the compression.

It should be noted that the intrinsic elasticity E_0 can be frequency dependent as a result of other relaxation processes in the surface such as chain reorientation. This is associated to a non-zero intrinsic surface viscosity. However, for soluble monolayers, these effects are usually negligible compared to the frequency dependences associated to the bulk-surface exchange process. Indeed, equations (2.6) generally describe well the experimental results for nonionic surfactants. In the case of ionic surfactants, the diffusion process is affected by the surface charges, and the theory is more complicated [19]. Let us also mention that the calculation of the elasticity at infinite frequency, E_0, is very delicate, and that the use of an equation of state for soluble monolayers (empirical relation between σ and Γ) can lead to large errors. The use of the Gibbs equation is more reliable, but requires very accurate surface tension measurements [32].

In the case of solid-like monolayers, it has been shown that if the detection is made with a Wilhelmy plate, the measured modulus depends on the orientation of the plate. This can be used to determine both shear and compression moduli [33]. However, the validity of the method is questionable in some cases [34].

Contrary to the excited waves techniques, there are only very few reported light-scattering studies of soluble monolayers in comparison with the relatively large amount of data on insoluble monolayers. With

Fig. 2.6 Surface viscoelastic compression moduli vs surfactant bulk concentration. Top: Real modulus E_r. Bottom: Imaginary modulus E_i. Circles: Measurements using excited capillary waves. Triangles: Measurements using excited longitudinal waves. Lines are from theory equations (2.6). Adapted from [32].

the light-scattering technique, the Boussinesq numbers are large (scales probed are small), and small viscosities can be measured. Data obtained making use of the different surface waves techniques reveal the important frequency variation of the viscosity, even for relatively expanded monolayers [35]. So far, the relaxation processes giving rise to the frequency dependences are mostly unknown, except for those associated with exchanges with the bulk in the case of soluble monolayers.

Methods involving large surface deformations include the oscillating bubble or drop method, which is becoming increasingly popular. One of its advantages is the use of very small amounts of fluid, and commercial instruments now available. In this method, the drop or the bubble is usually imaged with a video camera and the surface tension is calculated from a fit of the surface profile with the Laplace equation relating the pressure difference between the interior and the exterior to surface tension. Typical deformation amplitudes are of the order of 1 to 10%.[7] Still larger deformation amplitudes are obtained with overflowing cylinders, a method in which the surface velocity is measured with laser velocimetry and surface tension with a Wilhelmy plate. The use of large amplitudes will allow investigating in more detail the non linear behavior of monolayers, a topic little explored up to now.

Curvature elasticity can be determined by a variety of techniques involving bulk microstructures studies [20]. In the case of macroscopic surfaces, the elastic modulus K can be obtained through the study of the thermal fluctuations of the vertical position of the surface with either ellipsometry or X-ray reflectivity [21, 38]. This elasticity influences the behavior of the surface only at very small scales.

2.3 Bulk aggregates

Micelles

Micelles form when the gain in free energy obtained by hiding the surfactant hydrocarbon chains from water overcomes the entropy of monomers (Figure 2.7). The chemical potential of a surfactant molecule in water and in a micelle containing N molecules are respectively [39]:

$$\mu_1 = \mu_1^0 + k_B T \ln X_1 , \qquad \mu_N = \mu_N^0 + (k_B T/N) \ln (X_N/N) , \qquad (2.10)$$

where X_i is the mole fraction of surfactant in the form of monomers ($i = 1$) or micelles ($i = N$). The main chain contribution to $\mu_N^0 - \mu_1^0$ is the hydrophobic free energy associated to the transfer of surfactant hydrocarbon chains from water into the micellar core. In the following, we will consider the example of sodium dodecyl sulfate (SDS). In this case, the above hydrophobic free energy is $n\varepsilon_{CH2} - \varepsilon_{CH3}$, n being the number of CH_2 groups in the chain, $\varepsilon_{CH2} \sim 825\,\mathrm{cal\,mol^{-1}}$ (close to $k_B T \sim 600\,\mathrm{cal\,mol^{-1}}$) and $\varepsilon_{CH3} \sim 2100\,\mathrm{cal\,mol^{-1}}$ [40]. There is a second contribution from the hydrophobic attraction at the interface between water and hydrocarbon, which can be estimated as the product of a

[7] Recently, difficulties were pointed out in the case of irreversibly adsorbed layers (surface concentration increases during the expansion cycles) [36] and solid-like layers (shear stresses become important) [37].

hydrocarbon–water interfacial tension σ by a, the area per surfactant molecule at the micelle surface. The repulsive contributions are more difficult to evaluate: they include the electrostatic repulsion between the polar parts when these are charged, the hydratation repulsion, the steric repulsion, etc. As a first order approximation, the repulsion will be estimated as $\kappa^{-1}e^2/(2\varepsilon a)$, where κ^{-1} is the Debye–Hückel electrostatic screening length, e is the electron charge and ε is the dielectric constant. The polar head contribution to μ_N^0 is therefore:

$$\mu_{N\,\text{heads}}^0 = \sigma a + \kappa^{-1}e^2/(2\varepsilon a) \;,$$

minimum for $a = a_h = e/(\varepsilon\sigma\kappa)^{1/2}$. Numerical estimations using $\varepsilon \sim 40\varepsilon_0$ (ε_0: dielectric constant of vacuum), $\sigma \sim 50\,\text{mN/m}$, and $\kappa^{-1} \sim 5\,\text{Å}$ lead to $a_h \sim 60\,\text{Å}^2$; this value is in good agreement with the values measured in micelles and in monolayers, a remarkable result in view of the simplifications used. In the above simple model, the steric repulsion between the chains has been neglected. Mean-field theory shows that the free energy of the chains has a minimum at $a_c = 45\,\text{Å}^2$ for a 12-carbon tail, and that stretching the chains down to $a = 25\,\text{Å}^2$ costs as much as $3k_BT$ [20]. The optimum value a_0 of the area per surfactant molecule is therefore intermediate between a_c and a_h.

The surfactant area at the micelle surface is frequently close to that deduced from surface tension measurements at flat surfaces. It is described by equation (2.2). In practice, it should be different because the micelle surface is curved and the monolayer at the free surface is flat; moreover the chains are in contact with air in the latter case. Note that the area a deduced from these surface tension measurements is also about the same as those deduced from oil-water interfacial tension measurements. This confirms that the interactions between polar heads is dominant and that the contribution of curvature elastic energy is small.

If one assumes that the aggregates are spheres of radius R, a is related to N by: $a = 4\pi R^2/N = 3v/R$, v being the volume of the hydrocarbon tail. Using $v(\text{Å}^3) = 27.4 + 26.9n$, n being the number of carbon atoms of the chain, and $n = 12$, one finds $N \sim 60$, i.e. a value much larger than one, and close to that found experimentally (Table 2.2).

The distribution curve for the aggregates of size N can also be calculated: it has a maximum at $N = N_0$ and a width:

$$\delta N \approx \sqrt{\frac{9k_BT}{2\sigma\,(4\pi)^{1/3}\,(3v)^{2/3}}}\,N_0^{2/3}$$

for the above example $\delta N/N \sim 0.1$: the micelles are quite monodisperse.

It follows from equations (2.10) that:

$$X_N = N\left\{X_1 \exp\left[(\mu_1^0 - \mu_N^0)/k_BT\right]\right\}^N, \qquad (2.11)$$

with

$$\mu_N^0 - \mu_1^0 = 2\sigma a_0 - n\varepsilon_{\text{CH2}} - \varepsilon_{\text{CH3}}. \qquad (2.12)$$

Fig. 2.7 Schematic representation of a micelle (top) and a vesicle (middle) in water.

Table 2.2 Micelle aggregation number N, distribution width δN, dissociation rate constant k^-, and association rate constant k^+ at 25°C for various surfactants. After [41].

Surfactant	N	δN	$10^{-7} k^-$ s^{-1}	$10^{-9} k^+$ $M^{-1} s^{-1}$
Sodium hexyl sulfate	17	6	132	3.2
Sodium heptyl sulfate	22	10	73	3.3
Sodium octyl sulfate	27		10	0.77
Sodium nonyl sulfate	33		14	2.3
Sodium dodecyl sulfate (SDS)	64	13	1	1.2
Sodium tetradecyl sulfate	80	16.5	0.096	0.47
Sodium hexadecyl sulfate	100	11	0.006	0.13

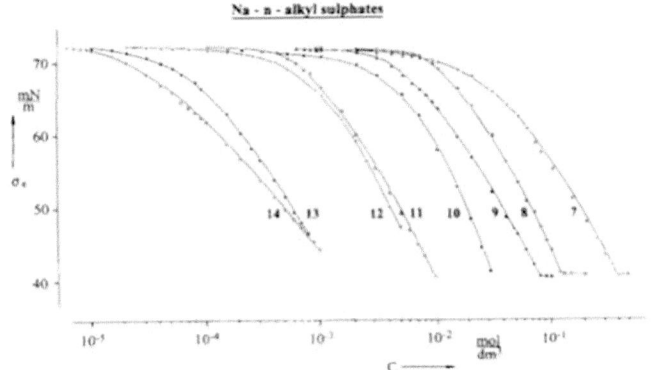

Fig. 2.8 Surface tension vs bulk surfactant concentration for a series of sodium alkyl sulfates. After [42].

Since N is large, the exponential factor in equation (2.11) is large but as long as X_1 is small enough, X_N is much smaller than X_1. X_N remains small until $X_1 \exp[(\mu_1^0 - \mu_N^0)/k_B T]$ is of order one, after which it increases abruptly; this X_1 is then the CMC and

$$\ln(\text{CMC}) = \left(\mu_N^0 - \mu_1^0\right)/k_B T. \quad (2.13)$$

The simple model described above explains why the transition at CMC is sharp. [8]

It follows from equations (2.12) and (2.13) that $\ln$ CMC is linear in the number of carbon atoms in the chain. This has been verified in many experiments (Figure 2.8) [42]. The slope of the curves $\ln$ CMC versus n gives a value of ε_{CH2} in agreement with calorimetric data for dissolution of alkanes in water, as discussed above. It follows that the CMC decreases by a factor of order 10 if the chain length is increased by two to three carbon atoms.

The chemical potential difference $\mu_N^0 - \mu_1^0$ also plays an important role in the exchanges of surfactant molecules between micelles and water. Indeed, the micelles are labile aggregates that constantly form and disappear. The corresponding chemical equilibrium has been studied by

[8] Note that for micelles in oil, the difference $\mu_N^0 - \mu_1^0$ is much smaller (there is no gain from the chains of about $n k_B T$), and the association is in general progressive with no well-defined CMC. It is known that the enthalpy and entropy variations accompanying a surfactant hydrocarbon chain out of water are both large and positive because of the ability of water to promote three-dimensional hydrogen-bonded structures around hydrocarbons groups [40]. However, these increments nearly compensate for each other and the unique properties of water do not play a dominant role in micellization. In non-aqueous solvents the entropy increments are negative, indicating that the solvent configuration around the chains is very different than in water.

Aniansson et al. [43]. For a surfactant species A, it can be written as:

$$A_N + A \leftrightarrow A_{N+1} .$$

When the rate constants k^+ and k^- are assumed to be independent of N, one finds $k^+ = k^-/\text{CMC}$. The association rate constant k^+ is usually close to, or only slightly lower than the value calculated for a diffusion controlled process $k^+ \sim 5 \times 10^8$–$3 \times 10^9 \,\text{Mol}^{-1}\text{s}^{-1}$. The dissociation rate constant k^- strongly depends on the surfactant chain length and CMC decreases by a factor close to 2–3 per additional CH_2 group [41]. A few examples are given in Table 2.2. For SDS for instance, the exit time is of the order of 10^{-7} s. For longer chain lengths, the exit time increases rapidly: for double chained lipids, it can reach values of the order of days. The same approach can be used to describe the exit step of molecules solubilized in surfactant aggregates [44].

Dilute solutions

Because of geometrical considerations, not all surfactant molecules can form spherical micelles. A useful model introduces a *surfactant packing parameter* [39], $v/a_0 l_c$, where v is the molecular volume, a_0 is the area per polar head and l_c is the extended chain length: $l_c(\text{Å}) = 1.265n + 1.5$ [40]. It can be predicted that the aggregates formed are:

- spherical micelles if $v/a_0 l_c < 1/3$
- elongated micelles if $1/3 < v/a_0 l_c < 1/2$
- lamellar phases if $1/2 < v/a_0 l_c < 1$

If $v/a_0 l_c > 1$, reverse aggregates can be formed in the presence of oil.

These very simple predictions are in good agreement with the observations. Sodium dodecyl sulfate (SDS) for instance forms spherical micelles in water; in its case $v \sim 350 \,\text{Å}^3$, $l_c \sim 17 \,\text{Å}$, $a_0 \sim 60 \,\text{Å}^2$ and $v/a_0 l_c \sim 0.34$. Phospholipid molecules are double chained amphiphiles that do not form micelles in water: the molecules directly associate into bilayers (Figure 2.7); typically $v \sim 1000 \,\text{Å}^3$, $a_0 \sim 70 \,\text{Å}^2$, $l_c \sim 23 \,\text{Å}$ and $v/a_0 l_c \sim 0.62$. AOT (sodium diethylhexyl sulfosuccinate) is also a double chain surfactant; but its polar head is less bulky than the phospholipid one: $v \sim 600 \,\text{Å}^3$, $l_c \sim 8 \,\text{Å}^2$, and $a_0 \sim 60 \,\text{Å}^2$ leads to $v/a_0 l_c \sim 1.25$, suggesting that AOT will form aggregates in oil; $v/a_0 l_c$ is too small for the formation of reverse micelles, however, the close packing of the surfactant polar heads leaves an empty volume in the center of the micellar core; this explains why micellization is observed only when trace water is added. The $C_i E_j$ (alkyl polyethyleneoxides) surfactants have slightly more flexible polar parts but do not form large monodisperse aggregates in oil in the absence of water.

The spontaneous curvature C_0 of the surfactant layer introduced in equation (2.7) is closely related to the surfactant parameter: for $v/a_0 l_c \sim 1$, $C_0 \sim 0$ whereas C_0 is either positive or negative when $v/a_0 l_c$ is smaller or larger than 1, respectively: the farther $v/a_0 l_c$ is from one, the largest

[9] Another related notion is that of HLB: *hydrophilic–hydrophobic balance* [45]. This notion is purely empirical and serves to distinguish surfactants according to their affinity to oil or water: high HLB corresponds to surfactant well soluble in water, and low HLB to surfactants more soluble in oil, the borderline being HLB ~ 7.

the absolute value of the spontaneous curvature. Let us recall however that the description in terms of equation (2.7) is only valid for surfactant layer thicknesses much smaller than the inverse curvatures and it is not suitable for spherical micelles for instance.[9]

It has to be noted that $v/a_0 l_c$, C_0, and HLB depend not only on the surfactant, but also on counterions for ionic surfactants: in the case of SDS, when sodium is replaced by lithium, the micelles are more spherical, because lithium is a highly hydrated ion and cannot penetrate in between polar heads, whereas sodium does. External conditions such as temperature and added salt can also change the surfactant parameter. In the case of nonionic surfactants, increasing temperature decreases the number of hydration water molecules linked to the polar heads and a_0 decreases, meaning that the micelles tend to elongate. When oil is present, an evolution toward reverse micelles is observed upon increasing temperature (see Section 2.4). In the case of ionic surfactants, adding salt decreases the electrostatic repulsion between polar heads, and a_0 is smaller, meaning that the micelles tend to elongate. Note that this effect is ion specific: counterions such as salicylate can promote the growth of cetyl trimethyl ammonium micelles, because they penetrate into the surfactant layer much more deeply than bromide ions. With this type of ions, the micelles may become worm-like and entangle as polymers [46]. Worm-like micelles can also be found with *gemini* surfactants (having two chains and two polar heads) which form very long micelles even without added salts [47]. The extreme sensitivity of packing parameter, spontaneous curvature, or HLB on external conditions requires that these notions should be used with care.

Vesicles being made of two surfactant monolayers (Figure 2.7) have by definition zero spontaneous curvature. They do not form spontaneously, as the lamellar phase is the system of lowest energy, but submicronic vesicles are easily obtained by sonication of the lamellar dispersions. When the solution contains two surfactants, vesicles may form spontaneously: this is the case of mixtures of cationic and anionic surfactants (sometimes called *catanionics*) [48]. In some cases, when the bilayers are very rigid, vesicles with original shapes (icosahedra [49] and nanodiscs [50]) can be obtained.

Positive values of $\bar{K}$ can be predicted for bilayers in which the spontaneous curvature of the monolayers are negative: the surfactant cannot form reverse systems in water, but can assemble into *frustated bilayers*. One can show that $\bar{K}_{\text{bilayer}} = 2\bar{K}_{\text{mono}} - 2hC_0 K_{\text{mono}}$ where h is the bilayer thickness. If $\bar{K}_{\text{mono}}$ is small and C_0 negative, $\bar{K}_{\text{bilayer}}$ is positive and favors sponge-like structures, i.e. structures with many holes in the bilayers. This is frequently observed in practice for instance when a short chain alcohol is added to a micellar system. This mainly increases v in the packing parameter and thus decreases C_0. The micelle shape evolves from spheres to cylinders and to lamellae; upon further addition of alcohol, a sponge phase is obtained [51].

In the case of long surfactants such as those of the Pluronics® series, which are made of sequences of oxyethylene separated by a sequence of oxypropylene, the spontaneous curvature is less well defined, because these surfactants are flexible and can easily change configurations (hairpin and other). They can in this way form micelles either in oil or in water [52] (Figure 2.9). This is another clear limit of the description using a surfactant parameter (or C_0 or HLB).

Bulk aggregates and concentrated solutions

The notion of packing parameter can no longer either be applied in concentrated solutions where interactions between aggregates become important and topological constraints arise. Spherical micelles interacting via repulsive forces can pack into ordered cubic phases. When the forces are attractive, a clouding phenomenon can be observed above a critical point, followed by a phase separation into two micellar phases. This also happens when temperature is raised in solutions of nonionic surfactants and the critical point is usually called *cloud point*. The coexistence curve is generally very asymmetric as in polymers solutions and for similar reasons: the size difference between solvent and solute molecules [3].

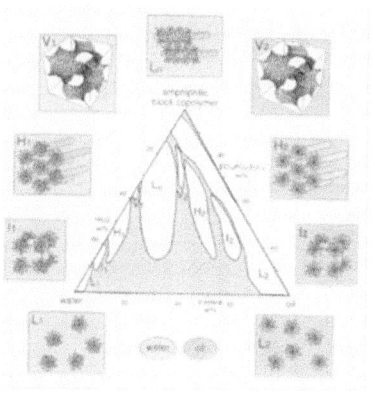

Fig. 2.9 Ternary phase diagram of a polymeric surfactant [52].

In general, surfactant with small packing parameters evolve from micellar phases (L_1) toward lyotropic liquid crystalline phases: hexagonal phases (H_α) made of long tubular micelles parallel to each other and then to lamellar phases (L_α). In some cases, cubic phases can be present: I_1, cubic arrangement of micelles between L_1 and H_α phases, or V_1, bicontinuous cubic arrangement between H_1 and L_α phases (Figure 2.9). Disordered bicontinous structures can also be found, these are the *sponge* phases L_3. When the solvent is oil, reverse structures are found, following the same order while increasing surfactant concentration: $L_2 \to I_2 \to H_2 \to V_2 \to L_\alpha$ (Figure 2.9).

In these structures, including the lamellar phases L_α, the hydrocarbon chains are disordered; for instance in micelles, the chain ends are located everywhere in the micelle interior, including the vicinity of the micelle surface. If the temperature falls below the melting temperature of the chains, the chains crystallize and micelles cannot form. The transition temperature is the Krafft temperature T_k and depends on the chain length and to some extent on the nature of the polar head group. For CTAB for instance, $T_k \sim 25°C$; because of the very low density difference between water and the CTAB crystals, phase separation is slow and the solution may remain turbid for hours. It should be noted that the surface tension exhibits the same concentration variation (Figure 2.2), and the onset of phase separation can be easily confounded with a CMC. At these temperatures, L_α phases are replaced by more ordered lamellar phases called L_β.

The structures discussed so far are not permanent structures. Surfactant molecules as well as the other constituents are constantly exchanging. These exchanges are of major importance in the understanding of the dynamic properties of the systems. For instance the sponge-phases, in which the surfactant bilayers are fully interconnected, have viscosities not much larger than that of water; this would be impossible in the absence of exchanges. The viscosity of entangled giant micelles is also much smaller than the viscosity estimated from structural data: this is because the micelles break and reform constantly.

In all these self-assembled structures, surfactant molecules are held together by weak non-covalent forces, and the equilibrium structures can be modified by the application of small stresses, for instance under moderate shear. This topic is still the object of active research. In particular, a large number of studies have been devoted to the rheology of worm-like micelles, which is very peculiar: shear bands can be seen close to solid walls due to micelle orientation [53].

2.4 Microemulsions

Microemulsions are dispersions containing both oil and water. The structure is made of oil and water microdomains, separated by surfactant monolayers. The most frequent microemulsion structures consist of droplets, either of water or of oil, surrounded by a surfactant layer and dispersed respectively in oil or in water (Figure 2.10).

Difference from emulsions (Section 2.6), microemulsions are thermodynamically stable because the surface tension σ is low enough to compensate the dispersion entropy as will be discussed in the following (σ below about 10^{-2} mN/m). Most common surfactants are unable to decrease the oil–water interfacial tension below about 1 mN/m; this is why a cosurfactant is frequently used. Because σ is so small, the curvature elastic energy, which is usually negligible compared to the surface energy, becomes very important in microemulsion systems.

The name microemulsion is sometimes restricted to droplets of a size sufficiently large that the physical properties of the dispersed oil or water (internal phase) are indistinguishable from those of the corresponding oil or water bulk phases; when the droplet core is too small, the medium is called *swollen micellar*. When the micelles are progressively swollen, the properties of the internal phase evolve smoothly toward those of a bulk phase, without any well defined transition.

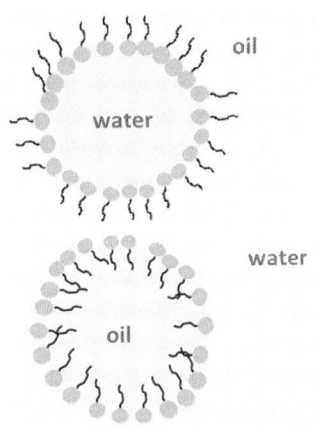

Fig. 2.10 Droplets in water/oil and oil/water microemulsions.

When oil is solubilized in micelles or water in reverse micelles, obviously the condition $R = 3v/a_0 < l_c$ for obtaining spherical aggregates no longer applies. For instance, oil can promote the evolution of cylindrical micelles towards spherical swollen micelles depending upon the amount of solubilized oil [54]. Similarly reverse tubular micelles can evolve towards spherical droplets upon addition of water [55]. The *packing parameter* $v/(a_0 l_c)$, or equivalently the spontaneous curvature C_0, however, determines the type of microemulsion. When the amount of solubilized oil or water is large, one observes either o/w microemulsions for $C_0 > 0$ and w/o microemulsions for $C_0 < 0$. When $C_0 \sim 0$, lamellar phases or bicontinuous microemulsions are observed.

C_0 can be varied in different ways. When a cosurfactant is used, C_0 depends on the surfactant/cosurfactant ratio in the surfactant layer

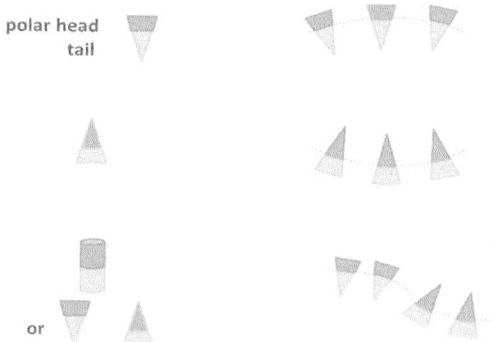

Fig. 2.11 Schematic representation of the packing of surfactant molecules with different packing parameters $v/(a_0 l_c)$ or spontaneous curvature C_0.

(Figure 2.11). The C_0 variation can be achieved by increasing the surfactant/cosurfactant ratio in the overall mixture with oil and water. But the cosurfactant being frequently soluble in oil and water, there is no control of the surfactant/cosurfactant ratio in the surfactant layer. C_0 is much better controlled in single surfactant systems. If the surfactant is ionic, the polar heads strongly repel each other in water and C_0 is positive, i.e. the layer is curved towards water. When salt is added, the electrostatic repulsion is screened; C_0 decreases and may become negative at high salt concentration. If the surfactant is nonionic, a similar trend is observed by rising the temperature, due to the release of the hydration water of the polar heads. When C_0 approaches zero, the structure become sponge-like or lamellar, depending on the value of the elastic constant K as will be seen in the following. However, this happens in a small range of C_0 values and droplet structures are the most frequent microemulsion structures.

When the composition of the medium is known, the droplet radius can be predicted quite accurately by using the following relation

$$R = \frac{3\phi}{ca}, \qquad (2.14)$$

where ϕ is the dispersed volume fraction, c is the number of surfactant molecules per unit volume, and a is the area per surfactant molecule. This relation expresses the fact that virtually all the surfactant molecules sit at the oil–water interface and that each of them occupies a well-defined area, independent of the composition.

In many cases, a continuous structural evolution from o/w to w/o via bicontinuous structures (Figure 2.12) can be observed. In order to describe this evolution, different types of space-filling models were proposed: Voronoi polyhedra [57] or cubes [22]. Cubes are particularly useful, because they allow the use of known results for Ising models and they reflect the fact that there is always a well-defined characteristic size in the microemulsion. A formula similar to equation (2.14) can be derived for the cube size ξ:

$$\xi = \frac{6\phi_o \phi_w}{ca}, \qquad (2.15)$$

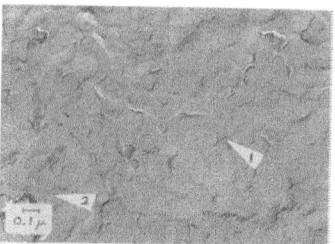

Fig. 2.12 Electron microscopy images of bicontinuous microemulsions displaying saddle-shaped structures (arrows 1 and 2). After [56].

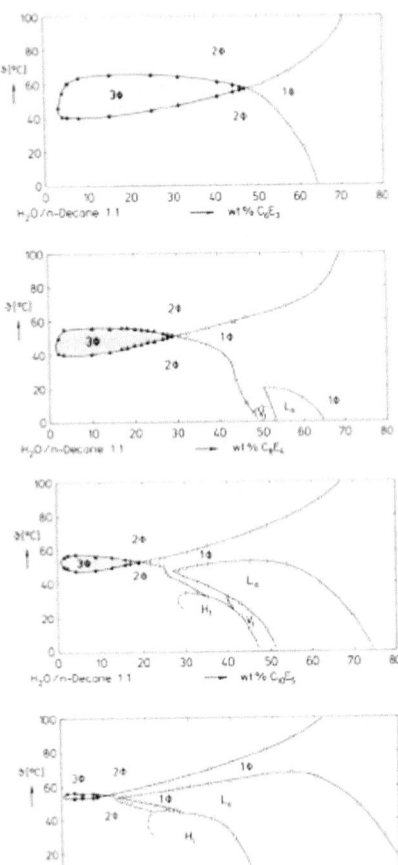

Fig. 2.13 Sections (water : oil = 1 : 1) through the phase diagrams of the system H_2O-n-decane-C_iE_j with increasing amphiphilicity of the surfactant. After [60].

where ϕ_o and ϕ_w are the oil and water volume fractions (in the limit of negligible surfactant volume fraction), respectively. There is usually some surfactant solubilized in the bulk water or oil. This amount corresponds to the limit concentration above which droplets are formed. It is very close to the critical micellar concentration (CMC) of the surfactant in either pure oil or water. It can be large in a few cases, as for nonionic alkyl polyoxyethylene ethers in oil. In these cases, the CMC has to be subtracted from c in equations (2.14) and (2.15), since R and ξ depend only on the amount of surfactant adsorbed at the oil–water interface.

The space-filling models show that the transitions between water or oil continuous and bicontinuous structures are percolation transitions and arise in the vicinity of ϕ_o and $\phi_w \sim 0.2$, respectively. This is as observed experimentally through the behavior of the electrical conductivity of w/o microemulsions (although ϕ_w is smaller in the case of strongly attracting droplets and electrical percolation may not occur in systems where the surfactant layer is very dense) [58].

The free energy can be evaluated simply for the cubic lattice. It is the sum of dispersion entropy, surface energy, and curvature energy. By neglecting the latter, the free energy per lattice site is:

$$F = k_BT(\phi_w \ln \phi_w + \phi_o \ln \phi_o) + 6\sigma\xi^2 \phi_o \phi_w, \quad (2.16)$$

an expression formally identical to the free energy of a regular binary solution. The problem is well known. A single-phase system is found for $\sigma < kT/\xi^2$. Typically, without surfactant, $\sigma_0 \sim kT/l^2 \sim 50\,\mathrm{mN/m}$, l being a molecular length. In order to get a large-scale dispersion, with ξ about 100 times larger than l, then $\sigma \approx 10^{-4}\sigma_0$. The microemulsion will therefore be thermodynamically stable only if σ is ultralow. σ can hardly be decreased below $10^{-3}\,\mathrm{mN/m}$; this means that the dispersion size will never exceed about 50 nm, as observed in practice.

The above model is very crude. It predicts that two-phase systems with symmetrical o/w and w/o microemulsions can coexist. This is never observed in practice, because of preferred curvature requirements. When the curvature energy is introduced within the frame of the cubic lattice, more realistic phase equilibria are found, which will be discussed in the next subsection [59]. In particular they predict that the maximum size in bicontinuous microemulsion is the persistence length ξ_k given by equation (2.9).

Bicontinuous microemulsions are found only when the bending elastic moduli are close to k_BT. Indeed, when the elastic modulus is larger than k_BT, ξ_k is large and the layers are flat over macroscopic distances; a lamellar phase is then obtained. Since this phase possesses long range orientational order, it differs from microemulsion phases which are disordered phases. There are indeed well-defined first order phase transitions between droplet microemulsion phases and lamellar phases when the surface layers are rigid (large K). These transitions are seen for instance when C_0 is varied by changing the temperature or another well-chosen parameter. The lamellar phases obtained when K approaches k_BT have large inter-lamellar distances. Indeed, this distance is determined by the repulsive forces between the surfactant layers across both oil and water. When $K \sim k_B T$, the repulsion is due to the large thermal undulations of the layers (entropic repulsion). Note that large inter-lamellar distances are also obtained with zwitterionic surfactants in which residual charges can create long-range electrostatic repulsions [20]. Large inter-lamellar distances of similar origin are also encountered in lamellar phases in oil.

When the elastic modulus is comparable to $k_B T$, ξ_k is microscopic, and a continuous inversion from w/o to o/w structures can be observed with intermediate bicontinous phases when C_0 is varied. When K is too small, ξ_k is of the order of a molecular length. The microemulsion structure begins to vanish; one obtains a molecular mixture, in which well defined surfactant layers are unable to form.

Lattice models were also proposed to describe microemulsions and analogies with existing Ising models were found [61,62]. For this purpose, a microscopic bending energy was used, with an interaction parameter J, related to the surfactant amphiphilicity, i.e. its probability to sit between oil and water sites. When J is large, lamellar structures are found; this is equivalent to large K. Indeed, K increases rapidly with surfactant length l ($K \sim l^3$, see page 58) as does the surfactant amphiphilicity. When J is small, the medium becomes completely structure-less, as a molecular mixture. This evolution has been observed with non ionic surfactants of variable molecular weight. Several phase diagrams of ternary oil–water–nonionic surfactant $C_i E_j$ are shown on Figure 2.13. In these phase diagrams, the oil and water amounts are equal [60]. When i and j are small, no liquid crystalline phases are seen in the phase diagrams, indicating that no well-defined surfactant layers are built (molecularly dispersed systems). At larger i and j, a lamellar phase L_α is formed at very low surfactant concentrations. For still larger i and j, the L_α phase forms at the expense of the microemulsion phase, as predicted by the theory. Precise determinations of characteristic sizes in the coexisting phases confirmed the theoretical predictions [63].

Swelling and phase diagrams

According to equation (2.14), when for instance oil is added to an o/w microemulsion, the droplet radius R increases. If the surfactant concentration is fixed, the total oil–water interfacial area is approximately constant. The minimization of the bending energy at constant total area leads to an optimal radius given by:

$$R_{\max} = \frac{2K + \bar{K}}{2K} R_0, \qquad (2.17)$$

with $R_0 = 1/C_0$. If $\bar{K} = 0$, $R_{\max} = R_0$; R_0 is in general lower, because in systems of spheres, $\bar{K} < 0$. Once $R_{\max}$ is reached during the swelling process, a further increase in the oil volume will increase the bending energy. An alternative possibility is to reject the oil as an excess phase, the bending energy cost rapidly exceeding the energy of the created

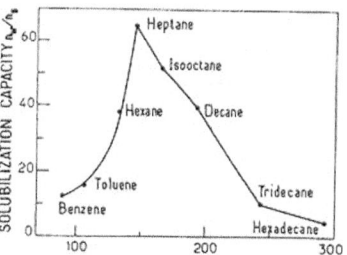

Fig. 2.14 Effect of the molecular volume of oil on the solubilization capacity of microemulsion AOT–oil–water at room temperature. The oil used was shown along each data point. Adapted from [64].

interface. This phase separation process was called *emulsification failure* [59]. The process leading to a w/o microemulsion in equilibrium with excess water is entirely similar. These two-phase equilibria are called, respectively, *Winsor I* and *II*.

In other cases, R_{max} cannot be reached. This happens when interactions between microemulsion droplets are attractive and increase with droplet radius. During the swelling process, the attraction thus increases. A phase separation of the liquid–gas type is then observed between a droplet-rich and a droplet-poor microemulsion phases. The maximum radius R'_{max} depends on the interaction potential. As for the liquid–gas transition, a critical point is found for a given microemulsion composition. The critical behavior of these systems has been extensively studied and non-classical (3D Ising) features have been evidenced in some cases.

Remarkably, R_{max} and R'_{max} have opposite variations with a number of parameters including oil and surfactant chain length and water salinity [64]. The role of oil chain length l_{oil} can be understood in terms of penetration in the surfactant layer; short-chain alkanes penetrate in the layer, increase v, and decrease the value of the packing parameter, hence R_{max} increases when l_{oil} increases. The limit alkane chain length is close to the surfactant chain length [65]. When the alkane penetration decreases, the interactions between surfactant chains become more attractive (as in a polymer–solvent mixture when the quality of the solvents decreases) and at some point lead to a phase separation, hence R'_{max} decreases when l_{oil} increases. The maximum droplet radius is largest at a particular chain length l^*_{oil} (Figure 2.14). In the two-phase region, the microemulsion coexists with excess water below l^*_{oil}, and with another microemulsion phase above l^*_{oil}. Close to l^*_{oil}, the formation of lamellar phases is sometimes obtained. The behavior observed when changing other parameters (water salinity, cosurfactant chain length, and cosurfactant/surfactant ratio) is similar and can be rationalized in the same way.

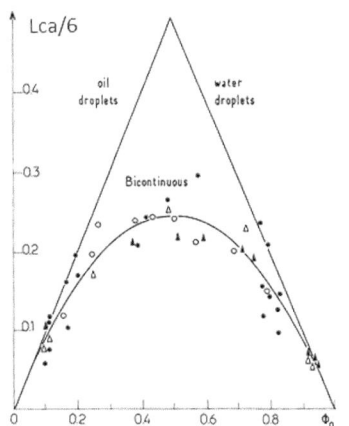

Fig. 2.15 Renormalized characteristic size in microemulsions (using $a = 60\,\text{Å}^2$) vs oil volume fraction. SDS-butanol-brine-toluene microemulsions (open symbols), SHBS-butanol-brine-toluene microemulsions (closed symbols). The straight lines are the predictions from equation (2.14) and the parabola from equation (2.15). Adapted from [66].

Bicontinuous structures can be swollen by both oil and water. When ξ reaches a limit value ξ_{max}, additional oil and water are rejected in excess phases and the microemulsion coexists with both oil and water (*Winsor III* equilibrium). The maximum swelling is controlled by the surfactant layer elasticity; the maximum average distance ξ_{max} between two oil or water domains is close to the persistence length of the layer ξ_k given by equation (2.9). A simple inspection of the phase diagrams such as those of Figure 2.13 confirms that the solubilization power is maximum in bicontinuous microemulsions; the surfactant concentration c in the phases in equilibrium with excess oil and/or water corresponds to the borderline of the single-phase region, from which one can estimate characteristic sizes using equations (2.14) and (2.15) (the smallest c creates the largest size; the bicontinuous microemulsions are found in the temperature range of the three-phase equilibria). Neutron and x-ray scattering determinations confirmed that ξ_{max} is smaller than $1/C_0$ (Figure 2.15) [66] and that ξ_{max} is close to ξ_k; the swelling is limited by thermal undulations in the surfactant layer [67].

Dynamic properties of droplet microemulsions

As micelles and other surfactant aggregates, microemulsion droplets constantly form and disappear in solution. Microemulsion droplets appear slightly polydisperse, but the polydispersity reflects shape fluctuations, and can be related to the elastic constants K and $\bar{K}$. Spin echo neutron scattering was used to investigate the shape fluctuations, allowing the determination of K and $\bar{K}$ [68]. These determinations were found to agree with independent determinations using droplet polydispersity measured by static neutron scattering experiments and surface roughness measured by ellipsometry [69].

Droplets lifetimes are very short although somewhat larger than those of micelles [41]. Because of fast molecular exchanges, encapsulation of substances in microemulsion droplets is not possible, in contrast to ordinary emulsions. Molecules can be transported from one drop to another without even going through the continuous phase. This involves collisions and transient merging of the droplets cores. The exchange process has been extensively studied in w/o microemulsions and the exchange rate and lifetime of the clusters formed during the collisions have been found to correlate with the increase in attractive interactions between droplets.

There is also a remarkable correlation between the nature of the interactions between droplets and the phenomenon of electrical percolation observed in w/o microemulsions. When the droplets volume fraction ϕ increases, a large electrical conductivity increase is observed for $\phi > \phi_p \sim 0.1$. This happens in systems in which the droplets interactions are sufficiently attractive [58].

Interfacial properties

Microemulsions can coexist with many different types of other phases: oil, water, lyotropic liquid crystals, or other microemulsions. Up to four different phases have been observed in equilibrium, and this number is certainly not a limit. A particularly important problem was the origin of the ultralow interfacial tensions between the microemulsion and the excess phases (Figure 2.16). The subject was very controversial in the 80's, but it is now well established that in most cases the tension σ is low because the surface pressure of the surfactant monolayer at the interface between the microemulsion and the excess phase is high and almost compensates the bare oil–water interfacial tension [70]. It was found that in these cases the interfacial tension σ is approximately $k_B T/L^2$, where L is either the droplet radius for microemulsions in equilibrium with excess oil or water (Winsor I or II equilibria) or ξ for microemulsions in equilibrium with both oil and water (Winsor III equilibrium) [56,71]. In this last case, the result applies to the largest tensions (higher branches of the curves such as those of Figure 2.16). In the case of drops, neglecting

Fig. 2.16 Interfacial tension between microemulsion and excess water phases versus brine salinity for different Winsor systems: brine-toluene-butanol-sodium dodecyl sulfate (SDS); brine-dodecane-butanol-sodium hexadecyl benzenesulfonate (SHBS); brine-toluene-butanol dodecyl trimethyl ammonium bromide (DTAB). The dispersion sizes are the smallest with DTAB and the largest with SHBS. Adapted from [70, 71].

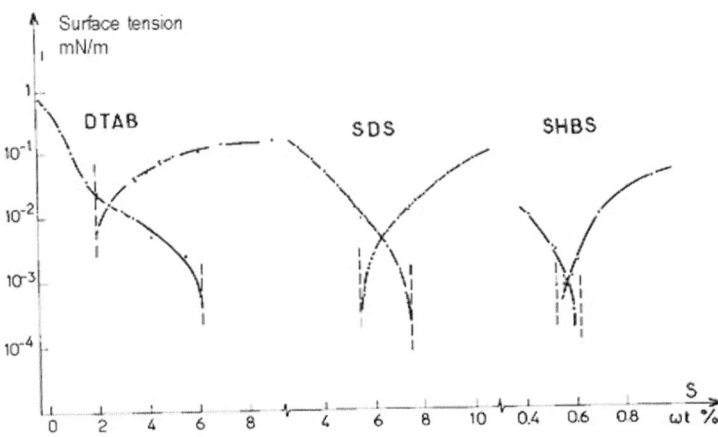

dispersion entropy contributions:

$$\sigma = \frac{2K + \bar{K}}{R^2} \qquad (2.18)$$

This expression was used to determine the elastic constants K and $\bar{K}$ which were found in agreement with other determinations [69]. It was also shown that the microemulsion phase does not wet the interface between oil and water phases, despite that the interfacial tensions are very low [72]. This is as predicted by theory, as well as an evolution from non-wetting to wetting behavior when the amphiphilicity of the surfactant decreases (evolution toward molecular mixtures) [63].

Interfacial structure has been probed with light reflectivity and ellipsometry [21]. The macroscopic interfaces are very rough, as are the microscopic ones. Indeed, interfacial tensions being ultralow, thermal fluctuations are very large; their mean square amplitude reaches values up to 1000Å. The short-wavelength fluctuations can be probed by ellipsometry, allowing the determination of the bending elastic modulus K.

The surfactant monolayer can also be disrupted at the macroscopic interface. The corresponding interfacial tensions are then the lowest; these low values have been ascribed to the proximity of critical end points in the phase diagrams [70]. They correspond to the lowest branches of the curves of Figure 2.16.

2.5 Non-equilibrium systems

Surfactants play important roles in systems where the surface–volume ratio is large such as drops, bubbles, and liquid films. When the surface tension is not ultralow, such systems are frequently away from thermodynamic equilibrium. Emulsions and foams, which are assemblies of drops and bubbles, respectively, belong to this category and will be described in Section 2.6.

When surfactant monolayers are present, surface tension gradients may arise during motion, giving rise to *Marangoni* forces, which can be quantified using the surface compression elastic modulus E shown in equation (2.4). When the modulus is high, the surfaces behave as *rigid* surfaces, when E is small, as *mobile* surfaces. The influence of exchanges between surfaces and bulk is sometimes treated using the effective surface modulus given by equation (2.6). This can however be incorrect, in particular when the motion is convective[10]. The motion can also involve viscous losses and the influence of the surface viscosities can be important. It is however difficult in general to disentangle the roles of E, η_E, and η_S.

[10] Equation (2.6) assumes diffusion controlled transport.

Drops and bubbles

The motion of very small drops or bubbles ($R < 1\,\mu$m) is Brownian-like, but when their size is larger than a few microns, they sediment or rise due to gravity. The rising velocity of a sphere of radius R and density ρ_i in a fluid of density ρ and viscosity η is (if $\rho > \rho_i$):

$$V = \frac{2R^2(\rho - \rho_i)g}{9\alpha\eta}, \quad (2.19)$$

where g is the acceleration of gravity and α is a numerical factor depending on viscous friction; $\alpha = 1$ if the sphere is solid and $\alpha = 2/3$ if the sphere is fluid [17] (this is because fluid is set in motion inside the sphere which reduces the drag force). If $\rho < \rho_i$, the sphere sediments and the velocity is given by the same expression, provided the sign of the density difference is changed.

In the case of drops or bubbles in water, the friction factor α is frequently found equal to one. This can be explained by the easy contamination of water and by the adsorption of surface active impurities at the surface of drops or bubbles. When the elastic modulus of surface compression and the viscosity are such that $E > \sigma/2$ or $\eta_E > 20\eta R$, a flat interface behaves as a rigid surface and the fluid stays immobile in the interior [17].

Surface elasticity also plays a role in the rupture of a drop in a laminar shear flow. The stress exerted on a drop by the flow is $\eta\dot{\gamma}$, where η is the continuous phase viscosity and $\dot{\gamma}$ is the velocity gradient; this stress is counteracted by the Laplace pressure $2\sigma/R$, where R is the drop radius. The ratio of both stresses is (half) the *capillary number*: $\text{Ca} = \eta\dot{\gamma}R/\sigma$. If Ca is larger than a critical value Ca_{cr}, the drop splits into two smaller drops (Taylor problem). Ca_{cr} depends on the type of flow and on the viscosity ratio η/η_i, η_i being the dispersed phase viscosity. When surfactant is present, it was found that Ca_{cr} is somewhat larger, especially in a concentration region close to the CMC [73]. The difference is attributed to the supplementary energy needed to counteract surface tension gradients, quantified by the surface compression elasticity. Indeed, there is a remarkable similarity between the surfactant concentration variations of Ca_{cr} and that of the compression elastic modulus E calculated at a

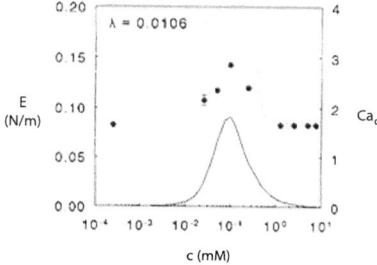

Fig. 2.17 Critical capillary number Ca_c (right vertical scale, points) for drop rupture in a shear flow as a function of added surfactant concentration. Calculated surface compression elastic modulus E at a frequency equal to the shear rate $\dot{\gamma}$ by using equation (2.6) (right scale, full line). Adapted from [73].

Liquid films on solid substrates

Wetting of films of a liquid (l) on a substrate (s) in a medium (f) depends on the value of the spreading parameter $S_{\text{wetting}} = \sigma_{\text{fs}} - (\sigma_{\text{fl}} + \sigma_{\text{ls}})$. Since wetting is a fast process, S_{wetting} cannot be evaluated using the equilibrium surface tensions. Figure 2.18 shows the plot of variation of S_{wetting} for a surfactant solution on a solid substrate versus surfactant concentration [74]. One sees that S_{wetting} is larger around CMC and is better correlated with surface elastic modulus than with surface tension γ (Figure 2.18).

Drops of surfactant solutions can undergo characteristic instabilities (Figure 2.18, bottom), that do not occur when the surfactant solution is used as a substrate and a simple liquid spreads on the top [75]. The origin of the instability is due to the fact that surfactant is depleted from the advancing wetting film and there is not enough surfactant in the film to replenish the surface; the spreading parameter increases with distance to the drop and, as a result, the film boundary is unstable. In the case where the substrate is the surfactant solution, the wetting film is in contact with the aqueous phase containing the surfactant, the monolayer replenishment is much faster, surface and bulk remain in equilibrium, and the spreading parameter can be evaluated using the equilibrium surface tensions.

Films can be also withdrawn from a liquid using a vertical solid substrate moving upward at a velocity V. In these conditions, a liquid layer of constant thickness h is formed, the capillary force σ/l_{cap} being balanced by the viscous force, of order $\eta h V$ (Landau–Levich problem). In the case of pure fluids and small enough velocities: $h = h_0 \text{Ca}^{2/3}$, Ca being the capillary number: $\text{Ca} = \eta V/\sigma$ and $h_0 \approx [\sigma/(\rho g)]^{1/2} = l_{\text{cap}}$ (*capillary length*) [17]. When a surfactant solution is used, the friction is increased (as in the cases discussed earlier with bubbles/drops) and h_0 is multiplied by a factor $4^{2/3}$ in the case of rigid surfaces. The intermediate case of finite surface elastic moduli has been addressed in reference [76]. Predictions for the case of finite surface viscosities have been recently tested with concentrated surfactant solutions for which the effective E is negligible and the corresponding surface viscosity was found compatible with known values [77].

The Landau–Levich law controls coating processes, and can be extended to films withdrawn on fibers [78]. Similar laws can also be found for films between tube walls and bubbles or drops traveling at a velocity V inside the tube (tube section smaller than the diameter of bubbles or drops). The film thickness is also proportional to $\text{Ca}^{2/3}$; this is the *Bretherton* law [79]. The motion of drops/bubbles is accompanied by a pressure drop, which depends on the surfactant concentration when surfactant solutions are used [80], as shown in Figure 2.19. Again the pressure drop exhibits a maximum close to the CMC, similar to what we

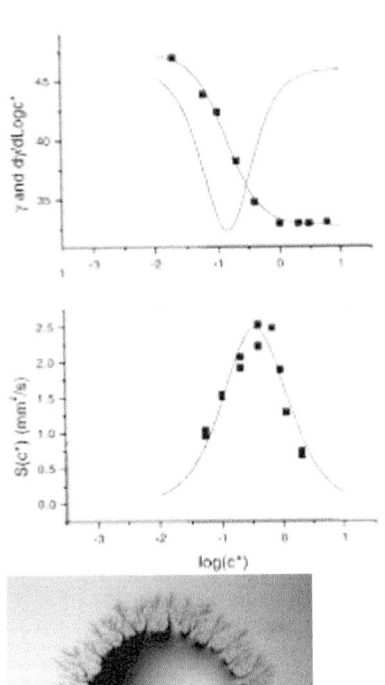

Fig. 2.18 Wetting of substrates by aqueous surfactant solutions. Surface tension (top) and reduced spreading velocity (middle) versus reduced surfactant concentration (c/CMC). Bottom: Image of the drop showing rim instabilities. After [74].

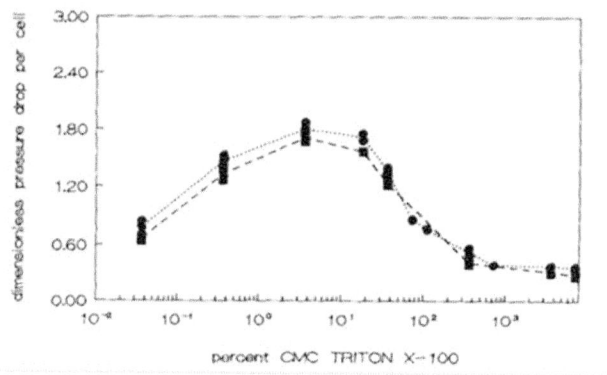

Fig. 2.19 Pressure drop vs surfactant concentration for an aqueous slug in contact with oil in a capillary, after [80].

saw in the threshold in drop rupture (Figure 2.17) and in the spreading parameter (Figure 2.18). This follows the trend of the effective surface elastic modulus variation predicted by equation (2.6); at small c, E increases with c and at large c, bulk-surface exchanges become fast and E decreases. A full hydrodynamic calculation (not relying on an effective elastic modulus) was used to obtain the theoretical curves shown in Figure 2.19.

The influence of the presence of surfactants on the Bretherton law is being actively investigated in connection with foam slippage at solid walls [81].

Freely suspended liquid films

In contrast to films on solid substrates, freely suspended films of pure liquids cannot be mechanically stable, because there is no force to balance the capillary force. If surfactant is added, surface tension gradients can be present and counteract the capillary force. If the film is withdrawn vertically from the solution at a velocity V, its thickness is in the case of rigid surfaces $2h_0 \text{Ca}^{2/3}$ (*Frankel* law) [82].

If left immobile, liquid films thin under the influence of gravity and capillary pressure; the menisci connecting the films with the holders are curved in such a way that the liquid pressure is smaller there than in the flat parts of the films. The thinning of films made out of surfactant solutions (*foam films*) has been extensively studied, in particular horizontal films, either formed between rising bubbles approaching the liquid surface or with devices in which the pressure can be controlled more easily: Sheludko cells [83], and porous plates [84].

Mysels distinguished several regimes in his early studies of vertical films [82]: *mobile* films where the film thickness does not remain uniform and sometimes surface turbulence can be observed and *rigid* films which drain much more regularly. Typical examples are films made from pure SDS solutions (mobile films) and mixed SDS-dodecanol solutions (rigid films). Dodecanol is nonionic and can be inserted between SDS molecules in the surface monolayer without changing the area per surfactant (60Å^2, whereas a hydrocarbon chain only occupies 20Å^2). This results in a much higher local density and in larger elasticities and viscosities [85].

The easiest way to estimate the velocity of film thinning V_f is to assume that the film surfaces are flat, parallel, and immobile, and that the fluid flows regularly from the center toward the Plateau borders. In such a simple case, an expression derived by Reynolds for the flow between two rigid plates brought together can be used:

$$V_f = -\frac{dh}{dt} = \frac{h^3}{3\eta r^2}\Delta P, \qquad (2.20)$$

where h is the film thickness, r is the film radius, and ΔP is the difference in pressure between the film center and border. When the film surfaces are not solid, surface flow may arise. Numerical calculations show that the influence of surface compression elasticity can be very important for thick enough films [86]. The surface elastic modulus that needs to be considered is the intrinsic modulus E_0; this is because surfactant is taken away along the surface to the Plateau borders and there is not enough surfactant in the foam film to replenish the surface ($h < l_{\text{ads}} = \Gamma/c$). When the film thickness further decreases, the surface velocity decreases and the role of the surface elasticity becomes negligible; for $h < 0.1$ mm and $E_0 \sim 10$ mN/m, V_f can be safely approximated by equation (2.20).

If films made of oil are immersed in a surfactant solution, the two sides of the film are in contact with the aqueous phase containing the surfactant and the surfactant monolayer replenishment during thinning is fast, because no surfactant depletion could occur. The effective surface elastic modulus to be considered in this second case is the low-frequency modulus, close to zero, and the corresponding thinning velocity is much larger than for a film of the surfactant solution immersed in oil. This argument has been invoked to explain the Bancroft rule, which states that the emulsion which forms is the one where the surfactant is in the continuous phase [45]. This is fully analogous to the differences described for wetting films on p.76: drops of surfactant solutions can undergo characteristic instabilities (Figure 2.18) that do not occur when the surfactant solution is used as a substrate and a simple liquid spreads on the top.

Coming back to the case of thick films, it can be shown that a *dimpling* instability frequently appears during the thinning of large circular horizontal films, in which the film is pinched off around its boundary and thicker in the center [87]. When the monolayer at the surface has a large surface elastic modulus, the dimple remains centro-symmetric and the velocity of thinning corresponds to the velocity of flat films with thicknesses equal to the smaller actual film thickness; see equation (2.20). When the surface elastic modulus is moderate (SDS solutions for instance), a second instability can take place in which the dimple loses its circular shape and is sucked away rapidly into the Plateau border. During this fast process the film frequently ruptures. The threshold for the second instability is predicted to depend upon the shear surface viscosity [88].

When the surfactant solution is above CMC, thinning becomes more complex. At the beginning of the 20th century, Johnnott and Perrin reported that foam film thinning occurs stepwise. Later on, it was shown that this was due to the presence of surfactant micelles which form layered structures below the film surfaces, and that thinning proceeds by expulsion of the layers, one after the other, into the Plateau border [89].

The thinning of films made of other types of concentrated solutions has been less extensively studied. The particular case of nonionic surfactant solutions above the cloud point can be mentioned. These solutions are cloudy because they slowly phase separate into a micelle-rich phase and a dilute phase, with surfactant concentration close to the CMC. In some cases, the surface tensions are such that droplets of the concentrated phase can enter into the interface between air and dilute phase and bridge to two sides of a thin film formed from the dilute phase [90]. This is one of the mechanisms of action of antifoams [91]. The thinning process leads then quickly to the rupture of the foam films once the film thickness is comparable to the size of the droplets of the concentrated phase (microns).

Surface forces

If the film thinning has been smooth enough so that no early rupture has occurred, a regime where interactions between the two sides of the film become significant can be attained; this is in a thickness region of about 50 nm and less. Typical interactions are van der Waals forces (attractive) and electrostatic forces (repulsive), called DLVO forces [92].

When the electrostatic repulsion is strong enough, an energy minimum can be found for thicknesses of about 10 nm. Short-range repulsive forces are also present and the different contributions to the force per unit area, called *disjoining pressure* Π_d can be written as:

$$\Pi_{d\ \text{vdw}} \sim -\frac{A}{6\pi\eta h^3}, \quad \Pi_{d\ \text{elect}} \sim Be^{-\kappa h}, \quad \Pi_{d\ \text{steric, hydration}} \sim Ce^{-h/l},$$
(2.21)

where A is the Hamaker constant, B and C are constants, κ^{-1} is the Debye-Hückel screening length, and l is the range of the short range forces, typically a few Å. If a lateral pressure ΔP is applied to the film (gravity, Laplace pressure), the thickness of the films decreases down to h_1 (Figure 2.20), and the corresponding equilibrium film is called *common black film*. If the pressure ΔP is larger than the electrostatic barrier, one may reach a very small film thickness where the water layer thickness is of order 1 nm: this is the so-called *Newton black film*.

In surfactant solutions containing micelles, structural oscillatory forces are also found, due to the local structuration into layers of micelles [84] (Figure 2.21). These forces are responsible for the step by step thinning of the liquid films.

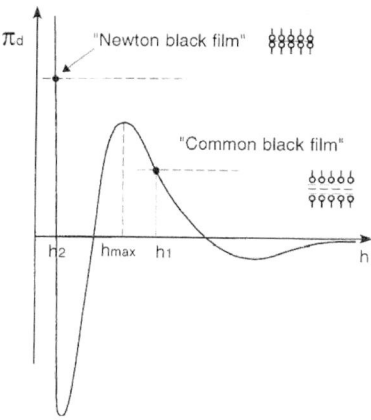

Fig. 2.20 Schematic representation of the variation of the disjoining pressure Π_d with film thickness h. The dotted lines correspond to different applied pressures ΔP and show the final equilibrium thicknesses of the film.

Film rupture

Different models for film rupture can be found in the literature. In the first type of model, worked out first by Sheludko [83] and later refined by Vrij [93], it is proposed that thermal thickness fluctuations can be amplified in some circumstances. If one writes (horizontal film, Figure 2.22):

$$h(x,y) = h_0 + \sum_q u_z(q) \exp(iq_x x + iq_y y),$$

then the mean-square amplitude of the fluctuation $u_z(q)$ is:

$$\langle u_z(q)^2 \rangle = \frac{k_B T}{2\sigma q^2 - \partial \Pi_d / \partial h}. \qquad (2.22)$$

In situations where Π_d decreases when h decreases, one can find q values for which the denominator of equation (2.22) vanishes. This happens for instance for $h < h_{\max}$ (Figure 2.20); if the short range forces are neglected, this process can be shown to lead to a film instability and rupture when a critical thickness h_c is reached.

When the film surfaces are covered by a dense surfactant monolayer, the short-range forces cannot be neglected. It has been shown in experiments done with a mica plate surface force machine that the fusion of bilayers adsorbed on the mica plates can only be achieved when these bilayers are formed from dilute surfactant solutions. Above CMC, the bilayers are compact and cannot be forced to fuse, even if pressures up to thousands of atmospheres are exerted between the mica plates [94]. Furthermore, the Vrij model predicts variations of the rupture time for rupture with film size, surface tension, and other parameters, variations that do not correspond to experimental observations. It is therefore likely that other mechanisms operate in the process of rupture of films covered by dense surfactant monolayers.

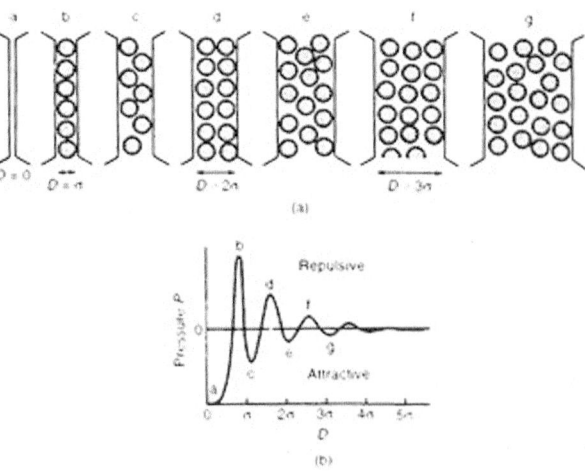

Fig. 2.21 Confinement of spheres between surfaces and oscillatory variation of the disjoining pressure (see later). After [92].

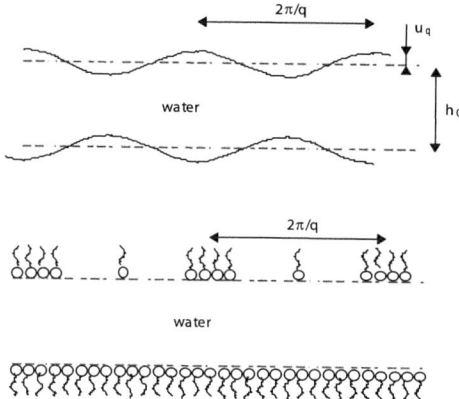

Fig. 2.22 Type of fluctuations leading to film rupture: thickness fluctuations (top); surface coverage fluctuation (bottom).

It has been proposed by other authors (Helm *et al.* for bilayer fusion [95], Exerowa *et al.* [96] for foam films, and de Gennes for emulsion films [97]) that the rupture occurs via thermal concentration fluctuations in the film surfaces. This process is controlled by compression elasticity rather than by surface tension and disjoining pressure given by equation (2.22), the mean-square amplitude of the density fluctuations in the surfactant layers being [15] (Figure 2.22):

$$\langle u_x(q)^2 \rangle = \frac{k_B T}{E q^2}. \qquad (2.23)$$

Once the surfactant layer is less dense at some point, the chains can tilt toward the opposite layer, and the hydrophobic attraction between opposite surfaces will be enhanced; this is the instability mechanism proposed by Helm *et al.* [95]. Of course, the state of the models is very preliminary, as is also the problem from the experimental side.

Correlations between pressure thresholds ΔP_c, rupture times τ_r, and compression elasticity E are lacking. The measurement of ΔP_c and τ_r for single films is not easy and the reproducibility of the results is poor. Indeed, they depend on many parameters, especially the velocity of thinning; if the velocity is large, the film may rupture for small applied pressures at which it would otherwise remain stable for long times if the thinning is slower. Even in identical conditions, ΔP_c and τ_r vary considerably and statistical averages need to be performed. This is why very few systematic studies have been made. Less controlled experiments can be performed with oil bubbles rising or water bubbles falling toward a flat water–oil interface. The coalescence times are larger for drops of the emulsion dispersed phase, in agreement with the Bancroft rule (Figure 2.23) [98].

2.6 Emulsions and foams

Emulsions are fine dispersions of oil in water (o/w) or of water in oil (w/o), with drop sizes usually in the range of microns, stabilized as

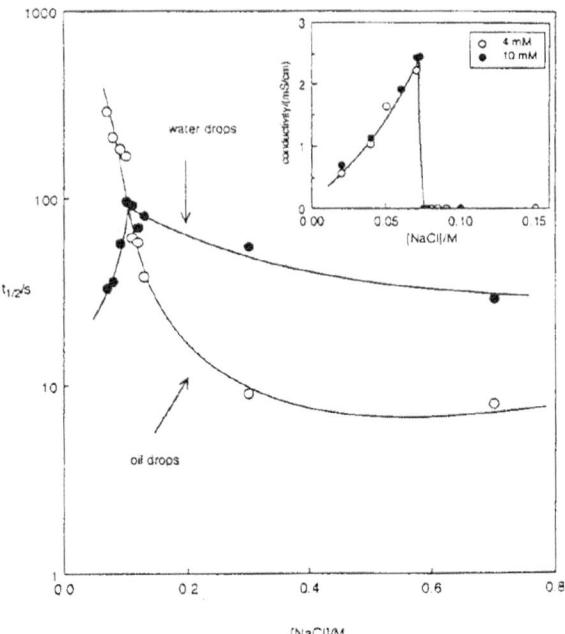

Fig. 2.23 Drop coalescence time vs added salt (NaCl) in water; heptane is used as oil and aerosolOT (AOT) as surfactant. The inset shows the conductivity of the corresponding emulsion (o/w at low salt concentration, w/o at large salt concentration). After [98].

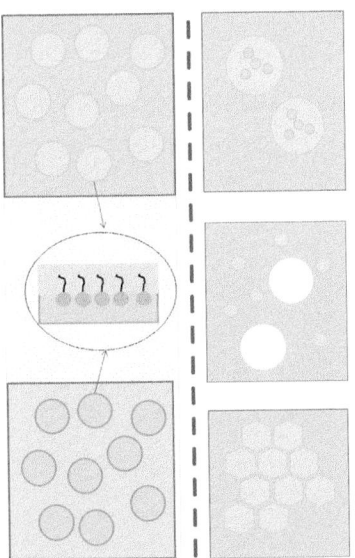

Fig. 2.24 Schematic representation of emulsion structures: o/w emulsion (upper left); w/o emulsion (lower left); multiple emulsion (lower right); aerated emulsion (middle right); gel emulsion (top right). Encircled: Enlarged view of a surfactant monolayer sitting at the oil-water interface.

microemulsions by surfactant monolayers (Figure 2.24). In some cases, multiple emulsions can be formed: o/w/o or w/o/w. Foams are coarser dispersions of air in a liquid, generally water, with bubble sizes usually in the range 100 μm–1 mm, also stabilized by surfactant monolayers. Some emulsions (frequently food emulsions) can be *aerated*, i.e. contain air bubbles, and are foams in which the liquid phase is an emulsion.

Emulsion and foam properties do not depend only on the nature of the liquid phases and of the stabilizing agents, but also on the disperse volume fraction ϕ. For instance, emulsions and foams have a finite shear modulus if ϕ is larger than about 0.6, and are solid-like. This is because the drops are no longer spherical at these volume fractions, they deform into polyhedra and elastic energy is stored at the expense of surface energy. These emulsions are sometimes called *biliquid foams*.

In order to make emulsions or foams, an energy is needed to create new surface area; this energy is equal to σA, σ being the surface tension and A the area created. The size of the dispersion is too large for the dispersion entropy to compensate the surface energy as in the case of microemulsions (p.68). As a consequence, emulsions and foams are thermodynamically unstable. However, as will be detailed in the following, metastable configurations can be attained, in which each drop or bubble takes a shape of minimal area: spheres for isolated drops/bubbles, polyhedra for large ϕ. The shape of these polyhedra is still an unsolved mathematical problem [99].

Bancroft rule and HLB

About 100 years ago, Bancroft proposed that when oil, water, and surfactant are mixed, the continuous phase of the emulsion that forms is the phase where the surfactant is more soluble. The justification of the rule came only in the late 60's when Lee and Hogdson attributed it to the role of surface tension gradients as discussed on page 77. So far, very few exceptions to the Bancroft rule were found [100]. In these exceptions, the surfactant concentration is very small, and it was proposed that the type of emulsion is then rather determined by the emulsification process: indeed the emulsions containing no surfactant were of the same type.

The wide applicability of the rule made it extremely useful to formulate emulsions. The main difficulty is that surfactant solubility can depend on the water pH and salinity and also on the temperature and the type of oil. The rule is mainly used for non-ionic surfactants that are insensitive to the presence of ions in water. The Bancroft rule was the basis of the empirical scale called the hydrophilic-lipophilic balance (HLB). Surfactants with HLB larger than 7 lead to o/w emulsions and HLB lower than 7 to w/o emulsions.

In the case of microemulsions, the Bancroft rule and the HLB concept have been rationalized in terms of spontaneous curvature of the surfactant layer: if the surfactant is more soluble in water, the interface will curve around the oil phase and vice versa.[11]

Emulsification and foaming procedures

There is a large variety of emulsification and foaming devices: simple shaking, mixing with rotor-stator systems, injection of liquid or gas through porous membranes, decompressing gas (foams), mixing with a porous piston (*mighty whipper*), turbulent mixing of liquid and gas jets, etc. [102]. When the emulsion or the foam is formed, the interfaces are stretched rapidly and ruptured in different flow conditions according to the method used: laminar with low-shear mixers, turbulent with high-shear machines, extensional with jets and porous membranes. It has been shown in the case of emulsions, where similar devices are used, that for a given energy input, the dispersion size varies by orders of magnitude [103, 104]. Indeed, different types of flow lead to different rupture mechanisms. For instance, in an extensional flow, one drop can be strongly elongated and then ruptured into several smaller droplets after a Rayleigh type of instability. This process leads to more monodisperse droplets than for rupture of drops in a shear flow. The surface compression elasticity also plays a role, although limited (see page 76): the energy used during the emulsification process is smaller if the modulus E is smaller. However, if E is decreased, coalescence can occur (see page 80) and a compromise needs to be found. Although no detailed studies of this kind exist for foams, it is likely that the general trends are the same.

There is a minimum concentration of surfactant molecules required for emulsification or foam production. This concentration is associated to a

[11] Several authors extended this argument to emulsions, although as early as 1941, Hildebrand commented that the curvature of emulsion droplets was too small for the argument to be valid. Deviations between spontaneous and actual curvature are associated to a bending energy with two elastic moduli, K and $\bar{K}$, which adds to the surface energy (see page 68). Kabalnov and coworkers applied these concepts to emulsions [101].

Fig. 2.25 Bottom: Picture of a foam made of two layers of bubbles, where one sees that the bubbles of one layer sit around the interstices between the bubbles of the layers just above or below, after [105].

surface coverage of the bubbles of the order of $1\,\mathrm{mg/m^2}$ and corresponds to the point where the curve $\sigma(c)$ shows a sharp decrease. This occurs at a bulk concentration of the order of CMC/10, and is associated to the onset of formation of Newton black films and to the appearance of very short-range repulsive forces between drops (see page 79). Below this concentration, coalescence occurs spontaneously, possibly via the Vrij–Sheludko mechanism, when the film thickness reaches a critical value h_c. The total surfactant concentration required includes the amount adsorbed at the surface of the bubbles and therefore depends on the method used; this amount is higher for high-energy mixers than for low-energy ones, such as whipping [106]. This is likely associated to the fact that high-energy mixing produces small bubbles, hence high surface area, and the amount of required surfactant molecules thus larger.

Important advances in the fundamental understanding of emulsions were made when procedures to produce monodisperse emulsions were discovered. The first method was proposed by Bibette in 1991 [107], and is analogous to crystallization fractionation: excess surfactant is added to a standard emulsion and produces depletion flocculation of the larger drops, which can then be separated from the rest of the emulsion [107]. When the procedure is repeated several times, the final emulsion can have a polydispersity as low as 10% or less. Rather monodisperse emulsions can also be produced with liquid jet devices and membranes. Another interesting method is the application of a sudden shear to a viscous emulsion [107]. Still better performances are obtained with microfluidic devices. Bubbles can also be produced in this way and are so remarkably monodisperse that they crystallize readily [105].

Some non-ionic surfactants produce o/w emulsions at low temperatures and w/o emulsions at high temperatures. When the temperature is close to the inversion temperature (PIT), Shinoda and Friberg showed that rather monodisperse emulsions can spontaneously be formed [108]. Miller and Rang found other spontaneous emulsification processes occurring when surfactant bilayers are present in one of the liquid phases [109]. In all these cases, the interfacial tension between oil and water is very low: $0.1\,\mathrm{mN/m}$ or less. The origin of these processes is still not yet fully elucidated.

Emulsion and foam stability

The time evolution of emulsions and foams at rest is governed by different processes occurring spontaneously, Ostwald ripening, sedimentation or creaming, and coalescence.

Ostwald ripening is the drop growth process occurring when the dispersed phase has a finite solubility into the continuous phase and can migrate between drops or bubbles of different sizes. Lifshitz and Slyozov predicted the corresponding time variation of drop/bubble radius in the case of dilute dispersions ($\phi \ll 1$):

$$R(t) = \left(\frac{8\sigma D c_{\text{eq}} v_m^2}{9 k_B T}\right)^{1/3} t^{1/3}, \quad (2.24)$$

where c_{eq} is the equilibrium concentration of the molecules of the dispersed phase in the continuous one, D is their diffusion coefficient there, and v_m is their molecular volume. It was found in experiments that the ripening process is either slower or larger than predicted by equation (2.24) [110, 111]. Faster velocities may occur at large drop/bubble volume fractions making exchanges between drops/bubbles easier. Indeed, at large ϕ, $R \sim t^{1/2}$ instead of $t^{1/3}$ [99].

Let us also mention the importance of the nature of the oil or gas used. Oils soluble in water such as short-chain alkanes give less stable o/w emulsions than less soluble ones such as long-chain alkanes or silicon oils, because transport across water films is slower in the last case. The stability of emulsions made with short-chain alkanes can be improved by adding small amounts of long-chain alkanes; since the oil composition in each drop cannot change (otherwise, the chemical potential would vary locally), the oil diffusion process is slowed down [112]. Similarly, gases soluble in water such as CO_2 give less stable foams than less soluble ones such as N_2, because CO_2 transport across water films is faster. The stability of CO_2 foams can be improved by adding small amounts of nitrogen [113].

During the shrinking/growth process of small drops or bubbles, the surface area is reduced/increased, and the surface elasticity opposes the evolution. Van Vliet and coworkers predicted an influence of the elasticity on the rate of ripening [114]. This was evidenced recently, both in emulsions and foams [115, 116]. It is also predicted that when $E > \sigma/2$, the ripening stops. However, ripening being slow, surfactant molecules have time to exchange with bulk and the effective elastic modulus can never reach values of order $\sigma/2$.

Gravity produces sedimentation or rising (creaming) of emulsion drops; see equation (2.19). The drop volume fraction increases with time, either in the bottom or at the top of the emulsion sample, where drops concentrate locally. Foam bubbles raise similarly, the process being called *foam drainage*.

When the dispersed volume fraction of the emulsion or the foam is large, the drops or bubbles are no longer spherical, they distort into polyhedra, the flattened regions being the liquid films. Gravity still produces drainage, a subject extensively studied in foams [99, 118, 119]. The liquid then flows through the interstitial spaces between drops or bubbles, which are either thin films, *Plateau borders* (PBs) made of connections of three films, and *nodes* made of connections of four PBs (Figure 2.26).

Two regimes were predicted corresponding respectively to mobile and rigid surfaces (small or large E). The transition between the two regimes has been observed with mixed solutions of sodium dodecyl sulfate (SDS) and dodecanol, upon increasing the amount of dodecanol (Figure 2.27)

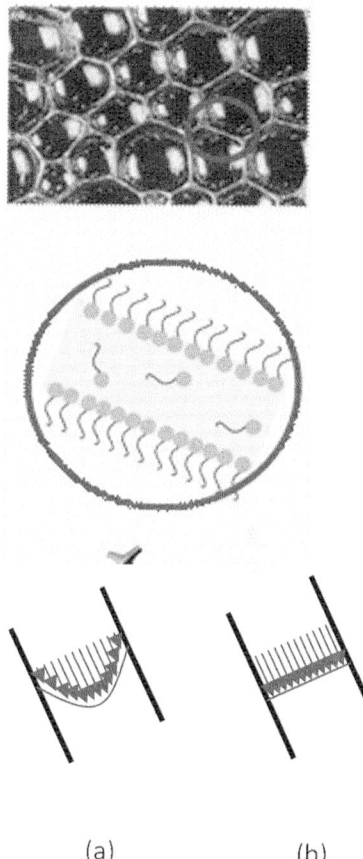

Fig. 2.26 Aqueous foam (top) where polyhedric bubbles are seen. Liquid film between bubbles covered by surfactant monolayers (middle). Schematic network of Plateau borders (bottom).

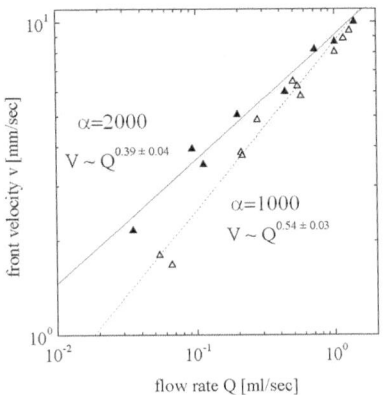

Fig. 2.27 Top: Schematic representation of Poiseuille flow (a) and plug flow (b) in Plateau borders. Bottom: Front velocity V versus flow rate Q for foams made with pure SDS-dodecanol solutions containing different SDS/dodecanol mass ratio α. Adapted from [117].

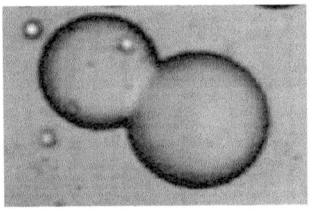

Fig. 2.28 Adhesive drops. After [107].

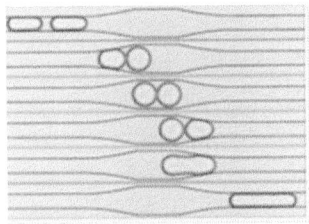

Fig. 2.29 Sequence of two droplets traveling and coalescing in a microfluidic device with channels of varying width (from 36 to $72\,\mu$m); the time elapsed between top and bottom pictures is 10 ms. After [123].

[117]. For small amounts of dodecanol, the bubble surface is mobile, and the flow in the PBs is quasi plug-like; for larger amounts of dodecanol, the surface becomes rigid and the flow is Poiseuille-like.

With time, the films separating the drops or the bubbles thin and eventually reach an equilibrium value (see Figure 2.20). If the electrostatic repulsion is screened for instance by adding salt, the electrostatic barrier is lowered, and it becomes easier to reach the second force minimum, which is quite deep, many $k_B T$ in terms of energy. In this case, emulsion drops may flocculate and *adhesive* emulsions are obtained (Figure 2.28). When the continuous phase (film phase) contains micelles, macromolecules, or particles, these objects will be expelled from the film when its thickness is comparable to their size. As a consequence, the film zone will be depleted and osmotic attraction between the surfaces will arise. This is the so-called *depletion force*, which can also lead to drop flocculation [107]. When the micelle concentration is large, the attractive depletion force turns into an oscillatory force, due to the local structuration of the film into layers of micelles, as seen on page 79, and several equilibrium film thicknesses can be reached depending on the externally applied pressure.

Coalescence

In dilute emulsions, transient films can be formed during collisions between drops. In concentrated emulsions and in foams, the drops or the bubbles are separated by films. Coalescence of drops and bubbles occur when these films are ruptured. Coalescence is therefore closely related to the film stability issues discussed in the section on page 80. Note that film thinning in emulsions and foams is different in nature, because emulsion droplets are much smaller than foam bubbles, and therefore much less sensitive to all the instabilities (dimpling) that may lead to film rupture in the early stages of a foam life.

It is very difficult to study coalescence in a controlled way. It is generally admitted that there is a threshold for coalescence; at rest, coalescence occurs above a certain drop/bubble volume fraction, or, if the volume fraction is fixed, above a certain drop/bubble size (reached for instance after a certain time because of Ostwald ripening). A more complete description can be found in [120]. However, most of the existing data correspond to emulsions and foams with broad size distributions and are therefore difficult to interpret.

No obvious correlation exists between coalescence and surface forces [121]. In some cases, larger repulsion even leads to less stable foams [122]. This illustrates the fact that in contrast with the case of solid colloidal particles, where the stability is well correlated with surface forces, drops and bubbles behave differently: the surfactant can move along film surfaces, and the interaction forces are probably strongly modified during the coalescence process.

It has been noted by Shinoda and Friberg that foams and emulsions made when lamellar phases are present can be very stable. When the lamellar phase is viscous, one obvious effect is the slowing down of the sedimentation/creaming process. A more subtle effect could be due to surface tensions balance; if the lamellar phase domains cannot enter the films, they will be trapped in the film borders [108].

Coalescence under shear is still less well understood. Shear has two competitive actions: it can fraction the drops or bubbles (Section 10) or favour their fusion; under shear, hydrodynamic forces create local increase in drop volume fraction, that depends on shear rate, and coalescence can be observed above a certain shear rate. Recently, microfluidic devices with channels of variable width have been designed to control coalescence [123]. In these devices, extremely monodisperse drops can be produced and directly observed in controlled flow conditions. It has been shown that coalescence occurs not when the drops come close to each other, but rather when they split apart (Figure 2.29). This has been attributed to a hydrodynamic instability [124], very different from that leading to coalescence at rest.

The problem of emulsion and foam destruction is very important in practice, and coalescence can be promoted by using additives:

- In the case of emulsions, adding surfactants soluble in the dispersed phase; in this way, surface tension gradients are suppressed and film thinning is faster [125].
- Adding particles in the continuous phase, able to spread or bridge the two surfaces of the film [91]. Such particles are spontaneously nucleated above the PIT (Section 2.24) in o/w emulsions made with non-ionic surfactants; the surfactant solubility in water decreases, phase separation between surfactant-poor and surfactant-rich phases occurs, and the droplets of the nucleated surfactant rich phase bridge and rupture the films between gas bubbles [90] or emulsion drops [126].

Conclusion

Surfactant molecules can self-assemble in bulk and create a rich variety of structures, either equilibrium structures (such as micelles, microemulsions, sponge phases, lyotropic liquid crystalline phases), and non-equilibrium structures (such as vesicles, emulsions and foams). The equilibrium behavior of surfactant systems is now rather well understood, but the non-equilibrium behavior is still unclear in most cases.

Surfactants also form self-assembled monolayers at liquid interfaces, which dynamic behavior is controlled not only by surface tension but also by surface viscoelastic parameters. A behavior intermediate between that of rigid and mobile surfaces is usually observed. Surface viscoelasticity is difficult to measure, hence the lack of understanding of many related phenomena. The most poorly understood issue is probably the problem of coalescence of bubbles and drops.

Many applications of surfactant systems make use of their very rich dynamic behavior and it is desirable to achieve a better understanding of the microscopic phenomena controlling the macroscopic behavior.

References

[1] Adamson, A. W. and Gast, A. P. (1997). *Physical Chemistry of Surfaces* (6th revised edn). Wiley, New York.
[2] Evans, F. and Wennerström, W. (1999). *The Colloidal Domain* (2nd edn). Wiley, New York.
[3] De Gennes, P. G. (1979). *Scaling Concepts in Polymer Physics*. Cornell University Press, Intaca, NY.
[4] Hill, R. M. (1998). *Curr. Opin. Coll. Interface Sci.*, **3**(3), 247–254.
[5] Lu, J. R., Hromadova, M., Thomas, R. K. *et al.* (1993). *Langmuir*, **9**(9), 2417–2425.
[6] Weissenborn, P. K. and Pugh, R. J. (1996). *J. Coll. Interface Sci.*, **184**(2), 550–563.
[7] Gaines, G. (1966). *Insoluble Monolayers at Liquid–Gas Interfaces*. Interscience, New York.
[8] Ward, A. F. H. and Tordai, L. (1946). *J. Chem. Phys.*, **14**(7), 453–461.
[9] Dukhin, S. S., Kretzschmar, G., and Miller, R. (1995). *Dynamics of Adsorption at Liquid Interfaces*. Elsevier, New York.
[10] Lee, S., Kim, D. H., and Needham, D. (2001). *Langmuir*, **17**(18), 5537–5543.
[11] Bonfillon, A., Sicoli, F., and Langevin, D. (1994). *J. Coll. Interface Sci.*, **168**(2), 497–504.
[12] Diamant, H. and Andelman, D. (1996). *J. Phys. Chem.*, **100**(32), 13732–13742.
[13] Hua, X. Y. and Rosen, M. J. (1988). *J. Coll. Interface Sci.*, **124**(2), 652–659.
[14] Lee, S., Kim, D. H., and Needham, D. (2001). *Langmuir*, **17**(18), 5544–5550.
[15] Langevin, D. (1992). In *Light Scattering by Liquid Surfaces* (ed. D. Langevin), pp. 161–201. Dekker, New York.
[16] Landau, L. and Lifshitz, E. (1959). *Theory of Elasticity*. Addison Wesley, Boston, MA.
[17] Levich, V. G. (1962). *Physicochemical Hydrodynamics*. Prentice Hall, Englewood Cliffs, NJ.
[18] van den Tempel, M. and Lucassen-Reynders, E. H. (1983). *Adv. Coll. Interface Sci.*, **18**(3–4), 281–301.
[19] Bonfillon, A. and Langevin, D. (1994). *Langmuir*, **10**(9), 2965–2971.
[20] Gelbart, W. M., Ben-Shaul, A., and Roux, D. (1994). *Micelles, Membranes, Microemulsions and Monolayers*. Springer, New York.
[21] Meunier, J. (1992). In *Light Scattering by Liquid Surfaces and Complementary Techniques* (ed. D. Langevin), pp. 333–364. Dekker, New York.

[22] de Gennes, P. G. and Taupin, C. (1982). *J. Phys. Chem.*, **86**(13), 2294–2304.
[23] Goodrich, F. C. (1981). *Proc. R. Soc. A: Math. Phys. Eng. Sci.*, **374**(1758), 341–370.
[24] Joly, M. (1972). In *Surface and Colloid Science* (ed. E. Matijevic), Volume 5. Interscience, New York.
[25] Miller, R., Wustneck, R., Kragel, J. *et al.* (1996). *Coll. Surf. A: Physicochem. Eng. Aspects*, **111**, 75–118.
[26] Schwartz, D. K., Knobler, C. M., and Bruinsma, R. (1994). *Phys. Rev. Lett.*, **73**(21), 2841–2844.
[27] Steffen, P., Wurlitzer, S., and Fischer, T. M. (2001). *J. Phys. Chem. A*, **105**(6), 8281–8283.
[28] Ortega, F., Ritacco, H., and Rubio, R. G. (2010). *Curr. Opin. Coll. Interface Sci.*, **15**(6), 237 – 245.
[29] Espinosa, G. and Langevin, D. (2009). *Langmuir*, **25**(20), 12201–12207.
[30] Brooks, C. F., Fuller, G. G., Frank, C. W. *et al.* (1999). *Langmuir*, **15**(7), 2450–2459.
[31] Miyano, K. (1992). In *Light Scattering by Liquid Surfaces and Complementary Techniques* (ed. D. Langevin), pp. 311–331. Dekker, New York.
[32] Jayalakshmi, Y., Ozanne, L., and Langevin, D. (1995). *J. Coll. Interface Sci.*, **170**(2), 358–366.
[33] Petkov, J. T., Gurkov, T. D., Campbell, B. E. *et al.* (2000). *Langmuir*, **16**(8), 3703–3711.
[34] Zang, D. Y., Rio, E., Langevin, D. *et al.* (2010). *Eur. Phys. J. E*, **31**(2), 125–134.
[35] Lemaire, C. and Langevin, D. (1992). *Coll. Surf.*, **65**(2–3), 101–112.
[36] Ritacco, H., Cagna, A., and Langevin, D. (2006). *Coll. Surf. A: Physicochem. Eng. Aspects*, **282**, 203–209.
[37] Yeung, A. and Zhang, L. C. (2006). *Langmuir*, **22**(2), 693–701.
[38] Gourier, C., Daillant, J., Braslau, A. *et al.* (1997). *Phys. Rev. Lett.*, **78**(16), 3157–3160.
[39] Israelachvili, J. N., Mitchell, D. J., and Ninham, B. W. (1976). *J. Chem. Soc., Faraday Trans. II*, **72**, 1525–1568.
[40] Tanford, C. (1973). *The Hydrophobic Effect*. Wiley, New York.
[41] Zana, R. and Lang, J. (1987). In *Surfactant Solutions* (ed. R. Zana), Volume 22, pp. 405–452. Dekker, New York.
[42] Lunkenheimer, K., Czichocki, G., Hirte, R. *et al.* (1995). *Coll. Surf. A: Physicochem. Eng. Aspects*, **101**(2–3), 187–197.
[43] Aniansson, E. A. G., Wall, S. N., Almgren, M. *et al.* (1976). *J. Phys. Chem.*, **80**(9), 905–922.
[44] Gauffre, F. and Roux, D. (1999). *Langmuir*, **15**(9), 3070–3077.
[45] Ruckenstein (1996). *Langmuir*, **12**(26), 6351–6353.
[46] Cates, M. E. and Candau, S. J. (1990). *J. Phys.: Cond. Mat.*, **2**(33), 6869–6892.
[47] Zana, R. (2002). *Adv. Coll. Interface Sci.*, **97**(1–3), 205–253.

[48] Kaler, E. W., Herrington, K. L., Murthy, A. K. *et al.* (1992). *J. Phys. Chem.*, **96**(16), 6698–6707.

[49] Dubois, M., Deme, B., Gulik-Krzywicki, T. *et al.* (2001). *Nature*, **411**(6838), 672–675.

[50] Zemb, T., Dubois, M., Deme, B. *et al.* (1999). *Science*, **283**(5403), 816–819.

[51] Porte, G., Appell, J., Bassereau, P. *et al.* (1989). *J. de Physique*, **50**(11), 1335–1347.

[52] Alexandridis, P., Olsson, U., and Lindman, B. (1998). *Langmuir*, **14**(10), 2627–2638.

[53] Salmon, J. B., Manneville, S., and Colin, A. (2003). *Phys. Rev. E*, **68**, 051503.

[54] Hoffmann, H. and Ulbricht, W. (1989). *J. Coll. Interface Sci.*, **129**(2), 388–405.

[55] Evans, D. F., Mitchell, D. J., and Ninham, B. W. (1986). *J. Phys. Chem.*, **90**(13), 2817–2825.

[56] Strey, R. (1994). *Coll. Polym. Sci.*, **272**(8), 1005–1019.

[57] Talmon, Y. and Prager, S. (1978). *J. Chem. Phys.*, **69**, 2984–2992.

[58] Cazabat, A. M., Chatenay, D., Langevin, D. *et al.* (1983). *Faraday Disc.*, **76**(76), 291–303.

[59] Safran, S. A. (1994). In *Micelles, Membranes, Microemulsions and Monolayers* (ed. W. M. Gelbart, A. Ben-Shaul, and D. Roux), pp. 427–484. Springer, New York.

[60] Kahlweit, M., Strey, R., and Firman, P. (1986). *J. Phys. Chem.*, **90**(4), 671–677.

[61] Widom, B. (1986). *J. Chem. Phys.*, **84**(12), 6943–6954.

[62] Gompper, G. and Schick, M. (1990). *Phys. Rev. B*, **41**(13), 9148–9162.

[63] Gradzielski, M., Langevin, D., Sottmann, T. *et al.* (1996). *J. Chem. Phys.*, **104**(10), 3782–3787.

[64] Hou, M. J. and Shah, D. O. (1987). *Langmuir*, **3**(6), 1086–1096.

[65] Binks, B. P., Kellay, H., and Meunier, J. (1991). *Europhys. Lett.*, **16**(1), 53–58.

[66] Guest, D., Auvray, L., and Langevin, D. (1985). *J. de Physique Lett.*, **46**(22), 1055–1063.

[67] Langevin, D., Guest, D., and Meunier, J. (1986). *Coll. Surf.*, **19**(2–3), 159–170.

[68] Hellweg, T. and Langevin, D. (1998). *Phys. Rev. E*, **57**(6), 6825–6834.

[69] Gradzielski, M., Langevin, D., and Farago, B. (1996). *Phys. Rev. E*, **53**(4), 3900–3919.

[70] Cazabat, A. M., Langevin, D., Meunier, J. *et al.* (1982). *Adv. Coll. Interface Sci.*, **16**(JUL), 175–199.

[71] Binks, B. P., Meunier, J., Abillon, O. *et al.* (1989). *Langmuir*, **5**(2), 415–421.

[72] Chatenay, D., Abillon, O., Meunier, J. *et al.* (1985). In *Macro and Microemulsions* (ed. D. O. Shah), Volume 272, pp. 119–132. American Chemical Society.

[73] Janssen, J. J. M., Boon, A., and Agterof, W. G. M. (1994). *Aiche J.*, **40**(12), 1929–1939.
[74] Cachile, M. and Cazabat, A. M. (1999). *Langmuir*, **15**(4), 1515–1521.
[75] Bergeron, V. and Langevin, D. (1996). *Phys. Rev. Lett.*, **76**(17), 3152–3155.
[76] Shen, A. Q., Gleason, B., McKinley, G. H. et al. (2002). *Phys. Fluids*, **14**(11), 4055–4068.
[77] Scheid, B., Delacotte, J., Dollet, B. et al. (2010). *Europhys. Lett.*, **90**, 24002.
[78] Quere, D. (1999). *Ann. Rev. Fluid Mech.*, **31**, 347–384.
[79] Bretherton, F. P. (1961). *J. Fluid Mech.*, **10**(2), 166–188.
[80] Stebe, K. J. and Maldarelli, C. (1994). *J. Coll. Interface Sci.*, **163**(1), 177–189.
[81] Denkov, N. D., Tcholakova, S., Golemanov, K. et al. (2009). *Soft Matter*, **5**(18), 3389–3408.
[82] Mysels, K., Shinoda, K., and Frankel, S. (1959). *Soap Films*. Pergamon, Oxford.
[83] Sheludko, A. (1967). *Adv. Coll. Interface Sci.*, **1**(4), 391–464.
[84] Bergeron, V. and Radke, C. J. (1992). *Langmuir*, **8**(12), 3020–3026.
[85] Maru, H. C. and Wasan, D. T. (1982). *Chem. Eng. Sci.*, **37**(2), 175–184.
[86] Sonin, A. A., Bonfillon, A., and Langevin, D. (1994). *J. Coll. Interface Sci.*, **162**(2), 323–330.
[87] Joye, J. L., Hirasaki, G. J., and Miller, C. A. (1994). *Langmuir*, **10**(9), 3174–3179.
[88] Joye, J. L., Hirasaki, G. J., and Miller, C. A. (1996). *J. Coll. Interface Sci.*, **177**(2), 542–552.
[89] Nikolov, A. D. and Wasan, D. T. (1989). *J. Coll. Interface Sci.*, **133**(1), 1–12.
[90] Bonfillon-Colin, A. and Langevin, D. (1997). *Langmuir*, **13**(4), 599–601.
[91] Garrett, P. (1993). *Defoaming, Theory and Industrial Applications*. Dekker, New York.
[92] Israelachvili, J. (1992). *Intermolecular and Surface Forces* (2nd edn). Academic Press, New York.
[93] Vrij, A. (1966). *Disc. Faraday Soc.*, **42**, 23–33.
[94] Helm, C. A., Israelachvili, J. N., and McGuiggan, P. M. (1989). *Science*, **246**(4932), 919–922.
[95] Helm, C. A., Israelachvili, J. N., and McGuiggan, P. M. (1992). *Biochemistry*, **31**(6), 1794–1805.
[96] Exerowa, D., Kashchiev, D., and Platikanov, D. (1992). *Adv. Coll. Interface Sci.*, **40**, 201–256.
[97] de Gennes, P. G. (2001). *Chem. Eng. Sci.*, **56**(19), 5449–5450.
[98] Aveyard, R., Binks, B. P., Fletcher, P. D. I. et al. (1992). In *Trends in Colloid and Interface Science VI*, pp. 114–117. Springer, Berlin.
[99] Weaire, D. and Hutzler, S. (1999). *The Physics of Foams*. Claren-

don Press, Oxford.
[100] Binks, B. P. (1993). *Langmuir*, **9**(1), 25–28.
[101] Kabalnov, A. and Wennerström, H. (1996). *Langmuir*, **12**(2), 276–292.
[102] Walstra, P. (1993). *Chem. Eng. Sci.*, **48**(2), 333–349.
[103] Karbstein, H. and Schubert, H. (1995). *Chem. Eng. Process.*, **34**(3), 205–211.
[104] Schubert, H., Ax, K., and Behrend, O. (2003). *Trends Food Sci. Technol.*, **14**(1–2), 9–16.
[105] van der Net, A., Drenckhan, W., Weaire, I. *et al.* (2006). *Soft Matter*, **2**(2), 129–134.
[106] Martin, A. H., Grolle, K., Bos, M. A. *et al.* (2002). *J. Coll. Interface Sci.*, **254**(1), 175–183.
[107] Bibette, J., Leal Calderon, F., Schmitt, V. *et al.* (2007). *Emulsion Science: Basic Principles* (2nd edn). Springer, Berlin.
[108] Shinoda, K. and Friberg, S. (1986). *Emulsion and Solubilisation*. Wiley, New York.
[109] Miller, C. A. and Rang, M. J. (2000). In *Emulsions, Foams, and Thin Films* (ed. K. L. Mittal and P. Kumar), pp. 105–120. Marcel Dekker, New York.
[110] Pena, A. A. and Miller, C. A. (2001). *J. Coll. Interface Sci.*, **244**(1), 154–163.
[111] Schmitt, V., Cattelet, C., and Leal-Calderon, F. (2004). *Langmuir*, **20**(1), 46–52.
[112] Webster, A. J. and Cates, M. E. (2001). *Langmuir*, **17**(3), 595–608.
[113] Weaire, D. and Pageron, V. (1990). *Phil. Mag. Lett.*, **62**(6), 417–421.
[114] Meinders, M. B. J., Kloek, W., and van Vliet, T. (2001). *Langmuir*, **17**(13), 3923–3929.
[115] Tcholakova, S., Mitrinova, Z., Golemanov, K. *et al.* (2011). *Langmuir*, **27**(24), 14807–14819.
[116] Georgieva, D., Schmitt, V., Leal-Calderon, F. *et al.* (2009). *Langmuir*, **25**(10), 5565–5573.
[117] Durand, M., Martinoty, G., and Langevin, D. (1999). *Phys. Rev. E*, **60**(6), R6307–R6308.
[118] Koehler, S. A., Hilgenfeldt, S., and Stone, H. A. (2000). *Langmuir*, **16**(15), 6327–6341.
[119] Saint-Jalmes, A. and Langevin, D. (2002). *J. Phys. Cond. Mat.*, **14**(40), 9397–9412.
[120] Langevin, D. and Rio, E. (2012). *Encyclopedia of Surface and Colloid Science, 2nd edition.* Volume 4. Taylor and Francis, London.
[121] Bergeron, V. (1997). *Langmuir*, **13**(13), 3474–3482.
[122] Espert, A., von Klitzing, R., Poulin, P. *et al.* (1998). *Langmuir*, **14**(15), 4251–4260.
[123] Bremond, N., Thiam, A. R., and Bibette, J. (2008). *Phys. Rev. Lett.*, **100**, 024501.

[124] Lai, A., Bremond, N., and Stone, H. A. (2009). *J. Fluid Mech.*, **632**, 97–107.
[125] Kim, Y. H. and Wasan, D. T. (1996). *Ind. Eng. Chem. Res.*, **35**(4), 1141–1149.
[126] Deminiere, B., Stora, T., Colin, A. *et al.* (1999). *Langmuir*, **15**(7), 2246–2249.

3 Liquid Crystals

Oleg D. Lavrentovich

Liquid Crystal Institute & Chemical Physics Interdisciplinary Program
Kent State University, Kent, Ohio 44242, USA

Liquid crystals (LCs) represent a state of matter with an orientational order of building units (which might be individual molecules or their aggregates) and complete or partial absence of the long-range positional order. The LCs are thus intermediate between regular solids with long-range positional order of atoms (or molecules, in which case there is also an orientational long-range order) and isotropic fluids in which the molecules show neither positional nor orientational order. The LC can flow, adopt the shape of a container, and form drops, as regular isotropic fluids do, but at the same time, the molecules in the liquid crystal sample are ordered. The direction of average orientation in LCs is called the director $\hat{n}$, with the properties $\hat{n} \equiv -\hat{n}$ and $|\hat{n}|^2$.

Molecular interactions responsible for orientation order in LCs are relatively weak. Most LC substances melt into the isotropic (I) phase in the range $(10\text{--}200)°$ C. As a result, the structural organization of LCs, most importantly, the spatial configuration of the director and thus the optical properties, are very sensitive to the external factors, such as electromagnetic fields, shear deformations, type of molecular orientation at bounding walls, etc. This sensitivity opened the doors for numerous applications of LCs, including revolutionary development of LC displays (LCDs); in these devices, weak voltage pulses reorient the director and thus change the optical appearance of the LC pixels.

Liquid crystals, discovered in 19th century,[1] represent nowadays one of the best studied classes of soft matter, along with colloids, polymer solutions and melts, gels, and foams. There is an extensive literature on physical phenomena in LCs [1–6], their chemical structure and materials parameters [7,8], display [9] and non-display [8] applications, and historical developments [10]. LCs are also prominently featured in the general texts on condensed and soft matter [11–13]. Phenomena observed in LCs and approached developed for their description become of heuristic value in other branches of science. Despite all the successes, there are still unsolved problems. Among these are the status of a biaxial nematic phase, the role of divergence elastic terms in the Oseen–Frank theory, the structure of many new phases formed by the molecules with the shape just a bit more complex than a simple rod or disk (the so-called N_x and B7 phases are just two examples), the expansion of the temperature stability range of the blue phases, the search for the new modes of electrooptic switching that would overcome the bottleneck of slow re-

---

[1] The first known article in the field has been published by Julius Planer of Lemberg University: Planer, J. (1861). Notiz über das cholestearin. *Annalen der Chemie und Pharmacie*, **116**, 25–27; for English translation, see "Note about Cholesteril". *Condensed Matter Physics*, **13**, 37002 (2010). Planer reported that cholesterol chloride shows iridescent colors within a certain temperature range. Nowadays the phenomenon is known as selective reflection of light on the periodic helicoidal structure of the cholesteric (also known as chiral nematic) phase of liquid crystals.

sponse time, etc. This review mainly covers the basic properties of the low-molecular weight thermotropic LCs, with an additional emphasis on the developments of the last decade, such as LC colloids.

3.1 Thermotropic and lyotropic systems

One distinguishes thermotropic and lyotropic LCs depending on the way the liquid-crystalline state (also known as *mesophase*) is produced.

The thermotropic LC states exist in a certain temperature range and are formed by materials with strongly anisometric molecules, either elongated ('calamitic' molecules), such as the molecule of octylcyanobiphenyl (8CB),[2] Figure 3.1, or disk-like ('discotic' molecules). Upon heating, many substances of this type yield the following phase sequence: solid crystal with a long-range orientational and positional order ↔ LC with a long-range orientational order and partial or no positional order ↔ I fluid with no orientational nor positional long-range order. An individual substance forming a thermotropic LC does not require any solvent to exhibit the mesophase, although in practical applications the LC materials are often mixed with each other to improve properties of practical interest (such as the temperature range of existence, viscosity, dielectric anisotropy), or doped with additives that are not necessarily mesomorphic, such as dyes, chiral molecules, polymers, solid inclusions, etc.

Lyotropic LCs form only in presence of a solvent, such as water or oil. Most commonly, lyotropic mesophases are formed by solutions of anisometric amphiphilic molecules, such as soaps, phospholipids, and surfactants [14]. Amphiphilic molecules have two distinct parts: a polar hydrophilic head and a nonpolar hydrophobic tail (generally, an aliphatic chain). This feature gives rise to a special 'self-organization' of amphiphiles in solvents. As the concentration of amphiphilic molecules increases, they aggregate into micelles in order to optimize the exposure of different parts of the molecules to the solvent. For example, micelles of surfactant molecules in water expose the hydrophilic groups at the periphery, while hiding the hydrophobic aliphatic chains in the interior. The packing geometry is very sensitive to the molecular structure, to the relative size of polar heads and aliphatic tails, etc. The shape of objects that form LCs varies from anisometric (disc-like, rod-like, or box-like) closed micelles to very long cylinders and bilayers with opposite orientation of amphiphilic molecules in each layer.

There is a distinct subgroup of lyotropic LCs, the so-called chromonic lyotropic liquid crystals (LCLCs), that embrace many dyes and drugs, proteins [15, 16], and even nucleotides [17]. Chromonic molecules are typically plank-like with a polyaromatic rigid flat core and polar groups at the periphery, Figure 3.2. When in water, the molecules self-assemble into aggregates. The assembly is typically face-to-face (H-aggregation), at least in the most studied materials such as disodium chromoglycate (DSCG) and Sunset Yellow [15, 16, 18]. The face-to-face aggregation results in rod-like formations with one, two, or more molecules in the

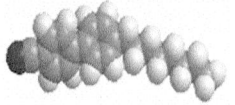

Fig. 3.1 Molecular structure and phase diagram of the typical thermotropic material octylcyanobiphenyl. Its phases have the following transition points: Solid crystal - 24°C - Smectic A - 34°C - Nematic - 42.6°C - Isotropic fluid.

[2] Cyanobiphenyls synthesized by George Gray of Hull University, UK, in the 1970's were the first materials that satisfied most of the requirements for the development of commercially viable liquid crystal displays based on the so-called twisted nematic effect patented earlier by Martin Schadt and Wolfgang Helfrich, both of Hoffmann-La Roche, Switzerland and by James Fergason of Kent State University, USA. Strong dielectric anisotropy of these materials allowed one to use the Frederiks effect of the optical axis reorientation in a relatively modest electric field.

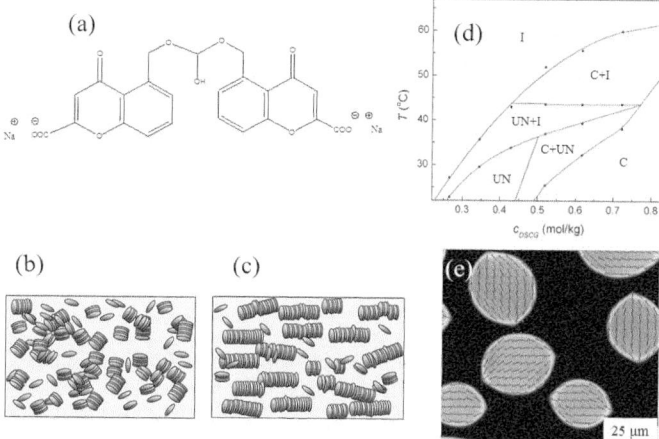

Fig. 3.2 Lyotropic chromonic liquid crystals. (a) Molecular structure of disodium chromoglycate (DSCG); (b) chromonic molecules arranged into columnar aggregates that are misaligned when the concentration of DSCG is too low; (c) the DSCG concentration is sufficiently high to form long aggregates that arrange into a UN; (d) phase diagram of DSCG in water; the isotropic, uniaxial nematic, and columnar phases are labeled I, UN, and C, respectively [16]; (e) PolScope texture of a biphasic state with coexisting UN (tactoids) and I (black) background; the bars show the local orientation of $\hat{n}$.

cross-section. The association energy E of two neighboring molecules within the aggregate is small, about $(5\text{–}15)\,k_B T$ [19–21]. The aggregates are polydisperse, with the length that depends on concentration and temperature. The underlying mechanism is analogous to the process resulting in worm-like micelles formed by surfactant molecules and the so-called 'living' polymerization. An aggregate would grow indefinitely if it were not for the entropy. The balance of the 'sticking' energy E (also called the scission energy, i.e. the energy needed to split an aggregate into two) and the entropy gained by producing more 'ends', results in a prediction that in the I phase, the average aggregation number scales with the volume fraction ϕ of the solute and absolute temperature as [22] $\langle n \rangle \propto \sqrt{\phi} \exp(E/2k_B T)$. The LCLCs are thus very different from both the thermotropic low-molecular weight LCs with molecules of covalently fixed shape, and from more familiar lyotropic LCs of Onsager-type [23], formed by dispersions of objects such as tobacco mosaic viruses [24], carbon nanotubes [25], or irreversibly polymerized long molecules, see, for example, [26]. The phase diagrams of LCLCs depend on both the temperature and concentration. The length of chromonic aggregates and thus their phase diagrams also depend on pH of the solutions [20], presence of additives such as salts [27, 28], and neutral particles [29, 30] that impose an osmotic pressure on the aggregates, facilitating their growth and closer packing.

Mesomorphic states might be also formed in solutions of certain polymers; one of the most famous examples is Kevlar, the super-strong polymer; some polymers can form thermotropic (solvent-free) LCs.

There are four basic types of LC phases, classified according to the dimensionality of the translational correlations of building units (individual molecules or aggregates): nematic (no translational correlations), smectic (1D correlations), columnar (2D correlations), and various 3D-correlated structures, such as cubic phases and blue phases. Upon heating, many thermotropic substances, such as 8CB, yield the following phase sequence: solid crystal → smectic (or columnar phase) → nematic

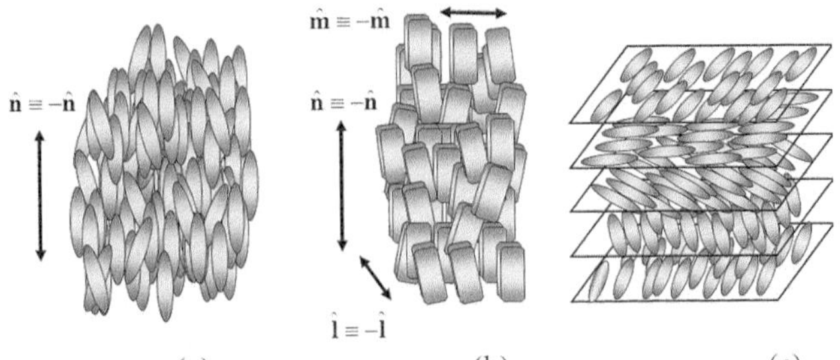

Fig. 3.3 Uniaxial nematic (a), biaxial nematic (b) and chiral nematic or cholesteric (c).

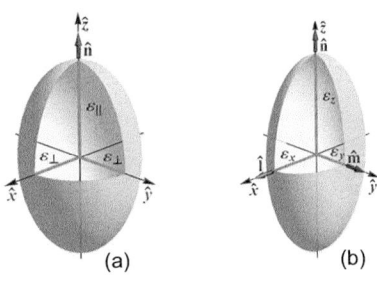

Fig. 3.4 Dielectric tensor for (a) uniaxial nematic and (b) biaxial nematic. Pictures by V. Borshch.

→ isotropic fluid (Figure 3.1). On cooling, the sequence is generally reversed, although it often includes phases that are absent during heating.

There are three different variations of the nematic phase: uniaxial nematic (UN), biaxial nematic (BN), and chiral nematic or cholesteric (Ch), Figure 3.3.

Uniaxial nematic

The UN is typically formed by the molecules of a rod-like or disk-like shape that can easily rotate around one of the axes, Figure 3.1. The average orientation of these axes of rotation determines the director $\hat{n}$ (Figure 3.3). Even when the molecules are polar, as 8CB in Figure 3.1, their head-to-head overlapping and flip-flops establish centrosymmetric arrangements in the UN bulk, so that $\hat{n}$ and $-\hat{n}$ are equivalent. The orientational order is never perfect, as at any instant of time, the individual molecular axes might point in a direction different from $\hat{n}$. All material properties of a UN are anisometric, with $\hat{n}$ defining the axis of rotational symmetry. The anisotropy of UN manifests itself in the anisotropy of diamagnetic susceptibility and dielectric permittivity. For example, the dielectric permittivity $\epsilon_\parallel$ of the UN measured parallel to $\hat{n}$ and permittivity $\epsilon_\perp$ measured along any direction normal to $\hat{n}$, are generally different. Figure 3.4(a) illustrates an anisotropic axisymmetric dielectric tensor for a so-called dielectrically positive UN with $\epsilon_a = \epsilon_\parallel - \epsilon_\perp > 0$. By tuning the molecular properties (direction of permanent dipoles, polarizability, and shape) in chemical synthesis, one can vary the absolute value and even the sign of ϵ_a; UN materials with $\epsilon_a = \epsilon_\parallel - \epsilon_\perp < 0$ are called dielectrically negative.

When the electric field $\boldsymbol{E}$ is applied, the director $\hat{n}$ tries to align along $\boldsymbol{E}$ if $\epsilon_a > 0$ and perpendicular to $\boldsymbol{E}$ if $\epsilon_a < 0$. The phenomenon is called the Frederiks effect.[3] It was first discovered for a magnetic field case by Repiova and Frederiks in 1927 [31]. A UN was confined between a flat glass and a convex lens, Figure 3.5, and observed between crossed polarizers. The set-up was placed between the poles of an electromagnet. In the absence of the magnetic field, the cell looked dark, since $\hat{n}$ was directed vertically, along the line of observation, Figure 3.5(a). Such an orientation is called *homeotropic*. It occurs whenever the molecular interactions at the boundary of the LC prefer a perpendicular orientation

---

[3] Vsevolod Frederiks, working in Leningrad in the Soviet Union, discovered the effect of field-induced reorientation of liquid crystal director in the late 1920's, first for the magnetic field. During the Stalin regime, in 1936 he was arrested on the ground of absurd political charges and sentenced to ten years in prison. He died in 1944. There are widely varying version of spelling of his last name in the literature: here we use the simplest version of Latin transliteration from Cyrillic.

of $\hat{n}$ (in real experiment, the wedge angle between the lens and the flat plate was small, much smaller than in Figure 3.4, thus $\hat{n}$ was practically perpendicular to both interfaces). When the magnetic field B was applied in the horizontal direction, it tilted $\hat{n}$ from its original vertical alignment. The tilt manifested itself as a series of bright and dark concentric rings that represented curves of equal difference of paths of ordinary and extraordinary rays. Importantly, the field-induced tilt was observed only for a portion of the wedge-like gap, of the thickness larger than some critical value d_c, Figure 3.5(b). Subsequent experiments with high fields by Frederiks and Zolina [32] concluded that the critical thickness separating the distorted and undistorted states is related to the acting magnetic field, as $Bd_c = $ const. A few years later [33], a similar relationship $Ed_c = $ const was found for the case of an electric field realigning a dielectrically anisotropic UN.

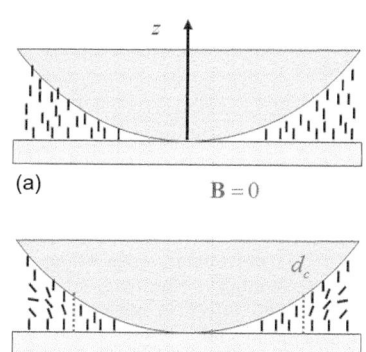

Fig. 3.5 Frederiks transition in a homeotropic UN confined between a lens and a flat glass, in the horizontal magnetic field: (a) field off; (b) field on; the thicker portion of the nematic wedge realigns along the magnetic field; the portion thinner than d_c remains unperturbed.

The underlying physical mechanism of both Frederiks field effects is the balance of (a) surface anchoring forces (that fix $\hat{n}$ at the bounding surfaces), and (b) orienting torque of the field. The latter in the magnetic case is equal to $\mu_0^{-1}\chi_a B^2 \sin\varphi \cos\varphi$ (here μ_0 is the magnetic constant, $\chi_a = \chi_\parallel - \chi_\perp$ is the diamagnetic anisotropy of the LC, calculated as the difference in susceptibility parallel and perpendicular to the director, and φ is the angle between B and $\hat{n}$), and, finally, (c) elasticity of the orientational order: to make the director varying from point to point, one needs to perform some work. Thus the diamagnetic reorienting torque needs to overcome the elastic restoring torque $\propto K\left(\partial^2\varphi/\partial z^2\right)$, where K is the Frank elastic constant, Figure 3.5. In modern setting, one typically uses a cell of a constant thickness; then the critical field that enables the Frederiks transition is determined from the balance of torques as

$$B_c = \frac{\pi}{d}\sqrt{\frac{K}{\mu_0^{-1}\chi_a}}. \quad (3.1)$$

A similar expression can be deduced for the dielectric case.[4] The UN is an optically anisotropic medium; the optic axis coincides with $\hat{n}$ (Figure 3.3a). The speed of light propagation in UN depends on how the polarization of light is directed with respect to the optic axis, giving rise to the phenomenon of *birefringence*. Light propagating through the UN with its polarization parallel to $\hat{n}$ experiences an extraordinary refractive index n_e (which means that its speed is c/n_e, where c is the speed of light in vacuum) while light polarized along a direction perpendicular to $\hat{n}$ experiences an ordinary refractive index n_o. Birefringence allows one to change the optical appearance of the material dramatically when it is viewed through a pair of crossed linear polarizers. For example, if the homeotropic UN slab with the optic axis $\hat{n}$ parallel to the direction of light propagation is placed between two polarizers, it would appear dark. If the director is tilted and the projection of this tilt on the cell plane is different from the direction of polarizers, the cell would appear bright. The pair of polarizers converts the difference in the speed of ordinary and extraordinary beams into the intensity modulation of the

[4] It is important to realize that $\hat{n}$ specifies only the average direction of orientation but not the *degree* of orientational order.

[5] Director reorientation is not the only way to cause an optical response. The electric field can be used to modify the refractive indices of the liquid crystal and thus cause an optical response (a change in retardance) even when the director remains in the original orientation. The effect is extremely fast (nanoseconds) [34].

transmitted light. The two facets of the dielectric tensor anisotropy of a UN, at optical frequencies and at low frequencies, allow one to use the material in a variety of electro-optical applications. Numerous variations of the electric Frederiks effect with different geometries of the initial director and electrode pattern form the basis of the modern multi-billion industry of UN devices.[5]

Biaxial nematics

In 1970, Freiser theoretically considered a biaxial nematic (BN) with orthorhombic symmetry in which physical properties are different along three principal directors, $\hat{n}$, $\hat{l}$, and $\hat{m} = \hat{n} \times \hat{l}$, such that $\hat{n} \equiv -\hat{n}$, $\hat{l} \equiv -\hat{l}$ and $\hat{m} \equiv -\hat{m}$, Figure 3.3(b). The corresponding tensors have all three diagonal components different from each other, Figure 3.4(b).

The discovery of the BN has been announced by Yu and Saupe in 1980 [35] for a surfactant-based lyotropic LC, in which the micelles became anisometric with three different principal axes in a certain temperature and concentration range. Another example of a BN in a lyotropic system was presented recently by van den Pol et al. [36]. It is argued [37] that the BP might be stabilized by polydispersity of the building units. An overview of theoretical and computer simulations of BN has been presented by Zannoni's group [38].

It is the search for the BN phase in the thermotropic solvent-free low-molecular weight materials that is attracting the most attention nowadays, fueled by a hope to achieve a much faster switching time for director reorientation as compared to the UN materials [39]. The appearance of BN phase was claimed for thermotropic LCs formed by ring-like cyclic molecular trimers [40], organosiloxane tetrapods [41], and bent-core molecules [42–44]. The case of ring-like molecules and tetrapods remains underexplored, most probably because of difficulties in chemical synthesis of these complex molecules. Bent-core molecules are easier to synthesize (the first bent-core LCs were prepared by Vorländer already in 1929, see the recent review by Jákli [45]). The recent studies of bent-core materials produce rather controversial results, as it has been shown by a number of groups that the biaxial features might originate from the unusual behavior of the standard UN [46] rather than from the true and stable BN order in the bulk. Among the potential effects that mimic biaxial features in the UN phase of bent-core materials, one finds: (a) formations of cybotactic smectic (of C type, with tilted molecules) clusters that can exist in a broad temperature range [47,48]; (b) surface effects, such as the formation of smectic layers with tilted molecules [49,50], in the latter case, the apparent biaxial features show dependence on the cell thickness [50]; (c) thermal expansion of the LC that results in a flow-induced tilt of the optic axis along two directions. The latter mimics an appearance of an optically biaxial state, sometimes with very long transient times [51]. The crucial difference is that in the flow-induced effect, there is just one local optic axis; this local axis tilts into two opposite directions in the top and bottom parts of the sample (see Section 3.5 below). In a true BN, there are two optic axes in each and every point of the medium. It is important to remember that experimental techniques based on an integral response (such as the measurements of optical retardance, X-ray, etc.) might result in misinterpretation of the type of orientational order. Interestingly, biaxial features in the nematic phase formed by micellar lyotropic LCs have also been claimed as being only transient [52]. One can thus conclude that the situation with the existence of the BN phase in low-molecular thermotropic and lyotropic LCs is not entirely clear. It appears that at present, the polymer nematic systems provide a more convincing case for the existence of BP, see [53] and references therein.

Chiral nematics

When the building unit (molecule or aggregate) is chiral, i.e. not equal to its mirror image, the nematic might show twisted structures. The most frequent reason for molecular chirality is presence of a carbon atom with four different ligands in the molecule. The simplest twisted structure forms when $\hat{n}$ is perpendicular to the twist axis and rotates continuously as one moves along the axis, forming a helicoid, Figure 3.3(c). The corresponding phase is called a cholesteric and denoted Ch or N^*. Historically the first LC textures observed have been the Ch textures of cholesterol derivatives, similar to Figure 3.6(a). Many molecules and objects of biological origin, including DNA, chitin, f-actin, and tobacco mosaic virus, are capable of forming Ch phase; for recent reviews, see [24, 54]. [6]

The pitch (period) of the helicoidal twist is much larger than the molecular size as molecular interactions responsible for the twist are weak as compared to the interactions responsible for the mesomorphic state itself. With a typical molecular size of a LC being 1–10 nm, the pitch P is often in the range of 0.1–1 μm, which overlaps with the range of the visible part of optical spectrum, producing a selective light reflection when light propagates along the axis. The blue color of planar texture in Figure 3.6(a) is not associated with absorption, as the material contains no dye; it results from the selective reflection of light, when the wavelength of light satisfies the Bragg condition, $\lambda = nP$, where n is the average refractive index of Ch. The Ch's are extremely sensitive to surface orientation, electric field, etc. In particular, the axis of the periodic Ch structure can be realigned to be in the plane of the cell. In Figure 3.6(b), it is achieved by applying the alternating-current (AC) electric field $\boldsymbol{E} = (0, 0, E)$ between the two glass plates [their inner surface are coated with transparent conducting layers of indium tin oxide (ITO)]. In the LC used in the experiment, the molecules prefer to align parallel to the electric field. Realignment along the field occurs in a number of steps. First, the Ch axis realigns along the x-axis [the substrates are rubbed beforehand along the x-axis to lift the degeneracy of orientation in the (x, y) plane], Figure 3.6(b). Note a complex character of director configuration near the bounding plates. As the field increases, the peri-

[6] An interesting conical version of chiral nematic was discovered by Berry et al in condensed suspensions of flagella isolated from bacteria [55]. Because of the helical shape of each flagellum, the parallel arrangement of flagella can be described by a local director that follows an oblique helicoid, making an acute angle (rather than a right angle) with the helicoid axis.

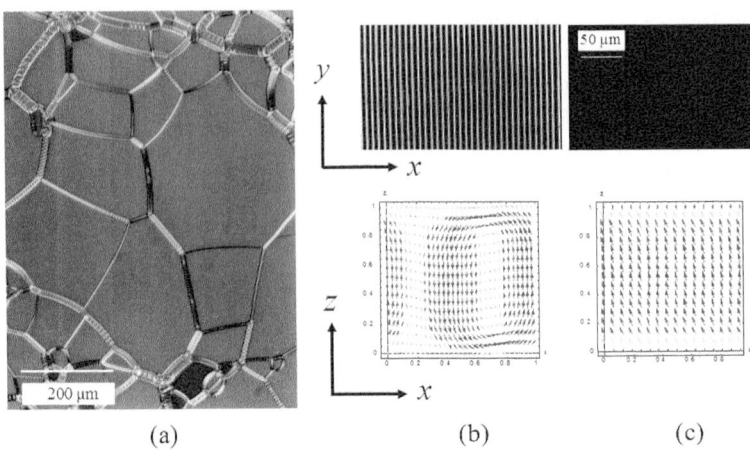

Fig. 3.6 (a) Planar texture of a Ch: the uniform color intensity is caused by Bragg reflection of light; bright and dark lines are the so-called 'oily streaks', representing parallel sets of dislocations and disclinations (sometimes containing periodic undulations and focal conic domains); the defects separate Ch monodomains. (b) Planar Ch cell, under the action of a 2 V voltage; the field is applied along the z-axis. (c) The same, the voltage is 5 V, $\hat{n}$ is practically vertical. In computer simulations of $\hat{n}$, the arrows indicate the direction of tilt. Simulations by S. Shiyanovskii [56].

odicity of the Ch diffractive-grating structure increases. At a sufficiently high field, the Ch unwinds completely into the homeotropic nematic texture, with $\hat{n}$ parallel to the z-axis practically everywhere, except in a thin layer near the bounding plates where the surface anchoring keep the local director tilted, Figure 3.6(c) [56].

Changes of Ch textures by electric field and by other means, such as mechanical pressure, are used in various display applications; the advantage of Ch over the UN is in the possibility to achieve optical contrast without polarizers and in bistable character of many structural transformations. The latter allows one to use a power source only to change the information displayed.

Blue phases

The uniaxially twisted Ch is not the only structure arising from chiral interactions. Imagine the director oriented along a single line, say, an axis z. In the vicinity of z, the chiral molecules tend to rotate helically along all the directions perpendicular to z, Figure 3.7. The ensuing double twist might be energetically more preferable than the uniaxial twist of a Ch phase. Within the framework of the Frank–Oseen elastic theory, the preference of double twist over the simple twist is usually associated with the so-called saddle-splay elastic term in the free energy expansion and the corresponding saddle-splay elastic constant K_{24} [57, 58]. As one moves away from the axis z, the double twist becomes more and more similar to the simple uniaxial twist, thus the energy preference of double twist gradually vanishes. The double twist should thus realize itself within domains of finite size. On the other hand, finite domains (say, cylinders) cannot fill the space continuously, without creating defects in director orientation. As a compromise between the local double twist and the global network of defects, a number of complex chiral LC phases form in certain chiral materials, collectively called the blue phases, BP-I, BP-II, and BP-III. Nowadays it is established that the defects are line disclinations [59]. The phases differ in the symmetry of disclination networks, with BP-III representing a disordered amorphous array of them.

The director gradients grow rapidly as one approaches the core of disclinations; at some distance, the orientational order should be strongly modified (roughly speaking, melted into an isotropic fluid). It suggests that the BPs can be stable only within a narrow temperature interval between the isotropic phase and a simple Ch phase. For many years, the PBs were indeed observed only within a few degrees of the melting transition. Recently, two approaches have been suggested to widen the BP temperature range.

The first approach is based on the polymer-stabilization of the BPs through their disclination network [61]. It can be traced back to the general idea that linear defects in ordered materials such as dislocations in metals and semiconductors can trap impurities at their cores. An impurity added to an ordered medium replaces a portion of this medium and it is thus natural to expect that the placement at the defect core replaces the most energetically costly region. Theoretically, Huang et al. [62] proposed that a stable BP can be formed by mixing the chiral nematic liquid crystal with isotropic liquids such as water, and adding surfactants to lower the interfacial energy between the phase separated components. The model has been expanded recently [63, 64]. Experimentally, attraction of a polymerizing material to the disclination core has been demonstrated for twisted nematic LC [60] and recently for BPs [65]. Dierking et al. explored a variety of additives, of low-molecular-weight, polymer, and colloidal nature, as stabilizers of the BPs, and achieved a significant ($\sim 10\,^\circ\mathrm{C}$) expansion of the BP temperature range on cooling [66]. The available literature on polymer-stabilized BP cites a hysteresis in their electro-optical performance [67].

The second approach is based on the search for molecules that yield beneficial material parameters such as elastic constants. A spectacular increase of the BPs temperature range was achieved with molecular dimers (two rigid mesogenic cores connected by a flexible aliphatic chain [68]); in the context of BP these flexible dimers were called 'bimesogens' [69]. In the original theory of the BP [57, 70], the central role belongs to the saddle-splay K_{24} elastic term. Kleman demonstrated that the double twist is accompanied by bend [58]. Empirically, it was noticed [71] that a small value of the bend constant K_{33} enhances the temperature stability of BPs, which is in accord with the Kleman's model [58] and with a recent extension of the original Meiboom theory [69, 72]. Furthermore, Castles et al. [73] proposed that the temperature range of BP phase might be expanded by flexoelectricity, as the latter reduces the elastic cost associated with splay and bend; the model is in qualitative agreement with experimental observations of a broad range BP phases in bimesogenic and bent-core materials.

An interesting variation of the BP, a K_{24}-stabilized 'double tilt' phase comprised of cylindrical elements with tilted director, has been theoretically predicted for non-chiral nematic materials by Chakrabarti et

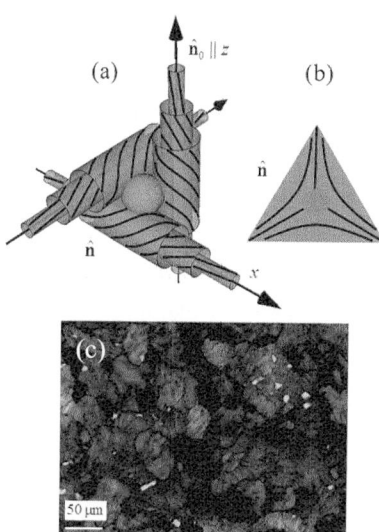

Fig. 3.7 (a) Three cylinders of double twist fill space and create a disclination line; (b) expanded view of the disclination line of strength $(-1/2)$. The core of the disclination can be filled with a stabilizing component (for example, a polymer or a colloidal particle) [60]. (c) Typical polarizing microscope texture of a BPI (Photo by J. Xiang).

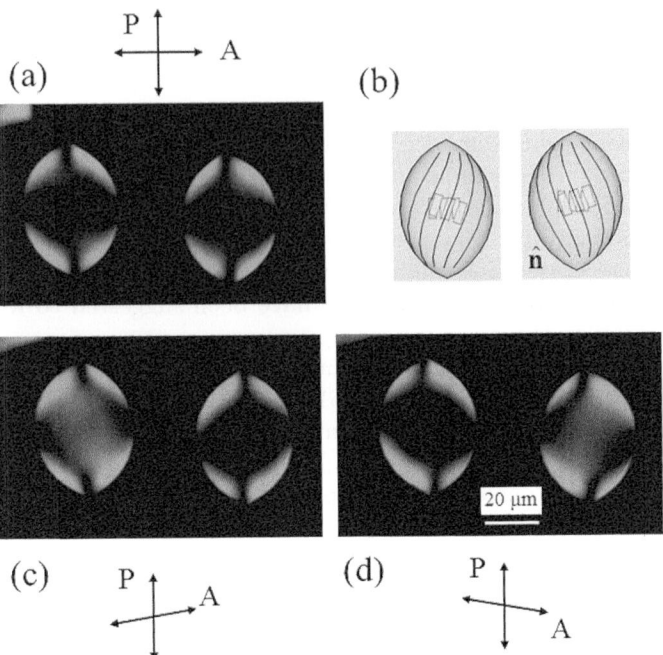

Fig. 3.8 Two tactoidal droplets of the UN phase of a chromonic lyotropic liquid crystal viewed between two crossed polarizers reveal the chiral structure despite the fact that there are no chiral molecules in the system; (a) polarizing microscopic texture between two crossed polarizers; (b) schematic director structure; (c, d) polarizing microscope textures for slightly uncrossed polarizer (P) and analyzer (A) revealing chirality and rotation of polarization direction of light transmitted through the droplet. Photos by L. Tortora.

al. [74]. A somewhat similar, but electric-field stabilized phase of ferroelectric domains with two dimensional splay and bend has been proposed by Alexander and Yeomans [75].

Macroscopic chirality in soft matter systems does not necessarily require chiral centers in the molecules. In LCs, chiral domains might form as a result of packing of achiral molecules with nontrivial shape [76]. The mirror symmetry can also be broken by an explicit action of boundary conditions (e.g. a UN slab placed between two parallel rubbed rigid plates, one of which is then twisted with respect to the other, the so-called Mauguin twist [1]) or as a result of a more subtle mechanism that involves a smallness of the elastic constant of twist as compared to the elastic constant of splay and bend. For example, a nonchiral UN confined to a droplet, can acquire a twisted structure because of this mechanism [30, 77, 78], Figure 3.8.

We stress that UN, BN, Ch, BP-I, BP-II and BP-III phases are liquid, with no correlations in molecular positions; if there is a spatial periodicity, it is the periodicity of orientations, not positions. BPs can be observed not only in the thermotropic materials, but also in concentrated DNA solutions [79]. An analog of the BP phases has been proposed for smectic LCs [80]. LCs demonstrate other phases stabilized by lattices of defects. One of the most interesting is the 'twist grain boundary' predicted by Renn and Lubensky [81], in which the Ch twist is made compatible with one-dimensional periodic modulation of density (smectic layering) by a lattice of screw dislocations; the phase has been observed experimentally [82, 83]. There is a growing body of data pointing to the possibility of existence of more than one UN phase with temperature-triggered transitions between them. The effect has been first reported for polymer main-chain polymers, in which the mesogenic units are linked by flexible chains [84]. More recently, studies of LC dimers such as $1'',7''$-bis(4-cyanobiphenyl-4'-yl)heptane, with the chemical formulae [85], NC-$(C_6H_4)(C_6H_4)$-$(CH_2)_7$-$(C_6H_4)(C_6H_4)$-CN (CB7CB), lead to the observation of a new UN with a periodic structure, called N_x [85–87]. The $N \to N_x$ transition is detectable in differential scanning calorimetry (DSC) and through textural changes: a typical UN Schlieren texture, Figure 3.18, transforms into N_x textures resembling those of layered phases, such as smectic A, Figure 3.20 in Section 3.5. Since X-ray scattering patterns show absence of positional order; the observed textures are probably associated with a periodic reorientations of the director field rather than with a periodic wave of density, as in smectic LCs.[7]

One can interpret the periodic structure of N_x phase in terms of an idea that some elastic constants become negative or within a more detailed theory put forward by Dozov, that predicts a mixed twist-bend modulated structure [90]. Both explanations need further exploration, as in the absence of experimental data on the detailed configuration of the director pattern it is hard to make a comparison with the theory.

Smectics

There are layered LC phases with a quasi-long-range 1D translational order of centers of molecules in a direction normal to the layers. This positional order is not exactly the long-range order as in a regular 3D crystal. As shown by Landau and Peierls, the fluctuative displacements of layers in 1D lattice diverge with the linear size of the sample. However, for typical smectics with period of the order of 1 nm, the effect is noticeable only on scales of 1 mm and larger; typical samples used in laboratory research and in applications are much thinner, $1-100\,\mu$m. In smectic A (SmA), the molecules within the layers show fluid-like arrangement, with no long-range in-plane positional order; it is a uniaxial medium with the optic axis $\hat{n}$ perpendicular to the layers, Figure 3.9(a).

In the lyotropic version of SmA, the lamellar L_α phase, the amphiphilic molecules arrange into bilayers. If the solvent is water, the exterior surfaces of the bilayer are formed by polar heads; the hydrophobic tails are hidden in the middle of the bilayer (note that membranes of many biological cells are organized in the similar way). The periodic structure of alternating surfactant and water layers gives rise to the L_α phase, Figure 3.9(b). The lamellar structure might retain its smectic order even when strongly diluted, being stabilized by thermal fluctuations of bilayers.

Other types of smectics show in-plane order, caused, for example, by a collective tilt of the rod-like molecules with respect to the normals to the layers (the so-called SmC), Figure 3.9(c). In chiral materials, the tilt of the molecules might lead to the helicoidal structure; the resulting chiral

---

[7] Recently, the N_x phase formed by molecular dimers with two rigid parts connected by a flexible aliphatic chain, such as CB7CB, has been experimentally proven to represent the twist-bend nematic N_{tb}. The flexible nature of the chain with the odd number of methylene groups enables local bend. The twist helps to spread the bend uniformly in space, as has been foreseen by R.B. Meyer, who predicted both N_{tb} and a splay-bend nematic [Meyer, R.B. (1976). Structural problems in Liquid Crystal Physics, In: Molecular Fluids. Les Houches Lectures, 1973 (eds. Balian, R. & Weill, G., Gordon and Breach)]. Although the N_{tb} state does not require the molecules to be chiral, its local organization is chiral. The period of director modulation is extraordinary small, only about 8 nm in N_{tb} phase of CB7CB [88,89]. The local director of N_{tb} forms an oblique helicoid that reveals itself through characteristic asymmetric periodic "arches" detected by Borshch et al by the freeze-fracture transmission electron microscopy. The splay-bend phase predicted by R.B. Meyer and I. Dozov, has not been experimentally observed so far.

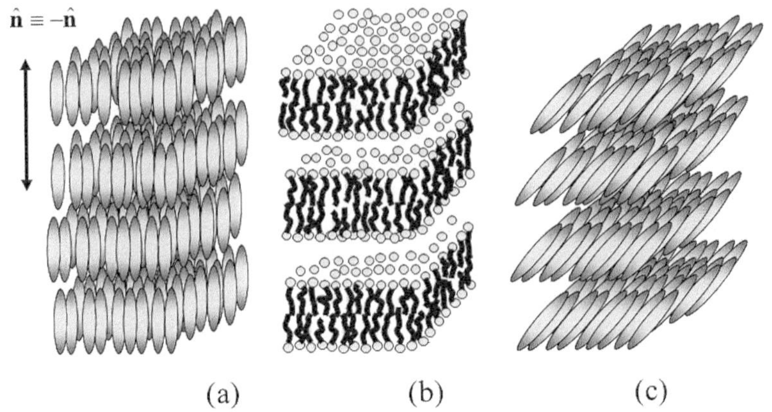

Fig. 3.9 Liquid crystals with 1D periodic structure: (a) thermotropic smectic A; (b) lyotropic lamellar phase formed by bilayers of surfactant molecules; (c) thermotropic smectic C.

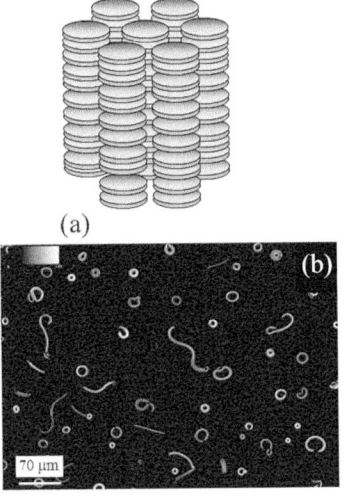

Fig. 3.10 (a) Columnar phase formed by disc-like molecules. (b) Nucleation of a columnar phase (bright) from an isotropic melt (dark background); Polscope texture; DSCG-water dispersion. Photo by L. Tortora.

SmC phase is of interest for applications in fast-switching electro-optical devices.

Columnar phases

Columnar phases are most frequently formed by hexagonal packing of cylindrical aggregates, as in the case of thermotropic materials formed by disc-like molecules. The positional order is two-dimensional only, as the intermolecular distances along the axes of the aggregates are not regular, Figure 3.10(a). Columnar phases form in thermotropic and lyotropic systems, including chromonics. Figure 3.10(b) shows nuclei of the chromonic columnar phase appearing from the isotropic phase in the shape of flexible filaments that close into loops to balance the energy of bend distortions and the surface energy [29].

3D-correlated structures

Three-dimensional correlated structures demonstrate a periodic structure along all three coordinates, but they are still different from the 3D crystals, as the periodicity is associated with repetition of the local director patterns rather than with a regular positional order of the molecular centers of mass; the molecules are free to move around adjusting to the local director orientation. BPs represent an example for thermotropic systems. An analog of a BP for smectic LCs has been theoretically predicted by DiDonna and Kamien [80]. For the case of lyotropics, a good example is a cubic phase, a 3D network formed by periodically curved layers of amphiphilic molecules; the molecules are free to move within the layers.

3.2 Order parameter

The concept of an order parameter has emerged in its modern form in the Landau model of phase transitions and has been later expanded to describe other features such as topologically stable defects in the ordered media. The order parameter of the LC can be related to the anisotropy of macroscopic properties such as magnetic susceptibility or dielectric permittivity. Measuring these anisotropies allows one to establish the degree of orientational order. The magnetic measurements are especially convenient compared with their electric counterparts, as in this case the local field acting on the molecules differs very little from the external field. In the UN phase, the components of the (symmetric) magnetic susceptibility tensor χ read in the frame in which the z-axis is parallel to the director $\hat{n}$, as

$$\chi = \begin{pmatrix} \chi_\perp & 0 & 0 \\ 0 & \chi_\perp & 0 \\ 0 & 0 & \chi_\parallel \end{pmatrix}. \tag{3.2}$$

The quantity $\chi_a = \chi_\parallel - \chi_\perp$ is called the anisotropy of magnetic susceptibility. In the usual UN of thermotropic type, $\chi_\parallel < 0$ and $\chi_\perp < 0$ (diamagnetism), and $\chi_a > 0$, so that $\hat{n}$ orients along the applied magnetic field, as was observed by Frederiks [31].

The degree of order in UN can be defined through its magnetic susceptibility. In the isotropic phase, $\chi_a = 0$; in the nematic phase, χ_a is determined by (a) molecular susceptibilities of individual molecules, and (b) degree of molecular order. For the latter, one can choose the temperature-dependent quantity $s(T) = (1/2) \langle 3\cos^2\theta - 1 \rangle$ where θ is the angle between the axis of an individual molecule and $\hat{n}$; $\langle \ldots \rangle$ means an average over molecular orientations; s is called the 'scalar order parameter'; $s = 1$ corresponds to an ideal order with all the molecules rigidly aligned along one direction, and $s = 0$ corresponds to absence of orientational order (in which case one deals with a regular isotropic fluids).

One can construct a traceless symmetric tensor $\mathbf{Q}$ with the components

$$Q_{ij} = Q\left(\chi_{ij} - \frac{1}{3}\delta_{ij}\mathrm{Tr}\chi\right), \tag{3.3}$$

where $\mathrm{Tr}\chi = \chi_{xx} + \chi_{yy} + \chi_{zz}$, such that $\mathrm{Tr}\mathbf{Q} = Q_{xx} + Q_{yy} + Q_{zz}$. Each component vanishes in the I phase, and it is proportional to χ_a in the UN:

$$\mathbf{Q} = Q\begin{pmatrix} -\chi_a/3 & 0 & 0 \\ 0 & -\chi_a/3 & 0 \\ 0 & 0 & 2\chi_a/3 \end{pmatrix}. \tag{3.4}$$

One can choose the constant Q in such a way that in an arbitrary coordinate system, where $\chi_{ij} = \chi_\perp \delta_{ij} + \chi_a n_i n_j$, the components are expressed as

$$Q_{ij} = s(T)\left(n_i n_j - \frac{1}{3}\delta_{ij}\right). \tag{3.5}$$

The tensor order parameter allows us to describe the BN as well:

$$Q_{ij} = s(T)\left(n_i n_j - \frac{1}{3}\delta_{ij}\right) + b(T)\left(l_i l_j - m_i m_j\right), \tag{3.6}$$

where b is the 'biaxiality parameter'; $b = 0$ in the UN.

For practical applications of LC, also very important is the anisotropy of dielectric permittivity at optical birefringence, i.e. the difference in the refractive indices for ordinary and extraordinary models of propagation, i.e. birefringence $\Delta n = n_e - n_o$.

Phenomenological theories

There are two main types of theories to describe the appearance of orientational order in soft matter systems during phase transitions: phenomenological models based on the Landau expansion and molecular-statistical theories focusing on modeling of molecular interactions in a thermal ensemble. Phenomenological theories are based on the Landau idea that the free energy near a phase transition can be represented as a power expansion in terms of a small amplitude order parameter. These theories rely on symmetry considerations; the main challenge is to connect the coefficients of the expansion to the physically meaningful parameters of molecular interactions. In the case of the I-nematic transition, the proper order parameter is the tensor $\boldsymbol{Q}$ introduced above. It has been used by de Gennes [1] to build the Landau expansion of the Gibbs free energy density:

$$g(T) = g_0 + \frac{1}{2}A(T)Q_{\alpha\beta}Q_{\beta\alpha} - \frac{1}{3}B(T)Q_{\alpha\beta}Q_{\beta\gamma}Q_{\gamma\alpha} \\ + \frac{1}{4}C(T)Q_{\alpha\beta}Q_{\alpha\beta}Q_{\gamma\delta}Q_{\gamma\delta} + \ldots, \quad (3.7)$$

where g_0 is the free energy density of the isotropic phase, A, B, and C are the temperature-dependent (and, generally, pressure-dependent) coefficients, and summation over repeated indices is implied. For a UN, assuming that A is positive in the high-temperature I phase and can be presented as a linear function of temperature, $A = a(T - T^*)$, while a, as well as B, and C are temperature-independent, one obtains an expression for the free energy density

$$g(T) = g_0 + \frac{1}{3}a(T - T^*)s^2 - \frac{2}{27}Bs^3 + \frac{1}{9}Cs^4 \quad (3.8)$$

that should be minimized with respect to s to obtain the stable states. Note the presence of a cubic term, which is not invariant under the transformation $s \to -s$. The UN states described by s and $-s$ are indeed different. For example, $s = 1/2$ describes a UN that is aligned relatively poorly along a single direction in space, while $s = -1/2$ describes a UN with molecules that are everywhere parallel to a plane but are arranged randomly within this plane, showing no in-plane preference. The cubic term implies that the isotropic-to-UN transition is of the first order. Numerous experiments indicate that the transition is of a weak first-order type, with a small latent heat and strong pretransitional effects over a wide temperature range. The closeness of the transition to the continuous (second-order) phase transitions partially justifies the very approach based on the Landau expansion. The equation $\partial g/\partial s = 0$ has a solutions $s = 0$ for the isotropic phase and a solution describing the UN, $s = (B/4C)\left[1 + \sqrt{1 - 24a\left(T - T^*\right)C/B^2}\right]$. The temperature at which the free energy densities of the two phases are equal, is $T_c = T^* + \left(B^2/27aC\right)$; the quantity T^* is the temperature at which the isotropic phase is no longer metastable when the system is cooled down. There is also a temperature $T^{**} = T_c + \left(B^2/216aC\right)$ above which the UN phase is absolutely unstable. Note that $T_c = T^* = T^{**}$ when $B = 0$.

Molecular-statistical theories

This theoretical approach starts with an appropriate model of molecular interactions. The key development in the field was the Onsager theory [23] that predicted an isotropic-to-nematic transition in colloidal systems using only a concept of 'excluded volume' that approximates repulsive interactions. The particles are depicted as rigid bodies that cannot simultaneously occupy the same portion of the physical volume. The phase behavior depends on the shape and concentration of the particles. The free energy contains only an entropy term, as there are no specific interactions except the steric interactions. For a system of hard elongated rods, there are two types of contributions to the entropy, a part associated with their translations in space and a part associated with their orientation. The orientational and translational degrees of freedom are coupled. For example, in a densely populated colloid, the particles might gain a translational freedom by aligning parallel to each other and thus decreasing the orientational entropy. The excluded volume depends on the mutual orientation of particles and is minimum when the particles are parallel to each other. Minimization of the excluded volume means that the volume available for the translation is maximized. In dilute solutions, different orientations of the rods would not alter their ability to move around and the system is in the isotropic state. At some concentration, one expects an isotropic-to-UN phase transition. The volume fraction at which the nematic phase emerges is inversely proportional to the aspect ratio length (L)/diameter (D) of the rods, and is given by an approximate relationship $\Phi \sim 4D/L$.

The original work by Onsager [23] also discussed the role of *electric double layers*. Typically, colloidal particles are charged, for example, because of the polar character of the surface groups. The electrostatic forces thus contribute to the repulsive forces between the particles. Since the system is electrically neutral, the solvent contains a number of counterions (see *Manning condensation*), most of which accumulate around the charged particles, screening the electrostatic repulsions. As a rough approximation, the electrostatic contribution to the repulsive forces can be accounted for by introducing an effective diameter of the rods that is somewhat larger than their bare diameter D.

Flory suggested [91] a modification of the Onsager theory by using a lattice model in which the interactions among the rods were still of steric nature; the tilted rods were modeled as a staircase-like chain of shorter units. He considered the effect of additive-induced condensation of the UN phase. Imagine a dispersion of the hard rods that is not dense enough to form a UN, to which one adds spherical particles. The spheres create an effective 'depletion attraction' among the rods. When two rods approach each other by a distance that is smaller than the diameter of the sphere, the spheres cannot penetrate this gap. As a result, the osmotic pressure on the opposite lateral sides of the rods is different. This difference creates an attractive force that moves the rods closer to each other and aligns them parallel to each other. This effect of 'depletion interactions' in a system comprised of different colloidal entities has been first described by Asakura and Oosawa [92] for particles in the solution of flexible macromolecules that can be approximated as spheres. Flory [91] extended it by demonstrating that the depletion attraction forces make it possible to condense the system of rods into a UN even when their original concentration is not high enough for the nematic to form. The depletion agent causes a phase separation in the system, with a condensed UN and an I phase. The rods tend to populate the UN portion while the depletion agent resides predominantly in the isotropic phase. These entropy mechanisms associated with the excluded-volume depletion interaction, also known as 'molecular crowding effects' in biophysics, play an important role in numerous systems ranging from the interior of living cells to the self-assembled nanoparticles in solutions. For example, depletion forces contribute to *protein crystallization* and condensation of aqueous solutions of DNAs into orientationally ordered domains. The entropy-driven transitions in mixtures of colloidal rods and spheres can lead not only to the formation of UN but also to the smectic (lamellar) and columnar phases with positional order, as demonstrated experimentally for mixtures of rod-like viruses and flexible polymers (or polymer spheres) [93].

Note that since in the Onsager theory the behavior is controlled by entropy, the properties of the corresponding systems are temperature independent (the temperature is simply a common factor for all the different entropic contributions to the free energy). These systems are called athermal. Many lyotropic LCs, say those formed by rod-like tobacco mosaic viruses, fall more or less into this category. However, the chromonic lyotropic liquid crystals do show a strong temperature dependence of their properties, since the chromonic molecules are assembled into aggregates by relatively weak dispersive forces rather than by covalent bonds; the building units are thus temperature-dependent themselves. The Onsager model is also less suitable to describe the thermotropic LCs in which the densities of particles are always high and the aspect ratio is moderate. For these materials, a better description is provided by the Maier–Saupe [1] molecular statistic theory.

The Maier–Saupe theory [1] focuses on the anisotropy of the attractive forces between molecules. It assumes that molecular interactions are of van der Waals type. The repulsive forces and excluded volume effects are not taken into consideration. The pairwise potential for two molecules located at r and r', is $U_{rr} \propto -\left(1/\left|r - r'\right|^6\right) P_2\left(\cos\gamma\right)$, where $P_2\left(\cos\gamma\right) = (1/2)\left(3\cos^2\gamma - 1\right)$ is the Legendre polynomial of the second order and γ is the angle between the two molecular axes $\boldsymbol{a}$ and $\boldsymbol{a}'$. The potential favors parallel alignment. Instead of calculating all pairwise interactions of a given molecule, the theory supposes that each molecule is submitted to some *mean potential* that is averaged over the positions and orientations of all other molecules, which writes $U = -bsP_2(\cos\theta)$, where b is a constant and θ is the angle between the molecular axis and the director. This mean potential is used to calculate the internal energy and entropy of the system and to obtain the values of the scalar order parameter s that corresponds to the minimum of the free energy. The theory predicts the first-order phase transition; when the two phases have the same energy, the scalar order parameter in the UN is relatively low (as compared to the Onsager and Flory theories), $s = 0.43$, which is close to many experimental values observed in thermotropic UN.

3.3 Elasticity

In real samples of LCs, the average molecular orientation changes from point to point because of the external fields, boundary conditions, presence of foreign particles, etc. In many cases of practical interest, such as the Frederiks transitions, the typical scale of distortions is much larger than the molecular scale; the deformations are weak in the sense that the scalar part of the order parameter, s, remains constant despite the spatial gradients of the director field $\hat{\boldsymbol{n}}(\boldsymbol{r})$.

Uniaxial nematic

There are three basic types of director distortions in UN, called splay, twist, and bend, Figure 3.11. The free energy density associated with these three types of (small) deformation writes in terms of the spatial director gradients $n_{ij} = \partial n_i/\partial x_j$ as

$$f_{\rm FO} = \frac{1}{2}K_{11}(\nabla\cdot\hat{\boldsymbol{n}})^2 + \frac{1}{2}K_{22}(\hat{\boldsymbol{n}}\cdot{\rm curl}\hat{\boldsymbol{n}})^2 + \frac{1}{2}K_{33}(\hat{\boldsymbol{n}}\times{\rm curl}\hat{\boldsymbol{n}})^2 \quad (3.9)$$

and is known as the Frank–Oseen energy density with Frank elastic constants of splay (K_1), twist (K_2), and bend (K_3); all three are necessarily positive definite; the dimensionality is that of a force. The energy density equation (3.9) is often supplemented with the so-called divergence terms:

$$f_d = f_{13} + f_{24} = K_{13}\nabla(\boldsymbol{n}\nabla\cdot\boldsymbol{n}) - K_{24}\nabla(\boldsymbol{n}\nabla\cdot\boldsymbol{n} + \boldsymbol{n}\times{\rm curl}\boldsymbol{n}). \quad (3.10)$$

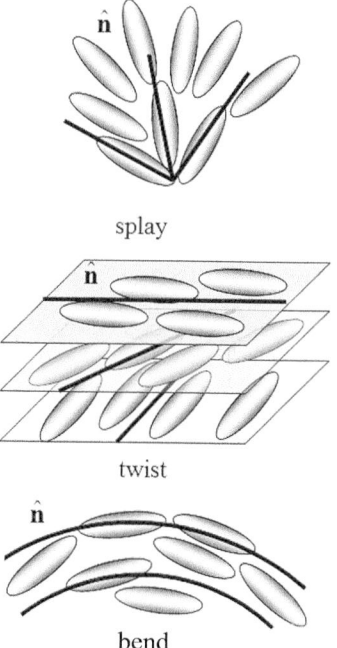

Fig. 3.11 Main types of director deformations, splay, twist and bend.

The saddle-splay K_{24} term can be re-expressed as a quadratic form of the first derivatives whereas the mixed splay-bend K_{13} term is proportional to the *second* derivatives $n_{i,jk}$ and thus might in principle be comparable to $f_{\rm FO} \sim n_{i,j}n_{k,l}$. The volume integrals of these terms can be re-expressed as the *surface* integrals by virtue of the Gauss theorem (but only when the elastic moduli K_{13} and K_{24} are constant which might not be the case at certain interfaces and at the core of defects). Therefore when one seeks for equilibrium director configurations by minimizing the total free energy functional $\int (f_{\text{FO}} + f_d)\, dV$, the K_{13} and K_{24} terms do not enter the Euler–Lagrange variational derivative for the bulk. However, they can contribute to the energy and influence the equilibrium director through boundary conditions at the surface, see [94] for a detailed review. Sometimes an argument is used that these terms should not be considered since their contribution is small in the systems with small surface/volume ratio. A good example to the contrary is the radial configuration (or point 'hedgehog'), $\hat{n} = r/r$ in the spherical coordinates, for which the free elastic energy is $F = 8\pi R (K_{11} - K_{24} + K_{13})$. The last expression shows that the 'surfacelike' divergence terms are as important as the bulk terms. The contribution of the divergence terms depends on the type of distortions. For example, the saddle-splay K_{24} term vanishes if the director depends only on one Cartesian coordinate. The important difference between the 'regular' Frank–Oseen terms such as splay, twist, and bend, and the divergence terms is that the former are all positive definite, while the latter are not. It implies that the divergence terms might cause spontaneous deformations. Ericksen [95] noted that if one wishes to maintain the uniform UN phase with $\hat{n} = \text{const}$, then, if $K_{13} = 0$, one should limit the values of the saddle-splay constant to $0 < K_{24} < K_{11}$ or $0 < K_{24} < K_{22}$ (whichever is smaller).

Usually, the K_{24} term is retained when the system experiences a topological change of the director field, for example, in the BPs [57, 70], in the consideration of topological point defects [96–99], in explanations of stripe structures observed in thin UN films with hybrid boundary conditions [96, 99, 100], Figure 3.12, showing spatially-periodic versions of the Frederiks transition. The K_{13} term is often neglected; very little is known about its value. Faetti and Riccardi [101] and then Yokoyama [102] used microscopic arguments to suggest $K_{13} = 0$. Later analysis by Pergamenshchik and Chernyshuk [103] suggested that the microscopic model does allow K_{13} to be non-zero. At present, there is no experimentally observed effect that can be directly linked to the K_{13}, although the thickness dependence of the period in experimentally observed stripe pattern in hybrid aligned thin films of UN was described better for $K_{13} = -0.2 K_{11}$ than for $K_{13} = 0$ [96].

The elastic constants can be estimated as the typical energy of molecular interactions responsible for the orientational order divided by the characteristic length, which is the molecular size: $K \sim U/l \sim k_B T/l \sim 4 \times 10^{-21}\, J/10^{-9} \sim 4\, \text{pN}$, which is a very good estimate for many thermotropic UN materials, as the experimental values are between 1 and 10 pN. Experimentally, for low-molecular weight nematic materials such as cyanobiphenyls, Figure 3.1, $K_{33} > K_{11} > K_{22}$. The temperature dependence of the elastic constants is predicted within the Landau–de Gennes theory to scale as $s^2(T)$ [1]. Near the UN–SmA transition, one expects a sharp increase in bend K_{33} and twist K_{22} elastic constants. For the polymer and lyotropic UNs, there are interesting relationships between the shape and properties of building units and the elastic con-

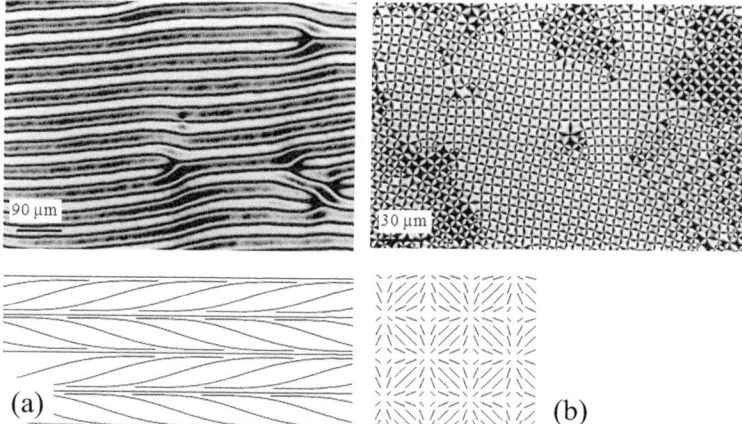

Fig. 3.12 Polarizing microscopy textures and schemes of director distortions in ground state of submicron hybrid-aligned thin films of pentylcyanobiphenyl (5CB) at glycerine substrate (upper surface is free): (a) Periodic stripe texture and (b) lattices of point defects-boojums formed in submicron films. Film thickness is 0.35 μm in (a) and 0.1 μm in (b).

stants.

In the models of lyotropic LC polymers, the molecules of covalently fixed length L and diameter D are considered in first approximation as rigid or semiflexible. If the rods are rigid, the excluded volume theory [104] predicts $K_{11} \propto \phi L/D$, $K_{11}/K_{22} = 3$, and $K_{33} \propto (\phi L/D)^3$, where ϕ is the volume fraction of the polymer in solution. One expects that the ratio K_{11}/K_{33} is much smaller than one and decreases as ϕ increases. In the model of semiflexible polymers [104–107], K_{22} and K_{33} are determined by the persistence length of polymer, rather than by the contour length: $K_{22} = (k_B T/D) [\phi (\lambda_P/D)]^{1/3}$ [104], $K_{33} = (4/\pi)(k_B T/D) \phi (\lambda_P/D)$ (in derivation of the last formula, the standard definition [108] is used for the persistent length $\lambda_P = B/k_B T$ through the bend modulus B which makes the bend elastic modulus twice as large as in the original work [107]). The splay elastic constant K_{11}, as explained by Meyer [109], should still depend on the contour length L rather than on λ_P: to preserve the constant density of the UN material under splay deformation, one must use the ends of molecules to fill the gaps between two polymers in splay configuration. Larger L implies a smaller number of molecular ends available, and thus a higher K_{11}: $K_{11} = (4/\pi)(k_B T/D) \phi (L/D)$. The splay-to-bend ratio is thus expected to follow the ratio of the contour length to the persistence length, $K_{11}/K_{33} = L/\lambda_P$. The latter relationship is also applicable to the case of self-assembled chromonic aggregates. Since the length of chromonic aggregates is not fixed by covalent bonds, it varies strongly with the volume fraction [110] and with the temperature. The average length can be estimated as [110, 111]

$$\bar{L} = L_0 \phi^{5/6} \left(\frac{\lambda_P}{D}\right)^{1/3} \exp \frac{E + \kappa \phi}{2 k_B T}, \qquad (3.11)$$

where $L_0 = 2\pi^{-2/3} \sqrt{a_z/D}$ is the characteristic size of the monomer, a_z is the period of molecular stacking along the aggregate, E is the scission energy, and κ is a constant associated with the enhancement of aggregation by the excluded volume effects, $\kappa \approx 4k_B T$ [110]. The experimental measurements [111] show that in the chromonic UN, the ratio K_{11}/K_{33} increases with temperature, which is consistent with shortening of the aggregates at higher temperatures.

In the presence of external field, the free energy density acquires additional terms. For example, for the magnetic field $\boldsymbol{B}$, the energy density, equations (3.9) and (3.10), should be supplemented by the term $-(1/2)\mu_0^{-1}\chi_a(\boldsymbol{B}\cdot\hat{\boldsymbol{n}})^2$. A similar term arises when the dielectrically anisotropic liquid crystal is placed in the electric field $\boldsymbol{E}$; here, however, one needs to be aware of the non-local character of dielectric coupling (the acting field depends on the local permittivity); for weak dielectric anisotropy, the dielectric energy density is $-(1/2)\epsilon_0\epsilon_a(\boldsymbol{E}\cdot\hat{\boldsymbol{n}})^2$. The magnetic and electric Frederiks transitions can be used to determine the elastic constants of the LC. In some cases, especially for UN made of polymer and lyotropic liquid crystals, this approach is not practical. Then one can rely on indirect measurements, exploring the contribution of various terms to the complex deformations. For example, Hudson and Thomas [112] proposed to use the director field around a disclination line that is typically comprised of splay and bend distortions, to extract the values of the K_{11}/K_{33} ratio. Williams [78] calculated the threshold of twist deformation in a nematic droplet that can also be used to estimate some of the combinations of elastic constants [113].

The dielectric coupling of the UN and the electric field is only one of the possible mechanisms of coupling between the director and the electric field. Other mechanisms involve flexoelectric polarization, surface polarization, and ionic effects. Flexoelectric polarization in liquid crystals, discovered by Meyer [114, 115], contributes a term $-\boldsymbol{E}\cdot\boldsymbol{P}_f$ to the free energy density, where the flexoelectric polarization is associated with the director gradients, $\boldsymbol{P}_f = e_1\hat{\boldsymbol{n}}(\nabla\cdot\hat{\boldsymbol{n}}) + e_3\mathrm{curl}\hat{\boldsymbol{n}}\times\hat{\boldsymbol{n}}$. Note an interesting correspondence between the flexoelectric polarization and the saddle-splay elastic K_{24} term. The surface polarization $\boldsymbol{P}_s$ (that arises, for example, because of different affinity of molecular ends to the bounding substrates), contributes a similar energy density $-\boldsymbol{E}\cdot\boldsymbol{P}_s$.

Smectic A

For the SmA phase, the elastic free energy density functional should be modified to take into account (a) certain restrictions that the layered structure imposes onto the director distortions and (b) elastic cost of changes in the thickness of the layers:

$$f = \frac{1}{2}K_{11}(\nabla\cdot\hat{\boldsymbol{n}})^2 + \frac{1}{2}B\gamma^2, \qquad (3.12)$$

where B is the Young modulus (layers compressibility modulus) and $\gamma = (d - d_0)/d_0$, the relative difference between the equilibrium period d_0 and the actual layer thickness measured along the director $\boldsymbol{n}$. The ratio of K_{11}/B defines an important scale

$$\lambda = \sqrt{\frac{K_{11}}{B}}, \qquad (3.13)$$

called 'the penetration length'; λ is of the order of the layer separation but diverges when the system approaches the SmA–nematic transition.

When the SmA layers undergo a large degree of reorientation (which is the case of focal conic domains, to be discussed later), the elastic free energy density is written in terms of the mean curvature and the Gaussian curvature of the layers:

$$f = \frac{1}{2}K_{11}\left(\frac{1}{R_1} + \frac{1}{R_2}\right)^2 + \bar{K}\frac{1}{R_1 R_2} + \frac{1}{2}B\gamma^2. \qquad (3.14)$$

As compared with equation (3.12), it is supplemented by the divergence saddle-splay term $\bar{K}$ that should satisfy the condition $-2K_{11} < \bar{K} \leq 0$ to ensure the stability of undistorted SmA; R_1 and R_2 are the local values of principal radii of curvature of the smectic layer, Figure 3.13. Here one uses the relationships [13]

$$\nabla \hat{\boldsymbol{n}} = \pm \left(\frac{1}{R_1} + \frac{1}{R_2}\right); \quad \nabla\left(\hat{\boldsymbol{n}} \cdot \nabla \hat{\boldsymbol{n}} + \hat{\boldsymbol{n}} \times \mathrm{curl}\hat{\boldsymbol{n}}\right) = \frac{2}{R_1 R_2}. \qquad (3.15)$$

Note that the saddle geometry shown in Figure 3.13(b) with radii of curvature of the opposite sign implies a reduced energy of splay deformation.

When the smectic A layers are perturbed weakly, one can write the free energy density in yet another form, using a new variable $u(x, y, z)$ for displacement of layers (or 'phase'), so that the displacement of layers from a position z_0 to z is written $z_0 = z - u(x, y, z)$. Then

$$f = \frac{1}{2}K_{11}\left(\frac{\partial^2 u}{\partial x^2} + \frac{\partial^2 u}{\partial y^2}\right)^2 + \frac{1}{2}B\left[\frac{\partial u}{\partial z} - \frac{1}{2}\left(\frac{\partial^2 u}{\partial x^2} + \frac{\partial^2 u}{\partial y^2}\right)^2\right]^2. \qquad (3.16)$$

The geometrical meaning of the terms in the last expression is illustrated in Figure 3.14 for the two-dimensional case. Note that the harmonic form of the compressibility term $(1/2) B (\partial u/\partial z)^2$ is written with

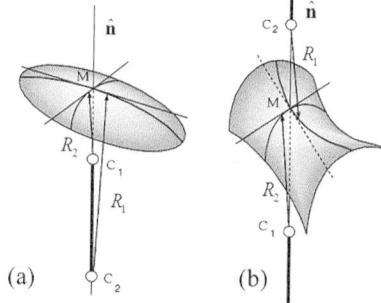

Fig. 3.13 Elements of curved surface in the vicinity of an (a) elliptic and (b) hyperbolic or saddle point M, with the two principal centers of curvature C_1, C_2 and two principal radii R_1, R_2. The point is located at the smectic layer; the director is normal to the surface.

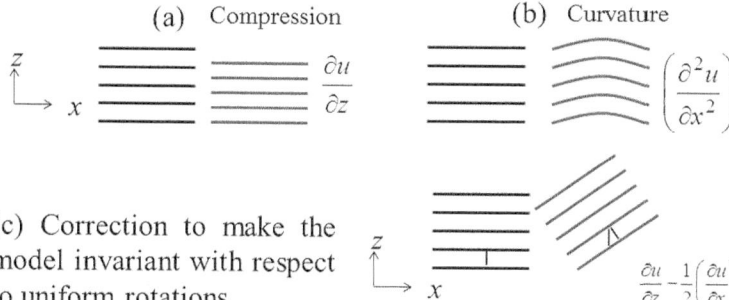

Fig. 3.14 Types of elementary distortions of a lamellar system such as smectic A or cholesteric with deformation scale much larger than the pitch. (a) Compression/dilation; (b) curvature; (c) correction to the compression term to make the model invariant with respect to the uniform rotations.

a nonlinear correction. The purpose of this correction is to make the compressibility term invariant with respect to uniform rotations. A uniform rotation of layers, say by an angle $\theta = \partial u/\partial x$, should not change the energy of the SmA, Figure 3.14(c). However, it increases the effective layer spacing measured along the fixed vertical axis, $d \to d/\cos\theta$. To account for an effective strain, one needs to correct the compression term by a quantity $d\left(1 - 1/\cos\theta\right)/d \approx -\theta^2/2 = -\left(\partial u/\partial x\right)^2/2$, which explains the correction. The correction terms make the elastic theory nonlinear and complicate analytical calculations. However, many elastic problems cannot be solved without these terms. For example, the behavior of layers above the threshold in the undulation instability of layers (the Helfrich–Hurault undulation) [1, 116–119] or the profile of layers around edge dislocations can be calculated only when the nonlinear term is retained [120–122]. Experimentally, the configuration of layers in lamellar systems can be used to deduce the value of penetration length $\lambda = \sqrt{K_{11}/B}$ [121].

The Helfrich–Hurault undulations [116, 117] in a system of flexible layers periodic in one dimension is an analog of the Frederiks transition in the UN. Figure 3.15(a) shows a system of cholesteric layers confined between the two glass plates. When there is no external electric field, the layers are parallel to the bounding plates. Once the field is applied along the vertical z-axis, the layers try to realign parallel to the field; however, surface anchoring and the compression term in the free energy density allow only a periodic tilt of the layers, mostly in the middle of the cell, Figure 3.15(b). The undulations can be triggered also by dilations [123, 124]. When the distance between the two glass plates is increased quickly, so that dislocations have no time to nucleate and to change the number of layers between in the gap, the only option for layers is to tilt, in order to fill the enlarged gap without voids. However, since the surface anchoring prevents a uniform tilt at the boundaries, the tilt regions of opposite directions alternate with the regions of large dilations; the amplitude of these sinusoidal-like deformations tends to zero near the bounding plates to satisfy the surface anchoring conditions. At higher stresses, experiments demonstrate that the simple sinusoidal deformation is replaced with zig-zag [119] and more complex patterns with parabolic defects [118].

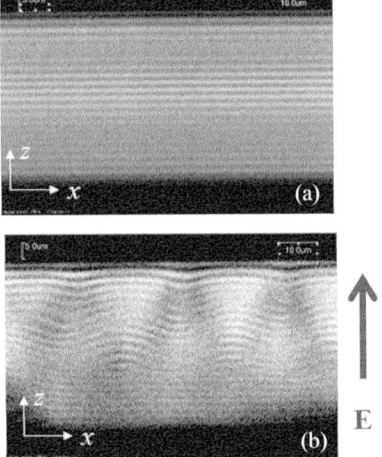

Fig. 3.15 Helfrich–Hurault instability in a Ch liquid crystal under the applied electric field. Flat horizontal layers (a) become undulated (b) when the field exceeds some threshold. Fluorescence confocal polarizing microscopy texture by B. Senyuk [118].

3.4 Surface anchoring

Observations of phase transitions in LCs reveal that the liquid crystalline phases appear as nuclei of complex internal structure and more or less round shape, see, for example, tactoids in Figures 3.2(e) and 3.8. The equilibrium of a LC of a given shape in an external field is determined by the minimum of the functional comprised of the bulk and surface terms,

$$F = \int_V (f_{\text{FO}} + f_d + f_{\text{field}})\,\mathrm{d}V + \int_S (\sigma_0 + f_{\text{anch}})\,\mathrm{d}S, \qquad (3.17)$$

where σ_0 is the 'bare' (director orientation-independent) surface tension of the liquid crystal boundary, and f_{anch} is the surface anchoring energy density. The nature of the surface anchoring term is in anisotropic molecular interactions between the liquid crystal and the adjacent medium. This anisotropy establishes one (sometimes more) preferred orientation(s) of $\hat{n}$ at the boundary, the so-called 'easy axis'. To change this orientation, one needs to perform some work. The surface anchoring density in UN is often represented by the simple potential $f_{\text{anch}} = (1/2) W_a \alpha^2$, with W_a being the surface anchoring coefficient and α the angle between the 'easy axis' and the actual director orientation at the surface. In practice, a convenient form is the one called Rapini–Papoular anchoring potential [125] $f_{\text{anch}} = W_2 (\hat{n} \cdot \hat{k})^2$, or more elaborated expressions such as $f_{\text{anch}} = W_2 (\hat{n} \cdot \hat{k})^2 + W_4 (\hat{n} \cdot \hat{k})^4$ [126].

The shape of the thermotropic UN droplets dispersed either in their own melt or in a foreign isotropic fluid, is nearly spherical. The reason is the high surface tension of the I–N interface and a relatively weak surface anchoring: $\sigma_0 \sim 10^{-5}\,\text{J/m}^2$, while $W_2 \sim (10^{-6}\text{--}10^{-7})\,\text{J/m}^2$ [127, 128], so that $W_2/\sigma_0 \sim 0.1\text{--}0.01$. The surface energy $F_S \propto \sigma_0 R^2 \sim 10^{-17}\,\text{J}$ of a thermotropic UN droplet of a radius $R \sim 1\,\mu\text{m}$ or larger overweighs the elastic energy of internal distortions $F_e \propto K (\nabla \hat{n})^2 R^3 \sim KR \sim 2 \times 10^{-18}\,\text{J}$.

The interplay of the surface and bulk effect during phase transitions in LCs is enriched when the materials are of lyotropic type. If the lyotropic N phase is formed by building units of a size in the range 10–100 nm, the interfacial surface energy is expected to be weaker than in the thermotropic case; experimentally, $\sigma_0 \sim (10^{-7}\text{--}10^{-5})\,\text{J/m}^2$ [129]. The theoretical expectation for a system of very long rigid rods is that [130, 131], $\sigma_0 = \chi (k_B T/LD)$, where k_B is the Boltzmann constant, T is the absolute temperature, L and D are the length and diameter of the rods, respectively, and χ is the numerical coefficient estimated by different models to be in the range 0.18–0.34. Large values L and D explain why the surface tension might be small. Smallness of σ_0 suggests that the UN droplets might not be able to maintain a spherical shape if the surface anchoring requires $\hat{n}$ to be distorted in the interior. And indeed, the pioneering observations by Bernal and Fankuchen [132] of the I–N phase transition in lyotropic UN formed by tobacco mosaic virus dispersed in water, revealed that the UN droplets with tangential director orientation are of a peculiar elongated shape with two cusped ends [132], Figure 3.8. These shapes were called tactoids [132, 133]; for more studies, see [26, 134–136].

The surface anchoring anisotropy in lyotropic LCs might be more pronounced than in their thermotropic counterparts. The experimental estimates range from $W_2/\sigma_0 \approx 4$ [25] for water dispersions of carbon nanotubes to $W_2/\sigma_0 \sim 10\text{--}100$ [$W_2 \sim (0.5\text{--}5) \times 10^{-5}\,\text{J/m}^2$ and $\sigma_0 \sim 5 \times 10^{-7}\,\text{J/m}^2$] for vanadium pentoxide dispersions [134, 137]. Theoretical consideration [138] for lyotropic systems of worm-like chains predicts a simple Rapini–Papoular type angular dependency of the surface tension, $f_{\text{anch}} \propto W_2 \cos^2 \alpha$, with $W_2/\sigma_0 \sim 0.5\text{--}1$ (depending on the flexibility of

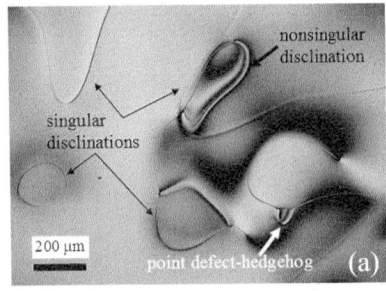

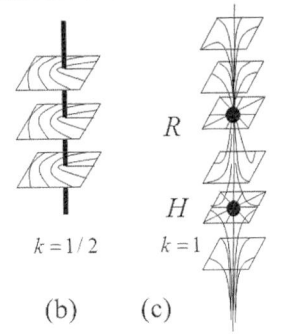

Fig. 3.16 (a) Thin (singular) and thick (nonsingular) defect lines in a thick UN sample; (b) director configuration for the thin disclination; (c) director configuration for the thick disclination with two point defects, a radial (R) and a hyperbolic (H) hedgehogs; each defect can be labeled with a topological charge N equal to either 1 or -1, see text.

[8] The disclinations inspired the term "nematic": George Friedel, observing under a microscope the line defects in what now is known as a nematic, derived the term from the Greek word "$\nu\varepsilon\mu\alpha$ = nema = thread".

the chains).

On the other hand, the Frank elastic constants in the lyotropic LCs are nearly of the same order as those in thermotropic LCs [111]. It is thus expected that the structure of nuclei in the I–UN phase transitions would be highly nontrivial, both in terms of their shape and the interior director structure, as the representative energies $\sigma_0 R^2$, $W_2 R^2$, and KR might vary in a much wider range than in the thermotropic LCs. Furthermore, in thermotropic systems with a relatively weak surface anisotropy, a Rapini–Papoular or even a simpler quadratic dependence $f_{\text{anch}} \propto (\alpha - \bar{\alpha})^2$ (where $\bar{\alpha}$ is the 'easy' angle) is often sufficient to describe the anchoring phenomena. It might not be true in the case of a lyotropic LC with a large W_2/σ_0; observation of the first-order anchoring transition of LCLCs in contact with solid substrates suggests that the surface potential should be different from the Rapini–Papoular form [126, 139].

Surface anchoring between a LC and a solid substrate is an important issue in practical applications of LCs. It determines the field-off state of the LC cell (pixel) and the dynamics of LC reorientation. Typically, one uses special alignment agents such as polyimides; these might be mechanically buffed to set an in-plane direction of orientation, see [140] for more details.

3.5 Topological defects

Uniaxial nematic

When a thick UN sample (say, $100\,\mu$m) is viewed under the microscope, it often shows a number of mobile flexible lines, the so-called disclinations. The disclinations are seen as thin and thick threads, Figures 3.16(a). Thin threads strongly scatter light and show up as sharp lines.[8] These are truly topologically stable defect lines, along which the nematic symmetry of rotation is broken. The disclinations are topologically stable in the sense that no continuous deformation can transform them into a state with the uniform director field, $\hat{\boldsymbol{n}}(\boldsymbol{r}) = \text{const}$. When one goes around the thin line, the director reorients by an angle $\pm\pi$ (we are reminded that the states described by $\hat{\boldsymbol{n}}$ and $-\hat{\boldsymbol{n}}$ are identical). The thin lines are assigned a 'strength' $k = 1/2$ and $k = -1/2$ depending on the sense of rotation. Thin disclinations are singular in the sense that the director is not defined along the line. Thick threads are line defects only in appearance; they are not singular disclinations. The director is smoothly curved and well-defined everywhere, except, perhaps, a number of point defects in the nematic bulk, also called 'hedgehogs'. Far away from the thick disclination axis, one can find that the director makes a rotation by so that the strength k can be put equal to 1 or -1. However, this assignment is not very meaningful, as the thick lines can be smoothly transformed into a uniform state with zero strength.

The question is how these linear and point defects get into the samples?

One of the mechanisms is through phase transitions. Very often, a sample is filled with a LC material in its I phase that is cooled down then to obtain, say, a UN phase. The I–UN transition is of the first order. The critical size R^* of N nuclei, determined by the gain in the bulk condensation energy and the loss in the surface energy of the I–N interface, is about 10 nm near the spinodal temperature T^* that marks the limit of metastability of the I phase upon cooling [141]. As these nuclei grow and coalesce, they can produce topological defects at the points of junction. The effect is known as the Kibble mechanism. In the model of early Universe proposed by Kibble [142], topological defects such as cosmic strings form when domains of the new 'phase' grow and merge. The order parameter is assumed to be uniform within each domain. When the domains with different 'orientation' of space–time merge, their junctions have a certain probability of producing defects. Similar effects are expected in condensed matter systems, ranging from superfluids to solids, and also in the I-to-UN phase transition.[9]

Chuang et al. [144] and Bowick et al. [145] performed the Kibble-mechanism-inspired experiments on the I–N transition and described the dynamics of defect networks. Mostly disclinations were observed, with hedgehogs seldom appearing. A probability of forming a disclination with a $\pm\pi$ director rotation is significant ($\sim 1/\pi$ when there are three merging domains [146]). For a hedgehog in a 3D space, as explained by Hindmarsh [147], one needs many more uncorrelated domains, which drastically reduces the probability of forming the defect.

As the UN domains grow larger that the de Gennes–Kleman anchoring length, $\lambda_{\text{dGK}} = K/W_2$, the surface anchoring sets a new mechanism of defect formations [13]. For a droplet of radius R much larger than λ_{dGK}, it is energetically preferable to maintain the surface orientation at the interface along the direction specified by the anisotropic molecular interaction. In that case, the director in the interior must be deformed, to satisfy the boundary orientation; the surface anchoring energy scales as $W_2 R^2$ and the bulk elastic energy scales as KR. Experimentally, for thermotropic UN in contact with their melt or with other fluids, W_2 is in the range 10^{-6} J/m^2 to 10^{-3} J/m^2, so that the de Gennes–Kleman length is often much larger than the molecular scale, being in the 'colloidal' range 10 nm–1 μm.

For example, if the surface anchoring favors normal orientation, one might expect a radial point defect-hedgehog in the interior of the droplet. In reality, experimental situation might be a bit more difficult, with the director conforming not only to the surface anchoring conditions but also to the anisotropy of elastic constants [97, 148, 149], but the overall picture is such that drops of size $R \gg K/W_2$ must contain a certain number of defects in equilibrium. Their type depends on the boundary conditions. For a director that is tangential to the interface, one might observe two surface point defects-boojums [113, 148, 150]. For tilted surface orientation, one of the most frequently met textures is the texture with an equatorial disclination ring [77], Figure 3.17. The latter can be observed in a UN nuclei growing from the I background, Figure

---

[9] The simplest illustration of the Kibble mechanism of defect formation in liquid crystals is the growth and merging of lyotropic chromonic tactoids such as those shown in Figure 2.2(e). If the merging tactoids have sufficiently different director orientation, their coalescence produces topological defects [143].

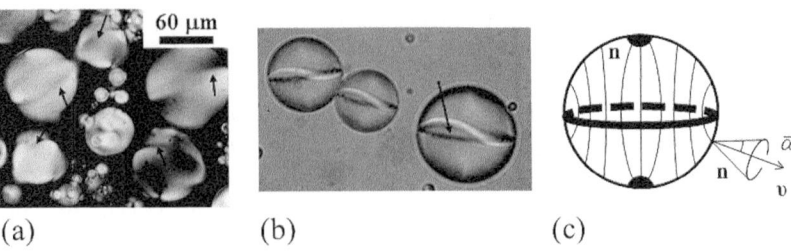

Fig. 3.17 Equatorial disclination rings in (a) droplets of UN growing from its I melt; (b) UN drops dispersed in glycerine-lecithine matrix; (c) director scheme inside the drop with tilted surface anchoring of the director.

3.17(a), for UN drops dispersed in other fluids, such as glycerine, Figure 3.17(b).

Defects in UN (and in other ordered media) can be characterized by certain numbers, such as the strength introduced above that can also be called topological charges. The topological charges play a key role in the classification of defects based on the homotopy groups, as demonstrated by Toulouse and Kleman [151, 152] and Volovik and Mineev [153]. The stability of a defect is guaranteed by the conservation of its topological charge. The laws of conservation of such charges regulate the decay and merger of defects, their creation, annihilation, and mutual transformations.

Analytically, the topological charge of a point defect in a 2D vector field $\boldsymbol{\tau} = (\tau_x, \tau_y)$ is defined as

$$N^{(2)} \equiv k = \frac{1}{2} \oint \left(\tau_x \frac{d\tau_y}{dl} - \tau_y \frac{d\tau_x}{dl} \right) dt = 0, \pm 1, \pm 2, \ldots, \quad (3.18)$$

where t is the natural parameter defined along the loop enclosing the defect point. For a three-dimensional system, one can assign charges k's to the linear defects or to point defects in the surface vector field $\boldsymbol{\tau} = \hat{\boldsymbol{n}} - \hat{\boldsymbol{v}}(\hat{\boldsymbol{n}} \cdot \hat{\boldsymbol{v}})$ which is the projection of the director onto the surface; $\hat{\boldsymbol{v}}$ is the normal to the interface. The number k, also known as the 'strength' of defect, shows how many times $\boldsymbol{\tau}$ rotates by 2π when one circumnavigates around the defect's core. The sign of k reflects whether the directions of rotation of $\boldsymbol{\tau}$ and direction of circumnavigation are the same ($k > 0$) or opposite ($k < 0$) to each other. In the case of disclinations in a UN, the configurations with opposite signs of k can be continuously transformed one into another.

For point defects in 3D vector fields [13],

$$N^{(3)} \equiv N = \frac{1}{4\pi} \oint \hat{\boldsymbol{n}} \left[\frac{\partial \hat{\boldsymbol{n}}}{\partial u} \times \frac{\partial \hat{\boldsymbol{n}}}{\partial v} \right] du\, dv = 0, \pm 1, \pm 2, \ldots, \quad (3.19)$$

where the coordinates u and v are specified on a sphere S^2 surrounding the defect. For a radial hedgehog $\hat{\boldsymbol{n}} = \hat{\boldsymbol{r}}$ in spherical coordinates, $N = 1$. N shows how many times one meets all possible orientations of the vector field while moving around a closed surface surrounding the point defect. Note that in UN, the sign of N is not defined: a substitution of $\hat{\boldsymbol{n}}$ with $-\hat{\boldsymbol{n}}$ obviously changes the sign of N in equation (3.19), but does not change the nematic state. The radial hedgehog can be described equally well with $N = -1$ and $N = 1$. The sign ambiguity leads to some spectacular effect described by Volovik and Mineev [153]: the result of merger of two hedgehogs in the presence of a singular disclination depends on the pathway toward the merger. Čopar and Žumer proposed to introduce a special characteristic, called defect rank, which reflects the geometrical structure of the point defect and not just its topology as N does [154].

An immediate useful feature of the introduced charges k and N is that one can predict the total charge of surface and point defects in a bounded UN, if one knows the Euler characteristic E of its boundary ($E = 2$ for sphere, $E = 0$ for torus), using the Poincaré and Gauss theorems of differential geometry:

$$\sum_{i=1}^{b} k_i = E\,; \qquad \sum_{j=1}^{b+q} N_j = \frac{E}{2}\,. \qquad (3.20)$$

Here b is the number of boojums at the surface and q is the number of hedgehogs in the bulk. Note that the last equation implies that the boojums are also characterized by 3D topological charges that adopt continuously varying values when the surface orientation of $\hat{n}$ changes with respect to $\hat{v}$ [77].[10] For example, each of the two boojums in a bipolar droplet can be assigned a 3D charge equal $1/2$, as each one represents $1/2$ of a radial hedgehog. For derivation and generalizations, see [77,155]. Note that by using the theorems above, one should be careful in dealing with the ambiguity associated with the non-polar character of $\hat{n}$. The ambiguity can be easily avoided, if the director field is replaced with the vector field, which is certainly permissible when the UN volume does not contain disclination lines [77]. Presence of topological defects in droplets leads to a number of interesting effects in the external fields, see for example [148, 156–160], and in the presence of additives such as surfactants [77, 161]. The first effect is used in electrically switchable 'privacy windows' and in developments of tunable microresonators [160] and generators of optical vortices [162]; the second is used in sensing applications [161, 163, 164]. Topological description of defects has recently received a renewed interest because of the studies of colloidal assemblies in LCs, see, e.g. [165–167], LC shells [167, 168], and LC in porous media [169].

In thin nematic samples (1–50 μm), the disclinations are often perpendicular to the bounding plates. Under a polarizing microscope with two crossed polarizers, one can see the ends of the disclinations as centers with emanating pairs of dark brushes, giving rise to the so-called *Schlieren texture*, Figure 3.18. The dark brushes display the areas where the director $\hat{n}(\mathbf{r})$ is either in the plane of polarization of light or in the perpendicular plane. Since the director rotates by an angle $\pm\pi$ when one goes around the end of the disclination at the surface, there are two dark brushes emanating from the core of a singular disclination in the Schlieren texture.

Sometimes one observes centers with four dark brushes; they correspond to point defects-boojums located at the surface, Figures 3.17 and

[10] In LCs that are more complex than UN, equations (3.20) explain the existence of structural analogs of the Dirac monopole, namely, a combination of a point defect with an attached line defect. For example, a spherical droplet of a SmC or a Ch with concentric packing of layers shows a radial point defect, $N = 1$, in the field of normals to the layers, accompanied by disclinations in the molecular orientation (Ch), either one radial line of strength $k = 2$, or two radial lines of strength $k = 1$ each.

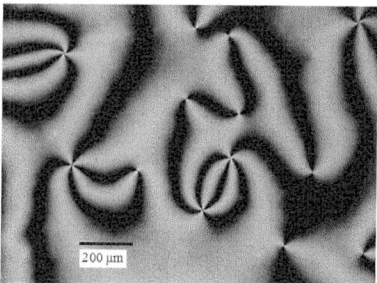

Fig. 3.18 Schlieren texture (long side approximately 3 mm) of a thermotropic nematic liquid crystal.

[11] Centers with two dark brushes, representing isolated disclinations, can be observed in Schlieren textures only when the director is strictly parallel to the bounding plates. Tilted surface alignment requires the disclinations to be connected by domain walls.

3.18. The director undergoes a $\pm 2\pi$ rotation around these four-brush centers ($k = \pm 1$). One can observe the difference between the two-brush and four-brush centers by gently shifting one of the bounding plates. Upon separation in the plane of observation, the centers with two brushes leave a clear singular trace—disclination, while the centers with four brushes do not.[11]

The intensity of linearly polarized light coming through a uniform UN slab depends on the angle β between the polarization direction and the optical axis (i.e. $\hat{n}$), projected onto the slab's plane [13]:

$$I = I_0 \sin^2 2\beta \sin^2 \left[\frac{\pi h}{\lambda} (n_{e,\text{eff}} - n_o) \right], \qquad (3.21)$$

where I_0 is the intensity of incident light, λ is the wavelength of the light, and $n_{e,\text{eff}}$ is the effective refractive index that depends on the ordinary index n_o, extraordinary index n_e, and the director orientation [13].

Equation (3.21) allows one to relate the number $|k|$ of director rotations by $\pm 2\pi$ around the defect core, to the number B of brushes:

$$|k| = \frac{B}{4}. \qquad (3.22)$$

As already indicated, k is called the strength of disclination and is related to a more general concept of a topological charge. Note that in some cases, for example in hybrid-aligned UN films with antagonistic surface anchoring at the two boundaries (say, normal and tangential), the sense of director rotation might change from counterclockwise to counter-clockwise and back, which would make the relationship (3.22) invalid [170].

When left intact, defectous textures relax into a more or less uniform state. There are, however, situations when the equilibrium state *requires* topological defects, such as the hybrid aligned films with periodic lattice of point defects, Figure 3.12(b) and nematic droplets, Figure 3.17.

Smectic A

In smectic phases, the layered structure leads to linear defects of positional order, dislocations, in addition to disclinations. There is also a special class of distortions known as focal conic domains (FCDs) that are associated with large-scale curvatures of layers.

Early in 20th century, Friedel and Grandjean [171] used a microscope to observe the intermediate states that appeared in some organic materials between a solid crystal and a regular fluid when the temperature was varied, Figure 3.19(a,b,c). The textures showed translucent regions with embedded dark lines. The lines were often shaped as perfect ellipses and one-branch hyperbolae, Figure 3.19(b,c). They appeared in pairs, situated in two orthogonal planes, Figure 3.19(a) in such a way that the apex of one partner was at the focus of the other. These shapes were all that Friedel needed to decipher the principal structural organization of the new state of matter that he called a 'smectic' [172] (nowadays known

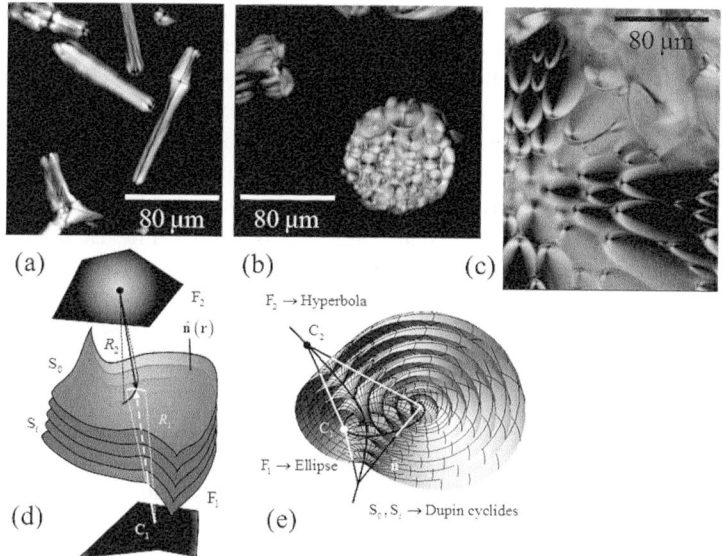

Fig. 3.19 Smectic A (dodecylcyanobiphenyl, 12CB) textures with focal conic domains: (a) rod-like nuclei (bâtonnets) of the smectic A phase emerging from the I melt and comprised of one or more focal conic domains; (b) spherical nuclei with numerous focal conic domains; their elliptic bases are located at the I–SmA interface; (c) thick ($120\,\mu m$) homogeneous smectic A sample filled with focal conic domains of the ellipse-hyperbola type; (d) family of parallel layers, the loci of the centers of curvature form two focal surfaces; (e) in a focal conic domain, the focal surfaces degenerate into focal lines; the corresponding flexible parallel layers adopt the geometry of Dupin cyclides.

as 'smectic A' or SmA). With no X-ray diffraction data available, he has predicted correctly that the smectic A is formed by flexible molecular layers of constant thickness.

To understand the principle of space filling by a smectic, consider a smoothly bent single molecular layer S_0, Figure 3.19(d). The layers' bend implies splay of the director $\propto |\nabla \hat{n}|$. Let us use S_0 as a seed layer to build an entire family of layers S_i that fill the 3D space, requiring that the distance between two neighboring layers, measured along the local normal $\hat{n}$ is everywhere the same, Figure 3.19(d). Each point of each layer is characterized by two principal radii of curvature, R_1 and R_2. The loci of the corresponding centers of curvature, C_1 and C_2, form two focal surfaces F_1 and F_2. The curvature of layers tends to infinity near C_1 and C_2, which means that the smectic order at the focal surfaces cannot be sustained and must be broken (for example, melted into an isotropic liquid). The energy associated with the broken order at focal surfaces is proportional to their area. The simplest and most efficient way to reduce this energy is to transform the 2D focal surfaces into 1D focal lines, Figure 3.19(e). Such a reduction is possible only for limited types of surfaces and focal lines. It was Pierre-Charles Dupin, in the 19th century, who discovered these special surfaces and called them cyclides [173]. The corresponding focal conics can be of three types: a pair of confocal ellipse and hyperbola, two parabolae [174], and a circle with a straight line passing through its center [13].[12]

The nanometer-thin smectic layers could not be seen directly under the microscope, but the elliptic-hyperbolic focal pairs are visible. First, they extend over distances of tens and hundreds of microns, serving as a frame for thousands of smectic layers folded as a single family of equidistant Dupin cyclides; Friedel called the region of space framed by the confocal pairs a FCD [172]. Second, the optic axis $\hat{n}(r)$, con-

[12] Dupin cyclides were first analyzed in context of physics by James Clerk Maxwell [On the cyclide. (1868) Q. J. Pure Appl. Math. 9, 111-126], who considered propagation of optical waves through an isotropic medium and stated that, if the light rays pass through two lines, these lines are parts of focal conics; the wave surfaces are equidistant Dupin cyclides to which the rays are normal. The liquid crystals were not classified yet, although the first article describing a liquid crystalline behavior of cholesteryl chloride had already been published by J. Planer, see note 1 on page 95.

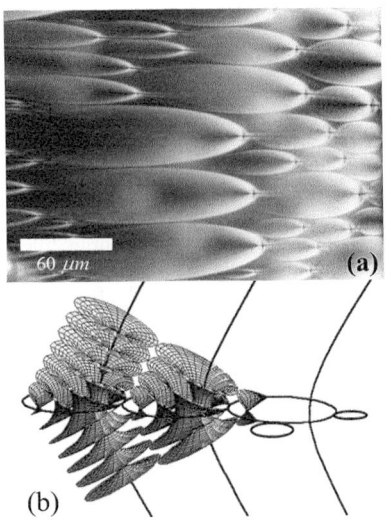

Fig. 3.20 A grain boundary in smectic A formed by an array of focal conic domains (a), schematically explained by layer surfaces in (b).

necting one focal line to the other, changes orientation from one location to another and diverges at the cores of focal conic defects, making them visible through deflection and scattering of light. Interestingly, Dupin cyclides were first analyzed in the context of physics and optics by Maxwell [175] who considered propagation of optical waves through an isotropic medium and stated that if the light rays pass through two lines, these lines are parts of focal conics; the wave surfaces are equidistant Dupin cyclides to which the rays are normal. The LCs were not known to Maxwell although the first article describing a liquid crystalline behavior of cholesteryl chloride had been already published [176].

To fill the 3D space up, the Dupin cyclides in smectics should form multi-lamellar parallel surfaces within each FCD and somehow find a way to exit one FCD and enter another, avoiding serious problems such as focal surfaces. Theoretical description of 3D space filling with Dupin cyclides is an interesting problem with many developments brought about by decades of studies [13, 177–180]. One of the most frequently met situations is relaxation of grain boundaries through an array of FCDs [181], Figure 3.20.

For more discussion of defects and textures in LCs, see the recent reviews, [13, 177, 182].

3.6 Polymer-liquid crystal hybrid materials

There are two currently very active fields of research that focus on hybrid materials in which the liquid crystals represent only one component and the other component is another soft material, such as a polymer or a colloid.

Combinations of polymers and LCs give rise to materials such as polymer dispersed liquid crystals (PDLCs) [156, 157, 183], polymer-stabilized liquid crystal textures (PSLCTs) [9, 184], LC elastomers (LCEs) [6], LC polymer networks [185], etc.

In PDLC, the LC is confined into closed inclusions within a polymer matrix. The material is formed by a phase separation, triggered by temperature quench, evaporation of a common solvent, or by polymerization of the monomers. The ground state of the inclusions is typically deformed by surface anchoring and the closed shape and often contains topological defects such as boojums and hedgehogs. An applied electric field realigns the LC molecules. For droplets of micron size, such a switching would result in a strong modulation of light scattering. If the materials are selected in such a way that $\epsilon_a > 0$ and the ordinary index of refraction of the liquid crystal matches the refractive index of the polymer matrix, the PDLC film becomes transparent in the field-on state. The composite film thus represents an electrically-switchable light-scattering panel and can be used in large area (square meters) 'privacy windows' that do not require polarizers, as usual LCDs do.

In PSLCTs, the polymer component does not enclose the LC domains but rather forms a 3D open network that helps to stabilize one of the desired states of the LC. Such a network is set into competition with the aligning force of the electric field and helps to achieve an effective reversible switching of the optical or other properties of the composites. In PSLCTs, the polymer threads forming the network represent bundles rather than individual polymer molecules, which is the main difference with the LCEs.

LCE [6] is a composite in which the polymer forms a crosslinked rubbery network. The main difference between a regular rubber and a LCE is that the polymer molecules in rubber are highly disordered, while in LCEs they show orientational order, even at rest [6]. LCEs can be considered as cross-linked main-chain or side-chain liquid-crystalline polymers. While the polymer chains are fixed by cross-linking nodes into a solid network, the mesogenic units form an orientationally ordered subsystem that preserves some degree of mobility.

3.7 Liquid crystal-colloid hybrid systems

In the consideration of surface anchoring and elasticity of LCs above, we already considered a suspension of LC droplets in an isotropic matrix. The distinctive feature of this system is that sufficiently large LC droplets contain a distorted director with a certain number of topological defects when their size is larger than the anchoring extrapolation length.

A complementary system is the one in which the LC is a dispersive medium and contains closed-shape inclusions of isotropic fluid or solids. At the interface between the colloidal particle and the LC host, the anisotropy of molecular interactions results in director distortions. For example, a sphere with a perpendicular orientation of $\hat{n}$ at its surface will tend to create a hedgehog-like configuration around itself. Whether or not such a radial configuration will be stabilized depends on the size of inclusion relative to the de Gennes–Kleman extrapolation length $\lambda_{\text{dGK}} = K/W_2$. When the typical size of particles R is much smaller than λ_{dGK}, one deals with the LC colloid in which $\hat{n}$ is not strongly pinned at the inclusions; if the initial director orientation is monocrystalline, it is likely to remain as such even if the small particles $R \ll \lambda_{\text{dGK}}$ are added. Although the director orientation does not change much, the added small particles can still impart new properties on the composition, modifying the effective elastic, viscous, dielectric, and optical parameters. The additives-modified molecular interactions can also change the overall director structure. A good example is an addition of chiral molecules to a UN that transform it into the Ch phase with a helicoidal director. One can call the system with $R \ll \lambda_{\text{dGK}}$ a LC colloid with weak intrinsic anchoring.

When the inclusions are much larger than λ_{dGK}, the situation changes dramatically, Figure 3.21. Since the anchoring energy prevails over the elastic energy, each particle tends to satisfy the boundary conditions imposed by the anisotropic surface anchoring potential, at the expense of the elastic distortions in the surrounding LC. The elastic distortions culminate in the appearance of topological defects in the neighborhood of each and every particle, as the locally distorted director field needs to match the uniform director configuration far away from the inclusion. The elasticity of distortions determines interactions between the embedded particles [186]. The corresponding long-range forces [186] have no analogs in regular colloids with an isotropic fluid as a dispersive medium. These systems, reviewed in [187], can be called LC colloids with strong intrinsic anchoring. Their description is challenging, as it requires one to treat the complex 3D configurations of the director field, taking into account the boundary conditions at all the interfaces [186]; for the recent theoretical models, see [188, 189]. The original approach to the description of LC colloids with strong anchoring was based on an analogy with electrostatics, as in both cases one deals with long-range forces [186]. Experimentally, it has been shown that the particles interact with each other as elastic 'dipoles' [186, 190, 191] Figure 3.21, and elastic 'quadrupoles' [192, 193], depending on the type of surface anchoring and director distortions around the particle. The anisotropic interactions in LC colloids with strong anchoring are being actively explored, see for example [166, 194–196].

A careful reader might notice that in many of the articles cited in this section, the size of the colloidal particle placed in a LC matrix is rather large, sometimes $10\,\mu$m or more. The standard nomenclature sets a limit on the use of the word 'colloid' for particles in the range $1\,$nm–$1\,\mu$m. The physical background is that a particle, to qualify for the status of a colloid, must be small enough to resist the gravity-driven sedimentation through the random Brownian motion. To compare the relative importance of the two factors, it is customary to consider the barometric formula that describes the probability of finding a sphere with an excess mass $m^* = (4/3)\pi R^3 \Delta \rho$, at some height z above the bottom of the container [197],

$$p(z) = \exp\left(\frac{-m^* g z}{k_B T}\right), \qquad (3.23)$$

where $\Delta\rho = \rho_p - \rho$ is the excess density of the dispersed particle as compared to density of the fluid, and $g \approx 9.8\,\mathrm{m/s^2}$ is the standard gravity. The last expression suggests that the thermal motion can raise the sphere up to the height $z_s = 3k_B T/4\pi R^3 g \Delta\rho$, called the sedimentation (or gravitation) length. The upper colloidal size is determined as $R_{\mathrm{upper}} = z_s$. For typical materials such as glass, polymers, dispersed in water at room temperature, with an excess density $\Delta\rho \sim 10^2\,\mathrm{kg/m^3}$, $R_{\mathrm{upper}} \approx 1\,\mu$m and the definition of colloid is satisfied as long as $R \leq 1\,\mu$m [197].

In a LC, sedimentation is opposed not only by thermal motion, but also by elastic repulsion from bounding walls [199], Figure 3.21. Consider a sphere with a radial anchoring and an accompanying hyperbolic hedgehog, placed in a semi-infinite UN bulk, bounded by a rigid flat wall at $z = 0$ with a strong surface anchoring parallel to the wall. The sphere

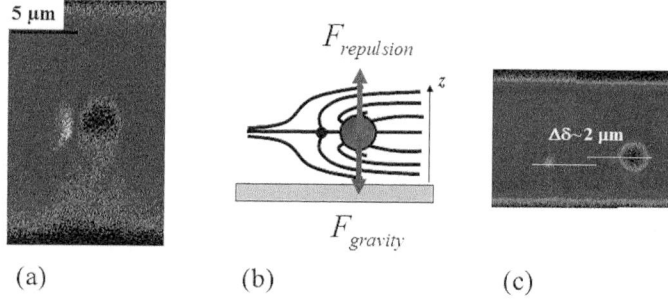

Fig. 3.21 (a) Colloidal particle with perpendicular surface anchoring in a UN is accompanied by the point defect-hedgehog; fluorescence confocal polarizing microscopy by O. Pishnyak. (b) Gravity is balanced by elastic repulsion from the plate, which allows the colloid to levitate in the nematic bilk. (c) Larger colloidal particle levitates at a higher level than the smaller colloid [198].

is repelled from the wall, as the uniform director field near the wall is incompatible with the director distortions around the sphere. The elastic potential of repulsion, in the first (dipole) approximation, neglecting in-plane anchoring effects, is

$$F_{\text{repulsion}} \approx A^2 \pi K \frac{R^4}{z^3}, \qquad (3.24)$$

where the dimensionless coefficient A (of the order of few units) depends on factors such as surface anchoring strength, elastic anisotropy of the material, etc. [186, 200, 201]. For $A = 1$, $K = 10\,\text{pN}$, $R = 1\,\mu\text{m}$, $z = 5\,\mu\text{m}$, and at room temperature, one finds the elastic energy to be much larger than the thermal energy: $F_{\text{repulsion}} \approx 2.4 \times 10^{-19}\,\text{J} \approx 60 k_B T$. When the elastic repulsion from the wall competes with gravity, the equilibrium height of the elastically levitating particle is $z_{\text{elastic}} = R \left(3\pi A^2 K/m^* g\right)^{1/4}$, or

$$z_{\text{elastic}} = \left(\frac{3}{2}A\right)^{1/2} \left(\frac{KR}{\Delta \rho g}\right)^{1/4}. \qquad (3.25)$$

For $A = 1$, $K = 10\,\text{pN}$, $R = 1\,\mu\text{m}$, and $\Delta \rho \sim 10^2\,\text{kg/m}^3$, one finds $z_{\text{elastic}} \sim 10\,\mu\text{m}$, one order of magnitude higher than the sedimentation distance $z_s \sim 1\,\mu\text{m}$ that one would measure for the same colloid if the LC were melted into the isotropic fluid. Note that $z_{\text{elastic}} \propto R^{1/4}$, increasing with the particle's size, an effect opposite to the dependence $z_s \propto 1/R^3$ for the isotropic medium. The increase of z_{elastic} with R has been demonstrated experimentally for spheres with normal anchoring [198], Figure 3.21. Therefore, the elastic nature of the LC allows the particles that are much larger than the classic colloidal limit, to levitate in the LC bulk and do not sediment under the force of gravity.

We conclude this section with a brief overview of the dynamics and transport of colloids in LCs. Some aspects (anisotropic viscosity revealed in Brownian motion, and transport through the backflow effect) have already been discussed above. Historically, the most popular technique of electrically-controlled transport of particles in an isotropic fluid is electrophoresis [202]. The particle needs to be charged; the charge is balanced by counterions in the medium. Under a uniform DC electric field, the particle moves with the velocity that depends linearly on the applied electric field,

$$v = \mu_1 E, \qquad (3.26)$$

where the electrophoretic mobility μ_1 is proportional to the particle's charge and inversely proportional to the fluid's viscosity. According to equation (3.26), an AC field with a zero time average produces no net displacement.

There is a growing interest in finding mechanisms for particle manipulation that use an AC electric field, as with the latter, it is much easier to produce steady flows and to avoid undesirable electrochemical reactions. One of the interesting developments was a discovery that a broken symmetry of the particle moving in an isotropic fluid can result in a nonlinear electrophoresis, with a quadratic dependence of the velocity on the field [203, 204].

Electrophoresis in LCs [205–210] is studied much less than in the isotropic fluids. A replacement of an isotropic fluid with a LC should bring first of all some anisotropy in the velocity of particles moving parallel and perpendicular to $\hat{n}_0$ because of the different Stokes drags. There are, however, some qualitative differences. Dierking et al. [207] has reported on the electromigration of microspheres in nematic LCs and noticed that the particles move under the AC field in the direction perpendicular to the field and parallel to $\hat{n}_0$; the velocity was linearly dependent on E. Ryzhkova et al. [206] found a nonlinear (cubic) term in the dependence of v on E, in addition to the classic linear term. A different type of electrophoresis in LC, with $v \propto E^2$, has been described recently [208, 209]. The effect is observed for colloidal particles with dipolar director configuration and is not observed for the quadrupolar particle. The mechanism is attributed to the broken symmetry of ionic currents around the particle: because the ionic motion in LC is anisotropic, the two ends of the elastic dipole with different director orientation impose a different effective viscosity for the moving ions; uncompensated ionic flow results in the propulsion of the particle. A reverse in the field polarity does not change the polarity of the dipolar configuration, thus $v \propto E^2$. Note that the latter dependence implies that one can transport even particles with zero charge. For a material with $\epsilon_a < 0$, the electric field $\boldsymbol{E} = E(1,0,0)$ can be applied normally to $\hat{n}_0 = (0,1,0)$, so that the director far away from the colloid is not perturbed. In this case, one can observe two different components of the electrophoretic velocity [209], Figure 3.22:

$$v_x = \mu_1 E_x + \mu_3 E_x^3, \qquad v_y = \beta E_x^2. \qquad (3.27)$$

Both components show a nonlinear dependence on the applied electric field.[13] The crucial point is that the quadratic component $v_y = \beta E_x^2$ is enabled exclusively by the broken dipolar symmetry of the director field around the colloid; it vanishes when the UN is heated into the I phase or when the colloid has a quadrupolar symmetry [208].

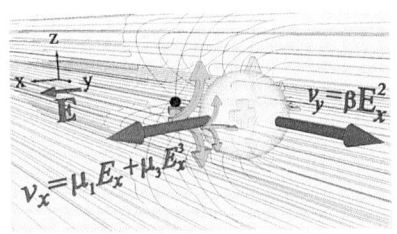

Fig. 3.22 Nonlinear electrophoresis of a positively charged sphere with normal anchoring in a UN liquid crystal with $\epsilon_a < 0$. One component of electrophoretic velocity is parallel to the applied electric field; it is an odd function of the field; the second component is perpendicular to the electric field and parallel to the overall director; it is quadratic in field. The black dot represents the core of a hyberbolic hedgehog that accompanies the sphere to compensate for its topological charge. Figure by I. Lazo.

---

[13] Thermotropic liquid crystal-enabled AC electrophoresis can be used for controlled transport of "microreactors", representing water droplets with various chemicals and drugs, see [211].

3.8 Dynamics

Anisotropic viscosity

Liquid crystals are fluids; they can flow preserving the orientational order. Flow imposes an orientational torque on $\hat{n}$. Most often, $\hat{n}$ tends to realign along the direction of flow. There is also a reverse effect: director distortions can cause the flow. This 'backflow' effect is of importance in liquid crystal displays. In the approximation of a constant scalar order parameter, the hydrodynamics of liquid crystals is described in terms of seven variables: (1) mass density, (2) three components of the velocity field, (3) energy density, and (4) two components of the director field $\hat{n}(\boldsymbol{r},t)$. In contrast to an isotropic fluid, the viscous stress tensor depends not only on the gradients of the velocity, but also on the director rotations. The viscous stress tensor has been expressed by Leslie [212] through the symmetric part $\boldsymbol{A}$ of the tensor of velocity gradients, $A_{ij} = A_{ji} = \frac{1}{2}(\partial v_i/\partial x_j + \partial v_j/\partial x_i)$, and the relative angular velocity $\boldsymbol{N} = d\hat{\boldsymbol{n}}/dt - [\boldsymbol{w} - \hat{\boldsymbol{n}}]$ of $\hat{\boldsymbol{n}}$ with respect to the fluid; here $\boldsymbol{w} = \frac{1}{2}\mathrm{curl}\boldsymbol{v}$ is the local rotation rate and $\boldsymbol{v}$ is the linear velocity of the fluid. The viscous stress tensor features six viscosity coefficients α_m (called the Leslie coefficients):

$$\sigma_{ij} = \alpha_1 n_i n_j n_k n_l A_{kl} + \alpha_2 n_j N_i \\ + \alpha_3 n_i N_j + \alpha_4 A_{ij} + \alpha_5 n_j n_p A_{pi} + \alpha_6 n_i n_p A_{pj}. \quad (3.28)$$

Only five of the Leslie coefficients are independent [213] because of the Onsager's reciprocal relations, $\alpha_2 + \alpha_3 = \alpha_6 - \alpha_5$.

The number of viscosities reduces to three when the director distortions are small. These three can be chosen as the effective viscosities for three idealized geometries of flow, also known as Miezowicz geometries, in which one assumes that the director is fixed (for example, by a strong magnetic field), Figure 3.23:

(a) When $\hat{\boldsymbol{n}} = (1,0,0)$ is perpendicular to both the flow direction and the velocity gradient, the UN behaves as an isotropic fluid with a viscosity η_a; however, director fluctuations coupled with certain values of the viscosity coefficients might destabilize the initial orientation.

(b,c) When $\hat{\boldsymbol{n}}$ is parallel to the flow (b) or parallel to the velocity gradient (c), the corresponding viscosities η_b and η_c are generally different from η_a and from each other; $\eta_b < \eta_a < \eta_c$ for a typical UN material composed of rod-like molecules. The result $\eta_b < \eta_c$ can be explained by assuming that the friction correlates with the cross-section of the molecules seen by the flow.

The anisotropic nature of viscoelastic properties of LCs leads to many new dynamic effects that are rather unusual from the point of view of the standard hydrodynamics of isotropic fluids. Consider, for example, a Brownian motion of a small colloidal particle. A sphere of a radius

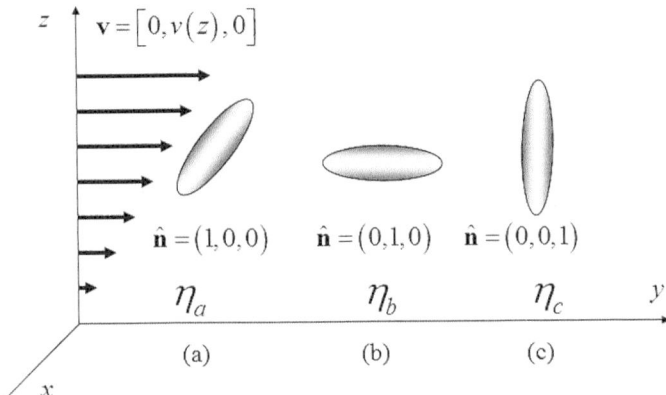

Fig. 3.23 Miezowicz geometries for effective viscosities of the UN.

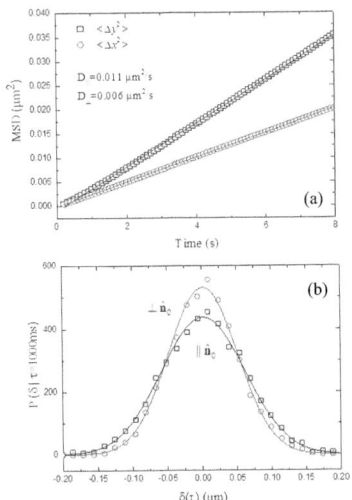

Fig. 3.24 (a) Mean-square displacement vs time lag of $2a = 5.08\,\mu\text{m}$ silica spheres dispersed in nematic mixture E7 in the directions parallel (y) and perpendicular (x) to the director $\hat{\bm{n}}_0$. (b) Probability distribution of particle displacement along and perpendicular to the director for time intervals of 1 s. Solid lines represent Gaussian fits [209].

R moving with the velocity v in an isotropic fluid of viscosity η, experiences a Stokes drag $F = 6\pi\eta R v$. For a LC, the Stokes drag should be anisotropic and depend on the direction of motion with respect to the director [214, 215]. Loudet et al. [216] described Brownian motion of spheres with tangential surface anchoring of the director and found that its self-diffusion should be described by two independent diffusion coefficients, associated with two different viscosities $\eta_\parallel$ and $\eta_\perp$, of the LC for motion parallel and perpendicular to the overall director $\hat{\bm{n}}_0$, respectively. In this case, the director configuration around the sphere resembles an elastic quadrupole [186]. Figure 3.24 illustrates a different case of perpendicular surface anchoring. The sphere creates dipolar distortions [186] that break the 'fore–aft' symmetry. However, this feature alone does not rectify Brownian diffusion of the sphere and does not result in unidirectional movement. The time dependence of the mean square displacements (MSD) along and perpendicular to $\hat{\bm{n}}_0$ follows the classic linear law, Figure 3.24(a), at least for the time scales larger than 100 ms, with an anisotropic ratio of self-diffusion coefficients $D_\parallel/D_\perp = \eta_\parallel/\eta_\perp = 1.7$ [209]. As these times are larger than the director relaxation time, the possible influence of director fluctuations on the linear time dependence of MSD vanishes. The time-average displacement of the particles is zero, as clear from the probability distribution P of particle displacements δ parallel and perpendicular to $\hat{\bm{n}}_0$, Figure 3.24(b).

Director dynamics and the Frederiks transition

The dynamics of director is usually described in the approximation that neglects the inertia effects (which is valid for low-frequency excitations [13]), as the balance of reorienting torque $[\hat{\bm{n}} \times \bm{h}]$ exerted by the external fields and elasticity of the LC on the director, and the frictional torque with the coefficients $\gamma_1 = \alpha_3 - \alpha_2 > 0$ and $\gamma_2 = \alpha_6 - \alpha_5$:

$$[\hat{\bm{n}} \times \bm{h}] - [\hat{\bm{n}} \times (\gamma_1 \bm{N} + \gamma_2 \bm{A} \cdot \hat{\bm{n}})] = 0. \quad (3.29)$$

Here, $\boldsymbol{h}$ is the so-called 'molecular field' [1]; its spatial components write, for the case of magnetically induced reorientation,

$$h_i = \mu_0^{-1} \chi_a (n_j B_j) B_i - \frac{\partial f_{\text{FO}}}{\partial n_i} + \frac{\partial}{\partial x_j}\left[\frac{\partial f_{\text{FO}}}{\partial (\partial n_i/\partial x_j)}\right], \quad (3.30)$$

where $i, j = 1, 2, 3$.

Viscous properties of UN define the response time of Frederiks transitions. Consider first a twist Frederiks transition, as in this case the director reorientation does not cause a backflow. The cell is planar, $\hat{\boldsymbol{n}} = (1, 0, 0)$ (the director is fixed at the top and bottom plates, say, by unidirectional rubbing of the polymer alignment layer), and the electric field is perpendicular to the director, $\boldsymbol{E} = (0, 1, 0)$; $\epsilon_a > 0$. The balance of dielectric, elastic, and viscous torques writes in the approximation of small angular perturbations of the director as

$$K_{22} \frac{\partial^2 \phi}{\partial z^2} + \epsilon_0 \epsilon_a E^2 \sin\phi \cos\phi = \gamma_1 \frac{\partial \phi}{\partial t}. \quad (3.31)$$

By considering a stationary case $\partial \phi/\partial t = 0$, one finds the field threshold above which the director tilt ϕ are nonzero, $E_c = (\pi/d)\sqrt{K_{22}/\epsilon_0 \epsilon_a}$. Suppose the electric field is very strong, $E \gg E_c$. Then the elastic torque can be neglected as compared to the dielectric one, and one finds that the dynamics of director realignment at the initial stage of director reorientation (small tilts) by the field follows an exponential dependence, $\phi \propto \exp(t/\tau_{\text{on}})$, with the characteristic time $\tau_{\text{on}} = \gamma_1/\epsilon_0 \epsilon_a E^2$. The last formula suggests that the UN can be switched rather quickly, if one uses a sufficiently high electric field. And indeed, as demonstrated by Clark et al. [217], for a UN with $\gamma = 0.078$ Pa s, $\epsilon_a \approx 11$, a high field of $E = 1.3 \times 10^8$ V/m leads to a rather small $\tau_{\text{on}} \approx 50$ ns. The bottleneck of the electrooptic response of UN is their relaxation to the field-off state. When the electric field is switched off, it is only the surface anchoring and the elastic nature of the orientational order that force $\hat{\boldsymbol{n}}$ to return to its original orientation. The balance of elastic and viscous torques leads to a slow realignment, $\phi \propto \exp(-t/\tau_{\text{off}})$, where $\tau_{\text{off}} = \gamma_1 d^2/\pi^2 K_{22}$, determined by the thickness d of the cell. For the typical $K_{22} = 6$ pN, $d = 5\,\mu$m, $\gamma = 0.1$ Pa s, one finds that the relaxation time is very slow, $\tau_{\text{off}} \approx 50$ ms. Even if the cell is very thin, $d = 1\,\mu$m, the relaxation time is still above 1 ms. The slow relaxation of the realigned director puts a serious limit on applications of UN.

In the twist Frederiks effect, the director reorientation does not lead to the materials flow. The effect, called the backflow, can be observed in a different geometry for a splay Frederiks transition, Figure 2.25. When the electric field is applied, it realigns the director mostly in the center of the cell, where the stabilizing elastic torque is the smallest, Figure 3.26(a). Reorientation induces flow, which is antisymmetric, $v(z) = -v(d-z)$. In the top part of the cell, $v_{\text{on}}(d/2 < z < d) < 0$, and in the bottom part, $v_{\text{on}}(0 < z < d/2) > 0$. As time elapses, the dielectric and elastic torques will balance each other and the fluid will come to rest. Note that in the field-on regime, the director in the central portion of cell is practically vertical. Near the bounding plates, the

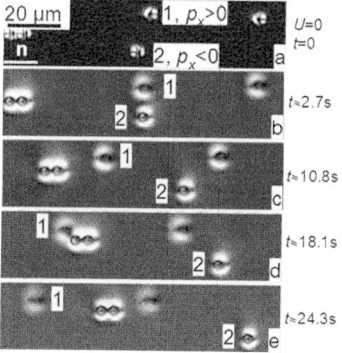

Fig. 3.25 Propulsion of colloidal particles in the nematic cell driven by voltage pulses and experiencing a splay Frederiks transition with backflow effect [199]. Particle labeled 1 moves from right to left near the bottom of the cell; particle 2 moves to the right near the top of the cell (the view is from the top). Textures by O. Pishnyak.

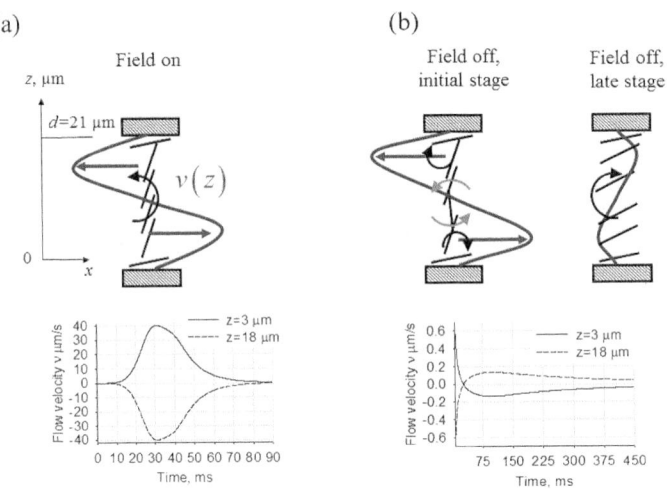

Fig. 3.26 Backflow effects in a planar cell and a splay Frederiks transition. (a) Director is realigned by the electric field counterclockwise, (b) direct flow at the beginning of switch OFF: director in the middle layer rotates counterclockwise and flips over, and (c) reverse flow observed during switch OFF: director in the middle layer rotates clockwise and relaxes to the initial state. Numerical simulations by Tang [199].

director distortions are strong because of the surface anchoring. When the electric field is removed, the strong elastic torque in these surface regions causes a clockwise rotation of the director, Figure 3.26(b). The director rotation induces a fluid flow, which might be so strong that the director in the middle part of the cell turns by an angle larger than $\pi/2$ (a 'kickback' effect). Subsequent relaxation of the director back to the original orientation along the x-axis produces a flow in the opposite direction that slowly decays until the cell reaches an equilibrium. The time dependence of the flow velocity for two regions, one near the top of the cell and another near the bottom, are illustrated in Figure 3.26. The net result of the cyclic driving of a nematic cell with voltage pulses is propulsion of the material from right to left in the bottom part of the cell and from left to right in the top part of the cell.

The backflow effect can be used to manipulate and drive small particles in the nematic fluid, Figure 3.25 (and alternatively, the tracing the trajectories of particles, one can deduce the properties of backflow) [199, 218, 219].

The backflow phenomenon is not desirable in display applications, as it often implies a longer relaxation time and lower contrast of the devices. Its inverse version, i.e. flow-induced director field, is currently explored in microfluidic and optofluidic devices controlled by pressure gradients [220]. Optofluidics studies control of light propagation by fluids; in this regard, the coupling between the director and flow is a very promising direction, as reorientation of optic axis(es) implies a dramatic change in optical appearance of the liquid crystal. Below we consider how the optical properties of the UN can be altered by a simple thermal expansion. The advantage is that the optical properties can be altered with no pressure pumps, which might be beneficial for lab-on-chip and sensing applications.

Flow induced by thermal expansion

In the description of UN dynamics, it is assumed that the UN density and volume are constant, independent of molecular reorientation and materials flow [1]. This 'incompressible flow' assumption is a natural extension of a similar simplification in the hydrodynamics of isotropic fluids. However, when the LC sample is heated or cooled, the volume and density do change. Thermally induced expansion and contraction cause LC flows, which in turn activate pronounced reorientation of the optical axis, Figure 3.27.

The mechanism of thermally induced flow is clear from the mass conservation equation, $\partial \rho/\partial t = -\rho \nabla \cdot \bm{v}$, which connects the time derivative of the fluid density ρ to the spatial gradients of its velocity. Consider a very long capillary extended along the x-axis, with open ends, Figure 3.27(a). The cross-section of the capillary is rectangular; the width of the capillary (measured along the x-axis) is much larger than its thickness d along the z-axis (but much smaller than the length of the capillary). The UN is aligned in a homeotropic fashion, along the z-axis, so that the texture appears black when viewed (along the z-axis) between to crossed polarizers.

The density of a NLC slab thermally expanding or shrinking along its axis x can be presented as $\rho(t) = \rho_0 (1 - \beta \xi t)$, where ρ_0 is the initial density of UN, β is the coefficient of thermal expansion, and $\xi = \partial T/\partial t$ is the rate of temperature change. The mass conservation equation then immediately yields $v_x \propto \beta \xi x$, i.e. a non-zero velocity along the axis x that depends linearly on the distance from the geometrical center $x = 0$. For a typical velocity of $10\,\mu\text{m/s}$, the corresponding Reynolds number $\text{Re} = \rho d v_x / |\alpha_2|$ is low, about 10^{-6}; here $|\alpha_2| = 0.1\,\text{kg}\,\text{m}^{-1}\,\text{s}^{-1}$ is the relevant viscosity [221]. Therefore, the velocity should satisfy the Stokes equation $\nabla p = \mu \nabla^2 \bm{v}$ (here ∇p is the pressure gradient and μ is the dynamic viscosity), as well as a no-slip condition $v_x = 0$ and no-penetration condition at the bounding walls. The solution for v_x then follows as:

$$|v_x| \approx 6\beta\xi x \frac{z}{d}\left(1 - \frac{z}{d}\right) \qquad (3.32)$$

(the z component of flow velocity is much smaller, $v_z = v_x (2z - d)/6x$, at distances of interest, $x \gg d$, $0 < z < d$). The flow along the x-axis realigns $\hat{\bm{n}}$ towards the x-axis. The viscous reorienting torque $\alpha_2 \partial v_x/\partial z$ is proportional to the shear rate $\partial v_x/\partial z$ that is vanishing at the walls and at the middle plane $z = d/2$, according to equation (3.32). The viscous torque is opposed by the elastic torque $K_3 \partial^2 \theta/\partial z^2$ that tends to keep $\hat{\bm{n}}$ along the z-axis. The balance of the two torques for small angles $\theta = \theta(z)$ between $\hat{\bm{n}}$ and the vertical z-axis determines the flow-induced director tilt:

$$\theta(z) = \beta\xi xz \frac{|\alpha_2|}{K_3}\left(1 - \frac{z}{d}\right)\left(1 - \frac{2z}{d}\right). \qquad (3.33)$$

Figure 3.27(b) shows schematically the velocity and the director tilt for the case of cooling. The director profile is of the bow type, with $\hat{\bm{n}}$ remaining homeotropic at $z = 0, d/2, d$, and tilted in the regions

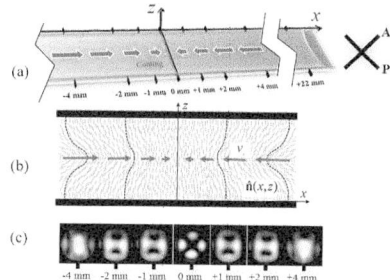

Fig. 3.27 (a) Long rectangular capillary with open ends, filled with a UN in a homeotropic texture and subject to temperature drop; (b) scheme of the thermally induced flows and director tilts of the UN in the vertical cross-section of the capillary; (c) conoscopic patterns observed in diffcrent portions of the capillary along its long axis. Textures by Y. K. Kim.

$0 < z < d/2$ and $d/2 < z < d$. Note that in the middle plane, $z = d/2$, the velocity is maximum, but the velocity gradient and thus tilt are zero. For a typical parameter of the UN and the cell, $|\alpha_2| = 0.28\,\mathrm{kg\,m^{-1}\,s^{-1}}$, $K_3 = 1.95 \times 10^{-11}\,\mathrm{N}$, $d = 50\,\mu\mathrm{m}$, $x = 10\,\mathrm{mm}$ and $\beta = 10^{-3}/^\circ\mathrm{C}$, one finds that even a modest rate of temperature change, $\xi = 0.5\,^\circ\mathrm{C/min}$, causes significant changes as the flow velocities reach the level of $10\,\mu\mathrm{m/s}$ and the tilt of $\hat{n}$ reaches tens of degrees.

Figure 3.27(c) illustrates how the director tilt manifests itself in the conoscopic patterns; the patterns are taken at a different distance from the geometrical center of the capillary. The central part shows a conoscopy pattern of a Maltese cross, which is a standard pattern of a homeotropic texture of a UN with $\theta = 0$. As one moves away from the center, progressive tilts of the optic axes result in conoscopic patterns with split isogyres. These are often associated with the biaxial nematic order, but in our case, the nematic remains uniaxial and the pseudo-biaxial features are caused by the coupling of the director tilt to the thermally induced flow.

3.9 Applications of liquid crystals

The most widely used effect in LCs is the Frederiks transition, i.e., the field induced reorientation of the director. The dielectric analog of the effect is used in many electrooptic devices (displays, optical shutters, etc.). The LC is usually sandwiched between two transparent electroconductive plates (for example, glass covered with indium tin oxide) coated with a suitable alignment layer. The voltage across the cell controls the director configuration and, thus, the optical properties of the cell. In one of the commercially most successful embodiments, the initial field-free state is the one in which $\hat{n}$ is perpendicular to the bounding plates (homeotropic orientation). The cell is sandwiched between two crossed light polarizing films. The light beam becomes linearly polarized after passing the first polarizer. This state of polarization is not changed by the liquid crystal and the light beam is thus blocked by the second polarizer. The UN material is of a negative dielectric anisotropy, $\epsilon_a < 0$. When the electric field is applied across the cell, the director deviates from the vertical axis. In this 'field-on' state, because of the finite birefringence, the cell allows the light beam to pass through the second polarizer. The direction of director tilt in the plane of the UN cell can be controlled by using specially patterned electrodes, thus the approach is known as the patterned vertical alignment (PVA) mode, Figure 3.28. The field can also be applied in the plane of the sample, comprising a layer of a UN with $\epsilon_a > 0$, thus forming the in-plane switching (IPS) mode. Finally, the polarization of light can be controlled by using the so-called twisted nematic (TN) cell, in which the electric field switched the twisted (chiral) nematic structure into a vertically aligned director configuration, thus altering the appearance of the UN cell between two polarizers, either crossed or parallel. The TN mode was the first one

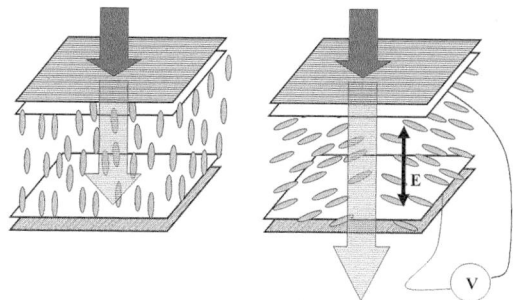

Fig. 3.28 A scheme of the display pixel based on the nematic with a negative dielectric anisotropy. The surfaces impose a vertical easy axis on the director. The vertical electric field causes the director to realign towards the horizontal plane, which make the pixel transparent to the linearly polarized light.

used in dielectrically driven commercially successful liquid crystal displays; earlier attempts to construct a 'dynamic light scattering' mode display using electrohydrodynamic motion of ions in the liquid crystal have failed because of the limited lifetime of the units and poor contrast. To create a full-color display using the PVA, IPS or TN techniques, one adds a pixellized color filter with red, blue, and green pixels to the glass plate facing the viewer. The panel that controls the voltage across the liquid crystal is pixelized, too, so that the intensity of light transmitted by each pixel and the corresponding color filter is controlled independently of other pixels. The most advanced scheme of electrical addressing the liquid crystal display is the so-called active matrix scheme based on the array of thin film transistors (TFTs). Note that the operating voltage needed in liquid crystal displays is relatively low, usually between 1 and 10 V. Much more information on the operation of LC displays can be found in the books by Yang and Wu [9, 184].

The flexibility of molecular architectures of liquid crystals allows many other types of applications, including the information displays based on the effects of light scattering rather than the changes of the polarization state. Historically, one of the first liquid crystal devices were cholesteric thermometers that changed colors as a function of temperature. In Ch, the helicoidal pitch is often in the range of 0.1–1 μm (the molecular interactions responsible for the chiral structure are relatively weak, thus the pitch is much larger than the size of an individual molecule). The well-aligned periodic director structure is thus capable of selective reflection in the visible part of spectrum, which brings a 'colored' appearance to the sample viewed in white light. The spectral range of this selective reflection is related to the value of the cholesteric pitch and thus can be changed by changing the pitch. Ch materials with temperature sensitive pitch are used in thermometers. Ch materials with electrically-controlled orientation of the helicoid axis are used in bistable light-reflective liquid crystal displays [184]. The Ch is switched by the electric voltage pulses between a well aligned (so-called planar) light-reflecting state and disoriented light-scattering state with numerous defects of the director configuration. By adding a black absorbing layer to the bottom of the Ch film, one creates a bistable Ch display with a bright colored appearance in the planar state and black appearance in the distorted state. The field is needed only to switch the Ch sample from one state to another, which greatly reduces energy consumption. Although only the planar state is truly stable, the distorted state is relatively stable, too, as its relaxation into the planar state is hindered by large energy barriers associated with the helix distortions. As the display works in the light-reflective mode, it does not need a special illuminating panel beneath the liquid crystal layer, as most other liquid crystals displays do. The Ch reflective displays are used in 'electronic books' and in the development of flexible displays in which the traditional glass panels are replaced with plastic films.[14]

[14]The unique optical Bragg reflection effect in Ch is widely used in optical security and counterfeit deterrence technologies, for example, in security labels that change the color of images when the viewing angle is changed, see, for example, Moia, F. "New color-shifting security devices". (2004) *Proc. SPIE 5310, Optical Security and Counterfeit Deterrence Techniques V*, 312, 312-320.

Smectic phases can also be used in effective display applications [222]. For example, electrically controlled FCD structures in SmA can be used in light-scattering displays similar to that based on Ch deformations. Chiral SmC* phase offers an especially interesting display mode, a truly bistable display, in which the director is switched between the two different states by a field of an appropriate polarity. Both states are truly stable in the absence of the field. The switching is fast (submillisecond) but the challenge is to align the SmC* uniformly over a significant area; it is more difficult than for the nematic-based devices.

Besides LCDs, the LCs are used in various sensing applications [223]. The effects are typically based on the response of LC surface or interfacial alignment to the chemical composition of the interfaces/substrates [164,224,225]. The ability of LCs to align uniformly is being actively explored for the development of organic semiconductors [8]. An extremely wide area of potential applications is offered by the coupling of LCs and propagation of light [226, 227]. Among these effects are light-induced chemical modifications such as trans-cis isomerization and the optical Frederiks transition [226].

Summary

Over the last five decades, liquid crystals transformed from a curious form of condensed matter into a key technological material, thanks to the progress in understanding of their elastic, electro-optical, and viscous properties. However, the intrinsic complexity of these materials still leaves plenty of room for further studies, not only of their applied nature, but also their fundamentals. In the field of thermotropic LCs, researchers continue to discover new types of structural organization, such as the phases formed by 'banana-shaped' molecules and molecules in which rigid mesogenic cores are linked by flexible chains. There is a continuous work to sharpen our understanding of even the 'old' problems, such as mechanisms of surface alignment, nature, and quantitative values of the elastic constants K_{13}, K_{24}, and $\bar{K}$. Even in the case of the electric Frederiks effect that is at the heart of modern LCD applications, the search continues as the field-induced director reorientation is generally very complex; in addition to the dielectric torque, it is controlled by a non-local character of the electric field in the anisotropic medium, finite electric conductivity, flexoelectricity, surface electric polarization, motion of ions, by a coupling of the director reorientation and the materials flows, appearance of topological defects, etc. Even if the coupling with the electric field is limited to the dielectric mechanism, the director dynamics are still hard to describe because of the dielectric dispersion effects [228]. The response of a LC to the applied electric field is not instantaneous and requires some time. As a result, the dielectric torque depends not only on the current instantaneous values of the field and state of director orientation but also on the prehistory of the system and might show a component that is linear in the field [229]. (rather than quadratic).

Many current research efforts are focused on composite systems such as LC colloids and polymer-LC composites. LCs are also explored within the context of metamaterials, i.e. artificially constructed composites of dielectric and metallic nature, with unusual properties such as negative refraction. New molecular architectures with LC properties are explored in the development of organic systems for energy applications. The advantage of the LC materials is that they allow one to achieve a well aligned functional molecular structure (i.e. a thin layer of an organic semiconductor), since the finite molecular mobility in the mesomorphic phases heals the structural defects. Over the next decade or so, one would expect that the emphasis in fundamental studies will gradually shift from the thermotropic LCs to their lyotropic counterparts, as the lyotropic type of orientational order is featured by many systems of biological significance, such as solutions of DNA, F-actin, etc.

Despite the long history of exploration, liquid crystals continue to surprise researchers and bring new challenges in our understanding of soft matter. This review considered only a limited list of LC properties; the choice is obviously biased by the research in the author's laboratory. For a deeper analysis, the reader is referred to the textbooks and reviews in the list of references. Current research in the field is published by specialized journals (*Liquid Crystals*, *Soft Matter* and *Liquid Crystals Reviews*) and in numerous general journals.

Acknowledgements

I owe my understanding of LCs to M. Kleman, M. V. Kurik and G. E. Volovik, collaboration with P. Boltenhagen, P. Bos, E. Gartland, A. Jákli, H.-T. Jung, B. Lev, N. V. Madhusudana, P. Palffy-Muhoray, P. Pasini, V. Pergamenshchik, J. Selinger, S. V. Shiyanovskii, A. Strigazzi, E. M. Terentjev, D.-K. Yang, C. Zannoni. I acknowledge the assistance my former and current graduate students and postdoctoral fellows: V. Bodnar, V. Borshch, W. R. Folks, M. Gu, T. Ishikawa, Y.-H. Kim, Y.-K. Kim, I. Lazo, Z. Li, Y. Nastishin, V. Nazarenko, H.-S. Park, O. Pishnyak, T. Schneider, B. Senyuk, V. Sergan, I. Smalyukh, L. Tortora, D. Voloshchenko, J. Xiang, and S. Zhou. I am sincerely thankful to all of them. Some of the work presented originated from the research grants supported by DOE DOEFG02-06ER 46331 and NSF, including the current grants DMR-1104850 and 1121288.

References

[1] de Gennes, P. G. and Prost, J. (1993). *The Physics of Liquid Crystals*. Clarendon Press, Oxford.
[2] Blinov, L. M. and Chigrinov, V. G. (1994). *Electrooptic Effects in Liquid Crystal Materials*. Springer, New York.
[3] Blinov, L. M. (2011). *Structure and Properties of Liquid Crystals*. Springer, Dordrecht.
[4] Oswald, P. and Pieranski, P. (2005). *Nematic and Cholesteric Liquid Crystals*. Taylor & Francis, Boca Raton.
[5] Oswald, P. and Pieranski, P. (2006). *Smectic and Columnar Liquid Crystals*. Taylor & Francis, Boca Raton.
[6] Warner, M. and Terentjev, E. M. (2003). *Liquid Crystal Elastomers*. Oxford University Press, New York.
[7] Kirzerow, H.-S. and Bahr, C. E. (2001). *Chirality in Liquid Crystals*. Springer, New York.
[8] Li, Q. E. (2012). *Liquid Crystals Beyond Displays: Chemistry, Physics, and Applications*. Wiley, Hoboken, New Jersey.
[9] Yang, D.-K. and Wu, S.-T. (2006). *Fundamentals of Liquid Crystal Devices*. Wiley, Chichester, England.
[10] Sluckin, T. G., Dunmur, D. A., and Stegemeyer, H. (2004). *Crystals That Flow*. Taylor & Francis, London.
[11] Chaikin, P. M. and Lubensky, T. C. (1995). *Principles of Condensed Matter Physics*. Cambridge University Press, Cambridge.
[12] Larson, R. G. (1999). *The Structure and Rheology of Complex Fluids*. Oxford University Press, New York.
[13] Kleman, M. and Lavrentovich, O. D. (2003). *Soft Matter Physics: An Introduction*. Springer, New York.
[14] Petrov, A. G. (1999). *The Lyotropic State of Matter*. Gordon and Breach, Amsterdam.
[15] Lydon, J. (2011). *Liq. Cryst.*, **38**, 1663–1681.
[16] Park, H. S. and Lavrentovich, O. D. (2012). In *Liquid Crystals Beyond Displays: Chemistry, Physics, and Applications* (ed. Q. Li), pp. 449–484. Wiley, Hoboken, NJ.
[17] Nakata, M. *et al.* (2007). *Science*, **318**, 1276–1279.
[18] Chami, F. and Wilson, M. R. (2010). *J. Amer. Chem. Soc.*, **132**, 7794–7802.
[19] Horowitz, V. R., Janowitz, L. A., Modic, A. L. *et al.* (2005). *Phys. Rev. E*, **72**, 041710.
[20] Park, H. S. *et al.* (2008). *J. Phys. Chem. B*, **112**, 16307–16319.
[21] Renshaw, M. P. and Day, I. J. (2010). *J. Phys. Chem. B*, **114**, 10032–10038.
[22] Gelbart, W. M. and BenShaul, A. (1996). *J. Phys. Chem.*, **100**, 13169–13189.
[23] Onsager, L. (1949). *Ann. NY Acad. Sci. J.*, **51**, 627–649.
[24] Dogic, Z. and Fraden, S. (2006). *Curr. Opin. Colloid & Interface Sci.*, **11**, 47–55.
[25] Puech, N., Grelet, E., Poulin, P. *et al.* (2010). *Phys. Rev. E*, **82**,

020702.

[26] Oakes, P. W., Viamontes, J., and Tang, J. X. (2007). *Phys. Rev. E*, **75**, 061902.

[27] Jones, J. W., Lue, L., Ormerod, A. P. *et al.* (2010). *Liq. Cryst.*, **37**, 711–722.

[28] Park, H. S., Kang, S. W., Tortora, L. *et al.* (2011). *Langmuir*, **27**, 4164–4175.

[29] Tortora, L. *et al.* (2010). *Soft Mat.*, **6**, 4157–4167.

[30] Tortora, L. and Lavrentovich, O. D. (2011). *Proc. Natl. Acad. Sci. USA*, **108**, 5163–5168.

[31] Repiova, A. and Frederiks, V. (1927). *J. Russian Phys. Chem. Soc.*, **59**, 183–200.

[32] Frederiks, V. and Zolina, V. (1929). *Trans. Amer. Electrochem. Soc.*, **55**, 85–96.

[33] Frederiks, V. and Tsvetkov, V. N. (1935). *Comptes Rendus Acad. Sci. USSR*, **4**, 131–142.

[34] Borsch, V., Shiyanovskii, S. V., and Lavrentovich, O. D. (2013). *Phys. Rev. Lett.*, **111**, 107802.

[35] Yu, L. J. and Saupe, A. (1980). *Phys. Rev. Lett.*, **45**, 1000–1003.

[36] van den Pol, E., Petukhov, A. V., Thies-Weesie, D. M. E. *et al.* (2009). *Phys. Rev. Lett.*, **103**, 258301.

[37] Belli, S., Patti, A., Dijkstra, M. *et al.* (2011). *Phys. Rev. Lett.*, **107**, 148303.

[38] Berardi, R., Muccioli, L., Orlandi, S. *et al.* (2008). *J. Phys. - Cond. Mat.*, **20**, 463101.

[39] Luckhurst, G. R. (2001). *Thin Solid Films*, **393**, 40–52.

[40] Li, J. F., Percec, V., Rosenblatt, C. *et al.* (1994). *Europhys. Lett.*, **25**, 199–204.

[41] Merkel, K. *et al.* (2004). *Phys. Rev. Lett.*, **93**, 237801.

[42] Madsen, L. A., Dingemans, T. J., Nakata, M. *et al.* (2004). *Phys. Rev. Lett.*, **92**, 145505.

[43] Acharya, B. R., Primak, A., and Kumar, S. (2004). *Phys. Rev. Lett.*, **92**, 145506.

[44] Seltmann, J. *et al.* (2011). *Chem. Mater.*, **23**, 2630–2636.

[45] Jákli, A. (2013). *Liq. Cryst. Rev.*, **1**, 65–82.

[46] Galerne, Y. (1998). *Mol. Cryst. Liq. Cryst.*, **323**, 211–229.

[47] Vaupotic, N. *et al.* (2009). *Phys. Rev. E*, **80**, 030701(R).

[48] Francescangeli, O., Vita, F., Ferrero, C. *et al.* (2011). *Soft Mat.*, **7**, 895–901.

[49] Le, K. V. *et al.* (2009). *Phys. Rev. E*, **79**, 030701.

[50] Senyuk, B. *et al.* (2011). *Mol. Cryst. Liq. Cryst.*, **540**, 20–41.

[51] Kim, Y.-K., Senyuk, B., and Lavrentovich, O. D. (2012). *Nature Commun.*, **3**, 1133.

[52] Berejnov, V., Cabuil, V., Perzynski, R. *et al.* (1998). *J. Phys. Chem. B*, **102**, 7132–7138.

[53] Brommel, F., Zou, P., Finkelmann, H. *et al.* (2013). *Soft Mat.*, **9**, 1674–1677.

[54] Bouligand, Y. (2011). In *Morphogenesis* (ed. P. Bourgine and

A. Lesne), pp. 49–86. Springer, Berlin Heidelberg.

[55] Berry, E., Hensel, Z., Dogic, Z. et al. (2006). *Phys. Rev. Lett.*, **96**, 018305.

[56] Subacius, D., Shiyanovskii, S. V., Bos, P. et al. (1997). *Appl. Phys. Lett.*, **71**, 3323–3325.

[57] Meiboom, S., Sethna, J. P., Anderson, P. W. et al. (1981). *Phys. Rev. Lett.*, **46**, 1216–1219.

[58] Kleman, M. (1985). *J. Phys.*, **46**, 1193–1203.

[59] Kitzerow, H. S. (2010). *Ferroelectrics*, **395**, 66–85.

[60] Voloschenko, D., Pishnyak, O. P., Shiyanovskii, S. V. et al. (2002). *Phys. Rev. E*, **65**, 060701(R).

[61] Kikuchi, H., Yokota, M., Hisakado, Y. et al. (2002). *Nature Mater.*, **1**, 64–68.

[62] Huang, C. Y., Stott, J. J., and Petschek, R. G. (1998). *Phys. Rev. Lett.*, **80**, 5603–5606.

[63] Fukuda, J. (2010). *Phys. Rev. E*, **82**, 061702.

[64] Ravnik, M., Alexander, G. P., Yeomans, J. M. et al. (2011). *Proc. Natl. Acad. Sci.*, **108**, 5188–5192.

[65] Higashiguchi, K., Yasui, K., Ozawa, M. et al. (2012). *Polymer J.*, **44**, 632–638.

[66] Dierking, I. et al. (2012). *Soft Mat.*, **8**, 4355–4362.

[67] Chen, K. M., Gauza, S., Xianyu, H. Q. et al. (2010). *J. Display Tech.*, **6**, 318–322.

[68] Emsley, J. W., Luckhurst, G. R., and Shilstone, G. N. (1984). *Mol. Phys.*, **53**, 1023–1028.

[69] Coles, H. J. and Pivnenko, M. N. (2005). *Nature*, **436**, 997–1000.

[70] Wright, D. C. and Mermin, N. D. (1989). *Rev. Mod. Phys.*, **61**, 385–432.

[71] Hur, S. T., Gim, M. J., Yoo, H. J. et al. (2011). *Soft Mat.*, **7**, 8800–8803.

[72] Fukuda, J. (2012). *Phys. Rev. E*, **85**, 020701.

[73] Castles, F., Morris, S. M., Terentjev, E. M. et al. (2010). *Phys. Rev. Lett.*, **104**, 157801.

[74] Chakrabarti, B., Hatwalne, Y., and Madhusudana, N. V. (2006). *Phys. Rev. Lett.*, **96**, 157801.

[75] Alexander, G. P. and Yeomans, J. M. (2007). *Phys. Rev. Lett.*, **99**, 067801.

[76] Hough, L. E., Spannuth, M., Nakata, M. et al. (2009). *Science*, **325**, 452–456.

[77] Volovik, G. E. and Lavrentovich, O. D. (1983). *Zh. Eksp. I Teor. Fiz. (Sov. Phys. JETP)*, **85**, 1997–2010.

[78] Williams, R. D. (1986). *J. Phys. A - Math. Gen.*, **19**, 3211–3222.

[79] Leforestier, A. and Livolant, F. (1994). *Liq. Cryst.*, **17**, 651–658.

[80] DiDonna, B. A. and Kamien, R. D. (2003). *Phys. Rev. E*, **68**, 041703.

[81] Renn, S. R. and Lubensky, T. C. (1988). *Phys. Rev. A*, **38**, 2132–2147.

[82] Goodby, J. W. et al. (1989). *Nature*, **337**, 449–452.

[83] Lavrentovich, O. D. et al. (1990). *Europhys. Lett.*, **13**, 313–318.
[84] Ungar, G., Percec, V., and Zuber, M. (1992). *Macromolecules*, **25**, 75–80.
[85] Cestari, M. et al. (2011). *Phys. Rev. E*, **84**, 031704.
[86] Panov, V. P. et al. (2010). *Phys. Rev. Lett.*, **105**, 167801.
[87] Henderson, P. A. and Imrie, C. T. (2011). *Liq. Cryst.*, **38**, 1407–1414.
[88] Chen, D., Porada, J. H., Hooper, J. B. et al. (2013). *Proc. Natl. Acad. Sci.*, **110**, 1593115936.
[89] Borshch, V., Kim, Y.-K., Xiang, J. et al. (2013). *Nature Comm.*, **4**, 2635.
[90] Dozov, I. (2001). *Europhys. Lett.*, **56**, 247–253.
[91] Flory, P. J. (1978). *Macromolecules*, **11**, 1138–1141.
[92] Asakura, S. and Oosawa, F. (1958). *J. Polym. Sci.*, **33**, 183–192.
[93] Adams, M., Dogic, Z., Keller, S. L. et al. (1998). *Nature*, **393**, 349–352.
[94] Lavrentovich, O. D. and Pergamenshchik, V. M. (1995). *Intl. J. Mod. Phys. B*, **9**, 2389–2437.
[95] Ericksen, J. L. (1966). *Phys. Fluids*, **9**, 1205–1207.
[96] Lavrentovich, O. D. and Pergamenshchik, V. M. (1994). *Phys. Rev. Lett.*, **73**, 979–982.
[97] Lavrentovich, O. D., Ishikawa, T., and Terentjev, E. M. (1997). *Mol. Cryst. Liq. Cryst.*, **299**, 301–306.
[98] Allender, D. W., Crawford, G. P., and Doane, J. W. (1991). *Phys. Rev. Lett.*, **67**, 1442–1445.
[99] Sparavigna, A., Lavrentovich, O. D., and Strigazzi, A. (1994). *Phys. Rev. E*, **49**, 1344–1352.
[100] Manyuhina, O. V., Cazabat, A. M., and Ben Amar, M. (2010). *Europhys. Lett.*, **92**, 16005.
[101] Faetti, S. and Riccardi, M. (1995). *J. Phys. II*, **5**, 1165–1191.
[102] Yokoyama, H. (1997). *Phys. Rev. E*, **55**, 2938–2957.
[103] Pergamenshchik, V. M. and Chernyshuk, S. B. (2002). *Phys. Rev. E*, **66**, 051712.
[104] Odijk, T. (1986). *Liq. Cryst.*, **1**, 553–559.
[105] Grosberg, A. Y. and Zhestkov, A. V. (1986). *Vysokomolekulyarnye Soedineniya Seriya A*, **28**, 86–91.
[106] Vroege, G. J. and Odijk, T. (1987). *J. Chem. Phys.*, **87**, 4223–4232.
[107] Taratuta, V. G., Lonberg, F., and Meyer, R. B. (1988). *Phys. Rev. A*, **37**, 1831–1834.
[108] Kornyshev, A. A., Lee, D. J., Leikin, S. et al. (2007). *Rev. Mod. Phys.*, **79**, 943–996.
[109] Meyer, R. B. (1982). In *Polymer Liquid Crystals* (ed. A. Ciferri, W. R. Krigbaum, and R. B. Meyer). Academic Press, New York.
[110] Vanderschoot, P. and Cates, M. E. (1994). *Europhys. Lett.*, **25**, 515–520.
[111] Zhou, S. et al. (2012). *Phys. Rev. Lett.*, **109**, 037801.
[112] Hudson, S. D. and Thomas, E. L. (1989). *Phys. Rev. Lett.*, **62**,

1993–1996.

[113] Lavrentovich, O. D. and Sergan, V. V. (1990). *Nuovo Cimento D - Cond. Mat.*, **12**, 1219–1222.
[114] Meyer, R. B. (1969). *Phys. Rev. Lett.*, **22**, 918–921.
[115] Patel, J. S. and Meyer, R. B. (1987). *Phys. Rev. Lett.*, **58**, 1538–1540.
[116] Helfrich, W. (1971). *J. Chem. Phys.*, **55**, 839–842.
[117] Hurault, J. P. (1973). *J. Chem. Phys.*, **59**, 2068–2075.
[118] Senyuk, B. I., Smalyukh, I. I., and Lavrentovich, O. D. (2006). *Phys. Rev. E*, **74**, 011712.
[119] Ishikawa, T. and Lavrentovich, O. D. (2001). In *Defects in Liquid Crystals: Computer Simulations, Theory, and Experiments* (ed. O. D. Lavrentovich, P. Pasini, C. Zannoni, and S. Žumer), pp. 271–301. Kluwer, Amsterdam.
[120] Brener, E. A. and Marchenko, V. I. (1999). *Phys. Rev. E*, **59**, R4752–R4753.
[121] Ishikawa, T. and Lavrentovich, O. D. (1999). *Phys. Rev. E*, **60**, R5037–R5039.
[122] Alexander, G. P., Kamien, R. D., and Santangelo, C. D. (2012). *Phys. Rev. Lett.*, **108**, 047802.
[123] Clark, N. A. and Meyer, R. B. (1973). *Appl. Phys. Lett.*, **22**, 493–494.
[124] Delaye, M., Ribotta, R., and Durand, G. (1973). *Phys. Lett. A*, **44**, 139–140.
[125] Jerome, B. (1991). *Rep. Prog. Phys.*, **54**, 391–451.
[126] Sluckin, T. J. and Poniewierski, A. (1986). In *Fluid Interfacial Phenomena* (ed. C. A. Croxton). Wiley, Chichester.
[127] Faetti, S. and Palleschi, V. (1984). *J. Chem. Phys.*, **81**, 6254–6258.
[128] Blinov, L. M., Kats, E. I., and Sonin, A. A. (1987). *Uspekhi Fizicheskikh Nauk*, **152**, 449–477.
[129] Chen, W. L. and Gray, D. G. (2002). *Langmuir*, **18**, 633–637.
[130] Doi, M. and Kuzuu, N. (1985). *Appl. Polymer Symp.*, **41**, 65–68.
[131] van der Schoot, P. (1999). *J. Phys. Chem. B*, **103**, 8804–8808.
[132] Bernal, J. D. and Fankuchen, I. (1941). *J. Gen. Physiol.*, **25**, 111–146.
[133] Zocher, H. (1925). *Z. Anorg. Allg. Chem.*, **147**, 91–110.
[134] Kaznacheev, A. V., Bogdanov, M. M., and Sonin, A. S. (2003). *J. Exp. Theor. Phys.*, **97**, 1159–1167.
[135] Prinsen, P. and van der Schoot, P. (2004). *J. Phys.-Cond. Mat.*, **16**, 8835–8850.
[136] Davidson, P. (2010). *Compt. Rend. Chimie*, **13**, 142–153.
[137] Kaznacheev, A. V., Bogdanov, M. M., and Taraskin, S. A. (2002). *J. Exp. Theor. Phys.*, **95**, 57–63.
[138] Jiang, Y. and Chen, J. Z. Y. (2010). *Macromolecules*, **43**, 10668–10678.
[139] Nazarenko, V. G. *et al.* (2010). *Phys. Rev. Lett.*, **105**, 017801.
[140] Sonin, A. A. (1995). *The Surface Physics of Liquid Crystals*. Gordon and Breach, Australia.

[141] PopaNita, V. and Sluckin, T. J. (1996). *J. Phys. II*, **6**, 873–884.
[142] Kibble, T. W. B. (1976). *J. Phys. A - Math. Gen.*, **9**, 1387–1398.
[143] Kim, Y.-K., Shiyanovskii, S. V., and Lavrentovich, O. D. (2013). *J. Phys.-Cond. Mat.*, **25**, 404202.
[144] Chuang, I., Durrer, R., Turok, N. et al. (1991). *Science*, **251**, 1336–1342.
[145] Bowick, M. J., Chandar, L., Schiff, E. A. et al. (1994). *Science*, **263**, 943–945.
[146] Vachaspati, T. (1991). *Phys. Rev. D*, **44**, 3723–3729.
[147] Hindmarsh, M. (1995). *Phys. Rev. Lett.*, **75**, 2502–2505.
[148] Candau, S., Leroy, P., and Debeauva, F. (1973). *Mol. Cryst. Liq. Cryst.*, **23**, 283–297.
[149] Lavrentovich, O. D. and Terentiev, E. M. (1986). *Zh. Eksp. I Teor. Fiz. (Sov. Phys. JETP)*, **91**, 2084–2096.
[150] Drzaic, P. S. (1999). *Liq. Cryst.*, **26**, 623–627.
[151] Toulouse, G. and Kleman, M. (1976). *J. Phys. Lett.*, **37**, L149–L151.
[152] Kleman, M., Michel, L., and Toulouse, G. (1977). *J. Phys. Lett.*, **38**, L195–L197.
[153] Volovik, G. E. and Mineyev, V. P. (1977). *Zh. Eksp. I Teor. Fiz. (Sov. Phys. JETP)*, **72**, 2256–2274.
[154] Copar, S. and Zumer, S. (2012). *Phys. Rev. E*, **85**, 031701.
[155] Lavrentovich, O. D. (1998). *Liq. Cryst.*, **24**, 117–125.
[156] Drzaic, P. S. (1995). *Liquid Crystal Dispersions*. World Scientific, Singapore.
[157] Doane, J. W., Vaz, N. A., Wu, B. G. et al. (1986). *Appl. Phys. Lett.*, **48**, 269–271.
[158] Bodnar, V. G., Lavrentovich, O. D., and Pergamenshchik, V. M. (1992). *Zh. Eksp. I Teor. Fiz. (Sov. Phys. JETP)*, **101**, 111–125.
[159] Prishchepa, O. O., Shabanov, A. V., and Zyryanov, V. Y. (2005). *Phys. Rev. E*, **72**, 031712.
[160] Humar, M., Ravnik, M., Pajk, S. et al. (2009). *Nature Photonics*, **3**, 595–600.
[161] Gupta, J. K., Zimmerman, J. S., de Pablo, J. J. et al. (2009). *Langmuir*, **25**, 9016–9024.
[162] Brasselet, E., Murazawa, N., Misawa, H. et al. (2009). *Phys. Rev. Lett.*, **103**, 103903.
[163] Miller, D. S. and Abbott, N. L. (2013). *Soft Mat.*, **9**, 374–382.
[164] Lin, I. H. et al. (2011). *Science*, **332**, 1297–1300.
[165] Musevic, I., Skarabot, M., Tkalec, U. et al. (2006). *Science*, **313**, 954–958.
[166] Senyuk, B. et al. (2013). *Nature*, **493**, 200–205.
[167] Nelson, D. R. (2002). *Nano Lett.*, **2**, 1125–1129.
[168] Fernandez-Nieves, A. et al. (2007). *Phys. Rev. Lett.*, **99**, 157801.
[169] Araki, T., Buscaglia, M., Bellini, T. et al. (2011). *Nature Mater.*, **10**, 303–309.
[170] Lavrentovich, O. D. and Nastishin, Y. A. (1990). *Europhys. Lett.*, **12**, 135–141.

[171] Friedel, G. and Grandjean, F. (1910). *Bull. Soc. Fr. Minéral.*, **33**, 409–465.

[172] Friedel, G. (1922). *Ann. de Phys.*, **18**, 273–315.

[173] Dupin, C. P. (1822). *Applications de géométrie et de méchanique la marine, aux ponts et chaussées, etc.* Bachelier, Paris.

[174] Rosenblatt, C. S., Pindak, R., Clark, N. A. *et al.* (1977). *J. Phys.*, **38**, 1105–1115.

[175] Maxwell, J. (1868). *Quart. J. Pure Appl. Math.*, **9**, 111–126.

[176] Planer, J. (1861). *Ann. Chem. Pharm.*, **118**, 25–27.

[177] Kléman, M (1982). *Points, Lines, and Walls.* Wiley, Chichester.

[178] Bragg, W. (1934). *Nature*, **133**, 445–465.

[179] Alexander, G. P., Chen, B. G. G., Matsumoto, E. A. *et al.* (2010). *Phys. Rev. Lett.*, **104**, 257802.

[180] Kleman, M. and Lavrentovich, O. D. (2009). *Liq. Cryst.*, **36**, 1085–1099.

[181] Kleman, M. and Lavrentovich, O. D. (2000). *Euro. Phys. J. E*, **2**, 47–57.

[182] Alexander, G. P., Chen, B. G. G., Matsumoto, E. A. *et al.* (2012). *Rev. Mod. Phys.*, **84**, 497–514.

[183] De Sio, L. *et al.* (2013). *Liq. Cryst. Rev.*, **1**, 2–19.

[184] Wu, S.-T. and Yang, D.-K. (2001). *Reflective Liquid Crystal Displays.* Wiley, Chichester.

[185] Liu, D. Q. and Broer, D. (2013). *Liq. Cryst. Rev.*, **1**, 20–28.

[186] Poulin, P., Stark, H., Lubensky, T. C. *et al.* (1997). *Science*, **275**, 1770–1773.

[187] Stark, H. (2001). *Phys. Rep.*, **351**, 387–474.

[188] Chernyshuk, S. B., Tovkach, O. M., and Lev, B. I. (2012). *Phys. Rev. E*, **85**, 011706.

[189] Pergamenshchik, V. M. and Uzunova, V. A. (2011). *Phys. Rev. E*, **83**, 021701.

[190] Noel, C. M., Bossis, G., Chaze, A. M. *et al.* (2006). *Phys. Rev. Lett.*, **96**, 217801.

[191] Smalyukh, I. I., Kuzmin, A. N., Kachynski, A. V. *et al.* (2005). *Appl. Phys. Lett.*, **86**, 021913.

[192] Smalyukh, I. I., Lavrentovich, O. D., Kuzmin, A. N. *et al.* (2005). *Phys. Rev. Lett.*, **95**, 157801.

[193] Kotar, J. *et al.* (2006). *Phys. Rev. Lett.*, **96**, 207801.

[194] Tkalec, U., Ravnik, M., Copar, S. *et al.* (2011). *Science*, **333**, 62–65.

[195] Lapointe, C. P., Mason, T. G., and Smalyukh, I. I. (2009). *Science*, **326**, 1083–1086.

[196] Wood, T. A., Lintuvuori, J. S., Schofield, A. B. *et al.* (2011). *Science*, **333**, 79–83.

[197] Lekkerkerker, H. N. W. and Tuinier, R. (2011). *Colloids and the Depletion Interactions.* Springer, Dordrecht.

[198] Pishnyak, O. P., Tang, S., Kelly, J. R. *et al.* (2009). *Ukr. J. Phys.*, **54**, 101–108.

[199] Pishnyak, O. P., Tang, S., Kelly, J. R. *et al.* (2007). *Phys. Rev.*

Lett., **99**, 127802.
[200] Chernyshuk, S. B. and Lev, B. I. (2011). *Phys. Rev. E*, **84**, 011707.
[201] Pergamenshchik, V. M. and Uzunova, V. A. (2009). *Phys. Rev. E*, **79**, 021704.
[202] Morgan, H. and Green, N. G. (2003). *AC Electrokinetics: Colloids and Nanoparticles*. Research Studies Press, Baldock, UK.
[203] Gangwal, S., Cayre, O. J., Bazant, M. Z. et al. (2008). *Phys. Rev. Lett.*, **100**, 058302.
[204] Bazant, M. Z. and Squires, T. M. (2010). *Curr. Opin. Coll. & Interface Sci.*, **15**, 203–213.
[205] Tatarkova, S. A., Burnham, D. R., Kirby, A. K. et al. (2007). *Phys. Rev. Lett.*, **98**, 157801.
[206] Ryzhkova, A. V., Podgornov, F. V., and Haase, W. (2010). *Appl. Phys. Lett.*, **96**, 151901.
[207] Dierking, I., Biddulph, G., and Matthews, K. (2006). *Phys. Rev. E*, **73**, 011702.
[208] Lavrentovich, O. D., Lazo, I., and Pishnyak, O. P. (2010). *Nature*, **467**, 947–950.
[209] Lazo, I. and Lavrentovich, O. D. (2013). *Phil. Trans. Royal Soc. A*, **381**, 20120255.
[210] Klein, S. (2013). *Liq. Cryst. Rev.*, **1**, 52–64.
[211] Hernandez-Navarro, S., Tierno, P., Ignés-Mullola, J. et al. (2013). *Soft Mat.*, **9**, 7999–8004.
[212] Leslie, F. (1968). *Arch. Ration. Mech. Anal. (Germany)*, **28**, 265–283.
[213] Parodi, O. (1970). *J. Phys. (Paris)*, **31**, 581–584.
[214] Ruhwandl, R. W. and Terentjev, E. M. (1996). *Phys. Rev. E*, **54**, 5204–5210.
[215] Stark, H. and Ventzki, D. (2001). *Phys. Rev. E*, **6403**, 031711.
[216] Loudet, J. C., Hanusse, P., and Poulin, P. (2004). *Science*, **306**, 1525–1525.
[217] Takanashi, H., MacLennan, J. E., and Clark, N. A. (1998). *Jap. J. Appl. Phys. Part 1*, **37**, 2587–2589.
[218] Pishnyak, O. P., Shiyanovskii, S. V., and Lavrentovich, O. D. (2011). *Phys. Rev. Lett.*, **106**, 047801.
[219] Mieda, Y. and Furutani, K. (2005). *Appl. Phys. Lett.*, **86**, 101901.
[220] Cuennet, J. G., Vasdekis, A. E., De Sio, L. et al. (2011). *Nature Photonics*, **5**, 234–238.
[221] Wang, H. Y., Wu, T. X., Gauza, S. et al. (2006). *Liq. Cryst.*, **33**, 91–98.
[222] Jákli, A. and Saupe, A. (2006). *One- and Two-dimensional Fluids*. Taylor & Francis, New York.
[223] Carlton, R. J. et al. (2013). *Liq. Cryst. Rev.*, **1**, 29–51.
[224] Luk, Y. Y. and Abbott, N. L. (2003). *Science*, **301**, 623–626.
[225] Shiyanovskii, S. V. et al. (2005). *Phys. Rev. E*, **71**, 020702.
[226] Khoo, I. C. (1995). *Liquid Crystals: Physical Properties and Nonlinear Optical Phenomena*. Wiley, New York.

[227] Vicari, L. E. (2003). *Optical Applications of Liquid Crystals.* Institute of Physics Publishing, Bristol and Philadelphia.
[228] Shiyanovskii, S. V. and Lavrentovich, O. D. (2010). *Liq. Cryst.*, **37**, 737–745.
[229] Gu, M. X., Shiyanovskii, S. V., and Lavrentovich, O. D. (2008). *Phys. Rev. Lett.*, **100**, 237801.

Foams

Denis Weaire and Stefan Hutzler
School of Physics, Trinity College Dublin
Dublin 2, Ireland

Introduction

The effective foundation of our subject was laid by Joseph Antoine Ferdinand Plateau in the 19th century. Plateau's inspiring devotion to this research, hindered or perhaps driven by self-inflicted blindness, would match that of the most single-minded of today's postdocs. He collated all of his results in his 1873 *magnum opus* [1], to which reference has been made ever since, despite its rarity. Today it may be readily found on the Internet, but it is hardly an easy read.[1] No later work was to match the achievement of Plateau's two volumes, but the curious may wish to consult attempts from various perspectives [2–4].

Numerous conference volumes followed the 1994 foundation of a regular European conference (now called EUFOAM) [5–10] and the proceedings of other workshops [11, 12] may also be consulted for recent developments. Foam found its place within the fashionable general topics of disordered materials, complexity, and soft matter. Weaire and Hutzler [13] summarised progress up to 2000, attempting to include all that was likely to survive further refinements. They pointed to dynamics as the main unfinished business. It is still so, but actively pursued (see the recent French multi-author book "Les Mousses" [14] which is currently being translated into English).

[1] K. Brakke provides a fully cross-linked translation into English at the following site: http://www.susqu.edu/brakke/PlateauBook/PlateauBook.html .

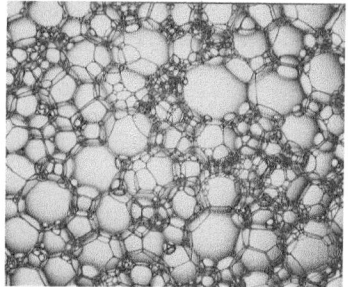

What is a foam?

For our purposes the prototypical foam consists of gas bubbles surrounded by liquid and closely packed together so that they are in contact with each other (see Figure 4.1). Paradoxically this mixture of two fluids is a solid, at least under a small enough stress – surface tension and gas pressure maintain it in equilibrium when deformed.

The corresponding mixture of two liquids is an emulsion, included here as being closely analogous. Again surface tension renders it stiff under low stress. Further variations include foams of anti-bubbles in which the role of gas and liquid are reversed: this topic remains rather esoteric, on account of instability.

Since a solid foam is usually a frozen liquid foam, it may have the same structure, but we do not explicitly cover it here.

In the familiar aqueous foam, the contacting bubbles are separated

Fig. 4.1 Photograph of a typical dry foam, a close-packing of bubbles of a wide distribution of sizes. Despite the apparent disorder of the arrangement, both local topology and geometry are well defined and obey Plateau's rules, stated in section 4.2. (Photograph: M. Boran.)

[2] Recent space experiments have been devoted to the question: what happens when we try to foam pure water?

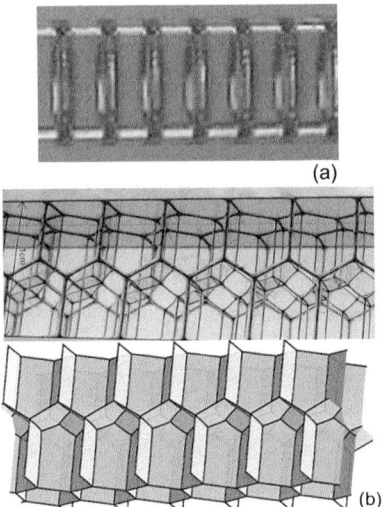

Fig. 4.2 A so-called bamboo foam consists of equally spaced parallel soap films confined in a tube (a). More complicated, but still perfectly ordered structures are obtained if the bubble diameter is less than the width of the confinement. This is shown in (b) for a foam in a tube with rectangular cross-section (left: photo, right: computer simulation).

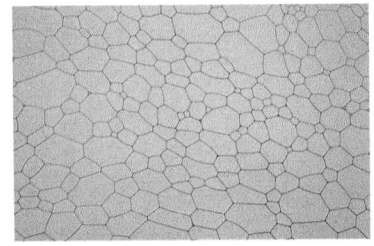

Fig. 4.3 Confining foam between two glass plates, with a spacing less than the smallest bubble diameter, leads to one particular type of experimental realisation of a 2D foam.

by a thin film of liquid. The addition of surfactants is required for this film to be stable, otherwise the foam would quickly collapse.[2] A foam stabilized in this way is still likely to be ephemeral over a timescale of minutes or hours, due to the coarsening process in which gas diffusion progressively eliminates shrinking bubbles.

Depending on the circumstances, a foam may contain a greater or lesser fraction of liquid. The volume fraction of gas ranges from about 0.65 for contacting spherical bubbles to practically unity when the bubbles are closely compacted and take polyhedral shapes. The bubbles that constitute the foam are usually polydisperse, although monodisperse samples are interesting enough that special care is sometimes taken to create them in the laboratory.

In most familiar examples from everyday life the bubbles are at least a few millimetres in diameter. In such a case, equilibrium of a substantial sample under gravity is only attained after most of the liquid has drained away: the liquid fraction will be less that one percent and the foam may be called *dry*. For bubbles of a size much less than a millimetre, gravity is less effective in competition with surface tension ("capillarity"): a small sample remains *wet*, even approaching the *wet limit* of contacting spherical bubbles.

Foams in confined geometries display fascinating structures and effects [15]. The most rudimentary, hardly to be considered a foam at all, is that of the *bamboo structure* (Figure 4.2a), nothing more than a sequence of flat films in a tube. The structures are considerably more intricate if the width of the confinement exceeds the bubble diameter, as shown in Figure 4.2b.

A two-dimensional foam is formed when bubbles are trapped between two closely spaced flat plates (Figure 4.3). In many ways it is analogous to its three-dimensional counterpart, and it presents the advantage of visibility. It is well described by idealised models that are strictly two-dimensional (but it is important to remember that the real system is not quite so). It also has high entertainment value for lectures and demonstrations.

Any process of shaking or shearing a foamable liquid will entrain air in bubbles that are progressively broken down into smaller ones (probably by the process of Rayleigh-Plateau instability) until some limit is reached, beyond which insufficient energy is being supplied to further refine the foam.

Another method of foam production is via the injection of gas from a needle or porous plug (sparging) will create the foam. Alternatively, the release of gas from solution (as in beer or champagne) creates a familiar foam or froth.

Applications of foams

We shall not try to catalogue the wide variety of foams that occur in nature and in industry in any detail. Most familiar cases are aqueous, but oil foams are important as well, as are plastic and metal foams, as precursors of solid foams. Foams can act as light, highly visible carriers of chemicals for cleaning and other purposes, using a minimum of liquid. The added property of adhesion to surfaces and clinging to them under gravity (because there is a finite yield stress) is also valuable.

Depending on the composition of their surfactants, foams may have a convenient life-span so that they do not persist in large quantities. Accidental foams, on the other hand, may be rather stable, as in the wind-blown spume at the seaside in winter, stabilized by natural surfactants from decaying vegetable matter. In waste water treatment, in scrubbers and other major chemical installations, in oil pipelines and elsewhere, the occurrence of persistent foam detracts from efficiency. Defoaming agents may be used to counter such foam formation, but such chemical recipes come at an economic and environmental cost – better to avoid foam formation in the first place by minimising the agitation that creates it.

Beverage companies have invested much research in the creation of the "head" that enhances the appearance and feel of their products. Many now use devices (*widgets*) that inject gas bubbles upon release of pressure in a can. Champagne producers insist upon glasses and bottles that are not pitted and scratched, as they will nucleate bubbles too rapidly. Sometimes a laser-etched single pit is made in an otherwise perfect glass, to produce an attractive rising stream of bubbles.

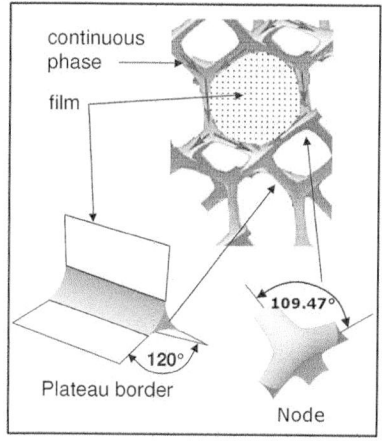

Fig. 4.4 Computer simulations of the key constituents of a dry and moderately wet foam. Three thin films meet in channels (Plateau borders), four of which in turn meet in nodes (junctions).

4.1 Statics and dynamics

On timescales of less than minutes a typical foam at rest may be regarded as being in static equilibrium, with consequences for local geometrical rules which we shall come to later. For now let us simply note the threefold elements of the structure, as shown in Figure 4.4: the soap films; the liquid-filled channels at their intersections (Plateau borders); the intersections of those channels (vertices or nodes). In the limit of a dry foam the Plateau borders may be regarded as lines, the vertices are points, and bubbles are polyhedral cells with curved faces and sharp edges and corners.

A nominally two-dimensional foam, an example of which is shown in Figure 4.3, has what may be called surface Plateau borders where films meet a confining plate. For the case of a dry foam this and other aspects of the full structure are often ignored in the idealisation of a planar structure of curved lines surrounding polygonal cells.

The equilibrium structure may be disturbed by shear or by the slow process of coarsening (Figure 4.5). In either case it is punctuated by abrupt local rearrangements, often called topological changes. In a 2D dry foam these generally occur when the length of a Plateau border is reduced to zero, provoking an instability, and the so-called T1 process (Figure 4.6). Physicists have seized on this as a "quantum of structural change" in terms of which theory and data analysis may proceed. In three dimensions similar things happen, but are slightly more compli-

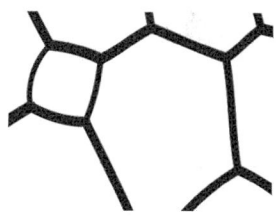

Fig. 4.5 Pressure differences in neighbouring cells drives the diffusion of gas from the smaller to the larger cells. This leads to the coarsening of a foam with time, as shown in the photographic sequence of Figure 4.16.

Fig. 4.6 Illustration of T1 or neighbour switching topological process in a 2D dry foam, such as that shown in Figure 4.3. Whenever a cell edge (line) shrinks to zero due to coarsening or applied shear, this results in a local rearrangement of the foam topology. The instability of four-fold vertices is a consequence of one of Plateau's rules (Section 4.2).

Fig. 4.7 The T1 process in 3D involves the disappearance of a small triangular face, followed by the creation of a new cell edge, or *vice versa*. This is a consequence of the instability of vertices with more than four edges in a 3D dry foam. Here the lines shown correspond to Plateau borders (lines, in the dry limit) in three dimensions and the numbers are indices of cell edges.

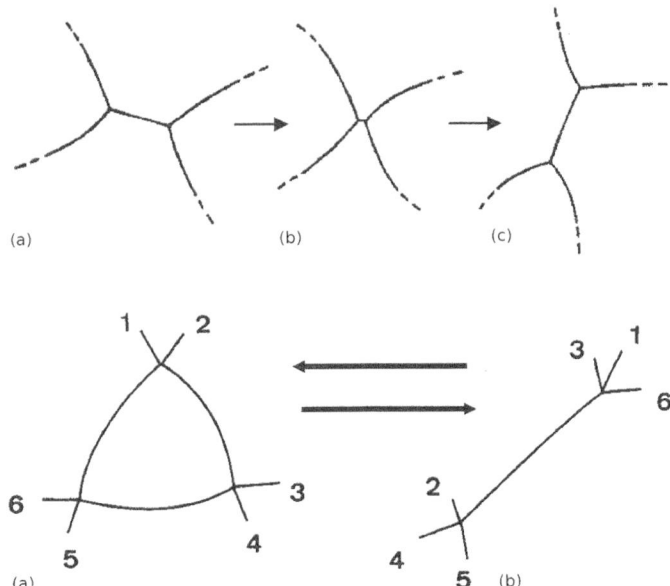

cated, as sketched in Figure 4.7.

Observing foams

Three-dimensional foams present some difficulties for observation, on account of light scattering: it may be possible to readily observe only a few layers of bubbles at the surface. Nevertheless, one historic and heroic experiment by the botanist E. Matzke explored the bulk structure of a foam [16]. He succeeded in compiling extensive statistics for the topological characteristics of the cells (distribution of cells with F faces). Detailed simulations [17] have been found to be in agreement with Matzke's findings.

Matzke's achievement has not been emulated: nowadays modern techniques of tomography (magnetic resonance imaging [18], X-ray tomography [19, 20]), are being adapted to this purpose. Figure 4.8 illustrates the important role that X-ray tomography can play in the identification of foam structure (also Section 4.3).

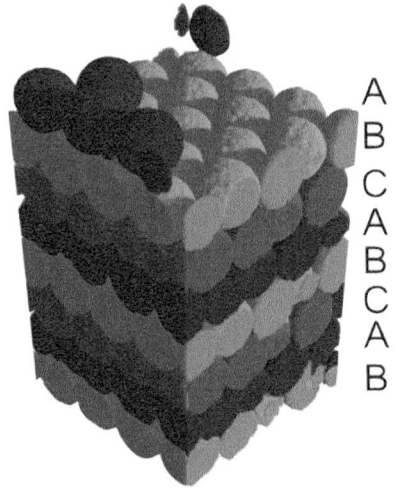

Fig. 4.8 X-ray tomography allows for the detailed 3D imaging of foam structure on the bubble scale. Shown is a section of an ordered monodisperse foam of bubble diameter $730 \pm 10 \mu m$. The bubbles are arranged in an fcc lattice with the corresponding sequence of layers ABCABCA... [20].

4.2 The rules of equilibrium

Before stating the simple rules that govern local equilibrium, we must first take a closer look at the forces involved and take note of some reasonable approximations.

In the general case of a wet foam, thin films meet in Plateau borders, as already described (see Figure 4.4). Each of the two surfaces of a film, taken here to be flat, is acted on by the gas pressure p_i of the bubble next to it (see Figure 4.9 for a sketch of the pressures involved).

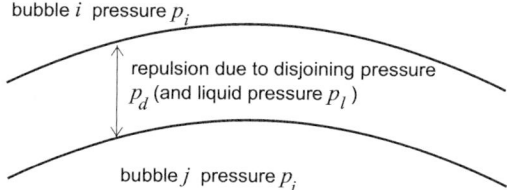

Fig. 4.9 Sketch of the pressures acting in a small film element, separating two bubbles. The disjoining pressure is due to a repulsive force (resulting from electrostatic and steric interactions) acting between the two surfaces of the thin film.

The gas pressure in each bubble is different, in general. However, it is generally close to atmospheric pressure p_0, as we shall see. (The laws of hydrostatics require that the liquid pressure is a function of height, but for now we ignore this variation, taking p_l to be constant over some small volume.) An additional force or pressure is needed to balance the forces on each liquid/gas interface. This is supplied by the disjoining pressure p_d, the mutual force between the two surfaces, attributable to molecular interactions. Hence for a flat film, the condition of equilibrium of its surfaces is approximately

$$p_l + p_d = p_0. \qquad (4.1)$$

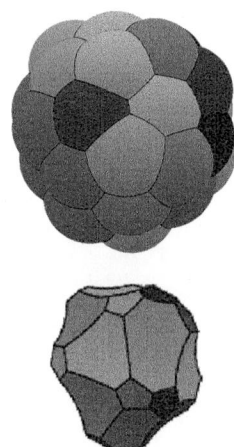

However, most films are not flat, but curved. They are not locally spherical, but have two principal curvatures, whose mean may be written as $1/r$, in terms of a single radius of curvature r. (Because the film is very thin we take its two surfaces to be parallel everywhere.) This brings another force into play in the local equilibrium of the film, associated with its surface tension 2σ (twice that of an individual surface), according to the Young-Laplace Law. This requires a pressure difference between the neighbouring bubbles to balance the force of surface tension on a small element, according to

$$p_i - p_j = 4\sigma/r \qquad (4.2)$$

It follows that the mean curvature $1/r$ is the same everywhere on a given film separating a pair of bubbles (see Figure 4.10 for an illustration of the curved surfaces).

Fig. 4.10 The films separating two bubbles are surfaces of constant mean curvature, as dictated by the Young-Laplace law, equation (4.2). The image illustrates such curvature for the case of a cluster containing 31 surface bubbles and one central bubble (shown *enlarged*) with 31 faces. Computations by S.J. Cox using the Surface Evolver, see also [21]. (Figure reproduced by permission of S.J. Cox.)

Note also the orders of magnitude involved, which justify some of our earlier remarks. The value of surface tension for a typical surfactant solution is roughly a third of that of pure water, i.e. about $25mN/m$. For bubbles of diameter of $1mm$ (whose faces have radii of curvature which are on order of magnitude less than this), as found in many applications, this results in pressure differences of about $100Pa$ – much less than the value of atmospheric pressure, $p_0 \simeq 10^5 Pa$. Bubble radii of about $10\mu m$ are required to obtain pressure differences that are comparable with atmospheric pressure. In consequence, most theories treat the gas as incompressible, since the pressure variations from cell to cell are not enough to induce significant volume changes.

The Young-Laplace law may also be applied to the single surface of a Plateau border, giving

$$p_i - p_b = 2\sigma/r \qquad (4.3)$$

with p_b the pressure in the Plateau border.

Topology and geometrical rules

Remarkably, the previous section contains all that is needed for a reasonable model of foam equilibrium. It dictates the rules that govern local curvature everywhere, in terms of the balance of surface tension forces and the pressures that act on an element of surface. To these must be added the appropriate constraints and boundary conditions, which may include the stipulation of constant liquid and cell volumes. Even in a simple case – say, a cluster of a several bubbles – to construct the solution surface looks forbiddingly difficult. However, the Surface Evolver software of Brakke [22][3] does so straightforwardly even for a very large system, as shown in Figure 4.11.

[3]The Surface Evolver is an interactive program for the computation of minimal surfaces. Since its launch by Ken Brakke in 1992 it has been extensively used for the simulation of foam structure. (Free downloaded from http://www.susqu.edu/brakke/)

In the case of a dry foam, which is often a good approximation to experimental samples, the liquid content is taken to be infinitesimal: films are represented by single surfaces, across which we may write the Young-Laplace law in terms of the adjacent gas pressures. They meet in lines (the reduced Plateau borders) and these come together at vertices.

In this dry limit, remarkably simple rules emerge, as first expounded by Plateau. They are dictated by equilibrium and stability.

- *Only three films can meet to form a line.*
- *Only four such lines can meet at a vertex.*

In both cases the local configuration is symmetric (with angles of 120 degrees and $\cos^{-1}(-1/3)$, respectively).

The Surface Evolver can be used in this limit, for an example see Figure 4.11, and the above rules are satisfied automatically in the process of minimising surface area. In the idealised 2D dry foam, lines meet at 120 degrees, similarly.

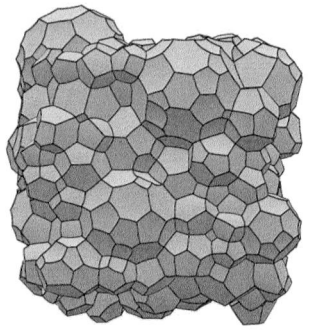

Fig. 4.11 Simulation of a polydisperse disordered foam with a wide range of bubble volumes. Using Ken Brakke's Surface Evolver software [22], A. Kraynik and colleagues performed a number of detailed studies of both and mono- and polydisperse random foams. [17, 23].

Equilibrium under gravity

Most foam physics is investigated under gravity. In equilibrium, it entails a vertical gradient of foam density or liquid fraction, with density decreasing with height. Commonly it becomes effectively dry (in the sense of having a liquid fraction less than, say, 1%) at a height of only a centimetre or so. Liquid *hold-up* – the property of retaining a finite liquid content – is due to capillarity, analogous to the rise of liquid in a capillary tube. Straightforward considerations of equilibrium, starting from the Pascal law of pressure, lead to an estimate of the variation of liquid fraction with height, as developed by Princen and Kiss [24].

A foam floating on underlying liquid has a liquid fraction which decreases to "dry" values over a vertical distance W, which depends crucially on the average diameter d of the bubbles:

$$W \simeq l_0^2/d. \qquad (4.4)$$

The quantity l_0^2 is given by

$$l_0^2 = \frac{\sigma}{\rho g}. \qquad (4.5)$$

Note that $\sqrt{2}l_0$ is the conventional capillary constant or capillarity length. For typical surfactants $l_0 \approx 1.5mm$.

The above rule, equation (4.4), entails an important distinction between foams with bubble diameters significantly less than or greater than the capillary length. Put simply, a sample of a few centimetres height, made up of very small bubbles and in contact with underlying liquid, will remain wet, while one of larger bubbles becomes entirely dry. We have recently suggested that one might refer to the number of wet bubble layers in a foam in equilibrium as the dimensionless *Princen number* [25]

$$P = \left(\frac{l_0}{d}\right)^2, \qquad (4.6)$$

after one of the pioneers of the modern study of foam drainage and rheology, Henricus Mattheus (Henry) Princen [26].

4.3 Structure of foam

For any given assortment of bubbles infinitely many structures are consistent with Plateau's local rules of equilibrium. What dictates the choice in practice? This is a much more awkward question than in those fields of condensed matter physics in which the rules of thermodynamics apply. In a foam $k_B T$ is many orders of magnitude smaller than the energy involved in a T1 process, which is of the order σd^2.[4] So in general the foam is trapped in a metastable state. This is, of course, not exceptional in soft matter.

Ideally, one may ask (as Lord Kelvin famously did): what is the lowest energy structure of a monodisperse dry foam? It is a fascinating question, but it has only limited relevance to real foams. Generally, we must "take foams as we find them", and seek to characterise them as best we can.

In the limit of a wet foam, the bubbles are spheres, and the close-packed crystal structures described in the context of hard sphere packings are of relevance. For bubbles of equal size, we may look for the face-centred cubic structure, for instance. It is indeed found [27, 29], see Figure 4.12. Structures of maximum packing density are to be expected when the foam forms under gravity. But bear in mind the admonition of the last section, which leads us to ask: how does it manage to order itself in this case? In our view there is as yet no clear answer to this.

Going to the opposite limit of a dry foam, the fcc structure becomes unstable, since it infringes Plateau's vertex rule. Kelvin showed that the natural alternative is body-centred cubic, known in this context as the Kelvin structure. This is not the structure of lowest energy (area) – since 1994 that status is provisionally reserved for the Weaire-Phelan structure

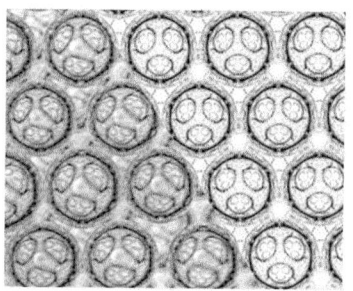

Fig. 4.12 Monodisperse bubbles of less than a few hundred μm in diameter spontaneously arrange into close-packed crystal structures. This image combines a photograph (bottom left corner) of an fcc packing of bubbles with diameter of about 200 μm with the result of a ray tracing simulation for an fcc packing of equal volume spheres (top right corner) [27, 28]. The dark lines (roughly circles) are not the perimeters of bubbles, which are larger and in contact, being affected by refraction. Within them can be seen refracted images of the next ordered layer (and so on...).

[4] At room temperature T and for bubbles with diameter $d = 1cm$, one obtains $k_B T/\sigma d^2 \simeq 10^{-15}$, where k_B is Boltzmann's constant. Even for bubbles as small as 20 μm in diameter, as in shaving foam, the ratio is still only 10^{-10}.

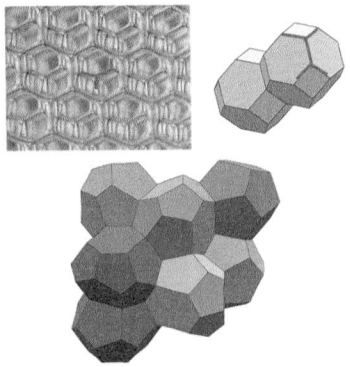

Fig. 4.13 Ordered arrangements of bubbles in a monodisperse dry foam. The so-called Kelvin structure (top row) is readily produced in the lab (see photograph, top left) while only fragments of the Weaire-Phelan structure have been observed experimentally. The image shown at the bottom is from a computer simulation by A. Kraynik using the Surface Evolver.

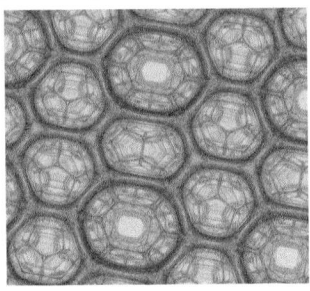

Fig. 4.14 Photograph of an experimentally produced Weaire-Phelan structure, viewed along its $\langle 100 \rangle$ direction. The entire sample (of which only a small section is shown) contains approximately 1500 bubbles, arranged into six layers. The bubble diameter is about 2mm.

[30]. Nevertheless, the Kelvin structure may be created experimentally relatively easy in various ways and remains stable, while the Weaire-Phelan (WP) structure has only been realised for the first time in the laboratory in 2011 [31]. This required the use of a container where the sides were templated with the WP structure. Such a procedure is not necessary for making a Kelvin foam since unlike the WP structure, the former is well adapted to fit against a planar surface.

Figure 4.13 shows computer simulations of both structures together with a photograph of a Kelvin foam as produced in the lab. A photograph of the Weaire-Phelan foam is shown in Figure 4.14.

Foams made by most ordinary methods are inevitably highly disordered, because they consist of a range of bubble sizes, so that their analysis takes on a statistical character.

Ordinary foams, however disordered, may first be characterized by the distribution $p(V)$ or $p(A)$ of the sizes (cell volume V in 3D or cell area A in 2D) of their cells. A second characteristic is the distribution $p(n)$ of the number of faces belonging to each cell, or of sides (edges) in two dimensions. This is more interesting, since it can vary according to the preparation and treatment of the sample, even though the size distribution is unaltered. In 2D its mean (for an infinite sample) is exactly 6 for dry foam, by Euler's theorem, and its second moment $\mu_{2,n}$ is a significant measure of disorder.

One may also define a second moment for $p(V)$ or $p(A)$, which may be used as the measure of polydispersity.

Using X-ray tomography for the study of foams structure, as mentioned in Section 4.1, makes it possible to also compute the radial distribution function $g(r)$ to probe for order within a sample. For ordered monodisperse foams, such as the one shown in Figure 4.8, this results in a sequence of narrow peaks, see Figure 4.15(top). A different variation of $g(r)$ may be found for the bulk of larger foam samples, where the foam is generally disordered. In this case $g(r)$ takes on the form known from the random (Bernal) packing of spheres, featuring a split second peak, as shown in Figure 4.15 (bottom) [20].

4.4 Key properties

Coarsening due to diffusion

A fresh foam sample typically comes into apparent equilibrium in some tens of seconds, but there is usually a slow process that progressively changes its detailed structure over longer times. It is due to the slow diffusion of gas between the cells, driven by their pressure difference. In general, smaller cells have higher pressures and so shrink and disappear. This continual loss of cells implies an increase in the average cell size. After an interesting transient period the foam reaches a scaling state, such that all the statistical measures of structure remain the same, save that the average cell size increases indefinitely. Figure 4.16 shows the evolution due to coarsening of a 2D foam.

Apart from sudden local rearrangements, the foam is effectively in equilibrium during this slow process, which plays the role of a changing constraint on the individual cell sizes in any detailed simulation.

How then should the average cell diameter d vary with time t in this scaling regime? Dimensional arguments and many experiments support the following law in both two or three dimensions:

$$d \propto t^{1/2}. \tag{4.7}$$

Looking more closely at the variation of individual cell area during coarsening in a two-dimensional dry foam, a neat relation emerges. It is given below and states that cells with less than six sides shrink, cells with more than 6 sides grow and six-sided cells remain constant in area (until they encounter a topological change which alters their number of sides).

$$\frac{dA_n}{dt} = \frac{2\pi}{3}\sigma\kappa(n-6) \tag{4.8}$$

where A_n is the area of an n-sided cell, σ is the surface tension, and κ is the permeability constant for the films (per unit transverse length).

This is known as the law of Von Neumann, or sometimes Mullins, although the latter was found in the slightly different context of grain growth. Von Neumann's law is exact in two dimensions. Its three-dimensional counterpart applies only to averages and is not linear, but nevertheless is of some theoretical interest [32, 33].[5]

Coalescence

Many foams collapse rapidly, due to the process of coalescence of cells brought about by film rupture. This can also lead to increase of cell size, and hence "coarsening", inviting some confusion. But film rupture is often confined mainly to top (free) surfaces, where films are thinnest.

Drainage

We return to the earliest process that takes place in a freshly made foam. Excess liquid drains out of it until equilibrium under gravity

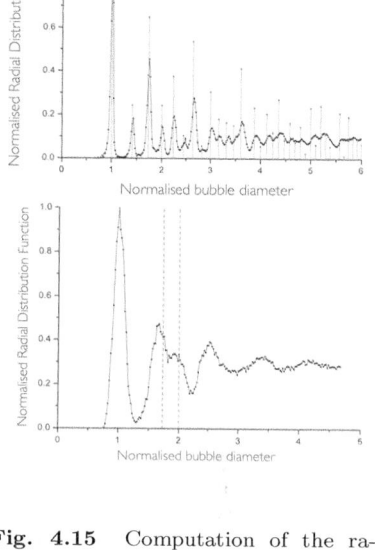

Fig. 4.15 Computation of the radial distribution function $g(r)$ for ordered (top) and (disordered) experimental foam samples. The peaks of $g(r)$ for the ordered sample coincide with the δ-peaks for an ideal fcc crystal. In the disordered sample $g(r)$ shows a characteristic split second peak with maxima close to $\sqrt{3} \simeq 1.73$ and 2, as in the Bernal packing of hard spheres.

[5] The coarsening process may be suppressed by incorporation of a small admixture of a relatively insoluble (hence non-diffusing) gas. This may be convenient for experiments, and it slows down the evolution of the head on some popular beers.

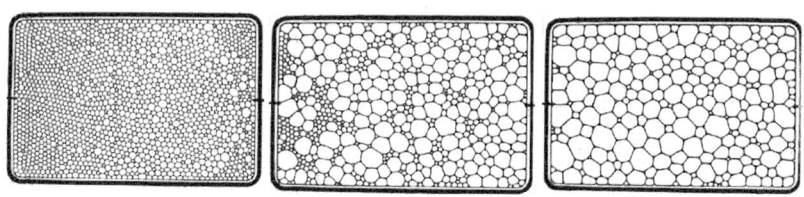

Fig. 4.16 Sequence of photographs showing the onset of coarsening of an initially almost monodisperse foam. The foam was produced with ordinary detergent and nitrogen gas, and trapped between two glass plates. The photographs were taken at intervals of 90 minutes, the dimensions of the Hele-Shaw cell are 120mm × 80mm × 1mm. (Reproduced by permission of the Liquid Interfaces group Orsay, University of Paris, France.)

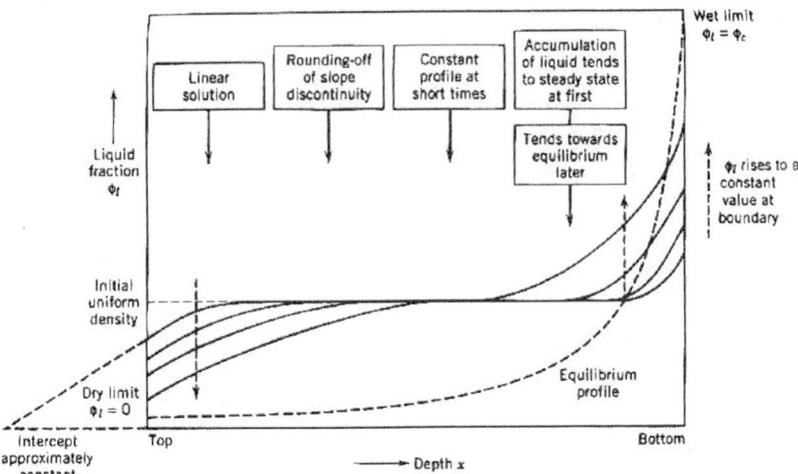

Fig. 4.17 Various regimes are encountered as an initially uniformly wet foam sample drains due to gravity. This needs to be taken into account when designing and performing foam tests.

is established (see Section 4.2). More generally, any departure from hydrostatic equilibrium results in liquid flow.

A model for such drainage may be framed in the form of a continuum theory, with quite extensive predictions, most of which are in accord with the main features of experimental results [34, 35].

Usually this treatment is confined to essentially one-dimensional drainage, as in a vertical column. The model centres on the Foam Drainage Equation, a partial differential equation for the local liquid fraction as a function of position (height) and time. It can take various forms according to the assumptions made about the nature of flow in the Plateau borders. The simplest form, for immobile borders, is:

$$\frac{\partial \alpha}{\partial \tau} + \frac{\partial}{\partial \xi}\left(\alpha^2 - \frac{\sqrt{\alpha}}{2}\frac{\partial \alpha}{\partial \xi}\right) = 0, \tag{4.9}$$

where the quantity $\alpha(\xi, \tau)$ is proportional to the liquid fraction at vertical position ξ and time τ. These quantities are all dimensionless, see [34, 35].

Much effort has been spent on the improvement of the basic model, and to take into account a finite mobility of the Plateau border surfaces [36–38], as required for many surfactant solutions (see further comments at the end of this section).

An analysis of the drainage equation (in any of its variants) shows there are various regimes of time, which need to be understood separately, as Figure 4.17 suggests.

Foamability tests

The development of reliable and meaningful methods for characterisation of foamability and foam stability is of practical importance whenever foams occur in an industrial context. Although a variety of such methods are in use (some as archaic as the simple shaking of a liquid in a closed vessel or the pouring of liquid, followed by the determination of the resulting foam volume [4]), most of them seem to suffer from serious drawbacks.

These foam tests often take the form of a set procedure, whose outcome results in the determination of some defined quantity. In the Rudin test [39] this is the time required for a specified amount of liquid to drain from a specified amount of foam, while in the NIBEM method [40] the foam/gas interface is monitored to determine a time for foam collapse.

While such foam tests may provide for a useful empirical comparison between different samples, their theoretical interpretation is hindered by the fact that generally several processes (e.g. drainage, coarsening, film-breakage) occur simultaneously and the test cannot distinguish between them.

A foam test that lends itself better to theoretical modelling dates back to Bikerman [4] and consists of the continuous production of foam in a tall column by sparging. Due to the breakage of films at the top, where the foam is driest, this results in a steady-state height of the foam, whose value depends on the flow rate of gas. It is possible to interpret this test within the framework of the drainage theory described above [34], and a theoretical expression for "Bikerman's unit of foaminess" Σ was recently derived [41].

Flowing foams are also of great relevance in the context of foam floatation and fractionation (see various articles in [42]). The former process describes the use of foams to separate mixtures of suspended substances, based on their different wettability properties, with applications in the mineral processing industries. In foam fractionation use is made of the different surface-activity of mixtures of solutions. The surface active component adsorbs to the surface of bubbles and can be concentrated by gathering the liquid of an overflowing foam column. A theoretical analysis of such a process involves foam drainage equations, such as equation (4.9), with appropriate boundary conditions [43, 44].

Rheology

"Foam flows" was the title of an early review article on foam rheology [45], containing many of the seeds of later understanding, marred only by the assumption of ordered structures in models. Disorder is of key importance in rheology and modern simulations have demonstrated its essential role, while posing further mysteries.

An understanding of foam rheology generally starts with the case of quasistatic shear, that is, very slow shear deformation. *Shear* sometimes implies simple shear (Figure 4.18) but there is a second type - extensional shear. Since we will be involved with finite shears, extensional and simple shear are not quite the same thing: we shall imply *simple* shear throughout this brief review.

For the moment we consider local properties, on a scale larger than a single cell, but not too much so. We assume that the initial sample is

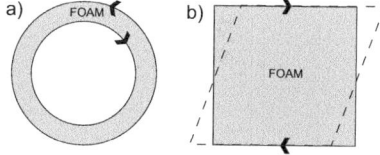

Fig. 4.18 Two standard geometries used in shear experiments. (a) Circular (Couette) geometry, (b) straight geometry.

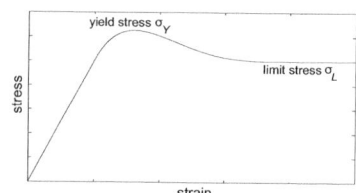

Fig. 4.19 Sketch of a typical stress - strain curve for foams. For low values of strain the foam is solid-like, with a finite shear modulus. For increasing strain, stress may increase up to a yield stress beyond which the foam flows liquid-like. Often this is accompanied by a decrease in stress to the yield-limit stress.

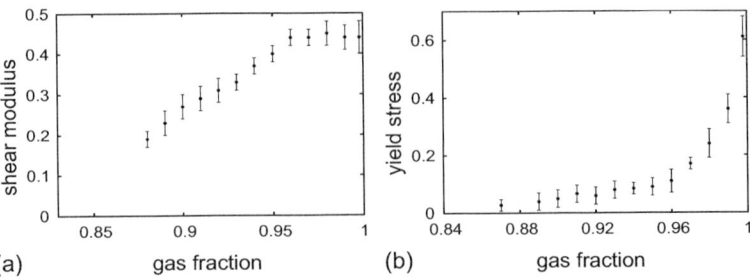

Fig. 4.20 The mechanical properties of foams, such as shear modulus and yield stress, are strongly dependent on liquid fraction. Data shown is from 2D computer simulations [47] using the software PLAT [48].

isotropic on this scale. Under slowly increasing shear deformation, the shear stress varies as in Figure 4.19. For imposed shear strain much less than unity, the foam is a linear elastic solid, to a good approximation. With increasing strain, T1 events (see Section 4.1) eventually introduce a plastic response. A maximum or yield stress is reached, and eventually a slightly lower limit stress. Deformation can continue without limit, because a self-healing steady state is reached. Note the hysteresis if the sign of the increment of shear stress is reversed.

The Stamenovic estimate of the shear elastic modulus G of a dry foam has proved to be a good one. It is given by

$$G = \frac{\sigma}{6} A_b / V_b, \tag{4.10}$$

where A_b and V_b are averages of cell surface area and cell volume, respectively, and σ is the surface tension.

Increasing liquid fraction reduces the shear modulus; it goes smoothly to zero at the wet limit, as demonstrated in Figure 4.20. This property is linked to disorder and was not anticipated in early work: today its precise variation close to the limit is an active topic of research [46].

The transition when the spherical bubbles come apart has been variously dubbed as *melting* or *rigidity loss* but is now generally related to the *jamming* transition of granular materials

The imposition of a finite *strain rate* entails additional local forces not considered in the quasistatic picture, which may be broadly characterised as *viscous*. The stress-strain relation must be generalised to include strain rate. The limit stress increases with strain rate $\dot{\gamma}$ and this is captured empirically in the Herschel-Bulkley relation

$$S = S_Y + c_v \dot{\gamma}^a. \tag{4.11}$$

Here c_v is the coefficient of the viscous contribution to stress, also called *consistency*.

The index a is found to take values substantially less than unity, as shown for example in the numerical data of Figure 4.21. (When it is assigned the value unity the model is associated with the name of Bingham.) The explanation of the observed values of a remains a challenge. It seems that it cannot simply be attributed to nonlinear local forces, but lies (at least in part) in subtle, complex dynamics. Again, disorder is an essential element of this observation. Finally, surfactant characteristics play an important role [49], as they control the type of viscous friction between the bubbles in a sheared foam (see also the comments at the end of this section).

Rheometry

We have sketched the empirical description of the deformation properties of a typical foam, being careful to stipulate that we speak of local properties. On some larger scale, deformation is generally inhomogeneous, for a variety of reasons. It may include regions of shearing flow and plug flow, as well as stagnant regions. Even cylindrical Couette geometry, commonly employed in commercial rheometers, necessarily entails shearing that is not homogeneous, even for a classical Newtonian liquid [50]. This is also true of pipe or capillary rheometers. Only the simplest shear experiment (awkward in practice) allows the possibility of homogeneous shear on a larger scale (and even then it may not be so, as we shall see). For a complex fluid, probed with any kind of rheometer, the analysis benefits from some direct visualization of the distribution of shear [51, 52], obtained for example with MRI. Regions of plug flow are made manifest by these techniques.

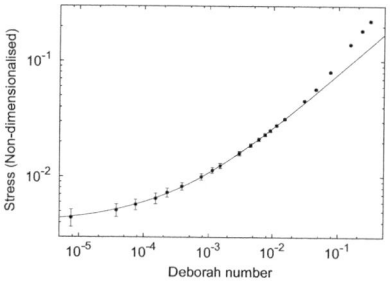

Fig. 4.21 Example of the variation of stress with Deborah number (which is proportional to strain rate) as obtained in simulations [53] using the bubble model for 2D foams [54]. The solid line is a fit to the Herschel-Bulkley relationship, equation (4.11), resulting in a Herschel-Bulkley exponent of $a \simeq 0.5$. A deviation occurs only at very high shear rates.

Shear localization

Localization can occur for reasons other that those imposed by geometry, and has been debated recently in the context of 2D foams. There is some evidence for shear banding in 3D foams [55], but most debate has centred on the 2D case, for which experimental and computational results have recently been summarised in [56]. Despite the advantages of direct visualization, shear localization at moving boundaries, as sketched in Figure 4.22, has yet to be entirely explained. One partial explanation attributes it to the effect of the drag exerted by fixed boundaries (the confining plates). A developed continuum model [56] is based on the Herschel-Bulkley relation and a drag force due to the confining plates. It generates neat formulae for the consequent localization, for which there is some validation, but open questions remain.

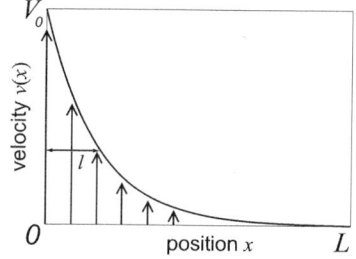

Fig. 4.22 The velocity profile in 2D foam shear experiments and simulations often features localization, expressed in terms of a localization length l.

Conductivity

Electrical conduction in a typical liquid foam takes place through the liquid. The liquid films are so thin as to make a negligible contribution to conduction, so it takes place through the network of Plateau borders and the junctions where they meet.

For a dry foam, the junctions are small enough that their precise size and shape insignificant, and the foam may essentially be regarded as a network of uniform conductors, much as in the elementary theory of drainage, described above. Indeed the interpretation of drainage data in terms of a liquid flow conductivity shows the direct analogy between liquid and electrical flow [57].

For a typical disordered foam, the conductivity is isotropic. It increases from zero, for liquid fraction $\phi = 0$ to σ_0, the liquid conductivity for $\phi = 1$.

The Lemlich formula is based on a simple ansatz for the current distribution:

$$\sigma = \frac{1}{3}\sigma_0 \phi. \tag{4.12}$$

It provides a good estimate of the foam conductivity σ for small values of ϕ (of the order of 1%), but fails at higher values.

The smooth variation of conductivity with liquid fraction that is observed in experiments is well fitted by various formulae, of which the best appears to be

$$\sigma/\sigma_0 = 2\phi(1 + 12\phi)/(6 + 29\phi - 9\phi^2), \tag{4.13}$$

as proposed by Feitosa et al. [58].

The Lemlich approximation may be given a good deal of theoretical rationalisation [59]. The more general formula, equation (4.13), is purely empirical.

Since conductivity depends strongly on ϕ and relatively little on the detailed structure or statistics of the foam, it is a convenient means of measurement of liquid fraction, and instruments have been commercialised for that purpose (e.g. by Sinterface Technologies, Teclis, Laval Lab).

The insights gained in studies of liquid foam conductivity have been useful in analyzing the *thermal* conductivity of some solid foams, for which conduction through the solid phase is only a minor contribution to the total.

Light scattering

One of the first appearances of foam in the scientific literature was concerned with its optical properties – its whiteness. The 17th century Robert Boyle correctly surmised that this was due to the multiple reflection or refraction of light by the disordered structure of the foam. Light incident on the surface undergoes many changes of direction within it with little absorption. At some stage this random walk must take it back out of the surface, at a random angle. Hence a black beer makes an almost white foam.

Such multiple light scattering (or "radiative transfer") crops up in several branches of physics and even in medicine. Chandrasekhar wrote a famously inscrutable book on its mathematical treatment, but it is nevertheless quite simple in some respects. Durian [60], among others, has pioneered its application to foams. From measurements of the fraction T of transmitted light through a foam sample of thickness L, he extracts a characteristic length l^*,

$$T \propto \frac{l^*}{L}. \tag{4.14}$$

This transport mean free path l^* indicates over which length an incoming beam of light is randomised in direction due to scattering events (l^* is typically a few bubble diameters). Since the scattering is due to the liquid interfaces in the foam, l^* may be used to measure the rate of change of bubble size due to coarsening (Section 4.4, which is otherwise difficult to assess in 3D foams.

The coarsening process raises another possibility to exploit the technique, in as much as it involves abrupt local rearrangements, the expanding structure remains the same only in an average statistical sense: in detail it changes continuously. The result is a corresponding change in the speckle (interference pattern) of a laser beam transmitted through the sample and the rate of the local transformations may be estimated from this change [61].

The role of surfactants

Typically, a foam is stabilized by a small concentration of surfactant, which coats all of its gas-liquid interfaces. Its primary effect is to lower the surface tension to about a third of that of water. (For a detailed acount of the role of surfactants see Chapter 2 of this Handbook.) The surface concentration of surfactant is reduced locally, then the surface tension must increase. This introduces an additional force which tends to "heal" any local thinning of a film. This and other related properties may be called the "Marangoni effect".

While the detailed chemistry of the interfaces does not play a role in determining foam structure, and many static foam properties, such as conductivity or shear modulus, except in affecting the value of the static surface tension, it is important for many dynamic processes.

Earlier in this section we mentioned the role of finite interface mobility (expressed in terms of a surface viscosity) for the details of foam drainage. If this is low, as for example in SDS foams, dissipation occurs mainly due to the flow of liquid through the Plateau border junctions. If it is high, as in protein foams, the flow through the Plateau borders themselves is the main source of dissipation.

The detailed mechanism of energy dissipation also plays a role in foam rheology. Here it was shown that the value of the surface dilational modulus, related to the expansion and contraction of a film surface, can control the value of the Herschel-Bulkley exponent [49].

4.5 Analogous systems

Particle-stabilized foams

The classic formulation for foam is liquid+gas+surfactant. In recent years great interest has arisen in the replacement of the surfactant molecules by small particles, giving rise to a remarkable class of foam with distinct and useful properties [62], especially in food and cosmetics. The understanding of these foams has been greatly advanced by the

Fig. 4.23 A delicious prospect? Example of a liquid foam stabilized with coated silica particles. Foam stability is controlled by the hydrophobicity of the coating [66]. (Photograph reproduced by permission of B. Binks.)

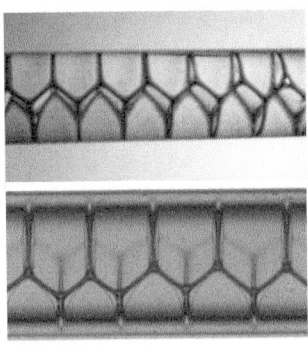

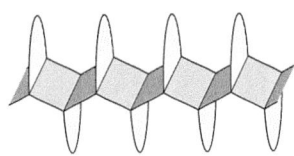

Fig. 4.24 The structure of confined monodisperse emulsions is identical to those of foams of comparable bubble size, as illustrated here for the so-called staircase structure, consisting of two equal volume bubbles per unit cell. The photograph on the left shows a silicon oil in water emulsion [68], stabilized by a surfactant, while the photograph in the centre features an air in water foam. The image on the right is the result of a computer simulation of this structure [15] using Ken Brakke's Surface Evolver.

group of B. Binks at the University of Hull, UK, and co-workers [63, 64].

It is hardly new, having been first investigated more than a century ago by Walter Ramsden [65]. Partly due to the self-effacing character of this distinguished researcher, there was little interest during the intervening period, but today it is certainly intense.

The particles in present use, for example fumed silica, are required to be partially hydrophobic so that they adsorb to the liquid-gas interface. The attachment energy is by about three orders of magnitude larger than that of surfactant molecules. Once the particles are attached, they remain there, unlike surfactant molecules. Both coarsening and film-rupture are greatly reduced in foams stabilized by particles (an example of which is shown in Figure 4.23), leading to sample life-times of months.

The details of stabilization mechanism are still under scrutiny and include the questions of role of particle shape, size and roughness. What is emerging, however, is the importance of the finite dilational elasticity of the particle stabilized films [63]. As a bubble increases, due to coarsening, so does the surface tension since there are no more particles available to adsorb from the bulk to the surface.

Emulsions

A lot of what was said in the previous sections about foams is equally applicable to emulsions. Indeed, some of the early work on foam rheology by Princen (e.g. determination of yield stress [67]) was carried out with oil-in-water emulsions, for reasons including increased stability of the samples with respect to coarsening and reduced drainage, due to smaller density differences of the two phases.

The analogy of foams and emulsions is particularly striking when using droplets with diameter of a few millimetres or larger. Figure 4.24 shows that the ordered arrangements, obtained when confining these in cylinders of similar width, are the same as when using foams [68]. Surface energy turns out to be the key parameter in both systems.

Also drainage in oil-in-water emulsions can be described in terms of the foam drainage equation 4.9, as introduced in Section 4.4 [57, 68]. Here the Plateau border surfaces may be considered as immobile, since the viscosity of the continuous phase (water) is much lower than that of the dispersed phase (oil).

Conclusion

For long the preserve of empirically-minded chemical engineers, foams have been intensively studied by physicists only for a few decades. Despite the sophistication of some of their experimental methods and simulation techniques, the essential description of a liquid foam and its basic properties remains quite simple and transparent.

It has fulfilled the promise of providing a test-bed for speculation and analysis that has much broader implications, across the range of soft matter physics and beyond. Few areas of research can readily combine insight and outreach so readily.

Looking on the negative side, it must be admitted that we are still a long way from a really comprehensive model of foam properties. Physicists like to "divide and conquer", singling out each property in turn, under the simplest of circumstances. But in real applications, many effects on many scales may be coupled. The simple act of pouring a glass of beer invokes the simultaneous nucleation, growth, drainage, flow, and collapse of a foam. And much of what goes on must be intimately linked to the local small-scale dynamics of surfactants.

Do we have enough understanding of the necessary ingredients to contemplate an integrated approach now, given the power of the modern computer to run massive simulations? Perhaps not, but it is certainly time to design more experiments that move from the idealisations of physics towards the hard realities of engineering. For example, the combination of drainage and rheology would be an important step forward. This might be combined with (fast) tomographic techniques to reveal in detail what goes on in this type of two-phase flow. Indeed the intervention of tomography may have an even more dramatic impact of these difficult problems than on static structure. The pay-off for such progress would be considerable if it provided chemical engineers with reliable tools for optimisation of major industrial processes, producing foam by design or accident.

Acknowledgements

This review has emanated from research conducted with the financial support of Science Foundation Ireland (08/RFP/MTR1083) and the European Space Agency (MAP AO-99-108: C14914/02/NL/SH and AO-99-075: C14308/00/NL/SH). Denis Weaire thanks the University of Hyderabad for hospitality during the period of completion of this chapter.

References

[1] Plateau, J.A.F. (1873). *Statique Expérimentale et Théorique des Liquides soumis aux seules Forces Moléculaires*. Gauthier-Villars, Paris.

[2] Lawrence, A. S. C. (1929). *Soap Films*. G. Bell & Sons Ltd., London.

[3] Manegold, E. (1953). *Schaum*. Straßenbau, Chemie und Technik Verlagsgesellschaft m.b.H., Heidelberg.

[4] Bikerman, J.J. (1973). *Foams*. Springer Verlag, Berlin.

[5] Zitha, P., Banhart, J., and Verbist, G. (ed.) (2000). *EUFoam 2000: Foams, emulsions and their applications*, Bremen, Germany. Verlag MIT.

[6] Vignès-Adler, M., Weaire, D., and Miller, R. (ed.) (2005). *EUFoam 2004: Foams, emulsions and applications*, Volume 263.

[7] Adler, M., Miller, R., and Weaire, D. (ed.) (2007). *EUFoam 2006: Foams, emulsions and applications*, Volume 309.

[8] Adler, M., Cilliers, J. J., Minster, O. *et al.* (ed.) (2009). *EUFoam 2008: Foams, emulsions and applications*, Volume 344.

[9] Denkov, N. D. and Kralchevsky, P. A. (ed.) (201). *EUFoam 2010: Foams, emulsions and applications*, Volume 382.

[10] Vaz, M. F., Rosa, M. E., and Teixeira, P. I. C. (ed.) (2013). *EUFoam 2012: Foams, emulsions and applications*.

[11] Sadoc, J.F. and Rivier, N. (ed.) (1999). *Proceedings of the NATO Advanced Study Institute on Foams, Emulsions and Cellular Materials*. Kluwer, Dordrecht.

[12] Weaire, D. and Banhart, J. (1999). *International Workshop on Foams and Films*. Verlag MIT, Bremen, Germany.

[13] Weaire, D. and Hutzler, S. (1999). *The Physics of Foams*. Oxford University Press, Oxford.

[14] Cantat, I., Cohen-Addad, S., Elias, F. *et al.* (2010). *Les mousses – Structure et dynamique*. Éditions Belin, Paris.

[15] Tobin, S.T., Barry, J.D., Meagher, A. *et al.* (2011). *Coll. Surf. A: Physicochem. Eng. Aspects*, **382**, 24–31.

[16] Matzke, E.B. (1946). *Am. J. Botany*, **33**, 58.

[17] Kraynik, A. M., Reinelt, D. A., and van Swol, F. (2003). *Phys. Rev. E*, **67**, 031403.

[18] Gonatas, C.P., Leigh, J.S., Yodh, A.G. *et al.* (1995). *Phys. Rev. Lett.*, **75**(3), 573–576.

[19] Lambert, J., Mokso, R., Cantat, I. *et al.* (2010). *Phys. Rev. Lett.*, **104**, 248304.

[20] Meagher, A. J., Mukherjee, M., Weaire, D. *et al.* (2011). *Soft Matter*, **7**, 9881 – 9885.

[21] Cox, S.J. and Graner, F. (2004). *Phys. Rev. E*, **69**, 031409.

[22] Brakke, K. A. (1992). *Exp. Math.*, **1**, 141–165.

[23] Kraynik, A.M., Reinelt, D.A., and van Swol, F. (2004). *Phys. Rev. Lett.*, **93**, 208301.

[24] Princen, H. M. and Kiss, A. D. (1987). *Langmuir*, **3**, 36–41.

[25] Weaire, D., Langlois, V., Saadatfar, M. *et al.* (2007). In *Granular and Complex Materials* (ed. T. Aste, T. D. Matteo, and A. Tordesillas), New Jersey, pp. 1–26. World Scientific Publishing.

[26] Hutzler, S. and Weaire, D. (2011). *Coll. Surf. A: Physicochem. Eng. Aspects*, **382**, 3–7.

[27] van der Net, A., Drenckhan, W., Weaire, D. *et al.* (2006). *Soft Matter*, **2**(2), 129–134.

[28] van der Net, A., Blonde, L., Saugey, A. *et al.* (2007). *Coll. Surf. A: Physicochem. Eng. Aspects*, **309**, 159–176.

[29] Höhler, R., Sang, Y.Y.C., Lorenceau, E. *et al.* (2008). *Langmuir*, **24**, 418–425.

[30] Weaire, D. and Phelan, R. (1994). *Phil. Mag. Lett.*, **69**(2), 107–110.

[31] Gabbrielli, R., Meagher, A.J., Weaire, D. *et al.* (2012). *Phil. Mag. Lett.*, **92**, 1–6.

[32] Monnereau, C., Pittet, N., and Weaire, D. (2000). *Europhys. Lett.*, **52**, 361 – 367.
[33] Lambert, J., Mokso, R., Cantat, I. *et al.* (2007). *Phys. Rev. Lett.*, **99**, 058304.
[34] Verbist, G., Weaire, D., and Kraynik, A.M. (1996). *J. Phys.: Cond. Matter*, **8**, 3715–3731.
[35] Weaire, D., Hutzler, S., Verbist, G. *et al.* (1997). *Adv. Chem. Phys.*, **102**, 315–374.
[36] Koehler, S.A., Hilgenfeldt, S., and Stone, H. A. (2000). *Langmuir*, **16**, 6327 –6341.
[37] Koehler, S.A., Hilgenfeldt, S., Weeks, E.R. *et al.* (2002). *Phys. Rev. E*, **66**, 040601.
[38] Koehler, S.A., Hilgenfeldt, S., Weeks, E.R. *et al.* (2004). *J. Coll. Interf. Sci.*, **276**, 439–449.
[39] Rudin, A.D. (1957). *J. Inst. Brewing*, **63**, 506 – 509.
[40] Somasundaran, P. (ed.) (2006). *Encyclopedia of Surface and Colloid Science, 2nd edition*. Volume 4. Taylor and Francis, London.
[41] Hutzler, S., Lösch, D., Carey, E. *et al.* (2011). *Phil. Mag.*, **91**, 537–552.
[42] Li, X. and Stevenson, P. (2012). In *Foam Engineering: Fundamentals and Applications* (ed. P. Stevenson), pp. 307 – 330. John Wiley & Sons, Ltd, New York.
[43] Neethling, S.J., Lee, H.T., and Cilliers, J.J. (2003). *J. Phys.: Cond. Matter*, **15**, 1563–1576.
[44] Hutzler, S., Tobin, S.T., Meagher, A.J. *et al.* (2013). *Proc. R. Soc. A*, **469**, 2154.
[45] Kraynik, A. M. (1988). *Ann. Rev. Fluid Mech.*, **20**, 325–357.
[46] van Hecke, M. (2010). *J. Phys.: Cond. Matter*, **22**, 033101.
[47] Hutzler, S., Weaire, D., and Bolton, F. (1995). *Phil. Mag. B*, **71**, 277.
[48] Bolton, F. (1996). *http://www.tcd.ie/Physics/Foams/plat.php*.
[49] Denkov, N. D., Tcholakova, S., Golemanov, K. *et al.* (2009). *Soft Matter*, **5**, 3389 – 3408.
[50] Khan, S.A., Schnepper, C.A., and Armstrong, R.C. (1988). *J. Rheol.*, **32**, 69–92.
[51] Coussot, P., Raynaud, J.S., Bertrand, F. *et al.* (2002). *Phys. Rev. Lett.*, **88**, 218301.
[52] Yoon, W.B. and McCarthy, K. (2002). *J. Texture Studies*, **33**, 431 – 444.
[53] Langlois, V.J., Hutzler, S., and Weaire, D. (2008). *Phys. Rev. E*, **78**, 021401.
[54] Durian, D.J. (1997). *Phys. Rev. E*, **55**, 1739–1751.
[55] Rodts, S., Baudez, J.C., and Coussot, P. (2005). *Europhys. Lett.*, **69**, 636.
[56] Weaire, D., Barry, J.D., and Hutzler, S. (2010). *J. Phys.: Cond. Matter*, **22**, 193101 (22pp).
[57] Péron, N., Cox, S.J., Hutzler, S. *et al.* (2007). *Eur. Phys. J. E*, **22**, 341–351.

[58] Feitosa, K., Marze, S., Saint-Jalmes, A. *et al.* (2005). *J. Phys.: Cond. Matter*, **17**, 63016305.

[59] Durand, M., Sadoc, J.F., and Weaire, D. (2004). *Proc. R. Soc. Lond. A*, **460**, 1269–1284.

[60] Durian, D.J., Weitz, D.A., and Pine, D.J. (1991). *Science*, **252**(5006), 686–688.

[61] Weitz, D.A., Zhu, J.X., Durian, D.J. *et al.* (1993). *Phys. Scripta*, **T49**, 610–621.

[62] Vignes-Adler, M. and Weaire, D. (2008). *Curr. Opin. Coll. Interface Sci.*, **13**, 141–149.

[63] Stocco, A., Drenckhan, W., Rio, E. *et al.* (2009). *Soft Matter*, **5**, 2215 2222.

[64] Cervantes-Martinez, A., Rio, E., Delon, G. *et al.* (2008). *Soft Matter*, **4**, 1531–1535.

[65] Ramsden, W. (1903). *Proc. R. Soc.*, **72**, 156 – 164.

[66] Binks, B.P. and Murakami, R. (2006). *Nature Mater.*, **5**, 865 – 868.

[67] Princen, H.M. (1985). *J. Coll. Interface Sci.*, **105**, 150–171.

[68] Hutzler, S., Péron, N., Weaire, D. *et al.* (2004). *Eur. Phys. J. E*, **14**, 381–386.

Physics of Granular Systems

5

Raphael Blumenfeld[1,2], Sam F. Edwards[2]
and Stephen M. Walley[2]

1. Earth Science & Engineering, Imperial College, London SW7 2AZ, UK
2. Cavendish Laboratory, J J Thomson Avenue, Cambridge CB3 0HE, UK

The science of granular matter has expanded rapidly in the last two decades from an activity in engineering applications to a fundamental research field in its own right. This has been accompanied by a surge in the literature, the review of which could fill several volumes. It is hopeless to attempt to cover this diverse work in one chapter. We therefore limit ourselves to a review of several current problems that plague the understanding of the fundamental physics of granular science.

We start with a review of a method to quantify the structure of static granular media, an open problem which we regard as essential to any static and quasi-static theory. We then review and discuss recent progress on, and the current understanding of, stress transmission in mechanically equilibrated granular assemblies. We then turn to recent advances in the ongoing attempt to construct a statistical mechanical theory of granular systems. Our main aim is not only to review the current state of affairs in these directions but also to clarify a number of misconceptions in the community, as well as to point out several outstanding problems.

This chapter is constructed as follows. After the introduction, we briefly outline in Section 5.2 a method to quantify the structure of granular matter on the grain scale and then use it to discuss quantitatively some structural and topological properties. In Section 5.3, we review the recent progress on the theory of stress transmission in static packings of grains. This aspect of granular science has been the source of controversy for two decades, with one camp advocating strain-based theories, especially elasticity and elasto-plasticity, while another camp is advancing a more recent isostaticity theory. Since elasticity and elasto-plasticity are relatively well known, we focus on a review of the less known isostaticity theory and include some recent results. We also discuss a proposal of a more general stato-elasticity theory that combines the two approaches and and briefly describe some of the issues involved in the construction of such theory. In this section, we also highlight results that dispel a number of misconceptions, such as the applicability of isostaticity theory to non-rigid grains and its usefulness to describe stress in realistic granular matter. Section 5.3 concludes with a critical discussion of the very premise that a complete stress theory of granular materials could be linear. In Section 5.4, we review the development of a statistical-mechanical formalism of granular matter. We relate this formalism to the structural description, discussed in Section 5.2, and demonstrate its usefulness. We also highlight several misconceptions and open problems that stand in the way of derivations of equations of state. We conclude in Section 5.5 with a summary and a brief discussion of some outstanding fundamental problems that require resolution.

5.1 Introduction to granular matter

Granular matter is all around us. The sand we walk on, the salt and pepper on our table, the cereals we pour into bowls, the variety of powders we use, and even the objects we organise on our desk, are examples of granular systems (GS). After water, it is the most ubiquitous form of matter on our planet. Granular materials also play extremely important roles in a wide range industries including agriculture, minerals, metals, pharmaceuticals, laundry products, and construction. In fact, almost all industrial materials pass, at some stage of their processing, through a particulate form, making the handling and processing of particulate and powdered materials essential to our society. An entire field of engineering is devoted to conveying and transporting granular and particulate materials.

To appreciate the significance of granular materials on our lives and our economy, consider the statistics presented in the margin note.[1] These statistics make it evident that even a small increase in our ability to predict the behaviour of granular materials can make a considerable economic impact. Due to their significance and ubiquity, granular materials have attracted attention for millennia. The large numbers of grains in natural systems have been used as metaphors for plentitude well before the concept of the atom. For example, already in the Bible we find passages such as: "... like the stars of the heavens, and like the sand which is on the seashore...." (Genesis 22:17). The significance of this form of matter has been acknowledged by the ancient Greeks, who had a goddess of sand beaches called Psamathe. Technological uses of granular materials by the ancient Greeks included filtration of water through sand and gravel as a method of treatment – a method that dates even further back to 2000 BC, as documented in Sanskrit (India) writings. Documented scientific investigations involving GS date back to Archimedes, who famously set out to estimate the number of grains of sand that the universe could contain [5]. In the 19th century, the science and engineering of granular materials was the subject of studies by luminaries, such as Faraday [6] and Reynolds [7]. Since then and until the late 1980's, granular materials were studied mainly in engineering contexts. However, the 1990's saw a surge of renewed interest in the fundamentals of granular physics.

The explosion of activity in the last two decades was triggered by a combination of the increasing technological significance of GS, the

[1] Approximately one-half of the products and at least three-quarters of the raw materials in the chemical industry are in particulate form [1] and, within this industry, particulate technology is estimated to engage over 60 billion US Dollars [2]. Approximately 1.3% of all the USA electrical power production goes towards grinding particles or ores [2]. Landslides cause annual property damage of up to 2 billion US Dollars and at least 25 fatalities in the USA alone [3]. The 1997-1998 landslide damages to the ten San Francisco counties alone was in excess of 140 million US Dollars [3]. Over 1,000 silos, bins, and hoppers fail in North America every year [4]. In Mexico, 5 million tons of corn are handled each year, 30% of which is lost due to poor handling systems. Over 10% of the world's energy supply goes towards handling particulates. About 50% of the world's energy resources are extracted from granular systems, such as soils and rocks.

increasing need for predictive models of GS, and a fundamental scientific interest in their complex behaviour. Examples of such behaviour are extensive. GS display both fluid- and solid-like properties under a range of external loading. When agitated externally by vibration, shaking or periodic excitations, they exhibit rich and spectacular phenomena, including unstable surface dynamics [8,9], pattern formation [6,10], and size segregation [11].

The broad relevance of GS makes this one of the most multi-disciplinary fields, involving a large number of communities: physics, mathematics, earth sciences, processing, civil and mechanical engineering [12–15], and even information theory [16]. Despote the recent intensive research effort, we are still a long way away from a fundamental understanding. This state of affairs not only handicaps effective modelling for practical applications but also encourages development of phenomenological models suited only for specific applications.

The term granular matter refers generically to a collection of macroscopic particles that are too massive to be affected by thermal fluctuations. For example, in a gravitational field, this means that the energy required to move a typical grain a distance comparable to its size is much larger than Nk_BT, where N is the number of molecules comprising the grain, k_B is the Boltzmann constant and T is the temperature. This means that thermal fluctuations cannot rearrange the configurations of GS and hence that, for all practical purposes, static granular assemblies can be considered to be at zero temperature.

Almost all the GS of interest to us are immersed in a fluid medium. When the surrounding fluid is liquid the system is called a colloidal suspension, or simply a suspension. More commonly the surrounding fluid is a gas such as air. The exceptions are GS in outer space, which exist in vacuum, but these are less relevant to the present human society. GS immersed in air are often referred to as dry, while those immersed in liquids are said to be wet. The main difference between the two types of systems is that the latter support attractive capillary forces between grains while the former have only compressive, or repulsive, forces on contact. The co-existence of two phases, grains and an immersing fluid, means that the dynamics of such GS involves not only direct interaction between grains but also interactions mediated by the fluid. In other words, movement of the grains affects the hydro- or aero-dynamics of the fluid, which in turn affects forces on other grains. This indirect interaction depends on the range of velocities; the faster the dynamics the stronger the hydrodynamics-mediated interactions. More complicated are wet GS, where three phases coexist: particles, liquid, and gas. These are normally modelled differently depending on the relative concentrations of the phases, from small bubbles in suspensions to small liquid necks between solid particles in air (see e.g. [17] and references within). In this presentation, we limit ourselves to a discussion of statics of dry systems only.

Although many GS of relevance to human society consist of grains of sizes that range between 100nm and a few millimeters, e.g. soil, the modelling of granular materials can be made scale-independent over a wide range of scales, as long as gravitational attraction does not overwhelm the physics. For example, modelling behaviour of dry granular matter in vacuum can apply equally well to grains of sand and meter-size boulders. The difference is usually that we normally encounter many more grains of sand in any specifically studied system than boulders.

The irrelevance of thermal fluctuations makes conventional modelling of molecular particles inapplicable to GS. Since grains are much bigger than the range of molecular interactions, then, for most practical purposes, hard grains can be regarded as interacting only upon contact. This simple feature is sufficient to give rise to a rich behaviour that sometimes resembles that of solids, sometimes of liquids, sometimes of gases, and quite often seems a simultaneous combination of these. Another characteristic of macroscopic grains is that they dissipate energy both on collision and when rubbing against one another. This undermines the many modelling approaches based on energy-conserving theories, which work well for conventional molecular systems. A well-known ramification of collisional energy dissipation is the phenomenon of clustering and apparent solidification of a flowing dense granular material [18]. This phenomenon can be understood on the granular level as follows. When two grains collide they lose kinetic energy and tend to remain in the vicinity of one another. This increases the likelihood that they collide again and lose even more energy and so on. This process increases the probability that the two grains eventually stick together, creating a two-grain cluster. Since this cluster is larger in size, it enhances the probability that other grains collide with it, leading to further loss of kinetic energy. This process is unstable and it can continue, giving rise to spontaneous generation of large clusters of grains that do not move relative to one another. In other words, this is a process of solidification. Other intriguing phenomena, such as segregation [11] under shaking and the Brazil Nut Effect [19], are also unique to GS. The rich and complex macroscopic behaviour of GS has even led to the suggestion that granular matter is a distinct state of matter.

The zero-temperature-like nature of static granular materials is reminiscent of more conventional systems of particles at very low temperatures - glassy systems. This similarity prompted several groups to study granular matter with analytical models derived for such systems. Glassy systems are defined by dynamics that are so sluggish that the system cannot 'explore' a sufficiently large number of states in phase space. This undermines well-established and powerful approaches, such as conventional thermodynamics and statistical mechanics. GS are similarly problematic in that, left to their own devices in a gravitational field, they do not rearrange readily and are confined to realising only a small number of configurations.

The main problem is that we have a limited 'tool box' of theoretical models to describe their complex behaviour - models that have been developed for systems with altogether different characteristics. The default modelling of many-body systems in science is by coarse-graining to the continuum, also known as homogenisation. This is in spite of the fact that the universe around us is discrete at any scale. Granular science is no exception. Whilst the behaviour and interactions of individual grains are well understood, it is often impractical to describe the behaviour of the distinct grains and it is preferable to have models that provide an approximate description of the bulk behaviour as a continuum.[2] Yet, the modern scientific attempts at coarse-grained modelling of many-grain systems have seen very little success for almost every aspect of granular physics. Frustratingly, fundamental theories that work well for conventional systems, such as elasticity theory and Navier-Stokes flow equations, fail for GS. Indeed, part of the problem in the field is the tendency to use conventional approaches outside their ranges of validity, at the expense of more focused work into construction of fresh theories from first principles.

But what does the science of granular matter consist of? And which part of such a science is fundamental? Part of the problem is that there is no general consensus on the answers to these issues. Different groups study completely different aspects of behaviour of granular materials. The range of phenomena investigated in this field is wide, including stress transmission in static systems, both dry and wet; yield and failure; jamming; fast and slow flows; pattern formations under various dynamics; interactions of granular systems with other systems; statistical mechanics of GS and the relations to glassy dynamics; and impact and penetration of object into granular matter. A number of reviews on the richness of GS can be found in [1, 14, 20, 21]. While a plethora of approaches is often an advantage in any research field, it also makes it difficult to bridge between models of phenomena that are controlled by a single parameter. For example, the static theory of stress transmission should be related to the theory of yield and failure, which in turn should be related, as the velocity increases, to the rheology of dense granular flow, which should then converge to the equations of fast flow of granular matter, modelled sufficiently well by the kinetic theory of gases. All these models should form a continuous spectrum of models controlled only by the flow velocity. We are nowhere near such an understanding and universal modelling.

The lack of consensus on most, if not all, of these issues is a clear indication that granular matter is far from being understood. These authors suspect that part of the problem is methodological. Namely, that the current approach, where groups work with no coordination on the many aspects of the field, is not the best route to such understanding and a more systematic and coherent approach is required. Broadly speaking, there are currently six main fundamental directions of activity in the field.

[2] One could argue that the problem with treating GS discretely has occurred already to Joseph son of Jacob: "And Joseph gathered corn as the sand of the sea, very much, until he left numbering; for it was without number" (Genesis 41:49). He is probably the first documented person to anticipate the necessity of descriptions of many-body systems beyond the discrete.

1. Development of a statistical mechanical formalism for derivation of constitutive relations and equations of state.
2. Development of a continuum stress theory for static systems in mechanical equilibrium.

3. Development of a theory for the failure and subsequent yield flow.
4. Construction of continuous equations for fully developed flow of intermediate and very dense granular fluids.
5. Development of a general theory to predict packing fractions of granular assemblies.
6. Understanding a range of complex pattern formation processes in agitated systems.

The first four of these directions address fundamental physical modelling in the traditional sense. They deal with basic mechanisms of relatively well-understood phenomena on the discrete grain scale and aim at developing coarse-grained continuum theories for these phenomena. As such, the first four directions are essential to the construction of predictive science of granular matter. The fifth direction is of immense importance to technology and it is studied intensively in the mathematical, physical, and engineering disciplines. These studies are mostly analytical and numerical. While this direction of activity is also fundamental, the problem seems to be not entirely well posed in the literature [22]. In particular, packing of granular systems is strongly sensitive to the packing procedure, shapes and sizes of grains and grain mechanical properties (e.g. friction and crushability), but a way to parameterise the entire parameter space of all possible packing processes is hopeless at this stage. Consequently, these communities focus on canonical and paradigmatic problems, which while potentially providing some insight into the problem, are not of much use for realistic systems. In general, we are still far from a roadmap that outlines a theory for packing of general granular materials. The sixth direction is important to a range of applications and belongs also in the studies of complexity. Patterns normally form in dynamic GS and often under well-defined external excitations. However, to make fundamental progress in this direction and understand the patterns that develop in excited GS, one needs to have, at the very least, an adequate theory for the dynamics of granular matter. Without that, a general theory of pattern formation is hopelessly difficult. It is for this reason that pattern forming processes in agitated GS can only be understood and modelled currently one case at a time, rather than as part of a general unified theory. The above six directions are not independent and there are many applications that involve several of them. Nevertheless, this division is useful for classification of the theoretical approaches required for modelling.

The large body of literature on GS makes it impossible to review all the activity in these directions. We therefore present in this chapter a personal view of the current state of the field, focusing on the issues that could underpin fundamental understanding of the physics that governs granular science. Specifically, we limit ourselves to discussing the first two directions: the nature of stress transmission in GS and the attempts to construct a statistical-mechanical-like entropic formalism. In the following, we not only review recent advances but also aim to clarify some existing misconceptions as well as point out the outstanding problems that stand in the way of the science of granular matter becoming more predictive.

5.2 Structural description

To construct any predictive theory of GS, one has to start from the granular level and coarse-grain, or homogenise, to the continuum. To this end, we must first possess a useful description of the 'micro'-structure on the granular level. Without such a quantitative description, any discussion of the structure remains abstract and qualitative. Many so-called structural descriptors consist of averages over specific structural quantities. Examples are density-density correlation functions, all diffraction methods, the mean coordination number (see below), and a host of other correlation functions. Common to all these descriptions is the volumetric averaging of a structural quantity either over the entire GS or over sufficiently large parts of it.

For a proper theory of GS we need an unambiguous *local* descriptor, a quantity that quantifies the structure at any arbitrary specific point within the space that the system occupies. The difference between such a local descriptor and the aforementioned averages is analogous to the difference between the local density and the mean density - while the latter is useful for some purposes, it is only with the former that a fundamental theory can be constructed.

A number of works in the literature have attempted to address the structural and statistical characteristics of granular packings [23]. A particularly useful local descriptor has been proposed recently for static systems in mechanical equilibrium, first in 2D [24,25] and then extended to 3D [26]. The rationale behind both descriptions follows three conceptual steps.

1. Construct a connectivity graph, whose nodes are the inter-granular contact points (or centroids of contact surfaces), such that the contact points around every grain form a polygon in 2D and a polyhedron in 3D.

2. Use this construction to tessellate the space occupied by the GS into elementary volumes, called 'quadrons'. These are generically quadrilaterals in 2D and non-convex octahedra in 3D.

3. Quantify the shape of every quadron by a shape tensor.

With this procedure we can quantify the micro-structure unambiguously at the scale of the quadrons. This scale is smaller than that of the grains, since several quadrons belong to the same grain, as will be seen below. It should be noted that although the description is unambiguous and well-defined, it is not unique. There are many ways to tessellate the space occupied by a GS. Nevertheless, this specific description has several advantages, which will be detailed below. We proceed to review this method first in detail for 2D systems and then only sketch the similar

[3]If the contacts are not points but small surfaces, then the contact 'point' is defined as the centroid of that surface.

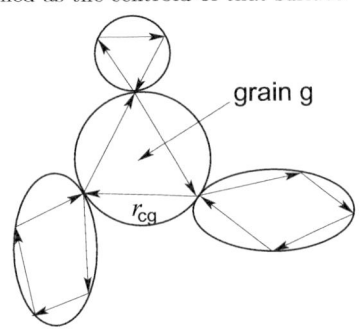

Fig. 5.1 The construction of the contact network of a 2D granular assembly under compressive boundary forces. The contact network is a directed graph made of vector edges r_{cg}.

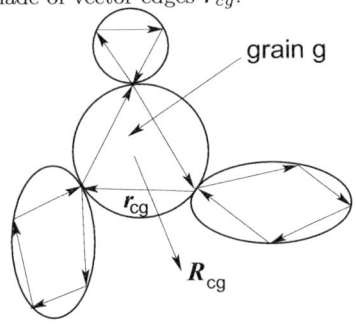

Fig. 5.2 The construction of the conjugate of the contact network, formed by the vectors R_{cg}.

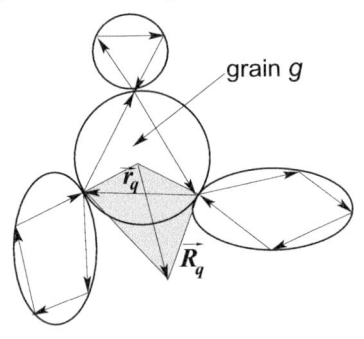

Fig. 5.3 The quadron q (shaded), whose diagonals are $R_{cg} = R_q$ and $r_{cg} = r_q$.

principles for 3D systems. For a detailed description of the extension to 3D, we refer the reader to [26].

Quantitative local structural description in 2D

The first step is to construct a graph that represents the local microstructure. This is done by connecting all the contact points around a grain to form a polygon.[3] The edges of the polygon are assigned directions such that they rotate clockwise around a grain, as shown in Figure 5.1. This forms a directed graph which we call the contact network. Every directed edge is then described by a vector r_{cg}. These vectors enclose loops (also known as cells) around the voids in the structure, rotating in the anti-clockwise direction around every loop.

We now define the centroid of a grain as the mean position vector of the contact points around it,

$$\boldsymbol{\rho}_g = \frac{1}{z_g} \sum_{g'=1}^{z_g} \boldsymbol{\rho}_{gg'} \;, \tag{5.1}$$

where z_g is the number of contacts (the coordination number) of grain g and the sum is over all the grains g' that are in contact with it. Similarly, we define the centroid of a cell, $\boldsymbol{\rho}_c$ as the mean position vector of the contact points between the grains that surround it. Let grain g reside on the perimeter of cell c. We define a vector $\boldsymbol{R}_{cg}$ that extends from the centroid of g to the centroid of c. The network formed by the vectors $\boldsymbol{R}$ is conjugate (or dual) to the contact network formed by the vectors $\boldsymbol{r}$ - every r_{cg} has one and only one corresponding vector R_{cg}, Figure 5.2. For GS in mechanical equilibrium under boundary forces only, when 'rattlers' (i.e. grains that do not carry forces in the absence of the body forces) are excluded, the polygons around cells must be convex. This means that the vectors r_{cg} and R_{cg} must cross one another.

We can now use the two conjugate networks to identify and quantify the local micro-structure. Consider the quadrilateral q, shown shaded in Figure 5.3, whose diagonals are the vectors r_{cg} and R_{cg}. Since this quadrilateral is indexed uniquely by given combination c and g, we re-index the vectors in the following by r_q and R_q, respectively. These quadrilaterals are the basic volume elements of the structure and they play a central role in the statistical mechanical formalism described here. They tessellate the plane of the granular assembly perfectly without gaps or overlaps. Due to their importance, these elements were given a name - 'quadrons'. The quadrons can be regarded as the quasi-particles of the granular statistical mechanics, as will become clear below.

Each quadron can be described by a local tensor that characterises its shape

$$(C_q)_{ij} = (r_q)_i (R_q)_j \quad \text{or formally} \quad C_q = \boldsymbol{r}_q \otimes \boldsymbol{R}_q \;. \tag{5.2}$$

The quadron shape tensor C_q is the quantitative descriptor of the local structure. The structure around grain g can be now described by

$$C_g = \sum_{q \in g} C_q \;, \tag{5.3}$$

where the sum is over all the z_g quadrons that belong to grain g. We now note that the volume (area) of the quadron can be obtained from the shape tensor, using the formula

$$\frac{1}{2}\left(C_q - C_q^T\right) = V_q \epsilon \; ; \quad \epsilon \equiv \begin{pmatrix} 0 & 1 \\ -1 & 0 \end{pmatrix}, \quad (5.4)$$

where C_q^T stands for the transposed of C_q. A more elegant form of this relation, which also generalises to 3D, as we shall see below, can be obtained by defining a slightly modified shape tensor

$$C_q' = (\epsilon \cdot r_q) \otimes R_q. \quad (5.5)$$

In terms of this tensor the quadron volume is simply

$$V_q = \frac{1}{2} Tr\left\{C_q'\right\}. \quad (5.6)$$

Since the trace is a linear operation, the volume associated with a grain is the sum of the volumes of its quadrons and therefore

$$V_g = \frac{1}{2} Tr\left\{C_g'\right\} = \frac{1}{2} Tr\left\{\sum_{q \in g} C_q'\right\}. \quad (5.7)$$

The volume of any region Γ in the assembly is therefore the sum of the volumes - grains and the voids that they enclose - of either the grain or the quadron shape tensors enclosed in it, $V_\Gamma = \sum_{g \in \Gamma} V_g = \sum_{q \in \Gamma} V_q$. The number of grains which the region Γ may contain, is arbitrary and, in particular, it could comprise the entire system. Note also that the shape tensor can be associated with every region, $C_\Gamma' = \sum_{g \in \Gamma} C_g' = \sum_{g \in \Gamma} \sum_{q \in g} C_q' = \sum_{q \in \Gamma} C_q'$.

Before we move on, we comment in passing that there may be a degenerate case, when the quadron is a triangle rather than a quadrilateral. This happens when a grain has exactly two contact points, in which case one of the triangles of the quadron - the one contained inside the grain - is 'squeezed' to a line. Then the quadron is comprised only of the triangle made by the two contact points and the cell centroid (see Figure 5.4). The above mathematical definition of the shape tensor, as well as the volume, remain valid.

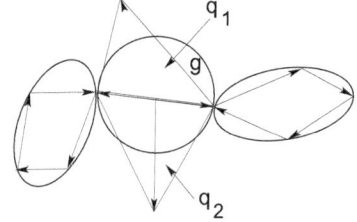

Fig. 5.4 Two degenerate quadrons of a two-contact grain, when the grain centroid happens to be on the vector r_q. The two quadrons of this grain are triangles.

Quantitative local structural description in 3D

The construction in 3D follows the same rationale. First, the space occupied by the GS is tessellated into basic volume elements, the quadrons, as shown in Figure 5.5 and as explained in detail in [26]. The 3D quadrons are generically non-convex octahedra. Then, the structure of every quadron is quantified by the shape tensor

$$C' = (\boldsymbol{\xi} \times \boldsymbol{r}) \otimes \boldsymbol{R}, \quad (5.8)$$

where the vectors $\boldsymbol{\xi}$, $\boldsymbol{r}$, and $\boldsymbol{R}$ are shown in Figure 5.5, $\boldsymbol{\xi} \times \boldsymbol{r}$ is a cross product, and $\otimes$ in equation (5.8) denotes an outer product. In

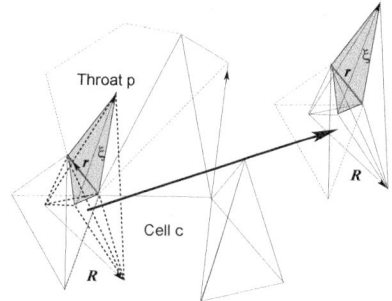

Fig. 5.5 The structural quantification in 3D, as detailed in [26]. The quadron is generically a non-convex octahedron, whose structure is quantified by the shape tensor in equation (5.8).

3D, degeneracy can occur when a grain has only two or three contact points, but the definition of C' remains unambiguous even then. The expression for the volume of the 3D quadron has a similar form to the 2D version, equation (5.6),

$$V = \frac{1}{3!} Tr\{C'\}. \tag{5.9}$$

The quadron method has several advantages:

1. It enables an *unambiguous* quantification of the local micro-structure by the shape tensors C (or C'). This is because all the quadrons are quadrilateral in 2D and octahedral in 3D (albeit occasionally degenerate). Other tessellations, which give rise to different polygonal, or polyhedral, shapes, make it impossible to assign equivalent tensors in a systematic manner.

2. The quadron tessellation has a distinct advantage over Voronoi-based methods [27] in that the quadron construction is based on the contact network and therefore takes account of the physical inter-granular connectivity. In contrast, most Voronoi-based tessellations disregard the physical contacts and are based on proximity of grains. This is crucial to derivations of physical properties that depend sensitively on the connectivity, such as mechanics and fluid transport through the porous granular material.

3. The quadrons are the natural 'quasi-particles' of the statistical mechanics formalism, as will be described later in chapter. In particular, the shape tensor plays a key role by allowing us to write an explicit volume function in the partition function

$$W_d = \frac{1}{d!} \sum_{all\ q} Tr\{C_d'^q\}, \tag{5.10}$$

where $d = 2, 3$ denotes the dimensionality of the system. The use of this method to the statistical mechanical formalism will be discussed in detail below.

4. The shape tensor C plays a key role in the theory of stress transmission in 2D isostatic granular materials. Specifically, its symmetric part comprises the coupling between the structure and the stress, which closes the stress equations (see below).

Structural / topological properties

Arguably, the most important characteristic of the structure of static granular matter is its connectivity. Given a large collection of grains, they can be assembled in mechanical equilibrium in an astronomically large number of configurations, each with its own connectivity network, as defined above. Each of these networks represents a 'graph' of a distinct topology. Thus, the topology of a configuration is one of the most important characteristics of a granular assembly. Here we will review some of the inter-relations between topological characteristics of a structure.

For simplicity, we discuss here only large assemblies of convex, but otherwise arbitrarily shaped, grains. The analysis can be extended readily to non-convex grains, as well as to finite size assemblies. The number of grains will be denoted in the following by N ($\gg 1$) and the assembly is assumed to be in mechanical equilibrium under a set of external compressive forces. For brevity, we disregard body forces, such as arising from gravity or other external fields. This also allows us to ignore 'rattlers', which, under this assumption, do not have any force carrying contact with other particles.

2D assemblies

Starting again from the contact network defined in Section 5.2, let us define the number of grains with exactly two contacts as N_2. The remaining $N - N_2$ grains are called in the following 'regular'. The contact network contains edges - the lines connecting contacts around a grain, and faces - the polygons that are made by the edges. These faces are of two types: the $N - N_2$ polygons that circulate grains (in the clockwise direction) and the N_c polygons that circulate cells (in the anti-clockwise direction), see Figure 5.6. Note that the N_2 two-contact grains make no intra-grain polygon. At the contact point between two regular grains, exactly four edges meet, with two belonging to each grain. At the contact between a two-contact grain and a regular grain, three edges meet and at the contact between two two-contact grains, only two edges meet. For reasons of stability, the latter contacts are extremely rare in mechanical equilibrium and we ignore these in the following. In other words, we assume that a contact between any two grains must involve at least one regular grain.

Every contact point is shared between two grains and, therefore, the number of contacts is $N\bar{z}/2$. The number of contacts connecting three edges is $2N_2$, and therefore the number of all the contacts connecting four edges is $N\bar{z}/2 - 2N_2$. Since every edge is shared between two contacts, it follows that the total number of edges is

$$E = \frac{1}{2}[4 \times (N\bar{z}/2 - 2N_2) + 3 \times 2N_2] = N\bar{z} - N_2 . \quad (5.11)$$

The total number of faces, grain and cell, is $F = N_c + N - N_2$. We now recall Euler's topological relation for a finite graph in 2D [28]

$$V - E + F = 1, \quad (5.12)$$

where V stands for the vertices of the graph, which in our case are the contact points, $V = N\bar{z}/2$. Using equation (5.12) and the above expressions, we can express the total number of cells, N_c, in terms of the number of grains and the mean coordination number:

$$N_c = \left(\frac{\bar{z}}{2} - 1\right) N + 1. \quad (5.13)$$

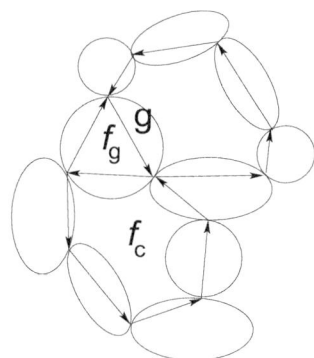

Fig. 5.6 The contact network consists of two types of polygonal faces: those inside grains, such as the triangle f_g inside grain g, and those inside cells (voids), such as the hexagon f_c inside cell c.

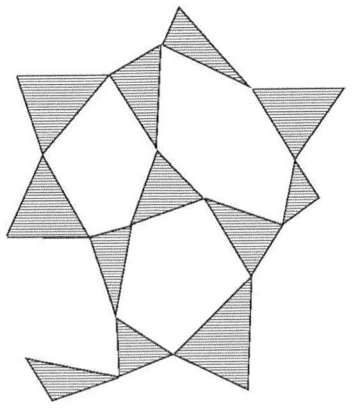

Fig. 5.7 Part of the surface of a cell as seen from a point within the void (see Figure 5.5). The boundary of the cell is made of triangles, which are the facets of the grain-representing polyhedra (shaded). These triangles join at their nodes, which are the intergranular contact points. The triangles enclose non-planar polygons – these are the 'throats' to adjacent cells.

Since $\bar{z} > 2$ in mechanical equilibrium, there are, to a good accuracy, $\frac{\bar{z}}{2} - 1 > 0$ cells per grain [the rightmost term on the right hand side of equation (5.13) is negligible for $N \gg 1$]. This relation is general and it is useful in assessing structural properties of granular assemblies in the plane.

To go beyond total numbers and averages, one needs to study statistical distributions of structural properties. It has been shown recently [29] that using the quadron characterisation, a number of such distributions collapse to a universal form, at least for a family of granular systems. This issue is still under investigation.

3D assemblies

The extension of the structural quantification method to 3D was described in detail in [26] and here we only review some aspects of it briefly. It is based on the same concept as in 2D: using the contact network as a basis to represent grains by polyhedra, tessellate the granular space into elementary quadron volumes and then quantify the shape of each quadron by a shape tensor. For simplicity, we will consider here only structures where each face of every grain-representing polyhedron opens to one cell. More generally, it may happen that two adjacent triangular faces may need to be joined and they both lie on the surface of one cell, but here we will ignore this complication. We also neglect cases of grains with fewer than four contact points, whose fraction is small in mechanically stable GS.

The contact network consists of: (i) edges between contact points; (ii) faces, F, of which there are two types: those belonging to the polyhedra that represent grains, F_g, and those made by the 'throats' between cells, F_t; (iii) cells, C, of which there are also two types: those made by the grain polyhedra, C_g, and those made by the polyhedra surrounding voids, C_v.

We will denote in the following quantities summed over the system with capital letters while lower case letters will be used for local variables. Before we continue, it is important to note that the polyhedral construction around a grain, with every face being a triangle, results in a unique relation between the number of faces, f_g, edges, e_g, and the number of contacts of that grain, z_g:

$$f_g = 2(z_g - 2), \quad (5.14)$$
$$e_g = 3(z_g - 2). \quad (5.15)$$

Looking out in all directions from a vantage point within the void, one then sees a surface made of triangles - the polyhedra facets - and non-planar polygons, which the triangles enclose (see Figure 5.7). The latter are the 'throats' to the adjacent cells. On the void's surface there are f_{gc} triangular facets and f_{tc} throats. The triangles connect at their nodes, which are the contacts of the 3D network. On the void's surface there emanate four edges from every such point - two to each triangle.

It follows that, on this surface, the planar graph has $3f_{gc}/2$ vertices, $3f_{gc}$ edges, and $f_{tc} + f_{gc}$ faces. Using Euler's relation for the 2D closed surface, we then find that the number of throats on the surface is

$$f_{tc} = 2 + \frac{f_{gc}}{2}. \tag{5.16}$$

We can now sum this expression over all the void cells. Every throat is shared between two cells and, therefore, in this summation, each throat is counted twice. It follows that the total number of throats is

$$F_T = \frac{1}{2}\sum_c f_{tc} = \frac{1}{2}\sum_c \left(2 + \frac{f_{gc}}{2}\right) = N_c + \frac{1}{4}\sum_c f_{gc}. \tag{5.17}$$

We now note that the right-most sum gives exactly all the faces of the grain polyhedra, F_g. We can then rewrite this term as a sum over the grains

$$F_g = \sum_c f_{gc} = \sum_g f_g = 2(\bar{z} - 2)N, \tag{5.18}$$

where in the last step we have used equation (5.14). We thus obtain for the total number of throats

$$F_T = N_c + \frac{\bar{z} - 2}{2}N. \tag{5.19}$$

From this expression we can now obtain the cell analogue of the mean coordination number; the mean number of throats per cell

$$\bar{e} \equiv \frac{F_T}{N_c} = 1 + \frac{\bar{z} - 2}{2}\frac{N}{N_c}. \tag{5.20}$$

This expression is yet further evidence that, unlike in 2D, Euler's relation is insufficient to determine the number of void cells or the number of throats independently in terms of the number of grains and the mean coordination number.

All the calculations in this subSection can be extended straightforwardly to systems containing grains with fewer than four contacts. A straightforward extension is also possible to GS whose network description necessitates having grain polyhedra with more than one triangular face residing on the same cell surface. These extensions have not been discussed in detail in the literature of granular science yet, mainly because the 3D structural representation discussed here is relatively recent.

5.3 Stress transmission

The way a material responds to stresses is arguably its most basic macroscopic characteristic. From a young age, we are taught that the state of matter of any material is strongly related to its response to external mechanical stresses: gases need to be held together by external pressure, liquids flow under stress, and solids first deform and then transmit the stress statically. Therefore, understanding stress transmission in GS is

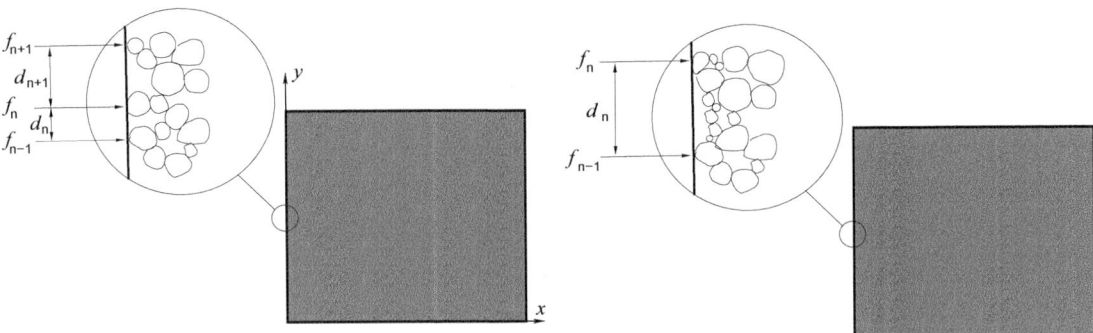

Fig. 5.8 A solid boundary plate, applying stresses on the granular medium, generates localised force sources at intervals d_n. When grains have a typical size (left) the intervals have a typical distance; when the grain sizes are broadly distributed (right) so are the distances between force sources.

essential to determining its state of matter. Furthermore, such understanding is key to an extraordinarily wide range of applications, from prediction of rock, mud, and snow avalanches to construction of stable structures in soils [30].

The main problem with modelling stresses in granular materials is that, although above some length scale they may appear uniform to the eye, the internal stress distribution is anything but. The stress field often exhibits a non-uniform structure in the form of a force networks, observed experimentally since the 1940's [31]. However, only in the 1990's did it became the focus of considerable attention [32, 33]. This phenomenon complicates attempts to construct a continuum stress theory since the filaments that carry the stresses, also known as arches, can be very thin and well separated. In 2D, this phenomenon is best known as 'force chains', which refers to the observation that localised loads on the boundary of the system appear to be supported by narrow force filaments.

It is often convenient to analyse GS loaded by compressive forces at the boundaries with specified boundary stresses. Consider, for example, a flat solid plate pressing the granular material, as sketched in Figure 5.8. The plate applies forces on protruding grains that are directly in contact with it (insets in Figure 5.8). These grains act as point-like force, or stress, sources, giving rise to the narrow stress arches in the medium. The distribution of distances between these localised sources, d_n, is an important factor in determining the medium's stress response. This distribution depends on the grain size distribution: when the grains have a typical size, as in Figure 5.8(left), the distance distribution is relatively narrow and it gives rise to a characteristic length-scale of the order of very few grains. When the grain size distribution is broad, the distribution of the distances d_n can also be broad, involving a large number of grains not in contact with the plate, as sketched in Figure 5.8(right). In either case, the total stress σ_{ix} over this boundary is the mean of the ith component of the force sources per unit boundary area.

Changing even slightly the external stress applied to the system or moving the boundaries, the force sources change and the network generated by them reorganises. This reorganisation involves arches disintegrating and new arches re-consolidating in other places, as has been demonstrated beautifully in [34]. It is important to note that the configuration of the stress network determines not only where failure occurs, but also the dynamics of the material as it deforms. As we will discuss in more detail below, this implies that determining the stress field must take into consideration the dynamics and the two must be modelled self-consistently.

Understanding stress transmission in GS is particularly important as granular matter transforms from fluid to solid - a process commonly called the jamming transition. In most cases of interest to soil mechanics, civil engineering and many other fields, the solid resulting from this transition has an irregular disordered internal micro-structure. We focus here on such systems, mainly because liquid-solid transitions in ordered systems (crystals) are well understood and covered in many textbooks. As will be discussed below, a stress theory for the jamming point is key to understanding stresses in granular matter in general. However, a prerequisite to constructing a general theory of stresses in granular materials is an understanding of the conditions for stability in these systems, which we proceed to discuss next.

To identify the criterion for stability of a GS, we need to examine how it becomes unstable under external force loading. A GS is said to yield and flow if, in response to a small external load, its internal grain configuration changes irreversibly. That is, when after rearrangement the grains do not return to their initial configuration upon removal of the load.

In the context of granular matter it is instrumental to ask the opposite question: under what conditions would a granular system remain stable under the application of an external force field? The set of boundary stresses that do not initiate failure flow are said to be below the yield threshold. A GS will remain stable when the inter-granular forces, which develop in response to the applied load, can be supported by the internal structure and the system remains in mechanical equilibrium. The significance of understanding and predicting the threshold to yield cannot be overemphasised [1, 3, 4]. Most of our buildings, bridges, and structures are built on soils, whose stability against failure is crucial. Lives and resources are lost due to poor predictability of avalanches and landslides. Even jogging on the beach involves making almost unconscious assumptions about the solidity of the sand under our feet. For clarity, we discuss in the following only assemblies of rigid non-compliant grains. In such systems there are no intra-granular deformations and structural changes can only take place by grain rearrangements.

The yield threshold concept is complementary to the jamming point. The former describes the liquefaction of a GS from a solid state while the latter describes the solidification, or jamming, from a fluid state. In the absence of hysteretic effects, both these processes should occur at the same state of the system. The jamming transition can be induced in a variety of ways: by bringing a flow to a halt under external stresses, by compression, and by increasing the density of colloidal suspensions and much work has focused on the transition, e.g. [35–41]. At the grain-scale, the emergence of mechanical stability at the jamming point is due to an increase in the number of force-supporting contacts between grains. This number per grain, $\bar{z}$, is known as the mean coordination number and is central to the understanding of granular mechanics. Studies of the relations between the coordination number distribution and the state of granular matter dates back to Bernal and Mason [42]. The criterion for stability is that the granular structure possesses sufficiently many inter-granular contacts supporting forces, $\bar{z} \geq z_c$, where z_c is a threshold value.

The jamming transition point corresponds to $\bar{z} = z_c$, at which state the GS is at the edge of stability, or marginally rigid [43, 44]. The value z_c depends on the dimensionality, on the size of the GS, and on granular characteristics, most notably, the inter-granular friction coefficient. To an extent, as will be detailed below, it also depends on the grain shapes. In fact, one can parametrise the stability of GS by the difference between the number of inter-granular force components, N_u (commonly called 'unknowns' or 'degrees of freedom'), and the number of conditions required to resolve for them, N_e. This quantity, which we call the determinacy parameter (DP) J, can be used to determine whether a finite region of N particles, within a larger system, is stable, marginally stable, or unstable,

$$J \equiv \frac{N_u - N_e}{N} \ . \tag{5.21}$$

The structure is stable, marginally rigid, or unstable, when J is positive, zero, or negative, respectively. z_c is evaluated by using the condition for marginal stability, $J = 0$. This is done in the following for GS in any dimensionality, $d \geq 2$, comprising either grains of arbitrary convex shape or spherical. For each of these cases we consider both finite and zero friction coefficients between grains. We also present an evaluation of z_c for two cases that are not discussed much in the literature: (i) slightly non-convex grains, where there is a finite probability of neighbouring grains touching at more than one contact point and (ii) cases when some of the balance conditions are redundant. In the following discussion we address GS loaded only by compressive boundary forces.

Conditions for marginal rigidity

Case I. Convex frictional grains in d dimensions

In mechanical equilibrium, each of the N grains of the GS must satisfy Newton's force and torque balance conditions. The force balance is a vector equation, giving d equations per grain. The number of torque balance conditions in d-dimensions is the same as the number of axes of rotation, $d(d-1)/2$. The total number of equations is then

$$N_e = N\left[d + \frac{d(d-1)}{2}\right] = \frac{Nd(d+1)}{2}. \tag{5.22}$$

The unknowns, which these equations need to solve for, are the forces at the inter-granular contact points. Let us denote the number of contacts by N_c. After taking Newton's third law of action and reaction into consideration, at every contact there are d components of the contact force vector to solve for and in total the number of unknowns is $N_u = dN_c$. If the grains are all convex, a grain can contact a neighbour only at one point.

Now, in the evaluation of the mean coordination number one counts over all the contacts around every grain and averages over the grains, which means that, except for boundary contacts, each contact is counted exactly twice. It follows that the mean coordination number is

$$\bar{z} = \frac{2N_c}{N} \tag{5.23}$$

and the number of unknowns is then

$$N_u = \frac{N\bar{z}}{2}d. \tag{5.24}$$

The numbers of unknowns and equations are therefore equal when

$$\bar{z} = z_c = d+1. \tag{5.25}$$

Thus, the condition that an assembly of frictional convex grains be in mechanical equilibrium under boundary compressive loads is that $\bar{z} \geq z_c$. When $\bar{z} = z_c$ the assembly is said to be marginally rigid, i.e. it is on the verge of becoming unstable. In two and three dimensions, $z_c = 3$ and 4, respectively.

Condition in equation (5.25) is accurate for infinitely large assemblies. In finite packs, the boundary grains naturally have lower coordination numbers, which must be taken into consideration. However, the correction due to the boundary effect, $\delta \bar{z}$, is proportional to the ratio of boundary to bulk grain numbers. In an assembly of $N = L^d$ grains, this ratio is of order L^{d-1}/L^d and the boundary correction decreases with size as $\delta \bar{z} \sim O\left(N^{-1/d}\right)$. To evaluate the boundary correction for moderate assembly sizes, a careful analysis is required, which must also take into consideration the number of forces acting on the boundary. This analysis is not difficult and we will give an example of such a calculation at the end of this section.

An alternative derivation of condition (5.25) is by considering, instead of unknowns, the 'determination' of forces via elimination of dynamic degrees of freedom upon becoming static. Consider a piling experiment, where the grains are deposited very gently onto a growing pile. On coming into contact with the pile, they come to rest without slipping and without disturbing the rest of the pile. In d dimensions, any solid, and in particular both the falling particle and the consolidated pile, has $d(d+1)/2$ degrees of freedom: motion along any of the d directions plus rotation around the $d(d-1)/2$ axes of rotation. Thus, prior to the settling, the combined system of grain and pile has $d(d+1)$ degrees of freedom. After the settling process has ended there is only one solid body - the new consolidated pile - and therefore only $d(d+1)/2$ degrees of freedom. It follows that $d(d+1)/2$ degrees of freedom have been used to satisfy $d(d+1)/2$ constraints. From these conditions we can solve for the forces between the recently-arrived grain and the pile. Again, using Newton's third law of action and reaction, at every contact there are only $d/2$ constraints to satisfy, giving $d+1$ unknowns per grain to solve for. Thus, when there is only one point of contact between grains, the inter-granular forces can be determined if there are on average $d+1$ contacts per grain. While the method of evaluating degrees of freedom versus constraints is equivalent to the one described previously, it seems to us that counting equations and unknowns is more intuitive and we shall therefore use it for our analysis.

Case II. Convex frictionless grains in d dimensions

To evaluate J for assemblies of frictionless grains, we note that the number of equations is exactly the same as in relation (5.22) in the previous case. Each grain needs to satisfy d force balance and $d(d-1)/2$ torque balance conditions. The main difference in the present case is that, at every contact point, the force must be normal to the $(d-1)$-dimensional plane tangent to both grains at the contact. Consequently, given the structure and the grain geometries, we know the directions of all the contact forces. The only unknowns are then the inter-granular force magnitudes. This leaves one unknown at every contact point and altogether N_c unknowns. The marginal rigidity condition then becomes

$$N_u = N_c = \frac{Nz_c}{2} = \frac{Nd(d+1)}{2}, \qquad (5.26)$$

which gives us the following condition for the marginal rigidity mean coordination number,

$$z_c = d(d+1). \qquad (5.27)$$

In two and three dimensions, $z_c = 6$ and 12, respectively.

Case III. Spherical frictionless grains in d dimensions

Consider now ideal systems of frictionless spheres. As in the previous case, the lack of friction dictates that only the magnitudes of the inter-granular force at the contacts are unknown, namely, $N_u = \bar{z}/2$. Since the grains are perfect spheres, the contact forces, which are oriented normal to the grain surfaces, apply no torque moments to the grains. This has a crucial effect on the above argument - once the force balance conditions are solved, the torque balance equations are satisfied automatically. In other words, the torque balance equations are redundant and $N_e = Nd$.

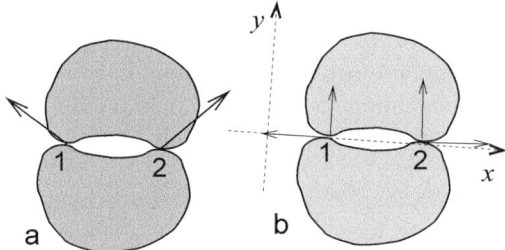

Fig. 5.9 (a) Two slightly non-convex grains, with a double contact at points 1 and 2; (b) decomposing the two contact forces to components, the magnitude of the x-components cannot be determined by the balance equations This, however, does not hinder the definition of the stress around this grain.

Equating N_e with the number of unknowns, N_u, we obtain the condition for marginal rigidity in these idealised systems,

$$z_c = 2d. \qquad (5.28)$$

In two and three dimensions, $z_c = 4$ and 6, respectively. Indeed, in numerical simulations of frictionless hard spheres and discs, of which there are many, these mean coordination numbers are observed to very good accuracy [36, 45]. It should be noted that in real experiments with spheres, the particles are never perfectly spherical. At the end of this Section we comment on the effect that this introduces to the above counting argument.

Case IV. Slightly non-convex frictional grains in d dimensions

When grains are non-convex, there is a finite probability that neighbouring particles would make contact at more than one point. If the non-convexity is small, as, for example, in kidney beans, then two objects can contact at most at d points in d dimensions. It is useful to understand how the above analysis can be extended to such assemblies, both because these are common in most particulate systems and because this analysis is also valid for objects with straight facets. In the latter systems, a facet-to-facet contact can be considered as a simultaneous contact of d points. In the following, we illustrate the extension to double contacts in two dimensions (Figure 5.9).

Consider an assembly of N grains, making N_c contacts. Suppose that a fraction of these, ω_2, are double, where each occurrence of a double contact is counted once. The number of equations is $N_e = Nd(d+1)/2$, as in case I, but the number of unknowns is different and needs to be evaluated carefully. A one-point contact transmits one inter-granular force vector, but the two points of the double contact have two such forces (Figure 5.9). Therefore, a double contact provides four unknowns in two dimensions. It is now useful to note that the main conclusion of the analysis in case I is that, for marginal rigidity, every contact must have, on average, $3/2$ unknowns. This means that, of the four unknowns, only three can be resolved simultaneously! A similar argument, leading to the same conclusion is the following. Consider a grain coming to rest gently against an already stationary pile and making a double contact with one of the grains of the pile. Each incoming grain has three degrees of freedom - translation in the x- and y-directions and rotation. Therefore, upon reaching a stable position, only three of the four unknown force components can be resolved. Thus, double contacts automatically exclude full static determinacy.

On the face of it, one might then conclude that such systems cannot be isostatic. But this need not necessarily be the case. Let us consider two such non-convex grains, a and b, in mechanical equilibrium and let us orient the coordinate axes such that one of them, say x, is aligned along the line joining the two contacts, as shown in Figure 5.9. In this coordinate system, the two x-components of the four force components must be equal and opposite. With three equations, we can then resolve three of the four unknowns and let us leave unresolved the magnitude of this x-component. The question is whether or not the indeterminacy of this component prevents determination of the stress field around these two grains. To answer it, consider a region Γ, surrounding a and b. The stress in this region can be written as

$$\sigma_{ij} = \frac{1}{V_\Gamma} \sum_{g \in \Gamma} \sum_{g'} r_{gg',i} f_{gg',j} , \qquad (5.29)$$

where $r_{gg',i}$ and $f_{gg',j}$ are, respectively, the ith and jth components of the position vector $\boldsymbol{r}_{gg'}$ at the contact between g and g' and the contact force there. Let us examine explicitly the terms in the sum that include the aforementioned x-component:

$$\delta\sigma_{i,x} = \frac{1}{V_\Gamma} \sum_{i=x,y} \left(r_{ab1,i} f_{ab1,x} + r_{b1a1,i} f_{b1a,x} + r_{ab2,i} f_{ab2,x} + r_{b2a1,i} f_{b2a,x} \right)$$

$$= \frac{1}{V_\Gamma} \sum_{i=x,y} \left(r_{ab1,i} - r_{b1a,i} - r_{ab2,i} + r_{b2a,i} \right) f_{ab1,x} = 0, \qquad (5.30)$$

where $r_{ab\alpha,i}$ is the i component of the vector from the centroid of grain a to its α-contact ($\alpha = 1, 2$) and the sum over the vectors vanishes because they form a closed loop. Thus, the terms containing the equal and opposite force component $f_{ab,x}$ cancel out and the magnitude of the unresolvable component does not affect the stress field even in the smallest region around the double contact. This argument, presented originally in [44] without a proof, leads to two conclusions. One is that the local stress field can be determined even if the inter-granular forces at double contacts are not completely determinate and, therefore that in this case the stress state can still be isostatic. The other is that the mean coordination number required for the marginally rigid state is higher due to the multiple contacts. Recalling that we count each occurrence of a double contact only once, the marginal rigidity value is

$$z_c = 3 + w_2 . \qquad (5.31)$$

Indeed, in a piling experiment, conducted in two dimensions with slightly non-convex grains, it has been found that the mean coordination number converges exactly to this value in the marginally rigid state [44].

General comments and extensions

The above considerations are based on a number of assumptions. One is that the granular assembly is sufficiently large so that the corrections due to the boundary grains, whose mean coordination is lower than that of the grains in the bulk, negligibly affect the total mean coordination number. However, it is straightforward to extend them and calculate the exact value of z_c when the ratio of boundary to bulk grains is taken explicitly into consideration. We will not carry out this exercise here as it does not add much to the conceptual understanding of marginal rigidity.

Another issue, which has not been discussed much in the literature, is the potential degeneracy of the equations. A good illustration of such degeneracy is assemblies of frictionless spheres (case III above). Recall that in that case all the contact forces are directed exactly toward the centres of the grains and, as a result, apply no torque moments on the grains, which in turn makes the torque balance equations redundant. Yet, there is no reason that such a situation could not arise in assemblies of other types of grains. When the directions of all the forces around a specific grain happen to coincide at a point, such as illustrated in Figure 5.10a, the torque balance equations for that grain are redundant. Suppose that there are N_r such grains in an assembly of frictional grains in d dimensions. Then the total number of equations is

$$N_e = (N - N_r)\frac{d(d+1)}{2} + N_r d, \tag{5.32}$$

while the number of unknowns remains

$$N_u = \frac{1}{2} N z d. \tag{5.33}$$

It follows that, for marginal rigidity, the mean coordination number is

$$z_c = d + 1 - (d-1)\frac{N_r}{N}. \tag{5.34}$$

Note that this value is *lower* than the value for the non-redundant system discussed in case I. A similar analysis for ideally frictionless grains of arbitrary shapes gives

$$z_c = d\left[d + 1 - (d-1)\frac{N_r}{N}\right]; \tag{5.35}$$

again, smaller than the value in the non-redundant case. In the case of frictionless spheres all the grains possess such redundancy, i.e. $N_r = N$, and relation (5.35) gives exactly the same condition as equation (5.28).

However, these calculations are idealised and taking into consideration the redundancy is problematic. In real systems in mechanical equilibrium, exact redundancy is possible only when a grain has exactly two force-carrying contacts, in which case they must be equal, opposite and align along the line between the contacts. For grains with three contacts or more, it is impossible to determine the force directions in general with

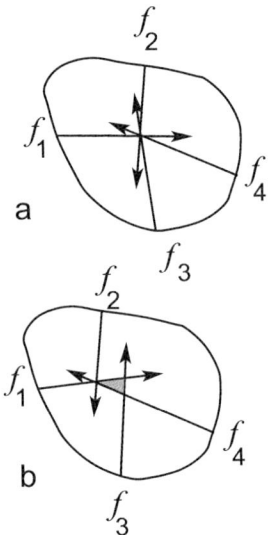

Fig. 5.10 (a) The redundancy of the torque balance equation in 2D - the contact force vectors coincide at a point. (b) In real measurements there is always a finite, however small, 'region of coincidence' (shaded), below which coincidence cannot be determined.

absolute accuracy and, in particular, whether those coincide exactly at a point. At most one can say that the forces apply very small torques to the grain because the 'region of coincidence' (the shaded region in Figure 5.10b) is small. Consequently, except for two-contact grains, the equations are never really redundant - they are at most negligible. It follows that the value of z_c depends to some extent on the level of accuracy of the measurement of forces; the more accurate the measurement, the more torque balance equations need to be included and the higher the value of z_c. This sounds strangely like an 'uncertainty principle' - the more accurate the measurement the more equations one needs to determine the stresses. This 'uncertainty principle' is also a manifestation of the inherent relation between the structure on the granular level and the stress field that the assembly supports - an issue that will be discussed below.

In conclusion, a sufficient condition for a granular assembly of convex grains in mechanical equilibrium to be marginally rigid is that it is generically statically determinate. For non-convex grains, a limited indeterminacy is also possible without affecting the marginally rigid state. The fundamental necessary condition for marginal rigidity is that the stress field (not the force field) be determined from statics alone. These conditions then translate into one requirement - that the mean coordination number has a specific value, z_c, which depends on the dimensionality, the grain characteristics (shape, friction), and on the size of the granular assembly. In the following we review the current state of affairs in the continuum modelling of the stress transmission in such marginally rigid assemblies.

As we shall see below, the equations of stress transmission in marginally rigid GS are different than those of conventional solids. It is important to remember that different stress transmission implies different responses to external loading and hence, by definition, *different states of solid matter*. The marginally rigid state lies between the fluid and the traditional solid states. However, we will see later on that the existence of this new state also affects the behaviour of GS that are close to it. The proximity to this state is often parameterised by the difference of the mean coordination number from z_c. However, a more accurate measure of it would be by the determinacy parameter J, given in expression (5.21).

How close a system is to the jamming transition, which generically takes place at the marginally rigid state, is important because in the vicinity of this state GS exhibit a range of phenomena associated with phase transitions. These include very long relaxation times, cooperative dynamics with appearance of long spatial correlations and absence of structural changes as the the transition is crossed. These phenomena are being studied and modelled from different angles and in several communities, interested in glass forming liquids, granular media, foams, colloids, gels, as well as in rigidity percolation. It is hopeless to review all this literature in one chapter and we only focus on attempts at developing a fundamental theory of stress transmission in GS at, and near, this marginally rigid state.

Isostaticity theory

The appearance of arches, force chains and generally nonuniform stress fields is one of the most spectacular phenomena in granular matter [31, 32]. Many beautiful experimental images of this phenomenon are in the literature, a large number of which originate from the Behringer laboratory [33, 34]. This phenomenon can be reproduced in numerical simulations on the granular scale [35, 36, 39–41, 46–59].

To understand the mechanical behaviour of consolidated granular matter it is essential to make the link between the transmission of forces on the granular level and a macroscopic description of a continuous stress field. While it is straightforward to write down the equations that govern the discrete inter-granular forces, the translation of these equations into a set of continuous differential equations for the stress field has proved elusive.

An empirical model to describe this phenomenon, named isostaticity theory (IT) was first proposed, in the mid 1990's [60–62]. The proposal underlying IT is that, to model stresses that are concentrated along narrow paths, the stress field equations must be hyperbolic. The reason is that solutions of hyperbolic equations 'propagate' along characteristic paths and these could be identified as the observed force chains.[4] This also means that, strictly speaking, IT is valid only for statically determinate, or isostatic, structures.

Statically determinate structures are those where internal forces can be resolved from statics alone, namely, from equations of force and torque balance only. A familiar example from high school textbooks is that of a ladder of a given weight standing on a rough floor and leaning against a frictionless wall. In this problem there are three force components that keep the ladder in place: the normal force applied to the ladder by the wall and the normal and tangential components of the force applied to the ladder by the floor. These three force components can be resolved by using two conditions of force balance and one of torque balance. Similarly, statically determinate granular assemblies contain a large number of internal forces, but they can all be resolved because there is exactly the right number of equations to solve for them. As discussed in the previous section, the only condition that a GS be statically determinate is that its mean coordination number $\bar{z}$ have a particular value, z_c. It is important to realise that static determinacy obviates the need for stress-strain constitutive relations. For example, in the case of the ladder we do not need to know anything about its elastic moduli, nor about those of the floor or the wall. This means that conventional approaches that resort to such relations, such as elasticity theory, use redundant information. Indeed, elasticity theory is hard pressed to explain the phenomenon of force chains and the highly non-uniform stress fields without resorting to elaborate mechanisms of anisotropic organisation of the bulk elastic moduli. The usefulness of IT to general granular materials has been debated much in the literature. In the next subsection, we will present an argument for the relevance of

---

[4] A familiar illustration of a hyperbolic equation is the wave equation in $1+1$ space and time dimensions,

$$\frac{\partial^2 f}{\partial x^2} - \frac{1}{c^2}\frac{\partial^2 f}{\partial t^2} = 0,$$

whose solutions are any functions of $x - ct$ (forward waves) and $x + ct$ (retarded waves). The latter are often discarded as unphysical because they violate causality. In the $x - t$ plane, these solutions propagate along the two characteristics $x \pm ct$ and are zero elsewhere.

IT to a stress theory of general GS.

The empirical theory of the 1990's was put on a firmer basis in 2002, when its fundamental stress equations were derived from first principles on the scale of very few grains [24,63]. The upscaling of those equations to the continuum posed a difficult coarse-graining problem, which was eventually resolved by employing a specialised technique [64] which relied on some special anti-correlations of the granular structure [24]. This coarse-graining technique lies outside the scope of this review and we shall not discuss it here. IT is well developed in 2D and we shall therefore focus on such systems. An extension of the first-principles derivation of the theory to 3D has not emerged in the literature yet.

Isostaticity theory in Cartesian coordinates

In 2D Cartesian coordinates, the stress field equations comprise four equations for the four components of the stress tensor σ. Three of these are balance conditions: two of force and one of torque.

$$\partial_x \sigma_{xx} + \partial_y \sigma_{xy} = g_x \quad \text{(force balance in } x\text{)} \quad (5.36)$$

$$\partial_x \sigma_{xy} + \partial_y \sigma_{yy} = g_y \quad \text{(force balance in } y\text{)} \quad (5.37)$$

$$\sigma_{xy} = \sigma_{yx} \quad \text{(torque balance)} \quad (5.38)$$

In these equations, $\partial_x, \partial_y \equiv \partial/\partial x, \partial/\partial y$ and the vector $\vec{g} = (g_x, g_y)$ is an external force field, which may or may not include body forces. For simplicity, we will ignore body forces in this presentation, but these can be readily included without qualitatively affecting the discussion.

The fourth equation must be a closure relation and it should include constitutive information on the specific medium. Yet, we have seen that IT cannot depend on the elastic properties of the medium. The question is then: what is the constitutive information that could be used for the closure relation? Casting our mind back to the case of the ladder, we can see that the only system-specific information we have is the angles that the ladder makes with the floor and the wall, as well as its length, in other words, the system structure. The ladder's elastic properties do not affect the calculation of the forces. Extending this insight to generally statically determinate GS, we conclude that the only constitutive properties one can use for closure must be related to the granular structure.

The initial work empirical offered such a closure [60], relating the stress components linearly to one another by coefficients that represented structural properties

$$Tr\{Q \cdot \sigma\} = 0 \quad \text{(stress − structure relation)}. \quad (5.39)$$

This form was later derived explicitly from first-principles in [24], where the coefficients were shown to relate directly to local micro-structural characteristics on the grain scale. The specific relation in [24] is

$$p_{xx}\sigma_{yy} + p_{yy}\sigma_{xx} = 2p_{xy}\sigma_{xy}, \quad (5.40)$$

where P is a symmetric fabric tensor. On the grain level, this tensor is derived from the shape tensor described in Section 5.2, $P_g = \left(C_g + C_g^T\right)/2$ [24, 63]. By defining

$$Q = \epsilon P \;\; ; \;\; \epsilon = \begin{pmatrix} 0 & 1 \\ -1 & 0 \end{pmatrix}, \tag{5.41}$$

the stress-structure relation (5.40) transforms into the form of equation (5.39). In the following we will use both P and Q.

The fundamental derivation elevated IT from an empirical to a first-principles theory. Furthermore, it provided a geometrical interpretation for the coefficients q_{ij} (or p_{ij}) that couple the structure to the stress. Specifically, the fabric tensors, P and Q quantify the local 'chirality' of grains, or how much they are rotated relatively to a global zero mean. Indeed, the closure relation was derived from considerations of local torque balance.

Equation (5.39) represents a coarse-grained version of the original equation of the same form on the granular level. It has been recognised already in the original paper [24] that the coarse-graining of this relation is not straightforward for systems of rough grains. The reason is that the values of the coefficients p_{ij} and q_{ij} fluctuate around a *zero mean*. This means that a simple area average of their values over increasing volumes tends to zero inversely proportionate to the linear size of the averaged area. Thus, the coarse-graining from P_g to the continuous P should be done very carefully. This difficulty was resolved eventually by the introduction of a specialised coarse-graining procedure that takes advantage of local anti-correlations between the fabric tensors of neighbouring grains [64]. Interestingly, the difficulty with the coarse-graining can be much more readily resolved in systems of frictionless grains [24]. While the coarse-graining problem and its resolution are quite interesting, it is not our aim to review it here.

For the set of equations (5.36)-(5.39) to be hyperbolic, the determinant of P (and equivalently of $\tilde{Q}$) must be negative [63]. It has been argued that realistic granular structures would organise themselves such that $det\{P\} < 0$ [63], but that there may occur regions wherein $det\{P\} > 0$. This means that pockets of elliptic behaviour may be observed even though the material is perfectly statically determinate on the granular scale. In the following, for simplicity, we do not consider such cases and assume that the equations are hyperbolic everywhere. When the fabric tensor is uniform, p_{ij} =const, the solutions of these equations in rectangular domain are not difficult - they are straight characteristics, very much like the solutions of the 1D wave equation. In other words, given a localised compressive stress at the boundary, the stress is supported by two 'stress chains' whose directions are determined uniquely by the fabric tensor. These stress chains act like arches or bridges, transferring the localised load sideways.

However, when the coefficients depend on spatial position, as is commonly the case in disordered granular matter, the equations become much less straightforward to solve. A comprehensive treatment of the isostatic stress solutions in these general cases has been presented in [65, 66] and we briefly review this analysis next.

To analyse the IT stress equations, it is useful to rewrite them first in a more convenient form. We review here the general case when all the p_{ij} are finite.[5] Defining

$$q_{ij} = \frac{p_{ij}}{p_{xx}} \quad \text{and} \quad \boldsymbol{u} = \begin{pmatrix} \sigma_{xx} \\ \sigma_{xy} \end{pmatrix}, \qquad (5.42)$$

and substituting equation (5.40) in equations (5.36) and (5.37), one obtains

$$\boldsymbol{u}_x + (A\boldsymbol{u})_y = \boldsymbol{g}, \qquad (5.43)$$

where

$$A = \begin{pmatrix} 0 & 1 \\ -q_{yy} & 2q_{xy} \end{pmatrix}. \qquad (5.44)$$

The eigenvalues of the matrix A, given by,

$$\lambda_{1,2} = q_{xy} \pm \sqrt{q_{xy}^2 - q_{yy}}, \qquad (5.45)$$

play an important role in the analysis of the solutions. It can be shown that since $\det(P) < 0$, the eigenvalues $\lambda_{1,2}$ are distinct and real and, therefore, that the system of stress field equations is indeed strictly hyperbolic. Such systems require some care in the formulation of the boundary conditions to ensure well-posedness, but this issue is well-addressed in the literature [67].

Diagonalising the matrix A is achieved via $A = Y\Lambda Y^{-1}$, with

$$Y = \begin{pmatrix} 1 & 1 \\ \lambda_1 & \lambda_2 \end{pmatrix}, Y^{-1} = \frac{1}{\lambda_2 - \lambda_1}\begin{pmatrix} \lambda_2 & -1 \\ -\lambda_1 & 1 \end{pmatrix}, \qquad (5.46)$$

$$\text{and } \Lambda = \begin{pmatrix} \lambda_1 & 0 \\ 0 & \lambda_2 \end{pmatrix}. \qquad (5.47)$$

The characteristic variables of equations (5.43) are ω_1 and ω_2, which can be expressed in vector form

$$\boldsymbol{w} = \begin{pmatrix} \omega_1 \\ \omega_2 \end{pmatrix} = Y^{-1}\boldsymbol{u} = \frac{1}{\lambda_2 - \lambda_1}\begin{pmatrix} \lambda_2\sigma_{xx} - \sigma_{xy} \\ -\lambda_1\sigma_{xx} + \sigma_{xy} \end{pmatrix}. \qquad (5.48)$$

These variables are particular combinations of the original stress components that take into consideration the local structure through the eigenvalues λ. We have finally arrived at our destination - in terms of these, the stress equations can now be cast in the form

$$\partial_x \boldsymbol{w} + \Lambda\partial_y \boldsymbol{w} = \boldsymbol{h} + (B - \Lambda_y)\boldsymbol{w}, \qquad (5.49)$$

where

$$\boldsymbol{h} = Y^{-1}\boldsymbol{g} \text{ and } B = -Y^{-1}(Y_x + Y_y\Lambda),$$

which can be seen to be singular

$$B = \frac{-1}{\lambda_1 - \lambda_2}\begin{pmatrix} \lambda_{1,x} + \lambda_1\lambda_{1,y} & \lambda_{2,x} + \lambda_2\lambda_{2,y} \\ -\lambda_{1,x} - \lambda_1\lambda_{1,y} & -\lambda_{2,x} - \lambda_2\lambda_{2,y} \end{pmatrix}. \qquad (5.50)$$

---

[5] When one of these terms vanishes, the analysis becomes much simpler and we will comment on these special cases at the end of this section.

In terms of the characteristic variables, the stress tensor is

$$\sigma = \omega_1 \begin{pmatrix} 1 & \lambda_1 \\ \lambda_1 & (\lambda_1)^2 \end{pmatrix} + \omega_2 \begin{pmatrix} 1 & \lambda_2 \\ \lambda_2 & (\lambda_2)^2 \end{pmatrix}. \tag{5.51}$$

This expression makes it possible not only to write explicitly the stress field everywhere inside the medium, as is illustrated below, but also to gain insight into the effect of the local structure on the solution in the following way.

The vector h, on the right hand side of equation (5.49), is independent of w and Λ is diagonal. This means that ω_1 and ω_2 are coupled only through the off-diagonal terms of the matrix B. When the fabric tensor P is uniform and independent of the local position, B vanishes altogether and the two equations for the characteristic variables are independent,

$$\partial_x \omega_i + \lambda_i \partial_y \omega_i = h_i \, , \quad i = 1, 2 \, . \tag{5.52}$$

This corresponds to the special case treated in most of the literature and it is the direct reason for theoretical derivations of straight stress chains [60–63].

This concludes the general solution. It only remains to specify the boundary conditions. For the problem to be well-posed one must be careful with their specification. As is well known for hyperbolic equations [67], the conditions (known as the Dirichlet conditions) for the stresses must be such that each characteristic variable is prescribed at exactly one boundary point. For example, if the domain is $0 \leq x \leq X$, $-\infty < y < \infty$ we can use a boundary condition of the form

$$\vec{u}(0, y) = \boldsymbol{f}(y). \tag{5.53}$$

As promised, we now need to consider the special case when p_{xx} vanishes, in which case we cannot define the parameters q_{ij}. From equation (5.40) we see that, when this happens, the stress components σ_{xx} and σ_{xy} are proportional to one another. Then they can be solved directly from equation (5.36). Substituting this solution in equations (5.37), one then solves for σ_{yy}. Similarly, the solutions are much simpler to obtain when either one of the other components of p_{ij} vanishes or when both p_{xx} and p_{yy} vanish. In all these cases, the resulting solutions are qualitatively the same as the one described above, namely, comprising stress chains. Note that p_{xx} and p_{xy} cannot vanish simultaneously without violating the assumption that $\det(P) < 0$.

Stress response to uniform fabric tensors

To gain insight into the theory, it is useful to first examine the solutions to equations (5.49) when the fabric tensor P is uniform. In this special case both B and Λ_y vanish and the equations decouple, as discussed above, into the two separate equations (5.52). The characteristic paths can be parameterised by an 'arc-length' variable s via

$$\frac{dx}{ds} = 1 \, , \quad \frac{dy}{ds} = \lambda_i \, , \quad i = 1, 2 \, . \tag{5.54}$$

The local direction of the ith characteristic path (i.e. the local gradient) is given by the eigenvectors of A, $(1, \lambda_i)$ and we have

$$\frac{d\omega_i}{ds} = h_i , \quad i = 1, 2 , \tag{5.55}$$

along the characteristic paths. For constant fabric tensors and in the absence of body forces, h_i are constant, which gives rise to straight characteristic paths.

For illustration, suppose that the granular medium occupies the semi-infinite plane $x \geq 0, -\infty \leq y \leq \infty$. The characteristic variables $\boldsymbol{w}$ can be calculated at each point (x', y') by integrating equations (5.55), each along its unique characteristic path,

$$\omega_i(x', y') = \omega_i(0, y' - \lambda_i x') + \int_0^{x'} h_i(s) ds , \quad i = 1, 2 \tag{5.56}$$

and the stresses σ_{xx} and σ_{xy} at (x', y') are given by

$$\boldsymbol{u}(x', y') = \begin{bmatrix} \sigma_{xx}(x', y') \\ \sigma_{xy}(x', y') \end{bmatrix} = Y\boldsymbol{w}(x', y') . \tag{5.57}$$

The remaining stress is determined through equations (5.38) and (5.40).

The physical interpretation of this solution is the following. A combination of stresses, applied at any point along the boundary, is supported solely by two characteristic paths that emanate from that point and 'propagate' into the medium. The stresses are localised along these paths, forming an arch. It is the formation of such arches that leads to observations of a minimum in the stresses under the apex of a conical pile of grains [68].

Since the stress equations are linear, the solution to an array of stress sources, distributed along the $x = 0$ boundary, is a superposition of the individual solutions to every source, with every characteristic path being traced back to its point of origin, using equation (5.56). For example, this means that, if the load is distributed along a finite region on the $x = 0$ boundary, say between $y_0 - w/2$ and $y_0 + w/2$, then from this region there will propagate into the medium two stress strips, each of width w.

It should be noted that, because q_{yy} is negative, the eigenvalues of A are always of opposite signs. This means that, in a Cartesian coordinate system in which the loaded boundary is parallel to the y-axis, the gradients of the stress paths have opposite signs, one going upwards and the other downwards. In particular, when $q_{yy} = -1$, the matrix A is symmetric, its eigenvectors are orthogonal and, therefore, the stress paths are orthogonal to one another.

Figure 5.11 shows the σ_{xx} component of the stress as a response to a Gaussian load applied at the $x = 0$ boundary around $y = 0$. The stress clearly localises on the characteristic paths, retaining a local Gaussian profile.

The sensitivity of the stress response to the local structure is best seen by considering the principal axes and principal stresses of the stress

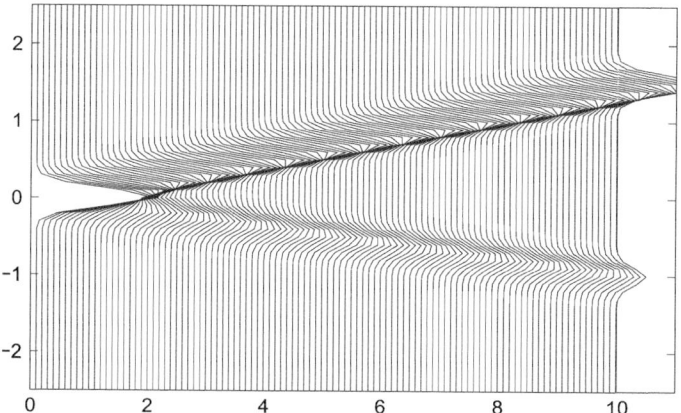

Fig. 5.11 Two stress chains of σ_{xx}, initiated by a narrow Gaussian σ_{xx} at the boundary. The chains remain straight throughout because the fabric tensor P is uniform and they continue all the way to the boundary.

tensor, which are, respectively, the eigenvectors and the eigenvalues of σ. Regardless of the particular distribution of the values of P in the medium, on the characteristic path, one principal axis is always oriented along the path and the other perpendicular to it. The corresponding principal stresses on the ith path are $\sigma_{max} = \omega_i\left(1 + \lambda_i^2\right)$ along the path and $\sigma_{min} = 0$ perpendicular to the path. The vanishing of the stress perpendicular to the path is intriguing - it underpins the fragility of stress paths to small side forces, as proposed on a phenomenological basis in [61]. This issue has not been addressed in any detail in the literature so far. However, a back-of-the-envelope linear stability analysis shows that the stress chain is unstable to such side forces pressing on it from the inside of the cone made by the two characteristic paths. In contrast, the characteristic path is *stable* to side forces pushing on it from the outside of the cone to the inside.

Stress response for position-dependent fabric tensors

While solving for media with constant P has its uses for gaining insight, such a uniform medium is hardly likely to arise in any natural, or even numerical, assembly of grains. The disorder and irregularity in realistic systems manifest in a position-dependent fabric tensor P. To understand the stress response in such systems, it is imperative to outline explicitly the general solutions to equations (5.49) when $P = P(\boldsymbol{r})$. The key to identifying the stress chains is the behaviour of the characteristic variables, given by the coupled equations (5.49):

$$\partial_x \omega_i + \lambda_i \partial_y \omega_i = h_i - \partial_y \lambda_i \omega_i + (B\boldsymbol{w})_i \qquad i = 1, 2 \ . \tag{5.58}$$

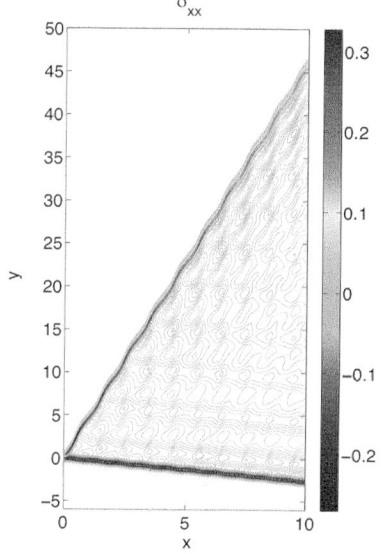

Fig. 5.12 The stress chain solution to the isostaticity stress equations when a narrow Gaussian loading is applied to the boundary of a medium with the slightly periodic ($\epsilon = 0.5$) fabric tensor equation (5.59). The characteristic paths are slightly wavy about straight lines and stresses 'leak' away from the chains and into the region between the paths - the 'cone of influence'. Source: reference [66].

As described above, stresses propagate along characteristic paths defined by equations (5.54), but unlike for the aforementioned uniform media, now the eigenvalues λ_i vary along the paths as functions of position. Since these eigenvalues define the local direction of the characteristics, then the implication is that the stress paths are no longer straight. This complicates the solutions – due to the coupling between the characteristic variables, equations (5.54) must be integrated explicitly to find the trajectories.

Gerritsen et al [66] demonstrated the effects of the coupling by solving the equations exactly for several variable fabric tensors, comprising of a constant part, modulated by a slight oscillation as a function of position,

$$P = \begin{pmatrix} 1 & 2 \\ 2 & -1 \end{pmatrix} + \varepsilon \cos(2\pi x) \begin{pmatrix} 1 & 1 \\ 1 & 1 \end{pmatrix}, \qquad (5.59)$$

where $\varepsilon < 1$ is a small parameter. The choice of this form is to illustrate the effect of the spatial periodicity on the stress solutions. Applying to this medium a narrow Gaussian-shaped σ_{xy} load around $(x=0, y=0)$ and solving the equations explicitly, gives rise to the stress chains shown in Figure 5.12. Note that we show the σ_{xx} component, which appears in the medium although we apply only σ_{xy} at the boundary. This mixing of the stress components is because the stresses are not the characteristic variables of the equations, but rather the quantities ω_i, which are combinations of the stress components. The resulting characteristic paths have a slight waviness about the predominant straight line, in response to the periodicity of the fabric tensor.

The right hand side of equations (5.58) can be grouped into three parts, each having a different effect on the stress field. The first, h_i, is independent of the stress field and can be regarded as an external source term, albeit a non-trivial one since it depends on position. The second group is $(B_{ii} - \partial_y \lambda_i) \omega_i$ and it is diagonal, which means that it does not give rise to any coupling between the two characteristic variables and paths. It is important to note that this term may give rise to amplification or attenuation of ω_i, depending on its sign. Over long paths stresses can either stay constant or attenuate, which means that in real systems the overall sign of this term should be negative when averaged over a sufficiently long stress path.

The third group consists of the off-diagonal terms of B and are the only reason for a coupling between the two characteristic variables ω_1 and ω_2. This coupling gives rise to 'leakage' from the stress paths into the region between the stress paths, which has been called the 'cone of influence' [65]. It should be noted that there can be no leakage of stresses to the region outside the cone of influence. It can also be seen that the larger the local gradient of P, the stronger the coupling between the ω_i and the stronger the leakage from that point. In fact, if the local gradient is very pronounced and localised, the leakage will take the form of a branch, as illustrated in Figure 5.13. The branches extend along the local characteristic directions.

It is important to remember that the linearity of the equations implies that the superposition principle applies. Namely, when the force on the boundary is not localised, the solution to the equations is a superposition of the individual solutions for every point along the boundary. This means that stress chains, originating at different points along the boundary, do not interact and pass through one another without change of shape or direction. This is an important conclusion that can be checked

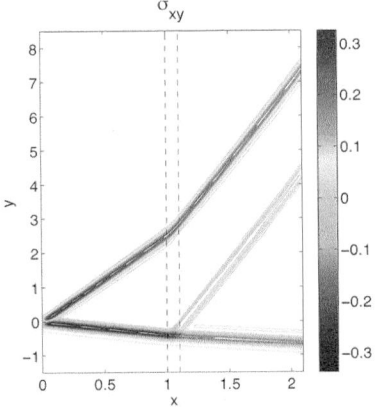

Fig. 5.13 Branching of a stress path of σ_{xx}, generated by a narrow Gaussian σ_{xx} at the boundary, due to a strong gradient in P in the thin mid-layer between the dashed lines. Source: reference [66].

against experimental results to test the theory.

Isostaticity theory in cylindrical coordinates

We cannot leave this Section without reviewing recent results on isostaticity theory in cylindrical coordinates. Cylindrical geometries, and in particular Couette flow, are common experimental setups [69]. In such experiments the granular material is often confined to an annulus between two circular cylinders of radii $r = r_{in}$ and $r = r_{out} > r_{in}$. Applying compressive stresses, σ_{rr}, at either of the boundaries and rotating the boundaries with respect to one other, generates a shear stress $\sigma_{r\theta}$. Along the cylindrical boundaries, forces are applied to individual grains and each such force can be regarded as a source of stress for the purpose of the analysis.

The symmetric stress tensor in the annulus is

$$\sigma = \begin{pmatrix} \sigma_{rr} & \sigma_{r\theta} \\ \sigma_{r\theta} & \sigma_{\theta\theta} \end{pmatrix}. \tag{5.60}$$

Setting the external force field to zero for simplicity, consider then a localised stress load applied at (r_{in}, θ_0) on the inner circular boundary. In cylindrical coordinates, the force and torque balance equations are

$$\partial_r \sigma_{rr} + \partial_\theta \sigma_{r\theta} - \sigma_{\theta\theta} = 0 \tag{5.61}$$
$$\partial_r (r\sigma_{r\theta}) + \partial_\theta \sigma_{\theta\theta} + \sigma_{r\theta} = 0 \tag{5.62}$$
$$\sigma_{r\theta} = \sigma_{\theta r} \tag{5.63}$$

To close the equations with a stress-structure condition, we can assume again a linear relation between the stress components

$$2p'_{r\theta}\sigma_{r\theta} - p'_{\theta\theta}\sigma_{rr} = p'_{rr}\sigma_{\theta\theta}. \tag{5.64}$$

The fabric components p'_{ij} can be expressed explicitly in terms of the components p_{kl} of the Cartesian fabric tensor in equation (5.40) and θ by comparing the two closure relations. Following a similar analysis to that above, one assumes $p'_{rr} \neq 0$ and defines $q'_{ij} = p'_{ij}/p'_{rr}$ and $\boldsymbol{u} = [\sigma_{rr}, \sigma_{r\theta}]^T$. This leads to the equation

$$\boldsymbol{u}_r + (A\boldsymbol{u})_\theta = D\boldsymbol{u}, \tag{5.65}$$

with

$$A = \frac{1}{r}\begin{pmatrix} 0 & 1 \\ -q'_{\theta\theta} & 2q'_{r\theta} \end{pmatrix}, \quad D = \frac{1}{r}\begin{pmatrix} -1 - q'_{\theta\theta} & 2q'_{r\theta} \\ 0 & -2 \end{pmatrix}.$$

The solution of this linear system of equations follows the same lines as used above. One transforms to equations for the characteristic variables $\boldsymbol{\omega}$ that have the form

$$\boldsymbol{\omega}_r + \Lambda\boldsymbol{\omega}_\theta = (B - \Lambda_\theta)\boldsymbol{\omega} + Y^{-1}DY\boldsymbol{\omega}, \tag{5.66}$$

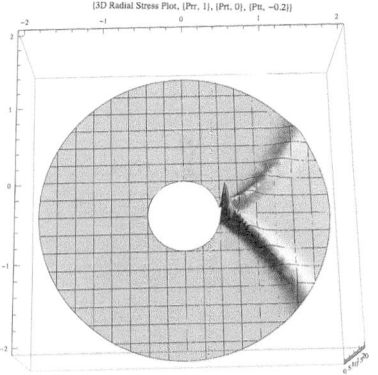

Fig. 5.14 The stress chain solutions in cylindrical granular media for a uniform fabric tensor. In contrast to the solution in rectangular systems, in this geometry, the stress chain solutions: (i) no longer follow straight lines but rather flare out; (ii) their widths are not constant and they broaden with the radius; (iii) the stress magnitudes attenuate along the chain.

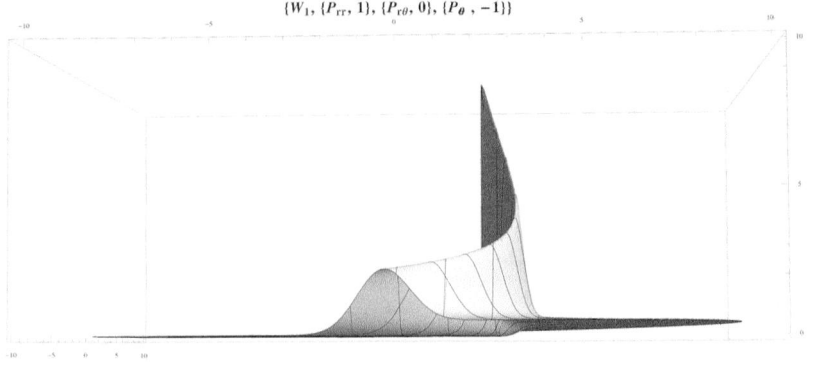

Fig. 5.15 The stress attenuation along the chain in cylindrical coordinates. A chain, initiated by a narrow Gaussian load at an inner radius, is shown here 'head-on', propagating to the outwards of the annulus. Its broadening and flaring out are clearly observed. Significantly, the stress attenuates as the chain approaches the external rim of the annulus.

where $\Lambda = \mathrm{diag}(\lambda_1, \lambda_2)$ is a diagonal matrix, λ_i are the eigenvalues of the matrix A, with corresponding eigenvectors $y_i = [1, r\lambda_i]^T$, forming the matrix Y, and

$$B = \frac{-1}{r(\lambda_1 - \lambda_2)} \begin{pmatrix} (r\lambda_1)_r + r\lambda_1\lambda_{1,\theta} & (r\lambda_2)_r + r\lambda_2\lambda_{2,\theta} \\ -(r\lambda_1)_r - r\lambda_1\lambda_{1,\theta} & -(r\lambda_2)_r - r\lambda_2\lambda_{2,\theta} \end{pmatrix}. \quad (5.67)$$

In spite of the similarities, there is a marked difference between the cylindrical and the Cartesian cases. The equations for ω_1 and ω_2 are coupled not only through the off-diagonal elements of B but also through the off-diagonal elements of $Y^{-1}DY$. While the matrix B vanishes when the fabric components p'_{ij} are everywhere constant, the off-diagonal terms of $Y^{-1}DY$ need not. This means that, unlike in Cartesian coordinates, the characteristic variables may be coupled even though the fabric tensor is uniform across the material. This is significant - it implies that in cylindrical coordinates one should always expect leakage, attenuation of stresses along stress chains, and branching, regardless of whether the fabric tensor is uniform or not.

Analysis [66] shows that, for constant fabric tensors, the characteristic paths flare out in a spiral, satisfying at every point the following relation between the angle and the radial position

$$\theta_i = \lambda_i \ln(r/r_{in}). \quad (5.68)$$

This is illustrated in Figure 5.14, where a narrow Gaussian-shaped load, σ_{rr}, is applied at the inner boundary of a medium of a constant fabric tensor $p'_{rr} = 1, p'_{r\theta} = 0$, and $p'_{\theta\theta} = -1$. Figure 5.15 shows how the characteristic variable attenuates along a chain.

It is also intriguing to note that the stress chains widen as they propagate from the inner to the outer boundary. This widening has been explained as follows. Consider a load source of finite width $w_{in} = r_{in}\delta\theta$ applied at r_{in}. It sends two stress chains of finite width into the system, ω_1 and ω_2, only the lowest of which, say ω_2, is shown in Figure 5.16. The widening of the stress chain can be understood by considering the two extreme w_2 characteristics, emanating from $r_{in}\theta_0$ and $r_{in}(\theta_0 + \delta\theta)$, respectively. For uniform fabric tensor, the system is invariant to rotations and therefore, if we rotate the system by $\delta\theta$, then the upper ω_2

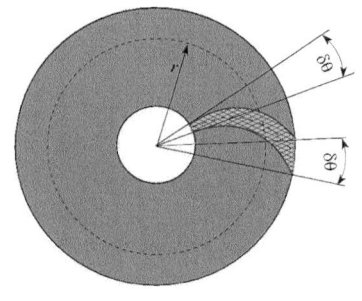

Fig. 5.16 The widening of a stress characteristic path is explained by considering the cylindrical symmetry and the geometry of the two extreme characteristic paths that flank the finite-width chain.

path must coincide with the lower one. This means that at r_{out} the separation between the characteristic paths is $w_{out} = r_{out}\delta\theta$. If we define the widening of stress paths as the width at a given radius, then it follows that the widening at radius r is proportional to r.

Isostaticity theory - general points

To conclude, isostaticity is a budding continuum theory that describes stress transmission in statically determinate systems in general, and in marginally rigid granular media in particular. Its main difference from elasticity is that it gives rise generically, but not always, to hyperbolic equations. The balance equations are closed by constitutive stress-structure relations (one in 2D, three in 3D and generally $d(d-1)/2$ in d dimensions), derived from a local torque balance theory. The closure in 2D is an expression that relates the stress components linearly, with the coefficients being the components of a fabric tensor. This fabric tensor characterises local fluctuations in the structure, explained in [24]. The upscaling, or homogenisation, of the fabric tensor is straightforward in systems of ideally frictionless grains [24], but it is far from straightforward in general systems of frictional grains [64]. Indeed, this suggests that there may be an inherent difference between the manner at which these types of systems transmit stresses. The upscaled fabric tensor is expected to be normally non-uniform, but experimental measurements of it are few and far between. It is therefore unknown currently at what length scales one can regard it as uniform.

The generic hyperbolic nature of the equations leads to stress chain solutions. A finite-width stress load at the boundary of a 2D Cartesian system of a uniform position-independent fabric tensor is supported by straight stress chains of the same width each, localised on two straight stress chains at an angle. The stress chain paths coincide with the characteristic variables of the hyperbolic equations, whose orientations are determined directly by the fabric tensor via the eigenvalues λ_i. These solutions are mathematically identical to those of the wave equation in time and one space dimension, albeit in two space dimensions.

However, realistic non-uniform fabric tensors give rise to rich and intriguing phenomena: the stress chains, whose local orientations can be determined directly from the local fabric tensor, are no longer straight; stress 'leaks' into the cone of influence, causing attenuation of stresses along the chain; and stress chains may branch where the gradients of the fabric tensor are large. Indeed, most of these phenomena have been observed in the experiments referenced above, providing support for this theory. Nevertheless, one should bear in mind that most relevant experiments are carried out on isostatic materials and therefore any comparison between the predictions of IT and observations should be made with caution.

Stresses in cylindrical geometries exhibit a more complex behaviour. Non-straight chains, leakage of stresses, and chain widening are phenomena that appear even for uniform fabric tensors. Position-dependent

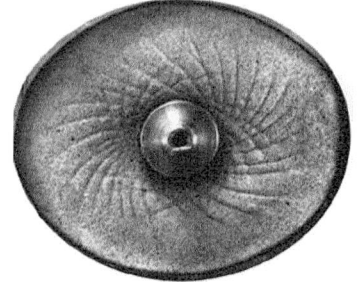

Fig. 5.17 Intersecting slip lines in a sand cylinder with an internal circular spigot and an external slightly elliptic boundary. The slip lines are generated by slowly twisting the inner boundary. Source: reference [70].

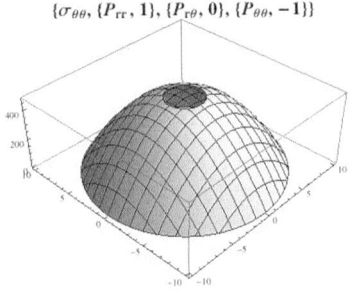

Fig. 5.18 The radial stress field in a cylindrical GS with a uniform fabric tensor under gravity. The gravity generates a uniform field of body forces inside the medium, all acting as sources for σ_{rr}. As a result the stress field fills the entire medium, with the stress increasing towards the inside of the medium.

fabric tensors give rise to even richer stress patterns, including strange branching configurations, which we do not discuss in this exposition.

It is interesting to note an experiment, carried out recently [70]. Subjecting a sand-filled cylinder with an inner circular spigot and an outer slightly elliptical boundary, to a slow shear by twisting slowly the inner spigot, the authors of [70] observed the emergence of an intersecting family of flaring out slip lines (Figure 5.17). The similarity between these slip line trajectories and the theoretical solutions discussed above (see Figure 5.14) begs the question of whether or not there is a relation between the two. It is tempting to conjecture that the slip lines originate from protruding grain force sources along the inner spigot, which propagate into the material following local failure at or near the stress chain paths. To substantiate this conjecture we need to have a first-principles failure and yield flow theory, which self-consistently relates slip line trajectories to the quasi-static stress field.

This Section demonstrates that the common wisdom - that isostatic granular systems must exhibit narrow stress chains - is misconceived. Such systems are likely to be disordered and therefore should exhibit stress attenuation and leakages, filling the medium with a highly complex pattern of stresses. One could argue that coarse-graining the fabric tensor to a sufficiently large length scale may make it uniform. However, in that case any stress chains are only as narrow as that homogenised length scale. Thus, the solutions discussed here have thrown new light on our understanding of the distributions of stress fields in isostatic systems in general.

Furthermore, one should recall that in gravitational field the weight of each grain acts as a load source. This means that the body forces induced by gravity generate a uniform background field that depends on the direction of gravity. For example, the self-gravity radial stress (σ_{rr}) developing in a two-dimensional disc is shown in Figure 5.18. Boundary stresses would superpose on this uniform field.

Beyond isostaticity theory

The relevance of IT to the description of real granular systems, consisting of frictional grains, has been the source of a fierce debate in the GS community. In particular, two aspects of the isostatic picture came under fire. One is that, while IT has been developed originally for perfectly rigid particles, real particles always have finite elastic moduli, however high. The other is that, while IT is specifically constructed to describe statically determinate systems, most real materials are not marginally rigid and the mean number of contacts per particle in mechanical equilibrium is often larger than z_c [71, 72]. In the following we address these two contentions.

The finite compliance of the particles means that contacts of grains in d dimensions are not points but $(d-1)$-dimensional surfaces. The implications of the 'broadening' of the contact have been discussed in detail in [73], where it has been shown that a gradual increase in the compliance of grains from the fully rigid state erodes the theory only gradually. Specifically, if the mean number of contact surfaces per grain is at the marginal stability value z_c, then the stress field is the same as predicted by IT if the particles were perfectly rigid and the contacts were points. The only difference is that there are elastic corrections to the basic isostatic field which tend to zero with an increase in the number of grains. It is not the aim of this exposition to detail that. Briefly, it is based on the observation that the unknown force distribution across every contact surface can be replaced by a single representative unknown force vector (see Figure 5.19). The magnitude and the position of this force can be determined uniquely from the surface force density. Since the number of contact surfaces per grain is z_c, this means that the number of equations for these forces is the same as the number of unknowns and the material is statically determinate for these forces. This, in turn, means that the IT equations are valid for this system and, consequently, that the stress field is isostatic. However, since the force densities across the contact surfaces are not known in the first place, then the exact location of the representative force vectors is also unknown. Nevertheless, by approximating the locations of the representative inter-granular forces at the centroids of their contact surfaces, the difference between the resulting approximate IT solution and the real solution decreases with system size. For example, for a 2D assembly of N grains, the difference between the approximate solution and the real one decreases as $N^{-3/4}$. The difference is also proportional to the inverse square root of the grains Young's modulus, $\sim E^{-1/2}$. It follows that, for assemblies of sufficiently large numbers of grains, IT is valid.

It should be noted that granular systems, almost by definition, cannot have too large contact surfaces - a medium can be considered an assembly of grains only if it is possible to identify distinct grains. This means that the linear size of contact surfaces must be smaller than the typical linear size of grains, if there is such a typical size. If there is no typical grain size, i.e. for broad size and shape distributions, then the linear size of a contact surface must be smaller than the linear size of the two grains in that particular contact. If this condition is not satisfied, the medium is not granular but rather falls into the class of generally porous. The distinction between the two classes of materials is not sharp, but the ratio of the linear sizes of the contact surface and the grain size is a good descriptor of such distinction.

The second contention is very important because it seems to undermine the applicability of IT to most real granular assemblies, which often contain more contacts than required for the marginally rigid state [43, 44]. To address this problem, it has been proposed in [63] that, for assemblies whose mean contact number is $\bar{z} > z_c$, over-connected regions form and that these regions are non-isostatic, transmitting stress differently than the isostatic regions. Thus, it was argued that most realistic granular materials are in fact two-phase composites – part isostatic and part non-isostatic. The stress transmission in the two-phase composites is such that the equations of IT apply within the isostatic regions, while

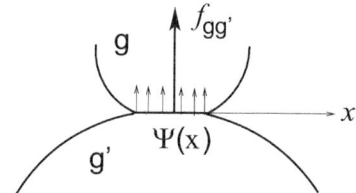

Fig. 5.19 A contact is normally a surface, across which an inter-granular force between grain g and g' is distributed with some force density. In this figure only the normal contact force density is shown, $\Psi(x)$. This force distribution can be replaced by a single force, $f_{gg'}$, which is the integral over the entire force distribution across the surface S, $f_{gg'} = \int_S \Psi(x)dx$. The location of this equivalent force vector is determined by the condition that it apply the same couple as the force distribution, namely, $x_0 f_{gg'} = \int_S x\Psi(x)dx$.

in the over-connected regions different equations apply, either those of elasticity theory or of elasto-plasticity. This picture is attractive since it suggests that the nature of the stress transmission in a granular material depends on its proximity to the marginally rigid state and it fits nicely with the identification of this state as a critical point [36, 38, 43, 44, 61] – the closer the system is to this point the longer the stress chains. Thus, in the two-phase picture, the closer an assembly is to the critical point the larger the isostatic regions it contains. This also gives a geometrical perspective to the problem – the stress transmission is likely to be different when either of the phases percolates between the system's boundaries.

This leads naturally to the idea that there is a continuous spectrum of stress responses in granular systems, from the pure isostatic response when $\bar{z} = z_c$ to the pure elastic response when $\bar{z} \gg z_c$. Such a continuum can be realised in the following thought experiment. Imagine a d-dimensional assembly of N ($\gg 1$) grains made of compliant elastic material, say rubber, possessing an appreciable Young's modulus. The assembly is in mechanical equilibrium under a set of very weak compressive boundary forces and it is marginally rigid, namely, its mean coordination number is exactly $\bar{z} = z_c = d + 1$. At this state the stress transmission is isostatic and the boundary forces are supported by arches that develop in the system (note that in 3D the stress arches would be more like 2D surfaces than 1D chains). The stress that develops in the medium can be then determined using IT. Now, increase all the boundary force magnitudes by a factor $A = 1 + \epsilon$, where $\epsilon \ll 1$. The contact force magnitudes increase by A and, correspondingly, the contact surfaces broaden and the grains centres change slightly. At this state, the stress field is still isostatic but there are corrections to the field proportional to $\sqrt{A}$ [73]. On further increase of A, there comes a moment when a new contact is made. The force transmitted by this contact cannot be determined and, around that contact, IT no longer gives the correct stress solution. At this stage the system may or may not be regarded as granular, depending on the local ratios of the linear sizes of the contact surface and the grain size. The region around the extra contact can then be regarded as an elastic enclave within an isostatic medium. As the boundary forces are increased further, more contacts are made and more elastic regions form, then grow and eventually join together. At some stage, the over-connected regions percolate between opposite boundaries and their stress transmission starts to dominate the behaviour of the medium. Finally, at sufficiently high loading on the boundaries, all the voids enclosed by the particles disappear and the medium is made fully of rubber. For this medium, elasticity theory must be valid. This shows that there is a continuous spectrum of stress states between the purely isostatic and the purely elastic states. These states, and the media that give rise to such states, have been called *stato-elastic*.

We can repeat the same thought experiment on perfectly rigid grains instead of rubber ones. Under the assumption that grains do not crush in the process, the compression of such a material increases the mean coordination number from $\bar{z} = z_c$ until it reaches a maximal value $\bar{z} = z_{max} > z_c$. One could plausibly regard this final state as the close random packing of this particular material under this process. At this state, the stress transmission is not isostatic, but it may not be fully elastic yet, as in the rubber grains case – the system is in between these two states. It is likely that most real granular systems are in such an intermediate state.

The question is then how to construct a *stato-elasticity* stress theory for such two-phase composites. This is not an easy task. In conventional composite materials, the different phases follow the same form of stress equations and only their constitutive properties, i.e. the elastic moduli, change as functions of position. In such cases, solving for the stress field normally involves solving in the different regions and matching the boundaries between regions to obtain continuous fields. The granular composites pose a much more difficult problem because the two phases follow different forms of stress equations. To complicate matters further, the elastic phase follows elliptic stress equations while the isostatic phase follows hyperbolic ones. This makes the matching of the stress solutions at the regions' boundaries cumbersome since these two types of equations require different types of boundary conditions. In particular, care needs to be taken when posing boundary conditions for the hyperbolic equations to avoid local ill-posedness. Modelling stresses in real materials along these lines is still in its infancy.

Limitations of linear stress theories

All the current theories of stress in granular systems, including IT and elasticity theory, are linear. Namely, they accommodate the superposition principle - the stress response to a given a set of boundary loads $\sigma_B(\boldsymbol{r}) = \sum_i \sigma_B(\boldsymbol{r} - \boldsymbol{r}_i)$ is the sum of the responses to the stress field generated by each boundary load independently. But how valid can a linear stress theory be for granular matter?

Let us consider first more closely the equations of IT. In principle, the stress solution σ depends on the particular set of boundary loads, $\sigma = \sigma(\sigma_B)$. Suppose that we change the load slightly to σ'_B. If the theory were linear, then the solution would be as before but with σ_B replaced by σ'_B. But we know that even the smallest change in the boundary loads of a granular systems leads to internal rearrangements. The grains shift and the network of forces on the granular scale reorganises with parts of it failing in some places and re-consolidating in others to accommodate the new loading. The change in the internal structure manifests in a change in the fabric tensor. This means that the fabric tensor depends implicitly on σ_B and on the local stress field $P = P(\sigma)$. It follows that the closure relation (5.40) is in fact *nonlinear*.

A similar problem plagues all linear theories of granular matter. For example, elasticity theory assumes a local stress-strain relation, but since the structure depends on the boundary loads, then so do the elastic moduli. It follows that the moduli are implicit functions of the solution and again the theory is nonlinear. Plasto-elasticity can also be shown to suffer from the same problem.

The nonlinearity is indicative of an incomplete approach to the problem. The above argument suggest that, to construct a physics-based theory, one has to include self-consistently the dynamics of the structural organisation under given boundary stresses. Ultimately, when applying boundary loads to a granular system, forces are transmitted through the medium and the structure rearranges until it consolidates with a particular configuration of force network. A key quantity governing this process is the yield surface. This is a surface in the space spanned by the stresses, such that for stresses larger than this surface, the material fails and below it the material is stable. Local structural changes take place until the yield surface is reached, at which moment the structure consolidates. A complete theory of stress in granular media must take this process into account. In our view, we are currently far from such a theory and a significant increase in activity in this direction is required.

5.4 Statistical mechanics

Application of statistical methods to analyse and classify granular materials has a long history [74]. Over two decades ago, Edwards and collaborators proposed to apply a statistical mechanical approach to explain the behaviour of large granular assemblies [75]. Much literature followed, e.g. [76–82] and references therein.

The aim of conventional statistical mechanics is to understand how the thermodynamic properties of macroscopic systems arise from the dynamics of basic elements, such as molecules, spins, etc. It is a first-principles theory to upscale from the molecular scale to the macroscopic continuum, establishing relations between the microscopic thermal fluctuations and responses and measurable quantities, such as energy, pressure, volume, and entropy. Similarly, the aim of this approach in granular matter is to understand the behaviour of systems of large assemblies of grains.

That such an approach should work is based on the apparent reproducibility of the macroscopic behaviour of granular materials in a wide range of situations. For example, joggers on the beach presume that the underlying ground will support them, hourglass egg timers are sold with the confidence that it always takes three minutes for the sand to flow from the top to the bottom compartment, buildings are built on soil that is expected to support them and normally does, and Brazil nuts are expected to rise to the top of mixed-nuts tins upon sufficient shaking. All these expectations are based on countless reproducible observations. While the configuration of the sand grains under the jogger's feet is different at every step it always supports the jogger's weight. This reproducibility can be also seen in controlled experiments [83], where the macroscopic properties of a granular bed, such as the density, are consistently the same after being vibrated by the same procedure. That the behaviour of granular matter is reproducible is not only an assumption underlying attempts to construct a predictive theory but it also suggests that, in spite of the astronomical number of possible configurations, there is a typical behaviour and therefore that the grain-scale seeming randomness is underpinned by limit statistics. This parallels the basis for constructing a statistical mechanical formalism in thermal systems. There is an astronomically large number of molecules of gas around us and these can be in uncountably many configurations. Yet, when we enter a room we expect that the oxygen molecules will be dispersed sufficiently homogeneously for us to breathe comfortably. We do not seriously consider the possibility of suffocation due to a stray configuration wherein all the oxygen molecules happen to be in the corner. We do not stick our noses inside the door to test this assumption every time we enter a new room. This is beyond the expectation of reproducibility - we stake our lives on this assumption! It is with a similar expectation of reproducibility that the jogger on the beach trusts that the sand underneath would support her weight and would not change into quicksand. Thus, the reproducibility suggests the existence of ensembles of steady-state (not equilibrium) configuration states that have similar features as the more conventional thermodynamic ensembles.

Since energy is hardly relevant to many of the fundamental issues of static granular matter, the suggestion in [75] was to apply the formalism of statistical mechanics, based on entropy alone. In this formalism, a configuration of the granular system is a micro-state and, as in conventional statistical mechanics, the entropy is the logarithm of the number of possible configurations.

However, the first question that arises is how useful is the concept of entropy for granular matter. This concept is key in conventional thermal systems because the micro-states are constantly shuffled by thermal fluctuations. In contrast, granular systems typically stay put unless excited by external means, such as shaking, vibrating the boundaries, shearing, rotating drums filled with particulates, etc. This is a significant difference. Not only does it cast doubt on the ability of granular matter to access a sufficiently large number of configurations dynamically but it also suggests that we cannot rely on the existence of *ergodicity*, a term coined by Boltzmann. An ergodic ensemble of systems, at least in physics, is one whose statistics are the same as the statistics of any one specific system, measured over time. This principle is extremely useful in thermal systems because it allows us to infer dynamical properties from analysis of ensembles and visa versa. The absence of ergodicity in granular ensembles denies us this important route to modelling the dynamics. Thus, in most cases, the statistical mechanics of granular matter is limited to ensemble statistics.

Since the micro-states cannot be accessed dynamically it is unclear how transferrable are the assumptions that underlie conventional statistical mechanics. Probably the most significant assumption is that of *uniform measure*, namely, that one can expect any relevant microstate to be realised by the system. Without this assumption one has to have a model for the different probabilities of micro-states, which is an insurmountable problem. Some numerical and experimental tests of the uniform measure assumption have been carried out on small systems [75,84]. However, due to the astronomically large number of possible configurations, it is impossible to sample sufficiently many systems, either numerically or experimentally, to resolve this issue conclusively. Additionally, it is unclear how the measure depends on the dynamics of generation of the sampled GS. For example, it is well recognised that different packing processes are likely to produce different measures. That said, since ergodicity does not apply anyway in most of the systems of interest, then the most practical approach is arguably to ignore the dynamics by which the phase space is explored, assume a uniform measure in the ensemble statistics and then test the resulting predictions against experimental measurements.

Mapping statistical mechanics from thermal to granular systems involves finding the parallels of the following.
(i) *The degrees of freedom.* These comprise the set of independent variables that describe all the possible micro-states - the configurations of the systems that belong to a particular ensemble. These variables define what is known as the phase space. In gases, for example, these are the positions and momenta of the molecules. In static granular systems, these are the positions and orientations of grains, however parameterised. We will quantify this set of variables more precisely below.
(ii) *The Hamiltonian.* In thermal systems, this function quantifies the energy of a micro-state, given a particular realisation of all the degrees of freedom. In the absence of dynamics and of potential energy between particles and in view of the irrelevance of thermal fluctuations in granular materials, the energy cannot play a role. It was therefore proposed that the volume determines the occurrence probability of micro-states [75].
(iii) *The temperature.* In thermal systems, the temperature is a measure of the fluctuations and it is defined as the derivative of the energy with respect to the entropy. A similar quantity, the compactivity, has been defined in GS for structural entropy.

The micro-canonical volume ensemble

This is the ensemble of all the configurations (micro-states) with a given volume V that N grains, of a given shape and size distribution, can assume. The entropy of this ensemble, $S(V)$, is defined as

$$S(V) = \ln\left[C \int \delta\left[W\left(\{u\}\right) - V\right] d\{u\}\right], \qquad (5.69)$$

where C is a system size dependent constant that makes the integral non-dimensional (analogous to the factor $1/N!h^{3N}$ in conventional statistical mechanics), $\{u\}$ is the set of all the structural degrees of freedom (SDF), and W is a volume function that depends on the SDF and which parallels the Hamiltonian in conventional statistical mechanics. This expression already assumes that the micro-states satisfy the constraints on the ensemble, e.g. that the systems are in mechanical equilibrium (see below for more details about the constraints). Alternatively, one can introduce into the integral in equation (5.69) a function Θ, consisting of a set of δ-functions that ensures that these constraints are indeed satisfied [75].

In terms of this, any macroscopic structural feature of the system, A, can be expressed as an expectation value

$$\langle A \rangle = \frac{\int A(\{u\}) \, \delta\left[W(\{u\}) - V\right] d\{u\}}{\int \delta\left(W[\{u\}] - V\right) d\{u\}}, \quad (5.70)$$

where the left-hand side is a measurable macroscopic quantity and the integrals on the right-hand side involve microscopic variables only. The denominator of equation (5.70) is the partition function and it is required for normalisation. One can express this expectation value (often called as the micro-canonical ensemble average) in terms of the entropy by replacing the partition function with the definition in equation (5.69).

Structural degrees of freedom

The above formalism is abstract without identifying explicitly the SDF, which we proceed to do next. For simplicity, we discuss only assemblies of rigid grains, where the inter-granular contacts can be regarded as points. This does not affect the generality of the description, which can be extended straightforwardly to contact surfaces. The volume of an assembly of grains is determined by the structure and this, in turn, depends on the locations of the inter-granular contact points.

There are two aspects to the structure: topological and geometric. The topology is determined by the graph whose nodes are the contact points, connected by lines, as explained below. The geometry is determined by the shapes of all possible grains that can be sculpted on any one such graph. These two aspects are independent of one another in some statistical ensembles and only very weakly dependent on others. To focus the discussion we now need to specify the ensemble. For our purpose here, we define it as all the possible configurations that a collection of $N \, (\gg 1)$ grains can make in mechanical equilibrium under a given set of boundary loads, discussed in more detail below. The grains are chosen from a given size and shape distribution and they possess specific physical properties, such as the inter-granular friction coefficients. The reason that friction is important is that, given a specific packing process, it affects strongly the final structure. Since the packing process is significant for the structural organisation, we also impose the constraint that all members of the ensemble are generated by an identical process. Finally, for convenience only, we also constrain the members of the ensemble to have the same mean coordination number $\bar{z}$.

Let us consider the topological structure of the contact network. Since $N \to \infty$, we can ignore boundary effects. Alternatively, we can assemble the pack on a $d+1$ hypersphere or impose periodic boundary conditions to do away with boundary corrections to the following description. In principle, the basic SDF for the topology is the vectors to every contact point between grains g and g', $\rho_{gg'}$, taken from an arbitrary origin. There are altogether $N_{cont} = N\bar{z}/2$ such contacts and therefore $N\bar{z}d/2$ independent variables. Rigid translation and rotation of the packing cannot affect its statistics and there are d axes of translation and $d(d-1)/2$ axes of rotation in d dimensions. Subtracting these from the total number of independent variables, we have that the number of SDF in d dimensions is

$$N_{\text{SDF}} = \frac{N\bar{z}d}{2} - \frac{d(d+1)}{2} \approx \frac{N\bar{z}d}{2}. \quad (5.71)$$

It is convenient to express the SDF in terms of the local vectors between contact points. To this end, we follow the structural description reviewed in Section 5.2. This description, proposed first for 2D assemblies [24,25] and extended in [26] to 3D, is especially useful to statistical mechanics, as we shall see below. The procedure follows the same rationale in 2D and in 3D and we start by describing it for 2D assemblies of convex grains.

Structural degrees of freedom of planar systems

Recalling that the volume function can be written as [see equation (5.10)]

$$W = \frac{1}{2} \sum_{q \in system} Tr\left\{(\epsilon \cdot \bm{r}_q) \otimes \bm{R}_q\right\}, \quad (5.72)$$

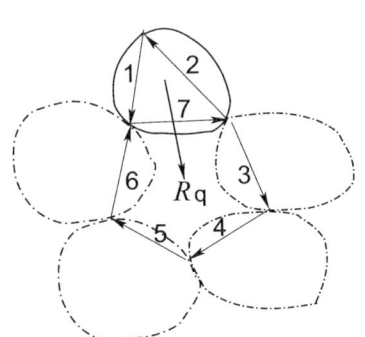

Fig. 5.20 The vectors $\bm{R}_q$ depend linearly on the vectors $\bm{r}_q$ (denoted only by q here, for brevity). For example, the shown vector $\bm{R}$ is a linear combination of $\bm{r}_q$, $q = 1, 2..., 6$, as expressed in equation (5.73) (only the indices of the $\bm{r}$ vectors are shown).

our aim is to determine how many, and which, of the variables that appear in the volume function are independent.

There are $N\bar{z}$ pairs of $\bm{r}_q$ and $\bm{R}_q$ vectors giving altogether $4N\bar{z}$ variables. However, these are not all independent. First, each $\bm{R}_q$-vector can be expressed as a linear combination of some $\bm{r}_q$-vectors. For example, the vector $\bm{R}$, shown in Figure 5.20, can be written in terms of the seven $\bm{r}_q$ around it as follows

$$\bm{R} = \frac{1}{5}(4\bm{r}_3 + 3\bm{r}_4 + 2\bm{r}_5 + \bm{r}_6) - \frac{1}{3}(2\bm{r}_2 + \bm{r}_1). \quad (5.73)$$

The general expression for $\bm{R}_q$, extending between grain g and nearby cell c, in terms of the vectors $\bm{r}_{cg}$ that circulate g and c, is

$$\bm{R}_q = \frac{1}{z_{cell}} \sum_{\substack{g'=1 \\ g' \neq g}}^{z_{cell}} (z_{cell} - g')\bm{r}_{cg'} - \frac{1}{z_g} \sum_{\substack{c'=1 \\ c' \neq c}}^{z_g} (z_g - c')\bm{r}_{c'g}, \quad (5.74)$$

where z_{cell} is the number of grains surrounding the cell (or the cell order) and z_g is the coordination number of the grain. The right-hand side of equation (5.74) involves $(z_{cell} + z_g - 2)$ vectors $\bm{r}_q$. However, since these form a loop, one of these can be expressed in terms of the others. This means that a $\bm{R}_q$ vector depends linearly on $(z_{cell} + z_g - 3)$ of the $\bm{r}_q$ vectors surrounding the cell and grain that it belongs to.

So, we know that the R_q vectors are not independent, which leaves only the r_q vectors as candidates for the SDF. However, there are twice as many components of these vectors as there are SDF. This is because the r_q vectors also form loops, each of which accounts for one dependent vector. Subtracting those loops gives that the number of SDF in 2D is $N\bar{z}$ – exactly the same as the number of quadrons, irrespective of the mean coordination number [25].

This leads us to two conclusions. One is that any independent subset of half of the components of the vectors r can be used as SDF to span the structural phase space – the other components can be expressed as linear combinations of those. The other conclusion is that, since $N_q = N_{sdf}$, then it is possible to use the quadron volumes as the SDF. As we will see below, this equivalence is unique to 2D, in 3D the number of quadrons is larger than the number of SDF. The implications of this latter observation will be discussed below.

Structural degrees of freedom of 3D systems

The construction of the structure phase space of 3D GS follows the same rationale as in 2D. This construction has been mentioned briefly previously and here we add some more details. The method consists of first tessellating the volume occupied by the GS into polyhedral volume elements based on the inter-granular contact points. We then quantify the shape of every volume element by a shape tensor, assigning numbers to the local micro-structure. Finally, we identify the SDF associated with these elements and the volume of every element in terms of these.

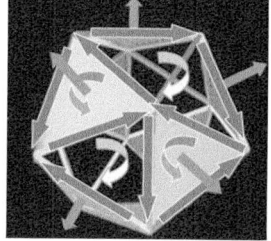

Fig. 5.21 A sketch of a cell (pore) in the 3D polyhedral structure. The cell is polyhedron-like, whose triangular facets (shaded) are the faces of the polyhedra that represent the grains around it. The remaining facet-like openings are non-planar and comprise the throats to neighbouring cells.

The construction of the 3D volume elements, which for consistency are also called quadrons, follows the same logic as in 2D. The first step is to connect the contact points around a grain in such a way that these form a polyhedron all of whose facets are triangles. The edges around each triangle are assigned directions, such that they circulate around the facet in the anti-clockwise direction when observed from a point inside the grain outwards. Generically (although not always, see below), every triangle opens to a specific cell (or pore), on the surface of which the grain resides. The contact points are the vertices of the polyhedra and each is shared by two neighbouring polyhedra (see Figure 5.5).

Consider now looking out from a vantage point inside a cell, c. What we see is a cell 'surface' made of the triangles of the polyhedra of the grains that surround the cell (Figure 5.7). These triangles enclose polygons on this surface, each of which being the 'throat' that leads from one cell to one of its neighbour. These polygons are non-planar. A sketch of such a cell is shown in Figure 5.21. Each triangle edge vector (arrows in the figure) is indexed r_{gcp}. This means it can be associated with grain g, it circulates a triangle that resides on the surface of cell c, and it is also one of the edges of polygon p. It is important to note that if c and c' are neighbouring cells sharing a throat p, then the triangles of grain g, which open to these two cells share an edge. These edges assume opposite directions, depending on whether it is r_{gcp} or $r_{gc'p}$,

$$r_{gcp} = -r_{gc'p}.$$

To simplify the description, we illustrate the method for tetrahedral structures, where every grain has exactly four contact points. This is an important class of structures because of its ubiquity in nature and in numerical applications. All cellular structures belong to this class, as well as all liquid and solid foams and Voronoi tessellations. In all these structures, exactly four edges always meet at a vertex, making the description below applicable to them all. Moreover, if one constructs the skeleton of a porous material, one finds that almost all vertices of the skeleton are also four-fold. In all these structures the polyhedra around the grains (or the polyhedra that 'dress' the vertices, see [26]) are tetrahedra. A moment's reflection will convince the reader that each triangular facet of each tetrahedron resides on the surface of one and only one cell. It follows that a tetrahedron is shared by four cells. The extension to more general polyhedral structures is complicated but possible; it will not be described here.

Having defined the polyhedral facets and their edges, we now define the centroids of the triangles as the mean position vectors of the contacts around it, ρ_{cg} (see Figure 5.5). We also define: the centroids of the polygons p as the mean position vectors of their vertices, ρ_{gcp}, the centroids of the grains as the mean position vectors of their contact points, ρ_g, and the centroids of the cells as the mean position vectors of their vertices, ρ_c.

It is now possible to show that the 3D space can be fully tessellated, using a similar procedure as in 2D, detailed in [26,85]. The basic volume elements, the quadrons, are generically octahedra, whose shapes can be quantified with the following three vectors, illustrated in Figure 5.5: r_{gcp}, $\xi_{gcp} = \rho_{gcp} - \rho_{cg}$, and $R_{gc} = \rho_c - \rho_g$. For every quadron we construct the shape tensor

$$(C_{gcp})_{ij} = (r_{gcp} \times \xi_{gcp})_i (R_{gc})_j \quad \text{or} \quad C_{gcp} = (r_{gcp} \times \xi_{gcp}) \otimes R_{cg} \tag{5.75}$$

where the term in the brackets on the right-most-hand side is a vector product between the vectors r_{gcp} and ξ_{gcp}. The volume of the quadron in terms of this tensor is

$$V_q = \frac{1}{3!} Tr\{C_q\}, \tag{5.76}$$

where we have shortened the indexing by identifying the indices gcp with a specific quadron q.

This description is particularly convenient because the volume associated with a grain can be then defined as the sum of the volumes of the quadrons that belong to this grain, $V_g = \sum_{cp} V_q$. This means that we can define a shape tensor associated with grain g

$$C_g = \sum_{cp} C_{gcp} = \sum_{q \in g} C_q, \tag{5.77}$$

such that

$$V_g = \frac{1}{3!} Tr\{C_g\}$$

and the volume of an arbitrary region Γ is

$$V_\Gamma = \frac{1}{3!} Tr\{C_\Gamma\} = \sum_{q \in \Gamma} V_q . \qquad (5.78)$$

This method allows us to quantify any region in the disordered granular structure, from the grain scale up, by a local shape tensor in 3D. The method is useful to many problems where system behaviour needs to be correlated with structural features. In particular, this description is useful for the statistical mechanical formalism in that it provides a straightforward volume function that depends on the SDF.

Prompted by the observations in 2D, it is instructive to compare the number of quadrons to the number of SDF. We have already seen above that of the latter there are $N_{sdf} = N\bar{z}d/2$. The number of quadrons per grain in tetrahedral structures is calculated as follows. There are three quadrons for every triangular face and four such faces, giving altogether $12N$ quadrons. Since the number of contacts per grain is $\bar{z} = 4$, there are altogether $N_{sdf} = 6N < N_q$.

It can be shown that, for non-tetrahedral structures, the number of quadrons is bounded from above by (but is close to)

$$N_{q,max} = 6(\bar{z} - 2)N. \qquad (5.79)$$

Comparing to the number of SDF we have

$$\frac{N_{q,max}}{N_{sdf}} = 4 - \frac{8}{\bar{z}} \geq 2. \qquad (5.80)$$

The inequality on the right hand side of equation (5.80) is due to the requirement of mechanical stability, which can be satisfied only for $\bar{z} \geq 4$. This relation shows clearly that the number of quadrons exceeds the number of SDF, which means that not all quadron volumes are independent. It follows that only a subset of N_{sdf} of the quadron volumes can be used as the variables for the integration of the partition function.

The canonical volume ensemble

The micro-canonical ensemble is not a useful description since in most experimental situations the total volume of the system fluctuates from packing to packing. This is reminiscent of thermal systems coupled to heat reservoirs, where the energy of the system fluctuates. One can then regard the entire system as composed of a reservoir (providing volume fluctuations) and the experimental pack. Assuming the probability of occurrence of a system i, p_i, follows the same Boltzmann-factor-like behaviour, but with the volume replacing the energy, we have the same result as in conventional statistical mechanics,

$$p_i = \frac{1}{Z_v} e^{-\frac{v_i}{X}} \qquad (5.81)$$

where Z_v is known as the structural (or volume) partition function, v_i is the volume of the ith packing (or structural configuration), and X

is the 'compactivity' - a scalar factor that characterises volume fluctuations analogous to the way that the temperature quantifies thermal fluctuations. This factor is defined as the derivative of the mean volume of an ensemble of such systems, $\langle V \rangle$ with respect to the entropy, $X = \partial \langle V \rangle / \partial S$, paralleling the definition of the temperature. The partition function is then

$$Z_v = \int e^{-\frac{W(\boldsymbol{u})}{X}} d^{N_{\text{SDF}}} \boldsymbol{u} , \qquad (5.82)$$

where $\boldsymbol{u}$ is a vector of all the SDF. As in the micro-canonical description, any structural feature, A, can be obtained as an expectation value over the partition function as long as it depends on the SDF,

$$\langle A \rangle = \int A(\boldsymbol{u}) e^{-\frac{W(\boldsymbol{u})}{X}} d^{N_{SDF}} \boldsymbol{u} . \qquad (5.83)$$

For example, the mean volume and volume fluctuations are

$$\langle V \rangle = \frac{1}{Z_v} \sum_q V_q e^{-V_q/X} = -\frac{\partial \ln Z_v}{\partial (1/X)} ; \qquad (5.84)$$

$$\langle V^2 \rangle - \langle V \rangle^2 = -\frac{\partial^2 \ln Z_v}{\partial (1/X)^2} . \qquad (5.85)$$

The quasi-particles of the volume ensemble

In conventional statistical mechanics the Hamiltonian represents the energy of entities and their interactions. These entities could be either real particles, such as atoms, molecules, etc. or they could be energy-carrying quasi-particles, such as spins, discrete frequencies, phonons, and so on. The nature of these entities depends on the system under study. Examples are: for gases these are the gas molecules, for magnetic systems - spins (Ising, XY, Heisenberg), and for black body radiation, superconductivity, and vibrations in solids - wave vectors and frequencies.

The question arises which are the most natural, or useful, such entities in static granular statistical mechanics. The natural tendency would be to assign this role to the grains. After all, they are the basic moving particles when the material flows and it is for them that Newton's equations apply, both kinematically and in mechanical equilibrium. Yet, to include only the volumes of the grain V_g in the volume function and treat them as SDF raises a conceptual problem since the number of grains is smaller than the number of SDF, $N < N_{\text{SDF}} = N\bar{z}d/2$. In contrast, there are sufficiently many quadrons to fit this bill. As shown above, in 2D their number equals exactly that of the SDF and in 3D $N_q > N_{\text{SDF}}$. This implies that one can choose the quadrons to be the quasi-particles of the granular statistical mechanics description. In 2D this means that we can integrate directly on their volumes, albeit with inclusion of an analogue of the 'density of states' (see below). In 3D, one can also write the partition function as an integral over *a subset N_{SDF} of the quadron volumes*, expressing the other quadrons in terms of those. This means that correlations between quadrons cannot be ignored.

To illustrate the use of quadron volumes as SDF, let us first consider a simple model in 2D - the ideal quadron gas approximation [25]. In this model, the quadrons are considered independent and the partition function can be written as

$$Z_v = \left[\int e^{-\frac{V}{X}} g(V) dV\right]^{N\bar{z}}, \qquad (5.86)$$

where $N\bar{z}$ is the number of quadrons. The function $g(V)$ is an analogue of the density of states. For illustration, suppose that the quadrons can assume one of two volumes, $V_0+\Delta$ or $V_0-\Delta < V_0+\Delta$, with probabilities p and $1-p$, respectively. The density of states is then

$$g(V) = p\delta(V - V_0 - \Delta) + (1-p)\delta(V - V_0 + \Delta). \qquad (5.87)$$

It should be noted that the distribution of the volumes associated with grains is much broader since the volume of every grain g is made of z_g quadrons, with each able to assume one of these two volumes. The partition function can be computed explicitly

$$Z_v = [\cosh(\Delta/X) + (1-2p)\sinh(\Delta/X)]^{N\bar{z}} e^{-\frac{N\bar{z}V_0}{X}}. \qquad (5.88)$$

The mean volume of this system per grain is

$$\langle V \rangle = -\frac{1}{N}\frac{\partial \ln Z_v}{\partial(1/X)} = \bar{z}\left[V_0 - \frac{\sinh(\Delta/X) + (1-2p)\cosh(\Delta/X)}{\cosh(\Delta/X) + (1-2p)\sinh(\Delta/X)}\Delta\right]. \qquad (5.89)$$

A plot of the mean volume per grain is shown in Figure 5.22. As the compactivity increases, a transition can be seen from $\bar{z}(V_0 - \Delta)$, when $X_0 \ll \Delta$, to $\bar{z}[V_0 - (1-2p)\Delta]$ when $X_0 \gg \Delta$.

The variance of the volume fluctuations per grain can be similarly calculated,

$$\langle \delta^2 V \rangle = -\frac{1}{N}\frac{\partial^2 \ln Z_v}{\partial(1/X)^2} = \frac{4p(1-p)\Delta^2}{[\cosh(\Delta/X) + (1-2p)\sinh(\Delta/X)]^2} \qquad (5.90)$$

and its dependence on the compactivity is also shown in Figure 5.22. This quantity also makes a transition with increasing compactivity from $2pe^{-2\Delta/X_0}$, when $X_0 \ll \Delta$, to $4p(1-p)$, when $X_0 \gg \Delta$. This is essentially the analogue of the Bragg-Williams phase transition in thermal systems.

The stress ensemble

Even before the debate on the applicability of the volume ensemble could be settled, another issue was tossed into the field, complicating matters further. It was proposed that the entropy of random granular assemblies must include also the different stress states [26, 81, 86]. This suggestion was promptly put to numerical tests, which seemed to support it [81, 87]. The idea is that one normally applies external stresses on a GS either via boundaries, such as plates or the walls of a vessel, or via body forces,

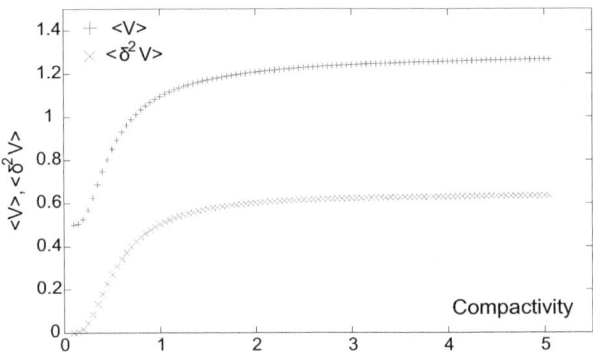

Fig. 5.22 The volume statistics of the binary ideal quadron gas model as a function of compactivity. As the compactivity increases, the mean volume per grain (+) undergoes a transition from $\langle V(X_0 \ll \Delta)\rangle = \bar{z}(V_0 - \Delta)$ to $\langle V(X_0 \gg \Delta)\rangle = \bar{z}[V_0 - (1-2p)\Delta]$. The expected value of the variance of the volume fluctuations, $\langle \delta^2 V\rangle$ (×), also changes with increasing compactivity from $2pe^{-2\Delta/X_0}$, when $X_0 \ll \Delta$, to $4p(1-p)$, when $X_0 \gg \Delta$.

e.g. gravity. For simplicity, we consider again only the former case. The boundary stress, $\sigma_{\alpha\beta}$, is an average of the total α force component on the boundary, per unit surface area in the β direction. For example, on a flat boundary whose normal is in the n direction, σ_{nn} would be the total force applied on it in that direction. In effect, the boundary stresses are transmitted to the system by direct forces on the grains that come into contact with the boundary. In other words, the boundary loading comprises a collection of individual forces applied to particular boundary grains. It follows that, to specify the boundary loading precisely requires identifying the boundary grains and specifying the position of each of these grains as well as the force vector applied to it.

Clearly, there are astronomically many combinations of such forces, giving rise to exactly the same boundary stress. In conventional materials this would not be a problem because above some coarse-grained scale, one expects this force distribution to be homogeneous. Therefore, in materials that transmit stresses uniformly, such as elastic media, the discreteness of the boundary forces blurs and hardly affects any macroscopic observable. This is not the case in granular materials. We saw that stresses in GS are not uniform and individual force chains emanate from individual forces at the particle scale. Thus, in these systems, the distribution of the boundary stresses strongly affects the force chain network inside the medium and hence the stress state.

It is this phenomenon that the stress ensemble attempts to capture. Every stress state is unique and the disorder of these 'micro-states' adds to the aforementioned disorder of the structure. The logarithm of the number of such stress micro-states is defined as the stress entropy. The phase space is then made up from the ensemble of all possible such micro-states, namely all the possible forces and boundary grains combinations that give rise to the same boundary stress.

Most convenient is to illustrate this ensemble in isostatic systems, in which, due to the static determinacy and the linearity of the equations, the number of stress micro-states is exactly the same as the number of possible combinations of individual forces on the boundary. Let us define the number of such boundary force sources at any one configuration as M. Significantly, the boundary forces $\boldsymbol{g}_i$, $(i=1,2,...,M)$ are the independent degrees of freedom - the inter-granular forces are determined from them by Newton's equations.

Limiting our discussion to 2D, for simplicity, the stress ensemble is described by the following partition function

$$Z_f = \int e^{-\sum_{ij} \frac{1}{X_{ij}} \mathcal{F}_{ij}} d\{\text{all boundary forces}\}. \quad (5.91)$$

Here the indices i, j run over the Cartesian components x, y and $\mathcal{F}_{ij}$ are the components of the *internal* force moment function, from which the stress σ_{ij} is derived,

$$\mathcal{F} = \sum_g V_g \sigma_g = \sum_{gg'} \boldsymbol{\rho}_{gg'} \otimes \boldsymbol{F}_{gg'} . \quad (5.92)$$

Here the sum runs over all pairs of grains in contact within the volume of interest, gg', $\boldsymbol{\rho}_{gg'}$ is the position of the contact point between these grains, measured from the centroid of grain g, $\boldsymbol{F}_{gg'}$ is the force that g' applies to g, and V_g is the volume associated with grain g, namely, the sum of the volumes of all the quadrons associated with it. The quantity $X_{ij} = \partial \mathcal{F}_{ij}/\partial S$ has been named 'angoricity' by Edwards and Blumenfeld [86] and it is a tensorial analogue of the temperature and the compactivity [88, 89]. S is the entire entropy associated with both the volume and the stress ensembles.

There are still several open questions concerning this ensemble, not all of them addressed in any detail in the literature. For example, when the granular system is not isostatic the number of stress states is not determined solely by the boundary forces; other variables need to be introduced to solve for the stress. These variables are also part of the phase space and should act as further degrees of freedom. Yet, so far there is no discussion, let alone identification of these variables. By specialising here to isostatic systems we circumvent this difficulty. Another issue is that, even in isostatic systems, the number of individual forces on the boundary, M, may fluctuate. This means that we also have to take into consideration the statistics of these fluctuations. This issue is also unaddressed so far in the literature.

Inter-dependence of the volume and stress ensembles

A significant misconception in the early literature on statistical mechanics of GS is the assumption that the stress and the volume ensembles are independent. Practically, all the earlier results used either of these ensembles alone. We proceed to show that this practice is incorrect, namely, that these two ensembles are coupled [90]. This can be based on the following three arguments, which, taken together, establish the inter-dependence of the volume and the stress ensembles.

1. *The volume ensemble alone does not capture all the entropy of mechanically stable granular systems.*
The volume ensemble consists of all the possible structural arrangements

under exactly the same boundary forces. However, no repeated experiment on a collection of many grains can reproduce the same precise forces on every boundary grain. Experimentally, only global boundary stresses can be controlled, i.e. the averages of boundary force components. Thus, the statistics of the boundary forces must be taken into consideration.

2. *The stress ensemble alone does not capture all the entropy of mechanically stable granular systems.*

The stress ensemble is composed of all the possible stress states that a fixed granular configuration can assume. In our isostatic systems, each state is achieved as a unique response to a specific combination of the boundary forces. The ensemble of boundary forces is subject to the constraint that the total boundary stresses are fixed. However, a little reflection will convince the reader that such a system cannot be realised experimentally in very large assemblies - any small change in the boundary forces is bound to change the internal structure.

3. *The volume and stress partition functions are inter-dependent, $Z \neq Z_v Z_f$.*

This statement is a consequence of the explicit mathematical dependence of both the volume function in Z_v and the force moment function in Z_f on the structural degrees of freedom. This argument, which follows the analysis in [90], is quantified in detail in the next subsection.

The implication of these arguments is that correct calculations of expectation values must be based on a combined ensemble of all structural arrangements and all boundary forces. Correspondingly, this means that the phase space consists of both the SDF and all the possible boundary forces. These arguments are independent of the dimensionality, holding both in 2D and in 3D. It is easier to illustrate this issue explicitly in 2D, where exact calculations are possible. Such an illustration is presented next, based on the work in [90].

Consider the ensemble of all N-grain systems ($N \gg 1$) in 2D, each system having the same mean contact number $\bar{z}$. All the systems are prepared by the same procedure, e.g. by pouring the same collection of grains into a box and compressing the system with the box boundaries or by shearing the system at a certain rate, etc. The systems are in mechanical equilibrium under M external compressive forces, acting on the boundary grains. We again emphasise that we disregard body forces, in the absence of which 'rattlers' can also be ignored, as they do not support forces and therefore do not affect the stress states in static piles. The discussion can be readily extended to include body forces and rattlers.

In the following, we calculate explicitly Z_v and Z_f for this ensemble and then we calculate the combined partition function Z, showing that the latter is not a product of the former two, $Z \neq Z_v Z_f$.

Evaluation of the volume partition function

As discussed previously, the number of SDF in each system is $N\bar{z}$, corresponding to the components of half of all the $\boldsymbol{r}$-vectors. Let us denote these as one long vector, $\boldsymbol{\rho} \equiv \left(r_{1x}, r_{2x}, ..., r_{N\bar{z}/2x}, r_{1y}, r_{2y}, ..., r_{N\bar{z}/2y}\right)$,[6] where r_{ni} is the ith component ($i = x, y$) of the nth vector $\boldsymbol{r}_n$. In terms of the $\boldsymbol{r}$'s, the volume function is exactly quadratic, $W = \frac{1}{2}\sum a_{\alpha\beta}^{qp} r_{iq} r_{jp} = \frac{1}{2}\boldsymbol{\rho} \cdot A \cdot \boldsymbol{\rho}$, where p, q run over quadrons, i, j run over vector components x, y, and A is a matrix whose elements are

$$(A)_{\alpha\beta}^{qp} = \frac{1}{X_0} \begin{cases} a_{xx}^{qp} & q, p \leq N\bar{z}/2 \\ a_{xy}^{qp} & q \leq N\bar{z}/2 \,, \, p > N\bar{z}/2 \\ a_{yx}^{qp} & p \leq N\bar{z}/2 \,, \, q > N\bar{z}/2 \\ a_{yy}^{qp} & q, p > N\bar{z}/2 \end{cases}$$

In the following, we assume that this matrix is not degenerate, i.e. that it has $N\bar{z}$ eigenvalues. This assumption is being investigated currently.

The volume partition function then has the following simple form

$$Z_v = \int e^{-\frac{1}{2}\boldsymbol{\rho} \cdot A \cdot \boldsymbol{\rho}} d^{N\bar{z}}\boldsymbol{\rho} \,. \tag{5.93}$$

[6] The vector $\boldsymbol{\rho}$ should not to be confused with the position vectors of the contact points $\boldsymbol{\rho}_{gg'}$.

It should be noted that the Gaussian form of the integral in equation (5.93) is exact, as opposed to a plethora of Gaussian approximations in conventional statistical mechanics of thermal systems. Here it is a direct consequence of the exact quadratic form of the volume function in the SDF $\boldsymbol{r}$.

The Gaussian nature of the integral makes it possible to evaluate the volume partition function exactly. Assuming a uniform measure of the degrees of freedom and noting that, for a grain size distribution with a maximum finite size, the contribution of very large $\boldsymbol{r}$ magnitudes vanishes, we obtain straightforwardly that

$$Z_v = \sqrt{\frac{(2\pi)^{N\bar{z}}}{\det A}} \,. \tag{5.94}$$

Evaluation of the stress partition function

To evaluate the stress partition function, we need to write it first in a convenient form. As discussed, Z_f sums over all the combinations of the M compressive forces on the boundary grains, $\boldsymbol{g}_1, \boldsymbol{g}_2, ..., \boldsymbol{g}_M$, subject to the condition that the total stresses on the boundaries are pre-specified. Since the structure of the system is presumed unchanged for all these combinations, the only permissible combinations are those that do not drive the system out of mechanical equilibrium, namely, the boundary stresses must be below the yield threshold.

The key to the analysis of the combined partition function (below) is via the loop force formalism, introduced in [24]. We therefore introduce the use of this formalism here. This formalism is based on parameterising the inter-granular forces, $\boldsymbol{F}_{gg'}$, in terms of loop forces, $\boldsymbol{f}_c$, such that every

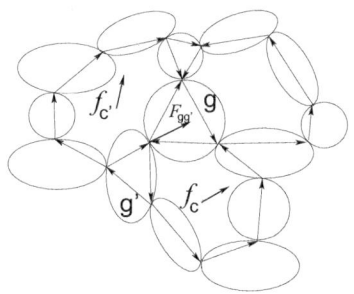

Fig. 5.23 The inter-granular forces are parameterised by loop forces. Here, the force that grain g' applies to grain g, $\boldsymbol{F}_{gg'}$, is parameterised by the two loop forces that share this contact, $\boldsymbol{F}_{gg'} = \boldsymbol{f}_{c'} - \boldsymbol{f}_c$ [24]. These loop forces automatically satisfy the force balance conditions. In isostatic systems, they are determined by the torque balance conditions alone.

inter-granular force can be written in terms of the difference between the loop forces associated with the two loops, c and c', that share the contact between grains g and g' (see Figure 5.23), $\boldsymbol{F}_{gg'} = \boldsymbol{f}_{c'} - \boldsymbol{f}_c$. The loop forces are reminiscent of Maxwell's construction [91]. The significant advantage of this parameterisation is that, by definition, the loop forces automatically satisfy Newton's force balance conditions on every grain. In 2D assemblies of arbitrarily-shaped frictional grains, there are $N(\bar{z}/2 - 1)$ loop forces (a consequence of the Euler relation), of which $N/2$ can be determined from the torque balance conditions. In the isostatic systems we discuss here, all the loop forces can be determined from the torque balance conditions.

The force moment function, equation (5.92), can be expressed in terms of the loop forces [24, 25],

$$\mathcal{F}_{ij} = \sum_{gc} f_{q_i} r_{qj} , \qquad (5.95)$$

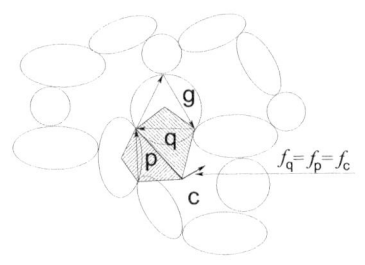

Fig. 5.24 The force associated with quadron q is defined as the loop force of the cell c that contains it, $\boldsymbol{f}_q = \boldsymbol{f}_c$ [90]. Quadrons q and p belong to the same loop and hence have the same loop force $\boldsymbol{f}_q = \boldsymbol{f}_p = \boldsymbol{f}_c$.

where $\boldsymbol{f}_{q_i}$ is the ith component of a loop force of the cell containing the quadron q (see Figure 5.23) and r_q is the $\boldsymbol{r}$ vector of quadron q. The advantage of this form, which will become clear as we use it below, is that it uses the same SDF, the vectors $\boldsymbol{r}_q$ that appear in the volume partition function.

Substituting equation (5.95) into the stress partition function (5.91) gives

$$Z_f = \int e^{-\sum_{q,i,j} \frac{1}{X_{ij}} f_{q_i} r_{qj}} \prod_{m=1}^{M} d^{2M} \{\boldsymbol{g}_m\} , \qquad (5.96)$$

where the integration runs over all the independent boundary forces $\boldsymbol{g}_m$. The loop forces depend linearly on the inter-granular forces, which in turn, depend linearly on the M boundary forces. Thus, each loop force is also a linear combination of the boundary forces. As illustrated in Figure 5.24, quadrons sharing the same cell have the same loop force. Consequently, only $N/2$ of the $N\bar{z}$ quadron forces are independent.

Following the same rationale as for the volume function, it is convenient to define a loop forces vector $\boldsymbol{\phi} \equiv (f_{1x}, f_{2x}, ..., f_{N/2x}, f_{1y}, f_{2y}, ..., f_{N/2y})$. Since the stress degrees of freedom are the boundary forces, we need to express the loop forces in terms of those boundary forces. This is done by first expressing the $N\bar{z}/2$ inter-granular forces in terms of the M boundary forces and then, using the definition of the loop forces, in terms of the contact forces. The result is

$$\phi_{ci} = C_{cmij} g_{mj} , \qquad (5.97)$$

where $i, j = x, y$; $c = 1, 2, \ldots, N/2$ runs over all cells, $m = 1, 2, \ldots, M$ runs over all boundary forces, and C is a matrix of $N \times 2M$. In terms of these, $\boldsymbol{f}_q = E\boldsymbol{\phi}$, where E is a matrix of $N\bar{z} \times N$. Defining further $B_{ij}^{qp} = (X^{-1})_{ij} \delta_{qp}$, with δ_{qp} being the delta function, we obtain the desired expression for Z_f

$$Z_f = \int e^{-\phi \cdot E^T \cdot B \cdot \rho} \prod_{m=1}^{M} d^{2M}\{g_m\} = \int e^{-g \cdot C^T \cdot E^T \cdot B \cdot \rho} \prod_{m=1}^{M} d^{2M}\{g_m\}. \tag{5.98}$$

The integrand is a simple exponential in the variables g_{mi} and therefore the integration is straightforward to carry out. The result, however, depends on SDF, namely the vectors r, which are not integrated on! This is a direct result of the incorrect treatment of the stress ensemble as independent.

The combined volume-stress partition function

To compute the total volume-stress partition function, we have to consider the combined volume-stress phase space, $dZ = dZ_v dZ_f$. The exponential dependence on the boundary forces and the Gaussian dependence on the SDF make the integration possible in principle. However, in practice, the involvement of the matrices A, B, C, and E makes the derivation cumbersome. For brevity, we define the matrix $Q = B^T \cdot E \cdot C$, in terms of which the partition function becomes

$$Z = \int e^{-\frac{1}{2}\rho \cdot A \cdot \rho - g \cdot Q^T \cdot \rho} \left(d^{N\bar{z}}\rho\right)\left(d^{2M}g\right) \tag{5.99}$$

This expression already demonstrates the above claim of inter-dependence of the partition functions. It clearly shows that Z is *not* the product of Z_v of equation (5.94) and Z_f of equation (5.98) – the vector of SDF, ρ, appears in both the volume and the stress integrals.

Having the correct partition function, we can start using it to obtain the important expressions – expectation values of measurable quantities. The most straightforward, and arguably the most usable, expectation value is the mean volume,

$$\langle V \rangle = \frac{X_0}{2Z} \int (\rho \cdot A \cdot \rho) \ e^{-\frac{1}{2}\rho \cdot A \cdot \rho - g \cdot Q^T \cdot \rho} \left(d^{N\bar{z}}\rho\right)\left(d^{2M}g\right). \tag{5.100}$$

We can 'complete the square' in the exponent by defining $\tilde{\rho} = \rho + A^{-1}Qg$ and express the variables in terms of $\tilde{\rho}$ and g. The result is

$$\langle V \rangle = \frac{X_0}{2Z} \int (\tilde{\rho} \cdot A \cdot \tilde{\rho} + gPg) \ e^{\frac{1}{2}(-\tilde{\rho} \cdot A \cdot \tilde{\rho} + gPg)} \left(d^{N\bar{z}}\tilde{\rho}\right)\left(d^{2M}g\right), \tag{5.101}$$

where $P \equiv Q^T \cdot A^{-1} \cdot Q$. This decouples the integrals and they can be computed separately to give a very compact result

$$\begin{aligned}\langle V \rangle &= \frac{X_0}{2}\left[\frac{\int \tilde{\rho} \cdot A \cdot \tilde{\rho}\ e^{-\frac{1}{2}\tilde{\rho} \cdot A \cdot \tilde{\rho}}d^{N\bar{z}}\tilde{\rho}}{\int e^{-\frac{1}{2}\tilde{\rho} \cdot A \cdot \tilde{\rho}}d^{N\bar{z}}\tilde{\rho}} + \frac{\int g \cdot P \cdot g\ e^{-\frac{1}{2}g \cdot P \cdot g}d^{2M}g}{\int e^{-\frac{1}{2}g \cdot P \cdot g}d^{2M}g}\right] \\ &= \frac{\bar{z}N + 2M}{2}X_0\ . \end{aligned} \tag{5.102}$$

This result is significant for a number of reasons. First, it is independent of the details of the connectivity matrix A and, therefore, of the statistics of the structure. This means that it holds for any preparation procedure, in spite of the fact that one of the constraints on the ensemble is that all its systems are prepared by the same procedure. Carrying out this calculation on ensembles whose systems are made by another procedure would give exactly the same result. Similarly, it is also independent of the statistics of the boundary forces – it only depends on their number. This means that this result is also independent of the statistics of the stress state configurations, which are determined by the boundary forces. Thus, this result is powerfully general.

Second, it reveals a striking equipartition principle: the mean volume, which is seemingly a structural property, is shared equally among *both* the $\bar{z}N$ SDF and the $2M$ force degrees of freedom, each getting on average a volume of $X_0/2$. This is the exact equivalent of the equipartition principle obtained for gases in thermal systems, $\frac{3k_B T}{2} N_{molecules}$.

Third, it expresses the compactivity X_0 in terms of the bulk quantities $\langle V \rangle$ and the total number of degrees of freedom. The former is directly measurable and the latter can be estimated in exactly the same way that we estimate the number of molecules to obtain the temperature in the classical expression for thermal gases. Thus, expression (5.102) makes it possible to measure the elusive compactivity in real systems from experimental measurements. In other words, we have obtained a 'compactometer' – the analogue of a thermometer.

It should not be surprising that the mean volume, a purely structural quantity, depends only on the compactivity X_0 and is independent of the angoricity X_{ij}. However, this is not the case with all structural quantities - they may not depend only on the compactivity. To demonstrate this, let us consider the expectation value of another structural quantity, the magnitude of the square of the vector $\boldsymbol{\rho}$,

$$\begin{aligned} \langle \boldsymbol{\rho} \cdot \boldsymbol{\rho} \rangle &= \langle | \tilde{\rho} - A^{-1} Q g |^2 \rangle \quad (5.103) \\ &= \frac{1}{Z} \int | \tilde{\rho} - A^{-1} Q g |^2 \; e^{\frac{1}{2}(-\tilde{\rho} A \tilde{\rho} + g P g)} \left(d^{N\bar{z}} \tilde{\rho} \right) \left(d^{2M} g \right) . \end{aligned}$$

The evaluation of this integral is straightforward, albeit involving a few diagonalisation steps. Defining the matrix that diagonalises P as Y and a new matrix $T = Y^T \cdot Q^T \cdot A^{-1} \cdot A^{-1} \cdot Q \cdot Y$, this integral can be evaluated exactly,

$$\langle \boldsymbol{\rho} \cdot \boldsymbol{\rho} \rangle = Tr\{A^{-1}\} + \sum_i^{2M} \frac{T_{ii}}{p_i} , \quad (5.104)$$

where p_i is the ith eigenvalue of the matrix P. To assess the dependence of the result on the angoricity and the compactivity, we note that Q is linear in the inverse angoricity and that P is therefore linear in the compactivity (since it is linear in the matrix A^{-1}) and it is a homogeneous function of order 2 in the inverse angoricity. From its definition, the matrix T is a homogeneous function of order 2 in the angoricity. Thus the second term in expression (5.104) is a homogeneous function of order zero in the angoricity. It follows that the expectation value of the purely structural quantity $\langle \boldsymbol{\rho} \cdot \boldsymbol{\rho} \rangle$ is not only linear in the compactivity X_0, as expected, but it also depends, unexpectedly, on the angoricity!

Another quantity of interest, studied much in the literature, is the mean magnitude square of the inter-granular forces, $\boldsymbol{F}$. Since these are simply related to the loop forces via $\boldsymbol{F}_{gg'} = \boldsymbol{f}_{c'} - \boldsymbol{f}_c$ (see above), we can calculate instead the mean magnitude square of $\langle \boldsymbol{f} \cdot \boldsymbol{f} \rangle$. The calculation of this integral follows the same lines as that of $\langle \boldsymbol{\rho} \cdot \boldsymbol{\rho} \rangle$ and it yields

$$\langle \boldsymbol{f} \cdot \boldsymbol{f} \rangle = \langle \boldsymbol{g} \cdot C^T \cdot E^T \cdot E \cdot C \cdot \boldsymbol{g} \rangle. \tag{5.105}$$

This expression can be written in terms of the matrices defined above, but this would not give more insight into the issue. What is interesting, however, is that this quantity, which is a measure of the mean magnitude square of the inter-granular forces, is not only a homogeneous function of order 2 in the angoricity tensor, as expected, but it is also surprisingly linear in the compactivity X_0.

These dependencies of structural properties on the angoricity tensor and pure force-based quantities on the compactivity are not only a direct result of the inter-dependence of the volume and stress partition functions, but also could not be obtained from an independent analysis of either. This conclusion, reached in [90], implies that many early results based on the assumption that these ensembles are independent, need to be revisited and possibly corrected.

Conclusion and discussion

To conclude this section, the entropy-based statistical mechanics of static granular matter is still under development. Although aspects of this formalism have been supported [77, 79, 80, 92], many fundamental issues remain open. Several such issues are: the identification of the proper degrees of freedom and counting their number; the difficulty in realising proper analogs of the thermometer for the structural and stress degrees of freedom - the compactometer and the angorometer. A particular handicap, compared with traditional statistical mechanics, is the absence of ergodicity - an equivalence between the statistics of an ensemble of static systems and those of the same system at different times under some dynamics. The latter problem makes it extremely difficult to infer from ensemble statistics on dynamical properties, such as diffusion (although attempts to form such a link have been made [78, 89]). The field also suffers from a number of misconceptions, e.g. mistaking the grain volumes for degrees of freedom, ignoring the inter-dependence of the volume and stress ensembles and misguided applications of ergodicity. All these difficulties have slowed down considerably progress in the field. Indeed, the useful results obtained so far by statistical mechanics are still few and far between, but they start to emerge.

These open issues notwithstanding, the statistical mechanical approach has enormous potential. It is the basis for derivation of equations of state, phase diagrams, and constitutive relations. It is also a springboard that can be used to construct a statistical mechanical description of the dynamics of granular matter. Static granular matter is commonly regarded in the community as the analogue of the zero-temperature limit of conventional classical particle systems. This is because, in most relevant situations, there is insufficient thermal energy to shuffle the systems from micro-state to micro-state. In conventional statistical mechanics this does not pose any conceptual problem because the system can exist in only one micro-state and the entropy vanishes. In granular systems, this is not the case - the 'zero energy' limit has an astronomically large number of configurations that the system may realise, both structural and stress-related. Therefore, the analogy to the zero-temperature state of thermal systems is not helpful. This also implies that one needs to consider carefully the degrees of freedom that describe the statistical mechanics of dynamical granular systems. As the velocities of the particles slow down and they eventually come to rest, the total phase space should reduce correspondingly to that of the static statistical mechanics, namely to a phase space spanned by structural and stress degrees of freedom. This issue, which may in fact be relevant to systems beyond the granular, has not been realised at all in the literature, let alone studied.

In this chapter, we have described a natural identification of the 'quasi-particles' of the structural statistical mechanics - the quadrons. They are the basic elements of the structure in the same way that real and quasi-particles are the basic elements for the purpose of energy evaluation in conventional systems. The quadrons not only make it possible to identify and treat conveniently the integration over the degrees of freedom but they also are fundamental to the theory of stress transmission in granular systems. Their advantages over traditional structural descriptions have been discussed and demonstrated.

Using this description, and using the corrected partition function in 2D, makes it possible to derive a number of useful results, including expectation values of several experimentally relevant quantities. We have reviewed the derivation of an equipartition principle, stating that the volume divides on average equally among the structural and the stress degrees of freedom, with each degree of freedom getting on average a volume of $X_0/2$. This allows one, in principle, to estimate the compactivity from experimental measurements. These are recent results that still need to be tested numerically and experimentally. Many more results need to be derived, especially in 3D. The equivalent derivation in 3D is more difficult because the volume function is cubic in the structural degrees of freedom, eliminating the fortuitous convenience of the 2D Gaussian form. Nevertheless, numerical computations of the relevant integrals are straightforward to carry out and, although the results may not be possible to present in a closed analytical forms as in 2D, they can still be calculated and plotted numerically, forming a test bed for experimental measurements.

5.5 Future directions

To conclude, the science of granular matter advances in several main directions: statistical mechanics of static and dynamic systems, dynamic and static response to stress, failure and yield, prediction of packing fractions, rheology and fully developed flow, and understanding the rich patterns that develop under different excitations. Of the many problems that granular science poses, we have chosen to focus on progress in two key areas: stress transmission in mechanically equilibrated GS and the current state of the statistical mechanical theory of static GS. We believe that progress in these areas is crucial to the development of the field in general. However, even reviewing these two directions comprehensively is well beyond the scope of one chapter and we therefore focused only on the very recent advances. It is unfortunate that the space here is too short to review progress on the third fundamental aspect of the field - the development of a theory of flow for dense granular materials.

Since constructing predictive science of granular matter must start from the granular level and involve coarse-graining to the continuum, we first reviewed a recent method to quantify the grain-level structure and discussed some of its uses. This method is promising in that it enables one to relate structural characteristics quantitatively to macroscopic behaviour. A step in this direction has been made recently for random systems in 2D [29], where it was shown that the quadron description makes it possible to identify 'universal' structural features, i.e. to collapse distributions of structural properties of different systems onto one general form.

Much more work is required on quantitative structural characterisation of generally disordered GS. This is crucial for quantifying relations between the disordered structural characteristics and macroscopic properties, such as transport properties, stress transmission, heat transport, catalysis, and reactive transport in the porous medium, etc. In the absence of such a fundamental description, any theory is bound to be either empirical / phenomenological or based on constitutive relations extracted from inappropriate models that are good for some applications but fail for others. Indeed, there are currently many such specific models, most of which depend on empirical parameters that need fitting. The larger the number of fitting parameters in a theory, the more specific, the less fundamental, and the less predictive it is.

We then turned to review the current understanding of stress transmission in mechanically equilibrated granular solids. Such an understanding is essential to clarifying the state of matter of these systems, as well as to the many applications that depend on the mechanics of granular materials. Moreover, in the long run, a correct theory of stress in static systems is a checkpoint for theories of flow of dense GS, which should converge to the static state in the limit when the local velocities vanish. This is a point that is not emphasised in the several models of plastic flow of granular matter. For example, slow shear of a granular material leads to dilation and to connectivity that is close to the jamming point. Then, as the material jams, the material should be very close to an isostatic stress state. Yet, this is not where plasticity models commonly land the system.

We reviewed the latest advances in the relatively recent theory of stress in isostatic systems, showing that, at least in 2D, it has matured into a well-established first-principles predictive theory. Nevertheless, much remains to be done, in particular in terms of applying it to different geometries and determining the constitutive fabric tensor, which relates the stress to the medium's local structure. The latter is still an outstanding problem.

We then went on to clarify several misconceptions, in particular the idea that isostatic solutions must always lead to long narrow stress chains. Most granular media are disordered and, as such, their local fabric tensors are not uniform. The solutions reviewed here show that such non-uniformity gives rise to leakage of stresses from the chains, to branching, and generally to smearing the sharp nature of the chains, even if the medium is purely isostatic. Moreover, once gravitational forces are taken into consideration, the weight of each grain acts as a load source, which smears the stress field even further depending on the relative magnitude of the resulting body forces and the external compressive loads.

The main insight into isostaticity theory comes from analysis of 2D systems. There is currently no first-principles parameter-free theory of isostaticity in 3D. Such a theory must provide the appropriate closure stress-structure conditions for the stress equations, based on an upscaling of the local 'micro-structure'. To develop a 3D isostaticity theory we need both a theoretical effort and clearer experimental measurements of stresses inside such granular materials. In particular, it is important to emphasise that, if the 3D isostaticity theory is also based on hyperbolic equations, then its characteristic solutions cannot be stress chains, as commonly assumed. Rather the non-uniformity of the stress field should manifest in *stress surfaces*. There are hardly any experimental efforts to identify such surfaces and work to clarify this issue is much needed.

Another important comment to make concerns the plethora of analyses of stress transmission in isostatic systems of frictionless grains. These analyses are often prompted by the relative convenience of simulating such systems numerically, but they may miss a vital point. As has been shown in [24], the coarse-grained structural constitutive characteristics in assemblies of frictionless particles are inherently different than their analogues in frictional ones. This suggests that the large-scale stress fields may also differ significantly. The detailed nature of this difference has yet to be studied and remains an open issue.

We then discussed the relevance of isostaticity theory to real granular materials. On the face of it, since most granular media are not isostatic, the theory seems inapplicable to any practical calculation, a fact that has been flagged up in the engineering community. Yet, stress chains are clearly observed experimentally whilst theories based on stress-strain relations are hard pressed to predict such non-uniform stress fields without resorting to esoteric constitutive material properties. This conceptual conundrum appears to be resolved by the idea that general granular materials are two-phase 'composites' - part isostatic and part over-connected, potentially elastic. This means that GS can span a spectrum of stress states from the purely elastic to the purely isostatic, as was demonstrated by the rubber grains thought experiment. The closer the material is to the marginally rigid state, the larger the isostatic regions and the longer the stress chains.

This conceptually attractive idea of a mixed elasto-static linear theory is yet to be developed. The construction of such a theory is fraught with mathematical problems. A complete stress solution over the entire two-phase medium needs to satisfy continuity of stresses across the boundaries of different regions. Yet, isostaticity is a hyperbolic theory, whose characteristics coincide with the stress chains, while elasticity is elliptic. The two types of equations require boundary conditions of different nature and care should be taken so that the stress equations within the isostatic regions are not ill-posed. This issue has to be tackled if we are to have a general linear theory of stress in granular materials.

However, there is an even more basic problem with the current attempts to construct a stress theory. The premise underlying these attempts is that the theory can be made linear. Yet, we have presented in this chapter what we believe is a compelling argument that a fully fundamental theory must be nonlinear. This is because both the structure and the stress depend on the boundary loading, which means that the fabric tensor depends indirectly on the stress field and therefore that the constitutive stress-structure is nonlinear. To the best of our knowledge, this problem is not addressed in the literature at all.

In the last section, we discussed the recent developments in the formulation of an entropy-based statistical mechanical theory for static GS. A proper statistical mechanical theory offers predictability, such as through equations of state and relations between the statistics of the structural characteristics and measurable bulk physical properties. We reviewed the basis for such a theory and clarified what is, and what is not, problematic with the basic assumptions of the formalism. We elucidated the identity and number of degrees of freedom, which seem easier to identify with the quadron structural description.

We then reviewed recent potentially controversial work, showing the flaw in the practice of focusing on either a volume ensemble or a stress ensemble, since these two ensembles are inherently inter-dependent. We reviewed the treatment of the combined ensemble, which makes it possible to derive explicit expressions for a number of useful expectation values in 2D. Intriguingly, this treatment also allows us to derive an equipartition principle for static systems - the volume of the system is shared on average equally among all the degrees of freedom, both structural and stress-based.

Several open questions remain: (i) the choice of the most useful structural degrees of freedom in 3D; (ii) the full phase space of static granular matter; (iii) the difficulty in realising proper analogues of the thermometer for the structural and stress degrees of freedom - the compactometer and the angorometer; (iv) the possible misuse of the assumption of uniform measure, i.e. that all the configurations are realised in the ensemble with equal probability, and (v) the absence of ergodicity. While touching on these difficulties, we have not discussed them in much detail, mainly because they are still not completely resolved. In particular, the absence of ergodicity is very handicapping – the severing of the link between ensemble and time statistics makes it impossible to infer from the former to the latter, a convenience that thermal statistical mechanics enjoys. In the absence of this bridge between static and dynamic systems it is very difficult to obtain results, e.g. on diffusion processes and dynamic phase behaviours.

This difficulty is also related to the absence of the equilibrium concept in GS. In thermodynamics, the thermal fluctuations are orders of magnitude faster and more frequent than the duration of any measurement. As a result, thermal systems are shuffled among a huge number of states during any one measurement. This means that experimental measurements probe representative, or typical, averages over many states. This is not the case in GS, which do not move frequently from state to state, even when excited artificially. Thus, any observed reproducible statistics are normally of steady-state nature rather than equilibrium.

This means that it is important to choose carefully the ensemble over which one models the statistics. In particular, since every ensemble is characterised by its constraints, these constraints should be clear and match the modelled GS. Examples of such constraints are: a fixed, or fluctuating, number of grains; fixed grain size and shape distributions; the mean coordination number; the total external stresses; the volume; and especially that all the systems in the ensemble be prepared by the same procedure.

The growing understanding of the statistical mechanics of static GS paves the way to another potentially very significant development - statistical mechanics of dynamic systems. Any attempt to develop such a theory must be consistent with, and converge to, statistical mechanics of static assemblies in the limit of vanishingly small grain velocities. To this end, understanding the relation between the static and dynamic theories is essential. This issue, which is also hardly addressed in the literature, poses several conceptual and practical problems. For example, what are the natural degrees of freedom of a potential dynamic theory? From the above discussion it is evident that these could not be only the traditional positions and momenta. The reason is that the entropy in this traditional phase space vanishes when the grain velocities vanish, while in GS this limit still entertains the extra entropy of static configurations. This suggests at least two possibilities. One is that statistical mechanics of dynamic granular matter possibly involves a larger phase space, of which the structural and stress degrees of freedom also are a part. Another possibility is that the conventional positions and momenta may not be the correct or most useful degrees of freedom. Support for the latter possibility is provided by observations that static GS enjoy an equipartition principle in a phase space made of structural and stress degrees of freedom, while dynamically agitated GS suffer from an absence of such a principle in the phase space consisting of positions and momenta [93].

The resolution of these difficulties requires a coherent collaborative work between theorists and experimentalists, both *in vivo* and *in silico*. The theoretical understanding of granular science is still in its infancy, while the numerical and physical experimental capabilities are considerable. As a result, the literature is full of spectacular experiments that involve a range of parameters and richness of behaviour. Yet, these types of experiments, while attractively visualisable and gratifyingly produces agreement between numerical simulations and real measurements, are of little help to theorists. To enhance fundamental understanding of GS, it is imperative to have measurements and simulations of simple and basic systems and processes. Examples are experiments providing detailed data and insight into stress and forces in static systems, well-characterised rheology in simple geometries and dependence of pack structures on the packing process. It is against such canonical measurements that theorists could test methods of coarse-graining and continuous modelling.

In conclusion, the science of granular matter is an active and exciting field of research, most of which is still unexplored and mysterious. If we are to resolve some of the many existing open questions, it would be only by a focused cooperative effort from diverse communities and especially by collaboration between theory, numerical simulations and experiments on the fundamental issues. This field is far from saturated and could benefit greatly from more researchers entering it.

References

[1] R.M. Nedderman, *Statics and Kinematics of Granular Matter*, (Cambridge University Press, Cambridge, 1992).
[2] B. J. Ennis, J. Green, and R. Davis, Particle Technol. **90**, 32 (1994).
[3] *Landslides Will Continue to Impact the US - Tumbling Rocks Cause Dollars and Lives*, Report of the US Dept. of the Interior, US Geological Society, released June 25th, 1999.
[4] T. M. Knowlton, J. W. Carson, G. E. Klinzing, and W.-C. Yang, Particle Technol. **90**, 44 (1994).
[5] Archimedes (287-212 BC), *The Sand Reckoner*, English translation by W.L. Heath, (Cambridge University Press, Cambridge, 1897)
[6] M. Faraday, Philos. Trans. R. Soc. London **52**, 299 (1831).
[7] O. Reynolds, Phil. Mag., Series 5, **20**, 469-481 (1885).
[8] P.G. de Gennes, C.R. Acad. Sci. IIb **321**, 501-506 (1995).
[9] H.K. Pak, R.P. Behringer, Phys. Rev. Lett. 71, 1832 (1993); H.K. Pak, R.P. Behringer, Nature **371**, 231 (1994).
[10] G. H. Ristow, Pattern formation in granular materials (Springer-Verlag Berlin-Heidelberg 2000) and references therein.
[11] M. Harrington, J. H. Weijs and W. Losert, Phys. Rev. Lett. **111**, 078001 (2013) and references therein.
[12] H.M. Jaeger, S.R. Nagel, R.P. Behringer, Phys. Today **49 (4)**, 32 (1996) and references therein.

[13] E. Ehrichs, H.M. Jaeger, G.S. Karczmar, J.B. Knight, V.Y. Kuperman, S.R. Nagel, Science **267**, 1632 (1995).
[14] Y. Jiang Y., and M. Liu, Eur. Phys. J. E **22**, 255-260 (2007).
[15] *Lecture Notes in Complex Systems*, pp 43-53, eds. T. Aste, A. Tordesillas and T. D. Matteo (World Scientific, Singapore 2008) and references therein.
[16] S. Roman, *Introduction to Coding and Information Theory*, (Springer, Berlin, 1996).
[17] N. Mitarai and F. Nori, Adv. Phys. **55**, 1-45 (2006).
[18] I. Goldhirsch and G. Zanetti, Phys. Rev. Lett. **70**, 1619 (1993).
[19] M. Bose M., U.U. Kumar, P,R. Mott, and V. Kumaran, Phys. Rev. E **72**, 021305 (2005) and references therein.
[20] P.G. de Gennes, Rev. Mod. Phys. **71**, S374-S382 (1999) and references therein.
[21] R. Behringer, H.M. Jaeger, S.R. Nagel, Chaos **9**, 509 (1999).
[22] S. Torquato, T. M. Truskett, and P. G. Debenedetti, Phys. Rev. Lett. **84**, 2064 (2000).
[23] G.W. Delaney, T. Di Matteo, and T. Aste, Soft Matter **6** (2010) 2992-3006 and references therein.
[24] R. C. Ball and R. Blumenfeld, Phys. Rev. Lett., 88 115505 (2002).
[25] R. Blumenfeld and S. F. Edwards, Phys. Rev. Lett. **90**, 114303 (2003).
[26] R. Blumenfeld and Sam F. Edwards, Eur. Phys. J. **E 19**, 23-30 (2006).
[27] A. Okabe, B. Boots, K. Sugihara and S. Nok Chiu, *Spatial Tessellations: Concepts and Applications of Voronoi Diagrams*, (John Wiley, New York, 2000).
[28] Euler's theorem, proven by him circa 1750. See, e.g. H. M. S. Coxeter, *Regular Polytopes*. (Dover, New York 1973).
[29] T. Matsushima and R. Blumenfeld, Phys. Rev. Lett. **112**, 098003 (2014).
[30] G.E. Barnes, *Soil Mechanics: Principles and Practice* (Palgrave Macmillan, London 2010).
[31] R.P. Seelig and J. Wulff, Trans. AIME **166**, 492-505 (1946).
[32] C. Liu, S.R. Nagel, D.A. Schecter, S.N. Coppersmith, S. Majumdar, O. Narayan and T.A. Witten, Science **269**, 513 (1995) and references within.
[33] T. S. Majmudar, M. Sperl, S. Luding and R. P. Behringer, Phys. Rev. Lett. **98** 058001 (2007).
[34] P. Yu and R. P. Behringer, Chaos **15**, 041102 (2005).
[35] S. Brand, R.C. Ball, and M. Nicodemi, Phys. Rev. E **83**, 031309 (2011).
[36] Ellenbroek W.G., Somfai E., van Saarloos W., van Hecke M. *Force response as a probe of the jamming transition*, in: R. Garcia-Rojo, H.J. Herrmann, S. McNamara (eds) "Powders and Grains 2005" (Balkema, Leiden, The Netherlands, 2005), p. 377-380.
[37] M.E. Cates, J.P. Wittmer, J.P. Bouchaud, and P. Claudin, Physica A **263**, 354-361 (1999).

[38] D. Lootens, H. van Damme, and P. Hebraud, Phys. Rev. Lett. **90**, 178301 (2003); L.E. Silbert, A.J. Liu, and S.R. Nagel, Phys. Rev. Lett. **95**, 098301 (2005).

[39] R. Arevalo, I. Zuriguel, and D. Maza, Phys. Rev. E **81**, 041302 (2010); M. Wyart, L.E. Silbert, S.R. Nagel, and T.A. Witten, Phys. Rev. E **72**, 051306 (2005).

[40] A.I. Campbell, and M.D. Haw, Soft Matter **6**, 4688-4693 (2010); P. Ballesta, A. Duri, and L. Cipelletti, Nature Phys. **4**, 550-554 (2008).

[41] A. Kumar, and J.Z. Wu, Appl. Phys. Lett. **84**, 4565-4567 (2004).

[42] J.D. Bernal, and J. Mason, Nature **188**, 908 (1960).

[43] R.C. Ball and R. Blumenfeld, Phil. Trans. Roy. Soc. Lond. A **360**, 731 (2003).

[44] R. Blumenfeld, S. F. Edwards and R. C. Ball, J. Phys.: Cond. Mat. **17**, S2481 (2005); R. Blumenfeld,S.F. Edwards, and R.C. Ball, J. Phys. - Cond. Mat. **7**, S2481–S2487 (2005).

[45] C. S. O'Hern, L. E. Silbert, A. J. Liu, and S. R. Nagel, Phys. Rev. E **68**, 011306 (2003).

[46] J.R. Melrose and R.C. Ball, Europhys. Lett. **32**, 535 (1995).

[47] C. Thornton, Kona Powder and Particle, **15**, 81 (1997).

[48] F. Radjai, D. Wolf, S. Roux, M. Jean, and J.J. Moreau, *Force networks in granular packings*, in: D.E. Wolf, P. Grassberger (eds) "Friction, Arching, Contact Dynamics" (World Scientific, Singapore, 1997), p. 169-179.

[49] C. Goldenberg, I. Goldhirsch, Phys. Rev. Lett. **89**, 084302 (2002).

[50] R.C. Hidalgo, H.J. Herrmann, E.J.R. Parteli, and F. Kun, *Force chains in granular packings*, in: F. Mallamace, H.E. Stanley (eds) "The Physics of Complex Systems: New Advances and Perspectives", (IOS Press, Amsterdam, 2004), p. 153-171.

[51] J.F. Peters, M. Muthuswamy, J. Wibowo, A. Tordesillas, Phys. Rev. E **72**, 041307 (2005).

[52] S. Ostojic, D. Panja, Phys. Rev. Lett. **97**, 208001 (2006).

[53] A. Tordesillas, *Stranger than friction: Force chain buckling and its implications for constitutive modelling*, in: T. Aste, T. Di Matteo, A. Tordesillas (eds) "Granular and Complex Materials", (World Scientific, Singapore, 2007), p. 95-109.

[54] A. Tordesillas, J. Zhang, R. Behringer, Geomech. Geoeng. **4**, 3-16 (2009).

[55] A. Rechenmacher, A. Abedi, O. Chupin, Geotechnique **60**, 343-351 (2010).

[56] E. Azema, F. Radjai, Phys. Rev. E **85**, 031303 (2012).

[57] S. Luding, Phys. Rev. E **55**, 4720 (1997).

[58] J. H. Snoeijer,T. J. H. Vlugt, M. van Hecke and W. van Saarloos, Phys. Rev. Lett. **92**, 054302 (2004).

[59] S. Ostojic, E. Somfai and B. Nienhuis, Nature **439**, 828 (2006).

[60] J.-P. Bouchaud, M. E. Cates and P. J. Claudin, J. Phys. I (France) **5**, 639 (1995).

[61] M. E. Cates, J. P. Wittmer, J.-P. Bouchaud and P. Claudin, Phys.

Rev. Lett. **81**, 1841 (1998).
[62] A. V. Tkachenko and T. A. Witten, Phys. Rev. E **62**, 2510 (2000).
[63] R. Blumenfeld, Phys. Rev. Lett. **93**, 108301-108304 (2004).
[64] R. Blumenfeld, Physica A **336**, 361 (2004).
[65] M. Gerritsen, G. Kreiss, R. Blumenfeld, Phys. Rev. Lett. **101**, 098001 (2008).
[66] M. Gerritsen, G. Kreiss, R. Blumenfeld, Physica A **387**, 6263-6276 (2008).
[67] H.-O. Kreiss, J. Lorenz, *Initial-boundary value problems and the Navier-Stokes equation*, (Academic Press, Boston, 1989).
[68] J. Smid, and J. Novosad, Proc. 1981 Powtech Conf., Inst. Chem. Eng. Symp. **63**, D3V 1-12 (1981).
[69] B. Utter, and R.P. Behringer, *Multiscale motion in the shear band of granular Couette flow*, in: Nakagawa, M. and Luding, S. (eds) "Powders and Grains 2009", (American Institute of Physics, Melville, NY 2009), p. 339-342, and references therein.
[70] A.P. Bobryakov, and A.F. Revuzhenko, J. Min. Sci. **45**, 99104 (2009).
[71] R.C. Ball, in: M. Lal, R. A. Mashelkar, B. D. Kulkarni, and V. M. Nai (eds) "Structures and Dynamics of Materials in the Mesoscopic Domain", (Imperial College Press, London 1999) .
[72] S.B. Savage, *Physics of Dry Granular Media*, in: H. J. Herrmann, J. P. Hovi and S. Luding (eds) "NATO ASI Series vol 25", (Kluwer, Amsterdam, 2002).
[73] R. Blumenfeld, *Stress transmission and isostatic states of non-rigid particulate systems*, in: M.C.T. Calderer and E.M. Terentjev (eds) "IMA Volume in Mathematics and its Applications" Vol. 141, (Springer-Verlag, Berlin, 2005), p. 235-246.
[74] W. O. Smith, P. D. Foote, and P. F. Busang, Phys. Rev. **34**, 1271-1274 (1929).
[75] S.F. Edwards, IMA Bulletin **25**, 94 (1989).
[76] C. C. Mounfield and S. F. Edwards, Physica A **210**, 279 (1994).
[77] A. Barrat, J. Kurchan, V. Loreto, and M. Sellitto, Phys. Rev. Lett. **85**, 5034 (2000).
[78] H.A. Makse and J. Kurchan, Nature **415**, 614 (2002).
[79] I.K. Ono, C.S. O'Hern, D.J. Durian, S.A. Langer, A.J. Liu, and S.R. Nagel, Phys. Rev. Lett. **89**, 095703 (2002).
[80] A. Coniglio, A. Fierro, M. Nicodemi, M.P. Ciamarra, and M. Tarzia, J. Phys. - Cond. Mat. **17**, S2557 (2005).
[81] S. Henkes, C. S. O'Hern and B. Chakraborty, Phys. Rev. Lett. **99**, 038002 (2007).
[82] S. Henkes and B. Chakraborty, Phys. Rev. E **79**, 061301 (2009).
[83] J.B. Knight, C.G. Fandrich, C.N. Lau, H.M. Jaeger, and S.R. Nagel, Phys. Rev. E **51**, 3957 (1995).
[84] N. Xu, J. Blawzdziewicz, and C. S. O'Hern, Phys. Rev. E **71**, 061306 (2005); G.-J. Gao, J. Blawzdziewicz, and C. S. O'Hern, Phys. Rev. E **74**, 061304 (2006); G. -J. Gao, J. Blawzdziewicz, C. S. O'Hern and M. Shattuck, Phys. Rev. E **80**, 061304 (2009).

[85] R. Blumenfeld, *On entropic characterization of granular materials*, in: T. Aste, A. Tordesillas and T. D. Matteo (eds) "Lecture Notes in Complex Systems", (World Scientific, Singapore, 2008), p. 43-53.

[86] S. F. Edwards and R. Blumenfeld, in: R. Garcia-Rojo, H. J. Herrmann, and S. McNamara (eds) "Powders and Grains, Stuttgart" (Balkema, Leiden, Netherlands, 2005), p. 3-5.

[87] L. A. Pugnaloni, I. Sánchez, P. A. Gago, J. Damas, I. Zuriguel and D. Maza, Phys. Rev. E **82**, 050301(R) (2010).

[88] S. F. Edwards and R. Blumenfeld, *Thermodynamics of granular materials*, in: A. Mehta (ed) "Physics of Granular Materials", (Cambridge University Press, Cambridge, 2007).

[89] R. Blumenfeld and S. F. Edwards, J. Phys. Chem. B **113**, 3981-3987 (2009).

[90] R. Blumenfeld, J. F. Jordan and S. F. Edwards, Phys. Rev. Lett. **109**, 238001 (2012).

[91] J. C. Maxwell, Phil. Mag., Ser 4, **27**, 250 (1864); J.C. Maxwell, Phil. Mag., **27**, 294-299 (1864); J. C. Maxwell, Trans. Roy. Soc. Edinb., **26**, 1 (1869).

[92] H.A. Makse and J. Kurchan, Nature **415**, 614 (2002).

[93] M. Alam, S. Luding, Gran. Matt. **4**, 139 (2002); H.-Q. Wang and N. Menon, Phys. Rev. Lett. **100**, 158001 (2008).

[94] P. Jop, Y. Forterre, O. Pouliquen, Nature **441**, 727 (2006); Y. Forterre, O. Pouliquen, Annu. Rev. Fluid Mech. **40**, 1 (2008); J.E. Andrade, Q. Chen, P.H. Le, C.F. Avila, and T.M. Evans, J. Mech. Phys. Solids **60**, 1122-1136 (2012).

Dynamics of Entangled Polymers

Ronald G. Larson[1] and Zuowei Wang[1,2]

[1]Department of Chemical Engineering, University of Michigan, Ann Arbor, Michigan 48109-2136
[2]School of Mathematical and Physical Sciences, University of Reading, Whiteknight, Reading RG6 6AX, UK

6

Long polymer molecules in the melt or concentrated solutions show unusually slow dynamics that are highly sensitive to molecular weight and polymer architecture. These dynamics are not only of great scientific interest, but are also essential to phenomena of practical importance such as the speed at which fibers can be spun, the strength of adhesives, and the resistance to cracking of plastics. The slow dynamics of these polymers has for decades been attributed to mysterious 'entanglements' that are thought of as being akin to those present in a tangle of string or fishing line. While defining an 'entanglement' precisely has proven to be difficult, the onset of the entanglement phenomenon is unambiguous. It reveals itself in a definite, though gradual, change in the dependence of viscosity, relaxation time, diffusivity, and other dynamic properties, on chain length or concentration [1, 2].[1]

Much has been written about the dynamics of entangled polymers, especially its role in diffusion and rheology. Many sophisticated experimental probes of these dynamics have been deployed and theoretical modeling is now very advanced. Monographs written at both introductory and advanced levels are available on this topic [2, 3, 5, 6]. In this chapter, we focus on the basic questions that faced the earliest researchers, and see how their answers stand up in light of what has been learned in the last 40 years. In the process, we hope to provide the novice with a basic understanding of the physical concepts without burdening him or her with the mathematics, or with detailed data that can easily be found in the books referred to above. At the same time, we hope to update the knowledgeable non-specialist with the latest developments in the area. For the expert, we hope to provoke new ways of thinking about an old, but incompletely solved, problem area.

The chapter is organized as follows. In Section 6.1, the state of entanglement is described, organized around the classical pictures of a 'tube' and a network of 'slip links'. The successes and limitations of these pictures are explored and new insights are gleaned from recent theoretical

---

[1]Given the space limitations, the data referred to will most often be rheological data, since these are of great practical importance and are most sensitive to dynamics on a wide range of time scales, especially at long times, where entanglement effects are most pronounced. We focus on basic rheological properties, such as the linear viscoelastic storage and loss moduli, G' and G'' and of strain-rate dependent shear viscosity and extensional viscosity. Briefly, G' and G'' are the contributions to the mechanical modulus in a small-amplitude oscillatory deformation, with G' being in phase with the deformation and G'' being 90° out of phase. These moduli are measures of dynamics of polymer molecules in the equilibrated state, i.e. free of significant distortion of chain conformation. The shear and extensional viscosities are the stress divided by the strain rate in each of these deformations and reflect the response of the polymers to large deformations. These rheological functions are described in detail in various monographs [e.g. 3–5].

developments and computer simulations. The focus is on the nature of an 'entanglement', both from a bottom-up molecular view, and from a phenomenological one. The effects on the entanglement structure of adding a solvent are discussed as well as the effects of imposing a large deformation. In Section 6.2, these ideas are applied to the prediction of the linear viscoelasticity of entangled polymer solutions and melts. The successes and limitations of existing theory will be described primarily in words and cartoons, and not as much with data plot and tables, which can be found elsewhere. Section 6.3 deals with nonlinear viscoelasticity, following the same approach as in Section 6.2. A summary and suggestions for future work can be found in theconcluding section on page 263.

6.1 Foundations of entangled polymer dynamics

Modeling entanglements using tubes and slip links

Over the period 1967–1979, Sam Edwards, Pierre-Gilles de Gennes, and Masao Doi laid the foundations for what remains the accepted coarse-grained theory for the dynamics of long polymer molecules in the melt or concentrated solution where the dynamics are dominated by intermolecular entanglements [2,7–12]. The basic approach is illustrated in Figures 6.1 and 6.2 [1]. At long time scales, chains surrounding a given 'test' chain primarily act to restrict the test chain's long-range lateral motion, while offering only short-range frictional resistance to motion along the coarse-grained contour of the test chain. A linear chain thus 'reptates' or slides as a whole along this coarse-grained contour; see Figure 6.2. At short time scales the chain undergoes vigorous short-range local fluctuations in all directions. These fast, localized, fluctuations do not feel the entanglements with other chains. At intermediate time scales, there are fluctuations of larger portions of the chain that are influenced by the entanglements and so are directed along the coarse-grain chain contour, but drag only a portion, and not the whole chain, in a given direction along the tube. Thus, the entanglements primarily affect the intermediate and long time scales of polymer relaxation, but not the short time scales.

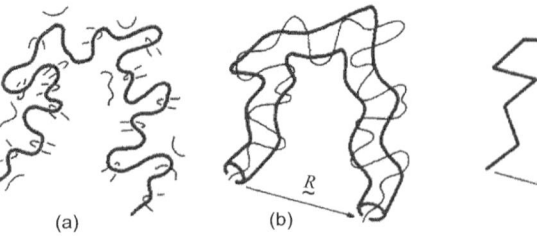

Fig. 6.1 (a) Illustration of a polymer molecule entangled with neighboring polymer molecules. (b) The entanglements with surrounding polymers are represented by a tube. (c) The primitive path of the tube, which is a random walk with the step size equal to the diameter of the tube. The contour length of the primitive path is much less than the contour length of the polymer. (Figure 6.7 in [1].)

On the intermediate and long time scales, the confining effect of the surrounding molecules has, since the earliest days, been depicted using a mesh of fixed obstacles, a 'tube' or a set of 'slip links'; see Figure 6.3. The tube and slip link pictures have been the most fertile ones for development of mathematical models. The tube picture emphasizes the collective effect of the many surrounding chains, which limits the short-time lateral motion of the chain to a distance characterized by the diameter of the tube. From experimental data, the tube diameter a is found to correspond to a factor of 10 to 30 times the diameter of the chain in the melt state, depending on the specific polymer [13], a value that increases as the chains are diluted with solvent. The slip-link picture, on the other hand, is suggestive of localized confinements, produced by one or perhaps a few surrounding chains, that limit lateral excursions of the test chain near each 'slip link', but allow rather wider excursions elsewhere.

Either a 'tube' or a set of 'slip links' has the same effect on the coarse-grained dynamics of the polymer chain, namely, the whole chain, or portions of it, must slide along the centerline of the tube, or along the slalom course marked out by the slip links. In either case this path, often called the 'primitive path', is much shorter than the contour length of the chain itself, by a factor of 10–30 or so for a melt, and by more than this in solutions. [2] If one were to somehow hold stationary the matrix environment, then at long time and length scales both tubes and slip links lead to the same coarse-grained dynamics. However, at shorter time scales, and in the usual cases where the matrix itself is allowed to be mobile, these two pictures can lead to different predictions. In addition, most polymers of practical and scientific interest are neither long enough nor densely entangled enough to reach the long-chain asymptote [5]. Hence, local behavior at the level of a few entanglements often has quantitatively significant effects on predictions. As long as the tube or slip-link models are viewed as qualitative or semi-quantitative models, these details may not matter much. But in recent years, much effort has been devoted toward making entanglement theories quantitative, so that they might even be used to infer precise molecular weights, molecular weight distributions, or the presence of long-chain branching, from rheological measurements [14–19]. In addition, a hunger has grown to build entanglement theories from a bottom-up analysis of motions of individual monomers, rather than relying on phenomenological notions such as that of a 'tube' [20].

For many years, the tube picture was more widely used than the slip

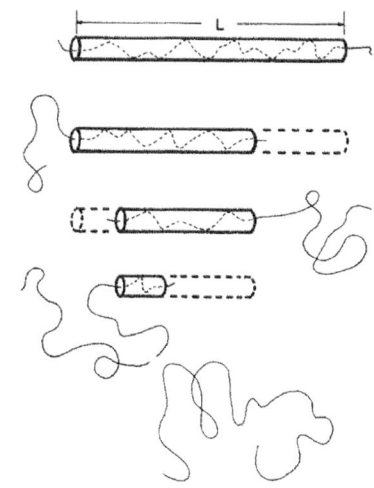

Fig. 6.2 A reptating polymer molecule disengages from its initial confining tube. (Figure 6.9 in [1].)

[2] The factor by which the tube diameter exceeds the diameter of the chain is related to the factor by which the tube length is shorter than the chain contour length, but they are not identical or even proportional to each other. This is because there is another length scale, the Kuhn step length of the chain, which is the distance over which the chain is effectively straight. It can be defined precisely as the ratio of the equilibrium mean-square end-to-end distance of the chain to its contour length, which affects both factors, as discussed below.

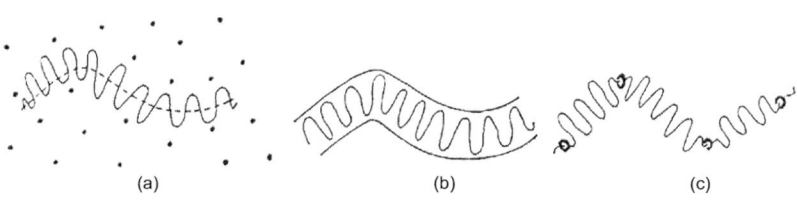

Fig. 6.3 Three models to describe entanglements: (a) the cage model, (b) the tube model and (c) the slip-link model. (Figure 1 in [10].)

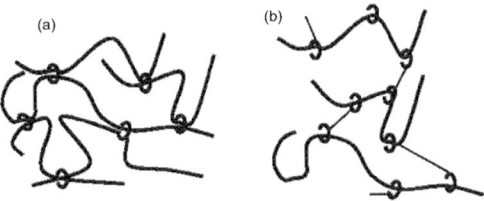

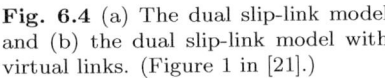

Fig. 6.4 (a) The dual slip-link model and (b) the dual slip-link model with virtual links. (Figure 1 in [21].)

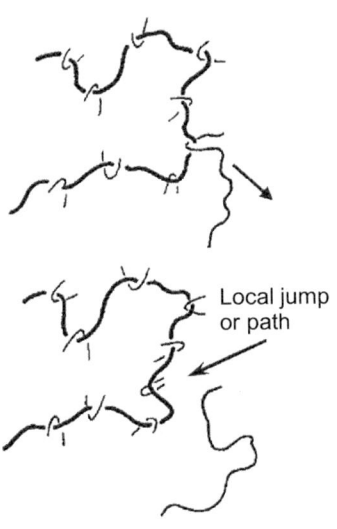

Fig. 6.5 Illustration of a constraint release event, as the end of a surrounding chain diffuses past the confined chain, releasing its confinement contribution. (Figure 7.11b in [5].)

link conception. An advantage of the tube model is that the centerline of the tube forms a continuous path, which can be parameterized by a one-dimensional tube coordinate. Coarse-grained motion along this coordinate can then be described by well-known diffusion equations [2]. Thus, in the early years of development of reptation theory, the focus was on the motion of a single chain, with surrounding chains blurred into a mean field that served to define the tube, which was taken to be immobile as long as the chain resided in it. However, after having established the basic validity of polymer reptation, research over the past 20–30 years has turned to adding refinements to make it more quantitative. Many of these refinements relate to the impermanence of the entanglement environment, owing to the motion of the 'matrix' chains that compose it, as discussed in the next section.

The mean-field picture can be clumsy or inaccurate when the matrix effects are highly heterogeneous, i.e. some chains or portions of chains can release their entangling effects much faster than others. Heterogeneities occur, for example, as a result of polydispersity in molecular weight where short chains are far more mobile than long ones. Hence the slip link picture has been in recent times gained increasing attention [21–29], since each slip link can be given a different lifetime and thus a wide range of entanglement lifetimes can easily be represented. Moreover, each slip link might represent an 'entanglement' between two specific chains, as illustrated in Figure 6.4. In this 'dual slip link' picture, the slip link remains intact until a chain end of *either* chain passes through the slip link and destroys it. While slip-link models require computer simulations to solve, this has become much less of a drawback as computers have become ever faster.

Despite these possible advantages, thinking of each slip link as an 'entanglement' is probably too literal an interpretation, since it suggests that only the motion of a specific chain end can release an entanglement on a second chain; see Figure 6.5. As we shall see, this localization of the entanglement interaction to just a pair of chains seems to be an adequate approximation for linear polymers, but is less applicable to polymers with long-chain branching. In truth, whether entanglements are best regarded as discrete interactions between pairs of chains, as a mean field of restrictions created by many chains, or as some mixture of the two, has still not been resolved.

The packing length and concentration effects

If 'entanglements' are discrete interactions between two chains, then when half the chains are removed and replaced with solvent, half the entanglement points on each remaining chain should disappear. And since only half the chains are left, there are only one quarter as many entanglement points remaining. Thus, the density of entanglements should scale as the square of the concentration of polymers. On the other hand, in the mean-field picture, as summarized by Milner [30], entanglements result when enough chains (around 20 in the melt) crowd into the same space (entanglement volume); see Figure 6.6. In a fixed volume, stubby, bulky, chain segments take up more room for themselves and leave less room for other chains, than do lithe, slender, ones. Bulkiness can be quantified by the *packing length* p, defined as $p \equiv M/(\langle R^2 \rangle \rho N_A)$, where M is the polymer molecular weight, $\langle R^2 \rangle$ is the mean-square end-to-end distance separating chain ends at equilibrium, ρ is the mass density, and N_A is Avogadro's number [13, 30, 31]. Although this formula for p involves the chain's molecular weight, this effect is cancelled out by dividing it by $\langle R^2 \rangle$, which is also proportional to molecular weight. Hence, the packing length p is really a measure of how much of the chain's mass or length is contained within a given radius around the chain contour at equilibrium. The ratio p will increase if the chain is flexible and bulky. Thus, if the polymer chain is viewed as a string of sausage links where each link can rotate freely, then the packing length is roughly given by d^2/l, where d and l are the diameter and length of each sausage link, respectively. The length of a sausage link, l, corresponds to the 'Kuhn step length' mentioned above. If one starts with a small spherical volume containing only one monomer of a chain and continuously expands this volume, then once the volume is large enough to encompass chain segments from enough other chains (around 20 different chains, as it turns out), an entanglement 'happens', according to the packing argument. The diameter of this spherical volume (entanglement volume) is then the entanglement length scale a, which is the tube diameter. The larger the packing length p, the larger the radius a of the volume needed to obtain an entanglement.[3] Fetters and coworkers have exhaustively marshaled the available data for polymers spanning a wide range of bulkiness, and have impressively confirmed the predictions of the packing model [13, 31]. From the tube diameter, we obtain the distance along the chain between effective entanglement points, which is typically given as the molecular weight between entanglements M_e.

When solvent is added to a polymer melt, what does the packing model imply for the dependence of the entanglement spacing a on polymer concentration? A complicating issue is the thermodynamics of polymer/solvent contacts [30, 32]. If the polymer tends to swell in the solvent, this will cause local deviations from a random walk configuration of the polymer, and might affect its packing and entanglement characteristics [30, 33]. Hence, it is simpler for present purposes to limit discussion to the case of a *theta solvent*, whose interactions with the

[3] Working out this argument yields the result that the tube diameter a is roughly 20 times the packing length p, or more precisely, $a = 19.36p$ at temperature $T = 413$ K and $a = 17.68p$ at 298 K [13]. Values for p range from around 1.5 to 10 Å, or so, depending on the polymer.

Fig. 6.6 Chain with the centers of gravity of many other chains, indicated by dots, lying within its pervaded volume. (Figure 7.1b in [5].)

polymer are tuned so that there is no significant departure of the polymer from the random walk configuration it has in the melt [34]. Taking a simple-minded extension of the packing argument, we might suppose that adding solvent expands the entanglement volume a^3 required to generate an entanglement enough that the same number of chains (around 20 chains) passes through this volume as in the case of the melt. But as a increases, more monomers *from each chain* are present in the volume. For a random-walk chain, the number of monomers from each chain present in the volume scales as a^2. So, increasing the entanglement volume a^3 leaves only a factor of a increase in the average volume available for each monomer. Hence, it follows that diluting a polymer melt down to a polymer volume fraction ϕ should, according to this simple argument, lead to an expansion of the tube diameter by a factor of $a \propto 1/\phi$ due to the increase of the available volume per monomer by this factor. This result is equivalent to uniformly coating each polymer molecule by its share of the solvent, thus fattening it into a thicker polymer [30]. Since the polymer takes on a random walk configuration, dilution by a factor of two, for example, will lead to entanglements that are spaced four times less densely along the polymer chain, and this yields the prediction that the entanglement density scales as the third power of concentration ϕ. This prediction sharply departs from experimental measurements [30].

Rejecting this simplistic theory view of entanglement dilution, Colby and Rubinstein [32] developed a theory based again on *binary contacts* between pairs of chains. Rather than ascribing to each entanglement a pairwise interaction between two particular chains, they assumed that a minimum number of 'binary contacts' are required to create an effective 'entanglement'. Since the number of 'binary contacts' per unit volume is proportional to ϕ^2 and the number of such contacts in a volume of a^3 is proportional to $\phi^2 a^3$, we immediately obtain the 'Colby–Rubinstein scaling law', $a \propto \phi^{-2/3}$. Because of its random walk configuration, a chain in a solvent has only a fraction proportional to $a^{-2} \propto \phi^{4/3}$ of the entanglements it has in the melt. The so-called 'dilution exponent' in the expression $a^2 \propto \phi^{-\alpha}$ is therefore $\alpha = 4/3$, while a simple slip-link picture, where each entanglement corresponds to an interaction between two chains, gives $\alpha = 1$, and the simplistic picture in which each chain is coated by its share of solvent gives $\alpha = 2$. Since the density of chains in solution is proportional to ϕ, the entanglement density should scale as $\phi^{1+\alpha}$, which yields $\phi^{7/3}$ for the Colby–Rubinstein scaling.[4] There is evidence [33, 35, 36], albeit contested [37], that the volume fraction scaling exponent is closer to 7/3 than to 2, which implies that α is closer to 4/3 than to unity [30]. However, the difference between these exponents is not large, and experimental results do not consistently favor one value over the other.

There are other difficulties and perplexities. Letting the number of 'binary contacts' (rather than the number of chains) define an entanglement volume leads to the result that as the polymer concentration drops, the number of distinct chains in the entanglement volume also decreases.

[4] Here the scaling is for entanglement density, not for tube diameter.

This can be seen by first remembering that if there is a constant number of chains in an entanglement volume, then $a \propto \phi^{-1}$. However, the Colby–Rubinstein scaling law is $a \propto \phi^{-2/3}$. Thus, as ϕ decreases, under Colby–Rubinstein scaling, the entanglement volume does not grow in size fast enough to keep constant the number of chains passing through that volume. Hence, the number of chains falls, and eventually, for sufficiently dilute systems, there is only one chain in the entanglement volume, and so no entanglements with other chains. The reason this occurs is that under Colby–Rubinstein scaling, the number of 'binary contacts' that are held constant includes the contacts of a chain with itself. For low enough concentrations, these are the only contacts left. Since an entanglement of a chain with itself is unlikely to have the same effect on chain dynamics as inter-chain entanglements have, this limit represents a breakdown in Colby–Rubinstein scaling. It is probably possible to repair this problem by subtracting self-contacts from the total number of 'binary contacts'. However, it is easy to show that the limit of a single chain in an entanglement volume is only achieved at very low polymer volume fractions, where all but extremely high molecular weight polymers would not be entangled anyway. Still, this observation highlights a basic problem: not all 'entanglements' are equal. One matrix chain might wrap like a noose around a test chain (see Figure 6.7), while another matrix chain just grazes the test chain. It is also easy to imagine that an 'entanglement' between two chains might be held in place by a third chain, which acts like a linchpin for the entanglement; see Figure 6.8. A slip link is likely to be a reasonable model for some entanglements, while for others the mean field picture is better.

What do computer simulations say about entanglements?

Given the difficulty in resolving the nature of entanglements using experiments and scaling arguments alone, researchers are turning toward computer simulations. Grest & coworkers [39] introduced a practical computational method of approximating the 'primitive paths' from molecular dynamics simulations of entangled long chains. In their method, after equilibrating a melt, all chain ends of the polymers are suddenly frozen in place, and the thermal motion of the monomers gradually turned off, allowing monomers on the same chain to overlap each other while preventing monomers on different chains from overlapping or passing through each other. The result is to 'shrink wrap' all polymers down to the shortest paths possible without them passing through each other; see Figure 6.7. This yields a network of 'primitive paths' (PPs) of line segments that bend at contact points with other 'primitive paths'. Following this work, multiple alternative methods have been developed for computing not only the PPs, but also the density of entanglements and the tube diameter [38, 40–44]. The 'Z1 Code' of Kröger and coworkers [41, 45, 46] is a particularly efficient method for extracting primitive paths and has recently been extended to extract time dependent tra-

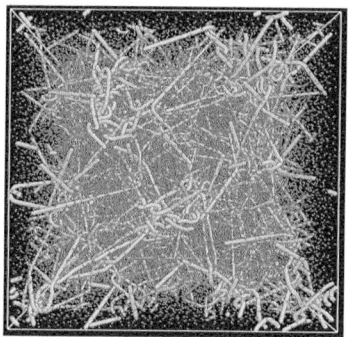

Fig. 6.7 Snapshot of primitive paths of entangled polymer chains in a melt. (Data from Zhou's molecular dynamics simulations [38].)

Fig. 6.8 Three-body entanglement: entanglement between the two upper chains is held in place by the third chain.

jectories of primitive paths and, from these, rheological properties. All methods for obtaining primitive paths from molecular dynamics simulations involve heuristic choices for mapping fine-scale coordinates of beads representing monomers onto coarse-grained tube coordinates. Moreover, the relatively short chain lengths able to be simulated so far restrict these methods to rather modest entanglement densities, and to the dense melt state. But all yield networks of zig-zag paths with bends at contact points with other paths, and whose path lengths are much shorter than the contour lengths of the original chains. Moreover, the density of entanglements inferred from the different methods are in rough agreement with each other [38] and with correlations with packing length derived from experimental data [39].

More importantly, a better picture of entanglements is gradually emerging from these computational studies. One discovery is that the 'primitive path' derived from microscopic computer simulations is usually not a random walk, but instead, neighboring steps of the primitive path are orientationally correlated. On long distance scales, the PPs still exhibit random-walk scaling, and so the mean square separation distance of the ends of the primitive path, $\langle R \rangle^2$, is proportional to the PP length L_{pp}. But the Kuhn step size of the PP is a factor of around 2.5 larger than the mean spacing between contact points, or 'topological contacts', along the primitive paths [42]. Thus, one infers that 2 or 3 topological contacts between primitive paths are required to generate a single entanglement, where an 'entanglement' is here defined by its contribution to the length of the primitive path; i.e. $\langle R^2 \rangle = L_{pp}^2/Z$. Here, Z is by definition the number of entanglement points along the polymer. The number of topological contacts (TCs) is around 2.5 times larger than this.

There are some interesting implications of this. First, the 'contact volume' of the melt necessary to contain, on average, a single topological contact between primitive paths is considerably smaller than the volume a^3 needed to contain a single entanglement. Since the primitive path shape is between a straight line and a random walk at distances below the entanglement spacing, the volume needed to contain a single contact point will be smaller than a^3 by a factor of between $(2.5)^{3/2}$ and $(2.5)^3$, i.e. a factor between 4 and 16. This smaller volume will contain fewer chains than the entanglement volume, in proportion to the cube root of this factor. Since the 'entanglement volume' a^3 contains around 20 distinct chains, there will be only around 8–12 chains in a 'contact volume'.

Secondly, since a single contact is insufficient to establish an 'entanglement', it would seem to follow that eliminating or creating one topological contact does not necessarily release or create an entanglement. Thus, the relationships between a slip link, a PP 'binary contact', and an 'entanglement' need to be clarified. It seems reasonable to follow the tube model by defining an 'entanglement' to be an inter-chain interaction that contributes $k_B T$ to the plateau modulus. More precisely, as discussed below, because of chain mass fluctuations along the PP,

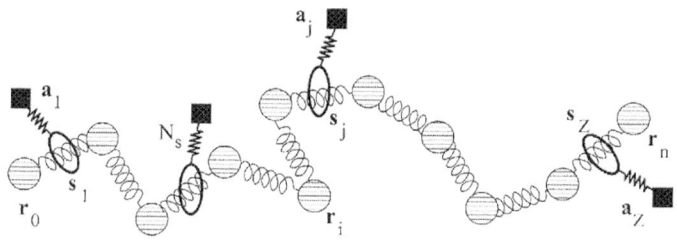

Fig. 6.9 Slip-spring model of entangled polymers: a standard Rouse chain of $N+1$ beads is constrained by a set of Z discrete virtual springs. (Figure 2 in [48].)

this contribution is reduced slightly to $(4/5)k_BT$. As remarked already, calculations of PP's from fine-grained molecular dynamics simulations indicates that there may be two to three times as many TC's as there are entanglements.

The number of slip links per entanglement, on the other hand, depends on the mobility allowed to each slip link. In the slip-link model of Doi and Takimoto [21], a slip link, once created, is frozen in space until it is destroyed, and each slip link delivers a contribution of $(4/5)k_BT$ to the shear stress [21, 27]. Hence, roughly speaking, there is one slip link per entanglement in this model. In the 'Primitive Chain Network' (PCN) model of Masubuchi et al. [23, 27], the contribution of each slip link to the stress is reduced from this by a factor of around 1.7 or so [28]. This is readily explained as the result of fluctuations in the slip-link positions, which are determined in the PCN model by a force balance among the four strands that are brought together at the slip link (two strands from each interacting polymer). Rubber elasticity theory suggests that these fluctuations should reduce the modulus by a factor of two [47]. The 'slip-spring' slip-link model of Likhtman [25], on the other hand, yields a contribution to the modulus, per slip link, that depends on the stiffness of the springs tethering each slip link to a fixed point in space; see Figure 6.9. Thus, the more freedom each slip has to move, i.e. the 'softer' the constraint imposed by a slip link, the lower the contribution it makes to the modulus.

Entanglements, slip links, and topological contacts

While these relationships allow the slip-link density of a given model to be related quantitatively to an 'entanglement', as defined by the tube model, it remains to relate either of these concepts more precisely to the notion of a 'topological contact' (TC) as obtained from a microscopic simulation. Note that a frozen slip link forces both chains to pass through a fixed point in space until a chain end from one chain or the other passes through it, while a 'topological contact' remains in place only as long as the two primitive paths exert force on one another. The fluctuations of the two participating chains and of surrounding chains might break the contact, temporarily or permanently; see Figure 6.10. One might thus think of a 'topological contact' as a barrier that blocks chain motion in one direction, while possibly allowing it in another direction. Noting that PP calculations based on fine-grained MD simulations

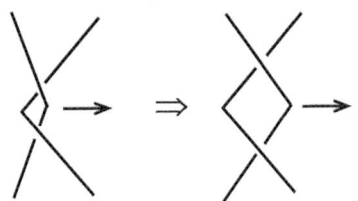

Fig. 6.10 Loss of topological contact when one chain moves away.

yield 2–3 TC's per entanglement, one can crudely judge that there are two or three 'gates' by which a chain might escape from a locally confined configuration. If chains are in 'loose' contact, relatively modest adjustment of the configuration of one or both of the chains, possibly through loss or creation of other contacts elsewhere along the chains, could cause one of the chains to move away from the topological contact without the other chain following it.

A TC thus differs conceptually from a slip-link in that the latter holds both chains together at the slip-link location, even if the slip-link location moves. Fluctuation of the slip link, as is allowed in the PCN model of Masubuchi et al. [23] or the slip-spring model of Likhtman [25], reduces the modulus, but retains a long-term association of the two chains. Breakage of a TC, on the other hand, might be either temporary or permanent, depending on whether the chains departing from that TC are driven together again, or are able to drift permanently apart, due to loss of other neighboring TC's that might hold the chains in close proximity. Thus, the dynamics of TC's are influenced by *many-body effects*; the influence of multiple chains can determine how long a given TC is retained. The dynamics of a slip link, on the other hand, are governed by a pair of chains.

To date, entangled chain dynamics has been described mainly by three basic types of models. (1) In *tube models*, the density of chain interactions is defined by the dimensionless length of the primitive path $Z \equiv L_{pp}^2/\langle R^2 \rangle$, which is used as the definition of the number of entanglements. (2) In *slip link* models, the density of interactions is specified by the number of slip links per chain, which is equal to or greater than Z, the number of entanglements in that polymer, with the exact number of slip links depending on the mobility of the slip links. (3) In *fine-grained* models, such as molecular dynamics or dissipative particle dynamics simulations, resolution of interactions is at the level of individual atoms or small groups of atoms that are not permitted to pass through each other. In fine-grained models, the density of interactions is not specified in advance, but can be calculated at any point during the simulation through an analysis of the primitive paths, yielding a measure of the density of topological contacts. For fine-grained simulations carried out so far, the number of topological contacts appears to be around a factor of two or three larger than Z. There is a fourth kind of modeling that is just beginning to be explored, in which chains are coarse grained, but are prevented from passing through each other through use of geometric constructions that identify chain crossings and then somehow block them [49–51]. Such models are computationally more expensive than coarse-grained tube models or slip-link models, but might be nearly as accurate as, and possibly considerably cheaper than, fine-grained simulation methods that resolve the chain at a level close to that of a monomer.

An interesting intermediate case is a lattice model described by the 'bond fluctuation' algorithm. In this model, monomers are centered on lattice points. When primitive paths are determined for this model, they turn out to be close to random walks, unlike the primitive paths generated from chains in continuous space [52]. The number of TC's is therefore close to the value of Z, the number of entanglements per lattice chain. For this lattice model, each entanglement can therefore be identified explicitly with a TC, located along the primitive path. Perhaps not coincidentally, the dilution exponent estimated for chains on this lattice appears to be very close to unity, indicating that entanglements on the lattice are well described as two-chain interactions. While the lattice chains certainly represent polymers less realistically than do continuous-space chains, these lattice results hint that entanglement interactions, and perhaps the dilution exponent, might have non-universal features that differ qualitatively from one kind of polymer to another.

Thus, the true molecular nature of entanglements and precisely how their density scales with polymer concentration remains frustratingly unresolved. Fortunately, rapid increases in computer speed and the development of better tools for analyzing molecular configurations and dynamics will likely lead to substantial progress in this area in coming years.

The effects of large deformations

Finally, when entangled polymers are subjected to large deformations that are imposed much more rapidly than chains can relax, the chain conformations can be greatly perturbed. Since the notion of an 'entanglement' remains ambiguous even in the well-equilibrated state where chains take on random walks, it should not be surprising that analysis of the effects of large deformations on entanglement structures and dynamics remains an open area of research. To date, work has been devoted mainly to extending models developed for dynamics of the undeformed state to the nonlinear regime, and we will discuss these in Section 6.3. However, molecular dynamics simulations are beginning to be used to explore the entanglement structure of deformed polymers. An example is a recent molecular dynamics study of confined polymer films where polymer chains are compressed in one direction, while retaining close to bulk dimensions in the other two directions [53]. Using an analysis of primitive paths, it was shown that compression of the polymer coils decreased their entanglement density. Some recent experiments on compressed polymers appear to support this computational result [54]. This is not surprising; the packing model discussed in Section 6.5 implies that reducing the volume of the coil leaves less room in that volume for other chains, and hence there will be fewer entanglements.

An analogous effect is expected in large nonlinear deformations, including deformations that stretch the material in a given direction, while shrinking it in a perpendicular direction. Since polymer melts and solutions basically retain the same volume in a deformation, in a fast deformation each coil is deformed into an ellipsoidal shape that retains the volume of the original coil. If all molecules deform more-or-less *affinely*—that is, in proportion to the macroscopic strain—then molecules that interpenetrate and entangle with a given coil before the deformation will

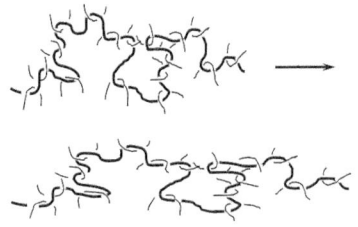

Fig. 6.11 Entanglements with a given polymer chain remain after the fast application of a large deformation.

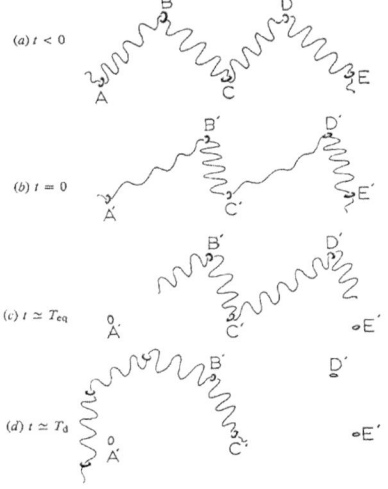

Fig. 6.12 Illustration of the relaxation process after a sudden deformation. (a) Initial equilibrium state, (b) immediately after the deformation: each part of the chain is either stretched or compressed, (c) after the first relaxation process: the primitive chain recovers its equilibrium arc length, but the conformation of the primitive chain is still in a non-equilibrium state, (d) the second relaxation process: the chain disengages from the original deformed slip-links and returns to the final equilibrium state in which it is entangled in new slip links. (Figure 3 in [10].)

continue to do so immediately after the fast deformation, even though the centers of mass of the coils are displaced away from each other in the stretching, or longitudinal, direction; see Figure 6.11. However, once the deformation ceases and the chains begin to relax, they will shrink back rapidly in the longitudinal direction while retaining nearly the same lateral dimensions. As explained by Graessley [1], this should release entanglements, as the volume pervaded by the polymer volume shrinks. Eventually, the polymer will re-expand in the lateral direction (along the short axis of the ellipse) and recover its original equilibrated volume and degree of interpenetration with other chains. This second relaxation process perpendicular to the stretch direction requires, however, that the chain escape from some remaining entanglements and create many more new ones. Hence this second process is expected to be much slower than the first process by which the chain shrinks along the stretch direction, since the first process does not require much maneuvering around entanglements and is hence expected to be fast.

In tube models, the tube diameter and the primitive path length are both determined by the entanglement density. At equilibrium, where the polymer takes on a random walk configuration, these two measures are related to each other through Z, the number of entanglements per chain, and the chain's radius of gyration. But, when the polymer is deformed, the entanglement constraints are likely no longer distributed isotropically, and the relationship between tube length and diameter becomes ambiguous. While the tube length is increased by the deformation, it is difficult to determine how the tube diameter should be chosen to reflect the effect of the deformation on entanglement density [55]. Marrucci and deCindio [56] suggested that the tube *volume* should remain constant during the fast deformation, implying that the tube diameter is compressed as a result of deformation. More recently, Marrucci and Ianniruberto [57] have proposed that deformation makes the tube cross-section become non-circular. The process by which the tube cross section recovers its original diameter and shape after deformation ceases must then be described somehow. It remains unclear how to do this and even what time scale is required for it to occur [58].

The slip-link model is more naturally extended to allow for large deformations. One needs only assume that the slip links are displaced affinely by the deformation. Chain retraction will then naturally lead to a decrease in the number of active slip links as slip links are abandoned [27]; see Figure 6.12. Even early slip link simulations yielded reasonable predictions of nonlinear behavior [21–23] and recent versions have shown remarkably accurate results [28].

To pursue these issues at a more microscopic level, it would be highly desirable to analyze distributions of topological contacts and primitive path lengths of chains during and after large deformations by molecular dynamics simulations. Such studies are now becoming feasible and represent the cutting edge of research that is likely to improve our understanding of entanglement polymers under flow.

6.2 Models of polymer dynamics in the linear regime

The 'standard tube model'

We now turn from the murky underpinnings of the tube model to a brief description of what might loosely be called the 'Standard Model', which is the most widely accepted theory of relaxation of linear and long-chain branched polymers.

Reptation

In Section 6.1, we already alluded to reptation, the coarse-grained sliding of a polymer chain along the contour of the tube. The time required for the chain to reptate completely out of the tube scales as the third power of the polymer length L. A power-law exponent of two is obtained from simple diffusion scaling, since the tube length is proportional to the length of the chain. The additional factor of L comes from the linear relationship between friction and chain length. Thus the reptation time τ_d is predicted proportional to L^3. At times much shorter than τ_d, but longer than the time τ_e it takes for the polymer to 'feel' its confinement in a tube, the polymer should behave as an elastic solid, or rubber, with modulus set by the density of entanglements, which on this time scale act somewhat like cross-links. Since the molecular weight per entanglement is given by M_e, and the modulus of a rubber is proportional to the thermal energy $k_B T$ times the number of effective cross links per unit volume, for entangled polymers an effective modulus of around $G_N^0 \approx \rho RT/M_e$ is obtained, where ρ is the melt density, $R = N_A k_B$ is the gas constant, and N_A is Avogadro's number. G_N^0 is called the 'plateau modulus', because for long polymers its value is obtained from frequency-dependent measurements which display a modulus that is relatively constant at the value G_N^0 over a wide range of frequencies;

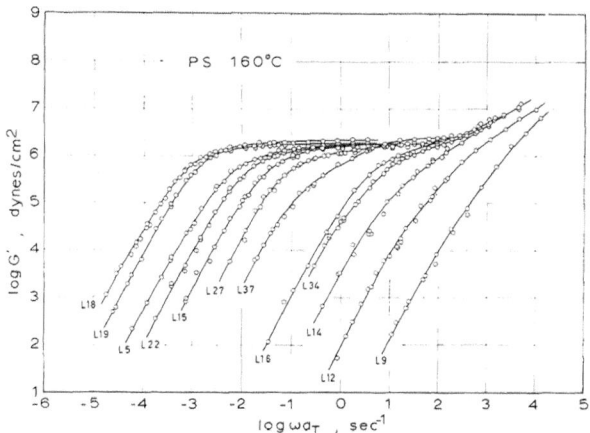

Fig. 6.13 Storage modulus, G', as a function of frequency reduced to 160°C for nearly monodisperse polystyrenes of molecular weight ranging from 580 000 to 47 000, from left to right. (Figure 2 in [59].)

see Figure 6.13. This range of frequencies lies between the inverse of the reptation time τ_d and the inverse of τ_e. It turns out that because the entanglement points allow sliding motion, which is not available for cross-links in the rubber/polymer network, the modulus is slightly smaller than G_N^0, and can be expressed as $G_N^0 = 4\rho RT/(5M_e)$. (This formula depends on how, exactly, the entanglement molecular weight M_e is defined; unfortunately there are competing conventions for this [60].) So, if polymer melts relax purely by reptation, then they should display, over a wide range of frequencies, a nearly constant modulus that is independent of molecular weight, and a longest relaxation time that scales with molecular weight to the third power.

Primitive path fluctuations

It has long been noted that the dependence of the longest relaxation time τ_d on the chain length deviates from a simple cubic power dependence, showing instead a stronger dependence roughly consistent with an exponent of around 3.3–3.6 [3], often labeled the '3.4 exponent'. The increase in the effective exponent over the reptation value of 3.0 is just what is expected from analysis of small-scale jiggling motions known as *primitive path fluctuations*. These jiggling motions within the tube depend only on the length of the chain segment undergoing the motion, and are not much affected by entanglements. For chains that are long enough, such small-scale motions should have essentially no effect on the time for reptation out of the tube, τ_d. However, their effect on τ_d is slow to die out as the chain length grows. The reason is that such chain fluctuations, when they include the chain ends, result in a nibbling away of the occupied portion of the tube; see Figure 6.14. Since the time to vacate the tube depends on the tube length to the third power, even a rather small fractional loss of tube length can appreciably accelerate the rate of reptation out of the tube. These fluctuations have their biggest relative accelerating effect when the chain is shortest, and diminish their relative importance as the chain increases in length. Hence they should increase the sensitivity of the reptation time to molecular weight, and thus *increase* the effective molecular weight exponent from 3.0 to something higher. Calculations that combine reptation with primitive path fluctuation yield a prediction that, over a range of around 10 to 100 entanglements per chain, the reptation time scales with molecular weight roughly to the 3.4 power [3], rather than 3.0, which is appropriate only for pure reptation [61]. Of course this higher exponent is really a pseudo-exponent, since it fits data only over a limited range of chain lengths, and for long enough chains, the exponent of 3.0 for pure reptation should be attained.

By combining reptation and primitive path fluctuations, almost quantitative predictions of the linear viscoelastic behavior of model linear polymer chains, including the dependence of reptation time on chain length, has been attained [15, 62]. The inclusion of primitive path fluctuations leads to a correction to the predicted reptation time in the

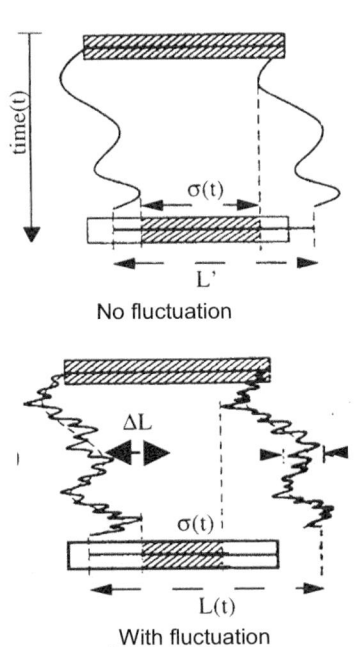

Fig. 6.14 Effect of primitive path fluctuations on accelerating the evacuation rate of the initial confining tube (shaded region). (Figure 6.9 in [2].)

form of $\tau_d(1 - 2C_1/\sqrt{Z} + C_2/Z + ...)$ where Z is the number of entanglements per chain, which can rather closely fit to the 3.4 power law by choosing proper values of the prefactors C_1 and C_2 [2, 63]. Nevertheless, this 'standard' explanation of the 3.4 power-law exponent is not universally accepted. One of the difficulties cited is that in experiments in which linear polymer chains are placed in a matrix of longer chains, the 3.4 power-law seems to revert to a 3.0 (or more precisely 3.1)-power-law dependence [64]. Since it is generally believed that the long matrix chains would not strongly affect the short-time fluctuations of a given test chain, these results call into question this standard explanation of the 3.4 power law. Another reason to question the standard explanation is that the primitive path fluctuations should also reduce the plateau modulus, because they nibble away the length of the tube and hence the number of effective entanglements. However, an analysis of the molecular weight dependence of the plateau modulus does not show the expected decrease in G_N^0 with decreasing polymer molecular weight [65]. The expected dependence of G_N^0 on M is small, however, and rather easily obscured by other small effects [62]. These observations suggest that the primitive path fluctuations only partly account for the 3.4 power-law. Contributions from other relaxation mechanisms, including constraint release as discussed below, also play an important role in determining the second- (reflected by the parameter C_2) and higher-order corrections to the reptation time and plateau modulus.

The 'standard model' of entanglement dynamics predicts that, for a given dimensionless molecular weight M/M_e, the rheological properties are universal, once the modulus is made dimensionless with respect to G_N^0, and the time or inverse frequency is made dimensionless using τ_e. However, rheological data show some significant deviations from this basic prediction. In particular, Fetters *et al.* [31] have found non-universal values of dimensionless ratios of various entanglement-dependent molecular weights. One of these is the critical molecular weight for entanglement M_c, defined as the molecular weight at which the zero shear viscosity transitions to a 3.4 power-law dependence on molecular weight. Another is M_r, the much higher molecular weight at which the 3.4 power-law dependence on molecular weight transitions to 'pure reptation' scaling with a 3.0 power-law exponent. The 'standard' tube model predicts that M_c/M_e and M_r/M_e should have universal values. However, Fetters *et al.* [31] have shown that the ratios M_c/M_e and M_r/M_e are not universal, but seem to correlate with the packing length p. This suggests that the number of Kuhn steps per entanglement plays a non-trivial role in entanglement dynamics. This, in turn, suggests that many entangled polymer melts are not in the highly flexible-chain limit; i.e. the entanglement spacing is not asymptotically large compared to the length scale at which chain stiffness becomes important.

Constraint release

As alluded to in the beginning of this Section, the matrix chains are themselves in motion and so the notion of a fixed 'tube' is built on shaky ground. This was made evident once rheological and diffusion experiments were conducted with mixtures of polymers of two widely different lengths. For such mixtures, after the short chains have relaxed, only the long chains remain entangled, and only with each other. The naive reptation theory discussed in Section 6.2 predicts that the plateau modulus after relaxation of the short chains should be proportional to the volume fraction of long chains ϕ_L, and therefore should be given by $\phi_L G_N^0$. Instead, however, the experiments reveal that the modulus after relaxation of short chains falls instead roughly to $\phi_L^2 G_N^0$ [66]. The roughly quadratic dependence on the volume fraction of the long chains not only rules out the naive model with fixed tubes, but is evocative of the simple slip link picture (Figure 6.4) in which 'it takes two to tango'. If the modulus is viewed as a sum of contributions from entanglements formed randomly by two chains, then the departure of all short chains from participation in the entanglements leaves only entanglements between pairs of long chains remaining, the frequency of which is proportional to the square of their concentration.

In fact, a serviceable model of the rheology of polydisperse entangled polymers can be crafted by assuming that the stress remaining at a given time after a deformation is just proportional to the sum of the remaining entanglements, and that each entanglement disappears as soon as the fastest-reptating chain abandons the entanglement [67–70]. Using random pairing of chains, and the single-chain reptation theory to obtain the departure times of each chain, the distribution of entanglement lifetimes is readily constructed, and the rheological response predicted.

This model, called 'double reptation', works best when the chains are broadly polydisperse in molecular weight, so that the limitations in the approach are covered over by overlaps of the contributions to the stress from the various components of the molecular weight distribution. These limitations are exposed more clearly in mixtures of chains of two greatly differing lengths, especially when the longer of these is present at low concentration. In that case, the reptation of all the short chains would, in double reptation, abruptly remove most of the stress. In reality, the relaxation of the short chains only frees the long chains to begin to move beyond the confines set by the entanglements with short chains. Before the long chains have moved far, they will again re-entangle with short chains, which must again move out of the way to allow further relaxation of the long chains. This combination of repeated short-chain disentanglement and long-chain drift can be described elegantly by treating the tube itself as a 'polymer' in a medium whose effective viscosity is set by the rate of disentanglement of the short matrix chains [1, 71]. The motion of such a 'polymer' in a viscous medium is described by the well-known *Rouse theory*. Hence, such motion is called 'tube Rouse' motion or 'constraint-release Rouse' motion. It is more accurate than the simple 'double reptation' picture in that it accounts for the gradual drift of long test chains resulting from repeated release of entanglements with short matrix chains. By analyzing reptation in a tube that is itself moving by Rouse dynamics, predictions have been made of the relaxation dynamics of mixtures of long and short chains that agree with multiple sets of experimental data [71, 72]. This agreement is attained only if the short chain is not greatly (more than a factor of 10 or so) shorter than the long chain.

If the short chain is very short, the relaxation of the short chain will occur before the long chain will even have reptated a single tube diameter in its own tube. In this case, the chain cannot be said to reptate in the original tube at all [73]. Instead, it makes more sense to think of the tube as expanding in diameter and shortening in its tortuous length as some of the chains defining its shape cease to present themselves as obstacles. Over long times, the test chain only feels a fraction of the entanglements that constrain it at short times, and so the long-time tube is effectively wider than the short-time one. Expansion of the tube diameter thus represents the time-scale dependence of the degree of entanglement felt by the test chain, and is called 'tube dilation' or 'dynamic dilution' [74], since the fast diffusing chains act as constraints at short times but as solvent on long times.

Both tube-Rouse motion and tube dilation have been accommodated into the mean-field tube model in which the matrix chains act collectively to define the tube location and diameter. This approach, accounting for reptation, primitive path fluctuations, and constraint release by tube-Rouse motion and tube dilation, constitutes a kind of 'standard model' for the relaxation of linear polymers, whether monodisperse, bidisperse, or polydisperse. Various computer codes have been created that implement these physical mechanisms of relaxation in somewhat different ways [16, 17, 19, 63, 75]. Many of them are quite successful in accounting for the linear viscoelastic properties of linear polymers.

The only 'fitting' parameters of the tube model mentioned thus far are the plateau modulus G_N^0, the entanglement molecular weight M_e, and the frictional time constant τ_e. In principle, the entanglement molecular weight and plateau modulus are related to each other by the formula $G_N^0 = 4\rho RT/(5M_e)$. Moreover, the plateau modulus G_N^0 is not very temperature sensitive and can be obtained rather easily from the plateau stress in linear viscoelastic measurements. Tabulations of values of G_N^0 can be found in [13]. If one uses the above formula to obtain M_e and the tabulated value of G_N^0, then only one fitting parameter τ_e remains. While this parameter can also be obtained from frictional measurements in unentangled melts, this is rarely done, in part because unentangled melts are so low in molecular weight that their frictional properties are typically affected by shifts in the glass transition temperature owing to the prevalence of chain ends. Moreover, a one-time fit yields a value for τ_e that is in principle valid for all molecular weights, molecular weight distributions, and long-chain branching structures for polymers of the same chemical make-up at the same temperature. We also remark that small adjustments of G_N^0 and M_e are frequently employed to improve fits of various versions of the tube (or slip link) model to experimental data. Since the tabulated values themselves are not sacrosanct, these adjustments are tolerable, as long as comparisons to multiple molecular weights and/or multiple chain architectures (i.e. different long-chain branching structures) are carried out using the same fixed values of the three tube-model parameters: G_N^0, M_e, and τ_e.

Finally, it is worth remarking that tube theories that include constraint release sometimes introduce one or more additional parameters to describe how effective the motion of a matrix chain is in reducing the entanglement density confining the test chains [5, 76]. Additional parameters can be avoided by following the 'double reptation' ansatz, which assumes that there is symmetry between the test and matrix chains whereby relaxation of a matrix chain yields loss of one entanglement on the test chain. This assumption is often carried over to the more sophisticated constraint-release Rouse model, which allows inclusion of constraint release effects with no additional fitting parameters. But other treatments introduce one or more additional parameters to describe constraint release effects.

The 'standard model' applied to long-chain branched polymers

Polymers with long-chain branching have long been of great fundamental and commercial interest. One of the earliest clues that the reptation idea was basically correct was the suppression of relaxation that occurs when a long branch is added to an otherwise linear chain, forming a 'three arm star'; see Figure 6.15. It was observed long ago [77] that high-molecular-weight star-branched polymers take much longer to completely relax than do linear polymers of comparable molecular weight. De Gennes [78] realized why: reptation cannot occur when a branch point anchors three or more polymer strands to the same point. The tube in this case has three legs, and the branch point connecting them will not spontaneously drag one of the arms into a tube surrounding one of the other arms. Thus, relaxation of each arm depends on its ability to fluctuate in its tube, gradually exploring by chance deeper and deeper into the tube and so forgetting larger and larger outer portions of the original tube. The process becomes very slow for deep fluctuations. Having fluctuated back to a certain point on the original tube and then wandered out again, forming a new outer portion of the tube, one must wait even longer for random events to cause the arm to fluctuate even deeper into the original tube, eventually back to the branch point itself.

Deep arm fluctuations are strongly disfavored entropically [81]. Only rare fluctuations will thread the tip of the arm through the entanglements lining the entire tube so that the arm escapes all of its entanglements [82]. An analysis of the chain configurations required leads to the prediction that the arm relaxation time τ_a should be *exponential* in molecular weight, which is a much steeper dependence even than the

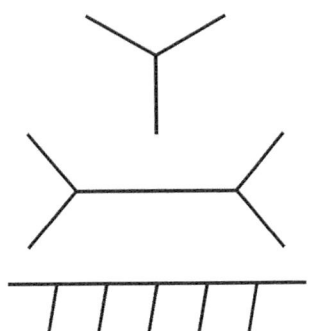

Fig. 6.15 Symmetric 3-arm star, H and comb polymers.

3.4 power law of linear chains. The exponential dependence of relaxation time and viscosity on arm molecular weight has been confirmed experimentally; see Figure 6.16. The 'entropic potential' U_{fluct} needed to describe these arm fluctuations was given by Doi and Kuzuu [81] and Doi and Edwards [2]. It is quadratic in the depth of the fluctuation along the tube. When this depth is measured as fractional distance ξ along the tube from the tip of the arm to the branch point, this quadratic potential is given by $U_{\text{fluct}} = \nu Z \xi^2$, where Z is the number of entanglements and ν is a numerical prefactor of $\nu = 1.5$. While this 'universal' form for the fluctuation potential is widely used, recent simulations and theory have suggested that the prefactor ν might decrease slowly with molecular weight, approaching a value close to unity for long arms [83]. Such a dependence on arm length might seem very weak, but it can have large effects, since the time scale of deep fluctuations depends exponentially on the potential.

Accounting for arm fluctuations gives only qualitatively correct predictions for star polymer relaxation; a quantitative treatment requires consideration of constraint release. This is because, in a melt of stars, many of the constraints, arising from entanglements with the tips of arms, are short-lived, and come and go rapidly compared to the time required for an arm to retract completely within the tube. Hence, as time goes by, ever-deeper fluctuations occur, but ever fewer of the entanglements composing the tube are still effective. Thus, a proper description of star arm relaxation requires treatment of constraint release, even when the star polymer is monodisperse and all arms are of the same length [74]. So widely varying are the times required for release of the various entanglements that a simple and reasonably successful model is to regard the tube as continuously dilating as the arm in it relaxes; see Figure 6.17. The relaxation of the arm is then described by a differential equation relating the fraction of chain relaxed to the logarithm of time. The instantaneous diameter of the tube that appears in this equation is set by the fraction of arm that has relaxed at that instant, since in a monodisperse melt of stars the surrounding matrix arms are just as relaxed as the test arm is. Integration of this equation then yields a rate of relaxation that accounts for the loss of entanglements that results from the relaxation itself. This 'self-consistent' picture of star arm relaxation,

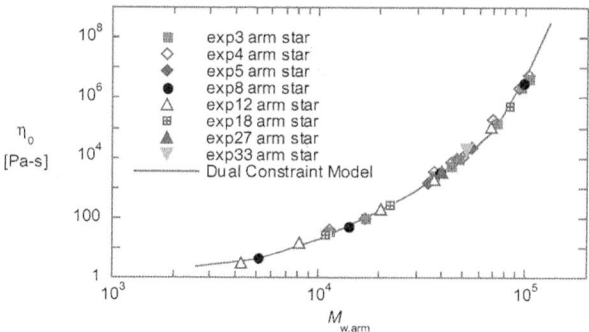

Fig. 6.16 Zero-shear viscosity versus arm molecular weight M_w of star polyisoprenes. Symbols are data from [79] at a reference temperature of 60°C. The line is the prediction of the dual constraint model with parameters $G_N^0 = 4.34 \times 10^6$ Pa from [80], and $\tau_e = 1.2 \times 10^{-5}$ s from a best fit to the data. (Figure 16 in [16].)

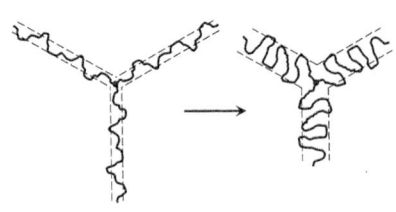

Fig. 6.17 Dynamic dilution in star polymers.

proposed by Ball and McLeish [74], was one of the first implementations of the idea of 'dynamic dilution'. In principle, it should apply to linear polymers [84], but is usually less important for them, since the relaxation of a linear polymer of given molecular weight is much less gradual than for a star. We should note, however, that the underpinnings of dynamic dilution, even for simple cases, is rather shaky [85], and the idea is more of an ansatz than a rigorous first-principles theory. As we see below, slip-link models capture the effects modeled by 'dynamic dilution' automatically with no need to insert an ansatz relating entanglement density to the fraction of the chain that has relaxed.

A star polymer, with a single branch point, cannot reptate, but each arm is free to fluctuate. How do polymers relax that contain a 'crossbar' terminated at each end by a branch point? This polymer, called an 'H polymer' (see Figure 6.15) has two branch points and the crossbar has no free ends, and so cannot relax by simple fluctuations as a star polymer can. In this case, McLeish & coworkers [86] proposed that each time an arm relaxes by primitive path fluctuations, the branch point can 'hop' a distance comparable to the entanglement spacing a. Since the effective diffusivity of the branch point is given by the square of the hopping distance divided by the hopping time, we obtain for the branch-point diffusivity the expression $D_b = p^2 a^2/\tau_a$, where τ_a is the arm relaxation time, and p is the ratio of the hopping distance to the tube diameter. The parameter p is expected to be around unity. The hopping of the branch points is very slow, if the arms are long, but over time will allow motion of the backbone. Since the backbone is trapped in a tube, it must move along that tube by reptation. But this is a renormalized reptation, since it is governed by the diffusivity of the branch points, which are set by the arm relaxation times. And these arm relaxation times are exponential in the arm molecular weight. Hence, the backbone reptation time depends quadratically on the molecular weight of the backbone, but exponentially on the molecular weight of the arms. Relatively recent experiments on monodisperse H polymers have confirmed this very strong dependence of H polymer relaxation on arm molecular weight [86]. However, one additional parameter, p, is required to make quantitative predictions. Not only is the value of this parameter only available through data fitting, but there are disquieting indications that its best-fit value can vary from one polymer melt to another [87].

The same basic model can be applied to 'comb' polymers, with multiple branch points along a linear backbone [88]; see Figure 6.15. As with the H polymer, the backbone of the comb relaxes by a renormalized reptation, whose rate is set by the relaxation times of the attached arms. Reasonably accurate predictions of the rheology of comb-branched polymers have been obtained using this idea [88].

Having introduced branch point motion into the tube model, it is possible to extend the model to apply to arbitrary mixtures of branched and linear polymer molecules, using a computer to keep track of the relaxations of each component of the mixture [17]. This is a potentially very powerful development, since commercial polymers with long-chain branching are invariably cocktails of polymers of various architectures: linear, star, H, comb, and even hyperbranched polymers with branches on branches. Moreover, the branches are polydisperse in length and in positioning along the backbone, and the backbone is itself polydisperse in length. Nevertheless, this complexity can be dealt with by using knowledge of the catalytic process by which the polymer is synthesized to create a computational ensemble of chains to represent the mixture produced by the catalyst [19, 89]. Several groups have created computer algorithms capable, in principle, of computing the linear viscoelastic properties of commercial polymers with long-chain branches [17–19, 75, 90]. Interest in these polymers is spurred on by the superior processing properties that long-chain branching can confer.

Successful application of these algorithms to predictions of the linear viscoelasticity of commercial long-chain-branched polyethylene has been reported [19, 90]. However, disturbingly, the algorithms that are the most successful in predicting the rheology of these complex mixtures are not always the most successful in predicting the rheological properties of simpler polymers, such as polydisperse linear chains or mixtures of star and linear polymers [90]. One of the persistent problems with these algorithms is in the choice of the value of the 'dilution exponent' α, discussed in the concluding section. Predictions of the rheology of mixtures of linear chains, or of lightly branched polymers, are generally in better agreement with experimental data when α is taken to be unity [19, 72, 90], while for stars and mixtures of stars and linear polymers, $\alpha = 4/3$ seems to provide better agreement [14, 91, 92]. The persistent inability of a single theory to provide accurate predictions for all molecular architectures suggests that something is amiss in the underpinnings of the 'standard' versions of the tube model. Perhaps even more troubling, the 'standard' tube model, whatever version is used, is really a pastiche of mechanisms (reptation, primitive path fluctuations, tube-Rouse motion, dynamic dilution, branch-point motion) combined in an *ad hoc* manner. Is there not a more elegant way to account for all of the relevant phenomena?

Slip link algorithms

Because of the complexities of the tube model, the idea of 'slip links' as embodiments of entanglements has been revisited in recent years [21–25]. With each slip link representing a localized 'entanglement' through which the chain is forced to pass, a coarse-grained model of the polymer chain then tracks only the locations of the slip links and the number of monomers along each polymer strand that spans two neighboring slip links. A diffusion equation for each strand can then be written, allowing the strand to exchange monomers with the two strands to which it is linked through the two slip links at its two ends. Solving these strand diffusion equations captures both reptation and primitive path fluctuations. Primitive path fluctuations occur when a chain end passes through a slip link, and the slip link is removed, or when a chain end

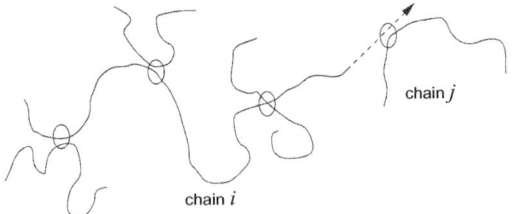

Fig. 6.18 Illustration of the slip-link model, in which an entanglement between two chains is represented by a slip link that is either created when the end of chain i fluctuates outward, randomly entangling partner chain j, or destroyed if either chain i or j fluctuates inward. (Figure 1 in [24].)

drags enough monomers beyond the terminal slip link to justify placing a new slip link on that chain end; see Figure 6.18. Reptation occurs naturally at long times as a collective result of monomer exchange among strands, which at short times captures high-frequency internal relaxation modes.

While this method elegantly captures much of the relevant single-chain physics, it does so at a high computational price. Each time step of the simulation must be small enough to track the movement of a few monomers from one strand to a neighboring strand. While much faster than a molecular dynamics simulation of all monomers would be, slip link simulations are typically much slower than computing reptation and fluctuation times from the equations of the tube theory. The simplest slip-link models introduce two parameters: the rate of monomer exchange and the average number of monomers between slip links. These parameters correspond, respectively, to the frictional time constant τ_e and the entanglement molecular weight M_e of the tube model. The plateau modulus G_N^0 emerges naturally from the choice of M_e, and the density of chains.

In addition to capturing reptation and primitive path fluctuations, constraint release is implemented easily by allowing a slip link to disappear or appear based on the motion of surrounding matrix chains. One way to do this is to explicitly pair the slip link created on the end of one chain with a 'partner' slip link created simultaneously at a position somewhere along another chain [21, 23, 24]; see Figure 6.4. The linkage of the two slip links is then maintained until one of them is destroyed by a chain end sliding through it, and then the partner slip link is destroyed as well. In a second method of including constraint release, the slip links are not explicitly paired, but for each loss of a slip link on the 'test' chain, another slip link on another chain can be randomly destroyed [25]. Similarly, the creation of a slip link on the end of the test chain can precipitate creation of a slip link randomly on another chain, but without retaining any pairing or memory of which slip link precipitated creation of the 'partner' slip link. This second method of incorporating constraint release is more consistent with a 'mean field' view of entanglements, in which no single, identifiable, matrix chain is responsible for the entanglement restriction that the test chain feels. One can make the non-local character of an entanglement more explicit by requiring, for example, that several slip links along a test chain be destroyed or created before a single matrix slip link is created or destroyed.

And there is no need to create slip links on matrix chains at the exact moment a slip link on the test chain is destroyed, as long as creation and destruction rates are kept in balance. Of course, creating and destroying slip links on matrix chains at a rate different from that for the test chain introduces an additional parameter – the relative frequency of matrix versus test-chain creation and destruction events.

Although the above describes the basic ingredients, many choices are needed to specify precisely a slip-link algorithm, and many of these choices are rather arbitrary. First, there is the decision alluded to above whether to retain an explicit pairing between slip links, or instead to create slip links in the matrix without pairing them with other slip links. There is also the relative rate of matrix vs test chain slip link creation and destruction. One can eschew explicit pairing of test and matrix chain slip links, but still insist on self-consistency in the rates of destruction by motions of the test chains and matrix chains, respectively [93]. In addition, there are choices regarding correlations between real-space positions of the chains and slip links. In the most detailed model, the positions of all slip links of all chains are tracked relative to each other in a simulation box [23]. In this case, motion of one chain end creates or destroys a slip link on a nearby chain. Alternatively, in a less detailed model, one can track only the positions of slip links relative to other slip links on the same chain [21]. In the latter case, the correlations with other 'partner' chains are created randomly, since no information is kept regarding the relative proximity of one chain to another. Finally, one can even discard all real-space positional information and only retain the relative ordering of slip links along a chain [24]. In this last case, stress relaxation is computed by simply counting the fraction of original slip links surviving after a given amount of time. If slip-link positions are tracked in real-space, then when a new slip link is created along a matrix chain, a decision must be made regarding whether to deform the matrix chain when putting the slip link on it or to simply insert it along a straight line connecting the two slip links between which it is placed. The former choice appears to be the better one [23]. One must also decide whether or not to let the slip links move. If they are allowed to move, one can choose to balance the force that one chain exerts on the slip link with the force produced by the partner chain, leading to deflection of the slip-link position to keep these forces in balance as each chain moves [23]. One can also decide to tether the slip link to a fixed point in space by means of a fictitious elastic spring that permits fluctuations in the slip-link position; see Figure 6.9 [25]. By choosing the spring constant of the tether appropriately, the limited lateral freedom a real chain has can be mimicked and so a kind of 'tube diameter' imposed. One must also decide when a chain end has grown long enough to justify creating a new slip link, or has retracted short enough to destroy one. Poorly thought-out choices can lead to wasted computation that, for example, adds and immediately deletes the same slip link over and over again.

Thus, while at first blush the slip-link models appear to be more 'elegant' than the ponderous versions of the tube model we have discussed, the number of rather arbitrary choices, and the large computational load for the slip-link models, are sobering realities. Perhaps even more distressing, there is not at this time a satisfactory slip-link model that can readily allow for branch point motion. Hence, the slip-link model is generally restricted to linear or star-branched polymers that can relax without branch-point motion. The problem that slip-link models have in handling polymers with two or more branch points is illustrated in Figure 6.19. Here, an entanglement between backbones of two different H polymers is depicted, which should be represented by a slip link. Suppose a slip link is released only when a chain end passes through it. Since for two entangled backbones, no chain end can pass through the slip link until a branch point passes through first, how will this slip link ever get released? One could allow any slip link, no matter where it is located, to be released at a rate determined by the average rate at which slip links are vacated by chain ends. However, this would no doubt release the slip link representing entanglements between two backbones much too rapidly, and would fail to account for the severe hindrance to relaxation that entanglements between two backbones actually present. Thus, while localizing entanglement points to slip links allows a wide range of constraint release time constants to be accounted for, it goes overboard by making some entanglement points on branched polymers virtually immortal, and so makes the polymer melt a virtual gel. Hence, to date, slip-link models have mostly been applied only to linear and star architectures, where branch point motion is non-existent, or not needed to describe rheological behavior.

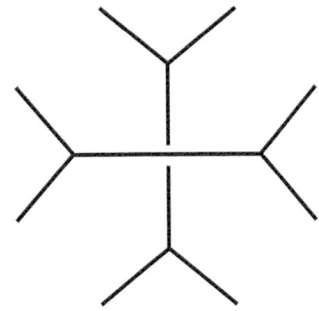

Fig. 6.19 Entangled H crossbars.

An exception is a schematic slip-link model by Shanbhag and Larson [94], which allows the branch point to hop one step along the path created by two of the branches whenever the third branch loses all of its slip links. This version of the slip-link model has shown success in predicting the behavior of star polymers and H polymers with relatively short arms. But it fails for H's with long arms, predicting terminal relaxation that is much too slow, evidently because requiring that the arm lose all its slip links before the branch point can hop is too severe a requirement. Moreover, it predicts that a branch point with functionality of four or greater is virtually immobile, since the likelihood of two arms simultaneously losing all slip links is essentially nil. A second model that permits branch-point motion is the Primitive Chain Network (PCN) model of Masubuchi et al. [95], which allows slip links to encompass two or more segments of the same polymer chain, so that a branch point can pass through a slip link. This model, while showing some promise, is also rather artificial, with arbitrary rules governing the probability of these events.

Thus, the slip link approach, which already requires multiple ambiguous choices of rules for creation, loss, and movement of slip links, must invoke still more such rules to be extended to even simple chains with mobile branch points. The tube model can be extended more readily to such cases, although also requiring selection of an empirical parameter

p^2. Since both the tube and slip-link models have significant advantages and disadvantages relative to each other, no doubt both will continue to be used for some time to come.

6.3 Models of polymer dynamics in the nonlinear regime

The above discussion was limited to linear viscoelastic phenomena, relevant for relaxation of polymers that are only slightly deformed from their equilibrium state. Nonlinear effects are produced by large deformations, and are sensitive to the rate, magnitude, and type of deformation. Two basic kinds of deformation are often considered, namely shear and extension, depicted in Figure 6.20. A shearing deformation involves layers sliding over each other, and the velocity gradient, i.e. the direction in which the velocity changes, is orthogonal to the direction of the velocity itself. In an extensional flow, the gradient of velocity is parallel to the velocity itself. Extensional flows tend to stretch material lines exponentially, while shearing only stretches them linearly in time. Flows that shape polymers into products are often mixtures of shear and extension. To predict the deformation of polymer molecules and the resulting stresses in general flows, one needs a nonlinear *constitutive equation*, which is a general relationship between the flow field and the polymer stress.

The modern tube model

Reptation

Doi and Edwards [12] produced a nonlinear constitutive equation for polymer dynamics based on reptation. The basic idea was that a chain, extended and oriented by a large deformation, can quickly relieve much of the stress by retracting back in its tube, leaving only the stress resulting from orientation of the tube remaining. This orientation can only relax by reptation of the chain out of the tube, a slow process. Using the reptation theory to predict this process, and assuming that the chain remains retracted at all times, Doi and Edwards developed the constitutive equation that bears their names. The theory explained the extreme shear thinning observed in typical polymer melts; i.e. the decrease in viscosity with increasing shear rate. A related thinning or 'softening' effect is observed if the shear is imposed suddenly in a single 'step'; see Figure 6.21. The Doi–Edwards theory explained that the shear thinning

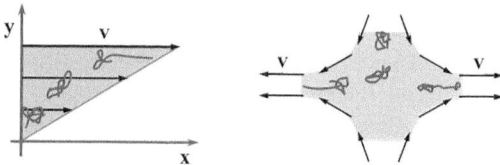

Fig. 6.20 Shear and extensional deformations.

or strain softening is due to the loss of stress that results from retraction of the chain in the tube in large deformation. The Doi–Edwards theory also yielded a reasonable value for the ratio of normal stress differences produced by shear. Normal stresses are anisotropic pressures that depend on direction relative to the shear direction. For ordinary fluids, pressure is independent of direction, but when polymers align, they generate a contractile force that is strongest in the direction parallel to the average orientation of the chains and weaker, but unequal, in the two perpendicular directions. Since the Doi–Edwards equation predicts the average orientation of the tube produced by shear, it can predict these anisotropic normal stresses rather well.

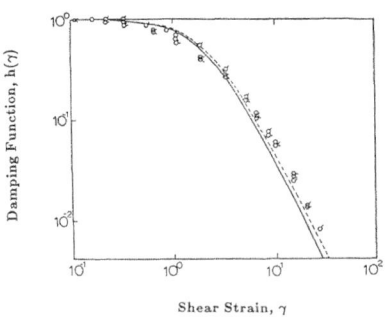

Fig. 6.21 Damping function $h(\gamma) = G(t,\gamma)/G(t)$, which describes the degree of strain softening of the melt, obtained from step-shear experiments on entangled solutions of polystyrene of molecular weight 3×10^6 in diethyl phthalate (symbols) (data of [96]) compared to the predictions of the Doi–Edwards theory with (solid line) without (dashed line) the independent alignment approximation. (Figure 6 in [10].)

Chain stretch

To go beyond these predictions, the Doi–Edwards theory needed to be extended to include additional physics. In the mid 1980's Marrucci and Grizzuti [97] incorporated the finite rate at which chains retract in their tubes. Since retraction within the tube does not require the chain to escape from any of its entanglement constraints, this retraction rate was taken to be the rate of relaxation of unentangled chains of the same length, given by the Rouse theory, which is much faster than the rate of reptation. The Rouse time scale for retraction should increase with the square of the molecular weight, while the time for escape from the tube is much longer and scales as the 3.4 power of molecular weight. The extension of the theory to include the retraction process allows it to predict the unusual behavior of polymer melts and concentrated solutions in steady extensional flows: the extensional viscosity first decreases with increasing extensional flow rate and then increases with further increases in extension rate. The initial decrease is the direct result of the chain retraction within the tube at low extension rates, which produces the extension thinning, just as retraction produces thinning in shear flows. The subsequent increase in viscosity occurs at extension rates fast enough that the chain does not have time to retract. Recent data, such as those in Figure 6.22 [98], confirm this peculiar non-monotonic dependence of extensional viscosity on extension rate and strongly support the notion that chains retract within their tubes much faster than they can reptate out of them. As shown in Figure 6.22, the predictions of tube models with time-dependent retraction are semi-quantitatively correct. However, the time constant associated with the 'fast' part of the relaxation after the step strain retraction process appears to be considerably longer, and might even be a stronger function of molecular weight, than expected for a Rouse time [99–102].

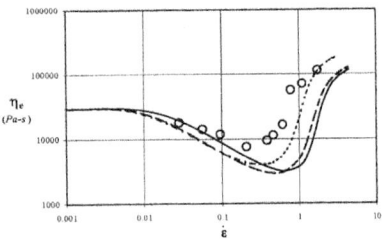

Fig. 6.22 Non-monotonic steady-state uniaxial extensional viscosity versus extension rate for a 6% solution of 10.2 million molecular weight polystyrene in diethyl phthalate at 21°C. Lines are the Doi–Edwards–Marrucci–Grizzuti (DEMG) model (solid line) and Mead–Larson–Doi (MLD) model (dashed and dotted lines). (Figure 12 in [98].)

Convective constraint release

Another outstanding issue is that in steady-state shearing flows, the shear thinning predicted by the original Doi–Edwards model is excessive, and would lead to flow instabilities if not moderated somehow [12]. One might expect that allowing for time-dependent, rather than instantaneous, retraction would bulk up the stress at high shear rates and so alleviate this problem, but, strangely, time-dependent retraction proves to be inadequate [97]. The reason is that the shear orients the tube in the flow direction, and since the flow gradient is orthogonal to the flow, there is little difference in velocity acting along the polymer chain once it becomes highly aligned. Hence, for the purpose of stretching entangled polymers, shear flow is self-defeating [103]. The more strongly it acts on the polymer, the more it aligns it, and hence the less variation in velocity is available to displace one part of the chain relative to another.

In fact, to stretch the polymer more effectively, one must paradoxically allow some additional mechanism for additional *isotropic* polymer *relaxation*, rather than higher flow rates. Unlike retraction along the tube, which merely contracts the polymer along the flow direction, isotropic relaxation includes relaxation perpendicular to the flow. And this allows the polymer conformation to expand into the gradient direction, giving the flow gradient a better grip on the chain. The result, oddly enough, is a more highly stretched chain than is possible when the chain remains highly aligned [103]. Since the tube is aligned by the flow, relaxation perpendicular to the flow direction requires that the polymer somehow escapes from the tube at a rate much faster than reptation would allow. There is in fact a mechanism that accomplishes this. As intuited years ago by Graessley [104], a flow or deformation separates neighboring chains from each other and thereby should destroy entanglement interactions. This occurs if the stretched chains have had time to retract back and so disentangle themselves from other chains [1]. Marrucci [105] called this process *convective constraint release* (CCR), and worked out a simple phenomenological model for its effect on the shear viscosity, finding that it alleviates the overly severe shear thinning of the original model. Mead *et al.* [103] first incorporated the idea into a full version of the tube model, and many other formulations of the idea have since followed, including a very sophisticated version by Milner *et al.* [106]. This latter version incorporates a parameter describing how effective the motion of a matrix chain is in reducing the entanglement density confining the test chain. A related parameter appears in some theories of constraint release for the linear viscoelastic regime; see the discussion on p. 248.

As a result of these efforts, tube models have become relatively successful in matching experimental rheology data for monodisperse linear polymers in both shear and extensional flows, both at steady-state and during transients [16, 103, 107–109]. Extensions of the tube model to mixtures of linear polymers have also been carried out, with substantial success. Development of nonlinear theories for long-chain branched polymers, such as stars or H's, is in a more primitive state, although rough models are available [110, 111]. While convective constraint release has now become part of the accepted 'standard' tube model for nonlinear polymer rheology, curiously, some recent slip link simulations of Schieber [112] show excellent agreement with experimental nonlinear shear rheology data of entangled polymers, without including any CCR contribution.

Unsolved problems

Strain-dependent tube diameter

Despite such successes, there are many unresolved issues in the nonlinear rheology of entangled polymers. Since a precise notion of what an 'entanglement' is remains elusive even for undeformed polymers, how these entanglement interactions are transformed by large deformations remains an even greater puzzle. For example, as discussed on p. 243, it has long been wondered if and how the tube diameter might change under deformation. Although the 'standard tube model' of nonlinear rheology has assumed that the tube diameter remains constant, it has been suggested that large deformations might cause the tube to change diameter or develop a non-circular cross section [57, 58, 113]. To date, there is no agreed-upon answer to this question, although it is likely that molecular dynamics simulations might soon give some new insight. It is likely that relaxation of the tube diameter is coupled to the chain retraction process after a step strain [58]. If so, then a correct analysis of tube diameter relaxation might shed light on the anomalously long values of the retraction time inferred from data on relaxation after a step strain, discussed above.

The magnitude of the retraction time

It is also worth noting that relaxation measurements have been carried out on individual optically visible long DNA molecules entangled with other such DNA molecules [114, 115]. Teixeira *et al.* [114] found that highly stretched DNA chains relax with two time constants, one of them consistent with the reptation time inferred from rheological measurements and the other corresponding to the 'retraction time', discussed above. As in the rheological studies on bulk samples, the 'retraction time' they measured on single molecules was much longer (by a factor of 10) than predicted from the Rouse theory. Robertson and Smith [115] measured the lateral force exerted by the matrix chains on a single entangled DNA molecule after it was displaced perpendicular to its orientation using optical tweezers. This force was found to relax with three time constants, one corresponding to the reptation time, the second to the fast Rouse time, and the third to a mysterious intermediate time scale. These results hint at some as yet unresolved 'matrix' time scale that may relate to the timescale for tube diameter relaxation.

Extreme strain softening and shear banding

In addition to the unresolved issues mentioned above, it was observed long ago by Osaki and Kurata [116] and others [117, 118] that polymers with more than 50 or so entanglements per chain displayed bizarre relaxation phenomena after a large, rapid 'step' strain. This relaxation consists of a collapse in stress to values lower than predicted by the tube theory even with no stretch. An early explanation of this phenomenon attributed it to a mechanical instability resulting from the extreme strain softening of the original Doi–Edwards equation [119]. The strain softening of the Doi–Edwards theory is so extreme that the stress is a non-monotonic function of strain, and at high strain *decreases* with increasing strain. Hence, a modestly deformed polymer sample can reduce its stress by dividing itself into two bands, one of which unloads most of its deformation onto the other band. Since the stress is small for both small and large deformations (owing to the extreme strain softening), this partitioning of strain is mechanically favored, and a small perturbation can prompt an instability that leads to a partitioning of this kind, which is analogous to gas–liquid phase separation.

Such banding can also in principle occur in steady shear, if the shear stress is a non-monotonic function of the shear rate [120]. The details, including the locations and numbers of the bands, are sensitive to details of the model and the flow geometry [121], but the instability is a robust result of a non-monotonic stress–strain curve. Bands can even form in case that the stress versus strain rate function is monotonic but very flat, if the stress is inhomogeneous [122]. This occurs in shearing flows with curved streamlines, such as a circular Couette flow. One possible form of bands is thin, highly sheared slip layers at the walls of the shearing device. Such slip layers have been inferred from microscopic observations using tracer particles [123]. If slip at the wall is suppressed, internal slip layers, corresponding to highly sheared layers, might form. Evidence of internal slip in highly entangled polymers for which wall slip is suppressed has been presented by Wang and coworkers [124–126]. The density of entanglements required to observe such phenomena might in some cases be lower than 50 entanglements per chain, but it is usually observed that weakly entangled polymers (less than around 10 entanglements per chain) do not show these effects.

As mentioned above, such phenomena are implicit in the original Doi–Edwards model [12]. But the precise manifestation of the phenomena and why some polymers seem more prone to it than others are not yet clear. A full explication of the phenomenon will probably require not only analysis of the constitutive equation but also solution of the full time-dependent numerical solution of the flow problem, using an appropriate version of the tube model for a constitutive equation. Efforts along these lines, using both simple constitutive equations, and a more sophisticated one based on the tube model, have been reported, respectively, by Kolkka *et al.* [120] and by Olmsted and coworkers [121, 122].

Re-entanglement

A long-standing un-explained observation in nonlinear polymer dynamics is the unusually long time required for polymers to 're-entangle' after they have been partially disentangled by shear. The long re-entanglement time is inferred from experiments in which the height of the overshoot in stress after start-up of steady shearing is measured after various rest periods between successive start-up shears. It is found that the time required after cessation of a previous shear for the peak to recover to its maximum height is more than an order of magnitude larger than the reptation time [127]. Robertson *et al.* [128] have found that the re-entanglement time scales with molecular weight to roughly the same power (around 3.2 ± 0.4) as the reptation time. Since the polymer chain conformation is expected to be completely renewed in a single relaxation time, the much longer time required for 're-entanglement' to take place remains a mystery.

Summary and prognosis

The now 'classical' model for entangled polymers is the tube model, which represents the entanglement effect as a confinement of each chain to a tube-like region. This simple idea has been remarkably successful and has delivered at least qualitatively correct predictions for many linear and nolinear rheological phenomena for linear and long-chain branched polymers. Its successes include an approximate 3.4 power law dependence of longest relaxation time with molecular weight in melts of linear chains and an exponential dependence in melts of stars, the prediction of strong shear thinning in simple shearing flows, thinning followed at high rates by thickening in extensional flows, qualitatively correct prediction of normal stress differences in shearing flows, and the extension of these predictions to entangled solutions of long polymers. Quantitative, or at least semi-quantitative agreement with experimental data can be achieved if the model includes the suitable mathematical descriptions of the mechanisms of reptation, primitive path fluctuations, constraint release, chain stretch and retraction, and convective constraint release. Moreover, the predictions that the tube model makes for many non-rheological properties have also proven to be remarkably reliable.

Despite the successes of the tube model, there are many persistent problems. A common malady is that the version of the tube model that works well for one type of architecture or one set of data, does not work as well for others. There are multiple algorithmic implements of the same basic physics available in the literature, and little clarity on which ones might be best overall, and why. A value of the 'dilution exponent' α of $4/3$ works well for predicting some of the data, while $\alpha = 1$ works better for others. These problems hint at unresolved issues or incomplete or erroneous physics contained in the current versions of the tube model. It is possible that the tube idea is inherently only qualitative and 'patches' to it can never make it accurate for a wide range of materials and deformation types.

Such difficulties and suspicions have driven the search for alternative coarse-grained approaches that are better at capturing the relevant physics. The most promising alternatives are slip-link models, which represent each entanglement explicitly using a constraint on each pair of entangling chains. Such models are able to capture multiple mechanisms of relaxation in a unified manner. Moreover, despite being pursued seriously starting only decade ago or so, slip-link models quickly yielded excellent agreement with linear and nonlinear rheological data in many cases. However, slip-link models, especially the most accurate ones, are computationally expensive. Even more seriously, slip-link models are difficult to extend to polymers with long chain branches, for which branch point mobility is important.

As summarized above, significant progress has been achieved to understand the dynamics of entangled polymers in both linear and nonlinear response regimes, but before theories and simulations can reliably give quantitative predictions of rheological properties, numerous open questions needed to be answered. Here we list some of the most critical ones:

- Linear viscoelasticity
    1. Can an 'entanglement' be precisely defined in terms of inter-chain interactions? Is it fruitful to define different degrees of inter-chain entanglement?
    2. What is the correct relationship between the entanglement molecular weight M_e and the plateau modulus G_N^0?
    3. Why are the ratios of M_c/M_e and M_r/M_e not universal? (cf. [31])
    4. Are polymer solutions equivalent to melts once the plateau modulus, entanglement spacing, and frictional time constants are re-scaled properly?
    5. What is the correct procedure for obtaining the primitive path from a microscopic simulation of entangled polymers?
    6. What is the proper entropic potential for primitive path fluctuations; i.e. is the potential given by the 'standard' quadratic form with a prefactor of $\nu = 1.5$, or does this prefactor decrease for long branches?
    7. Can constraint-release effects be adequately described by a combination of dynamic dilution and constraint-release Rouse dynamics?
    8. What is the correct dilution exponent α to use in tube theories?
    9. Can a generalized tube model be developed that will accurately predict the rheology of all entangled polymers, regardless of the distribution of architectures and molecular masses?
    10. Which slip-link models are the most realistic?
    11. Can slip-link models be applied to polymers with mobile branch points?

- Nonlinear viscoelasticity

    12. Is convective constraint release needed to explain shear rheology? (cf. [112])
    13. Does the tube diameter change or the tube cross section become elliptical in nonlinear deformations?
    14. Is the retraction time after a step strain given by the longest Rouse time?
    15. Is there an extra relaxation time in nonlinear rheology, besides the reptation time and the Rouse time? (cf. [115])
    16. Can a theory of slip at the wall and of 'internal slip' be developed?
    17. Can a model of slow re-entanglement after large nonlinear deformation be developed?

Acknowledgement

We acknowledge support from the National Science Foundation under grants DMR 0604965 and DMR 0906587. Any opinions, findings, conclusions, or recommendations expressed in this material are those of the authors and do not necessarily reflect the views of the NSF.

References

[1] Graessley, W. W. (1982). *Adv. Polym. Sci.*, **47**, 67–117.
[2] Doi, M. and Edwards, S. F. (1986). *The Theory of Polymer Dynamics.* Clarendon Press, Oxford.
[3] Dealy, J. M. and Larson, R. G. (2006). *Structure and rheology of molten polymers. From structure to flow behavior and back again.* Carl Hanser, Munich.
[4] Morrison, F. A. (2001). *Understanding Rheology.* Oxford University Press, New York.
[5] Graessley, W. W. (2008). *Polymeric Liquids & Networks: Dynamics and Rheology.* Taylor and Francis, New York.
[6] Doi, M. and See, H. (1996). *Introduction to Polymer Physics.* Oxford University Press, New York.
[7] Edwards, S. F. (1967). *Proc. Phys. Soc.*, **92**, 9–16.
[8] de Gennes, P. G. (1971). *J. Chem. Phys.*, **55**, 572–579.
[9] Doi, M. and Edwards, S. F. (1978). *J. Chem. Soc. Faraday Trans. II*, **74**, 1789–1801.
[10] Doi, M. and Edwards, S. F. (1978). *J. Chem. Soc. Faraday Trans. II*, **74**, 1802–1817.
[11] Doi, M. and Edwards, S. F. (1978). *J. Chem. Soc. Faraday Trans. II*, **74**, 1818–1832.
[12] Doi, M. and Edwards, S. F. (1979). *J. Chem. Soc. Faraday Trans. II*, **75**, 38–54.

[13] Fetters, L. J., Lohse, D. J., Richter, D. *et al.* (1994). *Macromolecules*, **27**, 4639–4647.

[14] Milner, S. T. and McLeish, T. C. B. (1997). *Macromolecules*, **30**, 2157–2166.

[15] Milner, S. T. and McLeish, T. C. B. (1998). *Phys. Rev. Lett.*, **81**, 725–728.

[16] Pattamaprom, C. and Larson, R. G. (2001). *Rheol. Acta*, **40**, 516–532.

[17] Larson, R. G. (2001). *Macromolecules*, **34**, 5229–5237.

[18] Park, S. J., Shanbhag, S., and Larson, R. G. (2005). *Rheol. Acta*, **44**, 319–330.

[19] Das, C., Inkson, N. J., Read, D. J. *et al.* (2006). *J. Rheol.*, **50**, 207–234.

[20] Likhtman, A. E. (2009). *J. Non-Newt. Fluid Mech.*, **157**, 158–161.

[21] Doi, M. and Takimoto, J. (2003). *Phil. Trans. R. Soc. Lond. A*, **361**, 641–650.

[22] Hua, C. C. and Schieber, J. D. (1998). *J. Chem. Phys.*, **109**, 10018–10027.

[23] Masubuchi, Y., Takimoto, J., Koyama, K. *et al.* (2001). *J. Chem. Phys.*, **115**, 4387–4394.

[24] Shanbhag, S., Larson, R. G., Takimoto, J. *et al.* (2001). *Phys. Rev. Lett.*, **87**, 195502.

[25] Likhtman, A. E. (2005). *Macromolecules*, **38**, 6128–6139.

[26] Nair, D. M. and Schieber, J. D. (2006). *Macromolecules*, **29**, 3386–3397.

[27] Masubuchi, Y., Watanabe, H., Ianniruberto, G. *et al.* (2008). *Macromolecules*, **41**, 8275–8280.

[28] Masubuchi, Y., Ianniruberto, G., Greco, F. *et al.* (2008). *J. Non-Newt. Fluid Mech.*, **149**, 87–92.

[29] Yaoita, T., Isaki, T., Masubuchi, Y. *et al.* (2008). *J. Chem. Phys.*, **128**, 154901.

[30] Milner, S. T. (2005). *Macromolecules*, **38**, 4929–4939.

[31] Fetters, L. J., Lohse, D. J., Milner, S. T. *et al.* (1999). *Macromolecules*, **32**, 6847–6851.

[32] Colby, R. H. and Rubinstein, M. (1990). *Macromolecules*, **23**, 2753–2757.

[33] Heo, Y. and Larson, R. G. (2008). *Macromolecules*, **41**, 8903–8915.

[34] Graessley, W. W. (2004). *Polymeric Liquids & Networks: Structure and Properties*. Taylor and Francis, New York.

[35] Adam, M. and Delsanti, M. (1984). *J. Phys. (Paris)*, **45**, 1513–1521.

[36] Colby, R. H., Fetters, L. J., Funk, W. G. *et al.* (1991). *Macromolecules*, **24**, 3873–3882.

[37] Tao, H., Huang, C., and Lodge, T. P. (1999). *Macromolecules*, **32**, 1212–1217.

[38] Zhou, Q. and Larson, R. G. (2005). *Macromolecules*, **38**, 5761–5765.

[39] Everaers, R., Sukumaran, S. K., Grest, G. S. *et al.* (2004). *Sci-*

ence, **303**, 823–826.
[40] Zhou, Q. and Larson, R. G. (2006). *Macromolecules*, **39**, 6737–6743.
[41] Foteinopoulou, K., Karayiannis, N. C., Mavrantzas, V. G. et al. (2006). *Macromolecules*, **39**, 4207–4216.
[42] Tzoumanekas, C. and Theodorou, D. N. (2006). *Macromolecules*, **39**, 4592–4604.
[43] Hoy, R. S. and Robbins, M. O. (2005). *Phys. Rev. E*, **72**, 061802.
[44] Shanbhag, S. and Larson, R. G. (2005). *Phys. Rev. Lett.*, **94**, 076001.
[45] Stephanou, P. S., Baig, C., Tsolou, G. et al. (2010). *J. Chem. Phys.*, **132**, 124904.
[46] Kröger, M., Ramírez, J., and Öttinger, H. C. (2002). *Polymer*, **43**, 477–487.
[47] Masubuchi, Y., Watanabe, H., Ianniruberto, G. et al. (2003). *J. Chem. Phys.*, **119**, 6925–6930.
[48] Ramirez, J., Sukumaran, S. K., and Likhtman, A. E. (2007). *Macromol. Symp.*, **252**, 119–129.
[49] Kumar, S. and Larson, R. G. (2001). *J. Chem. Phys.*, **114**, 6937–6941.
[50] Holleran, S. P. and Larson, R. G. (2008). *Rheol. Acta*, **47**, 3–17.
[51] Padding, J. T. and Briels, W. J. (2002). *J. Chem. Phys.*, **117**, 925–943.
[52] Shanbhag, S. and Larson, R. G. (2006). *Macromolecules*, **39**, 2413–2417.
[53] Meyer, H., Kreer, T., Cavallo, A. et al. (2007). *Eur. Phys. J. Special Topics*, **141**, 167–172.
[54] Rowland, H. D., King, W. P., Pethica, J. B. et al. (2008). *Science*, **322**, 720–724.
[55] Rubinstein, M. and Panyukov, S. (2002). *Macromolecules*, **35**, 6670–6686.
[56] Marrucci, G. and de Cindio, B. (1980). *Rheol. Acta*, **19**, 68–75.
[57] Marrucci, G. and Ianniruberto, G. (2005). *J. Non-Newt. Fluid Mech.*, **128**, 42–49.
[58] Mehtar, V. and Archer, L. A. (1999). *J. Non-Newt. Fluid Mech.*, **81**, 71–81.
[59] Onogi, S., Masuda, T., and Kitagawa, K. (1970). *Macromolecules*, **3**, 109–116.
[60] Larson, R. G., Sridhar, T., Leal, L. G. et al. (2003). *J. Rheol.*, **47**, 809–818.
[61] Doi, M. (1983). *J. Polym. Sci.: Polym. Phys. Ed.*, **21**, 667–684.
[62] Auhl, D., Ramirez, J., Likhtman, A. E. et al. (2008). *J. Rheol.*, **52**, 801–835.
[63] Likhtman, A. E. and McLeish, T. C. B. (2002). *Macromolecules*, **35**, 6332–6343.
[64] Liu, C. Y., Keunings, R., and Bailly, C. (2006). *Phys. Rev. Lett.*, **97**, 246001.
[65] Liu, C. Y., He, J., Keunings, R. et al. (2006). *Macromolecules*, **39**,

3093–3097.
[66] Rubinstein, M. and Colby, R. H. (1988). *J. Chem. Phys.*, **89**, 5291–5306.
[67] Tuminello, W. H. (1986). *Polym. Eng. Sci.*, **26**, 1339–1347.
[68] des Cloizeaux, J. (1990). *Macromolecules*, **23**, 4678–4687.
[69] Mead, D. W. (1994). *J. Rheol.*, **38**, 1797–1827.
[70] Tsenoglu, C. (1996). *J. Rheol.*, **40**, 633–661.
[71] Viovy, J. L., Rubinstein, M., and Colby, R. H. (1991). *Macromolecules*, **24**, 3587–3596.
[72] Park, S. J. and Larson, R. G. (2004). *Macromolecules*, **37**, 597–604.
[73] Doi, M., Graessley, W. W., Helfand, E. *et al.* (1987). *Macromolecules*, **20**, 1900–1906.
[74] Ball, R. C. and McLeish, T. C. B. (1989). *Macromolecules*, **22**, 1911–1913.
[75] van Ruymbeke, E., Bailly, C., Keunings, R. *et al.* (2006). *Macromolecules*, **39**, 6248–6259.
[76] Likhtman, A. E., Milner, S. T., and McLeish, T. C. B. (2000). *Phys. Rev. Lett.*, **85**, 4550–4553.
[77] Kraus, G. and Gruver, J. T. (1965). *J. Polym. Sci. A*, **3**, 105–122.
[78] de Gennes, P. G. (1975). *J. Physique*, **36**, 1199–1203.
[79] Fetters, L. J., Kiss, A. D., Pearson, D. S. *et al.* (1993). *Macromolecules*, **26**, 647–654.
[80] Pearson, D. S., Mueller, S. J., Fetters, L. J. *et al.* (1983). *J. Polym. Sci.; Polym. Phys. Ed.*, **21**, 2287–2298.
[81] Doi, M. and Kuzuu, N. Y. (1980). *J. Polym. Sci.: Polym. Lett. Ed.*, **18**, 775–780.
[82] Pearson, D. S. and Helfand, E. (1984). *Macromolecules*, **17**, 888–895.
[83] Khaliullin, R. N. and Schieber, J. D. (2008). *Phys. Rev. Lett.*, **18**, 188302.
[84] Marrucci, G. (1985). *J. Polym. Sci.: Polym. Phys. Ed.*, **23**, 159–177.
[85] McLeish, T. C. B. (2003). *J. Rheol.*, **47**, 177–198.
[86] McLeish, T. C. B., Allgaier, J., Bick, D. K. *et al.* (1999). *Macromolecules*, **32**, 6734–6758.
[87] Frischknecht, A. L., Milner, S. T., Pryke, A. *et al.* (2002). *Macromolecules*, **35**, 4801–4819.
[88] Daniels, D. R., McLeish, T. C. B., Crosby, B. J. *et al.* (2001). *Macromolecules*, **34**, 7025–7033.
[89] Park, S. J. and Larson, R. G. (2005). *J. Rheol.*, **49**, 523–536.
[90] Wang, Z. W., Chen, X., and Larson, R. G. (2010). *J. Rheol.*, **54**, 223–260.
[91] Milner, S. T., McLeish, T. C. B., Young, R. N. *et al.* (1998). *Macromolecules*, **31**, 9345–9353.
[92] Park, S. J. and Larson, R. G. (2003). *J. Rheol.*, **47**, 199–211.
[93] Schieber, J. D. and Khaliullin, R. (2008). In *Proc. XVth Intl. Congress on Rheology – The Society of Rheology 80th Annual*

Meeting, Parts 1 and 2, 1027, pp. 324–326.

[94] Shanbhag, S. and Larson, R. G. (2004). *Macromolecules*, **37**, 8160–8166.

[95] Masubuchi, Y., Ianniruberto, G., Greco, F. *et al.* (2006). *Rheol. Acta*, **46**, 297–303.

[96] Fukuda, M., Osaki, K., and Kurata, M. (1975). *J. Polym. Sci.: Polym. Phys. Ed.*, **13**, 1563–1576.

[97] Marrucci, G. and Grizzuti, N. (1988). *Gazz. Chim. Ital.*, **118**, 179–185.

[98] Bhattacharjee, P. K., Oberhauser, J., McKinley, G. H. *et al.* (2002). *Macromolecules*, **35**, 10131–10148.

[99] Archer, L. A. (1999). *J. Rheol.*, **43**, 1555–1571.

[100] Islam, M. T., Sanchez-Reyes, J., and Archer, L. A. (2001). *J. Rheol.*, **45**, 61–82.

[101] Sanchez-Reyes, J. and Archer, L. A. (2002). *Macromolecules*, **35**, 5194–5202.

[102] Inoue, T., Uematsu, T., Yamashita, Y. *et al.* (2002). *Macromolecules*, **35**, 4718–4724.

[103] Mead, D. W., Larson, R. G., and Doi, M. (1998). *Macromolecules*, **31**, 7895–7914.

[104] Graessley, W. W. (1965). *J. Chem. Phys.*, **43**, 2696–2703.

[105] Marrucci, G. (1996). *J. Non-Newt. Fluid Mech.*, **62**, 279–289.

[106] Milner, S. T., McLeish, T. C. B., and Likhtman, A. E. (2000). *J. Rheol.*, **45**, 539–563.

[107] Ianniruberto, G. and Marrucci, G. (2001). *J. Rheol.*, **45**, 1305–1318.

[108] Graham, R. S., Likhtman, A. E., Milner, S. T. *et al.* (2003). *J. Rheol.*, **47**, 1171–1200.

[109] Likhtman, A. E. and Graham, R. S. (2003). *J. Non-Newt. Fluid Mech.*, **114**, 1–12.

[110] McLeish, T. C. B. and Larson, R. G. (1998). *J. Rheol.*, **42**, 81–110.

[111] Verbeeten, W. M. H., Peters, G. W. M., and Baaijens, F. P. T. (2002). *J. Non-Newt. Fluid Mech.*, **108**, 301–326.

[112] Schieber, J. D., Nair, D. M., and Kitkrailaird, T. (2007). *J. Rheol.*, **51**, 1111–1141.

[113] Marrucci, G. and Hermans, J. J. (1980). *Macromolecules*, **13**, 380–387.

[114] Teixeira, R. E., Dambal, A. K., Richter, D. H. *et al.* (2007). *Macromolecules*, **40**, 2461–2476, 3514.

[115] Robertson, R. M. and Smith, D. E. (2007). *Phys. Rev. Lett.*, **99**, 126001.

[116] Osaki, K. and Kurata, M. (1980). *Macromolecules*, **13**, 671–676.

[117] Vrentas, C. M. and Graessley, W. W. (1982). *J. Rheol.*, **26**, 359–371.

[118] Morrison, F. A. and Larson, R. G. (1992). *J. Polym. Sci: Polym. Phys. Ed.*, **30**, 943–950.

[119] Marrucci, G. and Grizzuti, N. (1983). *J. Rheol.*, **27**, 433–450.

[120] Kolkka, R. W., Malkus, D. S., Hansen, M. G. *et al.* (1988). *J.*

Non-Newt. Fluid Mech., **29**, 303–335.
[121] Adams, J. M., Fielding, S. M., and Olmsted, P. D. (2008). *J. Non-Newt. Fluid Mech.*, **151**, 101–118.
[122] Adams, J. M. and Olmsted, P. D. (2009). *Phys. Rev. Lett.*, **103**, 067801.
[123] Archer, L. A., Larson, R. G., and Chen, Y. L. (1995). *J. Fluid Mech.*, **301**, 133–151.
[124] Tapadia, P. and Wang, S. Q. (2003). *Phys. Rev. Lett.*, **91**, 198301.
[125] Tapadia, P. and Wang, S. Q. (2006). *Phys. Rev. Lett.*, **96**, 016001.
[126] Boukany, P. E. and Wang, S. Q. (2007). *J. Rheol.*, **51**, 217–233.
[127] Stratton, R. A. and Butcher, A. F. (1973). *J. Polym. Sci. Polym. Phys.*, **11**, 1747–1758.
[128] Robertson, C. G., Warren, S., Plazek, D. J. *et al.* (2004). *Macromolecules*, **37**, 10018–10022.

Block Copolymers: Thermodynamics, Dynamics, and Small-Angle Scattering

Lee M. Trask, Nacú Hernandez and Eric W. Cochran
Department of Chemical & Biological Engineering
Iowa State University
Ames, IA 50011-2230, USA

The self-assembly of soft materials into well-ordered structures is a broad interdisciplinary topic with the potential to develop technologies featuring length scales ranging from Ångstroms to microns. Nature is replete with systems featuring specific interactions that promote the assembly of building blocks into a precise state of order over the entropically preferred random and isotropic state. Liquid crystals, lipid bilayers, surfactants, colloids, and self-assembled monolayers name just a few such examples. From recent advances in semiconductor processing to the chocolate industry to butterfly wings, self-assembly is pervasive in both the natural and artificial worlds. Block copolymers, comprised of two or more distinct sequences of dissimilar chemical repeat units, continue after more than four decades since their first commercialization to occupy a substantial proportion of the attention of the global research community as one of the most tunable and extensible classes of self-assembly.

The combination of small mixing entropy and unfavorable enthalpic interactions render most polymers immiscible with each other. Covalently linked, the blocks of a block copolymer are still thermodynamically incompatible over a broad range of experimentally accessible conditions, but are constrained in the length scale of the phases that they may form by the elastic energy generated as blocks separate. The so-called "mesophases" that form are typically highly periodic with well-defined crystallographic symmetry, having been chosen by nature as the optimized solution to interfacial contact between blocks and the elastic stretching energy required for the structure to fill space. The past 40 years have uncovered a multitude of such structures, ranging from one-dimensional (1D) lamellar structures to multiply periodic three-dimensional (3D) structures, to aperiodic networks (i.e. bicontinuous microemulsions) in certain homopolymer/block copolymer blends.

While commercial applications of block copolymers have been historically "low-tech" – e.g. the thermoplastic elastomers used as pressure sensitive adhesives, bitumen additives, sealants, etc. – the forefront of the research arena is replete with potential applications in areas ranging from drug delivery to separations to catalysis to photovoltaics to non-linear optics to photolithography . Each of these areas has received enough attention to merit numerous review articles, books, and book chapters.

In the present chapter we focus on providing what we perceive to be the fundamental basis for understanding what drives these applications, i.e. how block copolymers behave and how they are characterized. Section 7.1 summarizes the experimental data and various theories that comprise our understanding of block copolymer thermodynamics. Section 7.2 covers topics related to block copolymer dynamics – diffusion, viscoelasticity, shear-processing, and the kinetics of self-assembly. Finally, Section 7.3 supplies an overview of small-angle scattering techniques as applied to block copolymer characterization.

7.1 Phase behavior

Theory of microphase separation

Ordered structuring of block copolymers results from microphase separation as a result of unfavorable enthalpy of mixing compared to a much smaller favorable entropy of mixing. However, macrophase separation is not possible due to the covalent bonding of blocks. This microphase behavior of polymers has been studied at the theoretical level since the 1940's with the pioneering work of Flory [1] and Huggins [2]. This work focused on the behavior of long molecular chains on a lattice, with or without the presence of a solvent. Besides expressions for free energy and entropy, an important result of these studies was the recognition of a universal parameter now known as the Flory-Huggins segment-segment interaction parameter. For a pair of monomeric units, A and B, the driving force of separation is characterized by χ_{AB}, which describes the free energy cost per contact of segment A and segment B on adjacent cells of a lattice. This dimensionless parameter takes the form

$$\chi_{AB} = z/k_B T \left[\epsilon_{AB} - 0.5(\epsilon_{AA} + \epsilon_{BB})\right], \qquad (7.1)$$

where ϵ_{ij} is the energy of contact between a segment of i and a segment j and z is the lattice coordination number. χ_{AB}, which varies linearly with $1/T$, is usually positive for a given pair of monomers. This positive value signifies a repulsion between these two species. Along with the Flory-Huggins interaction parameter, the degree of polymerization, usually denoted as N, plays an important role in the phase behavior of a block copolymer system.

These two parameters play an important part in the theories developed for block copolymer systems because of the competing roles of enthalpy and entropy. Enthalpy usually drives a system to phase separate since χ_{AB} is usually positive, and this energy contribution only gets magnified as χ_{AB} grows. On the other hand, entropy drives a system toward disorder. Because longer polymers (larger N) result in fewer contacts between block species, entropic interactions scale as $1/N$. The competing role of these energy contributions results in $\chi_{AB}N$ being the parameter of interest in block copolymer systems. When considering this parameter, model systems fall into three regimes: weak segregation theory with $\chi_{AB}N \approx 10-20$, strong segregation theory with $\chi_{AB}N > 100$, and intermediate segregation theory with $\chi_{AB}N$ between these limits. Developed theories generally attempt to describe a specific regime.

Weak segregation theory

As mentioned above, at a $\chi_{AB}N \approx 10-20$ the level of theory is the so-called weak segregation theory or weak segregation limit. This regime exists near the order-disorder transition (ODT) where the interfaces are diffuse and the spatial variation of segment density is well-approximated with the superposition of only a few plane waves. Leibler laid out this theory in 1980 [3] building off previous work by de Gennes [4, 5]. For a monodisperse AB diblock copolymer melt with Flory-Huggins interaction parameter χ, degree of polymerization N, average composition f, and equal statistical segment lengths, the system is modeled using Landau-Ginzburg theory. In this theory, the free energy is expanded in terms of the composition order parameter defined by[1]

$$\psi(\mathbf{r}) = \langle \phi_A(\mathbf{r}) - f \rangle, \tag{7.2}$$

[1] In this context, the notation $\langle \ldots \rangle$ implies the average over the system volume.

where $\phi_A(\mathbf{r})$ is the local number density of monomer A. The free energy functional expansion near a second-order or weakly first-order phase transition becomes

$$F[\psi(\mathbf{q})] = \frac{1}{2!} \int_{\mathbf{q}} S^{-1}(\mathbf{q})\psi(\mathbf{q})\psi(-\mathbf{q})$$
$$+ \frac{1}{3!} \int_{\mathbf{q}} \int_{\mathbf{q}'} \mu(\mathbf{q}, \mathbf{q}', -\mathbf{q} - \mathbf{q}')\psi(\mathbf{q})\psi(\mathbf{q}')\psi(-\mathbf{q} - \mathbf{q}')$$
$$+ \frac{1}{4!} \int_{\mathbf{q}} \int_{\mathbf{q}'} \int_{\mathbf{q}''} \lambda(\mathbf{q}, \mathbf{q}', \mathbf{q}'', -\mathbf{q} - \mathbf{q}' - \mathbf{q}'')\psi(\mathbf{q})\psi(\mathbf{q}')\psi(\mathbf{q}'')\psi(-\mathbf{q} - \mathbf{q}' - \mathbf{q}'').$$
$$\tag{7.3}$$

In this equation, $\mathbf{q}$ are the wavevectors for the system; $\psi(\mathbf{q})$ is then the Fourier transform of $\psi(\mathbf{r})$, and $S(q = |\mathbf{q}|)$ is the Fourier transform of the structure factor (to be discussed in a later section). The coefficients in these expansions are found using de Gennes's random-phase approximation (RPA). The cubic term, μ, is found to be positive except for a symmetric system which has $\mu = 0$. The quartic term, λ, is positive for stability reasons.

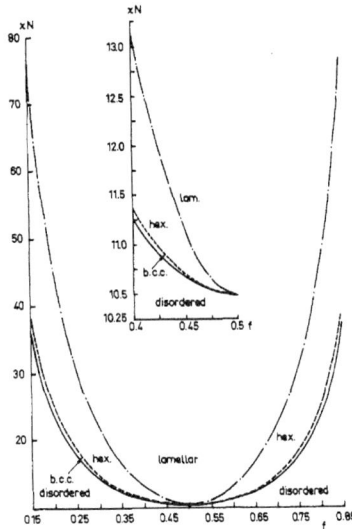

Fig. 7.1 Theoretical phase diagram for diblock copolymers calculated with weak segregation theory [3].

If equation (7.3) is transformed to real space with a term added to introduce fluctuations, the free energy takes the following form [6] (in dimensionless form)

$$F[\psi] = \int d\mathbf{r} \left[\frac{1}{2}\phi(r)[\tau + e(\nabla^2 + q_0^2)^2]\phi(\mathbf{r}) + \frac{\mu}{3!}\psi(\mathbf{r}) + \frac{\lambda}{4!}\psi(\mathbf{r}) \right]. \quad (7.4)$$

This energy equation can rescaled relative to the disordered phase for a symmetric diblock copolymer with the introduction of a variational functional A; the free energy then takes the form

$$F[A] = \tau A^2 + \frac{\lambda}{4} A^4. \quad (7.5)$$

The coefficient τ is found to be a reduced temperature variable that is a measure of distance from the spinodal point and takes the form $\tau = 2(\chi_S N - \chi N)/c^2$ where the constant c is 1.1019. λ is a function of f and N. For χN below the spinodal point value of $\chi_S N$, the disordered phase is stable. At the spinodal point, $A = 0$ and a second-order transition to the lamellar phase takes place. Continuing to higher χN, A becomes $\pm(2\tau/\lambda)^{1/2}$. Both these variational parameter values have a free energy of $F[A] = -\tau^2/\lambda < 0$, which is the energy of the stable lamellar phase.

For equation (7.3), Leibler used the leading harmonics of the various ordered phases to construct the first diblock copolymer phase diagram. By comparing the free energies of three traditional ordered phases with respect to the disordered state, he was able to construct a phase diagram. The phases he considered were the lamellar phase, cylinders in a hexagonal lattice, and spheres in a body centered cubic (bcc) lattice. Using χN and f as the phase space variables, the phase diagram was found and is shown in Figure 7.1. Leibler found that a critical point of the Landau theory occurs at $\chi N = 10.495$ and $f = 0.5$. This lamellar phase is predicted to have a period that scales as $D \sim N^{1/2}$. For compositions other than $f = 0.5$, the transition is predicted to be weakly first-order from the disordered state to the bcc state. This theory also predicts transitions between ordered states as the parameter χN is varied.

Structure factor

In order to construct his theory, Leibler [3] introduced a structure factor for the disordered state. If $W(q)$ is the determinant of the matrix of correlation functions for ideal independent copolymer chains and $\Sigma(q)$ is the sum of all elements, the structure factor is given by

$$S(q) = \frac{W(q)}{\Sigma(q) - 2\chi W(q)}. \quad (7.6)$$

Using the RPA, Leibler found the structure factor for a diblock copolymer to be

$$S_{11}(q) = N g(f, x), \quad (7.7)$$

$$S_{22}(q) = N g(1-f, x), \quad (7.8)$$

$$S_{12}(q) = \frac{1}{2}N\left[g(1,x) - g(f,x) - g(1-f,x)\right], \qquad (7.9)$$

where $x = q^2 R_g^2$ and $g(f,x)$ is the Debye function defined by

$$g(f,x) = \frac{2}{x^2}\left[fx + e^{-fx} - 1\right]. \qquad (7.10)$$

With these equations, structure factor expression (7.6) can be rewritten as

$$\frac{1}{S(q)} = \frac{F(f,x)}{N} - 2\chi, \qquad (7.11)$$

where

$$F(f,x) = \frac{g(1,x)}{g(f,x)g(1-f,x) - 0.25[g(1,x) - g(f,x) - g(1-f,x)]}. \qquad (7.12)$$

At large magnitudes of q, $S(q)$ tends toward zero like $1/q^2$. At small magnitudes of q, however, $S(q)$ tends toward zero like q^2. For both these cases, $S(q)$ is found to be independent of χ. At $q = 0$, $S(q) = 0$ since the system is incompressible.

This structure factor has a peak that diverges at the spinodal point, where the left-hand side of equation (7.11) is zero. While the peak is independent of segregation effects, the shape has a strong dependence on χN, as seen in Figure 7.2. Because of this relation, $S(q)$ can be used to find χ as a function of temperature. Small angle x-ray scattering can be used to find this information.

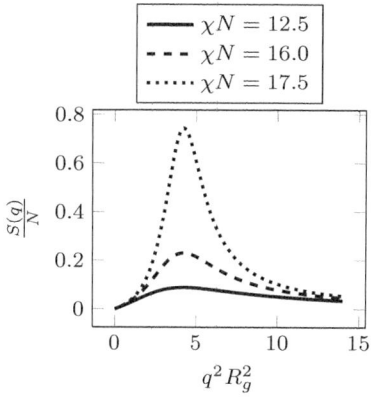

Fig. 7.2 Structure factor for diblock copolymer melt calculated by Leibler in [3] for $\chi N = \{12.5, 16, 17.5\}$. The spinodal point occurs at $\chi N = 18.2$.

Strong segregation theory

Strong segregation theory (SST), also known as strong segregation limit, corresponds to high values of χN, typically above 100. In this regime, the microphase density profiles resemble step functions – there are no interfacial boundaries with mixed compositions. The earliest works with this idea in mind appeared in the early 1970's [7–9]. However, these studies were limited at the time due to the lack to a theoretical framework for block copolymers. Two theories would emerge to tackle the problem of strongly segregated polymers.

In 1976, Helfand simplified self-consistent field theory (SCFT) with a narrow interphase approximation to predict an interfacial layer thickness of $a\chi^{-1/2}$ [10] where a is the statistical segment length. Three energy contributions were considered in their field theory calculations: (i) the enthalpy of contacts between the pure A and B microdomains, (ii) entropy loss due to chain stretching, and (iii) entropy of confining block junction points to the interface. They found the interfacial layer thickness d to be $2a/(\sqrt{6\chi})$ while the microdomain period scaled as $bN^{0.643}\chi^{0.143}$. Despite being limited by the computer resources of the time, these predictions matched experimental work on strongly segregated block copolymers [11].

In 1985, Semenov introduced an analytic theory for SST [12]. In his seminal work, Semenov found functional forms for the free energy in the asymptotic limit $\chi N \to \infty$. This work is based on the idea that the chain ends predominantly reside in the domain interiors. With this idea, the system resembles graft polymer brushes which greatly simplifies the theory. The domain contribution to the free energy was found to have the same scaling as for a Gaussian chain, despite the extreme chain deformations due to chain stretching. This contribution takes the form

$$\beta F_{\text{domain}} \sim \frac{d^2}{Na^2}. \tag{7.13}$$

The contribution from the observed nonuniform chain stretching is embodied as a constant prefactor. The domain free energy is countered by the interfacial free energy per chain, which takes the form

$$\beta F_{\text{interface}} \sim \gamma \sigma \frac{Nb\chi^{1/2}}{d}, \tag{7.14}$$

where the interfacial tension $\gamma \sim \chi^{1/2} a^{-2}$ and the area per chain $\sigma \sim Na^3/d$ leads to the last form in equation (7.14) [13]. Minimizing the sum of these two energy contributions with respect to d in the asymptotic limit $\chi N \to \infty$, the domain period is found to scale as

$$d \sim bN^{2/3}\chi^{1/6}. \tag{7.15}$$

This scaling behavior varies from that of Helfand's calculations because Helfand only carried out his numerical calculations to $d/bN^{1/2} \approx 3$, but $d/bN^{1/2} \gg 1$ needs to be satisfied. To verify that SCFT matches Semenov's predictions, Matsen and Whitmore [14] calculated SCFT boundaries between the traditional phases of lamellar (L), cylindrical (C), and spherical(S) up to just beyond $\chi N \approx 100$. Extrapolating their results to $\chi N \to \infty$, they found that the phase boundaries between phases occurred at $f_{L/C} = 0.3099$ and $f_{C/S} = 0.1104$, but SST predicted $f_{L/C} = 0.2991$ and $f_{C/S} = 0.1172$. Then in 2010, Matsen performed SCFT calculations to $\chi N = 512000$. For comparison, he modified SST by accounting for the exclusion zones in the corona of the cylindrical and spherical domains. Numerical calculations of this SST found phase boundaries of $f_{L/C} = 0.2990$ and $f_{C/S} = 0.1111$ which is in full agreement with his SCFT calculations.

Intermediate segregation and self-consistent field theory

Despite the successes of weak and strong segregation theories at predicting the traditional phases, these theories failed to account for more complex phases such the the double-gyroid phase. As a result, these theories were unified by Matsen and coworkers [15,16] and built on top of the foundations of self-consistent field theory (SCFT). These foundations of SCFT were first laid out by Edwards [17,18] using the techniques used in the reformulation of quantum mechanics with the use of path integrals. Helfand and Tagami [19] later represented Edward's work into a more standard model for numerical calculations.

SCFT has evolved into a complex theoretical framework to model all varieties of polymer systems [20]. As such, only a brief outline of SCFT of a monodisperse, incompressible AB diblock copolymer system is outlined using the canonical ensemble [15]. First the phenomenological Hamiltonian is constructed by considering all the potential terms of the system; this is then inserted into the configurational partition function

$$Z = \int \sum_{i=1}^{N} \widetilde{\mathcal{D}}\mathbf{r}_i \delta[1 - \hat{\phi}_A - \hat{\phi}_B] \exp\left(-\chi_{AB}\rho_0 \int \hat{\phi}_A \hat{\phi}_B\right), \quad (7.16)$$

where the sum is over all polymer chains, χ_{AB} is the Flory-Huggins interaction parameter, and ρ_0 is the reference polymer volume per chain. The functional integral is weighted over all configurations of the polymers, so that $\widetilde{\mathcal{D}}\mathbf{r}_i \equiv \mathcal{D}\mathbf{r}_i P[\mathbf{r}_i; 0, 1]$ with the weights defined by the chain stretching of continuous Gaussian chains

$$P[\mathbf{r}_i; 0, 1] \propto \exp\left(-\frac{3}{2Na^2} \int_{s_1}^{s_2} ds \left|\frac{d\mathbf{r}_i(s)}{ds}\right|^2\right), \quad (7.17)$$

where the dimension s is the coordinate that moves down the polymer chain backbone. The microscopic density operator for species A is defined as

$$\hat{\rho}_A(\mathbf{r}) = \frac{1}{\rho_0} \sum_{i=1}^{n} \int_0^{fN} ds\, \delta(\mathbf{r} - \mathbf{r}_i(s)). \quad (7.18)$$

The density operator for species B takes a similar form. At this point, density delta-function integrals, $1 = \int \mathcal{D}\delta[\phi_i - \hat{\phi}_i]$, are inserted into equation (7.16). These delta functions are then transformed into their Fourier space representation, $1 = \int \mathcal{D}\rho_i \mathcal{D}W_i \exp[-iW_i(\rho_i - \hat{\rho}_i)]$. This makes the field theory tractable by replacing $\hat{\phi}_i$ with ϕ_i. The configurational partition function now takes the form

$$Z = Z_0 \int \mathcal{D}\phi_A \mathcal{D}W_A \mathcal{D}\phi_B \mathcal{D}W_B \mathcal{D}P \exp(-\beta F), \quad (7.19)$$

where Z_0 is a normalization constant and

$$\beta F = \frac{N}{V} \int d\mathbf{r}[\chi N \phi_A \phi_B - W_A \phi_A - W_B \phi_B - P(1 - \phi_A - \phi_B)] - \ln Q, \quad (7.20)$$

$$Q \equiv \widetilde{\mathcal{D}}\mathbf{r}_i \exp\left(-\int_0^f ds W_A(\mathbf{r}_i) - \int_f^1 ds W_B(\mathbf{r}_i)\right). \quad (7.21)$$

From equation (7.20), F is minimized with respect to the variables to give the self-consistent equations

$$W_A(\mathbf{r}) = \chi N \phi_B + P(\mathbf{r}), \quad (7.22)$$

$$W_B(\mathbf{r}) = \chi N \phi_A + P(\mathbf{r}), \quad (7.23)$$

$$1 = \phi_A + \phi_B, \quad (7.24)$$

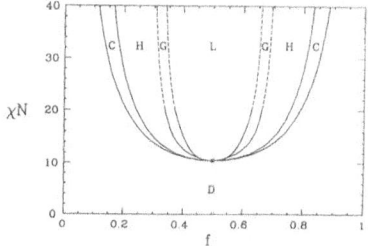

Fig. 7.3 Original spectral SCFT phase diagram for diblock copolymer. Diagram shows boundaries between the disorder (D), lamellar (L), gyroid (G), hexagonal cylinders (C), and body center cubic sphere (S) phases [15].

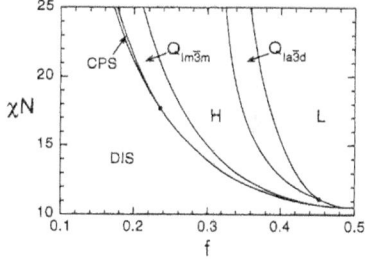

Fig. 7.4 Magnification of Figure 7.5 showing the weak segregation regime [16]. Updated phase diagram for diblock copolymer. The phases presented are the lamellar (L), bicontinuous $Ia\bar{3}d$ cubic (Q^{230}), hexagonal cylinders (H), body center cubic spheres (Q^{229}), close-packed spherical (CPS), and disordered (dis). Triple points are shown as dots.

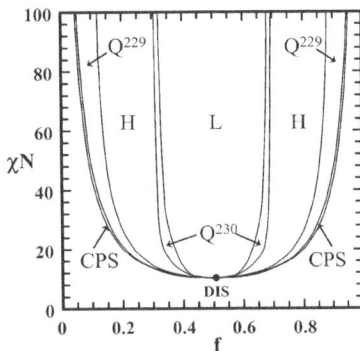

Fig. 7.5 Updated phase diagram for diblock copolymer. Diagram shows boundaries between the lamellar (L), bicontinuous $Ia\bar{3}d$ cubic ($Q^{230}(Ia\bar{3}d)$), hexagonal cylinders (H), body center cubic spheres ($Q^{229}(Im\bar{3}m)$), close-packed spherical (CPS), and disordered (dis) phases [16, 23].

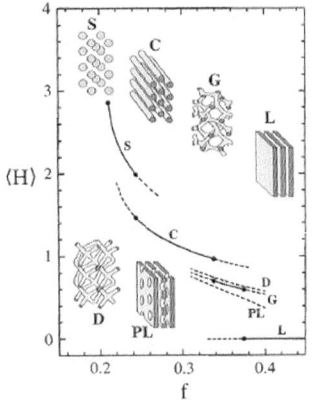

Fig. 7.6 Area-averaged mean curvature $\langle H \rangle$ as a function of species A composition, f. Structures are calculated using SCFT, where the stable structures are denoted with solid lines, the metastable structures are shown with dashed lines, and the transitions are demonstrated with dots [24].

$$\phi_A = -\frac{V}{Q}\frac{\mathcal{D}Q}{\mathcal{D}W_A}, \quad (7.25)$$

$$\phi_B = -\frac{V}{Q}\frac{\mathcal{D}Q}{\mathcal{D}W_B}. \quad (7.26)$$

The single molecule partition function can be expanded into a product of transition probabilities and transition weights. From here, the Feynman-Kac [21, 22] formula transforms equation (7.21) to

$$Q = \frac{1}{V}\int d\mathbf{r}\, q(\mathbf{r}, s = N). \quad (7.27)$$

The so-called propagators, $q(\mathbf{r}, s)$, come from the modified diffusion equation

$$\frac{\partial}{\partial s}q(\mathbf{r}, s) = \frac{a^2}{6}\nabla^2 q(\mathbf{r}, s) - W_i(\mathbf{r})q(\mathbf{r}, s)\,, \quad (7.28)$$

where $i \equiv A$ for $s \in [0, f]$ and $i \equiv B$ for $s \in [f, 1]$. The density operators can be rewritten in terms of the propagators so that

$$\rho_A(\mathbf{r}) = \frac{V}{Q}\int_0^f ds\, q(\mathbf{r}, s)q^\dagger(\mathbf{r}, s), \quad (7.29)$$

$$\rho_B(\mathbf{r}) = \frac{V}{Q}\int_f^1 ds\, q(\mathbf{r}, s)q^\dagger(\mathbf{r}, s). \quad (7.30)$$

These equations define the governing equations for this system, but must be solved to self-consistency. As mentioned, many systems can be examined with this approach. Furthermore, many numerical methods exist for solving these equations [20].

These methods have been used extensively since their inception. Matsen used these methods to calculate phase diagrams as a function of composition and Flory-Huggins interaction parameter [15]. Figure 7.3 was the first diagram found, but this was improved in 1996 [16]. Unlike the diagram found by Leibler, this work predicted the existence of the double-gyroid phase, Q^{230}. This phase was found to not extend to the critical point. Instead this phase terminates at a triple point found to be at $\chi N = 11.14$ with $f = 0.452$ or $f = 0.548$. The triple points of this system are shown in Figure 7.4. Matsen's work suggested that the gyroid phase became unstable at large χ_N, but numerical issues prevented the convergence of the equations; in2006, Cochran et al. showed that the Q^{230} phase window persisted to at least $\chi N = 100$ as shown in Figure 7.5 [23].

Previous theories showed that the phase transitions are a result of the AB interfaces trying to maximize curvature as the polymer composition becomes asymmetric. In this way, the stretching of each block is kept to a minimum. From this idea, the area-average ($\langle H \rangle$) of the mean curvature $H = 1/2(C_1 + C_2)$, with C_i being the principal curvatures at a given point on a surface of the phase, for each structure was thought to be a controlling parameter to determine which phase transitions to another. Figure 7.6 shows the structures predicted with the idea of maximizing interfacial curvature. However, this idea predicts the the stability of the perforated layer structure as well as the stability of the double diamond (OBDD) structure while these structures are not predicted by SCFT or found experimentally. Matsen and Bates [25] explain that while the area-average curvature $\langle H \rangle$ controls the order of the phase transitions, it is really the standard deviation σ_H of this variable that determines the phase selection. Another requirement for selection is to minimize packing frustration – the system aligns so no one chain has to stretch excessively. With these considerations, the stability of the perforated layer structure and the double diamond (OBDD) are prevented due to packing frustration. The interfacial curvatures for various phases are shown in Figure 7.7.

Block copolymer systems

As mentioned in previous sections, there exist several stable phases for the diblock copolymer. These are the lamellar, cylinders in a hexagonal array, the double gyroid, the orthorhombic O^{70} network [26–28], and spheres in a body centered cubic lattice. Further complexity can be added to the phase space by modifying the polymer's structure, chemistry, or environment. Below we outline a few of these techniques. The most straightforward is the addition of additional blocks of chemically different species. This can occur any number of times and also form branched structures, called dendrimers. Chemistry can also play a role in altering the phase behavior of a block copolymer – systems that incorporate stiff, rod-like blocks can exhibit interesting behavior. The last alteration discussed is the modification of the environment by confining the diblock copolymer to a thin film. These systems lead to a rich landscape of polymer phase behavior.

ABC triblock copolymers

When a third, chemically different block is added to a diblock copolymer, the parameters in phase space increase greatly. Instead of the two characteristic parameters of Flory-Huggins interaction strength, χN, and composition, f, there is now three interaction parameters (χ_{AB}, χ_{AC}, χ_{BC}) along with two independent compositions. Moreover, the sequence of the block positions can play an important role in morphology. Figure 7.8 shows traditional triblock copolymer phases that greatly extend the phase space of the diblock copolymer.

When the interaction parameters and composition fractions are nearly equal, all three species form a lamellar phase as shown in Figure 7.8a. However, as the interaction strength between a pair becomes more repulsive, the polymers move to try to minimize phase boundary surface areas. Since there are two tethers for the middle block, maneuvering is limited. Due to the minimization of surface areas for certain boundaries, several structures emerge that are reminiscent of diblock copolymer phases – lamellar, cylindrical, spherical, and gyroid structures are present.

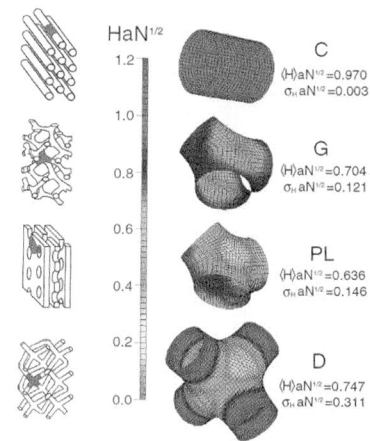

Fig. 7.7 Interfacial curvature for various phases (shown on left). The shading scale indicates the level of local curvature on each heat map [24].

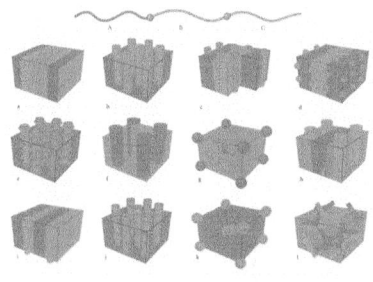

Fig. 7.8 Different morphologies for a ABC triblock copolymer. Shading scheme is the same as the molecule at top [29].

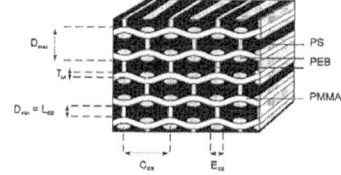

Fig. 7.9 Schematic of the *cmm knitting* pattern morphology, a type of "decorated" lamellae [30].

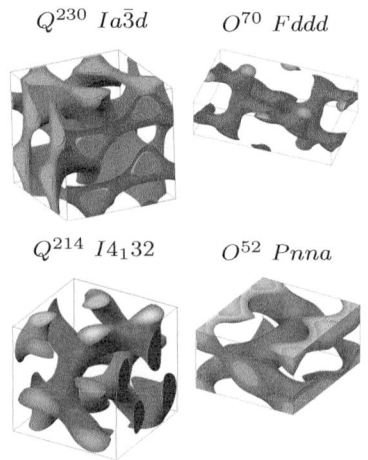

Fig. 7.10 Network structures known to form in ABC triblock copolymers.

Besides these traditional phases, more complex structures can form with triblock copolymers. One of the more complex structures, the so-called "knitting pattern" morphology with cmm plane-group symmetry, was found by Stadler and coworkers over a range of nearly symmetric poly(styrene-b-butadiene-b-methylmethacrylate) compositions and is shown in Figure 7.9.

ABCs are now notorious for an important component of their phase behavior differentiating them from ABs: they form multiply continuous network structures over a significant portion of the phase space. A total of four different network morphologies have been discovered in ABCs: the core-shell gyroid (Q^{230}) [31], the alternating gyroid ($I4_132$, Q^{214}) [26], the $Fddd$ O^{70} orthorhombic network [26], and the (metastable in ABCs, but stable in ABCA's) $Pnna$ O^{52} orthorhombic network [32, 33], illustrated in Figure 7.10.

Rod-coil block copolymers

Rod-coil diblock copolymers have been the focus of attention recently due to applications in nanolithography, nanopatterning, templating, and tailored thermoplastics [29, 34]. Due to the flexibility of the molecular backbone, the applications of available materials are generally limited by the self-assembly of Gaussian coil polymers. However, adding rigid structure through π-conjugation along the polymer backbone, helical secondary structure, or the introduction of aromatic groups allows for a wider array of conformations. These allow ordering on the 10-100 nm length scales. Many studies of these rod-coil block copolymers include investigations of functional biological [35] or semiconducting polymers [36] for use in organic electronics [37] and biological molecule materials [38].

The phase space of these rod-coil systems can be quite complex due to added anisotropic interactions of the rod blocks. When these anisotropic interactions dominate, the stiffness of the rod blocks leads to a variety

Fig. 7.11 Different smectic structures for a rod-coil diblock copolymer. (A) PHIC-PS layers forming a zigzag structure. The dark domains are PS due to RuO$_4$ staining. (B) The so-called arrowhead morphology for very high volume fractions of rod block. (C) and (D): Lamellar and cylindrical structures as the rod and coil species have a strong negative interaction strength [39].

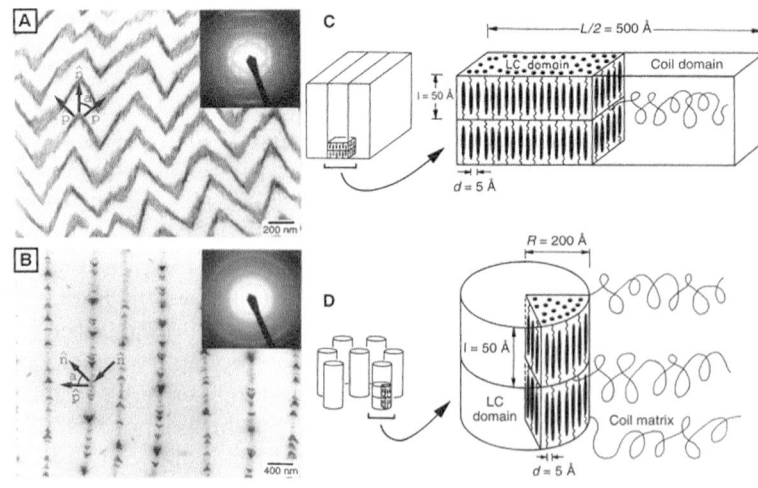

of liquid crystalline phases.

Micrographs in Figure 7.11 show an example of a rod-coil block copolymer system of poly(hexylisocyanate-b-styrene) (PHIC-PS). The smectic C phase is shown in Figure 7.11(A) as the different blocks alternate in a zigzag manner. This phase forms when the rod blocks are much longer than the PS blocks; as such the rods tilt away from the normal axis to the domain interface. Micrograph (B) shows the so-called arrowhead phase. This phase results from a very high rod block volume fraction. These arrowheads form from PS blocks and alternate orientation by 180° between adjacent PS-rich layers. If instead, the Flory-Huggins interaction parameter is very large for the rod and coil species, then the lamellar and cylindrical structures shown in Figure 7.11(C) and 7.11(D) form. Note that the rods align themselves parallel with the rod-coil domain interface.

If solvent is added to a rod-coil block copolymer system that interacts favorably with the coil species, another unique structure can form [40]. These structures are hexagonally ordered arrays of hollow spherical micelles. Figure 7.12 shows how to form these as well as fluorescence photomicrographs of these structures. Due to the difficulty of filling space as the rod blocks align themselves, the rod blocks are unable to fill the center of the spheres. By changing molecular weight or composition, the diameter, periodicity, and wall thickness can change. This allows for applications such as catalysis and selective separation media.

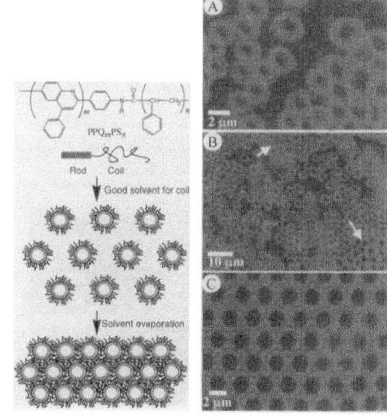

Fig. 7.12 (Left) Molecular structure of hollow micelles formed from rod-coil diblock copolymer in a selective solvent with favorable interactions with the coil species. (Right) Fluorescence photomicrographs of rod-coil diblock copolymers exhibiting hollow micelles in a two-dimensional array [40].

Thin films

Due to interest in membranes, coatings, light-emitting diodes and thin film transistors, application of block copolymers in thin films has become an emerging area of research. In a thin film environment, several characteristics of the polymer can change. These include the glass transition temperature, diffusion coefficients through the polymers, as well as the viscosity.

Thin films of poly(styrene-b-methylmethacrylate) (PS-PMMA) confined to one surface were studied by Russell [42] and were found to form lamellar structures. These lamellar sheets were found to be parallel to the silicon surface with PMMA preferring to locate on the surface. Russell concluded that this was a result of the lower surface energy of the PS. Due to this preferential segregation the film thickness becomes quantized – an initial thickness that doesn't match one of these quantized values leads to the formation of islands. With time, these islands will combine in an effort to minimize the surface area. Figure 7.13 shows a profile of these islands for poly(styrene-b-butyl methacrylate) (PS-PBMA) [41]. Formation of islands at the surface has been studied by Coulon and coworkers [43,44] for lamellar phase films using atomic force microscopy. Using a polymeric flow model, the kinetics of growth were also investigated [43]. Flows associated with domain growth were found to be confined to a bilayer which is disrupted at dislocation lines; this process of domain growth is enhanced by increased temperatures.

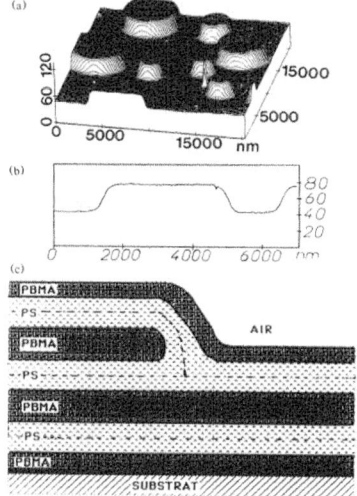

Fig. 7.13 Images of PS-PBMA diblock copolymer film. (a) AFM image of islands forming on free surface. (b) Section of one island with a diameter of 4 μm. (c) Assumed domain edge structure. [41]

The interfacial thickness of PS-PMMA was found to be 50 ± 3 Å, which is considered a typical interfacial thickness [45]. This thickness does not depend on copolymer molecular weight.

The surfaces of films that are confined to one surface were studied by Turturro et al. [46] and Henkee et al. [47] using transmission electron microscopy. They found non-equilibrium structures perpendicular to the free surface that contained lamellar and cylindrical microdomains when solvent is cast at high evaporation rates. Slowing the evaporation rates or annealing the samples will result in structures moving to become parallel to the two interfaces. Moreover, the surface itself was found to induce a lamellar structure, even when the bulk copolymer would be in the disordered state [48, 49]. This is a result of the constraints imposed by a finite film thickness. Because of the lack of a disordered state, no order-disorder transition is defined in these films. Instead, a transition from partially ordered to fully ordered occurs on lowering the temperature.

Confining diblock copolymers between two surfaces was investigated by Lambooy et al. [51] and Russell et al. [50]. Symmetric PS-PMMA was sandwiched between a silicon substrate and silicon oxide with the PMMA surface in contact with the silicon oxide. The period of these structures deviated from bulk in a cyclic manner as shown in Figure 7.14. Kellogg et al. [52] varied the wall-block interactions for this system in order to study the effect of surface energy on phase ordering. The PMMA was found at the interface of both walls if the walls were bare. However, if the walls were coated with random PS-PMMA copolymer, the wall's interactions were less selective and the diblock copolymer oriented in a perpendicular structure so both species were found at the wall.

It has been proposed that the block with a smaller statistical segment length is the species that preferentially segregate to the wall interface [53]. This is because of an unequal entropic energy penalty from unequal conformational perturbations resulting from the wall. As such, the species that is more flexible (smaller segment length) is pushed to the wall interface. It should be noted that enthalpic interactions can be increased by changing the wall's interactions with each species. In this way, the species found at the wall can be altered.

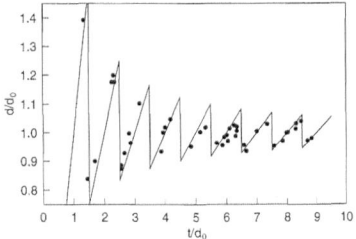

Fig. 7.14 Period, d, of lamellar formed in PS-PMMA diblock copolymer film as a function of t, the initial film thickness. The normalization, d_0, is the period of the lamellar phase in bulk [50].

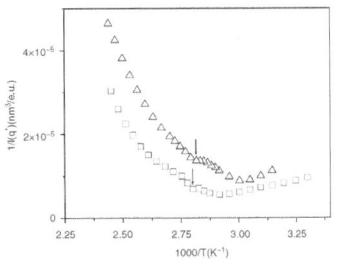

Fig. 7.15 $I(q^*)$ as a function of inverse temperature for two symmetric poly(styrene-b-isoprene) diblock copolymers. Arrows show the order-to-disorder transition temperatures.

Phase states

Despite the successes of the theories outlined above, there are cases where the theory fails to accurately describe the phase behavior, predict structures, or to calculate the structure factor that matches experimental data near the order-disorder transition (ODT). The experimentally measured $S(q^*)$ for two nearly symmetric poly(styrene-b-isoprene) diblock copolymers is shown in Figure 7.15 which gives an example of this discrepancy. Whereas weak segregation theory predicts the inverse structure factor to scale as $1/T$, this study shows otherwise. There exists curvature at $T > T_{ODT}$ that is not predicted by theory. Furthermore, the peak intensity is finite at the transition. Experiments [54] also found

7.1 Phase behavior

the heat of fusion at the T_{ODT} suggesting a first-order transition. Also, experimental work on phase diagrams showed it's possible to have a direct transformation between the lamellar phase and the disordered state. These results prompted modifications to the existing theories.

Fluctuation effects

One of the first studies of the effects of fluctuations was performed by Fredrickson and Helfand [55]. They found that fluctuations caused a first-order transition instead of the second-order transition predicted by Leibler. Using the Hartree approximation for a system of identical statistical segment length polymers that are incompressible, the free energy density has the form [6]

$$F_H(A) = \tau_R A^2 + \frac{u_R}{4} A^4 + \frac{w_R}{36} A^6, \qquad (7.31)$$

where the prefactors are temperature dependent and A is a variational parameter. Figure 7.16 shows the qualitative dependence of $F_H(A)$.

When $\chi N \ll \chi_S N$ such as in high-temperature regimes, a single minimum at $A = 0$ occurs, which represents the stable disordered phase. This indicates stability in only the disordered state. As χN approaches $\chi_S N$, where χ_S is the stability limit (the spinodal point) of the lamellar phase in the disordered region, two additional minima form in the potential curve. When $\chi N = \chi_T N$, the fluctuation induced first-order transition (MST) occurs. All minima have $F_H(A) = 0$. As χN increases away from the spinodal point, the lamellar phase is stable. This is a result of the formation of two symmetric minima with $A \neq 0$. The disordered state with $A = 0$ becomes metastable.

In addition, it was found that fluctuation effects occur in both the disordered and ordered phases in the vicinity of the ODT. For the disordered phase, they modified the structure factor as

$$\frac{S(q)}{N} = \frac{1}{\epsilon + F(f,x) - F(f,x*)}, \qquad (7.32)$$

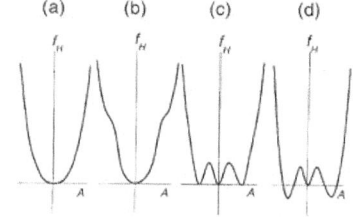

Fig. 7.16 Qualitative behavior of the free energy density in the Hartree approximation. (a) When temperatures are high, $\chi N \ll \chi_S N$ resulting in a single minimum at $A = 0$. This indicates stability in only the disordered state. (b) At $\chi_S N$, the stability limit of the lamellar phase in the disordered region, two additional minima form in the potential curve. These remain metastable for $\chi_S N < \chi N < \chi_T N$. (c) When $\chi N = \chi_T N$, the fluctuation induced first-order transition (MST) occurs. (d) Symmetric minima with $A \neq 0$ occur below the the MST, while the $A = 0$ minimum becomes metastable. These symmetric minima are stable lamellar phases [6].

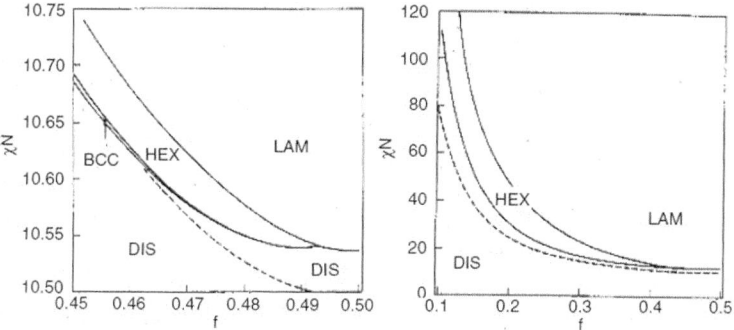

Fig. 7.17 Phase diagrams calculated with the Hartree approximation for different values of $\bar{N}$. The left is for $\bar{N} = 10^9$, whereas the right is with $\bar{N} = 10^4$. Notice that the smaller $\bar{N}$ causes the approximation to break down as the phase diagram is missing the body centered cubic spheres phase. The dashed curves represent the classical spinodal curve calculated by Leibler [55].

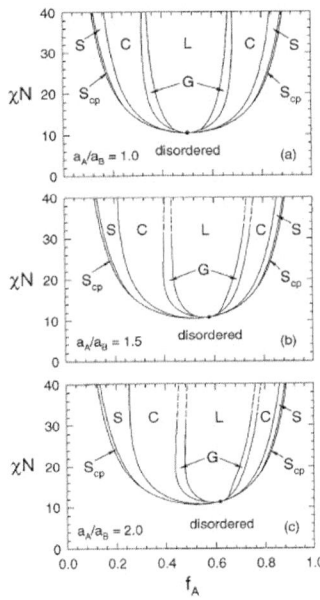

where

$$\epsilon = F(f, x*) - 2\chi N + \frac{c^3 d\lambda}{(\epsilon N)^{1/2}}, \qquad (7.33)$$

with c and λ being composition dependent coefficients; $d = 3x^*/2\pi$. Combining these equations, we finally obtain

$$\frac{S(q)}{N} = \frac{1}{F(f,x) - 2\chi N + \frac{c^3 d\lambda}{(N)^{1/2}} \frac{\sqrt{S(q*)}}{\sqrt{N}}}, \qquad (7.34)$$

where $\bar{N} = Nb^2/v^2$. b is the statistical segment length, and v is the volume. Since the last term is independent of q, fluctuations only affect the peak height. Ultimately, equation (7.34) predicts a nonlinear relation between $1/S(q)$ and $1/T$. This is in agreement with the experimental work shown above. For a symmetric diblock copolymer, these fluctuations result in a transition at $\chi\bar{N} = 10.495 + 41.022\bar{N}^{-1/3}$ and the peak scattering intensity reaches a maximum as $0.12328\bar{N}^{1/3}$. However, the Hartree approximation is only rigorously valid for $\bar{N} > 10^9$ – polymers of very high molecular weights. Figure 7.17 shows an example where this approximation breaks down for $\bar{N} = 10^4$. This phase diagram is missing the bcc phase.

Fig. 7.18 Phase diagrams calculated for a diblock copolymer while varying the ratio of the statistical segment lengths, a_A/a_B, of species A and B. The phases are labeled as L (lamellar), G (gyroid), C (cylindrical), S (spherical), and S_{cp} (close-packed spheres) [56].

Conformational asymmetry

Self-consistent field theory simulations usually consider studies where the different species have the same statistical segment length. However, varying this parameter from equality can have pronounced effects on the phase diagram of the polymer. The main causes for a segment length inequality are differences in monomer volume and also backbone flexibility of a particular species. Matsen and Bates [56] calculated the the free energy of several traditional and complex phases while varying the ratio of the statistical segment lengths of the two species of a diblock copolymer. They found the phase diagrams shown in Figure 7.18.

A trend is observed when this ratio increases: the phase boundaries shift toward compositions with greater quantity in the large segment length specie. Moreover, the boundaries between ordered phases has a larger shift than the shift in the critical point. Shifts in the ordered-ordered boundaries result from an increased stretching ability of the A blocks (as b_A/b_B increases); with a smaller segment length, this is not true for the B blocks. Because of this, the B blocks can relax at the expense of the A blocks. This forces interfaces to curve toward the A blocks. While the effect is pronounced for the ordered-ordered phase boundaries, there is a smaller effect on the critical point because the B blocks in the spherical domains must be pulled into a disordered state. The energy penalty incurred is independent of the segment lengths. Therefore, the effect on the critical point remains small. Apart from phase boundaries, the asymmetry also causes a change in the domain spacing.

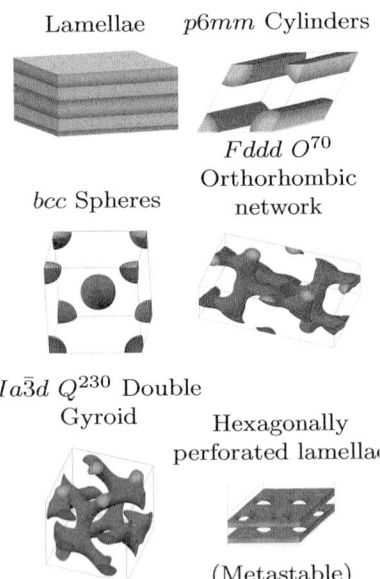

Fig. 7.19 Equilibrium phases in diblock copolymer melts.

Phase diagrams

There have been many block copolymer systems studied over the years, for example: poly(isoprene-b-styrene) (PI-PS), poly(styrene-b-methyl methacrylate), poly(ethleneoxide-b-isoprene) (PEO-PI), poly(ethylene--b-ethylene propylene) (PE-PEP), poly(styrene-b-2-vinylpyridine) (PS-P2VP), poly(ethylene propylene-b-ethylethylene) (PEP-PEE), poly(ethylene-b-ethylethylene) (PE-PEE), poly(diethylhexyloxy-p-phenylcncvinylene-b-methyl methacrylate) (DEH-PPVb- PMMA), and poly(styrene-b-isoprene-b-ethyleneoxide) (PS-PI-PEO) to name a few. There are many interesting features of these systems that over the past decades has lead to a reasonably complete picture of phase behavior in diblocks and a robust foundation in linear triblocks.

For diblock copolymers, the commonly observed phases are $Im\bar{3}m$ (bcc) spheres, $p6mm$ cylinders, lamellae, (metastable) hexagonally perforated lamellae and the bicontinuous $Ia\bar{3}d$ Q^{230} double gyroid. Recently, theory and experiment have both shown the orthorhombic $Fddd$ O^{70} network to be an additional equilibrium phase although it occupies a very small region of the phase diagram [27,58]. Real-space depictions of these phases appear in Figure 7.19. Experimental phase diagrams have been asymmetric about $f = frac12$, indicating a strong correlation between phase behavior and the asymmetry in the statistical segment lengths. As seen in the phase diagram for PEO-PI, Figure 7.20, the gyroid phase exists on both sides of $f = 0.5$. This phase is absent in the PE-PEP diagram, Figure 7.22 below, but stable for a wide range of parameter values of the PI-PS and PEO-PI systems. This leads to the conclusion that the asymmetry of the segment lengths plays an insignificant role in the formation of the gyroid phase while fluctuation effects could be instrumental in its formation. On the other hand, the PE-PEP data did indicate the HPL phase, and were presented [59] well before its acceptance as a metastable phase [60] that precedes the transition to Q^{230} [61].

Figure 7.21 shows the phase diagram for PI-PS as a function of χN and f_{PI} along with the mean-field prediction as a dash-dot curve [62]. The circles indicate transitions for ordered to ordered states (open circles) and the order-disorder transitions (closed circles). The lines presented in the figure are to guide the eye and may or may not represent the true phase boundary. Figure 7.22 shows the phase diagrams reported by Bates et al. [59], with solid lines delimiting the experimental data. This figure shows three systems with three different pairs of species with three different fluctuation parameters, $\bar{N}$. These include PE-PEP ($\bar{N} = 27 \times 10^3$), PEP-PEE and PE-PEE ($\bar{N} = 3.4x10^3$), and PI-PS ($\bar{N} = 1.1 \times 10^3$). It can be seen that decreasing $\bar{N}$ leads to the formation of the double gyroid phase – fluctuations stabilize the phase.

The PEO-PI phase diagram [57] in Figure 7.20 shows is built from experiments on 25 diblock copolymers with a composition ranging from $0.05 < f_{PEO} < 0.8$. Based on scattering data, viscoelasticity, and TEM results, the authors identified five ordered phases. These are the crys-

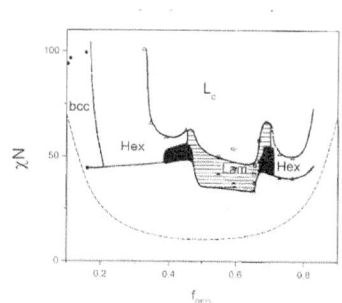

Fig. 7.20 Phase diagram for the system PEO-PI comprising 25 diblock copolymers samples. The phases are: crystalline lamellar (L_c), amorphous lamellar (Lam), Hex, hexagonal packed cylinders (Hex), gyroid (shadowed areas), and body centered cubic spheres (bcc). The dashed line gives the spinodal line in the mean-field prediction. [57].

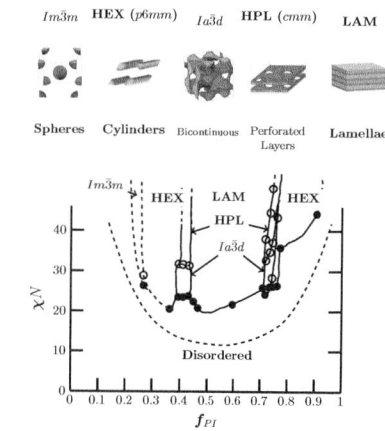

Fig. 7.21 Phase diagram for PI-PS diblock copolymer system [62].

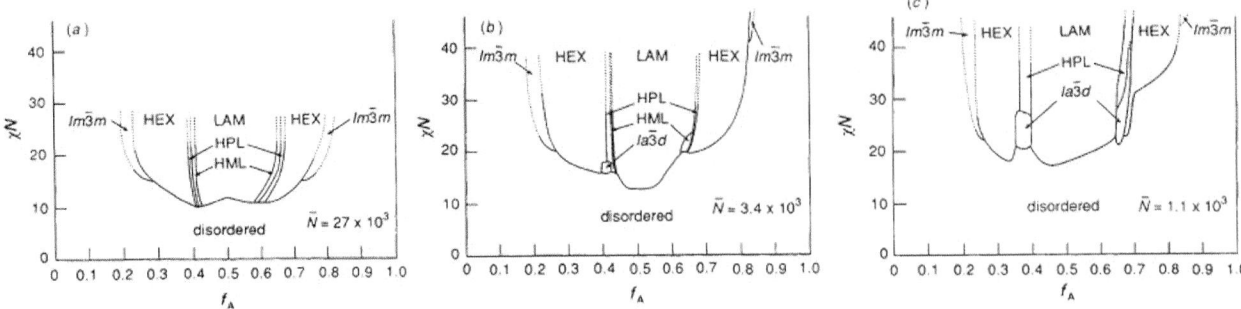

Fig. 7.22 Phase diagram for three diblock copolymer systems with varying fluctuation parameter, $\bar{N}$: (a) PE-PEP with $\bar{N} = 27 \times 10^3$; (b) PEP-PEE and PE-PEE with $\bar{N} = 3.4 \times 10^3$; and (c) PI-PS with $\bar{N} = 1.1 \times 10^3$. The gyroid phase develops on both sides of the phase diagrams as $\bar{N}$ decreases, leading to the conclusion that fluctuations stabilize the phase [59].

talline lamellar, amorphous lamellar, cylinders in a hexagonal array, the gyroid phase, and spheres in a body centered cubic lattice. Despite finding a perforated layered structure, this structure was not present when the polymers were cooled. As such, this was not considered a stable phase (and not present in the phase diagram), but merely facilitated the lamellar transformation to the gyroid phase.

The phase states of the diblock systems can be drastically changed with the addition of a chemically distinct species C block to a linear diblock copolymer. Phases not present in traditional ABA tripolymer systems have been experimentally observed with ABC triblocks. Instead of the usual two control parameters, these have two independent compositions, three Flory-Huggins interaction parameters, and three sequence possibilities (ABC, ACB, BAC), which leads to a rich – and nearly overwhelming – phase landscape. Bailey & coworkers suggested a classification scheme that is particularly useful in making sense of the vast array of reported ABC morphologies which considers the sequencing of interaction parameters [63]. A phase-separated linear ABC triblock copolymer will naturally form A/B and B/C interfaces to partition the dissimilar segments – interactions between the A and C blocks is "optional" and thus will only occur when there is an enthalpic benefit for doing so, at the expense of additional elastic energy. A "non-frustrated" (F^0-type) ABC is sequenced such $\chi_{AB} \approx \chi_{BC} < \chi_{AC}$, i.e. the energetics of the interfaces dictated by chemical connectivity are also more favorable than the optional interface. A singly-frustrated system (F^1-type) is sequenced such that one of the mandated interfaces is significantly more energetic than the optional interface, $\chi_{AB} \gg \chi_{BC} \approx \chi_{AC}$. In a doubly-frustrated system (F^2-type), *both* chemically mandated interfaces are significantly more energetic than the optional interface, $\chi_{AB} \approx \chi_{BC} \gg \chi_{AC}$.

As an example of an F^1 type system, Figure 7.23 shows the phase diagram of PS-PI-PEO as a function of temperature and f_{PEO}: $\chi_{IO} \gg \chi_{SI} \approx \chi_{SO}$. This series of triblock copolymers were synthesized along the $f_I \approx f_S$ isopleth by growing poly(ethylene oxide) off of the hydroxyl-

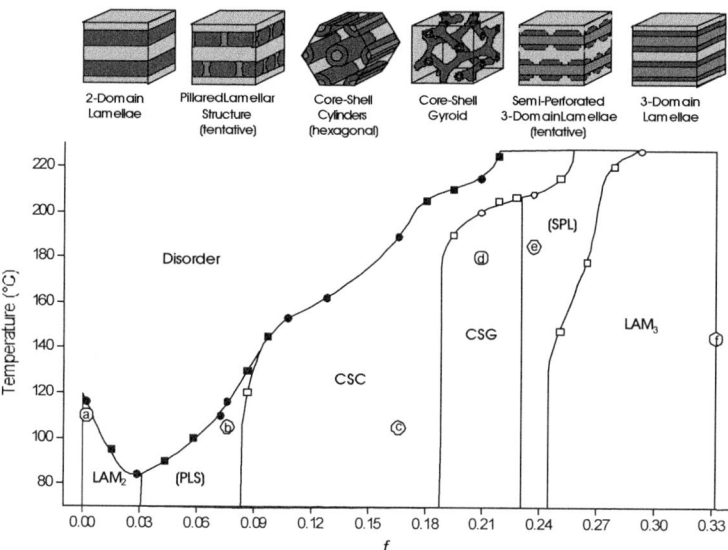

Fig. 7.23 Phase diagram of PS-PI-PEO where f_{PEO} is varied from 2.9 to 33.2% [63], illustrating the progression of phase states in a Type I frustrated system: $\chi_{IO} \gg \chi_{SI} \approx \chi_{OS}$.

terminus of a 18000 g/mol molecular weight PS-PI-OH diblock copolymer to span the composition range $0 < f_O \lesssim \frac{1}{3}$ [63]. This investigation revealed six phases that appear at the top of the figure. These are the two domain lamellar, core-shell cylinders in a hexagonal array, core-shell gyroid, three domain lamellar, along with two tentatively assigned HPL-like structures. These phases were thought to form largely because of the very unfavorable interaction between the PI and PEO blocks. The PS blocks in the middle of the polymer act as screening agents between PI and PEO and, therefore, morphologies are formed with only two interfaces.

Changing the block sequence to ISO presents a non-frustrated F^0 system, with completely different phase behavior as shown in Figure 7.24(i).

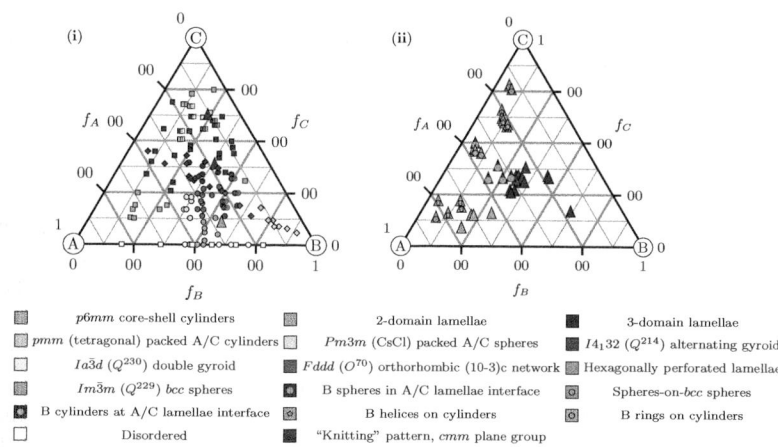

Fig. 7.24 Phase behavior for ABC block copolymers. **(i)** Unfrustrated chain sequences, $\chi_{AC} \gg \chi_{AB} \approx \chi_{BC}$. Data compiled from [26, 64–70]. **(ii)** Type II frustration, $\chi_{AC} \ll \chi_{AB} \approx \chi_{BC}$. Data compiled from [71–76].

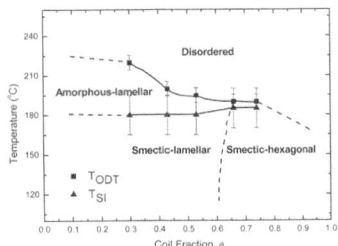

Fig. 7.25 Experimental phase diagram for poly(diethylhexyloxy-p-phenylenevinylene-b-methyl methacrylate), a rod-coil diblock copolymer system. [77].

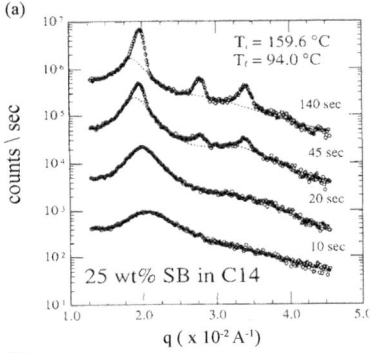

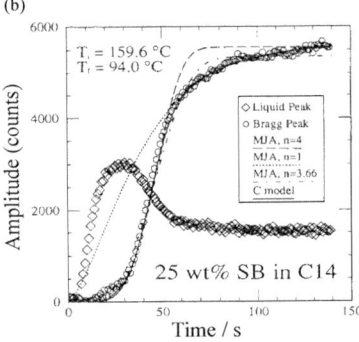

Fig. 7.26 (a): Time-resolved small-angle x-ray scattering of a 52 kDa 30 wt% styrene poly(styrene-b-butadiene), decomposed into contributions between crystalline (solid) and liquid-like (dashed) order. (b): Fits of the kinetic data to the Avrami models and the Cahn nucleation model.

On the other hand, the poly(styrene-b-ethylene butylene-b-methyl methacrylate) system favored by the late Stadler is a F^2 system, with the A/C interface highly favorable over either the A/B or B/C; these polymers are notorious for forming the "decorated" phases, i.e. classical diblock copolymer morphologies with "B" block "decorations".

Recently a phase diagram has been found for an experimental rod-coil diblock copolymer system. The phase diagram is shown in Figure 7.25 as a function of temperature and coil block composition fraction. This study focused on poly(diethylhexyloxy-p-phenylenevinylene-b-methyl methacrylate) (DEH-PPV-PMMA). To find the states, the coil length was varied as the rod length was kept fixed. Cooling the system for different compositions resulted in two main regimes: for coil fractions below 53%, polymers transitioned from a disorder to amorphous-lamellar and then to a smectic-lamellar state; above 63% coil composition, the transition was from disordered to a hexagonal smectic structure.

7.2 Block copolymer dynamics

Commercially produced block copolymers have been used in a variety of applications, ranging from lithography, oil additives, food supplements, to the production of thermoplastic elastomers, which currently represents their largest use. For over the past 30 years block copolymers have attracted attention for their capabilities of forming highly ordered structures in the nanometer scale, making them suitable for a wide range of nanotechnological applications [78]. The overall degree of polymerization (N), the composition (f_i), the Flory-Huggins interaction parameter (χ_{ij}), and the architecture of macromolecules are the essential parameters that control the thermodynamics and the structure of block copolymers (BCPs). These parameters also strongly influence the dynamics of BCPs, in particular, the segregation strength $\chi_{ij}N$ plays a leading role [79].

The study of the dynamics of block copolymer melts has been instrumental in the elucidation of principles governing the formation of morphologies with long-range order [80]. Segmental diffusion rates and viscoelastic properties are coupled inextricably to the ordered phase and the characteristics of the grains and defects thereof within a macroscopic specimen. Scientists have used several experimental methods to study block copolymer dynamics: dynamic mechanical spectroscopy, static birefringence, nuclear magnetic resonance (NMR), dynamic light scattering, and forced Rayleigh scattering. Some of these techniques play an important role in the study of the phase transition kinetics in BCPs [81] and will be described later in this section.

Kinetics of microphase separation

The mechanism and rate by which ordered structures form in block copolymer melts and solutions are strongly coupled to the initial state and external processing conditions. Fundamental to the ordering process is the chain diffusivity, which should be homogeneous in very weakly-segregated systems but strongly anisotropic in strongly-segregated systems. Transition mechanisms are therefore quite different for disorder-to-order transitions (DOTs) as compared to order-to-order transitions (OOTs).

Some of the first insights into the dynamics of the disorder-to-order transition were provided by Almdal and Bates in their discovery through dynamic shear experiments that time-temperature superposition failed in a disordered diblock copolymer melt in a temperature window ranging from the order-disorder transition temperature T_{ODT} through roughly $T_{ODT} + 50\,°C$ below a critical frequency ω_c [83]. In this regime, the onset of liquid-like terminal behavior shifted to progressively lower frequencies, indicative of the formation of structure within the disordered state. Subsequent small-angle neutron scattering (SANS) experiments showed that these structural features were in fact density fluctuations that increased in amplitude as $T \to T_{ODT}$ [84]. Composition patterns suggested by the SANS data were reminiscent of spinodal decomposition in other binary mixtures. In the absence of externally aligning fields these fluctuations are isotropic and ordered phases formed upon a temperature quench will form a number of isotropically oriented grains. Reminiscent of polymer crystallization, the nucleation rate depends on the ordering temperature, $T_o < T_{ODT}$. Floudas et al. [85] tracked the formation of lamellae from disorder and found that Avrami kinetics for heterogeneous nucleation and three dimensional growth described the data for shallow quenches, i.e. $T_{ODT} - T_o < 3\,°C$, while a different mechanism dominates at deeper quenches. Harkless et al. [86] examined a *bcc*-sphere forming poly(styrene-*b*-butadiene) melt using time-resolved high resolution small-angle x-ray scattering with rapid quenches deep into the ordered regime $T_{ODT} - T_o > 50\,°C$ and found a two-stage progression of the final morphology. The first stage was characterized by the rapid formation of spheres with liquid-like order followed by ripening onto the final lattice over a much larger time scale, Figure 7.26. [86]

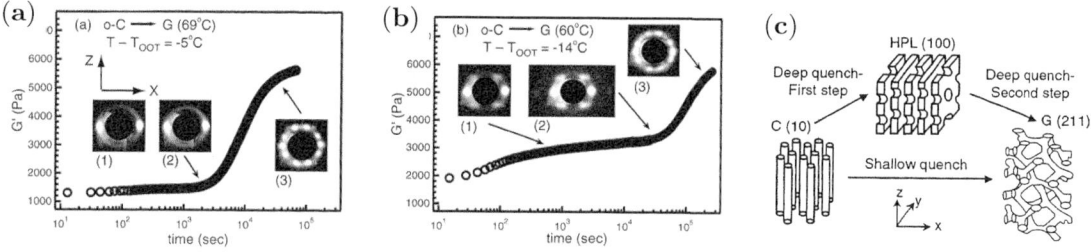

Fig. 7.27 (a) Evolution of the elastic modulus G' of a preoriented cylindrical 44 kDa 25 wt% styrene poly(styrene-*b*-isoprene) diblock quenched 5 °C below the cylinder-to-gyroid transition. Small-angle x-ray scattering shows a progression from oriented cylinders to an oriented Q^{230} state. (b) The same experiment, with a 14 °C quench, showing two plateaus in G' corresponding to the HPL intermediate state and the final Q^{230} state, with associated SAXS overlays. (c) Wang and Lodge's schematic summary of dependence of temperature quench depth on the ordering transition. From [82].

The first stage again obeyed Avrami kinetics, while the latter stage behaves as a pure exponential curve. The nucleation model by Cahn [87] which models nucleation events as occurring on two-dimensional defects successfully accounted for both regimes, suggesting that the presence of such defects dominates the nucleation process at deeper quenches. Time constants for ordering of this polymer ranged from 30-200 s.

Order-to-order transitions necessarily occur via a different process than DOTs, and in general proceed much more slowly. Epitaxial relationships between neighboring phases have been fully documented for the cylinder-to-sphere [88], cylinder-to-gyroid [89], cylinder-to-lamellae [90], and lamellae-to-gyroid transitions [91]. Similar to the DOT, the kinetics and pathways of OOTs are highly sensitive to the quench depth. Shallow quenches, with weaker driving forces, evidently proceed more readily to the final equilibrium state through a nucleation and growth process, while in deeper quenches (i.e. beyond the stability limit of the initial state) spinodal decomposition dominates and can lead to long-lived metastable states. For instance, hexagonally perforated layer (HLP) morphology in diblock copolymers readily forms as a metastable structure within the the cylinder-to-gyroid transition. Wang and Lodge [82] studied this transition in detail and found that at shallow temperature quenches the Q^{230} phase formed directly, following Avrami kinetics, with a time constant on the order of 1–3 hours, see Figure 7.27. For deep quenches, however, HPL formed quickly (1–10 min) with a much longer (2 hours – 2 days) decay into the final Q^{230} state.

Diffusion in block copolymers

A fundamental component of ordering kinetics in block copolymers is the nature of diffusion in these materials, which may be organized as usual into two categories: self-diffusion, which reflects the translational mobility of a single chain within a matrix of other identical chains, and collective diffusion, which for block copolymers reflects the ability of mesodomains to reconfigure themselves during the annealing process where non-equilibrium polycrystalline mesophases gradually ripen and defects are gradually annihilated.

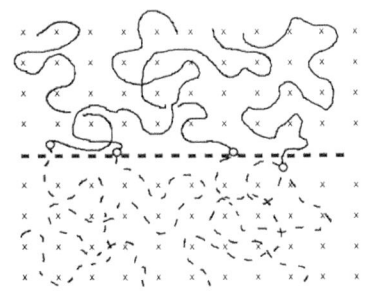

Fig. 7.28 Illustration of an entangled BCP melt where the thick dashed lines represent the interface and the x's represent the entanglement constraints, from [92].

Self-diffusion

The self-diffusivity of diblock copolymer chains has been studied extensively over the past 25 years, typically with a combination of forward recoil spectrometry, field gradient NMR [93], dynamic light scattering [94, 95], and forced Rayleigh scattering [96]. The most prominent distinction between homopolymer self-diffusion and block copolymer self-diffusion is that the spatial variations in the density pattern map directly to energy barriers which retard the mobility of chain segments in areas populated by dissimilar segments. Clearly then in ordered phases one should expect that the well-defined interfaces partitioning domains should be more permissive to parallel motion but more restrictive of the passage of chains across domain boundaries. This is indeed the case, at least for entangled melts, however even in the disordered phase density fluctuations form temporary barriers to single chain diffusion and thus restrict it significantly [97]; this effect persists to well above T_{ODT} ($\approx 100\,°C$). Chain entanglement also plays an important role, e.g. Eastman and Lodge found that Rouse chains (i.e. chains below the critical molecular weight) exhibited no distinguishable anisotropy in the diffusion coefficient [98]. Further experiments by Lodge and Dalvi [96] more carefully investigated differences between the diffusivities between Rouse-like and entangled diblock copolymer melts. In the case of non-entangled BCP melts the anisotropy is small since polymer chains can move through the interface without causing the two blocks to mix, while for entangled BCP melts, see Figure 7.28, the anisotropy is large, as polymer chains are not able to reptate without mixing the two blocks [99]. Moreover in entangled systems reptation of chains carries a thermodynamic penalty that depends on the degree of incompatibility of the chains (χN), since the A chains will be moving through the B chains [79], yielding an exponential dependence $D/D_0 = \exp\left[-\alpha\chi N\right]$ as illustrated in Figure 7.29.

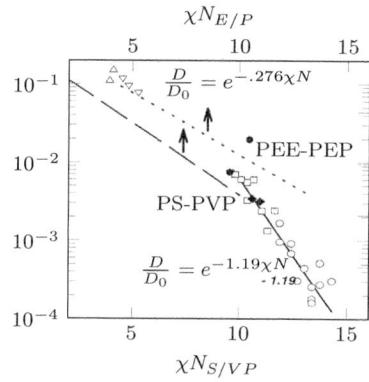

Fig. 7.29 Comparison of the block copolymer self-diffusivity D of entangled poly(ethyl ethylene-b-ethylenepropylene) (PEE-PEP), [96], and poly(styrene-b-vinyl pyridine) (PS-PVP), [100], compared to the free diffusivity ($\chi N \to 0$), D_0.

Collective diffusion

Collective diffusion in block copolymer melts may be most simply construed as the modes and rates over which the composition pattern in an ordering melt evolves over time. A fundamental origin for modeling such diffusive behavior begins with a Ginzburg-Landau equation of motion:

$$\frac{\partial}{\partial t}\psi(\mathbf{r},t) = -\int_V d\mathbf{r}'\Lambda(\mathbf{r},\mathbf{r}')\nabla\frac{\partial F[\psi]}{\partial \psi(\mathbf{r}',t)} + \theta(\mathbf{r},t), \quad (7.35)$$

$$\frac{\partial}{\partial t}\psi(\mathbf{q},t) = \Lambda(q)\frac{\partial F[\psi]}{\partial \psi(\mathbf{q},t)} + \theta(\mathbf{q},t), \quad (7.36)$$

where $\psi(\mathbf{r}) = \rho(\mathbf{r}) - \langle\rho(\mathbf{r})\rangle$, i.e. the same order parameter employed by Leibler's RPA [3]; $F[\psi]$ is a free energy functional describing the energy cost of a particular composition pattern, e.g. the Landau expansion employed by Leibler [3]; Λ is an Onsager coefficient, and θ is a white noise term.

Experiments directly probing collective dynamics are significantly less abundant than those treating single chain diffusivity. Here dynamic light scattering is the method of choice, which gives a measurement of the dynamic structure factor:

$$S^2(q,t) \propto G(q,t) - 1, \quad (7.37)$$

where $G(q,t)$ is the autocorrelation function of the scattered intensity. $S(q,t)$ contains a superposition of exponential decays $S(q,t) \propto \sum a_i e^{-\Gamma_i t}$ with amplitudes a_i and decay rates Γ_i. Anastasiadis et al. [95] identified an internal stress relaxation process with $\Gamma_I = \tau_I^{-1}$ associated with the relative motion of individual blocks within the chain and two collective diffusive relaxation modes with q^2-dependent rates. Subsequent experiments have shown repeatedly the existence of three diffusive modes:

- A "cluster" mode, which is a slow mode (1–10 s), in which clusters of ≈ 1 nm^3 in size reorganize [101].
- A "heterogeneity" mode, corresponding to the dynamics described by (7.35), also known as the pattern relaxation mode, with a decay rate of $\Gamma_H = q^2 D(1 - 2\kappa\chi N)$; D is the single-chain diffusivity and κ is a heterogeneity factor. This mode is unique to compositionally heterogeneous materials ($\kappa > 0$), and is the most pronounced in DLS of ordered block copolymer melts.
- An "intermediate" mode, whose characteristic time corresponds closely with the ω_c identified by Almdal et al. [83] below which time-temperature superposition fails.

The "heterogeneity" mode has been studied extensively from a theoretical perspective, although moreso as a tool for the exploration of phase space than in the study of collective diffusion. Most notable in the latter regard is the 1989 work of Kawasaki and Sekimoto [102], who treated reptating diblock copolymer chains in the lamellar phase, developed expressions for the Onsager coefficients $\Lambda(q)$, and in turn equations of motion for the deformation of lamellae. In terms of using the heterogeneity mode as a tool for the investigation of the kinetics of microphase separation, the dynamic density functional theory implemented by Fraajie [103] employs diffusive dynamics to evolve an initially homogeneous melt into ordered structures, e.g. as depicted in Figure 7.30.

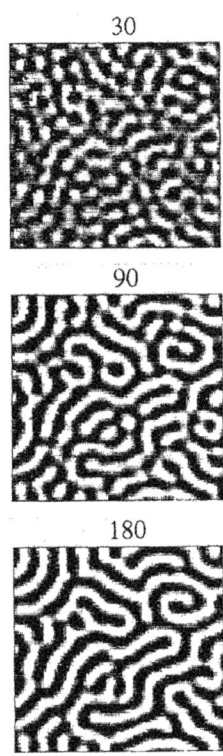

Fig. 7.30 Snapshots of the evolution of the lamellar morphology in a symmetric AB diblock copolymer. $\tau = tD/a^2$ is the time nondimensionalized by the single-chain diffusivity D and the statistical segment length a. Reproduced from [103].

Viscoelastic properties and melt rheology

Viscoelastic properties of block copolymers can be determined by the effects of variables like the chemical composition, molecular weight, block sequence, block architecture, among others. Far above the T_{ODT}, the rheology of block copolymers is quite similar to that of a typical homopolymer, exhibiting the classical glassy, rubbery, and terminal states across the dynamic spectrum. The terminal response of a typical homogeneous polymer fluid may be obtained by the evaluation of the limit of the Rouse moduli as the frequency $\omega \to 0$, which reveals that the scaling behavior of G' and G'' is as ω^2 and ω^1, respectively. It is in this low-frequency regime beyond the importance of local chain dynamics where the morphology of the block copolymer is influential. Even prior to ordering, density fluctuations give rise to spontaneous heterogeneities that elevate the moduli relative to the otherwise equivalent homopolymer.

Ordered block copolymer melts have even more pronounced influences in the terminal regime yielding responses ranging from nearly liquid-like to solid-like. Various relationships have been established in diblock copolymers for the characteristic response of each of the four phases, summarized best by Kossuth et al. [104] in Figure 7.31. It has been shown with calculation [105] and experiment [106] that polycrystalline lamellar phases exhibit scaling behavior as $G^* \propto (i\omega)^{1/2}$. The hexagonal C phase features scaling as $G^*(\omega) \propto (i\omega)^{1/3}$ as demonstrated by Ryu et al. [107]. Zhao et al. [108] demonstrated that both Q^{229} and Q^{230} have a large elastic plateau modulus ($G' \propto \omega^0$) that persist to low frequency. This result was initially surprising because the elasticity of the gyroid network was thought to be a consequence of three-dimensional connectivity. However, the Q^{229} spherical phase is completely discontinuous but nonetheless triply periodic. It was speculated that the three-dimensional quality of a morphology was solely responsible for the solid-like viscoelastic response.

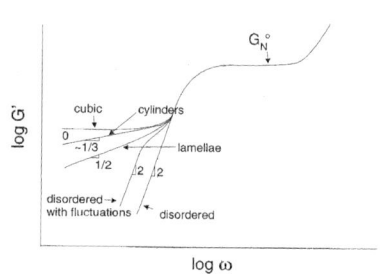

Fig. 7.31 Master curve of linear viscoelastic modulli in block copolymers. From [104].

Melt rheology of block copolymers

Melt rheology of BCPs has helped to further the understanding of the relationship between structure and stress under a variety of different flows. Rosedale and Bates [109] demonstrated how the terminal response of a PEP-PEE diblock copolymer melt changed from that of a liquid to that of a solid just by lowering the temperature from the disordered state to the ordered state, and they were able to conclude that these phenomena were universal for all symmetric diblock copolymers. It was also discovered that BCP's physical properties when in the disordered state were those as the homopolymer. These findings proved to be fundamental for the field of rheology [80].

Molecular architecture has a large effect on a number of dynamic and thermodynamic properties of block copolymers: phase morphology, polymer mobility, viscosity, molecular weight, composition, and concentration. The simplest architecture is the (A-B) block copolymer structure, followed by the triblock copolymer architecture (A-B-C), star-block copolymer architecture where three or more segments grow from a central hub, among other less common structures.

Effect of shear on block copolymers

It has been recognized as early as 1970 that flow fields can be particularly effective in altering the state of statically annealed block copolymers. The first such studies were conducted by Keller et al. in the extrusion of Kraton ® poly(styrene-b-butadiene-b-styrene) (SBS) triblock copolymers [110]. In these studies it was discovered that the extruded polymer, which forms styrene cylinders in a butadiene matrix, was nearly perfectly anisotropic with the cylinder axis oriented parallel to the flow direction. Hadziioannou et al. [111] later found similar results with styrene-isoprene based di- and triblocks in both the cylindrical and lamellar morphologies. More recently, shear alignment techniques have also been applied to block copolymer thin films [112].

The effects of shear orientation on ordered block copolymer mesophases, and also disordered polymers near the disorder-to-order transition, are controlled by a number of pertinent characteristics relating the shear field to the system undergoing flow:

Viscoelastic contrast: Refers to the difference in moduli among the different mesodomains. Softer domains more readily dissipate an applied stress, and the orientation of the morphology with respect to the shear field is governed by which best allows the soft component to deform freely while minimizing the deformation of the harder domains, i.e. the modulus of the oriented phase is minimized with respect to the shear field. Krishnan *et al.* observed peculiar flow-induced behavior primarily due to this effect in bicontinuous microemulsions of poly(ethyl ethylene)/poly(dimethyl siloxane) (PEE/PDMS) in which the mixture phase separated during flow to allow channels of relatively inviscid PDMS to dissipate the applied stress [113]. In the shear orientation of diblock copolymer lamellae, viscoelastic contrast dominates in the high frequency/high shear rate regime [114, 115].

Static equilibrium phase state: Shear experiments conducted near phase boundaries have shown that order-to-order transitions, order-to-disorder transitions, and disorder-to-order transitions readily occur. Koppi *et al.* demonstrated that at a critical shear rate in large-amplitude oscillating shear (LAOS) experiments with lamella-forming diblock PEP-PEE that disordered melts slightly above T_{ODT} will order, ostensibly due to the suppression of concentration fluctuations [116]. Similar behavior was observed in subsequent experiments with a cylinder-forming composition [117]. Winter *et al.* employed milder shear conditions after shallow temperature quenches from above T_{ODT} and monitored the microphase separation kinetics and found that these conditions suppress the ordering process, i.e. T_{ODT} is lowered under these conditions [118]. Hajduk *et al.* made similar conclusions in a PEP-PEE-PEP triblock copolymer [119]. Jackson *et al.* discovered a "Martensitic" transition of $p6mm$ SBS cylinders to $p3m$ cylinders, also dictated by the shear rate [120]. Harada *et al.* were able to produce either cylinders or lamellae by simply changing processing conditions in the wet extrusion of triblocks and pentablocks comprised of poly(cyclohexyl ethylene) (PC) and polyethylene (PE) [121].

The geometry of the morphology is obviously also tremendously influential. 1D and 2D morphologies, i.e. cylinders and lamellae have clearly defined "slip planes" (as well as *bcc* spheres). Network morphologies, on the other hand, have no mechanism by which the original morphology may remain intact during large amplitude shear. Cochran and Bates subjected an O^{70} forming triblock copolymer to LAOS and found an intermediate state which quickly annealed to the O^{52} structure upon the cessation of shear [32]. Eskimergen *et al.* showed that Q^{230} in diblock copolymers quickly formed cylinders during shear and then slowly annealed back into the gyroid phase after shear [122]. The *bcc*-phase, which contains slip planes, can either be aligned into twinned structures [123], transformed into the fcc phase [124], or disordered entirely [125] depending upon the processing conditions and also the presence of solvent.

Deformation time scale and amplitude: Koppi *et al.* [106] found that near the ODT and using low shear frequencies, the lamellae morphology arranged parallel to the velocity gradient direction and perpendicular to the flow direction, higher frequencies produced samples where the lamella morphology has the unit normal perpendicular to the flow and velocity gradients. They also found that at temperatures above the ODT, parallel layer orientation is obtained no matter to what shear frequency the sample is subjected to [106]. Low frequencies/low shear rates can be expected to couple to the relatively slow relaxation dynamics of the morphological domains [126], higher rates to the dynamics of the local mesostructure, while higher shear rates will excite primarily single-chain relaxation dynamics [127]. Gupta *et al.* have also reported experiments explicitly designed to elucidate the role of the strain amplitude [128].

Block architecture: Block connectivity has been shown to be yet another important variable in the response of ordered polymers to shear [129]. In AB diblocks, chain ends dangle freely into their domain interiors, and may accordingly slip past one another when subjected to shear. ABA diblocks contain a distribution of bridging and looping B chains, while ABABA pentablocks contain even more domain-bridging chains of both the A and B components. For example, lamellar CECEC pentablock copolymers can form the "transverse" orientation where the lamellar normal is parallel to the flow direction, whereas this orientation is unstable in AB and ABA sequences [130].

Computational studis of dynamics

Several computational methods have been employed to investigate the dynamics and statistics of polymers on the coarse-grained scale. They are divided into two main groups: those that operate only in systems with a relatively large fraction of lattice sites left free in order to allow movement of molecules, and those where all lattice sites are occupied by molecular elements. In the latter group, there are only two simulation methods that do not require empty spaces to allow molecular mobility and that are capable of simulating dense polymer melts. These are: the bond breaking algorithm [131] and the cooperative motion algorithm (CMA) [132], including a recently formulated method for low molecular weight liquids called the Dynamic Lattice Liquid model (DLL) [133]. CMA is the most widely used method because of its ability to simulate well-defined monodisperse polymers in the static limit [132].

Molecular Dynamics (MD) is another computational technique used for the investigation of the dynamic properties of polymer systems (a detailed explanation of MD simulations of polymers can be found in [133]).

Cooperative motion algorithm

The CMA has the ability to simulate polymer melts of dense systems, were the lattices are completely occupied by monomers and each of them is connected to a nonbreakable bond that forms structures that represent the different polymer architectures [132]. Each monomer (often referred as a 'bead') can only occupy one lattice site for it to satisfy the excluded-volume condition [81]. Moreover, according to Pakula and Floudas, this algorithm makes use of cooperative dynamics in which monomer rearrangements need to satisfy local continuity of the system, where no empty lattice sites are generated. This can be accomplished when monomers rearrange and replace one of its neighbors along closed trajectories. These trajectories are randomly generated around the simulated systems, with the sum of displacements equal to zero (local continuity). During these rearrangements the chains are subject to conformational transformation, additionally local energy is altered as the monomers contact new neighbors [132]. Monitoring position and/or orientations of system elements in time allows the study of the system's dynamics, see Figure 7.32 [81].

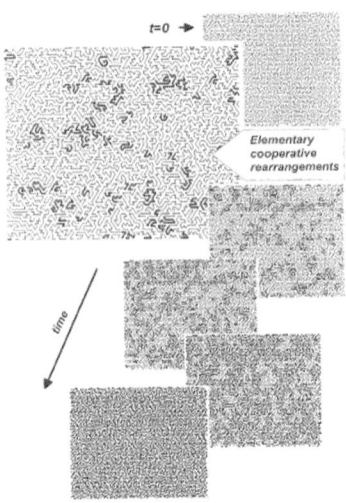

Fig. 7.32 A dense linear chain system simulated using CMA, from $t = 0$ until t, where t is time when all chains have been rearranged at least once (shown by the thicker lines), from [81].

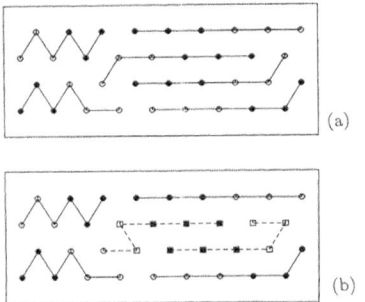

Fig. 7.33 (a) The configuration of copolymer melt where the filled and open circles represent the different types of monomers of the system. (b) The new configuration of the melt. The squares represent the movements of the monomers and the dashed lines the moved bonds, from [132].

Simulation of BCPs using the CMA. The CMA for diblock copolymers involves two monomer chains connected to form a copolymer of length $N = N_A + N_B$ [132]. The two monomers are partially compatible and described by the interaction parameter ϵ_{ij}. According to Pakula [133] the energy of mixing is given only by the interaction of different types of monomers so it is assumed that $\epsilon_{AA} = 0$, $\epsilon_{BB} = 0$, and $\epsilon_{AB} = 1$. The effective energy of one monomer E_m will depend on the local structure and will be given by the sum of ϵ_{AB} over z nearest neighbors. Equilibrium states are generated when systems' chains are moved at a given temperature (still following the cooperative dynamics, see Figure 7.33), this alters local energy as monomers contact new neighbors.

Every attempt to move a chain element is identified as a Monte Carlo step and the probability of motion is related to the interaction energy of the monomer in the attempted position. At temperature T, the Boltzmann factor $p = \exp\left(-\Delta E/k_B T\right)$ is compared with a random number r, where $0 < r < 1$. If the Boltzmann factor is greater than the generated random number $(p > r)$, rearrangement happens and the motion of another monomer is attempted. At low temperatures the different types of monomers tend to separate from each other in order to minimize local energy by minimizing the number of neighboring A-B contacts [81]. Simulations run over a wide rang of temperatures and provide thermodynamic, dynamic, and structural properties of the system, see Figure 7.34.

7.3 Small-angle X-ray and neutron scattering

Small-angle scattering techniques have been instrumental over the past decades in the development of the current state-of-the-art in block copolymer technologies and their fundamental understanding. Here we provide a brief treatment of the origin of the scattering phenomenon, followed by an overview of the various experimental techniques commonly applied to block copolymer melts, solutions, and thin films.

Basics of scattering

When an incident plane-wave of frequency ω, i.e. X-rays or neutrons, interacts with matter, it does so through a number of disparate mechanisms; the origin of the scattering phenomenon is dependent on the nature of the wave and that of the particle with which it interacts (i.e. electrons, protons, or neutrons), but in general case of elastic scattering each particle emits a scattered wave with spherical symmetry, an s-wave. The amplitude is given generally by

$$A_s = f \frac{A_0}{r} e^{i(\omega t - \mathbf{k} \cdot \mathbf{r} - \varphi)} , \qquad (7.38)$$

where $\mathbf{k}$ is the wavevector, which points in the direction of propagation with a modulus $|\mathbf{k}| = 2\pi/\lambda$; f, the scattering factor, is the ratio between the incident and scattered amplitude; and φ is the scattering phase shift which is often combined with f to form the complex scattering factor, $fe^{-i\varphi}$. In the case of x-rays interacting with electrons, $\varphi = \pi$ and so the scattered wave is precisely out of phase with the incident wave.

For most purposes each atomic particle may be treated as a point-like scatterer, and the observed scattered radiation is simply the superposition of the s-waves (spherical waves) generated by all scatterers. Consider the situation illustrated in Figure 7.35 where two such arbitrarily positioned scatterers are excited by incident radiation with wavevector $\mathbf{k}_{in}$. The two particles are separated by $\mathbf{r}_{12} = \mathbf{r}_2 - \mathbf{r}_1$ and each emit an s-wave $A_{s,i}$ with amplitudes given by equation (7.38). At some position $\mathbf{R}$ with $|\mathbf{R} - \mathbf{r}_1| \approx |\mathbf{R} - \mathbf{r}_2| \gg \mathbf{r}_{12}$, the intensity of the x-rays given by the superposition of A_{s1} and A_{s2} are recorded by a detector at a scattering angle θ. The quantities δ_1 and δ_2 are determined by the projection of $\mathbf{r}_1$ onto $\mathbf{k}_{in,2}$, and the projection of $\mathbf{r}_2$ onto $\mathbf{k}_{s1}$, respectively. δ_1 represents the extra distance in radians A_{s1} must travel to the detector compared to A_{s2}, while δ_2 is the extra distance $A_{in,2}$ must travel to reach $\mathbf{r}_2$ compared to A_{s1} to reach $\mathbf{r}_1$. The quantity $\delta = \delta_2 - \delta_1$ is then the phase shift of A_{s2} relative to A_{s1}:

$$\begin{aligned} \delta = \delta_1 - \delta_2 &= \mathbf{k}_{in,1} \cdot \mathbf{r}_{12} - \mathbf{k}_{s1} \cdot \mathbf{r}_{12} \\ &= (\mathbf{k}_{in,1} - \mathbf{k}_{s,1}) \cdot \mathbf{r}_{12} \qquad (7.39) \\ &= \mathbf{q} \cdot \mathbf{r}_{12} . \end{aligned}$$

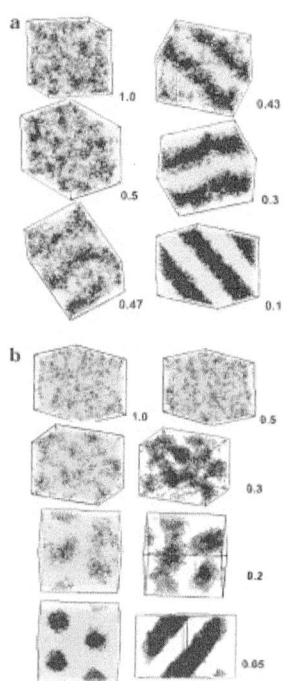

Fig. 7.34 Snapshots of lamellar (a) and cylindrical (b) architectures simulated using the CMA. (a) corresponds to symmetric and (b) to asymmetric diblock copolymer melts above and below their order-disorder transition temperature, from [134].

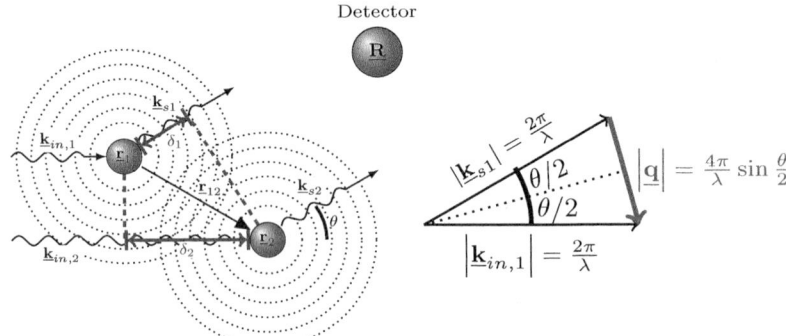

Fig. 7.35 Illustration of the origin of the static structure factor.

[2] Another common notation often favored by crystallographers is $s \equiv \frac{2}{\lambda} \sin \theta$, where the scattering angle is defined as 2θ.

The vector $\mathbf{q}$ is known as the momentum transfer or scattering vector[2] and as shown in Figure 7.35, $|\mathbf{q}| = \frac{4\pi}{\lambda} \sin \frac{\theta}{2}$. Arbitrarily defining $\mathbf{r}_1 \equiv 0$ as the origin, the net amplitude observed by the detector at $\mathbf{R}$ is then

$$A_s(\mathbf{R}) = A_{s1} + A_{s2}$$
$$= \frac{f}{|\mathbf{R}|} e^{i(\omega t - \mathbf{k}_s \cdot \mathbf{R})} + \frac{f}{|\mathbf{R} - \mathbf{r}_{12}|} e^{i(\omega t - \mathbf{k}_s \cdot \mathbf{R})} \underbrace{e^{-i(\mathbf{k}_{in} - \mathbf{k}_s) \cdot \mathbf{r}_{12}}}_{\text{phase shift}} \quad (7.40)$$
$$\approx f \frac{e^{i(\omega t - \mathbf{k}_s \cdot \mathbf{R})}}{R} \left(e^0 + e^{-i\mathbf{q} \cdot \mathbf{r}_{12}} \right),$$

where the approximate relation is nearly exact since $|\mathbf{r}_{12}| \ll |\mathbf{R}|$. For an ensemble of n point scatterers, the scattered wave function is

$$A_s(\mathbf{q}) = A_0 \frac{e^{i(\omega t - \mathbf{k}_s \cdot \mathbf{R})}}{R} \sum_i^n f_i e^{-i\mathbf{q} \cdot \mathbf{r}_i} . \quad (7.41)$$

The $\mathbf{q}$-dependent quantity is known as the static structure factor:

$$S(\mathbf{q}) \equiv \sum f_i e^{-i\mathbf{q} \cdot \mathbf{r}_i} . \quad (7.42)$$

When the scattering angle is small such that $qa \ll 1$ (where a is a characteristic interatomic distance), the medium may be treated as a continuum:

$$S(\mathbf{q}) = \int_V d\mathbf{r}\, f(\mathbf{r}) e^{-i\mathbf{q} \cdot \mathbf{r}} . \quad (7.43)$$

In either case, $S(\mathbf{q})$ is the component of the amplitude of the scattered wave that contains the structural information of the sample, and is simply the Fourier transform of the field of scattering factors.

The scattered intensity of the radiation observed at the detector is proportional to $|S(\mathbf{q})|^2$:

$$I(\mathbf{q}) \propto \int_V d\mathbf{r}'\, f(\mathbf{r}') e^{i\mathbf{q} \cdot \mathbf{r}'} \int_V d\mathbf{r}''\, f(\mathbf{r}'') e^{-i\mathbf{q} \cdot \mathbf{r}''}$$
$$= \int_V d\mathbf{r}'\, d\mathbf{r}''\, f(\mathbf{r}') f(\mathbf{r}'') e^{-i\mathbf{q} \cdot (\mathbf{r}' - \mathbf{r}'')} . \quad (7.44)$$

Scattering factors for X-ray

The X-ray's oscillating electric field $\mathbf{E}_0(\mathbf{r}, t)$ is responsible for generating the scattered radiation through which measurements are possible. Electrons (and protons) experience an acceleration whose amplitude is proportional to that of the incident field and the particle's charge to mass ratio (Z/m). Since protons are 1843 times more massive than electrons, they experience negligible acceleration compared to that of electrons. Note that the acceleration will be out of phase with $\mathbf{E}_0$ by π radians. Figure 7.36 illustrates the scenario for an isolated electron with incident beam of amplitude E_0. The beam is non-polarized, and so may be viewed as two beams, $\mathbf{E}_0^\perp$ (propagating along z) and $\mathbf{E}_0^\parallel$ (propagating along y). The amplitude of the wave emitted by the electron is proportional to the sin of the angle between the incident electric field vector and the point of observation: ϕ for $\mathbf{E}_0^\parallel$ and $\pi - \theta$ for $\mathbf{E}_0^\perp$. The scattered wave induced by $\mathbf{E}_0^\parallel$ observed at $\mathbf{R}$ is:

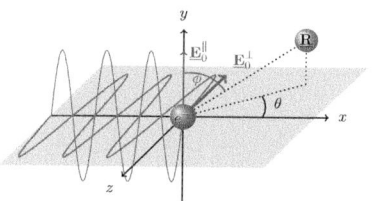

Fig. 7.36 Scattering of an X-ray by an electron.

$$\mathbf{E}^\parallel = -\frac{\mu_0}{4\pi} \frac{e}{m_e} \frac{\sin\phi}{|\mathbf{R}|} E_0^\parallel = -r_e \frac{\sin\phi}{|\mathbf{R}|} E_0^\parallel, \quad (7.45)$$

where $\mu_0 = 4\pi \times 10^{-7}$ and e is the charge of an electron. $r_e = 2.818 \times 10^{-15}$m is known as the classical radius of an electron. The scattered wave due to $\mathbf{E}_0^\perp$ is

$$\mathbf{E}^\perp = -r_e \frac{\cos\theta}{|\mathbf{R}|} E_0^\perp. \quad (7.46)$$

The observed intensity is the square of the electric field amplitude, and so the mean intensity observed from an unpolarized incident beam is given by the Thompson formula:

$$I_e(\theta) = r_e^2 I_0 \frac{1 + \cos^2\theta}{2}. \quad (7.47)$$

Remembering that a scattered wave has spherical symmetry, in SI units, the intensity $I_e(\theta)$ represents the power (in Watts) per steradian (unit solid angle) per second, emitted by a single electron subject to an incident beam transmitting I_0 (measured in Watts per square meter). $I_e(\theta)$ is also called the differential scattering cross-section of an electron, $I_e \equiv d\sigma/d\Omega$.

I_e as calculated by the Thompson formula can be further decomposed into two contributions: coherent and incoherent scattering. Coherent scattering arises from collective interference of waves of precisely the same wavelength and as such its intensity varies predictably as a function of the scattering vector. Incoherent scattering involves radiation of various wavelengths and has no such relationship. It arises in x-ray scattering through Compton scattering, which may be understood as a consequence of electron-photon collisions.

For atomic electrons, quantum mechanics specifies a charge density distribution $\rho_e(\mathbf{r})$ about the nucleus and so the waves emitted by the electron cloud will interfere coherently, and may be evaluated using the equation (7.43),

$$f_e = \int d\mathbf{r} \, \frac{\rho_e(\mathbf{r})}{e} e^{-i\mathbf{q}\cdot\mathbf{r}} \,. \tag{7.48}$$

f_e is called the scattering factor for an electron; it is unity at $\mathbf{q} = 0$ and decays to zero as $\theta \to \pi$. The coherent intensity from an atomic electron is

$$I_{\text{coh}} = f^2 I_e \tag{7.49}$$

and the incoherent part is

$$I_{\text{Compton}} = I_e - I_{\text{coh}} = (1 - f^2) I_e \,. \tag{7.50}$$

The total scattering from all Z atomic electrons is additive, so that the atomic scattering factor is $f_a = \sum_{i=1}^{Z} f_i$.

For small-angle scattering, $\theta \to 0$, so $f_a \to Z$. The static structure factor in equation (7.42) then becomes

$$S(\mathbf{q}) = \sum_{\text{atoms}} r_e Z_i e^{-i\mathbf{q}\cdot\mathbf{r}_i} \approx \int d\mathbf{r} \, \underbrace{b(\mathbf{r})}_{r_e \langle Z(\mathbf{r}) \rangle} e^{-i\mathbf{q}\cdot\mathbf{r}} \,, \tag{7.51}$$

where $b(\mathbf{r})$ is the electron scattering length density and where the brackets denote the thermal average.

Scattering factors for neutrons

Neutron scattering arises from interactions between an incident neutron and the atomic nucleus. The nature of this interaction is complex, although the Fermi approximation introducing a weak pseudo-potential $V(\mathbf{r}) = \sum_j b_j \delta(\mathbf{r} - \mathbf{r}_j)$ was used by Max Born to conclude that the probability of an incident neutron wave with wavevector $\mathbf{k}_{in}$ to become a scattered plane wave with wavevector $\mathbf{k}_s$ is proportional to $\left| \int d\mathbf{r} \, V(\mathbf{r}) e^{-i\mathbf{q}\cdot\mathbf{r}} \right|^2$. Van Hove used the Born approximation to show in 1954 that the number of neutrons scattered per incident neutron is given by:

$$I(\mathbf{q}, \epsilon) = \frac{1}{h} \frac{|\mathbf{k}_s|}{|\mathbf{k}_{in}|} \sum_{j,k}^{n} b_j b_k \int_{-\infty}^{\infty} dt \left\langle e^{-i\mathbf{q}\cdot(r_j(t) - r_k(0))} \right\rangle e^{-i\epsilon t} \,. \tag{7.52}$$

Using the identity $\int d\mathbf{r} \, \delta(\mathbf{r} - \mathbf{r}') \times e^{-i\mathbf{q}\cdot\mathbf{r}} = 1$ the double sum of the Van Hove equation may be transformed:

$$\sum_{j,k} b_j b_k \left\langle e^{-i\mathbf{q}\cdot(r_j(t) - r_k(0))} \right\rangle = \sum_{j,k} b_j b_k \int d\mathbf{r} \, \delta \left\langle \mathbf{r} - \mathbf{r}_k(0) + \mathbf{r}_j(t) \right\rangle e^{-i\mathbf{q}\cdot\mathbf{r}}$$

$$= nb^2 \int d\mathbf{r} \, G(\mathbf{r}, t) e^{-i\mathbf{q}\cdot\mathbf{r}} \,, \tag{7.53}$$

where the quantity $G(\mathbf{r}, t) \equiv \frac{1}{n} \sum_{j,k} \langle \delta \left(\mathbf{r} + \mathbf{r}_j(t) - \mathbf{r}_k(0) \right) \rangle$ is called the time-dependent pair-correlation function and represents the probability of finding two particles separated by $\mathbf{r}$ where the positions of the particles are measured at times differing by t. The Fourier transform is the dynamic structure factor, $S(\mathbf{q}, \epsilon)$, and for SANS experiments which employ thermal neutrons, $\epsilon \to 0$ and thus $S(\mathbf{q}, \epsilon) \to S(\mathbf{q})$. Thus equation (7.52) for SANS reduces to:

$$\begin{aligned} I(\mathbf{q}) = \frac{d\sigma}{d\Omega} &= \frac{1}{h} \frac{|\mathbf{k}_s|}{|\mathbf{k}_{in}|} \sum_{j,k} \left\langle e^{-i\mathbf{q}\cdot(\mathbf{r}_j(t)-\mathbf{r}_k(0))} \right\rangle \\ &= \frac{1}{h} \frac{|\mathbf{k}_s|}{|\mathbf{k}_{in}|} \left\langle \left| \sum_j b_j e^{-i\mathbf{q}\cdot\mathbf{r}} \right|^2 \right\rangle \\ &\approx \frac{1}{h} \frac{|\mathbf{k}_s|}{|\mathbf{k}_{in}|} \left\langle \left| \int d\mathbf{r}\, b(\mathbf{r}) e^{-i\mathbf{q}\cdot\mathbf{r}} \right|^2 \right\rangle, \end{aligned} \quad (7.54)$$

where

$$b(\mathbf{r}) \equiv \frac{\sum b_i}{v_0} = \frac{\text{Sum over atomic scattering lengths}}{\text{Molecular volume}} \quad (7.55)$$

is the scattering length density.[3] As in x-ray scattering, neutron scattering generates both coherent and incoherent radiation. The nature of the incoherent scattering arises since nearly all materials are comprised of a distribution of nuclear isotopes, and the combined nuclear spin of the incident neutron and nucleus will be randomly oriented and can take a variety of values. Each isotope and spin state has a different scattering length, and thus the thermally averaged scattering length of an element is observed in experiment. The thermal average of equation (7.54) may be further manipulated to see which averages are of interest:

[3]Scattering length densities may be calculated online via: http://www.ncnr.nist.gov/resources/sldcalc.html. The NIST website which enumerates atomic neutron scattering lengths for many isotopes has the URL: http://www.ncnr.nist.gov/resources/n-lengths/.

$$\left\langle \left| \int d\mathbf{r}\, b(\mathbf{r}) e^{-i\mathbf{q}\cdot\mathbf{r}} \right|^2 \right\rangle = \int d\mathbf{r}d\mathbf{r}'\, \langle b(\mathbf{r})b(\mathbf{r}')\rangle e^{-iq\cdot(\mathbf{r}-\mathbf{r}')}. \quad (7.56)$$

The b-values on different nuclei have no spatial correlation and so

$$\langle b(\mathbf{r})b(\mathbf{r}')\rangle = \begin{cases} \langle b(\mathbf{r})^2 \rangle & \mathbf{r} = \mathbf{r}' \\ \langle b(\mathbf{r}) \rangle^2 & \mathbf{r} \neq \mathbf{r}' \end{cases} \quad (7.57)$$

Equation (7.56) may then be further decomposed:

$$\int d\mathbf{r}d\mathbf{r}'\, \langle b(\mathbf{r})b(\mathbf{r}')\rangle e^{-iq\cdot(\mathbf{r}-\mathbf{r}')} = \underbrace{\int d\mathbf{r}d\mathbf{r}'\, \langle b(\mathbf{r})\rangle^2 e^{-iq\cdot(\mathbf{r}-\mathbf{r}')}}_{\text{coherent}} + \underbrace{\int d\mathbf{r} \left[\langle b(\mathbf{r})^2 \rangle - \langle b(\mathbf{r})\rangle^2 \right]}_{\text{incoherent}} \quad (7.58)$$

And thus the coherent scattering from a SANS experiment is:

$$I_{coh}(\mathbf{q}) = \frac{d\sigma_{coh}}{d\Omega} = \frac{1}{h} \frac{|\mathbf{k}_s|}{|\mathbf{k}_{in}|} \underbrace{\left| \int d\mathbf{r}\, \langle b(\mathbf{r})\rangle e^{-i\mathbf{q}\cdot\mathbf{r}} \right|^2}_{|S(\mathbf{q})|^2}. \quad (7.59)$$

Laue scattering (Bragg diffraction)

Small-angle scattering is of instrumental importance to the characterization of microphase-separated block copolymer melts. Block copolymer phases are periodic and so many of the techniques employed by crystallographers to deduce the structure of proteins and small crystalline molecules may be adapted to the small-angle limit appropriate for the length scales common to polymeric structures.

Bravais lattices

The general equations presented in Section 7.3 may be further reduced for materials with periodic long-range order. The morphology of such materials is comprised of a unit cell which repeats itself on an infinite lattice. In real space, the origin of any particular unit cell is described by the lattice vector $\mathbf{d}_{hkl}$ where the integers h, k, and l are known as the *Miller indices*:

$$\mathbf{d}_{hkl} = h\mathbf{a} + k\mathbf{b} + l\mathbf{c} . \tag{7.60}$$

The contents of each unit cell are identical, and so any periodic system is said to possess at least translational symmetry, which allows the scattering length density to be expressed as:

$$b(\mathbf{r}) = b(\mathbf{r} + \mathbf{d}_{hkl}) \tag{7.61}$$

The lattice vectors $\mathbf{a}$, $\mathbf{b}$, and $\mathbf{c}$ form the basis set of the vector space associated with the material. Conventionally, the lengths $a = |\mathbf{a}|$, $b = |\mathbf{b}|$, and $c = |\mathbf{c}|$ are used in conjunction with the Euler angles

$$\gamma \equiv \cos^{-1}\left(\frac{\mathbf{a}\cdot\mathbf{b}}{ab}\right); \tag{7.62}$$

$$\beta \equiv \cos^{-1}\left(\frac{\mathbf{a}\cdot\mathbf{c}}{ac}\right); \tag{7.63}$$

$$\alpha \equiv \cos^{-1}\left(\frac{\mathbf{b}\cdot\mathbf{c}}{bc}\right), \tag{7.64}$$

to define the shape of the unit cell, which is always a parallelepiped, or the lattice system. Conversely in Cartesian coordinates the lattice vectors may be expressed in terms of their lattice constants using:

$$\mathbf{a} = \begin{pmatrix} a \\ 0 \\ 0 \end{pmatrix}; \quad \mathbf{b} = \begin{pmatrix} b\cos\gamma \\ b\sin\gamma \\ 0 \end{pmatrix}; \tag{7.65}$$

$$\mathbf{c} = \begin{pmatrix} c\cos\beta \\ c\left(\frac{\cos\beta\cos\gamma-\cos\alpha}{\sin\gamma}\right) \\ c\sqrt{1-\cos^2\beta - \left(\frac{\cos\beta\cos\gamma-\cos\alpha}{\sin\gamma}\right)^2} \end{pmatrix}.$$

In 2D there are four lattice systems (see Table 7.2), while in 3D there are seven (see Table 7.3, categorized based on the values of the Euler angles and cell lengths.

7.3 Small-angle X-ray and neutron scattering

Table 7.1 Lattice centering systems.

Centering	# Lattice points	Lattice systems	Symmetry operations
Primitive (P)	1	All	—
Face centered (C)	2	Monoclinic, Tetragonal, Orthorhombic	$b(\mathbf{r}) = b\left(\mathbf{r} + \frac{1}{2}(\mathbf{a}+\mathbf{b})\right)$
Face centered (F)	4	Orthorhombic, Cubic	$b(\mathbf{r}) = b\left(\mathbf{r} + \frac{1}{2}(\mathbf{a}+\mathbf{b})\right)$ $= b\left(\mathbf{r} + \frac{1}{2}(\mathbf{b}+\mathbf{c})\right)$ $= b\left(\mathbf{r} + \frac{1}{2}(\mathbf{a}+\mathbf{c})\right)$
Body centered (I)	2	Cubic	$b(\mathbf{r}) = b\left(\mathbf{r} + \frac{1}{2}(\mathbf{a}+\mathbf{b}+\mathbf{c})\right)$

The lattice is said to be *primitive* if it contains only a single lattice point per unit cell. For certain crystallographic systems, such as fcc or bcc, the primitive unit cell may not be expressed as a simple parallelepiped. Rather, the conventional lattice must employ a centering system which introduces additional lattice points to the unit cell related by translational symmetry operations which are tabulated in Table 7.1. Combining a particular lattice system with a centering system yields the Bravais lattice. There are a total of 20 different types of Bravais lattices: one lattice for 1D structures (e.g., block copolymer lamellae), five lattices for 2D structures (e.g. the cylindrical phase), and 14 lattices for 3D phases. These are illustrated in Table 7.2 and Table 7.3.

In infinitely periodic systems the structure factor may be factored using equation (7.61) to account for the translational symmetry. Recall that the origin of any particular unit cell is $\mathbf{d}_{hkl}$ and denote the volume it encloses as Ω_{hkl}:

$$\begin{aligned} S(\mathbf{q}) &= \int_V d\mathbf{r}\, b(\mathbf{r}) e^{-i\mathbf{q}\cdot\mathbf{r}} \\ &= \sum_{h,k,l=-\infty}^{\infty} \int_{\Omega_{hkl}} d\mathbf{r}\, b(\underbrace{\mathbf{r}-\mathbf{d}_{hkl}}_{\mathbf{r}'\in\Omega_{000}} + \mathbf{d}_{hkl}) e^{-i\mathbf{q}\cdot(\mathbf{r}'+\mathbf{d}_{hkl})} \\ &= \underbrace{\left[\sum_{hkl} e^{-i\mathbf{q}\cdot\mathbf{d}_{hkl}}\right]}_{L(\mathbf{q})} \underbrace{\left[\int_{\Omega_{000}} d\mathbf{r}'\, b(\mathbf{r}') e^{-i\mathbf{q}\cdot\mathbf{r}'}\right]}_{P(\mathbf{q})} . \end{aligned} \quad (7.66)$$

The term $P(\mathbf{q})$ [$F(\mathbf{q})$ is also a common notation] is known as the *form factor* for the material and contains all of the structural information pertaining to the contents of the unit cell. The term $L(\mathbf{q})$ [often denoted $S(\mathbf{q})$ in the crystallographic literature] is also known as the structure

Table 7.2 Two-dimensional Bravais lattices.

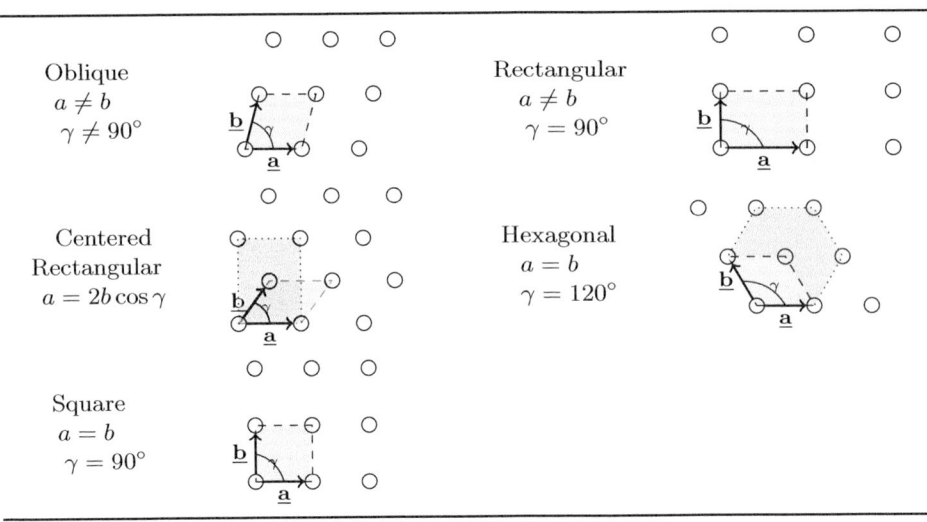

Oblique	Rectangular
$a \neq b$	$a \neq b$
$\gamma \neq 90°$	$\gamma = 90°$
Centered Rectangular	Hexagonal
$a = 2b \cos \gamma$	$a = b$
	$\gamma = 120°$
Square	
$a = b$	
$\gamma = 90°$	

factor and thus is unfortunately ambiguous to the definitions given by equations (7.51) and (7.59). $L(\mathbf{q})$ describes the Bravais lattice of the material and thus all of the pertinent length scales for block copolymer mesophases. The Laue condition is evident from the form of $L(\mathbf{q})$: each term of the summation has unit modulus with a phase angle of $-i\mathbf{q} \cdot \mathbf{r}$, and so given the infinite summation limits, these phase angles will be isotropically distributed and $L(\mathbf{q}) \to 0$ unless $\mathbf{q} \cdot \mathbf{d}_{hkl} = 2\pi n$. Thus for an arbitrary choice of $\mathbf{q}$ there is no observed intensity since the scattering from each unit cell produces destructive interference that asymptotically annihilates the wave as the scattering volume becomes large. For certain values of $\mathbf{q}$, however, the interference is perfectly constructive. Let $\underline{\mathbf{R}} = (\mathbf{a}, \mathbf{b}, \mathbf{c})^T$ such that $\mathbf{d}_{hkl} = (h, k, l) \cdot \underline{\mathbf{R}}$ and define $\underline{\mathbf{K}} \equiv (\mathbf{a}^*, \mathbf{b}^*, \mathbf{c}^*)$ such that $\mathbf{q}_{hkl} = (h, k, l) \cdot \underline{\mathbf{K}}^T$. Then to satisfy the Laue condition

$$\mathbf{q} \cdot \mathbf{d}_{hkl} = (h, k, l) \cdot \underline{\mathbf{R}} \cdot \underline{\mathbf{K}} \cdot \begin{pmatrix} h^* \\ k^* \\ l^* \end{pmatrix} = 2\pi n \to \underline{\mathbf{K}} = 2\pi \underline{\mathbf{R}}^{-1} . \quad (7.67)$$

The columns of $\underline{\mathbf{K}}$ form the basis vectors of the reciprocal lattice such that any reciprocal lattice point may be expressed in terms of the reciprocal lattice vectors and Miller indices as $\mathbf{q}_{hkl} = h\mathbf{a}^* + k\mathbf{b}^* + l\mathbf{c}^*$. The reciprocal-space vectors are often parametrized in lengths and Euler angles analogously to the real-space lattice: $a^* = |\mathbf{a}^*|$, $\gamma^* = \cos^{-1}\left(\frac{\mathbf{b}^* \cdot \mathbf{c}^*}{|b^* c^*|}\right)$, etc. Relationships between real- and reciprocal-space quantities appear in Table 7.4 The modulus of a lattice vector in terms of these lattice constants is:

Table 7.3 Three-dimensional Bravais Lattices.

Triclinic
$a \neq b \neq c \quad \alpha \neq \beta \neq \gamma \neq 90°$

Monoclinic
$a \neq b \neq c \quad \alpha \neq 90° \quad \beta = \gamma = 90°$

P C

Orthorhombic
$a = b = c \quad \alpha = \beta = \gamma \neq 90°$

P C I F

Tetragonal
$a = b \neq c \quad \alpha = \beta = \gamma = 90°$

P C

Rhombohedral (Trigonal)
$a = b \neq c \quad \alpha = \beta = \gamma = 90°$

R

Hexagonal
$a = b \neq c \quad \alpha = \beta = 90°$
$\gamma = 120°$

P

Cubic
$a = b = c$
$\alpha = \beta = \gamma = 90°$

P I F

$$|\mathbf{q}_{hkl}| = \frac{2\pi}{|\mathbf{d}_{hkl}|} = 2\pi \sqrt{\begin{aligned}&(ha^*)^2 + (kb^*)^2 + (lc^*)^2 \\ &+ 2hka^*b^* \cos\gamma + 2hla^*c^* \cos\beta + 2klb^*c^* \cos\alpha\end{aligned}}$$
(7.68)

Table 7.4 Relationships between real- and reciprocal-space lattice parameters.

$$\mathbf{a}^* = \frac{\mathbf{b} \times \mathbf{c}}{\det \underline{\mathbf{R}}} \quad a^* = \frac{bc}{V} \sin \alpha \quad \cos \alpha^* = \frac{\cos \beta \cos \gamma - \cos \alpha}{\sin \beta \sin \gamma}$$
$$\mathbf{b}^* = \frac{\mathbf{a} \times \mathbf{c}}{\det \underline{\mathbf{R}}} \quad b^* = \frac{ac}{V} \sin \beta \quad \cos \beta^* = \frac{\cos \alpha \cos \gamma - \cos \beta}{\sin \alpha \sin \gamma}$$
$$\mathbf{c}^* = \frac{\mathbf{a} \times \mathbf{b}}{\det \underline{\mathbf{R}}} \quad c^* = \frac{ab}{V} \sin \gamma \quad \cos \gamma^* = \frac{\cos \alpha \cos \beta - \cos \gamma}{\sin \alpha \sin \beta}$$
$$V^* = \det \underline{\mathbf{K}} = \frac{1}{V} = a^* b^* c^* \sqrt{1 - \cos^2 \alpha - \cos^2 \beta - \cos^2 \gamma - 2 \cos \alpha \cos \beta \cos \gamma}$$

Symmetry and systematic extinctions

The Bravais lattices are further categorized into space groups according to the crystallographic symmetry operations present in the unit cell. In general, a symmetry operation may be described by a linear operator in addition to a translation:

$$b(\mathbf{r}) = b\left(\{\underline{\mathbf{W}}|\mathbf{w}\}\mathbf{r}\right) \tag{7.69}$$

$$\{\underline{\mathbf{W}}|\mathbf{w}\}\mathbf{r} = \underline{\mathbf{W}} \cdot \mathbf{r} + \mathbf{w} \tag{7.70}$$

where $\{\underline{\mathbf{W}}|\mathbf{w}\}$ is known as the Seitz symbol. Unit cell coordinates are most conveniently described using the fractional coordinate system,

$$\mathbf{r} = \underline{\mathbf{R}} \cdot \mathbf{x}; \quad \mathbf{x} = (x, y, z); \quad x, y, z \in [0, 1] \tag{7.71}$$

In this way the linear operator $\underline{\mathbf{W}}$ and translation vector $\mathbf{w}$ are decoupled from the lattice constants. There are a total of eight unique kinds of symmetry operations mapping a point $\mathbf{x}$ to $\mathbf{x}'$:

Identity: Trivial mapping an object to itself, which is denoted as 1.
$$\underline{\mathbf{W}} = \underline{\mathbf{I}} = \begin{pmatrix} 1 & 0 & 0 \\ 0 & 1 & 0 \\ 0 & 0 & 1 \end{pmatrix}, \mathbf{w} = 0.$$

$$\mathbf{x}' = \mathbf{x}. \tag{7.72}$$

Inversion: Denoted as $\bar{1}$, inversion reflects an object through a point $\mathbf{o}$ known as the inversion center or center of symmetry. $\underline{\mathbf{W}} = -\underline{\mathbf{I}}$, $\mathbf{w} = 0$.

$$\mathbf{x}' = -\mathbf{x}. \tag{7.73}$$

Inversion reverses the chirality of the objects upon which it acts.

Translation: Shifts an object by a fractional vector $\mathbf{t}$. $\underline{\mathbf{W}} = \underline{\mathbf{I}}$, $\mathbf{w} = \mathbf{t}$.

$$\mathbf{x}' = \mathbf{x} + \mathbf{t}. \tag{7.74}$$

Rotation: A linear operator $\mathcal{R}_N$ that leaves a single line called the rotation axis, described by unit vector $\mathbf{u}$, fixed while all other points are rotated by an angle θ about this axis. In crystallography rotations are designated N-fold rotations, where $\theta = 2\pi/N$, and are denoted symbolically by N, i.e. three for a 3-fold (120°) rotation.

Rotations preserve the chirality of the structures upon which they operate. $\underline{\mathbf{W}} = \underline{\mathcal{R}}_N$, $\mathbf{w} = 0$.

$$\underline{\mathbf{u}}_\times \equiv \begin{pmatrix} 0 & -u_3 & u_2 \\ u_3 & 0 & -u_1 \\ -u_2 & u_1 & 0 \end{pmatrix}, \quad (7.75)$$

$$\underline{\mathcal{R}}_N = \cos\left(\frac{2\pi}{N}\right)\underline{\mathbf{I}} + \sin\left(\frac{2\pi}{N}\right)\underline{\mathbf{u}}_\times + \left(1 - \cos\left(\frac{2\pi}{N}\right)\right)\mathbf{u}\,\mathbf{u}, \quad (7.76)$$

$$\mathbf{x}' = \underline{\mathcal{R}}_N \mathbf{x}, \quad (7.77)$$

where $\mathbf{u}\,\mathbf{u}$ is the dyad product of $\mathbf{u}$.

Screw Axis: An N-fold rotation in conjunction with a translation along $\mathbf{u}$. Screw axes are denoted N_M, where N and M are positive integers such that $\underline{\mathbf{W}} = \underline{\mathcal{R}}_N$ and $\mathbf{w} = \mathbf{t}_{N_M} = \frac{M}{N}\mathbf{u}$.

$$\mathbf{x}' = \underline{\mathcal{R}}_N \mathbf{x} + \mathbf{t}_{N_M}. \quad (7.78)$$

Rotoinversion: Denoted $\bar{N}$, the combination of an N-fold rotation ($N > 3$) and inversion through a point on the rotation axis. $\underline{\mathbf{W}} = -\underline{\mathcal{R}}_N$, $\mathbf{w} = 0$.

$$\mathbf{x}' = -\underline{\mathcal{R}}_N \mathbf{x}. \quad (7.79)$$

Reflection: Denoted m, a reflection operates across a mirror plane. Reflection is in fact simply rotoinversion for $N = 2$, where the rotation axis is normal to the mirror plane and the inversion center resides at the intersection of the axis and plane. $\underline{\mathbf{W}} = -\underline{\mathcal{R}}_2$, $\mathbf{w} = 0$.

$$\mathbf{x}' = -\underline{\mathcal{R}}_2 \mathbf{x}. \quad (7.80)$$

Glide Reflection A reflection combined with a translation along a glide axis which is parallel to the mirror plane. Glides are designated as a lower case letter: a, b, c, d, or n. $\underline{\mathbf{W}} = -\underline{\mathcal{R}}_2$, $\mathbf{w} = \mathbf{t}$.

$$\mathbf{t} = \begin{cases} \left(\frac{1}{2}\ 0\ 0\right) & a\text{-glide}(\perp \mathbf{b} \text{ or } \mathbf{c}) \\ \left(0\ \frac{1}{2}\ 0\right) & b\text{-glide}(\perp \mathbf{a} \text{ or } \mathbf{c}) \\ \left(0\ 0\ \frac{1}{2}\right) & c\text{-glide}(\perp \mathbf{a} \text{ or } \mathbf{b}) \\ \left(\frac{1}{2}\ \frac{1}{2}\ 0\right) & n\text{-glide}(\perp \mathbf{c}) \\ \left(\frac{1}{4}\ \frac{1}{4}\ 0\right) & d\text{-glide}(\perp \mathbf{c}) \\ \left(\frac{1}{4}\ \frac{1}{4}\ \frac{1}{4}\right) & d\text{-glide(cubic/tetragonal systems only)} \\ \left(\frac{1}{2}\ 0\right), \left(0\ \frac{1}{2}\right) & g\text{-glide (2D)} \end{cases}$$

$$\mathbf{x} = -\underline{\mathcal{R}}_2 \mathbf{x}' + \mathbf{t}. \quad (7.81)$$

Each of the seven 3D crystal lattice systems intrinsically contain a minimal group of symmetry operations comprised of rotation axes and mirror planes oriented along or normal to standardized directions within the unit cell. This group is known as the minimal point group for the lattice (a formal mathematical group of operations containing an identity operation and satisfying closure, associativity, and invertability). Other supergroups of this minimal set with higher symmetries are also possible. In a scattering experiment, the intensity data $\propto |P(\mathbf{q})|^2$ have less symmetry than the crystalline lattice; the point group to which the intensity data belong is the Laue class of the crystal and is equal to that of the reciprocal lattice with the addition of an inversion center. Point groups are usually designated using the Hermann-Maugin notation, where for each lattice system there is a defined set of unique symmetry axes; the point group symbol is constructed by assigning the appropriate symmetry operator symbol for each axis. For example, in orthorhombic lattices the symmetry axes are 100, 010, and 001 (i.e. parallel to the $\mathbf{d}_{100}$, $\mathbf{d}_{010}$, $\mathbf{d}_{001}$ or $\mathbf{q}_{100}$, $\mathbf{q}_{010}$, $\mathbf{q}_{001}$ direct/reciprocal lattice vectors) and so the $m2m$ point group contains mirror planes perpendicular to the 001 and 100 directions and a two-fold rotation axis along the 010 axis. Point groups pertaining to each of the 12 lattice systems (one in 1D, four in 2D, seven in 3D) are tabulated in Table 7.5.

The point group of a crystal with the addition of the symmetry operations with translation (centered lattices, screw axes, and glide planes) forms the space group of the crystal. In 1D there two space groups, or "line groups" (centrosymmetric and non-centrosymmetric); in 2D there are 17 and are called either "plane" or "wallpaper" groups; in 3D there are 230. An exhaustive tabulation of the properties of these space groups appears in the International Tables for Crystallography, Vol. A1 [135]. A space group may be referred to either by its unique index as it appears in [135] or its Hermann-Maugin symbol, which is identical to that of the underlying point group with symbols for rotation axes and mirror planes substituted for screw axes and glides as needed, prepended by a letter corresponding to the lattice centering. For the 3D space groups the lattice centering indicator is uppercase whereas it is lowercase for the wallpaper groups. Thus a maximum of four symbols is needed to completely describe all of the symmetry operations within the unit cell. All unique combinations of these four operations form the group of symmetry operations, unique mappings of a point within the asymmetric unit to all other equivalent positions. A coordinate $\mathbf{x} = (x\,y\,z)$ is said to be a general position if all symmetry operations within the group except the identity transform $\mathbf{x} \to \mathbf{x}'$ such that $\mathbf{x} \neq \mathbf{x}'$. Certain positions within the cell may be unaltered by a non-trivial operation, for example $\mathbf{x} = (x, 0, 0)$ is invariant to any rotation axes along the 100 direction; such positions are called special positions.

For example, space group #230 is identified by the Hermann-Maugin symbol $Ia\bar{3}d$. Thus the underlying point group is $m\bar{3}m$, a cubic point group, and the "I" indicates that the lattice is body-centered. There are a-glides normal to the three principal axes, d-glides normal to the

Table 7.5 Point-group symmetries of the 12 crystal lattice systems in 1, 2, and 3 dimensions. 1- and 2D lattices are designated by parentheses. Centrosymmetric groups are shown in boldface.

Crystal system	Minimal symmetry	Point group symmetry	Laue symmetry	Symmetry directions[a]
Triclinic, (1D)	1	$1, \bar{1}$	$\bar{1}$	000
Monoclinic	2 or m	$2, m, \mathbf{2/m}$	$\mathbf{2/m}$	010 or 001
(Oblique)	1	$1, 2$	2	(Rotation)
Orthorhombic	222 or $mm2$[b]	$222, mm2$[b], **mmm**	**mmm**	$1°: 100$ $2°: 010$ $3°: 001$
(Rectangular)	$1m1$	$m, \mathbf{2mm}$	$\mathbf{2mm}$	$1°: $ (Rotation) $2°: 10$ $3°: 01$
Tetragonal	41 or $\bar{4}1$	$4, \bar{4}, \mathbf{4/m},$ $422, 4mm,$ $\bar{4}2m, \mathbf{4/mmm}$	$\mathbf{4/m},$ $\mathbf{4/mmm}$	$1°: 001$ $2°: 100, 010$ $3°: 110, 1\bar{1}0$
(Square)	411	$\mathbf{4}, \mathbf{4mm}$	$\mathbf{4}, \mathbf{4mm}$	$1°: $ (Rotation) $2°: 10, 01$ $3°: 11, 1\bar{1}$
Rhombohedral / Trigonal	31 or $\bar{3}1$	$3, \bar{3},$ $32, 3m, \mathbf{\bar{3}m}$	$\bar{3},$ $\bar{3}m$	$1°: 001$ $2°: 1\bar{1}0, 01\bar{1}, \bar{1}01$
Hexagonal	61 or $\bar{6}1$	$6, \bar{6}, \mathbf{6/m},$ $6mm, \bar{6}2m,$ $622, \mathbf{6/mmm}$	$\mathbf{6/m},$ $\mathbf{6/mmm}$	$1°: 001$ $2°: 100, 010, \bar{1}10$ $3°: 1\bar{1}0, 120, \bar{2}10$
(Hexagonal)	311 or 611	$3m, 31m, \mathbf{6},$ $\mathbf{6mm}$	$\mathbf{6},$ $\mathbf{6mm}$	$1°: $ (Rotation) $2°: 10, 01, 1\bar{1}$ $3°: 1\bar{1}, 12, \bar{2}1$
Cubic	231	$23, \mathbf{2/m\bar{3}},$ $432, \bar{4}3m, \mathbf{m\bar{3}m}$	$\mathbf{m\bar{3}},$ $\mathbf{m\bar{3}m}$	$1°: 100, 010, 001$ $2°: 111, \bar{1}11, 1\bar{1}1, \bar{1}\bar{1}1$ $3°: 1\bar{1}0, 110, 01\bar{1}, 011$

[a]The point group symmetry of each lattice is represented by one symmetry operator symbol for each unique axis. In groups where the operator is the identity the "1" is typically suppressed.
Example: The cubic point group $m\bar{3}m$ has three mirror planes on the faces of the unit cell (normal to 001, 010, and 100), four 3-fold rotoinversion axes along the cell diagonals, and four mirror planes normal to the face diagonals ($1\bar{1}0$, 110, $01\bar{1}$, and 011).
[b]Also $m2m, 2mm$.

face diagonals, and three-fold rotoinversion axes along the body diagonals. This information is sufficient to deduce the 96 general positions pertaining to the space group.

Space groups that contain symmetry operations with a translational element exhibit a phenomenon in scattering experiments known as systematic absences. While $L(\mathbf{q})$ enforces the Laue condition according to the lattice system, centered lattices, screw axes, and glide planes analytically extinguish certain scattered waves from within the form factor. To illustrate how this occurs, consider a crystal with space group $C2$. The complete set of general positions is:

$$\begin{array}{cc} (x,y,z) & (\bar{x},y,\bar{z}) \\ (x+\tfrac{1}{2},y+\tfrac{1}{2},z) & (\bar{x}+\tfrac{1}{2},\bar{y}+\tfrac{1}{2},\bar{z}) \end{array} \quad . \tag{7.82}$$

The form factor for this crystal may be simplified as follows:

$$P(\mathbf{q}_{hkl}) = \int_{\text{unit cell}} d\mathbf{r}\, b(\mathbf{r}) e^{-i\mathbf{q}_{hkl}\cdot \mathbf{r}} \tag{7.83}$$

$$= V \int_{\text{ASU}} d\mathbf{x}\, b(\mathbf{x}) \sum_{i\in \substack{\text{General}\\ \text{Positions}}} e^{-i(h\,k\,l)\cdot \mathbf{K}^T \cdot \mathbf{R}^T \cdot \mathbf{x}_i}$$

$$= V \int_{\text{ASU}} d\mathbf{x}\, b(\mathbf{x}) \\ \cdot \begin{pmatrix} e^{-2\pi i(h\,k\,l)\cdot (x\,y\,z)} + e^{-2\pi i(h\,k\,l)\cdot (\bar{x}\,y\,\bar{z})} + \\ e^{-2\pi i(h\,k\,l)\cdot (x+\tfrac{1}{2}\,y+\tfrac{1}{2}\,z)} + e^{-2\pi i(h\,k\,l)\cdot (\bar{x}+\tfrac{1}{2}\,\bar{y}+\tfrac{1}{2}\,\bar{z})} \end{pmatrix}$$

$$= V \int_{\text{ASU}} d\mathbf{x}\, b(\mathbf{x}) \\ \cdot \left(\left(1 + e^{-i\pi(h+k)}\right) \begin{pmatrix} e^{-2\pi i(h\,k\,l)\cdot (x\,y\,z)} \\ + e^{-2\pi i(h\,k\,l)\cdot (\bar{x}\,y\,\bar{z})} \end{pmatrix} \right)$$

$$= 0 \text{ for } h+k = \text{odd},$$

where $\int_{ASU}$ is the volume integral over the asymmetric unit of the dimensionless unit cell. This simple example can be extended to the 230 space groups to generate extinction rules, tabulated in Table 7.6, that can be used to unambiguously assign a space group to a specimen on the basis of peaks absent from the measured reciprocal space lattice.

Form factor extinctions

In addition to systematic extinctions which rigorously annihilate scattering peaks due to symmetry operations, the form factor may also cause the apparent disappearance of peaks otherwise allowed by the space group. Such extinctions depend on the contents of the asymmetric unit of the material, and accordingly will depend on the material composition as well as its space group. Awareness of these so-called form factor extinctions may prevent erroneous space group assignments or explain the absence of peaks otherwise expected of a material in a given phase.

As an illustrative example, both AB and ABC block copolymers in the lamellar phase will produce form factor extinctions when the volumetric block composition is perfectly symmetric. For these 1D structures with domain spacing a, the form factor amplitude becomes:

$$|P(\mathbf{q})|^2 = \left| \int_0^a dx\, b(r) e^{-2\pi i \tfrac{h}{a} x} \right|^2 = \left| \int_0^1 dx\, b(x) e^{-2\pi i h x} \right|^2 \propto I(\mathbf{q})\ . \tag{7.84}$$

Table 7.6 Systematic absences from lattice centerings and translation elements.

Symmetry element	Extinction condition	Symmetry element	Extinction condition	Symmetry element	Extinction condition
A	$k+l = $ odd	B	$h+l = $ odd	C	$h+k = $ odd
		F	h,k,l $\begin{Bmatrix}\text{mixed even}\\\text{and odd}\end{Bmatrix}$	I	$h+k+l = $ odd
$a \perp 010$	$k=0$, $h=$ odd	$b \perp 100$	$h=0$, $k=$ odd	$c \perp 100$	$h=0$, $l=$ odd
$a \perp 001$	$l=0$, $h=$ odd	$b \perp 001$	$l=0$, $k=$ odd	$c \perp 010$	$k=0$, $l=$ odd
		$c \perp 110^a$	$h+k=0$, $l=$ odd	$c \perp 1\bar{1}0^{a,b}$	$h=k$, $l=$ odd
$n \perp 100$	$h=0$, $k+l=$ odd	$n \perp 010$	$k=0$, $h+l=$ odd	$n \perp 001$	$l=0$, $h+k=$ odd
$d \perp 100$	$h=0$, $k+l \mod 4 \neq 0$	$d \perp 010$	$k=0$, $h+l \mod 4 \neq 0$	$d \perp 001$	$l=0$, $h+k \mod 4 \neq 0$
		$n \perp 1\bar{1}0^b$	$h=k$, $2h+l=$ odd	$d \perp 1\bar{1}0^b$	$h=k$, $2h+l \mod 4 \neq 0$
$2_1, 4_2 \parallel 100$	$k=l=0$, $h=$ odd	$4_1, 4_3 \parallel 100$	$k=l=0$, $h \mod 4 \neq 0$	$2_1 \parallel 010$	$h=l=0$, $k=$ odd
		$2_1, 4_2, 6_3 \parallel 001$	$h=k=0$, $l=$ odd	$4_1, 4_3 \parallel 001$	$h=k=0$, $l \mod 4 \neq 0$
		$3_1, 3_2, 6_2, 6_4 \parallel 001$	$h=k=0$, $l \mod 3 \neq 0$	$6_1, 6_5 \parallel 001$	$h=k=0$, $l \mod 6 \neq 0$

[a] Hexagonal Lattices.
[b] Rhombohedral, Tetragonal, and Cubic lattices.

Lamellae are centrosymmetric so this reduces to

$$|P(\mathbf{q})|^2 = \left| \int_0^{\frac{1}{2}} dx\, b(x) \left[\cos(2\pi h x) + \cos(-2\pi h x) \right] \right|^2 . \qquad (7.85)$$

For strongly segregated, compositionally symmetric lamellae, the scattering length density $b(x)$ may be approximated as a square wave:

$$\underbrace{b(x) = \begin{cases} b_A & x \in [0, \frac{1}{4}] \\ b_B & x \in (\frac{1}{4}, \frac{1}{2}] \end{cases}}_{\text{AB lamellae}} \quad \text{or} \quad \underbrace{b(x) = \begin{cases} b_A & x \in [0, \frac{1}{6}] \\ b_B & x \in (\frac{1}{6}, \frac{1}{3}] \\ b_C & x \in (\frac{1}{3}, \frac{1}{2}] \end{cases}}_{\text{ABC lamellae}} . \quad (7.86)$$

With these definitions, equation (7.85) may be integrated analytically to yield:

$$|P(\mathbf{q})|^2 \propto I(\mathbf{q}) \propto \begin{cases} \sin^2 \frac{h\pi}{2} & \text{AB lamellae} \\ \sin^2 \frac{h\pi}{3} & \text{ABC lamellae} \end{cases} . \qquad (7.87)$$

which indicates that even Miller indices will be extinct for compositionally symmetric AB lamellae and indices divisible by three will vanish for compositionally symmetric ABC lamellae (cf. Figure 7.38 on page 313.)

The Ewald sphere in small-angle scattering

The physics of scattering/diffraction is no different for small molecules in which atomic resolution is sought versus small-angle experiments where the pertinent length scales are 1–2 orders of magnitude greater. An important practical distinction between crystallography and small-angle scattering arises that is responsible for the very different instrumentation used for each of the techniques. To illustrate, consider a perfect cubic crystalline lattice illuminated with $\lambda = 1.54$Å radiation (Cu-K$_\alpha$ x-rays) with $a = b = c \approx 3.3$ Å and $a^* = 2\pi/\lambda \approx 1.9$Å^{-1}, oriented such that the $\mathbf{q}_{100}$ lattice vector is colinear with the incident wavevector $\mathbf{k}_{in}$. Ewald's sphere, or the sphere of reflection, is the sphere whose center corresponds to the origin of the incident wavevector (of modulus $2\pi/\lambda$) placed such that it terminates at the origin of the reciprocal lattice, $\mathbf{q}_{000}$, as illustrated in Figure 7.37(b). Constructed in this manner, Ewald's sphere represents the domain of elastically scattered waves; only reciprocal lattice points that intersect the sphere will generate a reflection. As Figure 7.37(a) shows, in this example diffraction will only be measurable at four discrete reciprocal lattice points corresponding to $\mathbf{q}_{\bar{3}20}$, $\mathbf{q}_{\bar{3}\bar{2}0}$, $\mathbf{q}_{\bar{3}02}$, and $\mathbf{q}_{\bar{3}0\bar{2}}$ irrespective of the orientation of the detector relative to the crystal. The other lattice points are observable only by altering the orientation of the crystal relative to the incident beam.

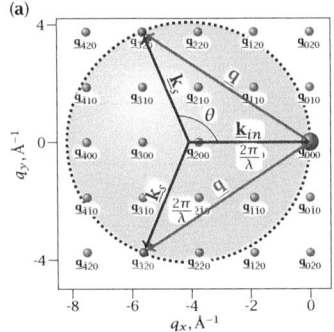

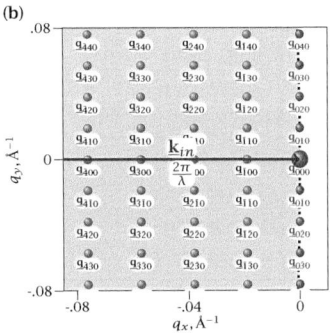

Fig. 7.37 Ewald sphere for: (a) small-molecule crystallography vs (b) small-angle scattering. Only reciprocal lattice points are measureable given the beam orientation.

[4]Effectively, all experimentally detectable reflections in the yz plane may be recorded simultaneously without reorienting the specimen in the beam.

Consider on the other hand, a single crystal of ordered block copolymer with cubic symmetry whose unit cell dimension is $a = 33$nm. The Ewald sphere is shown again for this system in Figure 7.37(b); its dimensions are precisely the same as in the former scenario, but now the reciprocal lattice points are much more densely packed such that the surface of the sphere is nearly planar for many of the lower-order lattice points in the yz plane.[4]

Scattering from ordered block copolymer melts

Small-angle scattering has been an indispensable tool for the characterization of the phase behavior of block copolymer melts. For diblock copolymers, the use of SAS for the assignment of morphology and finding the thermal transitions between various phases is usually straightforward. The vast majority of such specimens form only one of four equilibrium phases: body-centered spheres ($Im\bar{3}m$), hexagonally packed cylinders ($p6mm$), double gyroid ($Ia\bar{3}d$), or lamellae ($p\bar{1}$). In a typical scenario, the polymer is ordered yet polycrystalline with a nearly isotropic grain distribution. Figure 7.38(a) shows a SAXS pattern characteristic of such a material, a compositionally symmetric poly(isoprene-b-styrene-b-ethylene oxide) (ISO) [26, 136]. This sample scatters nearly isotropically and displays four distinct Bragg rings whose intensity varies only slightly in the azimuthal direction, and accordingly the structural information provided by this experiment resides largely in the 1-D function $I(|\mathbf{q}|)$. Figure 7.38(b) presents the same data in this fashion, obtained by numerically evaluating the integral $I(q) = \int_0^{2\pi} d\theta\, qI(q\cos\theta, q\sin\theta)$. The conclusion that this specimen assumes the lamellar morphology may be reached through analysis of the locations of the peaks in the 1D spectrum. Let the scattering vector modulus $q^* \approx 0.0287 \text{Å}^{-1}$ denote the center of the lowest-order observable peak in the spectrum; evidently the locations of the remaining peaks are $2q^*$, $3q^*$, $4q^*$, and $5q^*$. Assignment of the morphology in this case reduces to determining the Bravais lattice that allows Bragg peaks at these locations in conjunction with the identification of any systematically absent peaks to determine the space group symmetry. Referring to Table 7.7, the assignment of lamellae is unambiguous. Unambiguous assignment of morphology is in general possible from 1D spectra for well-resolved cubic, hexagonal, and 1D structures.

The assignment of morphology from scattering alone is not always possible, however, particularly in polymers with poorly formed long-range order, those that form structures depart from the four classical diblock copolymer phases, or specimens that form recently- or un-discovered phases that have yet to be fully characterized. Orthorhombic network morphologies such as $Fddd$ (also known as O^{70}) or $Pnna$ (O^{52}), for example, are prevalent in a number triblock copolymer systems [26, 32] and over a small region of phase space even in diblock copolymer systems [27, 28, 137]. 1D scattering spectra in such morphologies are insufficient to deduce the space-group symmetry or even the Bravais lattice due to the presence of multiple length scales in the lattice constants.[5] Figure 7.39(c), for example, shows a 1D SAXS spectrum characteristic of the $Fddd$ network morphology. The 1D pattern is complex, and moreover a number of distinct Bragg peaks are indistinguishable from each other, as elucidated by the original 2D spectrum in polar coordinates.

[5]In the orthorhombic network, assuming $a = b/2 = c\sqrt{3}/6$ as in [26]; actual lattice constants may differ slightly depending on the material and experimental conditions.

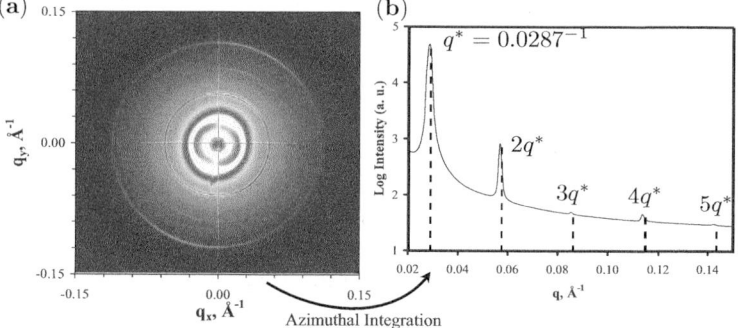

Fig. 7.38 SAXS of a polycrystalline poly(isoprene-b-styrene-b-ethylene oxide) triblock copolymer, ISO-3l from [26, 136] (compositionally symmetric, 3-domain lamellae). (a) Raw intensity data as collected by the detector. (b) 1D SAXS spectrum obtained by azimuthal integration of the 2D data in (a).

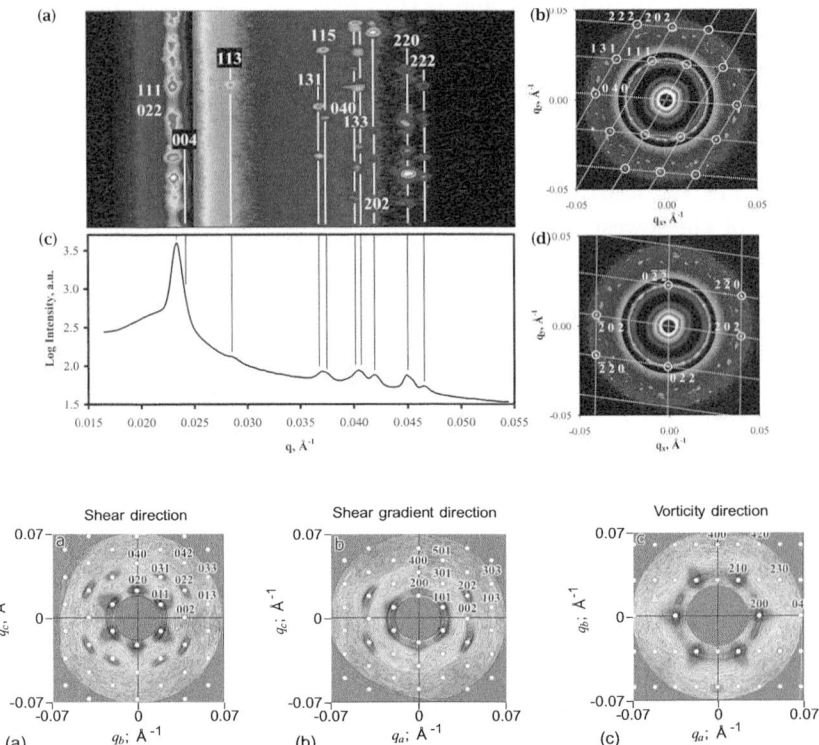

Fig. 7.39 SAXS of a polycrystalline poly(isoprene-b-styrene-b-ethylene oxide) triblock copolymer, ISO-3l from [26, 136] (compositionally symmetric, 3-domain lamellae). (a) Raw intensity data as collected by the detector. (b) 1D SAXS spectrum obtained by azimuthal integration of the 2D data in (a).

Fig. 7.40 SAXS images of single-crystalline poly(cyclohexylethylene-b-ethylene-b-ethyl ethylene) collected from three orthogonal axes: (a) the shear direction, (b) the shear gradient direction, and (c) the vorticity direction. Data from [32].

Scattering from block copolymers under externally applied fields: Quasi-single crystalline behavior.

As discussed earlier, the application of externally applied fields such as mechanical (e.g. shear), electronic, and even magnetic (in partially liquid-crystalline systems) to block copolymer melts and thin films is well known to produce materials nearly single-crystalline in character. The utility of SAS experiments with such materials is extended beyond the characterization of morphology and thermodynamic phase boundaries to include *in situ* characterization of dynamic processes such as mechanical deformation [138, 139] or nucleation and growth of ordered phases [86, 140].

In the absence of such fields, an isotropic grain distribution leading to azimuthally symmetric scattering patterns is the standard result from an ordered block copolymer melt; only the scattering vector moduli may be recovered, and an assignment of the space group must be unambiguous based on these data alone. While in practice this is often sufficient, newly discovered or as-of-yet undiscovered phases with multiple lattice constants or systematic extinctions require further information. For this reason the single-crystalline nature of polymers aligned through external fields may be highly desirable. In their study of polyolefin-based triblock copolymers, for instance, Cochran and Bates used large amplitude reciprocating shear to produce a single-crystalline poly(cyclohexylethylene-

Phase	Space group	Peaks
Lamellae	$p\bar{1}$	$1, 2, 3, 4\ldots$
Hex	$p6mm$	$\underbrace{1}_{q_{10}}, \underbrace{\sqrt{3}}_{q_{11}}, \underbrace{\sqrt{4}}_{q_{20}}, \underbrace{\sqrt{7}}_{q_{21}}, \underbrace{\sqrt{9}}_{q_{30}}, \underbrace{\sqrt{12}}_{q_{22}}, \ldots$
bcc spheres (Q^{229})	$Im\bar{3}m$	$\underbrace{\sqrt{2}}_{q^*=q_{110}}, \underbrace{\sqrt{4}}_{q_{200}}, \underbrace{\sqrt{6}}_{q_{211}}, \underbrace{\sqrt{8}}_{q_{220}}, \underbrace{\sqrt{10}}_{q_{310}}, \underbrace{\sqrt{12}}_{q_{222}}, \ldots$
Double gyroid (Q^{230})	$Ia\bar{3}d$	$\underbrace{\sqrt{6}}_{q^*=q_{211}}, \underbrace{\sqrt{8}}_{q_{220}}, \underbrace{\sqrt{14}}_{q_{321}}, \underbrace{\sqrt{16}}_{q_{400}}, \underbrace{\sqrt{20}}_{q_{420}}, \underbrace{\sqrt{22}}_{q_{332}}, \ldots$
Single gyroid (Q^{214})	$I4_132$	$\underbrace{\sqrt{2}}_{q^*=q_{110}}, \underbrace{\sqrt{6}}_{q_{211}}, \underbrace{\sqrt{8}}_{q_{220}}, \underbrace{\sqrt{10}}_{q_{310}}, \underbrace{\sqrt{12}}_{q_{222}}, \underbrace{\sqrt{14}}_{q_{321}}, \ldots$
Orthorhombic network (O^{70})	$Fddd$	$\underbrace{\sqrt{\tfrac{16}{12}}}_{q^*=q_{004},q_{111},q_{022}}, \underbrace{\sqrt{\tfrac{24}{12}}}_{q_{113}}, \underbrace{\sqrt{\tfrac{40}{12}}}_{q_{131},q_{115}}, \underbrace{\sqrt{\tfrac{48}{12}}}_{q_{040},q_{133}}, \underbrace{\sqrt{\tfrac{52}{12}}}_{q_{202}}, \underbrace{\sqrt{\tfrac{60}{12}}}_{q_{220}}, \underbrace{\sqrt{\tfrac{64}{12}}}_{q_{135},q_{044},q_{222}}, \underbrace{\sqrt{\tfrac{76}{12}}}_{q_{224}} \ldots$

Table 7.7 Scattering positions for commonly observed block copolymer morphologies[5]

b-ethylene-b-ethyl ethylene), Figure 7.40. The sheared polymer was sectioned into cubes allowing SAXS patterns to be collected after further annealing along each of the three unique shear axes. The resultant patterns exhibited systematic extinctions that allowed the unique assignment of the $Pnna$ space group. However, in this case it was discovered that the shear itself had induced an order-to-order transition from the equilibrium $Fddd$ network structure to the $Pnna$ morphology; these results accentuate the care that must be taken in the assessment of block copolymer phase behavior.

As another example, Figure 7.41 reproduces SAXS data from work conducted by Bang et al. which elucidates the mechanism responsible for an $fcc \rightarrow bcc$ transformation in a sphere-forming poly(styrene-b-isoprene) solution. Here the researchers modified the shear cell of a standard rheometer to allow the passage of the x-ray beam through not only the edge of the sample (yielding scattering perpendicular to the direction of flow), but also in the radial direction, giving access to the structure of the material normal to the vorticity direction. Such access to complementary views of reciprocal-space yielded not only an unambiguous assessment of the space group but also the orientation of the unit cell with respect to the flow coordinate system. Real-time monitoring of the morphology provided concrete evidence of the epitaxial transition between the two observed states in this system.

Lim et al. used a combination wide-angle x-ray scattering (WAXS) and SAXS to obtain a complementary view of crystalline and block copolymer morphology in their determination of structure-property relationships and deformation mechanisms in triblock and pentablock copolymers comprised of poly(cyclohexylethylene) (C) and polyethylene (E),

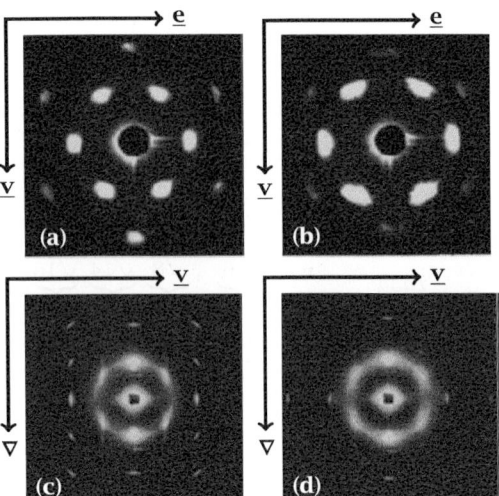

Fig. 7.41 SAXS patterns from [124] of a 30 wt% solution of poly(styrene-b-isoprene) copolymer in diethyl phthalate. Patterns (a) and (c) show that the solution self-assembles into spheres on an $fcc + hcp$ lattice under steady shear in a parallel plate configuration at 30 °C, while patterns collected after shearing at 40 °C show that the structure spontaneously transforms to bcc. **v** denotes the flow direction, **e** the vorticity direction, and ∇ the shear gradient direction.

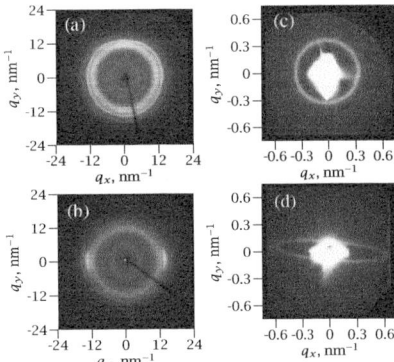

Fig. 7.42 (a,b): 2D WAXS and (c,d): SAXS images from a CECEC pentablock copolymer comprised of poly(cyclohexyl ethylene) (C) and polyethylene (E) blocks (a,c) prior to and after (b,d) *in situ* tensile elongation. Reproduced from [139]

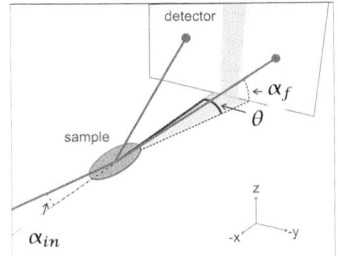

Fig. 7.43 Typical GI-SAXS configuration. Reproduced from [141].

i.e. CEC and CECEC [139]. In this work, all of the copolymers formed lamellae with an isotropic polyethylene crystallite distribution, Figure 7.42(a) (WAXS), and block copolymer grain distribution, Figure 7.42(c) (SAXS). Samples were subjected to tensile elongation, and then scattering showed how both the crystalline and block copolymer domains responded to the deformation, i.e. the elongation/reorientation of the lamellae and alignment of the polyethylene chains.

Scattering from block copolymer thin films: Grazing-incidence small-angle X-ray scattering (GI-SAXS)

Grazing-incidence small-angle x-ray scattering (GI-SAXS) is a relatively new technique first developed by Levine *et al.* in 1989 as a method for studying the growth of thin gold films on glass substrates [142]. Over the past decade, the technique has become an increasingly common method for the characterization of block copolymer thin films, i.e. films of less than ≈ 100 nm deposited on smooth planar substrates. x-rays become surface sensitive when incident upon the film at an angle near its critical angle, α_c, at which the phenomenon of total external reflection occurs. In contrast to lower energy electromagnetic radiation, the refractive index of materials with respect to x-rays is less than unity, i.e. $n = 1 - \delta(+i\beta); \delta > 0$. It can be shown that when the incident angle $\alpha_i < \sqrt{2\delta} = \alpha_c \propto \lambda\sqrt{Z}$, 100% of the incident radiation is reflected at angle equal to the incident angle, provided that the imaginary component β of the refractive index is 0 and the surface is perfectly smooth. In practice neither condition is met perfectly and a fraction of the radiation penetrates a short depth, from 10's of Å for $\alpha_i < \alpha_c$ to 1000's of Å for α_i near and even slightly above α_c.

In GI-SAXS of polymer thin films, the appropriate range of penetration depths may typically be achieved with incident angles ranging from about half of the critical angle to a few multiples thereof, where the

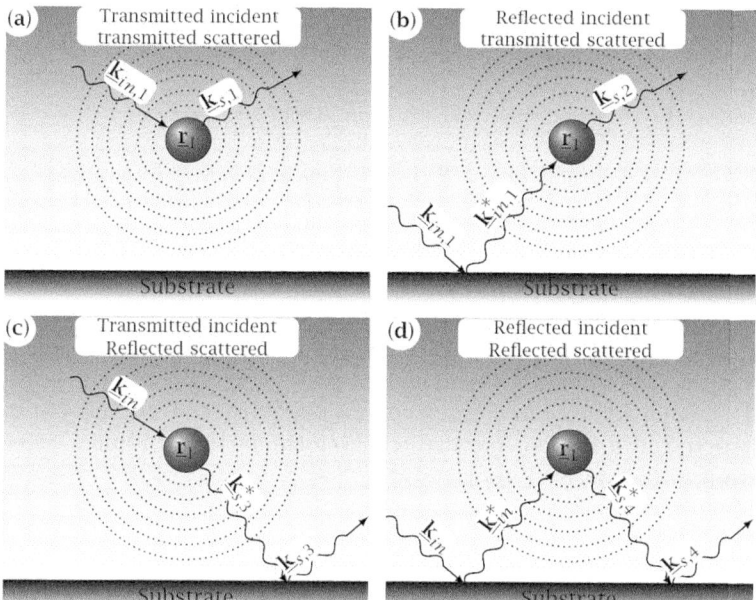

Fig. 7.44 Scattering events in GI-SAXS, where the incident and/or scattered wave may be either transmitted through or reflected (by the substrate) back into the polymer film.

critical angle is on the order of 0.1°. The configuration of a typical experimental setup appears in Figure 7.43. Perfect Bragg scattering from structural features of the polymer film occurs due to inhomogeneities in the lateral (xy-plane) directions only, and thus a limitation of GI-SAXS is that the structural details in the z-direction are more difficult to deduce and may need to be complemented with other techniques. Due to the Z-dependence of α_c, for polymer thin films $\alpha_{c,polymer} < \alpha_{c,substrate}$ for essentially all commonly employed substrate materials, e.g. silicon wafers. When the penetration depth exceeds the film thickness, both the incident and scattered x-rays may potentially be reflected from the substrate, which complicates the analysis of the scattered radiation. A common simplifying assumption known as the kinematic approximation states that each x-ray is scattered by at most a single scattering center; each incident ray has the possibility of reflecting off of the substrate prior to scattering, and each scattered ray may also potentially be reflected. The so-called distorted wave Born approximation (DWBA) treats the total scattering as the superposition of scattering from two potentials such that the total scattered intensity from reflection and transmittance events are simply additive. Under the kinematic approximation there are thus four possibilities for the scattered wave as depicted in Figure 7.44.

The incident and scattered wavevectors are:

$$\mathbf{k}_{in}^{t} = \begin{pmatrix} \cos\alpha_{in} \\ 0 \\ -\sqrt{n^2 - \cos^2\alpha_{in}} \end{pmatrix} \quad \mathbf{k}_{f}^{t} = \begin{pmatrix} \cos\alpha_{f}\cos\theta \\ \cos\alpha_{f}\sin\theta \\ \sqrt{n^2 - \cos^2\alpha_{f}} \end{pmatrix} \quad (7.88)$$

$$\mathbf{k}_{in}^{r} = \begin{pmatrix} \cos\alpha_{in} \\ 0 \\ \sqrt{n^2 - \cos^2\alpha_{in}} \end{pmatrix} \quad \mathbf{k}_{f}^{r} = \begin{pmatrix} \cos\alpha_{f}\cos\theta \\ \cos\alpha_{f}\sin\theta \\ -\sqrt{n^2 - \cos^2\alpha_{f}} \end{pmatrix} \quad (7.89)$$

where the superscripts t and r refer to transmitted and reflected waves, respectively. The scattered intensity will be a superposition of the appropriately weighted combination of all four modes of scattering from the substrate – dependent on the fours scattering vectors $\mathbf{q}^{ij} \equiv \mathbf{k}_s^i - \mathbf{k}_i^j; i,j \in \{t,r\}$. For ordered thin films, the DWBA scattered intensity may be constructed in a fashion parallel to the formulation of Laue scattering, equation (7.66) derived above:

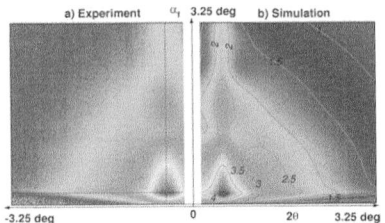

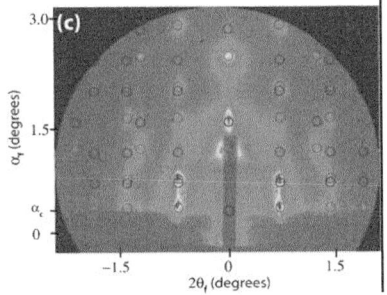

Fig. 7.45 (a): Experimental GI-SAXS pattern collected from gold sputtered on a TiO$_2$ substrate. (b): *IsGISAXS* simulation of the same. Both images are plotted using the same logarithmic color axes. Reproduced from [143].

$$I(\theta, \alpha_f) \propto \left| \sum_{i,j \in \{t,r\}} A_{in}^i A_f^j P(\mathbf{q}^{ij}) \mathcal{L}(\mathbf{q}^{ij}) \right|^2, \quad (7.90)$$

where $A_{in,s}^{t,r}$ represents the amplitude of the transmitted/reflected incident/scattered wave, $P(\mathbf{q}^{ij})$ is the form factor of the unit cell as usual, and

$$\mathcal{L}(\mathbf{q}^{ij}) = \sum_{h=-\infty}^{\infty} e^{-i\mathbf{q}^{ij}\cdot\mathbf{d}_{h00}} \sum_{k=-\infty}^{\infty} e^{-i\mathbf{q}^{ij}\cdot\mathbf{d}_{0k0}} \sum_{l=0}^{N} e^{-i\mathbf{q}^{ij}\cdot\mathbf{d}_{00l}}, \quad (7.91)$$

i.e. the structure factor analogous to which leads to the Laue condition for bulk scattering, differing in that a finite N unit cells are stacked in the z direction. Consequently $\mathcal{L}(q_{hkl}^{ij}) = 0$ when h or k are non-integers, while any value of l leads to measurable scattering. Models of $A(\mathbf{q})$ for transmitted/reflected incident/scattered waves may be constructed from the dynamical theory of Parratt [145, 146], or solely for the purpose of calculating peak positions, may be set to unity [144].

Fig. 7.46 GI-SAXS pattern of a rhombohedral R$\bar{3}$m nanoporous thin film with DWBA-predicted peak locations from both transmitted and reflected beams calculated using NANOCELL. Reproduced from [144]

A number of software tools have been developed over the past decade to assist researchers in the prediction and analysis of GI-SAXS patterns.

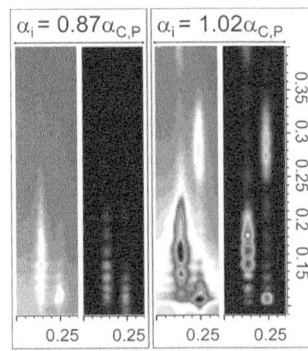
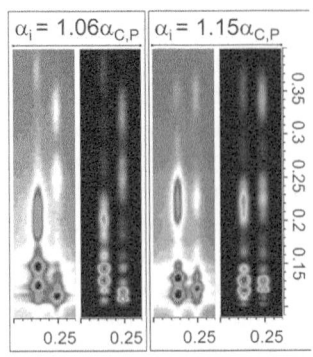

Fig. 7.47 GI-SAXS patterns (shaded) of sphere-forming poly(styrene-b-vinyl pyridine) diblock copolymers at various incident angles, compared side-by-side with DWBA predictions (black/white). $\alpha_{C,P}$ is the critical angle of the polymer thin film, which is 5 unit cells thick. Reproduced from [141].

In 2002 Lazzari described a Fortran 90 code called *IsGISAXS* which implements the DWBA to treat arrays of ordered and disordered islands on a surface, Figure 7.45 [143]. Lee *et al.* presented calculations and data for GI-SAXS of ordered diblock copolymer thin films [147]. Tate *et al.* described and made available (by individual request) software written in Mathematica ® called **NANOCELL** which computes Bragg peak locations for ordered crystalline films with arbitrary space group and incident angle, Figure 7.46 [144]. Stein *et al.* used the DWBA to determine the lattice spacing and packing symmetry of multilayered poly(styrene-*b*-vinyl pyridine) thin films, Figure 7.47 [141].

Scattering from non-periodic polymers

Block copolymers in solution

An important application of small-angle scattering arises in systems comprised of a suspension of micelles in solution. There is no translational order in such a system; however, the shape of the micelle has an associated form factor which may be used to analyze scattering data to yield average characteristic dimensions and to validate that the proposed shape of the micelle is in fact the correct one. Since the system is not periodic, there is no Laue condition, and equation (7.66) reduces to $I(\mathbf{q}) \propto |P(\mathbf{q})S(\mathbf{q})|^2$, where $P(\mathbf{q})$ is the form factor describing the scattering from a single micelle and the structure factor $S(\mathbf{q})$ accounts for particle-particle correlations. The structure factor approaches unity in the limit of dilute systems since interparticle correlations disappear.

The field of amphiphilic block copolymers in water has relied heavily on such techniques. Here SANS becomes particularly important due to the ability to alter the scattering contrast between the polymeric components and the solvent. A typical amphiphilic system is comprised of a solvent (s, e.g. water), a miscible block (A), and an immiscible block (B). The form factor for a micelle in such a system may be constructed by first proposing a function that describes the scattering length density $b(\mathbf{r})$ within the micelle. B blocks collapse into solid domains, and may modeled as having constant b_B within their domain. A blocks extend into the solvent domain as a polymer brush, and thus a model appropriate to describing the segment density distribution thereof, e.g. the Fermi-Dirac distribution [148, 149], should be employed for solvated domains. As an example, consider a spherical micelle with a "B" interior of radius R_B and an "A" exterior. The scattering length density function may be written as:

$$b(\mathbf{r}) = b(r) = \begin{cases} b_B & r < R_B \\ b_A \left[\exp\left(\frac{r-R_B-l_0}{s}\right)\right]^{-1} & r > R_B \text{ (Fermi-Dirac)} \end{cases} \quad (7.92)$$

where b_A and b_B refer to the bulk scattering length density of A and B homopolymer and l_0/s are adjustable parameters that controls the curvature/skewness of the Fermi-Dirac distribution. The form factor is simply the Fourier transform of equation (7.92):

$$P(\mathbf{q}) = \int_V d\mathbf{r}\, b(\mathbf{r}) e^{-i\mathbf{q}\cdot\mathbf{r}} \qquad (7.93)$$

and the intensity from a scattering experiment with perfectly monodisperse micelles from an instrument with perfect q-resolution is simply $I(\mathbf{q}) = \frac{n}{V^t}|P(\mathbf{q})|^2$, where $\frac{n}{V^t}$ is the number of micelles per volume. Polydispersity effects may be accounted for by treating the total observed intensity as the superposition of intensities from a mixture of micelles whose concentrations are governed by a distribution function $P(R_B)$ (e.g. Won et al. [149] fit SANS data using log-normal distribution function) such that:

$$I^{\text{polydisperse}}(\mathbf{q}) = \int_0^\infty dR_B\, I(\mathbf{q}, R_B) P(R_B). \qquad (7.94)$$

Detector resolution limits may also be accounted for as described by Pedersen et al. [150], by introducing another distribution function $P(\mathbf{q}, \mathbf{q}') = \frac{1}{\sqrt{2\pi}\sigma} \exp\left[-\frac{1}{2}\left(\frac{|\mathbf{q}|-|\mathbf{q}'|}{\sigma}\right)\right] d\mathbf{q}'$ which describes the fraction of scattering events at $\mathbf{q}'$ detected at $\mathbf{q}$, such that:

$$I^{expt}(\mathbf{q}) = \int d\mathbf{q}'\, I^{\text{polydisperse}}(\mathbf{q}') P(\mathbf{q}, \mathbf{q}'). \qquad (7.95)$$

The integrals of equations (7.93)-(7.95) may be evaluated numerically subject to the collection of adjustable parameters and the application of non-linear regression to fit equation (7.95) to experimental data yields the size distribution information to the micelles. It is straightforward to adapt this procedure to other solvated structures through the construction of the segment density function analogous to equation (7.92).

Scattering from disordered block copolymer melts via the Random Phase Approximation.

Leibler's 1980 paper [3] treating the Random Phase Approximation (RPA) of de Gennes [4] to diblock copolymer melts is one of the most frequently cited in the field; of interest to the present topic is the derivation of the structure factor $S(q)$ for a disordered block copolymer melt, equation (7.11), plotted in Figure 7.2. Scattering in disordered block copolymers arises from the so-called "correlation-hole effect", where the local polymer composition is homogeneous but density fluctuations occur over a length scale dictated by the polymer composition, molecular weight, and degree of segregation.

The RPA is easily extended to arbitrary chain architectures. In 2003 Cochran et al. illustrated the use of the RPA in conjunction with experimentally measured $\chi_{ij}(T)$ correlations to estimate the order-to-disorder transition temperature of linear block copolymers by calculating the spinodal temperature T_s at which $S(q^*) \to \infty$ at some critical wavelength corresponding to q^* [151]. These authors also showed the utility

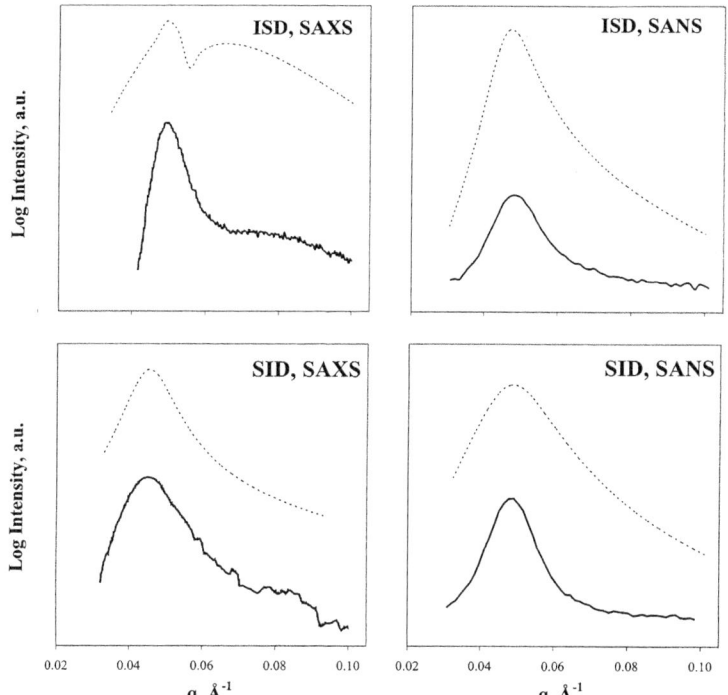

Fig. 7.48 Experimental and RPA scattering from disordered melts of poly(styrene-b-isoprene-b-dimethylsiloxane) (SID), ISD, poly(isoprene-b-styrene-b-ethylene oxide) (ISO), and ISO. Reproduced from [151].

of the RPA structure factor in correctly reproducing the scattering from disordered multiblock copolymer melts, Figure 7.48, even capturing the seemingly anomalous behavior of poly(isoprene-b-styrene-b-dimethylsiloxane) which exhibits two maxima in $I(q)$ in its disordered state.

Here we present a further generalization that allows the application of the RPA to polymers of arbitrary chain architectures, including hyperbranched structures such as dendrimers. The RPA computes the static structure factor by evaluating the linear response of a non-interacting homogeneous polymer melt to a small external potential that introduces segment-segment interactions and incompressibility. Consider the polymer to be comprised of n chemically distinct block types; there may be an arbitrary number of each block and blocks may be interconnected in any fashion so long as there are no loops, e.g. structures such as those illustrated in Figure 7.49. As Figure 7.49 depicts, the chain architecture may be redrawn as a directed acyclic graph. The chain is broken into "levels", where the root level, "Level 0", contains exactly one unique block and may be attached to an arbitrary number of children, which are designated to be in "Level 1". In this manner a block at level l may be uniquely indexed with a multiplet $\mathcal{M} \equiv m_0 m_1 \cdots m_l$, i.e. every block is child m_l of its parent block $m_0 m_1 \cdots m_{l-1}$. The polymer chain is divided into statistical segments of volume v_0. A block of molar mass $M_\mathcal{M}$ and mass density $\rho_\mathcal{M}$ occupies $N_\mathcal{M} = \frac{M_\mathcal{M}}{v_0 \rho_\mathcal{M} N_{av}}$ such sites, where N_{av} is Avogadro's number. The statistical segment length $a_\mathcal{M}$ is de-

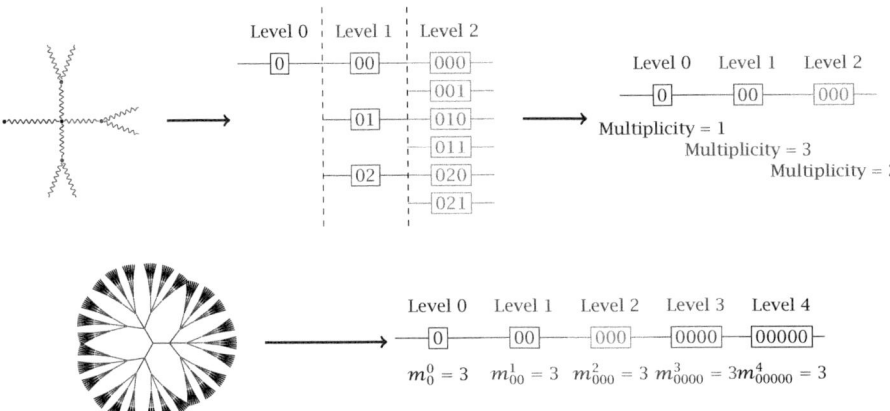

Fig. 7.49 Conceptualization of how to describe an arbitrarily branched block copolymer chain architecture using a tree diagram. Identical branches may be further simplified by tracking their multiplicity at each level of the tree. In this manner even a 5th generation dendrimer becomes tractable.

fined such that $R_{g,\mathcal{M}}^2 = \frac{N_\mathcal{M} a_\mathcal{M}^2}{6}$, i.e. $a_\mathcal{M}^2 = \frac{M_\mathcal{M}}{N_\mathcal{M}} \left(\frac{h_0^2}{M}\right)_\mathcal{M}$, where the ratio $\left(\frac{h_0^2}{M}\right)_\mathcal{M}$ is the unperturbed end-to-end distance squared per molar mass for the block (tabulated for many common polymers in the *Polymer Handbook* [152] and by Fetters [153]) and $R_{g,\mathcal{M}}$ is the radius of gyration for block $\mathcal{M}$. Let $\varrho_i(\mathbf{r})$ denote the microscopic number density for segments of type i, and $\boldsymbol{\varrho}$ be the vector of such operators. In the disordered state, the thermal average of the microscopic density is $f_i = \langle \varrho \mathbf{r} \rangle = \frac{N_i}{N}$, where N is the total number of segments in the polymer and N_i is the number of segments of type i. Define an order parameter:

$$\psi(\mathbf{r}) = \langle \varrho(\mathbf{r}) - \langle \varrho(\mathbf{r}) \rangle \rangle . \qquad (7.96)$$

$\psi(\mathbf{r}) = 0$ in the disordered (homogeneous) state, and varies periodically for ordered phases. To calculate the structure factor of the homogeneous melt, we first find the linear response function $R_{ij}(\mathbf{q})$ that describes the change in $\psi_i(\mathbf{r})$ upon the application of a small potential $\boldsymbol{\mu}^{ext}(\mathbf{r})$ and then invoke the fluctuation-dissipation theorem which states that $\underline{\mathbf{S}}(\mathbf{q}) = k_B T \underline{\mathbf{R}}(\mathbf{q})$. Define $\underline{\mathbf{R}}(\mathbf{q})$ such that it satisfies

$$\psi(\mathbf{r}) = -\int d\mathbf{r}' \, \underline{\mathbf{R}}(\mathbf{r} - \mathbf{r}')\boldsymbol{\mu}^{ext}(\mathbf{r}'), \text{ or in Fourier-space} \qquad (7.97)$$

$$\psi(\mathbf{q}) = \underline{\mathbf{R}}(\mathbf{q}) \cdot \boldsymbol{\mu}^{ext}(\mathbf{q}). \qquad (7.98)$$

The potential $\boldsymbol{\mu}^{\text{ext}}(\mathbf{q})$ is now replaced with an effective potential $\boldsymbol{\mu}^{\text{eff}}(\mathbf{q})$ that adds two terms: one to account for (mean-field) binary interactions χ_{ij} between dissimilar segments and another to enforce the constraint of incompressibility:

$$\boldsymbol{\mu}^{\text{eff}} = \boldsymbol{\mu}^{\text{ext}}(\mathbf{q}) + k_B T \underline{\chi}(T)\psi(\mathbf{q}) + \lambda\boldsymbol{\epsilon} , \qquad (7.99)$$

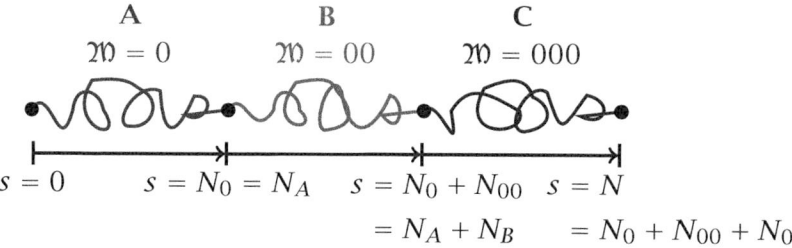

Fig. 7.50 Chain contour coordinates of a linear ABC triblock copolymer.

where λ is a Lagrange multiplier and ϵ is a vector of 1's of length n. Lambda is chosen to satisfy $\epsilon \cdot \psi(\mathbf{q}) = 0$:

$$0 = \epsilon \cdot \psi(\mathbf{q}) = \epsilon \cdot \underline{\mathbf{R}} \cdot \boldsymbol{\mu}^{\text{ext}} + k_B T \underline{\mathbf{R}} \cdot \underline{\chi} \cdot \psi + \lambda \epsilon \cdot \underline{\mathbf{R}} \cdot \epsilon \quad (7.100)$$

$$\lambda = -k_B T \frac{\epsilon \cdot \underline{\mathbf{R}}(\boldsymbol{\mu}^{\text{ext}} \cdot \underline{\chi} \cdot \psi)}{\epsilon \cdot \underline{\mathbf{R}} \cdot \epsilon} . \quad (7.101)$$

Substitution of equation (7.100) into (7.99), and finally into (7.97) yields an expression for the response function:

$$\underline{\mathbf{R}}(\mathbf{q}) = \left(\underline{\widetilde{\mathbf{R}}}^{-1}(\mathbf{q}) + k_B T \underline{\chi}\right)^{-1} ; \quad \underline{\mathbf{S}}(\mathbf{q}) = \left(\underline{\widetilde{\mathbf{S}}}^{-1}(\mathbf{q}) + \underline{\chi}\right)^{-1} , \quad (7.102)$$

where $\underline{\widetilde{\mathbf{S}}}(\mathbf{q})$ is the matrix of structure factors whose elements $\widetilde{S}_{ij}(\mathbf{q})$ are the pair-correlation functions for the disordered incompressible melt in the absence of binary interactions. For Gaussian coils, these correlation functions are the well-known Debye functions. To construct the form of $\underline{\widetilde{\mathbf{S}}}$ for an arbitrary chain architecture, first consider the linear ABC triblock copolymer depicted in Figure 7.50. Define the chain contour variable s which ranges from 0 at the beginning of block A and reaches $s = N$ at the end of block C. Intra-block correlations are computed via integrals of the form:

$$g_{\mathcal{M},\mathcal{M}}(\mathbf{q}) = \int_0^{N_{\mathcal{M}}} ds \int_0^{N_{\mathcal{M}}} ds' \, e^{-x_{\mathcal{M}}|s-s'|} = \frac{2}{x_{\mathcal{M}}}(N - h_{\mathcal{M}}(\mathbf{q}))$$

$$h_{\mathcal{M}}(\mathbf{q}) = \frac{1}{x_{\mathcal{M}}}(1 - \ell_{\mathcal{M}}(\mathbf{q})) \quad \ell_{\mathcal{M}}(\mathbf{q}) = e^{-xN_{\mathcal{M}}} \quad (7.103)$$

where $x_{\mathcal{M}}(\mathbf{q}) \equiv q^2 a_{\mathcal{M}}^2$. Similarly, inter-block correlations, e.g. for blocks A and C in the triblock illustrated by Figure 7.50, are given by:

$$g_{A,C} = \int_0^{N_A} ds \int_{N_A+N_B}^{N} ds' \exp\left[-q^2 \left(a_A^2(N_A - s)\right.\right. \quad (7.104)$$
$$\left.\left. + a_C^2(s' - N_A - N_B) + a_B^2 N_B\right)\right] = h_A h_C \ell_B$$

More generally, it is evident that interblock correlation functions between blocks $\mathcal{I}$ and $\mathcal{J}$ may be constructed as a product of the h-functions for each block and the ℓ-functions of the blocks spanning $\mathcal{I}$ and $\mathcal{J}$ in the chain architecture. More formally, define $\{\mathcal{B}\}_i$ as the set of blocks belonging to chemical species i, and $\{\mathcal{P}\}_{\mathcal{I}\mathcal{J}}$ as the set of blocks on the Path which connects $\mathcal{I}$ and $\mathcal{J}$. The generic *inter*block correlation function is:

$$g_{\mathcal{I},\mathcal{J}} = h_{\mathcal{I}} h_{\mathcal{J}} \prod_{\mathcal{K} \in \{\mathcal{P}\}_{\mathcal{I},\mathcal{J}}} \ell_{\mathcal{K}} . \qquad (7.105)$$

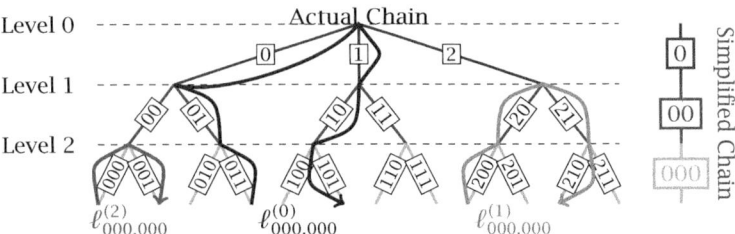

Fig. 7.51 Chain contour coordinates of a linear ABC triblock copolymer.

The elements of the non-interacting structure factor matrix $\widetilde{S}_{ij}$ are then simply the sum of all correlation functions for blocks in $\{\mathcal{B}\}_i$ and $\{\mathcal{B}\}_j$:

$$\widetilde{S}_{i,j}(\mathbf{q}) = \begin{cases} \displaystyle\sum_{\mathcal{I} \in \{\mathcal{B}\}_i} g_{\mathcal{I},\mathcal{I}}(\mathbf{q}) & i = j \\ \displaystyle\sum_{\mathcal{I} \in \{\mathcal{B}\}_i} \sum_{\mathcal{J} \in \{\mathcal{B}\}_j} g_{\mathcal{I},\mathcal{J}}(\mathbf{q}) & i \neq j \end{cases} \qquad (7.106)$$

Finally, the RPA prediction of $I(\mathbf{q})$ is realized with the pre- and post-multiplication of the structure factor matrix $\mathbf{S}(\mathbf{q})$ with the vector of pure component scattering length densities $\mathbf{b}$:

$$I(\mathbf{q}) = \mathbf{b} \cdot \underline{\mathbf{S}} \cdot \mathbf{b} \qquad (7.107)$$

Simplifications for symmetrically branched architectures. In a chain with n_b blocks there are a total of $\frac{1}{2}n_b(n_b - 1)$ pair correlations functions to evaluate in the formulation of equation (7.106). For branched polymers such as the dendrimer depicted in Figure 7.49 with symmetry in their chain architecture, this number may be substantially reduced through the exploitation of that symmetry. Sibling blocks whose descendants are identical are said to be *multiples* of each other, and may be replaced by a single linear segment on the chain architecture tree diagram. Let $L_{\mathcal{M}}$ be the level index of block $\mathcal{M}$, and $m_{\mathcal{M}}^{L_{\mathcal{M}}}$ represent the multiplicity of block $\mathcal{M}$, i.e. the number of times it appears on its parent block. For example, the 5th generation dendrimer of Figure 7.49 is composed of five unique blocks, each of which has $m_{\mathcal{M}}^{L_{\mathcal{M}}} = 3$. That is, each block m at level $L_{\mathcal{M}}$ repeats itself three times from its parent level. Define $m_{\mathcal{M}}^i$ to be the number of times block $\mathcal{M}$ appears as viewed from the origin of level i. Thus the total number of times the block appears in the chain is

$$m_{\mathcal{M}}^0 = m_{\mathcal{M}}^{L_{\mathcal{M}}} \prod_{\mathcal{I} \in \{\mathcal{A}\}_{\mathcal{M}}} m_{\mathcal{I}}^{L_{\mathcal{I}}}, \qquad (7.108)$$

where $\{\mathcal{A}\}_{\mathcal{M}}$ is the set of blocks with an $\mathcal{A}$ncestral relationship with $\mathcal{M}$. For example, the terminal block of our 5th generation dendrimer appears a total of $m^0_{00000} = m^4_{00000} m^3_{0000} m^2_{000} m^1_{00} m^0_0 = 3^5$ times. Let $\{\mathcal{U}\}_i$ designate the set of $\mathcal{U}$nique blocks of type i. With the preceding nomenclature in place, the pair correlation functions may now be expressed in terms of the *unique* blocks of the simplified chain and combinatorial multipliers that account for all possibles paths spanning blocks $\mathcal{I}$ and $\mathcal{J}$:

$$\widetilde{S}_{ii} = \sum_{\mathcal{I} \in \{\mathcal{U}\}_i} m^0_\mathcal{I} g_{\mathcal{I},\mathcal{I}} \qquad (7.109)$$

$$\widetilde{S}_{ij} = \sum_{\mathcal{I} \in \{\mathcal{U}\}_i} \sum_{\mathcal{J} \in \{\mathcal{U}\}_j} h_\mathcal{I} h_\mathcal{J} \ell^*_{ij} \qquad (7.110)$$

$$\ell^*_{\mathcal{I},\mathcal{J}} = \begin{cases} \left(m^{L_\mathcal{J}+1}_\mathcal{I} m^0_\mathcal{J} \ell^{L_\mathcal{J}}_{\mathcal{I},\mathcal{J}} + \sum_{k=0}^{L_\mathcal{J}-1} m^{k+1}_\mathcal{I} m^{L_\mathcal{J}-k}_\mathcal{J} \binom{m^k_\mathcal{J}}{2} \ell^{(k)}_{\mathcal{I},\mathcal{J}} \right) & \begin{array}{l} L_j < L_i, \\ \mathcal{J} \in \{\mathcal{A}\}_\mathcal{I} \\ \text{(in the simplified chain)} \end{array} \\ \left(m^{L_\mathcal{J}}_\mathcal{I} m^{L_\mathcal{J}}_\mathcal{J} m^1_\mathcal{J} \ell^{L_\mathcal{J}}_{\mathcal{I},\mathcal{J}} + \sum_{k=0}^{L_\mathcal{J}-1} m^k_\mathcal{I} m^k_\mathcal{J} m^{L_\mathcal{J}-1-k}_\mathcal{J} \binom{m^k_\mathcal{J}}{2} \ell^{(k)}_{\mathcal{I},\mathcal{J}} \right) & \begin{array}{l} L_j < L_i, \\ \mathcal{J} \notin \{\mathcal{A}\}_\mathcal{I} \\ \text{(in the simplified chain)} \end{array} \end{cases}$$

$$(7.111)$$

$$\ell^{(k)}_{\mathcal{I},\mathcal{J}} \equiv \prod_{\mathcal{M} \in \{\mathcal{P}\}^{(k)}_{\mathcal{I},\mathcal{J}}} \ell_\mathcal{M}(\mathbf{q}) = \left\{ \begin{array}{l} \text{The product of } \ell \text{ functions over} \\ \text{the path of blocks connecting} \\ \mathcal{I} \text{ and } \mathcal{J}, \text{ passing through level} \\ k. \text{c.f. Figure 7.51} \end{array} \right\} . \qquad (7.112)$$

References

[1] Flory, P. J. (1941). *J. Chem. Phys.*, **9**, 660.
[2] Huggins, M. L. (1941). *J. Chem. Phys.*, **9**, 440.
[3] Leibler, L. (1980). *Macromolecules*, **13**, 1602–1617.
[4] de Gennes, P. G. (1970). *J. Phys.*, **31**, 235–238.
[5] de Gennes, P.G. (1979). *Scaling Concepts in Polymer Physics*. Cornell University Press, New York.
[6] Fredrickson, G. H. and Binder, K. (1989). *J. Chem. Phys.*, **91**, 7265–7275.
[7] Meier, D. J. (1969). *J. Polym. Sci. Part C: Polym. Symp.*, **26**, 81–96.
[8] Leary, D. F. and Williams, M. C. (1970). *J. Polym. Sci. Part B: Polym. Lett.*, **8**, 335–340.
[9] Leary, D. F. and Williams, M. C. (1973). *J. Polym. Sci. Part B: Polym. Phys.*, **11**, 345–358.
[10] Helfand, E. and Wasserman, Z. R. (1976). *Macromolecules*, **9**, 879–888.
[11] Hasegawa, H., Tanaka, H., Yamasaki, K. *et al.* (1987). *Macromolecules*, **20**, 1651–1662.
[12] Semenov, A. N. (1985). *Zh. Eksp. I Teor. Fiz. (Sov. Phys. JETP)*, **88**, 1242–1256.

[13] Bates, F. S. and Fredrickson, G. H. (1990). *Annu. Rev. Phys. Chem.*, **41**, 525–557.
[14] Matsen, M. W. and Whitmore, M. D. (1996). *J. Chem. Phys.*, **105**, 9698–9701.
[15] Matsen, M. W. and Schick, M. (1994). *Phys. Rev. Lett.*, **72**, 2660–2663.
[16] Matsen, M. W. and Bates, F. S. (1996). *Macromolecules*, **29**(4), 1091–1098.
[17] Edwards, S.F (1965). *Proc. Phys. Soc.*, **85**, 613.
[18] Edwards, S.F (1966). *Proc. Phys. Soc.*, **88**, 265.
[19] Helfand, E. and Tagami, Y. (1972). *J. Chem. Phys.*, **56**, 3592–3601.
[20] Fredrickson, G.H. (2006). *The Equilibrium Theory of Inhomogeneous Polymers (International Series of Monographs on Physics)*. Oxford University Press, New York.
[21] Feynman, R. P. (1948). *Rev. Mod. Phys.*, **20**, 367–387.
[22] Kac, M. (1949). *Trans. Amer. Math. Soc.*, **65**, 1–13.
[23] Cochran, E. W., Garcia-Cervera, C. J., and Fredrickson, G. H. (2006). *Macromolecules*, **39**(7), 2449–2451.
[24] Matsen, M. W. and Bates, F. S. (1996). *Macromolecules*, **29**, 7641–7644.
[25] Matsen, M. W. and Bates, F. S. (1997). *J. Chem. Phys.*, **106**, 2436–2448.
[26] Epps, T. H., Cochran, E. W., Bailey, T. S. *et al.* (2004). *Macromolecules*, **37**(22), 8325–8341.
[27] Tyler, C. A. and Morse, D. C. (2005). *Phys. Rev. Lett.*, **94**, 208302.
[28] Takenaka, M., Wakada, T., Akasaka, S. *et al.* (2007). *Macromolecules*, **40**, 4399–4402.
[29] Bates, F. S. and Fredrickson, G. H. (1999). *Physics Today*, **52**, 32–38.
[30] Breiner, U., Krappe, U., Thomas, E. L. *et al.* (1998). *Macromolecules*, **31**, 135–141.
[31] Shefelbine, T. A., Vigild, M. E., Matsen, M. W. *et al.* (1999). *J. Amer. Chem. Soc.*, **121**, 8457–8465.
[32] Cochran, E. W. and Bates, F. S. (2004). *Phys. Rev. Lett.*, **93**(8), 087802.
[33] Bluemle, M. J., Fleury, G., Lodge, T. P. *et al.* (2009). *Soft Matter*, **5**, 1587.
[34] Segalman, R.A. (2005). *Mater. Sci. Eng.: Reports*, **48**, 191–226.
[35] Minich, E. A., Nowak, A. P., Deming, T. J. *et al.* (2004). *Polymer*, **45**, 1951–1957.
[36] Dai, C. A., Yen, W. C., Lee, Y. H. *et al.* (2007). *J. Amer. Chem. Soc.*, **129**, 11036–11038.
[37] Higgins, A.M., Martin, S.J., Thompson, R.L. *et al.* (2005). *J. Phys. - Cond. Mat.*, **17**, 1319–1328.
[38] Van Hest, J.C.M. (2007). *Polymer Reviews*, **47**, 63–92.
[39] Muthukumar, M., Ober, C. K., and Thomas, E. L. (1997). *Science*, **277**, 1225–1232.

[40] Jenekhe, S. A. and Chen, X. L. (1999). *Science*, **283**, 372–375.
[41] Maaloum, M., Ausserre, D., Chatenay, D. *et al.* (1992). *Phys. Rev. Lett.*, **68**, 1575–1578.
[42] Russell, T. P. (1990). *Materials Science Report*, **5**, 171–271.
[43] Ausserré, D., Chatenay, D., Coulon, G. *et al.* (1990). *J. Phys.*, **51**, 2571–2580.
[44] Ausserré, D., Raghunathan, V. A., and Maaloum, M. (1993). *J. Phys. II*, **3**, 1485–1496.
[45] Anastasiadis, S. H., Russell, T. P., Satija, S. K. *et al.* (1990). *J. Chem. Phys.*, **92**, 5677–5691.
[46] Turturro, A., Gattiglia, E., Vacca, P. *et al.* (1995). *Polymer*, **36**, 3987–3996.
[47] Henkee, C. S., Thomas, E. L., and Fetters, L. J. (1988). *Journal of Materials Science*, **23**, 1685–1694.
[48] Anastasiadis, S. H., Russell, T. P., Satija, S. K. *et al.* (1989). *Phys. Rev. Lett.*, **62**, 1852–1855.
[49] Menelle, A., Russell, T. P., and Anastasiadis, S. H. (1992). *Phys. Rev. Lett.*, **68**, 67–70.
[50] Russell, T. P., Lambooy, P., Kellogg, G. J. *et al.* (1995). *Physica B*, **213**, 22–25.
[51] Lambooy, P., Russell, T. P., Kellogg, G. J. *et al.* (1994). *Phys. Rev. Lett.*, **72**, 2899–2902.
[52] Kellogg, G. J., Walton, D. G., Mayes, A. M. *et al.* (1996). *Phys. Rev. Lett.*, **76**, 2503–2506.
[53] Donley, J.P. and Fredrickson, G.H. (1995). *J. Polym. Sci. Part B: Polym. Phys.*, **33**, 1341–1351.
[54] Stühn, B. (1992). *J. Polym. Sci. Part B: Polym. Phys.*, **30**, 61013–1019.
[55] Fredrickson, G. H. and Helfand, E. (1987). *J. Chem. Phys.*, **87**, 697–705.
[56] Matsen, M. W. and Bates, F. S. (1997). *J. Polym. Sci. Part B: Polym. Phys.*, **35**, 945–952.
[57] Floudas, G., Vazaiou, B., Schipper, F. *et al.* (2001). *Macromolecules*, **34**, 2947–2957.
[58] Yamada, K., Nonomura, M., and Ohta, T. (2006). *J. Phys. - Cond. Mat.*, **18**(32), L421.
[59] Bates, F. S., Schulz, M. F., Khandpur, A. K. *et al.* (1994). *Faraday Discussions*, **98**, 7–18.
[60] Matsen, M. W. (1996). *J. Chem. Phys.*, **104**, 7758–7764.
[61] Wang, C.-Y. and Lodge, T. P. (2002). *Macromol. Rapid Comm.*, **23**(1), 49–54.
[62] Khandpur, A. K., Foerster, S., Bates, F. S. *et al.* (1995). *Macromolecules*, **28**, 8796–8806.
[63] Bailey, T. S., Pham, H. D., and Bates, F. S. (2001). *Macromolecules*, **34**, 6994–7008.
[64] Mogi, Y., Mori, K., Matsushita, Y. *et al.* (1992). *Macromolecules*, **25**(20), 5412–5415.
[65] Mogi, Y., Mori, K., Kotsuji, H. *et al.* (1993). *Macro-*

molecules, **26**(19), 5169–5173.

[66] Matsushita, Y., Tamura, M., and Noda, I. (1994). *Macromolecules*, **27**, 3680–3682.

[67] Mogi, Y., Nomura, M., Kotsuji, H. *et al.* (1994). *Macromolecules*, **27**(23), 6755–6760.

[68] Hückstädt, H., Göpfert, A., and Abetz, V. (2000). *Polymer*, **41**(26), 9089 – 9094.

[69] Abetz, V. and Goldacker, T. (2000). *Macromol. Rapid Comm.*, **21**(1), 16–34.

[70] Chatterjee, J., Jain, S., and Bates, F. S. (2007). *Macromolecules*, **40**(8), 2882–2896.

[71] Krappe, U., Stadler, R., and Voigt-Martin, I. (1995). *Macromolecules*, **28**(13), 4558–4561.

[72] Brinkmann, S., Stadler, R., and Thomas, E. L. (1998). *Macromolecules*, **31**(19), 6566–6572.

[73] Stadler, R., Auschra, C., Beckmann, J. *et al.* (1995). *Macromolecules*, **28**(9), 3080–3097.

[74] Balsamo, V., von Gyldenfeldt, F., and Stadler, R. (1999). *Macromolecules*, **32**(4), 1226–1232.

[75] Ott, H., Abetz, V., and Altstädt, V. (2001). *Macromolecules*, **34**(7), 2121–2128.

[76] Hückstädt, H., Goldacker, T., Göpfert, A. *et al.* (2000). *Macromolecules*, **33**(10), 3757–3761.

[77] Ho, C-C, Lee, Y-H, Dai, C-A *et al.* (2009). *Macromolecules*, **42**, 4208–4219.

[78] Rehse, S., Mecke, K., and Magerle, R. (2008, May). *Phys. Rev. E*, **77**, 051805.

[79] Fredrickson, G. H. and Bates, F. S. (1996). *Annu. Rev. Mater. Sci.*, **26**, 501–550.

[80] Colby, R. H. (1996). *Curr. Opin. Coll. & Interface Sci.*, **1**(4), 454–465.

[81] Pakula, T. and Floudas, G. (2000). In *Block Copolymers* (ed. F. Balta-Calleja and Z. Roslaniec). Marcel Decker, New York-Basel.

[82] Wang, C.-Y. and Lodge, T. P. (2002). *Macromolecules*, **35**(18), 6997–7006.

[83] Almdal, K., Rosedale, J. H., and Bates, F. S. (1990). *Macromolecules*, **23**(19), 4336–4338.

[84] Bates, F.S., Rosedale, J. H., and Fredrickson, G. H. (1990). *J. Chem. Phys.*, **92**(10), 6255–6270.

[85] Floudas, G., Pakula, T., Fischer, E.W. *et al.* (1994). *Acta Polymerica*, **45**(3), 176–181.

[86] Harkless, C. R., Singh, M. A., Nagler, S. E. *et al.* (1990). *Phys. Rev. Lett.*, **64**, 2285–2288.

[87] Cahn, J. W. (1956). *Atca Metallurg.*, **4**(5), 449–459.

[88] Sakurai, S., Kawada, H., Hashimoto, T. *et al.* (1993). *Macromolecules*, **26**(21), 5796–5802.

[89] Sakurai, S., Momii, T., Taie, K. *et al.* (1993). *Macro-*

molecules, **26**(3), 485–491.

[90] Hajduk, D. A., Takenouchi, H., Hillmyer, M. A. *et al.* (1997). *Macromolecules*, **30**(13), 3788–3795.

[91] Schulz, M. F., Bates, F. S., Almdal, K. *et al.* (1994). *Phys. Rev. Lett.*, **73**, 86–89.

[92] Dalvi, M. C., Eastman, C. E., and Lodge, T. P. (1993, Oct). *Phys. Rev. Lett.*, **71**, 2591–2594.

[93] Fleischer, G., Fujara, F., and Stuehn, B. (1993). *Macromolecules*, **26**(9), 2340–2345.

[94] Vogt, S., Anastasiadis, S. H., Fytas, G. *et al.* (1994). *Macromolecules*, **27**(15), 4335–4343.

[95] Anastasiadis, S. H., Fytas, G., Vogt, S. *et al.* (1993). *Phys. Rev. Lett.*, **70**, 2415–2418.

[96] Lodge, T. P. and Dalvi, M. C. (1995). *Phys. Rev. Lett.*, **75**, 657–660.

[97] Dalvi, M. C. and Lodge, T. P. (1994). *Macromolecules*, **27**(13), 3487–3492.

[98] Eastman, C. E. and Lodge, T. P. (1994). *Macromolecules*, **27**(20), 5591–5598.

[99] Hamersky, M. W., Hillmyer, M. A., Tirrell, M. *et al.* (1998). *Macromolecules*, **31**(16), 5363–5370.

[100] Yokoyama, H. and Kramer, E. J. (1998). *Macromolecules*, **31**(22), 7871–7876.

[101] Stepanek, P. and Lodge, T. P. (1996). *Macromolecules*, **29**(4), 1244–1251.

[102] Kawasaki, K. and Sekimoto, K. (1989). *Macromolecules*, **22**(7), 3063–3075.

[103] Fraaije, J. G. E. M. (1993). *J. Chem. Phys.*, **99**, 9202–9212.

[104] Kossuth, M. B., Morse, D. C., and Bates, F. S. (1999). *J. Rheol.*, **43**(1), 167–196.

[105] Kawasaki, K. and Onuki, A. (1990, Sep). *Phys. Rev. A*, **42**, 3664–3666.

[106] Koppi, K. A., Tirrell, M., Bates, F. S. *et al.* (1992). *J. Phys. II France*, **2**(11), 1941–1959.

[107] Ryu, C. Y., Lee, M. S., Hajduk, D. A. *et al.* (1997). *J. Polym. Sci. Part B: Polym. Phys.*, **35**(17), 2811–2823.

[108] Zhao, J., Majumdar, B., Schulz, M. F. *et al.* (1996). *Macromolecules*, **29**, 1204–1215.

[109] Rosedale, J. H. and Bates, F. S. (1990). *Macromolecules*, **23**(8), 2329–2338.

[110] Keller, A., Pedemonte, E., and Willmouth, F.M. (1970). *Colloid & Polym. Sci.*, **238**, 385–389. 10.1007/BF02085559.

[111] Hadziioannou, G., Mathis, A., and Skoulios, A. (1979). *Colloid & Polymer Science*, **257**, 136–139. 10.1007/BF01638138.

[112] Bang, J., Jeong, U., Ryu, D. Y. *et al.* (2009). *Advanced Materials*, **21**, 4769–4792.

[113] Krishnan, K., Almdal, K., Burghardt, W. R. *et al.* (2001). *Phys. Rev. Lett.*, **87**, 098301.

[114] Kannan, R. M. and Kornfield, J. A. (1994). *Macromolecules*, **27**(5), 1177–1186.

[115] Patel, S. S., Larson, R. G., Winey, K. I. *et al.* (1995). *Macromolecules*, **28**(12), 4313–4318.

[116] Koppi, K. A., Tirrell, M., and Bates, F. S. (1993, Mar). *Phys. Rev. Lett.*, **70**, 1449–1452.

[117] Bates, F. S., Koppi, K. A., Tirrell, M. *et al.* (1994). *Macromolecules*, **27**(20), 5934–5936.

[118] Winter, H. H., Scott, D. B., Gronski, W. *et al.* (1993). *Macromolecules*, **26**, 7236–7244.

[119] Hajduk, D. A., Tepe, T., Takenouchi, H. *et al.* (1998). *J. Chem. Phys.*, **108**(1), 326–333.

[120] Jackson, C. L., Barnes, K. A., Morrison, F. A. *et al.* (1995). *Macromolecules*, **28**(3), 713–722.

[121] Harada, T., Bates, F. S., and Lodge, T. P. (2003). *Macromolecules*, **36**, 5440–5442.

[122] Eskimergen, R., Mortensen, K., and Vigild, M. E. (2005). *Macromolecules*, **38**, 1286–1291.

[123] Almdal, K., Koppi, K. A., and Bates, F. S. (1993). *Macromolecules*, **26**(15), 4058–4060.

[124] Bang, J., Lodge, T. P., Wang, X. *et al.* (2002). *Phys. Rev. Lett.*, **89**, 215505.

[125] Sebastian, J. M., Lai, C., Graessley, W. W. *et al.* (2002). *Macromolecules*, **35**(7), 2707–2713.

[126] Meins, T., Hyun, K., Dingenouts, N. *et al.* (2012). *Macromolecules*, **45**(1), 455–472.

[127] Chen, Z.-R. and Kornfield, J. A (1998). *Polymer*, **39**(19), 4679 – 4699.

[128] Gupta, V. K., Krishnamoorti, R., Kornfield, J. A. *et al.* (1996). *Macromolecules*, **29**(4), 1359–1362.

[129] Wu, L., Lodge, T. P., and Bates, F. S. (2005). *J. Rheol.*, **49**(6), 1231–1252.

[130] Vigild, M. E., Chu, C., Sugiyama, M. *et al.* (2001). *Macromolecules*, **34**(4), 951–964.

[131] Olaj, O. F. and Lantschbauer, W. (1982). *Makromol. Chem. Rapid Comm.*, **3**, 847–858.

[132] Weyersberg, A. and Vilgis, T. A. (1993). *Phys. Rev. E*, **48**(1), 377.

[133] Pakula, T. (2004). In *Simulation Methods for Polymers* (ed. M. Kotelyanskii and D. N. Theodorou), p. 602. Marcel Dekker, Inc., New York-Basel.

[134] Pakula, T., Karatasos, K., Anastasiadis, S. H. *et al.* (1997). *Macromolecules*, **30**(26), 8463–8472.

[135] Hahn, Th. (ed.) (2010). *International Tables for Crystallography*. Volume A: Space-group symmetry. Wiley.

[136] Cochran, E. W. (2004). Ph.D. thesis, University of Minnesota – Twin Cities, MN.

[137] Kim, M. I., Wakada, T., Akasaka, S. *et al.* (2008). *Macro-

molecules, **41**(20), 7667–7670.

[138] Ugaz, V. M. and Burghardt, W. R. (1998). *Macromolecules*, **31**(24), 8474–8484.

[139] Lim, L. S., Harada, T., Hillmyer, M. A. *et al.* (2004). *Macromolecules*, **37**, 5847–5850.

[140] Rangarajan, P., Register, R. A., Adamson, D. H. *et al.* (1995). *Macromolecules*, **28**(5), 1422–1428.

[141] Stein, G. E., Kramer, E. J., Li, X. *et al.* (2007). *Macromolecules*, **40**(7), 2453–2460.

[142] Levine, J. R., Cohen, J. B., Chung, Y. W. *et al.* (1989). *J. Appl. Crystallogr.*, **22**, 528–532.

[143] Lazzari, R. (2002). *J. Appl. Crystallogr.*, **35**, 406–421.

[144] Tate, M. P., Urade, V. N., Kowalski, J. D. *et al.* (2006). *J. Phys. Chem. B*, **110**, 9882–9892. PMID: 16706443.

[145] Parratt, L. G. (1954, Jul). *Phys. Rev.*, **95**, 359–369.

[146] Tolan, Metin (1999). *X-Ray Scattering from Soft-Matter Thin Films: Materials Science and Basic Research*. Volume 148, Springer Tracts in Modern Physics. Springer.

[147] Lee, Byeongdu, Park, Insun, Yoon, Jinhwan *et al.* (2005). *Macromolecules*, **38**(10), 4311–4323.

[148] Kittel, C. (1995). *Introduction to Solid State Physics* (7 edn). Wiley.

[149] Won, Y.-Y., Davis, H. T., Bates, F. S. *et al.* (2000). *J. Phys. Chem. B*, **104**, 7134–7143.

[150] Pedersen, J. S., Posselt, D., and Mortensen, K. (1990). *J. Appl. Crystallogr.*, **23**, 321–333.

[151] Cochran, E. W., Morse, D. C., and Bates, F. S. (2003). *Macromolecules*, **36**(3), 782–792.

[152] Brandrup, J., Immergut, Edmund H., Grulke, Eric A. *et al.* (1999; 2005). *Polymer Handbook*. (4th edn). Wiley.

[153] Fetters, L. J., Lohse, D. J., Richter, D. *et al.* (1994). *Macromolecules*, **27**(17), 4639–4647.

Elastomers and Rubberlike Elasticity

8

James E. Mark[1] and Burak Erman[2]

[1]Department of Chemistry and the Polymer Research Center, University of Cincinnati, Cincinnati, OH 45221-0172, USA
[2]Department of Chemical and Biological Engineering, Koc University, Rumelifeneri Yolu, Sariyer 80910, Istanbul, Turkey

Introduction

Elastomers have long held great interest in polymer science and engineering because of their remarkable properties. Specifically, the features of such a rubberlike material that are so striking are its ability to undergo very large deformations with essentially complete recoverability. In order for a material to exhibit such unusual behavior, it must consist of relatively long polymeric chains that have a high degree of flexibility and mobility, and are joined into a network structure. The requirement of flexibility and mobility is associated with the very high deformability. As a result of an externally imposed stress, the long chains alter their spatial arrangements or configurations quite rapidly due to their high mobilities [1, 2]. Not surprisingly, this type of elasticity has been a subject of continuing interest [3–6]. The requirement of linking the chains into a network structure is associated with the solid-like features of an elastomer, where the chains are prevented from flowing relative to each other under external stresses. As a result, a typical rubber or elastomer may be stretched up to about ten times its original length. Upon removal of the external force, it rapidly recovers its original dimensions, with essentially no residual or non-recoverable strain. As a result of these unique mechanical properties, elastomers find important usages ranging from automobile tires to heart valves, and to gaskets in jet planes and space vehicles [7].

The above general features of elastomeric materials have long been known and, in fact, the area of rubberlike elasticity has had one of the longest and most distinguished histories in all of polymer science [1,2,8].[1]

The rubberlike materials described in this chapter are so designated since they exhibit the high deformability and recoverability so reminiscent of natural rubber [2]. They are thus frequently called 'rubbers', and the terms rubberlike materials, rubbers, and elastomers are used essentially interchangeably in the literature. Since high flexibility and mobility are required for rubberlike elasticity, elastomers generally do

[1]Quantitative measurements of the mechanical and thermodynamic properties of natural rubber and other elastomers go back to 1805, and some of the earliest studies have been carried out by luminaries such as Joule and Maxwell. Also, the earliest molecular theories for polymer properties of any kind were, in fact, addressed to the phenomenon of rubberlike elasticity.

not contain stiffening groups such as ring structures and bulky side chains [2,5]. These characteristics are evidenced by the low glass transition temperatures T_g exhibited by these materials. These polymers also tend to have low melting points, if any, but some do undergo crystallization upon sufficiently large deformations. Examples of typical elastomers which do undergo strain-induced crystallization include natural rubber, butyl rubber, and high-*cis* polybutadiene, with recent attention focusing on copolymers of high-*trans* polybutadiene [9]. In contrast, poly(dimethylsiloxane) (PDMS) [–Si(CH$_3$)$_2$O–], poly(ethyl acrylate), styrene-butadiene copolymer, and ethylene-propylene copolymer generally do not undergo strain-induced crystallization. Other examples can be found in the general literature, including handbooks [10,11].

Some polymers are not elastomeric under normal conditions but can be made so by raising the temperature or adding a diluent ('plasticizer'). Polyethylene is in this category because of its high degree of crystallinity. Polystyrene, poly(vinyl chloride), and the biopolymer elastin are also of this type, but because of their relatively high glass transition temperatures also require plasticization to be elastomeric [5].

A final class of polymers is inherently nonelastomeric. Examples are polymeric sulfur, because its chains are too unstable, poly(*p*-phenylene) because its chains are too rigid, and thermosetting resins, because their chains are too short [5].

There is much current interest in hydrogels, particularly those robust enough to be used in biomedical applications (as described below).

8.1 Preparation and structure of networks

Networks are formed by joining individual chains into a three-dimensional structure. The conventional method of network formation is by random crosslinking with a suitable crosslinking agent. More modern techniques involve a more-controlled 'end-linking' of chains, in order to obtain networks of known structure. Polymerization into a crosslinked system by adding a small amount of multifunctional reactants into the polymerizing system, or by joining the chains by physical forces such as hydrogen bonding are also utilized.

Random chemical crosslinking

In random crosslinking, chemical reactions are used that attack a pair of chains, at essentially random locations [1,5]. Examples are the addition of sulfur atoms to the double bonds of diene elastomers and the attack of free radicals from peroxide thermolysis or high-energy radiation on side chains (frequently unsaturated) or on the chain backbone itself.

The crosslinks formed in this way are generally highly stable. These networks, however, still present the problem that the crosslinking reaction used is highly uncontrolled since it is not known how many crosslinks are introduced or where they lie along the chain trajectories. It is thus very difficult to use these networks for quantitative purposes such as the development of structure-property relationships. It is virtually impossible to obtain independent measures of their degree of crosslinking (as represented directly by the number density of crosslinks or the number density of network chains, or inversely by the average molecular weight M_c between the crosslinks that mark off a network chain) [5].

Highly specific chemical end-linking

If networks are formed by end-linking functionally-terminated chains instead of haphazardly joining chain segments at random, then the nature of this very specific chemical reaction provides the desired structural information [5, 12–16]. Thus, the functionality of the crosslinks is the same as that of the end-linking agent, and the molecular weight M_c between crosslinks and its distribution is the same as those of the starting chains prior to their being end-linked.

Because of their known structures, such end-linked elastomers are now the preferred materials for the quantitative characterization of rubberlike elasticity. Additionally, bimodal elastomers prepared by these end-linking techniques have very good ultimate properties, and there is currently much interest in preparing and characterizing such networks [15, 17–20] and developing theoretical interpretations for their properties [21, 22]. Both subjects are discussed below.

Polymerizations with multi-functional monomers

Another way of making a network is by a copolymerization in which at least one of the comonomers has a functionality ϕ of three or larger. This is one of the oldest ways of preparing networks but has been used mostly to prepare materials so heavily crosslinked as to be relatively hard thermosets rather than highly deformable elastomers [5, 23]. Important work in this area involves the use of sol fractions and statistical arguments to obtain information on the structures of the sol phases and gel (elastomeric) phases [24, 25].

Physical aggregation

Preparation of elastomeric networks is also possible by causing physical aggregation of some of the chain segments [5, 26]. Examples are the adsorption of chain segments onto filler particles, formation of polymer microcrystallites, condensation of ionic side chains onto metal ions, chelation of ligand side chains to metal ions, and microphase separation of glassy end blocks in an elastomeric triblock copolymer[2] [27]. These materials are not only of commercial interest, but are also now much used in quantitative studies of rubberlike elasticity. The nature and extent of the crosslinking can change with temperature, presence of diluent, and degree of deformation in an uncontrolled manner, and the 'crosslinks' are frequently so large as to complicate greatly the theoretical analysis of any experimental results.

[2] For example, Kraton ® thermoplastic elastomer.

Networks prepared under unusual conditions

Several indirect approaches are undertaken in preparing networks of simpler topology with reduced degrees of network chain entangling [28]. Some of the techniques employed involve (i) separating the chains prior to their crosslinking by either dissolution [29] or stretching [30], (ii) simultaneously forming two different but interpenetrating systems, (iii) trapping cyclics into the network, (iv) simultaneously crosslinking one component highly and the other lightly, and (v) using platelets such as exfoliated clays as planar junctions.

Crosslinking in solution

This approach produced elastomers that had fewer entanglements, as indicated by the observation that such networks came to elastic equilibrium much more rapidly than elastomers crosslinked in the dry state. When a network is crosslinked in solution and the solvent then removed, the chains collapse in such a way that there is reduced overlap in their configurational domains. It is primarily in this regard – namely, decreased chain-crosslink entangling – that solution-crosslinked samples have simpler topologies.

These elastomers also exhibited stress-strain isotherms in elongation that were closer in form to those expected from the simplest molecular theories of rubberlike elasticity. Specifically, there were large decreases in the Mooney-Rivlin $2C_2$ correction constant described below.

In these procedures, removal of the solvent has the additional effect of putting the chains into a 'supercontracted' state. Experiments on strain-induced crystallization carried out on such solution crosslinked elastomers indicated that the decreased entangling was less important than the supercontraction of the chains, in that crystallization required larger values of the elongation than was the case for the usual elastomers crosslinked in the dry state. Other recent work in this area has focused on the unusually high extensibilities of such elastomers [31].

One novel approach involves the use of collapsed crosslink regions that can unfold under stress and then refold reversibly upon retraction of the elastomer [32]. This is essentially biomimicry of some body tissues, such as the muscle protein titin, in which such behavior gives these biomaterials an impressive combination of deformability and strength.

Crosslinking in the deformed state

In this approach a first network is generally introduced in the undeformed state, the resulting elastomer is elongated, and a second series of crosslinks are introduced in the stretched state. Release of the stress permits the network to retract, but the second network of this 'double-network' structure prevents retraction down to the original dimensions. The most interesting feature of the retracted network is the fact that it is anisotropic in structure and properties. In some cases, double networks have shown increases in orientability and strain-induced crystallization, and improved fatigue resistance [33]. In fact, some results show that there may be less of a compromise between failure properties in general and the modulus [34], which may be due in part to the decreased hysteresis observed for some of these elastomers. Theoretical treatment of such materials goes back to 1960 [35], and research has continued on them up to the present time [33]. Better molecular understandings of these observations are being sought with, for example, extensive studies of residual strains and birefringence [36].

Interpenetrating networks

If two types of chains have different end groups, then it is possible to end-link them simultaneously into two networks that *interpenetrate* one another [37]. Such a network could, for example, be made by reacting hydroxyl-terminated PDMS chains with tetraethoxysilane (in a condensation reaction), while reacting vinyl-terminated PDMS chains mixed into them with a multifunctional silane (in an addition reaction). Interpenetrating networks can be very unusual with regard to both equilibrium and dynamic mechanical properties [38]. For example, such materials can have considerable toughness and unusual damping characteristics.

Slide-ring gels

Slide-ring gels are topological networks with some similarities to linear rotaxane assemblies [39]. In these gels, a number of cyclic molecules are threaded onto a linear polymer chain and then trapped by placing bulky capping groups at the two ends of the chain. Some of the cyclics are then fused together to form mobile crosslinks. In the case of two fused cyclics, the result is a figure-eight structure: ∞ [40–42]. These crosslinks are called slide rings, and act like pulleys for the chains threading through them. The sliding is thought to thus equalize any network tensions cooperatively. This gives the gels unusual mechanical properties, in particular very high deformabilities and degrees of swelling. There has been considerable interest in the theoretical treatment of such materials [43, 44].

Double networks

Double networks belong to a class of hydrogels that consist of two independently crosslinked networks, one consisting of a rigid polyelectrolyte and the other a flexible uncharged polymer [45–47]. The term 'double networks' is an unfortunate choice of terminology since this name has long been applied to the completely different elastomeric materials described above as networks in which a second set of crosslinks is introduced while the first network is in the deformed state.

Double networks are used as water-swollen gels, and exhibit the best mechanical properties when the first network is highly crosslinked and the second only lightly crosslinked. The molar ratio of the second component to the first should be a factor as high as ten or more, which makes them rather different from most interpenetrating networks (IPNs) [37]. The enhanced mechanical properties of these double-network hydrogels are thought to be a result of the second network preventing cracks from growing to the point of producing catastrophic failure of the material. More specifically, this dissipation of crack energy may be facilitated by the second network appearing as clusters in voids occurring in what is apparently an inhomogeneous matrix of the first network. There is expected to be considerable entangling of the two types of chains in these domains. In some cases, a third component is used to form a triple network with, for example, uncrosslinked polymer being added to reduce surface friction [48]. Theoretical modeling has been carried out in attempts to better understand the properties of these double-network hydrogels [49, 50].

Nanocomposite gels

Nanocomposite gels use exfoliated platelets of clay as the crosslinking junctions [51, 52]. Ends of the polymer chains absorb strongly on the surfaces of the platelets and enough of the chains attach to *different* platelets to provide the bridges that constitute a network structure. The fact that the crosslinks are planar sheets of considerable dimensions and junction functionality somehow yields unusual mechanical properties, including very good toughness. Other types of nanoparticles introduced into hydrogels are also of obvious interest [53].

Network structure

The elastic activity of a network depends directly on the molecular structure. Suitable reference materials are perfect networks that have no dangling chains (that are connected to the network structure at only one end) and no loops where the two ends of a chain terminate at the same junction. The structure of a perfect network may be defined by two variables: the cycle rank ξ and the average junction functionality ϕ [5, 26]. Cycle rank is defined as the number of chains that must be cut in order to reduce a network structure to that of a tree [54], which is a giant molecule containing no closed structures or cycles. The number of junctions μ, the number of chains ν, and the molecular weight M_c, of chains between two junctions, may be obtained from ξ and ϕ by the following three relations:

$$\nu = \xi / (1 - 2/\phi) \tag{8.1}$$

$$\mu = 2\nu/\phi \tag{8.2}$$

$$M_c = \frac{(1 - 2/\phi)\, \rho N_A}{\xi/V_0}, \tag{8.3}$$

where ρ is the density, V_0 is the reference volume of the network, and N_A is Avogadro's number. The number of junctions μ and the number

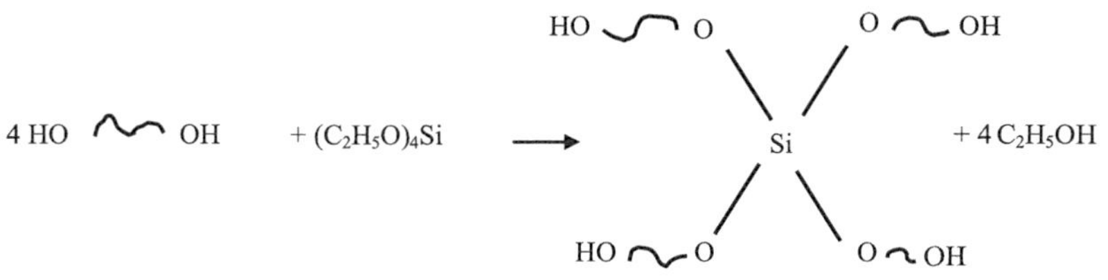

Fig. 8.1 End-linking by a condensation reaction between hydroxyl groups at the ends of a polymer chain and the alkoxy groups on a tetrafunctional end-linking agent.

of chains ν, are typically given as number *densities*, by dividing them by the volume of the unswollen network. They are the standard two direct measures of the degree of crosslinking. The molecular weight M_c, of the network chains between two crosslinks, is an inverse measure, and typically has a value ranging from 5000 to 15000 g/mol. This corresponds to using approximately one out of a hundred repeat units for a crosslink [5].

Effects of network structure on elastomeric properties

Chain lengths and chain-length distributions

The length of a network chain between two crosslinks decreases with increasing number of crosslinks as can be seen from equations (8.1) and (8.3). As will be elaborated further in the following sections, the modulus of a network varies inversely with network chain length [3, 5].

Elastomeric networks are generally prepared in an uncontrolled manner such as peroxide curing [1, 2]. In such crosslinking, segments close together in space are linked irrespective of their locations along the chain trajectories. This results in a highly random network structure in which the distribution of chain lengths between crosslinks is essentially unknown. New synthetic techniques are now available, however, for the preparation of 'model' polymer networks of known structure. More specifically, if networks are formed by end-linking functionally-terminated chains of known lengths and length distributions instead of haphazardly joining chain segments at random, then the nature of this very specific chemical reaction provides the desired structural information [3, 5, 16]. Thus, characterizing the uncrosslinked chains with respect to molecular weight M_n and molecular weight distribution, and then carrying out the specified reaction to completion gives elastomers in which the network chains have these characteristics, in particular a molecular weight M_c between crosslinks equal to M_n, a network chain-length distribution the same as that of the starting chains, and crosslinks having the functionality of the end-linking agent. A widely investigated system is end-linked PDMS using tetraethyl orthosilicate. The reaction is shown in Figure 8.1 [55].

One reason for this choice is the fact that the polymer is readily available with either hydroxyl or vinyl end groups, and the reactions these groups participate in are relatively free of complicating side reactions. In the application of interest here, a mixture of two hydroxyl-terminated polymers is being crosslinked, one consisting of very short chains and the other – of much longer chains. An alternative approach involves the addition reaction between vinyl groups at the ends of a polymer chain and the active hydrogen atoms on silicon atoms in the [–Si(CH$_3$)$_2$O–] repeat units in an oligomeric poly(methyl hydrogen siloxane). These ideas can obviously be extended to higher modalities (trimodal, etc.).

The most novel elastomer of known structure has a *multimodal* distribution of network chain lengths. In the *bimodal* case, it consists of a combination of unusually short network chains (molecular weights of a few hundred) and the much longer chains typically associated with elastomeric behavior (molecular weights of ten or twenty thousand). One use of elastomers having such a bimodal distribution is possibly to improve the ultimate properties of an elastomer.

'Bimodal' elastomers prepared by these end-linking techniques generally do have very good ultimate properties [3,5]. In addition to this experimental work, there are now theoretical studies addressing the novel mechanical properties of bimodal elastomers [5].

The distribution of network chain lengths in a bimodal elastomer can be extremely unusual, in that there is simultaneously a large *number* fraction of short chains and a large *weight* fraction of long chains.

Although there have been attempts to evaluate the mechanical properties of trimodal elastomers, this has not been done in any organized manner. The basic problem is the large number of variables involved, specifically three molecular weights and two independent composition variables (mol fractions); this makes it practically impossible to do an exhaustive series of relevant experiments. For this reason, mechanical property experiments that have been carried out have generally involved arbitrarily chosen molecular weights and compositions. Not surprisingly, only modest improvements have been obtained over the bimodal materials in these cases.

Recent computational studies [56] suggest that a trimodal network prepared by incorporating small numbers of very long chains into a bimodal network of long and short chains could significantly improve the ultimate properties.

Junction functionality

End-linking of chains with linkers of a known functionality is used to control the structure in this way. Increasing the junction functionality decreases the fluctuation amplitudes of the junctions in the undeformed state, and such a network with suppressed junctions behaves close to an affine network under deformation. However, the affineness diminishes under increasing extension. Trifunctional and tetrafunctional PDMS networks prepared in this way have been used to test the molecular theories of rubber elasticity with regard to the increase in non-affineness of the network deformation with increasing elongation [57]. The ratio $2C_2/2C_1$ (which is a measure of the increasing non-affineness as the elongation increases) decreases with increase in crosslink functionality. This is due to the fact that crosslinks connecting four chains are more constrained than those connecting only three. There is therefore less of a decrease in modulus brought about by the fluctuations which are enhanced at high deformation and give the deformation its non-affine character. The decrease in $2C_2/2C_1$ with decrease in network chain molecular weight is due to the fact that there is less configurational interpenetration in the case of short network chains. This decreases the firmness with which the crosslinks are embedded and thus the deformation is already highly non-affine even at relatively small deformations.

Entanglements

An increase in the functionality of junctions suppresses their fluctuations and therefore fluctuations in chain dimensions are decreased. Also, the entanglements along the chains that come from the neighboring chains may further act as if they were additional (albeit temporary) junctions. Consequently, the modulus of such networks may exceed that of an affine network.

The idea of entanglement effects on network modulus originates from the trapped-entanglement concept of Langley [58], and Graessley [59] stating that some fraction of the entanglements which is present in the bulk polymer before crosslinking becomes permanently trapped by the crosslinking process and acts as additional crosslinks. These trapped entanglements, unlike the chemical crosslinks, have some freedom, and the two chains forming the entanglement may slide relative to one other. The two chains may therefore be regarded to be attached to each other by means of a fictitious 'slip-link'. An excellent review article on this topic is by Heinrich et al. [60].

Model networks may also be used to provide a direct test of molecular predictions of the modulus of a network of known degree of crosslinking. Experiments on model networks [3, 61] have given values of the elastic modulus in good agreement with theory. Others [12, 62] have given values significantly larger than predicted [3,5], and the increases in modulus have been attributed to contributions from 'permanent' chain entanglements. There are disagreements and the issue has not yet been resolved. Since the relationship of modulus to structure is of such fundamental importance, there has been a great deal of research activity in this area.

Dangling chains

Since dangling chains represent imperfections in a network structure, their presence should have a detrimental effect on ultimate properties such as the tensile strength, as gauged by the stress at rupture. This expectation was confirmed by an extensive series of results obtained on PDMS networks which had been tetrafunctionally crosslinked using a variety of techniques.

Additional, more quantitative information has been obtained using the very specific chemical reactions used to form ideal elastomers, but now modified to prepare intentionally non-ideal networks containing known numbers and lengths of dangling-chain irregularities. If more chain ends are present than reactive groups on the end-linking molecules, then dangling ends will be produced and their number is directly determined by the extent of the stoichiometric imbalance. Their lengths, however, are of necessity the same as those of the elastically-effective chains. This constraint can be removed by separately preparing monofunctionally-terminated chains of any desired length and then attaching them as parts of the network.

Values of the ultimate strength of the networks containing the dangling ends were found to be lower than those of the more nearly perfect networks, with the largest differences occurring at high proportions of dangling ends (low $2C_1$), as expected [63]. The values of the maximum extensibility showed a similar dependence, also as expected.

Trapped cyclics

If cyclic molecules are present during the end-linking of chains, some of them will be trapped because of having been threaded by the linear chains prior to the latter being chemically bonded into the network structure [3, 64]. The fraction trapped is readily estimated from solvent extraction studies. Some typical results, in terms of the fraction trapped as a function of degree of polymerization of the cyclic [65] showed that, as expected, very small cyclics don't get trapped at all, but almost all of the largest cyclics do.

These cyclics can change the properties of the network in which they're 'incarcerated'. For example, when PDMS cyclics are trapped in a thermoplastic material, they can act as a plasticizer that is in a sense intermediate to the usual external (dissolved) and internal (copolymerized) varieties. Interesting changes in mechanical properties have been observed in materials of this type [66].

It may also be possible to use this technique to prepare networks having no crosslinks whatsoever [67]. Mixing linear chains with large amounts of cyclic and then *di*functionally end-linking them could give sufficient cyclic interlinking to yield an 'Olympic' or 'chain-mail' network [68]. Computer simulations [69] could be particularly useful with regard to establishing the conditions most likely to produce such novel structures.

Effects on ultimate properties

Of particular interest here are the upturns in the elastic modulus frequently exhibited by unfilled elastomers at very high elongations [5], and the commercially important increases in ultimate strength associated with them. This increase is important since it corresponds to a significant toughening of the elastomer. Its molecular origin, however, had been the source of some controversy. It had been widely attributed to the 'limited extensibility' of the network chains. However, the increases in modulus had generally been observed only in networks that could undergo strain-induced crystallization, which could account for the increase in modulus, primarily because the crystallites thus formed would act as additional crosslinks in the network.

This type of reinforcement resulting from strain-induced crystallization was identified by the fact that the higher the temperature, the lower the extent of crystallization and the worse the ultimate properties. The effects of increase in swelling were found to parallel those for increase in temperature, as was expected, since diluent also suppresses network crystallization [70]. On the other hand, in those cases where the upturns are due to limited chain extensibility, increase in temperature has relatively little effect on the upturns [3].

Attempts to observe upturns in the modulus arising due to non-Gaussian effects in non-crystallizable networks were made with some of the end-linked, non-crystallizable model PDMS networks described above. These networks have high extensibilities, presumably because of their very low incidence of dangling-chain network irregularities. They have particularly high extensibilities when they are prepared from a mixture of very short chains (around a few hundred $g\,mol^{-1}$) with relatively long chains (around $18000\,g\,mol^{-1}$), giving the already mentioned bimodal distributions of network chain lengths [17]. Apparently the very short chains in such networks are important because of their limited extensibilities, and the relatively long chains because of their ability to retard the rupture process. Stress-strain measurements on such bimodal PDMS networks exhibited upturns in modulus that were much less pronounced than those in crystallizable polymer networks. Furthermore, they are independent of temperature and are not diminished by incorporation of solvent. These characteristics are what are to be expected in the case of limited chain extensibility [5]. Thus these results permit interpretation of properties such as the elongation at the upturn in the modulus and the elongation at rupture ('maximum extensibility').

In the case of elastomers capable of undergoing strain-induced crystallization, such as *cis*-1,4-polybutadiene networks [71], the higher the temperature, the lower the extent of crystallization and, correspondingly, the lower the ultimate properties. The effects of increase in swelling parallel those for increase in temperature, since diluent also suppresses network crystallization. For non-crystallizable networks, however, neither change is as important.

The weakest-link theory states that the shortest chains in a network are primarily responsible to elastomer rupture, and this idea was tested by preparing end-linked networks containing increasing amounts of short chains [5]. Remarkably, these elastomers showed no significant decreases in ultimate properties with such increases in the numbers of short chains, in striking disagreement with the suggested mode of elastomer failure. Networks are apparently much more resourceful than given credit for in this theory. Apparently, the strain is continually being reapportioned during the deformation, in such a way that the much more easily deformed long chains bear most of the burden of the deformation. The flaw in the weakest-link theory is thus the implicit assumption that all parts of the network deform in exactly the same way, i.e. 'affinely', whereas the deformation is actually markedly non-affine [3, 5].

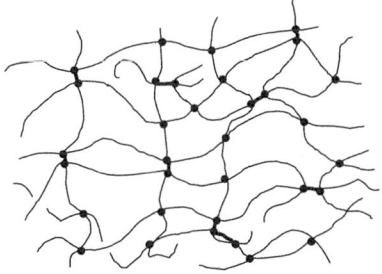

Fig. 8.2 A network having a bimodal distribution of network chain lengths. The short chains are arbitrarily shown by heavier lines than the long chains, and the dots represent the crosslinks, typically resulting from the end-linking of functionally-terminated chains.

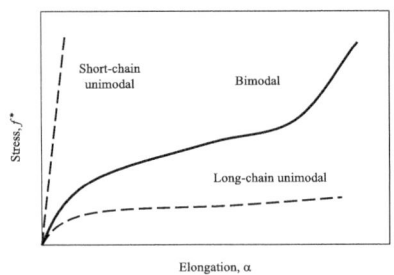

Fig. 8.3 Typical dependence of nominal stress against elongation for two unimodal networks having either all short chains or all long chains, and a bimodal network having some of both.

The weakest-link issue having been resolved, it became of interest to see what would happen in the case of bimodal networks having such overwhelming numbers of short chains that they could not be ignored in the network's response. This is illustrated in Figure 8.2. As described below, there is a synergistic effect leading to mechanical properties that are better than those obtainable from the usual unimodal distributions!

Such bimodal distributions of network chain lengths have been shown to give significant improvements in mechanical properties. This is illustrated in Figure 8.3 (after [72]), in which data on PDMS networks are plotted in such a way that the area under a stress-strain isotherm corresponds to the energy required to rupture the network. If the network is all short chains, it is brittle, which means that the maximum extensibility is very small. If the network is all long chains, the ultimate strength is very low. In neither case is the material a tough elastomer. As can readily be seen from the plot, the bimodal networks are much improved elastomers in that they can have a high ultimate strength without the usual decrease in maximum extensibility. Apparently, the observed increases in modulus are due to the limited chain extensibility of the short chains, with the long chains serving to retard the rupture process.

8.2 Elasticity experiments and theories

Mechanical properties

Mechanical properties determine the unique usefulness of elastomeric materials and have been studied extensively [1–3,5]. The relationship of primary interest is the stress-strain isotherm, and most experiments have evolved simple elongation, largely because of the simplicity of the techniques involved. Measurements to determine the stress-strain isotherm under simple tension are typically carried out to the rupture point of a sample, thus yielding as well the two ultimate properties of the elastomer, namely its ultimate strength and maximum extensibility. In addition to providing practical information, such results are also much used to obtain estimates of the degree of crosslinking, and to test and compare the predictions of the various molecular theories. Results obtained with other deformations [73–75], such as biaxial extension, compression, shear, and torsion are, however, particularly valuable for gauging the generality of a particular theory.

Elongation results

Experimental results for elongation of unswollen and swollen networks are represented in a manner consistent with the Mooney-Rivlin equation [2, 76, 77]:

$$[f^*] = 2C_1 + 2C_2\alpha^{-1} , \qquad (8.4)$$

where $2C_1$ and $2C_2$ are constants independent of elongation α. A typical Mooney-Rivlin plot is presented in Figure 8.4. The straight line represents the Mooney-Rivlin line, and the curved line below the Mooney-

Rivlin line is representative of experimental data. Typically, the reduced force in simple tension exhibits a decrease represented by the nonzero slope $2C_2$. At high elongations experimental data show an upturn, thus deviating significantly from the Mooney-Rivlin plot. Also, compression data depart significantly from the simple Mooney-Rivlin plot [3,5].

There are numerous other deformations of interest, including biaxial extension, shear, and torsion [2]. The equation of state for compression ($\alpha < 1$) is the same as that for elongation ($\alpha > 1$), and the equations for the other deformations may all be derived by proper specifications of the deformation ratios [1,2]. Some of these deformations are considerably more difficult to study than simple elongation and, unfortunately, have therefore not been as extensively investigated.

Measurements in biaxial extension have involved direct stretching of a sample sheet in two perpendicular directions within its plane, by two independently-variable amounts. In the equi-biaxial case, the deformation is equivalent to compression. A good account of such experimental results [78] has been given by the simple molecular theory, with improvements at lower extensions upon use of the constrained-junction theory [3,5].

Biaxial extension studies can also be carried out by the inflation of sheets of the elastomer [2]. Such results on unimodal and bimodal networks of PDMS show upturns in the modulus at high biaxial extensions, as expected. Also of interest, however, are the pronounced maxima preceding the upturns. This represents a challenging feature to be explained by molecular theories addressed to bimodal elastomeric networks in general. Equi-biaxial extension results have also been obtained on unimodal and bimodal networks of PDMS. Upturns in the modulus were found to occur at high biaxial extensions, as expected.

In shear measurements on unimodal and bimodal networks of PDMS [79], the bimodal PDMS networks showed large upturns in the pure-shear modulus at high strains which were similar to those reported for elongation and biaxial extension. Very little work has been done on elastomers in torsion (twisting a cylindrical sample around its long axis). The same types of bimodal PDMS networks showed rather different behavior in torsion [75]. Specifically, no unambiguous upturns in modulus were observed at large deformations. It has not yet been established whether this is due to the inability, to date, of reaching sufficiently large torsions, or whether this is some inherent difference in this type of deformation. There are also some torsion results on stress-strain behavior and network thermoelasticity [2].

Tear tests have been carried out on bimodal PDMS elastomers [80–82], using the standard 'trouser-leg' method. Tear energies were found to be considerably increased by the use of a bimodal distribution, with documentation of the effects of compositional changes and changes in the ratio of molecular weights of the short and long chains. The increase in tear energy did not seem to depend on tear rate [80], an important observation that seems to suggest that viscoelastic effects are not of paramount importance in explaining the observed improvements. Rheovibron vis-

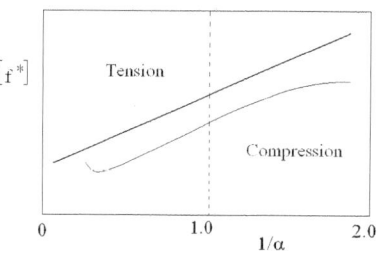

Fig. 8.4 Reduced stress as a function of reciprocal extension ratio for tension and compression, in the Mooney-Rivlin representation.

coelasticity results have been reported for bimodal PDMS networks [83]. To provide guidance for the desired interpretations, measurements were first carried out on unimodal networks consisting of the types of chains used in combination in the bimodal networks. One of the important results was documentation of the dependence of crystallinity on the network chain-length distribution.

Measurements on permanently set PDMS networks in compressive cyclic deformations [84] showed that there appeared to be less permanent set or creep in the case of the bimodal elastomers. This is consistent in a general way with some early results for polyurethane elastomers [85]. Specifically, cyclic elongation measurements on unimodal and bimodal networks indicated that the bimodal ones survived many more cycles before the occurrence of failure from fatigue. The number of cycles to failure is approximately an order of magnitude higher for the bimodal network, at the same value of the modulus [3]!

In addition to mechanical properties, some important non-mechanical properties are mentioned in the following sections.

Swelling

In a swelling experiment the network is typically placed into an excess of solvent, which it imbibes until the dilational stretching of the chains prevents further absorption [5, 86]. This equilibrium extent of swelling can be interpreted to yield the degree of crosslinking of the network, provided the polymer-solvent interaction parameter χ [1] is known. Conversely, if the degree of crosslinking is known from an independent experiment, then the interaction parameter can be determined. Most studies of networks in swelling equilibrium give values for the crosslink density or related quantities that are in satisfactory agreement with those obtained from mechanical property measurements [1, 2]. More specifically, estimates in crosslink densities agree within the limits of the accuracy of the swelling measurements and the accompanying theory, and in particular within the uncertainties occasioned by unanticipated changes in the modulus with deformation [3, 5].

A more interesting area involving swollen networks or 'gels' is their abrupt collapse (decrease in volume) upon relatively minor changes in temperature, pH, solvent composition, etc. [3, 5, 87]. Although the collapse is quite slow, being controlled by diffusion of solvent in and out of a large, monolithic piece of gel, it is rapid enough in fibers and films to make the phenomenon interesting with regard to the construction of switches and related devices.

Sorption and extraction of diluents

In these experiments, the rate at which a suitable diluent is absorbed into a network of known and controlled 'pore size' and the rate at which it can subsequently be extracted are determined [86, 88]. Of obvious interest is the dependence of the extraction efficiency on the molecular weight

M_c of the network chains (as a measure of pore size), the molecular weight M_d of the diluent, the structure of the diluent (linear, branched, or cyclic), and whether or not the diluent had been present during the end-linking process.

One way of obtaining a network swollen with diluent is to form the network in a first step, and then absorb an unreactive diluent into it. Alternatively, the same diluent can be mixed into the reactive chains prior to their being end-linked into a network structure. In either case, oligomeric and polymeric diluents are of greatest interest, and they must be functionally inactive for them to 'reptate' through the network rather than being bonded to it. Both types of networks can then be extracted to determine the ease with which the various diluents can be removed, as a function of M_d and M_c.

The efficiency of diluent removal was found to decrease with increase in M_d and with decrease in M_c, as expected. High-molecular-weight diluents are extremely hard to remove at values of M_c of interest in the preparation of model networks, a circumstance complicating the analysis of soluble polymer fractions in terms of degrees of perfection of the network structure. The diluents added after the end-linking were the more easily removed, possibly because they were less entangled with the network structure, and this could correspond to differences in chain conformations of the diluent.

There is a complication, however, which can occur in the case of networks of polar polymers at relatively high degrees of swelling [5]. The observation is that different solvents, at the same degree of swelling, can have significantly different effects on the elastic force. This is apparently due to a 'specific solvent effect' on the unperturbed dimensions, which appear in the various molecular forms on the elastic equations of state. The effect is not yet well understood. It is apparently partly due to the effect of the solvent's dielectric constant on the Coulombic interactions between parts of a chain, but probably also to solvent-polymer segment interactions that change the conformational preferences of the chain backbone [89].

Other characterization techniques

Optics and spectroscopy

An example of a relevant optical property is the birefringence of a deformed polymer network [26]. This strain-induced birefringence can be used to characterize segmental orientation, both Gaussian and non-Gaussian elasticity, and to obtain new insights into the network chain orientation necessary for strain-induced crystallization [2, 5, 90, 91]. Infrared dichroism has also been particularly helpful in this regard. Of predominant interest is the orientation factor $S \equiv (1/2)(3 < \cos^2 \chi > -1)$, which can be obtained experimentally from the ratio of absorbances of a chosen peak parallel and perpendicular to the direction in which an elastomer is stretched [3, 92]. One representation of such results is the effect of network chain length on the reduced orientation factor $[S] \equiv S/(\lambda^2 - \lambda^{-1})$, where λ is the elongation. A comparison is made among typical theoretical results in which the affine model assumes the chain dimensions to change linearly with the imposed macroscopic strain, and the phantom model allows for junction fluctuations that make the relationship nonlinear. The experimental results were found to be close to the phantom relationship. Combined techniques, such as Fourier-transform infrared spectroscopy combined with rheometry, are also of increasing interest [93].

Other optical and spectroscopic techniques are also important, particularly with regard to segmental orientation. Some examples are fluorescence polarization, deuterium nuclear magnetic resonance (NMR), and polarized infrared spectroscopy [5, 26]. Also relevant here is some work indicating that microwave techniques can be used to image elastomeric materials, for example with regard to internal damage [94].

Microscopies

A great deal of information is now being obtained on crystallinity, filler dispersion, and other aspects of elastomer structure and morphology by scanning probe microscopy, which consists of several approaches [95–97]. One approach is that of scanning tunneling microscopy (STM), in which an extremely sharp metal tip on a cantilever is passed along the surface being characterized while measuring the electric current flowing through quantum mechanical tunneling. Monitoring the current then permits maintaining the probe at a fixed height above the surface, and display of probe height as a function of surface coordinates then gives the desired topographic map. One limitation of this approach is the obvious requirement that the sample be electrically conductive.

Atomic force microscopy (AFM), on the other hand, does not require a conducting surface. The probe simply responds to attractions and repulsions from the surface, and its corresponding downward and upward motions are directly recorded to give the relief map of the surface structure. The probe can be either in contact with the surface, or adjacent to it and sensing only Coulombic or van der Waals forces.

There are numerous other types of scanning probe microscopy, including electrochemical STM and AFM, frictional force microscopy, surface force compliance, magnetic force microscopy, electric force microscopy, scanning thermal microscopy, and near-field scanning optical microscopy. Some of these various techniques not only generate topographic relief images, but also provide the opportunity to transport separate individual atoms and molecules into nanoscale arrangements.

An example of an application to elastomers is the characterization of binodal and spinodal phase-separated structures occurring in model PDMS networks [98]. Another application of these microscopy devices involves attaching probes to the two ends of a single polymer chain and then stretching it to determine its equilibrium and dynamic mechanical properties. This is generally referred to as 'single molecule elasticity' [99–103]. Some rather sophisticated equipment is required, such as 'optical tweezers', and sensitive force-measuring devices. Most of the effort thus far has involved biopolymers, and mechanically-induced transitions between their various conformations. Although such studies are obviously not relevant to the many unresolved issues that involve the interactions of chains within an elastomeric network, they are certainly of interest in their own right.

NMR

Nuclear magnetic resonance (NMR) has been much used to study the characteristics of the network chains themselves, particularly with regard to orientations and molecular motions, and their effects on the diffusion of small molecules [104]. Aspects related to the structures of the networks include the degree of crosslinking [105], the distributions of crosslinks [106] and stresses [107], and topologies [108]. Another example is the use of NMR to clarify some issues in the areas of aging and phase separation [109].

Most elastomers require reinforcing fillers to function effectively, and NMR has been used to characterize the structures of such composites as well. Examples are the adsorption of chains onto filler surfaces, the immobilization of these chains into 'bound rubber', and the imaging of the filler itself [110, 111].

Small-angle scattering

The technique of this type that is of greatest utility in the study of elastomers is small-angle neutron scattering, for example, from deuterated chains in a non-deuterated host [112, 113]. One application has been the determination of the degree of randomness of the chain configurations in the undeformed state, an issue of importance with regard to the basic postulates of elasticity theory described below. Of even greater importance is determination of the manner in which the dimensions of the chains follow the macroscopic dimensions of the sample, i.e. the degree of affineness of the deformation. This relationship between the microscopic and macroscopic levels in an elastomer is one of the central problems in rubberlike elasticity.

Some small-angle x-ray scattering techniques have also been applied to elastomers. Examples are the characterization of fillers precipitated into elastomers and the corresponding incorporation of elastomers into ceramic matrices, in both cases to improve mechanical properties [5].

Brillouin scattering

The application of Brillouin scattering to the characterization of elastomers [114, 115] is an interesting extension of earlier work on polymers in general [116–118]. It should be quite useful for looking at glassy-state properties of elastomers at very high frequencies.

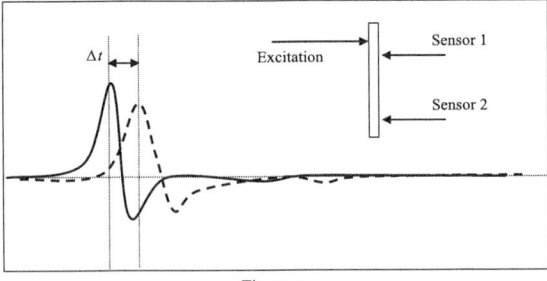

Fig. 8.5 Pulse propagation results, showing measurement of the time Δt required for a pulse to pass downward through an elastomer from sensor 1 to sensor 2.

Pulse propagation

A relatively new technique for the non-invasive, non-destructive characterization of network structures involves the acoustic pulse-propagation measurements [119,120]. In this technique, the delay Δt in a pulse passing through the network is used to obtain information on the network structure, for example the chain length between crosslinks or between entanglements. This is illustrated in Figure 8.5.

Elasticity theories

In this section we review the expressions for the elastic free energy for different theoretical models. In later sections, we derive the force deformation relations for these models. The simplest molecular theories of rubberlike elasticity are based on network chains that are Gaussian, and are therefore referred to as Gaussian theories. In these theories, the probability $W(r)$ for a distance r between the two ends of a network chain is given by the Gaussian function [1,5]:

$$W(r) = \left(\frac{3}{2\pi \langle r^2 \rangle_0}\right)^{3/2} \exp\left(-\frac{3r^2}{2\langle r^2 \rangle_0}\right). \tag{8.5}$$

Here, $\langle r^2 \rangle_0$ represents average of the squared end-to-end vectors, and the subscript zero indicates that the chain is in the unperturbed or the so-called theta-state [1]. It is now well established that chains in the bulk undiluted state are in this unperturbed state. For chains having fewer than 50 bonds, such as the short chains in a bimodal network, for example, the actual distribution departs markedly from the Gaussian limit [121]. Here we discuss molecular theories of networks that are based on the Gaussian picture of the individual network chains.

The elastic free energy A_{el} of a Gaussian chain is related to the probability distribution $W(r)$ by the thermodynamic expression [5]

$$A_{el} = C(T) - kT \ln W(r) \tag{8.6}$$

where $C(T)$ is a function only of temperature T, and k is the Boltzmann constant. Substituting equation (8.5) into equation (8.6) leads to

$$A_{el} = A^*(T) + \left(\frac{3kT}{2\langle r^2 \rangle_0}\right) r^2 \tag{8.7}$$

where $A^*(T)$ is a function of temperature alone. Equation (8.7) represents the elastic free energy of a single Gaussian chain with ends fixed at a separation r. The average force required to keep the two ends at this separation is obtained from the thermodynamic expression [5, 122]

$$f = \left(\frac{\partial A_{\text{el}}}{\partial r}\right)_T = \left(\frac{3kT}{\langle r^2 \rangle_0}\right) r \qquad (8.8)$$

where the second equality is obtained by using equation (8.7). The subscript T denotes differentiation at fixed temperature.

Equation (8.8) shows that the single chain behaves like a linear spring, obeying Hooke's law with a spring constant equal to $3kT/\langle r^2 \rangle_0$.

This approach is directly extendable to the chains making up a network structure because the total elastic free energy of the network is the sum of the elastic free energies of individual chains. Specifically, the difference in the total elastic free energy ΔA_{el} of the network in the deformed and undeformed states is obtained by summing equation (8.7) over the ν chains of the network [1–3, 5]:

$$\Delta A_{\text{el}} = \frac{3kT}{2\langle r^2 \rangle_0} \sum_\nu \left(r^2 - \langle r^2 \rangle_0\right) \qquad (8.9)$$

$$= \frac{3kT}{2}\left(\frac{\overline{r^2}}{\langle r^2 \rangle_0} - 1\right) \qquad (8.10)$$

where $\langle r^2 \rangle = \sum r^2/\nu$ is the average square end-to-end vectors of chains in the deformed network. Substituting

$$\langle r^2 \rangle = \langle x^2 \rangle + \langle y^2 \rangle + \langle z^2 \rangle \qquad (8.11)$$

into equation (8.10) and using the fact that chain dimensions are isotropic in the undeformed state, i.e.

$$\langle r^2 \rangle_0 = \langle x^2 \rangle_0 + \langle y^2 \rangle_0 + \langle z^2 \rangle_0$$
$$= 3\langle x^2 \rangle_0 = 3\langle y^2 \rangle_0 = 3\langle z^2 \rangle_0 \qquad (8.12)$$

yields

$$\Delta A_{\text{el}} = \frac{\nu kT}{2}\left[\frac{\langle x^2 \rangle}{\langle x^2 \rangle_0} + \frac{\langle y^2 \rangle}{\langle y^2 \rangle_0} + \frac{\langle z^2 \rangle}{\langle z^2 \rangle_0} - 3\right] \qquad (8.13)$$

The x, y, and z coordinates appearing in the above equations represent laboratory-fixed coordinates.

The ratios of mean-squared dimensions appearing in equation (8.13) are microscopic quantities. In order to express the elastic free energy of a network in terms of the macroscopic state of deformation, an assumption has to be made relating microscopic chain dimensions to macroscopic deformation.

The relationship of chain dimensions to macroscopic state of deformation has been the focus of a wide body of research. Two classical theories, the affine and the phantom network models, that relate the microscopic deformation to the macroscopic one, are described first. We then discuss more recent theories.

The affine network model

The classical theories comprising the phantom and the affine network models were introduced between 1941 and 1943. The term 'classical' is used in describing the phantom and affine models because of their conceptual simplicity and their status as suitable reference systems for all later treatments of the molecular theory. Below, the characteristic features of the classical theories are outlined.

One of the earlier assumptions regarding microscopic deformation in networks is that the junction points in the networks move affinely with macroscopic deformation. The affine network model was developed by Kuhn [123], and Wall and Flory [124]. According to the model, chain end-to-end vectors deform affinely, which gives

$$\Delta A_{\text{el,affine}} = \frac{1}{2}\nu kT \left(\lambda_x^2 + \lambda_y^2 + \lambda_z^2 - 3\right) \tag{8.14}$$

where the λ's are *macroscopic* deformation ratios along the three coordinate axes.

Theoretical developments after 1975 (see below) have not supported the affine network model [54, 125, 126], mainly because the assumption of embedding each junction securely into the continuum of the network volume was considered to be unrealistic. Even more important, neutron-scattering results [127] provide clear evidence for junction fluctuations of the nature assumed in the phantom network model described below.

The phantom network model

The theory of James and Guth, which has subsequently been termed the 'phantom network theory', was first outlined in [128], followed by a mathematically more rigorous treatment [129]. Subsequent work has been carried out by Eichinger [130], Graessley [131], Flory [54], Pearson [132], and Kloczkowski *et al.* [133]. The most important physical feature is the occurrence of junction fluctuations, which occur asymmetrically in an elongated network in such a manner that the network chains sense less of a deformation than that imposed macroscopically. As a result, the modulus predicted in this theory is substantially less than that predicted in the affine theory.

The relations resulting from the phantom network model, and their derivations, are given in many literature sources [3,5,54]. The final result of relevance here is

$$\Delta A_{\text{el,phantom}} = \frac{1}{2}\xi kT \left(\lambda_x^2 + \lambda_y^2 + \lambda_z^2 - 3\right). \tag{8.15}$$

The elastic free energies for the affine and the phantom network models differ only in the front factor, i.e. the factor $(1/2)\xi kT$ appearing in equation (8.15) replaces the front factor $(1/2)\nu kT$ in equation (8.14).

The affine and phantom network models, based on the Gaussian chain, act as the two limiting idealized models of amorphous networks in the absence of intermolecular interactions such as interactions between network chains. Expressions for the elastic free energy of more realistic models than the affine and the phantom network models are given in the following sections.

Extension of the classical models to non-Gaussian chains

The affine and phantom network models are based on the Gaussian distribution of end-to-end distances of the network chains. This distribution, however, is not suitable for very short chains or for any chains stretched to near the limits of their extensibility [2,5,134]. In these cases the modulus shows a marked upturn at high elongations, a result that may be of practical as well as fundamental importance.

As a result of the non-Gaussian behavior of network chains, the distribution $W(r)$ given by equation (8.5) is modified to

$$W(r) = \left(\frac{A\beta}{rl^2}\right)(\beta^{-1}\sinh\beta)^n \exp\left(-\frac{\beta r}{l}\right) \quad (8.16)$$

where A is a constant that normalizes the distribution, n is the number of repeat units of a network chain, l is the length of each repeat unit, and β is the inverse Langevin function $\beta = L^*(r/\langle r^2 \rangle_0^{1/2})$ which is expressed in series as

$$\beta = 3\rho + \frac{9}{5}\rho^3 + \frac{297}{175}\rho^5 + \dots \quad (8.17)$$

where $\rho = r/\langle r^2 \rangle_0^{1/2}$. Equation (8.16) now takes the form [2]

$$W(r) = \frac{3A}{\langle r^2 \rangle_0 l}\left[1 + \frac{3}{5}\rho^2 + \frac{99}{175}\rho^4 \dots\right] \\ \times \exp\left\{-\frac{3}{2\langle r^2 \rangle}r^2\left[1 + \frac{3}{10}\rho^2 \frac{33}{175}\rho^4 + \dots\right]\right\} \quad (8.18)$$

With this modified distribution of the end-to-end vector of a network chain, the evaluation of the elastic free energy follows the same lines as given above for the affine model. When the chains are sufficiently long, the parameter $\rho \to 0$, and the Gaussian affine model is recovered.

The constrained-junction theory

The constrained-junction model is one of the modern theories in the field. The term 'modern' refers to theories of rubberlike elasticity introduced after 1975 mainly to account for the disagreement between experiment and the predictions of the phantom or affine network models. All the theories in this category may essentially be regarded as corrections to the phantom network model. Corrections come in the form of contributions of entanglements to the elastic free energy.

The constrained-junction model assumes that the fluctuations of junctions are diminished below those of the phantom network due to the presence of entanglements [135]. It is based on initial work carried out by Ronca and Allegra [125], and subsequently developed by Flory and Erman [3,26]. These studies have shown that the fluctuations in a phantom network are substantial. For a tetrafunctional network, the mean-square fluctuations of junctions amount to as much as half of the mean-square end-to-end vector magnitude of the network chains. According to the model, each junction fluctuates in a domain whose size is dictated by the strength of the constraints acting on it. Such constraints are illustrated in the upper portion of Figure 8.6. The circle depicts the spherical volume in which the given junction fluctuates. The other junctions that share this volume are represented by crosses. If the junction is highly constrained, the size of this volume becomes small. The strength of the constraints is measured by a parameter κ, defined as

$$\kappa = \frac{\langle (\Delta R)^2 \rangle}{\langle (\Delta s)^2 \rangle} \tag{8.19}$$

where $\langle (\Delta R)^2 \rangle$ and $\langle (\Delta s)^2 \rangle$ denote, respectively, the mean-square dimension of junction fluctuations in the phantom network and in the entanglement domain. If the range of fluctuations decreases to zero due to entanglements, κ becomes infinitely large. If the effect of entanglements is nil, then the junction is unconstrained and can move in an indefinite space, and $\kappa = 0$. The elastic free energy of the constrained-junction model is given by the expression

$$\Delta A_{\text{el}} = \frac{1}{2}\xi kT \left(\sum_{t=1}^{3} \left(\lambda_t^2 - 1 \right) \right. \\ \left. + \frac{\mu}{\xi} \sum_{t=1}^{3} \left[B_t + D_t - \ln\left(1 + B_t\right) - \ln\left(1 + D_t\right) \right] \right) \tag{8.20}$$

where

$$B_t = \kappa^2 \frac{\lambda_t^2 - 1}{\left(\lambda_t^2 + \kappa \right)^2}, \qquad D_t = \lambda_t^2 B_t / \kappa . \tag{8.21}$$

This theory has been shown to give results which agree quantitatively with stress-strain swelling data [136], birefringence, and segmental orientation [137], in uniaxially stretched networks and with experimental data obtained in shear and multiaxial states of stress [138].

The Ronca-Allegra theory [125], and Flory-Erman theory [26,136], are both based on the idea that effects of constraints are local and decrease with increasing strain and swelling. The basic difference between the two theories is that in the Ronca-Allegra theory the fluctuations of junctions become exactly affine as the undeformed state is approached, whereas in the Flory theory they are close to but below those of the affine state.

The constrained-chain theory

This particular refinement of the constrained-junction model is based on re-examination of the constraint problem and evaluation of some neutron scattering estimates of actual junction fluctuations [127]. It was concluded that the suppression of the fluctuations was overestimated in

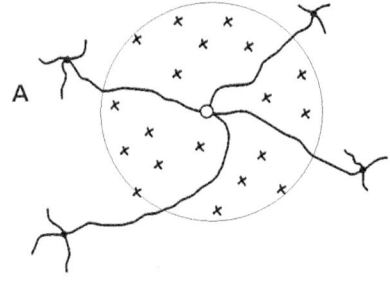
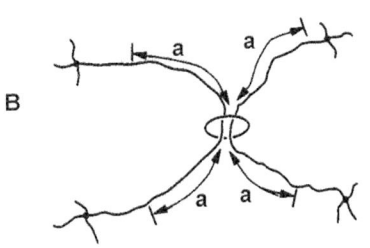

Fig. 8.6 (A) A central tetrafunctional junction surrounded by four other topologically neighboring junctions and a number of spatially neighboring junctions. (B) Schematic drawing of a slip link, with its possible motions along the network chains specified by the distances a, and its locking into position as a crosslink.

the theory, presumably because the entire effect of the inter-chain interactions was arbitrarily placed on the junctions. The theory was therefore revised to make it more realistic by placing the effects of the constraints on the centers of mass of the network chains [139]. This modification also provided improved agreement between theory and experiment.

The diffused-constraints theory

This theory attempts at even greater realism, by distributing the constraints continuously along the network chains. In its application to stress-strain isotherms in elongation [140], it has the advantage of having only a single constraint parameter and the values it exhibits upon comparing theory and experiment seem more reasonable than in the earlier models. Applications to strain birefringence, on the other hand, yield values of the birefringence that are much larger than those in the constrained-junction and constrained-chain theories.

Network models based on the Edwards approach

The above theories on the effects of constraints on fluctuations start from a detailed molecular model of the real network [26]. The fundamental postulate of the theory states that only strain-dependent contributions to the elastic free energy are of importance, and that these contributions vanish at infinitely large extensions or swelling. Contributions from trapped entanglements, for example, are categorically eliminated when this postulate is accepted. According to this approach, whether the trapped entanglements contribute to the stress can only be determined by experiments. A complete statistical mechanical theory that does not depend on physical assumptions of the type stated above has been outlined by Freed [141], and a mathematical theory of elasticity of networks with internal constraints has been worked out by Edwards and collaborators [142–144]. According to these theories, a set of internal constraints, including knots and entanglements, is assumed to be fixed during the formation of the network. These constraints are conserved under deformation, thus contributing to the elastic properties.

Edwards and colleagues formulated the contributions from entanglements by two different models: (i) the tube model and (ii) the slip-link model. In the tube model [145], the topological contributions are applied to every monomer of the network chain, confining its fluctuations to a tube. This potential is independent of network deformation. In a more recent work [146] the strength of tube constraints was made proportional to the deformation. This model is referred to as the 'nonaffine tube model'.

In the slip-link model [147–149], the chain length along the contour of the tube has been replaced by a slip-link. This is illustrated in the lower portion of Figure 8.6. In tube models the intermolecular potential acting on a given network chain is formed collectively by many neighboring chains. In slip-link models each entanglement is formed by a pair of chains. Thus, according to the mechanism in the slip-link theory, a link joins two different chains which may slide a distance a along the contour of the chains. The effect of entanglements enter as further contributions and is proportional to the number of slip links. The slip-link and constraint models have been compared by Vilgis and Erman [150].

Recently, Rubinstein and Panyukov [151] reviewed the constraint models and proposed a 'slip-tube' model by combining the ideas of the tube and slip-link models.

Computer simulations

In addition to the analytical formulations outlined in the preceding sections, rubber elasticity has been widely studied by computer simulations. The major applications of computer simulations are either in the form of Monte Carlo or molecular dynamics simulations. These simulations either characterize the detailed statistical properties of single chains that constitute the building blocks of rubbers, or collective statistical properties of the networks. PDMS has been a subject of particular interest [152].

Single-chain simulations

Monte Carlo simulations have been used to calculate thermoelastic results through the temperature coefficient of the unperturbed dimensions. In the case of networks of the protein elastin, such results were used to evaluate alternative theories for the molecular deformation mechanism for this bioelastomer [153].

Stress-strain isotherms have also been calculated with this approach. Examples are unimodal networks of polyethylene and PDMS, polymeric sulfur and selenium, short n-alkane chains, natural rubber, several polyoxides [154], elastin , and bimodal networks of PDMS [155]. It is possible to include excluded volume effects in such simulations [156]. In the case of the partially-helical polymer polyoxymethylene, the simulations were used to resolve the overall distributions into contributions from unbroken rods, once-broken rods, twice-broken rods, etc. [154]. It was also shown how applying stresses to the ends of chains of this type can be used to bias the distributions in the direction of increased helical content and increased average end-to-end distances [154]. In this sense, imposition of a stress has the same effect on the helix-coil equilibrium as a decrease in temperature.

There have also been simulations on chains of cis-1,4-polyisoprene and cis-1,4-polybutadiene with regard to the extent to which they can be stretched, the effects of temperature on cis-1,4-polybutadiene chain retractions, and their bonding of PDMS onto silica particles. Stress-strain relationships have also been simulated for networks, both unfilled cis-1,4-polyisoprene and PDMS filled with silica particles [157, 158].

The trapping process of cyclics in networks has been simulated using Monte Carlo methods based on a rotational isomeric state model for the cyclic chains [65]. The first step was generation of a sufficient number of cyclic chains having the known geometric features and conformational preferences, and the desired degree of polymerization. In the present application, a chain having an end-to-end distance less than a threshold value was considered to be a cyclic. The coordinates of each 'cyclic' chain thus generated were stored for detailed examination of the chain's configurational characteristics, in particular the size of the 'hole' it would present to a threading linear chain. The trapping process was simulated using a torus centered around each repeat unit in the cyclic [65]. Any torus found to be 'empty' was considered to provide a pathway for a chain of specified diameter to pass through, threading it and then incarcerating it once the end-linking process has been completed. The simulation results thus obtained gave a good representation of the experimental trapping efficiencies.

Chains in networks

One of the first studies of chains in networks is by Gao and Weiner [159] who performed extended simulations of short chains with fixed (affinely-moving) end-to-end vectors. The first extensive molecular dynamics simulations of realistic networks were performed by Kremer and collaborators [160]. These calculations were based on a molecular dynamics method that has been applied to study entanglement effects in polymer melts [161]. The networks obtained by crosslinking the melts were then used to study the effect of entanglements on the motion of the crosslinks and the moduli of the networks. The moduli calculated without any adjustable parameters were close to the phantom network model for short chains, and supported the Edwards tube model for long ones. Similar molecular dynamics analyses were used to understand the role of entanglements in deformed networks in subsequent studies [162]. There has also been interest in networks in which the junctions alternate regularly in their functionalities, with regard to chain dimensions and fluctuations and small-angle neutron scattering form factors [163].

There is now considerable interest in using simulations for characterizing crystallization in copolymeric materials. In particular, Windle and coworkers [164] have developed models capable of simulating chain ordering in copolymers composed of two comonomers, at least one of which is crystallizable. Typically, the chains are placed in parallel, two-dimensional arrangements. Neighboring chains are then searched for like-sequence matches in order to estimate extents of crystallinity. Chains stacked in arbitrary registrations are taken to model quenched samples. Annealed samples, on the other hand, are modeled by sliding the chains past one another longitudinally to search for the largest possible matching densities. The longitudinal movement of the chains relative to one another, out of register, approximately models the lateral searching of sequences in copolymeric chains during annealing.

One example of such a study involved modeling random and semi-blocky poly(diphenylsiloxane-co-dimethylsiloxane) copolymers. In this example, the chains were placed alongside one another in a two-dimensional array, with black squares representing dimethylsiloxane (DMS) units and white squares representing diphenylsiloxane (DPS) units. 'Like' squares neighboring each other in the same row were then viewed as coalescing into blocks the lengths of which are under scrutiny. It is thus possible to identify crystallizable DPS regions as distinct from a non-crystallizable DMS component, or units of the crystallizable DPS component that were not long enough to participate in the crystallization [165]. A value of the degree of crystallinity of a simulated sample can then be determined by counting the units involved in the matching sequences with respect to the total number of units of all the chains. The crystallites thus identified presumably act as crosslinking sites and reinforcing domains, providing the additional toughness the semi-blocky copolymers are known to have over their random counterparts.

Phenomenological theory

The phenomenological approach to elasticity theory does not consider the molecular constitution of networks. It is based on continuum mechanics and symmetry arguments rather than on molecular concepts [2, 166]. It attempts to fit stress-strain data with a minimum number of parameters, which are then used to predict other mechanical properties of the same material.

The elastic free energy given by the elementary and the more advanced theories in this area are symmetric functions of the three extension ratios λ_x, λ_y and λ_z. One may also express the dependence of the elastic free energy on strain in terms of three other variables, which are in turn functions of λ_x, λ_y and λ_z. In phenomenological theories of continuum mechanics, where only the observed behavior of the material is of concern rather than the molecular mechanism, these three functions are chosen to be

$$\begin{aligned} I_1 &= \lambda_1^2 + \lambda_2^2 + \lambda_3^2 \\ I_2 &= \lambda_1^2\lambda_2^2 + \lambda_1^2\lambda_3^2 + \lambda_1^2\lambda_3^2 \\ I_3 &= \lambda_1^2\lambda_2^2\lambda_3^2 \ . \end{aligned} \tag{8.22}$$

Here, I_1, I_2 and I_3 are referred to as the three strain invariants. The term invariant designates the property in which the value of the expression is independent of the choice of the coordinate frame. The elastic free energies of the phantom and the affine network models now take the simple form

$$\Delta A_{\text{el}} = FkT\left(I_1 - 3\right) \tag{8.23}$$

where F equates to $\xi/2$ and $\nu/2$ for the phantom and the affine network models, respectively. In general the elastic free energy of the network is expressed as

$$\Delta A_{\text{el}} = \Delta A_{\text{el}}\left(I_1, I_2, I_3\right). \tag{8.24}$$

The most general form of the elastic free energy may be written as a power series

$$\Delta A_{\text{el}} = \sum_{i,j,k=0}^{\infty} C_{ijk} (I_1 - 3)^i (I_2 - 3)^j (I_3 - 1)^k \qquad (8.25)$$

where C_{ijk} are the phenomenological coefficients. The simple case of the phantom and affine networks is obtained as the first term of the series

$$\Delta A_{\text{el}} = C_{100} (I_1 - 3). \qquad (8.26)$$

The elastic free energy of the so-called Mooney-Rivlin solid is obtained from equation (8.26) as

$$\Delta A_{\text{el}} = C_{100} (I_1 - 3) + C_{010} (I_2 - 3) \qquad (8.27)$$

where C_{100} and C_{010} correspond to $2C_1$ and $2C_2$ of the Mooney-Rivlin form, respectively. This form of the elastic free energy has been used widely in the treatment of data obtained from simple tension experiments on elastomers. It should be noted, however, that while the Mooney-Rivlin expression represents the behavior of rubbers under tension satisfactorily, it fails utterly in compression.

8.3 Stress-strain-temperature relationships

Here, we describe the stress-strain relations, or the equations of state, of networks by considering the homogeneous deformation of a cube of sides L_0 in the reference state and L_x, L_y, and L_z in the deformed state. More general states of deformation are discussed in several textbooks [2, 3, 5]. The reference state is defined as the state in which the network has been formed. The deformation ratios are defined as the ratio of the final to the reference state values of a dimension, such as $\lambda_x = L_x/L_0$, etc. The state of the network during formation is of particular importance in obtaining the correct expression for the stress. If the network is prepared in the presence of a diluent, the reference volume $V_0 = L_0^3$ equals the sum of the dry volume V_d of bulk polymer and the volume of solvent V_s.

Also, $\lambda_x = \alpha_x(v_{2c}/v_2)$, where α_x represents the ratio of the length along the x direction in the deformed state to that in the swollen but undistorted state, with similar definitions for the y and z directions. The quantity v_{2c} is the volume fraction of polymer during network formation and v_2 is the polymer volume fraction during the experiment (and is assumed to remain constant during network deformation). This approximation fails, obviously, if the sample is highly compressed or if the deformation is carried out while the sample is immersed in the solvent. The volume of a dry network remains approximately constant during deformation. This condition is that of incompressibility, which approximately holds for rubbers. The incompressibility condition is given in terms of deformation components by

$$\alpha_x \alpha_y \alpha_z = 1 \qquad (8.28)$$

and
$$\lambda_x \lambda_y \lambda_z = V/V_0 \tag{8.29}$$

The equation of state relating the components of the true stress tensor t_i to deformation is given for an incompressible material by the thermodynamic relation

$$t_i = \frac{2}{V} \lambda_i^2 \frac{\partial \Delta A_{\text{el}}}{\partial \lambda_i^2} , \qquad i = 1, 2, 3 \tag{8.30}$$

where the ith component t_i of the true stress tensor is defined as the force along the ith coordinate direction per unit deformed area. It should be noted that the three components t_i ($i = 1, 2, 3$) constitute the diagonal elements of the stress tensor $\boldsymbol{t}$, corresponding to the normal stresses acting on the three faces of the prism. The off-diagonal components correspond to shear stresses. In the particular case when the shear stresses equate to zero, the normal stresses are referred to as the principal stresses. In the following sections only simple tension or compression, which is the simplest state of principal stresses, is considered. The reader is referred to the literature [2, 3, 5] for applications to other states of stress.

Equation (8.30) allows the derivation of the three principal stresses from the elastic free energy of the network. This free energy follows either from the molecular theories discussed above or from the phenomenological formulation.

The quantity of greatest interest is the reduced stress $[f^*]$, defined by

$$[f^*] \equiv \frac{f v_2^{1/3}}{A_d \left(\alpha - \alpha^{-2} \right)} . \tag{8.31}$$

For uniaxial deformation, $\lambda_1 = \lambda$, $\lambda_2 = \lambda_3 = \lambda^{-1/2}$, and the force f is obtained by differentiating the elastic free energy given in equation (8.30) with respect to λ. For the affine and phantom network models, the force is obtained as

$$f = 2 \frac{FkT}{L_0} \left(\lambda - \frac{1}{\lambda^2} \right) \tag{8.32}$$

where F equates to $\xi/2$ and $\nu/2$ for the phantom and the affine network models, respectively. For the constrained-junction model, the uniaxial force f is obtained by differentiating equation (8.20) as specified in equation (8.30):

$$f = \left(\frac{\xi k T}{L_0} \right) \left(\lambda - \frac{1}{\lambda^2} \right) \left[1 + \frac{\lambda K(\lambda) - \lambda^{-2} K \left(\lambda^{-1} \right)}{\lambda - \lambda^{-2}} \right] \tag{8.33}$$

where L_0 is the length of the sample, and

$$K(\lambda) = B \left[\frac{\dot{B}}{B+1} + \kappa_0^{-1} \frac{\lambda^2 \dot{B} + B}{B + \kappa_0 \lambda^{-2}} \right], \quad \dot{B} = B \left[\frac{1}{\lambda^2 - 1} - \frac{2}{\lambda^2 + \kappa_0} \right] \tag{8.34}$$

The reduced force from equation (8.31) becomes [5,167]

$$[f^*] = \left(\frac{\xi kT}{V_d}\right) v_2^{1/3} \left[1 + \frac{\mu}{\xi}\frac{\alpha K(\lambda_x) - \alpha^{-2} K(\lambda_y^{-1})}{\alpha - \alpha^{-2}}\right]. \quad (8.35)$$

The reduced force may be interpreted as the shear modulus of the network [5]. According to equation (8.35), the reduced force consists of a term due to contributions from the phantom network and another due to the constraints. The contribution of constraints, proportional to the term μ/ξ in the brackets in equation (8.35) decreases as the network is stretched or swollen, an important feature in the comparisons of theory and experiment.

Stress-temperature relationships

The major component of elasticity of a network arises from the 'entropic elasticity' of the individual chains. This was the basic assumption of the early molecular theories of rubber elasticity [3]. A closer consideration of the statistics of the single chain shows that the rotational isomeric states allowable to each skeletal bond of the chain are not of the same energy, and stretching a chain or changing the temperature may move them from one isomeric minimum to another one. This results in an energetic contribution to the elasticity of a chain. Thus the total force acting on a network may be written as the sum of an entropic contribution, f_s, and an energetic contribution, f_e

$$f = f_s + f_e. \quad (8.36)$$

Force-temperature ('thermoelastic') relations lead to a quantitative assessment of the relative amounts of entropic and energetic components of the elasticity of the network.

In uniaxial deformation, the energetic contribution to the total elastic force [2,3,5,168] is given by the thermodynamically exact relation

$$\frac{f_e}{f} \equiv -T \left[\frac{\partial \ln(f/T)}{dT}\right]_{L,V}. \quad (8.37)$$

The subscripts L and V denote that differentiation is performed at constant length and volume. To carry out the differentiation indicated in equation (8.37), an expression for the total tensile force f is needed. One may use the expression given by equation (8.32) for the phantom network model. Applying the right-hand side of equation (8.37) to equation (8.32) leads to

$$\frac{f_e}{f} = \frac{T d \ln \langle r^2 \rangle_0}{dT}. \quad (8.38)$$

Equation (8.38) is important because the right-hand side relates to a microscopic quantity, $\langle r^2 \rangle_0$, and the left-hand side is the thermodynamic ratio of the energetic component of the force to the total force, both macroscopic quantities. It should be noted that equation (8.38) is obtained by using a molecular model. Experimentally, the determination of the force at constant volume is not easy. For this reason, expressions for the force measured at constant length and pressure p, or constant α and p, are used. These expressions are

$$\frac{f_e}{f} \equiv -T\left[\frac{\partial \ln(f/T)}{\mathrm{d}T}\right]_{L,p} - \frac{\beta T}{\alpha^3 - 1} \tag{8.39}$$

$$\frac{f_e}{f} \equiv -T\left[\frac{\partial \ln(f/T)}{\mathrm{d}T}\right]_{\alpha,p} - \frac{\beta T}{3} \tag{8.40}$$

where β is the thermal expansion coefficient of the network. It should be noted, however, that both of these equations are derived on the basis of the equation of state for simple molecular models and therefore are not quantities based purely on experimental data. Values of the energetic contribution for some typical elastomers are given elsewhere [5].

8.4 Swelling and gel collapse

As stated above, swelling of a network in a well-characterized solvent is a convenient means of obtaining information on the structure of networks, or conversely, of characterizing the polymer-solvent interactions when the network structure is known. In recent years, much emphasis has been placed on swelling of networks and their phase transitions under different activities of the network-solvent system. Large-scale volume transitions triggered by small changes in environmental variables directed attention to possible uses of swollen gels in the field of responsive-materials technologies. The transition involves the gel exuding solvent, for example upon decrease in temperature. The resulting shrinkage ('syneresis') is widely known as 'gel collapse', and is illustrated in Figure 8.7. The presence of charges on network chains facilitates the volume phase transitions in swollen gels.

The change in free energy of a network upon swelling is taken as the sum of the change in the elastic free energy ΔA_{el} and the change in free energy of mixing ΔA_{mix} and the contributions from ionic groups ΔA_i:

$$\Delta A = \Delta A_{\mathrm{el}} + \Delta A_{\mathrm{mix}} + \Delta A_i \tag{8.41}$$

where ΔA_{el} may be taken as any of the expressions resulting from a model, ΔA_{mix} is the free energy of mixing, and ΔA_i is the contribution of the ionic groups on the chains. The total chemical potential $\Delta \mu_1$ of solvent in the swollen network is obtained for the constrained-junction model as

$$\begin{aligned}\frac{\Delta \mu_1}{RT} &= \ln(1-v_2) + v_2 + \chi v_2^2 + \frac{1}{\lambda}\frac{\rho V_1}{M_c}\left[1 + \frac{\mu}{\xi}K\left(\lambda^2\right)\right] \\ &\quad -i\nu\left(\frac{V_1}{V_0 N_A}\right)\left(\frac{v_2}{v_{20}}\right), \end{aligned} \tag{8.42}$$

where v_2 is the volume fraction of polymer, χ is the Flory interaction parameter, V_1 is the molar volume of solvent, M_c is the molecular weight

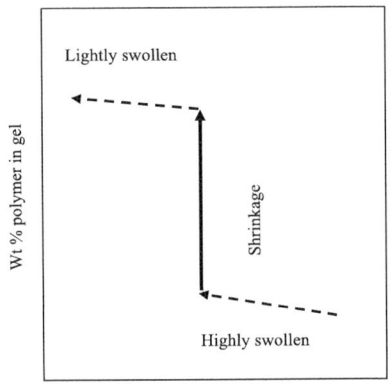

Fig. 8.7 A gel exuding solvent upon decrease in temperature, with the shrinkage ('syneresis') generally described as 'gel collapse'.

of a network chain, and λ is the extension ratio, defined for the swelling case as

$$\lambda = \left(\frac{V}{V_0}\right)^{1/3} = \left(\frac{n_1 V_1 + x V_1 n_2}{V_0}\right). \tag{8.43}$$

Here, x is the number of repeat units in one network chain, n_1 is the number of solvent molecules, n_2 is the total number of network chains in the system, i is the number of ionic groups on the chains, ν is the number of chains, and v_{20} is the volume fraction of chains during the formation of the network.

Equating the chemical potential to zero gives a relationship between the equilibrium degree of swelling and the molecular weight M_c. The relation for $M_{c,\text{ph}}$ is obtained for a tetrafunctional phantom network model as

$$M_{c,\text{ph}} = -\frac{\frac{1}{2}\rho V_1 \left(v_2/v_{20}\right)^{1/3}}{\ln\left(1 - v_2\right) + v_2 + \chi v_2^2 - i\nu \left(V_1/V_0 N_A\right)\left(v_2/v_{20}\right)} \tag{8.44}$$

where v_2 denotes the equilibrium degree of swelling.

For the affine network model, the molecular weight between crosslinks $M_{c,\text{af}}$ is obtained as

$$M_{c,\text{af}} = -\frac{\rho V_1 \left(v_2^{1/3} - (v_2/2v_{20})\right)}{\ln\left(1 - v_2\right) + v_2 + \chi v_2^2 - i\nu \left(V_1/V_0 N_A\right)\left(v_2/v_{20}\right)}. \tag{8.45}$$

Equations (8.44) and (8.45) form the basis of determining the molecular weights of network chains [1]. Equation (8.45) is the celebrated Flory-Rehner expression, improved by the incorporation of ionic group effects and the degree of swelling during network formation. The behavior of a swollen network is found to obey the constrained-junction model [136], which predicts results between those of the phantom network model given by equation (8.41) and the affine network model given by equation (8.45). Alternatively, the chemical potential expression may be solved for v_2, leading to a value for the degree of swelling of the network. The result shows that the degree of swelling increases as the chain length between crosslinks increases. The dominant forces that operate in swollen uncharged gels are van der Waals forces, hydrogen-bonding forces, hydrophobic forces, and forces resulting from chain entropies.

For known network structure, equation (8.44) or (8.45) may be solved for the polymer-solvent interaction parameter χ. Indeed, swelling serves as a convenient means of determining values of this parameter.

When the network chains contain ionic groups, there will be additional forces that affect their swelling properties. Translational entropy of counter-ions, Coulomb interactions, and ion-pair multiplets are forces that lead to interesting phenomena in ion-containing gels. These phenomena were studied in detail by Khokhlov and collaborators [169–171]. The free energy of the networks used by this group is

$$\Delta A = \Delta A_{\text{mix}} + \Delta A_{\text{el}} + \Delta A_{\text{trans}} + \Delta A_{\text{Coulomb}} \tag{8.46}$$

where ΔA_{trans} and $\Delta A_{\text{Coulomb}}$ are the contributions to the elastic free energy of the networks from the translational entropy of the counterions and the free energy of Coulomb interactions. Several interesting features of gels are obtained through the use of equation (8.46). A network chain of a polyampholyte gel contains both positive and negative charges. The liquid phase in the swollen polyampholyte gel may contain additional counterions. The theoretical and experimental literature on such gels was reviewed recently by Nisato and Candau [171]. In gels containing ions, the gel swells excessively when the ion-containing groups are fully dissociated, because of the tendency of the free counterions to occupy as much space as possible. In the other extreme case, called the ionomer regime, counterions are condensed on oppositely-charged monomer units, forming ion pairs followed by formation of multiplets. This decreases the osmotic pressure of the gel and results in its collapse. The conditions for ion-pair formation and physical and chemical factors leading to gel swelling and collapse have been discussed by Philippova et al. [170].

Theory versus experiment

The constrained-junction theory successfully describes most of the features of numerous investigations that have been made on stress-strain relationships involving a variety of types of deformations [1, 2, 5, 149, 172, 173]. Specifically, the decrease in modulus $[f^*]$ with increase in elongation is viewed as due to the deformation becoming more non-affine as the stretching of the network chains decreases chain-junction entangling; this in turn increases junction fluctuations. The observation that the decrease in $[f^*]$ is less in the case of swollen networks results from the swelling diluent decreasing some of this entangling even at low elongations. The observation that the isotherms become approximately horizontal in compression is also successfully predicted by the theory [5, 174]. The upturn in $[f^*]$ at very high elongations is a non-Gaussian effect not accounted for in the theory and must be treated separately. Interpretation of the upturn in $[f^*]$ in terms of the limited extensibility of the network chains first requires demonstration that reinforcement from strain-induced crystallization is not a significant part of the effect [175]. Failure to test for this by swelling or increase in temperature has caused a great deal of confusion in the literature.

Experiment and theory are in at least approximate agreement regarding swelling, in that estimates of degree of crosslinking from equilibrium swelling are generally in fair agreement with values obtained from mechanical properties [1, 2, 176].

A major unresolved issue concerns the observation that the swelling or dilation modulus frequently goes through an unanticipated maximum with increase in degree of swelling [177, 178]. Modifications of existing theories to reproduce this feature quantitatively have not been very successful. In a worst-case scenario, this discrepancy could undermine one of the crucial assumptions in swelling theory, namely that the mixing and elastic contributions to the swelling free energy are separable and simply additive [178].

The reduced birefringence shows some properties that parallel the reduced stress or modulus. In other respects, however, it can be quite different, showing for example significant increases with increase in degree of swelling. Nonetheless, experimental results and those from the constrained-junction theory seem to be in at least fair agreement.

Experimental results on the elongation and swelling dependence of the reduced orientation seem to be well reproduced by theory [137]. Preparing networks by crosslinking a polymer in solution [179] can give very useful orientation results, but it is essential in such cases to account for changes in reference state in the interpretation of the data [180].

The most important elasticity result of this type is obtained by small-angle neutron scattering. In particular, such studies yield information on how the radius of gyration of the network chains transforms with elongation [5]. Preliminary comparisons between theory and experiment are quite encouraging, but many issues remain unresolved in this relatively new approach to characterizing elastomeric behavior.

8.5 Elastomers with internal microstructure

Liquid-crystalline elastomers

In the case of polymers, liquid crystallinity ('mesomorphism') frequently occurs in very stiff chains such as the Kevlars^{TM} and other aromatic polyamides [181]. It can also occur with flexible chains and it is these flexible chains in the elastomeric state that are relevant here. Some reasons why such liquid-crystalline elastomers are of particular interest are the fact that: (i) they can be extensively deformed, (ii) the deformation produces alignment of the chains, and (iii) alignment of the chains is central to the formation of liquid-crystalline phases. A comprehensive treatise of physical principles behind these and other effects is given in [182].

A liquid-crystalline phase lies between the limits of completely ordered (crystalline) and completely disordered (isotropic) phases. In the case of *nematic* phases, the disorder that gives rise to the liquidity or fluidity is the sliding of chain segments relative to one another, to place them out of register. There are also a variety of *smectic* liquid-crystalline phases [183], in which layers of molecules or chain sequences occur in layers that are disordered relative to one another. In contrast, *cholesteric* phases have layers of nematic arrangements that are stacked in rotated arrangements, and a similar stacking occurs in the case of the *discotics*.

There can be liquid-crystalline arrangements that involve the side chains attached to the chain backbones. The three possibilities are groups that can form liquid-crystalline phases occurring in the backbone [184], in the side chains, and in both (in what are called 'combined'

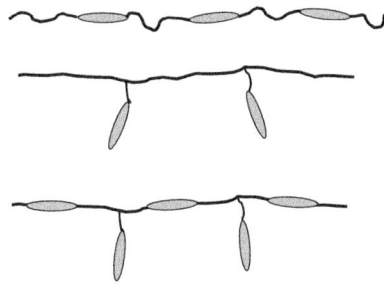

Fig. 8.8 Liquid-crystalline polymers in which the mesomorphic sequences occur in the chain backbones, the side-chains, or in both ('combination' structures).

structures). These arrangements are illustrated in Figure 8.8. Polymer chains that are very stiff can also appear in *rigid-rod networks*. Such networks deform in response to a stress by rearranging these rigid chains relative to one another instead of the usual unwinding of a flexible chain from a compact state to one that is more extended in the direction of the strain.

Main-chain liquid-crystalline elastomers

The well-studied polymer in this category is poly(diethylsiloxane) (PDES) [–Si(OC$_2$H$_5$)$_2$O–]. It forms a nematic mesophase, and relevant studies have focused on a wide range of properties. For example, stretching a PDES elastomer aligns its chains in a way that changes the temperature range over which the liquid-crystalline phase is stable. Several symmetric polysiloxane elastomers having longer side chains also show liquid-crystalline behavior, and these range from poly(di-n-propylsiloxane) to poly(di-n-decylsiloxane). The temperatures at which the nematic phase disappears (the 'isotropization temperatures') show a very interesting monotonic increase with increase in the number of methylene groups in the side chains [185]. Beyond poly(di-n-decylsiloxane), complications arise, probably because the side chains are sufficiently long to crystallize themselves. Some polysiloxanes with cyclic groups in the backbone also form mesophases as do some of the polyphosphazenes [186].

The property these materials exhibit of greatest relevance here are their stress-strain isotherms. Some obtained for PDES [187] as a function of degree of mesomorphic structure are shown schematically in Figure 8.9. The curves exhibit yield points similar to those shown by partially crystalline polymers, and the overall shapes of the curve differ greatly with increase in the amounts of mesophase present, either present initially or induced by the deformation. As is generally the case, formation of a second phase leads to irreversibility in the stress-strain isotherms, however, the yielding plateau in liquid-crystalline elastomers is due to the so-called polydomain-monodomain transition [188], i.e. the alignment of initially randomly oriented liquid-crystalline domains under stress.

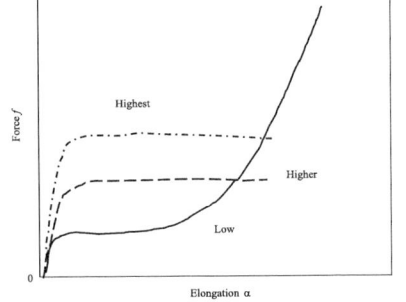

Fig. 8.9 Sketch showing stress-strain isotherms for a main-chain liquid-crystalline elastomer such as poly(diethylsiloxane), as a function of the amount of mesophase present.

Side-chain liquid-crystalline elastomers

Of considerable interest is the orientation of the mesogenic groups and the chain backbones to which they are attached. Frequently studied backbones include siloxanes and acrylates, but a variety of other structures have also been studied, including amphiphilics. The side chains in these structures can rearrange into positions either parallel or perpendicular to the deformed chain backbone [189]. The outcome depends particularly on the nature and length of the flexible spacer connecting the mesogenic groups to the chain backbone. As expected, the physical properties can become strongly anisotropic.

Some liquid-crystalline materials of this type could be oriented by imposing an electric or magnetic field [190]. The chains could also be aligned when these liquid-crystalline elastomers are deformed (generally in elongation but also in some cases in compression), and then crosslinked into network structures. The mesogenic behavior of such networks obviously depends strongly on their structures, in particular the degree of crosslinking, and composition (in the case of copolymers). The phase transitions exhibited also depend significantly on spacer length, with closer coupling between the mesogenic groups and the polymer backbone tending to make the system more sensitive to mechanical deformations.

The thermoelastic behavior of these materials has also been reported in many experiments, summarized in [182]. Such experiments often resolve the nematic-to-isotropic transition into entropic and enthalpic contributions and provide values of the corresponding energetic and entropic parts of the elastic force [191].

Considerably less work has been done on discotic liquid-crystalline elastomers and the same seems to be true for smectic elastomers even though some of them have the additional interesting property of being chiral [192]! Some liquid-crystalline elastomers are ferroelectric (possess spontaneous electric polarization) or piezoelectric (become electrically polarized when mechanically strained, and become mechanically strained when exposed to an electric field) [193]. Finally, some liquid-crystalline elastomers exhibit interesting photonic effects such as non-linear optical properties [194].

Theory

The unusual properties cited above have been the focus of numerous theoretical investigations [182]. The simplest physical picture of nematic elasticity can be visualized by the lattice theory, in which semi-flexible chains are assumed to be embedded in an isotropic lattice of solvent molecules. For example, calculations based on a lattice model show an abrupt jump in the segmental orientation function at a finite elongation, and this corresponds to the isotropic-nematic transition. Below the transition, the sample behaves close to that of isotropic Gaussian elasticity. At a fixed uniaxial force, the transition is marked by a sudden elongation, resulting from the abrupt alignment of the segments along the direction of the force during the transition. Above the transition point the segments are highly oriented along the direction of stretch, and the network becomes nematic, and its elastomeric behavior changes correspondingly.

Bioelastomers

There are a number of crosslinked proteins that are elastomeric, and investigation of their properties may be used to obtain insights into elastic behavior in general. For example, elastin [195–197], which occurs in mammals, illustrates the relevance of several molecular characteristics to the achievement of rubberlike properties. First, a high degree of chain flexibility is achieved in elastin by its chemically irregular structure, and by choices of side groups that are almost invariably very small. Since strong intermolecular interactions are generally not conducive to good elastomeric properties, the choices of side chains are also almost always restricted to non-polar groups. Finally, elastin has a glass transition temperature of approximately 200 °C in the dry state, which means it would be elastomeric only above this temperature. Nature, however, apparently also knows about 'plasticizers'. Elastin, as used in the body, is invariably swollen with sufficient aqueous solutions to bring its glass transition temperature below the operating temperature of the body.

Elastin chains are crosslinked *in vivo* in a highly specific manner, using techniques very unlike those usually used to cure commercial elastomers. The crosslinking occurs through lysine repeat units, the number and placement of which along the chains are carefully controlled in the synthesis of elastin in the ribosomes. An analogous reaction has been carried out commercially on perfluoroelastomers, which are usually very difficult to crosslink because of their inertness. Nitrile side groups placed along the chains are trimerized to triazine, thus giving similarly stable, aromatic crosslinks.

Another bioelastomer, found in some insects, is resilin [198]. It is an unusual material because it is thought to have a relatively high efficiency in storing elastic energy (i.e. very small losses due to viscous effects). A molecular understanding of this very attractive property could obviously have considerable practical as well as fundamental importance.

Filled elastomers

This topic is part of the general subject of polymer nanocomposites, which are of increasing interest, particularly with regard to the development of molecular theories [199]. Relevant here is the fact that elastomers, particularly those that cannot undergo strain-induced crystallization, are generally compounded with a reinforcing filler. The two most important examples are the addition of carbon black to natural rubber and to some synthetic elastomers, and silica to polysiloxane ('silicone') rubbers and other elastomers [200]. The advantages obtained include improved abrasion resistance, tear strength, and tensile strength [201]. Disadvantages include increases in hysteresis (and thus heat buildup) and compression set (permanent deformation). The mechanism of the reinforcement obtained is only poorly understood in molecular terms. Some elucidation might be obtained by incorporating the fillers in a more carefully controlled manner, particularly using sol-gel technology, as described below.

Sol-gel generation of ceramic-like phases

In the most important example, hydrolysis of an alkoxysilane such as tetraethoxysilane or tetraethylorthosilicate (TEOS),

$$\mathrm{Si(OEt)_4 + 2H_2O \rightarrow SiO_2 + 4EtOH} \tag{8.47}$$

is used to precipitate very small, well-dispersed particles of silica into a polymeric material [55, 202]. A variety of substances, generally bases, catalyze this reaction, which occurs quite readily near room temperature. Another example would be the catalyzed hydrolysis of a titanate, in the *in-situ* precipitation of titania. Silica, titania, and related ceramic-like fillers thus produced within an elastomer have been shown to give good reinforcement in a variety of deformations.

Methods for carrying out sol-gel reactions within elastomers

The most common approach is to generate the filler after curing. In this technique, the polymer is first cured or vulcanized into a network structure using any of the well-known crosslinking techniques. The network is then swelled with the organosilane or related molecule to be hydrolyzed, after which it is exposed to water at room temperature, in the presence of a catalyst, for a period of time (up to several hours for thick samples) The swollen sample can be either placed directly into an excess of water containing the catalyst, or merely exposed to the vapors from the catalyst-water solution. Drying the sample then gives an elastomer that is filled, and thus reinforced, with the ceramic particles resulting from the hydrolysis and condensation reactions.

The polymer used typically has end groups, such as hydroxyls, that can participate in the hydrolysis-condensation reactions. Such end groups provide better bonding between the two rather disparate phases, but bonding agents may also be introduced for this purpose. It is thus possible to mix hydroxyl-terminated chains (such as those of PDMS) with excess TEOS, which then serves simultaneously to tetrafunctionally end-link the PDMS into a network structure and to act as the source of silica upon hydrolysis [3].

In the previous two techniques, removal of the unreacted TEOS and the ROH alcohol by-product causes a significant decrease in volume [203], which could be disadvantageous in some applications. One way of overcoming this problem is by precipitating the particles into a polymer that is inert under the hydrolysis conditions, for example, vinyl-terminated PDMS. After removal of volatiles, the elastomer in this mixture can be subsequently crosslinked, with only the usual, very small volume changes occurring in any curing process.

Approximately spherical particles

Because of the nature of the *in-situ* precipitation, the particles are well dispersed and are essentially unagglomerated (as demonstrated by electron microscopy). The mechanism for their growth seems to involve simple homogeneous nucleation, and since the particles are separated by polymer, they don't have the opportunity to coalesce. The particles are typically relatively monodisperse, with most of them having diameters in the range 100–200 Å. A variety of catalysts work well in the typical hydrolysis reactions used, including acids, bases, and salts. Basic catalysts give precipitated phases that are generally well-defined particles,

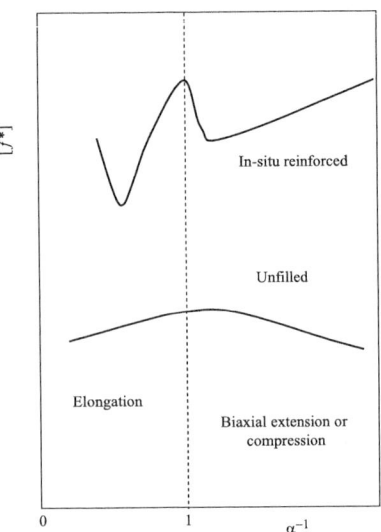

Fig. 8.10 Stress-strain isotherms for *in-situ* reinforced elastomers in elongation (region to the left of the vertical dashed line, with $\alpha^{-1} < 1$), and in biaxial extension (compression) (to the right, with $\alpha^{-1} > 1$).

whereas the acidic catalysts give more poorly-defined, diffuse particles.

The reinforcing ability of such *in-situ* generated particles has been amply demonstrated for a variety of deformations, including uniaxial extension (simple elongation), biaxial extension (compression), shear, and torsion. In the case of uniaxial extension, the modulus $[f^*]$ frequently increases by more than an order of magnitude, with the isotherms generally showing the upturns at high elongations that are the signature of good reinforcement [204]. Some typical results in elongation and biaxial extension are shown schematically in Figure 8.10. As is generally the case in filled elastomers, there is considerable irreversibility in the isotherms, which is thought to be due primarily to some irrecoverable sliding of the chains over the surfaces of the filler particles upon being strained.

Some fillers other than silica, for example titania, do give stress-strain isotherms that are reversible, indicating interesting differences in surface chemistry, including increased ability of the chains to slide along the particle surfaces.

These *in-situ* generated silica fillers also give increased resistance to creep or compression set in cyclic deformations. The *in-situ* filled PDMS samples showed very little compression set, and these samples also exhibited increased thermal stability.

The low-temperature properties of some of these peculiarly-filled materials have also been studied by calorimetric techniques. Of particular interest is the way in which reinforcing particles can affect the crystallization of a polysiloxane, both in the undeformed state and at high elongations. Finally, a number of studies using x-ray and neutron scattering have also been carried out on filled elastomers. The results on the nature of the particles are generally consistent with those obtained by electron microscopy.

Fillers with controlled interfaces

By choosing the appropriate chemical structures, chains that span filler particles in PDMS-based composites can be designed so that they are either durable, are breakable irreversibly, or are breakable reversibly [205, 206].

There has been some interest in generating silica-like particles using templates, as occurs in Nature in the biosilicification processes. Various particle shapes have been obtained; platelet forms would be of particular interest with regard to their abilities to provide reinforcement and decrease permeability, as occurs with the layered fillers described below.

Other inorganic particles

Polyhedral oligomeric silsesquioxanes (POSS)

These fillers are cage-like silicon-oxygen structures, and have been called the smallest possible silica particles [207, 208]. The most common structure has eight silicon atoms, each carrying a single organic group. The

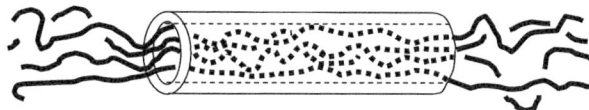

Fig. 8.11 Threading polymer chains through a cavity by polymerizing monomer absorbed into a zeolite.

particles on which none of the groups are functionally reactive can be simply blended into elastomers such as PDMS using the usual mixing or compounding techniques. In this case, the inert groups are chosen to improve miscibility with the elastomeric host matrix. POSS molecules having one reactive functional group can be attached to a polymer as side chains. Those with two reactive groups can be incorporated into polymer backbones by copolymerization, and those with more than two can be used for forming crosslinks, thus generating network structures.

Porous particles

Some fillers such as zeolites are sufficiently porous to accommodate monomers, which can then be polymerized. This threads the resulting chains through the cavities, with unusually intimate interactions between the reinforcing phase and the host elastomeric matrix [209]. This is illustrated in Figure 8.11. Unusually good reinforcement is generally obtained. Also, because of the constraints imposed by the cavity walls, these confined polymers frequently show no glass transition temperatures or melting points. PDMS chains have also been threaded through cyclodextrins, to form *pseudo*-rotaxanes [210].

Layered fillers

Exfoliating layered particles such as the clays, mica, or graphite are being used to provide very effective reinforcement of elastomers at loading levels much smaller than in the case of solid particles such as carbon black and silica [211]. Other properties can also be substantially improved, including increased resistance to solvents, and reduced permeability and flammability.

Miscellaneous fillers

There are a variety of miscellaneous fillers that are of interest for reinforcing elastomers. Examples are ground-up silica xerogels [212], carbon-coated silica [213], and functionalized silica particles [214].

Organic glassy particles deformable into ellipsoids

It is also possible to obtain reinforcement of an elastomer by polymerizing a monomer such as styrene to yield hard glassy domains within the elastomer. Roughly spherical polystyrene (PS) particles are formed, and good reinforcement is obtained. It is possible to improve bonding onto the filler particles, for example by including some trifunctional $R'Si(OC_2H_5)_3$ in the hydrolysis (where R' is an unsaturated group). The

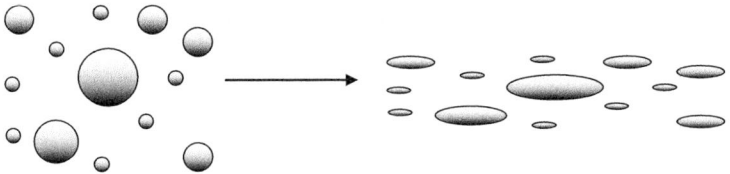

Fig. 8.12 Deformation of spherical filler particles into prolate (needle-shaped) ellipsoids.

R' groups on the particle surfaces then participate in the polymerization, thereby bonding the elastomer chains to the reinforcing particles.

The PS domains have a relatively low glass transition temperature ($T_g \approx 100\,°\mathrm{C}$) and are totally amorphous. Thus, it is possible to convert the essentially spherical PS particles into ellipsoids [215]. First, the PS-elastomer composite is raised to a temperature well above the T_g of PS. The composite is then deformed, and cooled while in the stretched state. The particles in it are thereby deformed into ellipsoids, and retain this shape when cooled. This is illustrated in Figure 8.12. Uniaxial deformations of the composite give prolate (needle-shaped) ellipsoids, and biaxial deformations give oblate (disc-shaped) ellipsoids. In these anisotropic materials, elongation moduli in the direction of the stretching were found to be significantly larger than those of the untreated PS-elastomer composite, whereas in the perpendicular direction they were significantly lower. Such differences were to be expected from the anisotropic nature of the systems. In the case of non-spherical particles in general, degrees of orientation are also of considerable importance. One interest here is the anisotropic reinforcements such particles provide, and there have been simulations to better understand the mechanical properties of such composites [216]. Of, course, the ellipsoidal particles can be removed by dissolving away the host polymer matrix and then redispersing them randomly into another elastomer prior to its crosslinking.

Nanotubes and metal particles

Carbon nanotubes are also of considerable interest with regard to both reinforcement and possible increases in electrical conductivity [217, 218]. There is considerable interest in characterizing the flexibility of these nanotubes, in minimizing their tendencies to aggregate, and in maximizing their miscibilities with inorganic as well as organic polymers.

Incorporating reinforcing particles that respond to a magnetic field is important with regard to aligning even spherical particles to improve mechanical properties anisotropically. Considerable anisotropy in structure and mechanical properties can be obtained [219]. Specifically, the reinforcement is found to be significantly higher in the direction parallel to the magnetic lines of force.

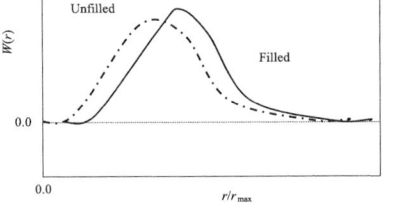

Fig. 8.13 Sketches of radial distributions for two network chains as a function of the fraction of their full extension. The presence of filler particles is seen to increase the end-to-end separation of the chains.

Simulations on fillers

Monte Carlo computer simulations have been carried out on a variety of filled elastomers [220] in an attempt to obtain a better molecular interpretation of how such dispersed phases reinforce elastomers. The approach taken enabled estimation of the effect of the excluded volume of the filler particles on the network chains and on the elastic properties of the networks. In the first step, distribution functions for the end-to-end vectors of the chains were obtained by applying Monte-Carlo methods to rotational isomeric state representations of the chains [221]. Conformations of chains which overlapped with any filler particle during the simulation were rejected. The distributions move to higher values of the end-to-end separation, as is illustrated in Figure 8.13. The resulting perturbed distributions were then used in the three-chain elasticity model [2] to obtain the desired stress-strain isotherms in elongation. These isotherms showed substantial increases in stress and modulus with increase in filler content and elongation that are in at least qualitative agreement with experiment. This is illustrated in Figure 8.14.

In the case of non-spherical filler particles, it has been possible to simulate the anisotropic reinforcement obtained, for various types of particle orientations [222]. Different types and degrees of particle agglomeration can also be investigated.

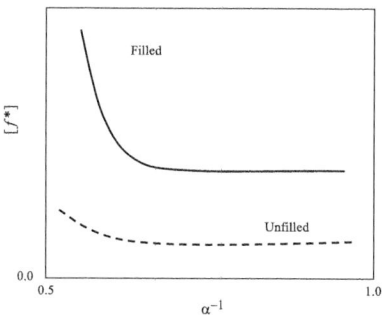

Fig. 8.14 Mooney-Rivlin isotherms from the distributions described in Figure 8.13, showing the resulting increases in modulus due to the chains being stretched by the presence of the filler particles.

Acknowledgement

James Mark wished to acknowledge the financial support generously provided by the National Science Foundation through Grant DMR-0314760 (Polymers Program, Division of Materials Research).

References

[1] Flory, P. J. (1953). *Principles of Polymer Chemistry*. Cornell University Press, Ithaca, NY.

[2] Treloar, L. R. G. (1975). *The Physics of Rubber Elasticity* (3rd edn). Clarendon Press, Oxford.

[3] Erman, B. and Mark, J. E. (1997). *Structures and Properties of Rubberlike Networks*. Oxford University Press, New York.

[4] Mark, J. E., Ngai, K. L., Graessley, W. W. et al. (2004). *Physical Properties of Polymers* (3rd edn). Cambridge University Press, Cambridge.

[5] Mark, J. E. and Erman, B. (2007). *Rubberlike Elasticity. A Molecular Primer* (2nd edn). Cambridge University Press, Cambridge.

[6] Erman, B. and Mark, J. E. (2010). In *Comprehensive Polymer Science* (ed. A. R. Khokhlov and L. Leibler). Elsevier, Amsterdam.

[7] Gent, A. N. (ed.) (2001). *Engineering with Rubber. How to Design Rubber Components* (2nd edn). Hanser, New York.

[8] Morawetz, H. (1985). *Polymers: The Origins and Growth of a Science*. Wiley-Interscience, New York.

[9] Zhou, D. and Mark, J. E. (2004). *J. Macromol. Sci. – Pure Appl. Sci.*, **41**, 1221.

[10] Brandrup, J., Immergut, E. H., and Grulke, E. A. (ed.) (1999). *Polymer Handbook* (4th edn). Wiley, New York.

[11] Mark, J. E. (ed.) (2007). *Physical Properties of Polymers Handbook* (2nd edn). Springer-Verlag, New York.
[12] Gent, A. N., Liu, G. L., and Mazurek, M. (1994). *J. Polym. Sci., Polym. Phys. Ed.*, **32**, 271.
[13] Shibayama, M., Takahashi, H., Yamaguchi, H. et al. (1994). *Polymer*, **35**, 2944.
[14] Trautenberg, H. L., Sommer, J.-U., and Goritz, D. (1995). *J. Chem. Soc. Faraday Trans.*, **91**, 2649.
[15] Besbes, S., Bokobza, L., Monnerie, L. et al. (1995). *Macromolecules*, **28**, 231.
[16] Hedden, R. C., Saxena, H., and Cohen, C. (2000). *Macromolecules*, **33**, 8676.
[17] Mark, J. E. (1994). *Acc. Chem. Res.*, **27**, 271.
[18] Roland, C. M. and Buckley, G. S. (1991). *Rubber Chem. Technol.*, **64**, 74.
[19] Oikawa, H. (1992). *Polymer*, **33**, 1116.
[20] Subramanian, P. R. and Galiatsatos, V. (1993). *Macromol. Symp.*, **76**, 233.
[21] Kloczkowski, A., Mark, J. E., and Erman, B. (1991). *Macromolecules*, **24**, 3266.
[22] Bahar, I., Erman, B., Bokobza, L. et al. (1995). *Macromolecules*, **28**, 225.
[23] Labana, S. S. and Dickie, R. A. (ed.) (1984). *Characterization of Highly Cross-linked Polymers.* American Chemical Society, Washington, DC.
[24] Dusek, K. (1986). *Adv. Polym. Sci.*, **78**, 1.
[25] Bahar, I., Erman, B., and Baysal, B. (1986). *Macromolecules*, **19**, 1703.
[26] Erman, B. and Mark, J. E. (1989). *Ann. Rev. Phys. Chem.*, **40**, 351.
[27] Holden, G. (2000). *Understanding Thermoplastic Elastomers* (2nd edn). Hanser, Munich.
[28] Mark, J. E. (1993). *New J. Chem.*, **17**, 703.
[29] Johnson, R. M. and Mark, J. E. (1972). *Macromolecules*, **5**, 41.
[30] Greene, A., Smith, Jr, K. J., and Ciferri, A. (1965). *Trans. Faraday Soc.*, **61**, 2772.
[31] Urayama, K. and Kohjiya, S. (1998). *Eur. Phys. J. B*, **2**, 75.
[32] Kushner, A. M., Gabuchian, V., Johnson, E. G. et al. (2007). *J. Am. Chem. Soc.*, **129**, 14110.
[33] Santangelo, P. G. and Roland, C. M. (2003). *Rubber Chem. Technol.*, **76**, 892.
[34] Santangelo, P. G. and Roland, C. M. (1995). *Rubber Chem. Technol.*, **68**, 124.
[35] Flory, P. J. (1960). *Trans. Faraday Soc.*, **56**, 722.
[36] Mott, P. H. and Roland, C. M. (2000). *Macromolecules*, **33**, 4132.
[37] Sperling, L. H. (1981). *Interpenetrating Polymer Networks and Related Materials.* Plenum, New York.
[38] Yoo, S. H., Cohen, C., and Hui, C.-Y. (2006). *Polymer*, **47**, 6226.

[39] Gibson, H. W., Bheda, M. C., and Engen, P. T. (1994). *Prog. Polym. Sci.*, **19**, 843.
[40] Karino, T., Okumura, Y., Ito, K. *et al.* (2004). *Macromolecules*, **37**, 6177.
[41] Granick, S. and Rubinstein, M. (2004). *Nature Materials*, **63**, 586.
[42] Johnson, J. A., Turro, N. J., Koberstein, J. T. *et al.* (2010). *Prog. Polym. Sci.*, **35**, 332.
[43] Graessley, W. W. and Pearson, D. S. (1977). *J. Chem. Phys.*, **66**, 3363.
[44] de Gennes, P.-G. (1999). *Physica A*, **271**, 231.
[45] Gong, J. P., Katsuyama, Y., Kurokawa, T. *et al.* (2003). *Adv. Mater.*, **15**, 1155.
[46] Webber, R. E., Creton, C., Brown, H. R. *et al.* (2007). *Macromolecules*, **40**, 2919.
[47] Yu, Q. M., Tanaka, Y., Furukawa, H. *et al.* (2009). *Macromolecules*, **42**, 3852.
[48] Kaneko, D., Tada, T., Kurokawa, T. *et al.* (2005). *Adv. Mater.*, **17**, 535.
[49] Jang, S. S., Goddard, III, W. A., Yashaar, M. *et al.* (2007). *J. Phys. Chem. B*, **111**, 14440.
[50] Brown, H. R. (2007). *Macromolecules*, **40**, 3815.
[51] Haraguchi, K. and Takehisa, T. (2002). *Adv. Mater.*, **14**, 1120.
[52] Wang, Q., Mynar, J. L., Yoshida, M. *et al.* (2010). *Nature Lett.*, **463**, 339.
[53] Yanagioka, Y., Toney, M. F., and Frank, C. W. (2009). *Macromolecules*, **42**, 1331.
[54] Flory, P. J. (1976). *Proc. R. Soc. London, A*, **351**, 351.
[55] Mark, J. E., Allcock, H. R., and West, R. (2005). *Inorganic Polymers* (2nd edn). Oxford University Press, New York.
[56] Erman, B. and Mark, J. E. (1998). *Macromolecules*, **31**, 3099.
[57] Mark, J. E., Rahalkar, R. R., and Sullivan, J. L. (1979). *J. Chem. Phys.*, **70**, 1794.
[58] Langley, N. R. (1968). *Macromolecules*, **1**, 348.
[59] Graessley, W. W. (1982). *Adv. Polym. Sci.*, **47**, 67.
[60] Heinrich, G., Straube, E., and Helmis, G. (1988). *Adv. Polym. Sci.*, **85**, 33.
[61] Sharaf, M. A. and Mark, J. E. (1995). *J. Polym. Sci., Polym. Phys. Ed.*, **33**, 1151.
[62] Venkatraman, S. J. (1993). *Appl. Polym. Sci.*, **48**, 1383.
[63] Andrady, A. L., Llorente, M. A., Sharaf, M. A. *et al.* (1981). *J. Appl. Polym. Sci.*, **26**, 1829.
[64] Clarson, S. J., Mark, J. E., and Semlyen, J. A. (1987). *Polym. Comm.*, **28**, 151.
[65] DeBolt, L. C. and Mark, J. E. (1987). *Macromolecules*, **20**, 2369.
[66] Huang, W., Frisch, H. L., Hua, Y. *et al.* (1990). *J. Polym. Sci., Polym. Chem. Ed.*, **26**, 1807.
[67] Rigbi, Z. and Mark, J. E. (1986). *J. Polym. Sci., Polym. Phys. Ed.*, **24**, 443.

[68] de Gennes, P. G. (1979). *Scaling Concepts in Polymer Physics*. Cornell University Press, Ithaca, NY.
[69] Iwata, K. and Ohtsuki, T. (1993). *J. Polym. Sci., Polym. Phys. Ed.*, **31**, 441.
[70] Mandelkern, L. (2003). *Crystallization of Polymers*. Volume I. Cambridge University Press, Cambridge.
[71] Su, T.-K. and Mark, J. E. (1977). *Macromolecules*, **10**, 120.
[72] Llorente, M. A., Andrady, A. L., and Mark, J. E. (1981). *J. Polym. Sci., Polym. Phys. Ed.*, **19**, 621.
[73] Mohsin, M. A., Berry, J. P., and Treloar, L. R. G. (1986). *Brit. Polym. J.*, **18**, 145.
[74] Xu, P. and Mark, J. E. (1991). *Makromol. Chemie*, **192**, 567.
[75] Wen, J. and Mark, J. E. (1994). *Polym. J.*, **26**, 151.
[76] Mooney, M. (1948). *J. Appl. Phys.*, **19**, 434.
[77] Rivlin, R. S. (1948). *Phil. Trans. R. Soc., Part A*, **241**, 379.
[78] Obata, Y., Kawabata, S., and Kawai, H. (1970). *J. Polym. Sci., Part A-2*, **8**, 903.
[79] Wang, S. and Mark, J. E. (1992). *J. Polym. Sci., Polym. Phys. Ed.*, **30**, 801.
[80] Yanyo, L. C. and Kelley, F. N. (1987). *Rubber Chem. Technol.*, **60**, 78.
[81] Smith, T. L., Haidar, B., and Hedrick, J. L. (1990). *Rubber Chem. Technol.*, **63**, 256.
[82] Shah, G. B. (2004). *J. Appl. Polym. Sci.*, **94**, 1719.
[83] Andrady, A. L., Llorente, M. A., and Mark, J. E. (1991). *Polym. Bulletin*, **26**, 357.
[84] Wen, J., Mark, J. E., and Fitzgerald, J. J. (1994). *J. Macromol. Sci., Macromol. Rep.*, **A31**, 429.
[85] Kaneko, Y., Watanabe, Y., Okamoto, T. *et al.* (1980). *J. Appl. Polym. Sci.*, **25**, 2467.
[86] Mazan, J., Leclerc, B., Galandrin, N. *et al.* (1995). *Eur. Polym. J.*, **31**, 803.
[87] Tanaka, T. (1978). *Phys. Rev. Lett.*, **40**, 820.
[88] Mark, J. E. (1989). *J. Appl. Polym. Sci., Symp.*, **44**, 209.
[89] Yu, C. U. and Mark, J. E. (1974). *Macromolecules*, **7**, 229.
[90] Mark, J. E. (2003). *Prog. Polym. Sci.*, **28**, 1205.
[91] Bokobza, L. and Nugay, N. (2001). *J. Appl. Polym. Sci.*, **81**, 215.
[92] Besbes, S., Cermelli, I., Bokobza, L. *et al.* (1992). *Macromolecules*, **25**, 1949.
[93] Leblanc, J. L. (2005). *Rubber Chem. Technol.*, **78**, 54.
[94] Fitzgerald, W. C., Davis, M. N., Blackshire, J. L. *et al.* (2002). *J. Corrosion Sci. Eng.*, **3**, 15.
[95] Bonnell, D. A. (ed.) (2001). *Scanning Probe Microscopy and Spectroscopy. Theory, Techniques, and Applications* (2nd edn). Springer, Berlin.
[96] Stefanis, A. D. and Tomlinson, A. A. G. (2001). *Scanning Probe Microscopies. From Surface Structure to Nano-Scale Engineering*. Trans Tech Publications Ltd., Uetikon-Zurich.

[97] Allcock, H. R., Lampe, F. W., and Mark, J. E. (2003). *Contemporary Polymer Chemistry* (3rd edn). Prentice Hall, Englewood Cliffs, NJ.
[98] Viers, B. D. and Mark, J. E. (2007). *J. Macro. Sci., Pure Appl. Chem.*, **44**, 131.
[99] Chu, S. (1991). *Science*, **253**, 861.
[100] Smith, S. B., Cui, Y., and Bustamante, C. (1996). *Science*, **271**, 795.
[101] Li, H., Rief, M., Oesterhelt, F. *et al.* (1998). *Adv. Mater.*, **3**, 316.
[102] Ortiz, C. and Hadziioannou, G. (1999). *Macromolecules*, **32**, 780.
[103] Janshoff, A., Neitzert, M., Oberdorfer, Y. *et al.* (2000). *Angew. Chem. Int. Ed.*, **39**, 3213.
[104] Litvinov, V. M. (2001). *Macromolecules*, **34**, 8468.
[105] Garbarczyk, M., Grinberg, F., Nestle, N. *et al.* (2001). *J. Polym. Sci., Polym. Phys.*, **39**, 2207.
[106] Litvinov, V. M. and Dias, A. A. (2001). *Macromolecules*, **34**, 4051.
[107] Schneider, M., Demco, D. E., and Blumich, B. (2001). *Macromolecules*, **34**, 4019.
[108] Beshah, K., Mark, J. E., and Ackerman, J. L. (1986). *Macromolecules*, **19**, 2194.
[109] Blumler, P. and Blumich, B. (1991). *Macromolecules*, **24**, 2183.
[110] Litvinov, V. M., Barthel, H., and Weis, J. (2002). *Macromolecules*, **35**, 4356.
[111] Garrido, L., Mark, J. E., Wang, S. *et al.* (1992). *Polymer*, **33**, 1826.
[112] Higgins, J. S. and Benoit, H. (1994). *Neutron Scattering from Polymers*. Clarendon Press, Oxford.
[113] Roe, R.-J. (2000). *Methods of X-Ray and Neutron Scattering in Polymer Science*. Oxford University Press, Oxford.
[114] Anders, S. H., Krbecek, H. H., and Pietralla, M. (1997). *J. Polym. Sci., Polym. Phys. Ed.*, **35**, 1661.
[115] Sinha, M., Mark, J. E., Jackson, H. E. *et al.* (2002). *J. Chem. Phys.*, **117**, 2968.
[116] Patterson, G. D. (1977). *J. Polym. Sci., Polym. Phys. Ed.*, **35**, 455.
[117] Sakurai, T., Matsuoka, T., Koda, S. *et al.* (2000). *J. Appl. Polym. Sci.*, **76**, 978.
[118] Anders, S. H., Eberle, R., Peetz, L. *et al.* (2002). *J. Polym. Sci., Polym. Phys. Ed.*, **40**, 1201.
[119] Gent, A. N. and Marteny, P. (1982). *J. Appl. Phys.*, **53**, 6069.
[120] Sinha, M., Erman, B., Mark, J. E. *et al.* (2003). *Macromolecules*, **36**, 6127.
[121] Erman, B. and Mark, J. E. (1988). *J. Chem. Phys.*, **89**, 3314.
[122] Callen, H. B. (2000). *Thermodynamics and an Introduction to Thermostatistics*. Wiley, New York.
[123] Kuhn, W. (1946). *J. Polym. Sci.*, **1**, 380.
[124] Wall, F. T. and Flory, P. J. (1951). *J. Chem. Phys.*, **19**, 1435.
[125] Ronca, G. and Allegra, G. (1975). *J. Chem. Phys.*, **63**, 4990.

[126] Flory, P. J. (1985). *Polym. J.*, **17**, 1.
[127] Oeser, R., Ewen, B., Richter, D. et al. (1988). *Phys. Rev. Lett.*, **60**, 1041.
[128] James, H. M. and Guth, E. (1942). *Ind. Eng. Chem.*, **34**, 1365.
[129] James, H. M. and Guth, E. (1953). *J. Chem. Phys.*, **21**, 1039.
[130] Eichinger, B. E. (1972). *Macromolecules*, **5**, 496.
[131] Graessley, W. W. (1975). *Macromolecules*, **8**, 865.
[132] Pearson, D. S. (1977). *Macromolecules*, **10**, 696.
[133] Kloczkowski, A., Mark, J. E., and Erman, B. (1989). *Macromolecules*, **22**, 1423.
[134] Nagai, K. (1963). *J. Chem. Phys.*, **39**, 924.
[135] Erman, B. (2010). *Curr. Opin. Solid. State Mats. Sci.*, **14**, 35.
[136] Erman, B. and Flory, P. J. (1982). *Macromolecules*, **15**, 806.
[137] Queslel, J. P., Erman, B., and Monnerie, L. (1985). *Macromolecules*, **18**, 1991.
[138] Erman, B. (1981). *J. Polym. Sci., Polym. Phys. Ed.*, **19**, 829.
[139] Erman, B. and Monnerie, L. (1992). *Macromolecules*, **25**, 4456.
[140] Kloczkowski, A., Mark, J. E., and Erman, B. (1995). *Macromolecules*, **28**, 5089.
[141] Freed, K. (1971). *J. Chem. Phys.*, **55**, 5588.
[142] Edwards, S. F. and Freed, K. F. (1970). *J. Phys. C*, **3**, 739.
[143] Edwards, S. F. (1971). *Brit. Polym. J.*, **3**, 140.
[144] Deam, R. T. and Edwards, S. F. (1976). *Philos. Trans. R. Soc. London, Ser. A*, **280**, 317.
[145] Edwards, S. F. (1967). *Proc. Phys. Soc. (London)*, **92**, 9.
[146] Rubinstein, M. and Panyukov, S. V. (1997). *Macromolecules*, **30**, 8036.
[147] Ball, R. C., Doi, M., and Edwards, S. F. (1981). *Polymer*, **22**, 1010.
[148] Edwards, S. F. and Vilgis, T. A. (1986). *Polymer*, **27**, 483.
[149] Edwards, S. F. and Vilgis, T. A. (1988). *Rep. Prog. Phys.*, **51**, 243.
[150] Vilgis, T. A. and Erman, B. (1993). *Macromolecules*, **26**, 6657.
[151] Rubinstein, M. and Panyukov, S. (2002). *Macromolecules*, **35**, 6670.
[152] Wu, S. and Mark, J. E. (2007). *J. Macro. Sci., Polym. Revs*, **47**, 463.
[153] DeBolt, L. C. and Mark, J. E. (1987). *Polymer*, **28**, 416.
[154] Curro, J. G., Schweizer, K. S., Adolf, D. et al. (1986). *Macromolecules*, **19**, 1739.
[155] Curro, J. G. and Mark, J. E. (1984). *J. Chem. Phys.*, **80**, 4521.
[156] Cifra, P. and Bleha, T. (1995). *J. Chem. Soc. Faraday Trans.*, **91**, 2465.
[157] Hanson, D. E. and Martin, R. L. (2009). *J. Chem. Phys.*, **130**, 0649031.
[158] Hanson, D. E. (2000). *J. Chem. Phys.*, **113**, 7656.
[159] Gao, J. and Weiner, J. H. (1994). *Macromolecules*, **27**, 1201.
[160] Duering, E. R., Kremer, K., and Grest, G. S. (1994). *J. Chem.*

Phys., **101**, 8169.

[161] Kremer, K. and Grest, G. S. (1992). *J. Chem. Soc. Faraday Trans.*, **88**, 1707.
[162] Everaers, R. and Kremer, K. (1996). *Macromolecules*, **28**, 7291.
[163] Skliros, A., Mark, J. E., and Kloczkowski, A. (2009). *Macromol. Theory Simul.*, **18**, 537.
[164] Hanna, S. and Windle, A. H. (1988). *Polymer*, **29**, 207.
[165] Madkour, T. M., Azzam, R. A., and Mark, J. E. (2006). *J. Polym. Sci., Polym. Phys.*, **44**, 2524.
[166] Ogden, R. W. (1987). *Polymer*, **28**, 379.
[167] Mark, J. E. and Erman, B. (1998). In *Polymer Networks* (ed. R. F. T. Stepto). Blackie Academic, Chapman & Hall, Glasgow.
[168] Flory, P. J. (1961). *Trans. Faraday Soc.*, **57**, 829.
[169] Khokhlov, A. R. (1992). In *Responsive Gels: Volume Transitions I* (ed. K. Dusek), p. 125. Springer, Berlin.
[170] Philippova, O. E., Pieper, T. G., Sitnikova, N. L. *et al.* (1995). *Macromolecules*, **28**, 3925.
[171] Nisato, G. and Candau, S. J. (2002). In *Polymer Gels and Networks* (ed. Y. Osada and A. R. Khokhlov), p. 131. Marcel Dekker, New York.
[172] Gottlieb, M. and Gaylord, R. J. (1983). *Polymer*, **24**, 1644.
[173] Brereton, M. G. and Klein, P. G. (1988). *Polymer*, **29**, 970.
[174] Eichinger, B. E. (1983). *Ann. Rev. Phys. Chem.*, **34**, 359.
[175] Mark, J. E. (1979). *Polym. Eng. Sci.*, **19**, 409.
[176] Tan, Z., Jaeger, R., and Vancso, G. J. (1994). *Polymer*, **35**, 3230.
[177] Gottlieb, M. and Gaylord, R. J. (1984). *Macromolecules*, **17**, 2024.
[178] Neuberger, N. A. and Eichinger, B. E. (1988). *Macromolecules*, **21**, 3060.
[179] Dubault, A., Deloche, B., and Herz, J. (1987). *Macromolecules*, **20**, 2096.
[180] Erman, B. and Mark, J. E. (1989). *Macromolecules*, **22**, 480.
[181] Donald, A. M. and Windle, A. H. (1992). *Liquid-Crystalline Polymers*. Cambridge University Press, Cambridge, UK.
[182] Warner, M. and Terentjev, E. M. (2007). *Liquid Crystal Elastomers* (2nd edn). Oxford University Press, Oxford.
[183] Kramer, D. and Finkelmann, H. (2007). *Macromol. Rapid Commun.*, **28**, 2318.
[184] Beyer, P., Terentjev, E. M., and Zentel, R. (2007). *Macromol. Rapid Commun.*, **28**, 1485.
[185] Godovsky, Y. K., Makarova, N. N., Papkov, V. S. *et al.* (1985). *Makromol. Chem.*, **6**, 443.
[186] Kojima, M., Magill, J. H., Franz, U. *et al.* (1995). *Makromol. Chem. Phys.*, **196**, 1739.
[187] Papkov, V. S., Turetskii, A. A., Out, G. J. J. *et al.* (2002). *Int. J. Polym. Mat.*, **51**, 369.
[188] Fridrikh, S. V. and Terentjev, E. M. (1999). *Phys. Rev. E*, **60**, 1847.
[189] Zentel, R. and Benalia, M. (1987). *Makromol. Chem.*, **188**, 665.

[190] Finkelmann, H., Kock, H.-J., Gleim, W. *et al.* (1984). *Makromol. Rapid Commun.*, **5**, 287.
[191] Shenoy, D. K., Thomsen, D. L., Keller, P. *et al.* (2003). *J. Phys. Chem.*, **107**, 13755.
[192] Semmler, K. and Finkelmann, H. (1995). *Makromol. Chem. Phys.*, **196**, 3197.
[193] Wendorff, J. H. (1991). *Angew. Chem., Int. Ed. Eng.*, **30**, 405.
[194] Warner, M. and Terentjev, E. M. (2003). *Macromol. Symp.*, **200**, 81.
[195] Ross, R. and Bornstein, P. (1971). *Sci. Am.*, **224**, 44.
[196] Sandberg, L. B., Gray, W. R., and Franzblau, C. (ed.) (1977). *Elastin and Elastic Tissue.* Plenum, New York.
[197] Andrady, A. L. and Mark, J. E. (1980). *Biopolymers*, **19**, 849.
[198] Jensen, M. and Weis-Fogh, T. (1962). *Phil. Trans. R. Soc. London, Ser. B*, **245**, 137.
[199] Hall, L. M., Jayaraman, A., and Schweizer, K. S. (2010). *Curr. Opin. Solild. State Mats. Sci.*, **14**, 38.
[200] Rodgers, B. (ed.) (2004). *Rubber Compounding. Chemistry and Applications.* Marcel Dekker, New York.
[201] Heinrich, G., Kluppel, M., and Vilgis, T. A. (2002). *Curr. Opinion Solid State Mats. Sci.*, **6**, 195.
[202] (2001). In *Filled and Nanocomposite Polymer Materials* (ed. R. J. Hjelm, A. I. Nakatani, M. Gerspacher, and R. Krishnamoorti), Volume 661. Materials Research Society, Warrendale, PA.
[203] Novak, B. M. (1993). *Adv. Mats.*, **5**, 422.
[204] Mark, J. E. and Ning, Y.-P. (1984). *Polym. Bulletin*, **12**, 413.
[205] Schaefer, D. W., Vu, B. T. N., and Mark, J. E. (2002). *Rubber Chem. Technol.*, **75**, 795.
[206] Vu, B. T. N., Mark, J. E., and Schaefer, D. W. (2003). *Composite Interfaces*, **10**, 451.
[207] Laine, R. M., Choi, J., and Lee, I. (2001). *Adv. Mats.*, **13**, 800.
[208] Phillips, S. H., Haddad, T. S., and Tomczak, S. J. (2004). *Curr. Opinion Solid State Mater. Sci.*, **8**, 21.
[209] Kickelbick, G. (2003). *Prog. Polym. Sci.*, **28**, 83.
[210] Okumura, H., Okada, M., Kawaguchi, Y. *et al.* (2000). *Macromolecules*, **33**, 4297.
[211] Vaia, R. A. and Giannelis, E. P. (2001). *MRS Bull.*, **26**(5), 394.
[212] Deng, Q. Q., Hahn, J. R., Stasser, J. *et al.* (2000). *Rubber Chem. Technol.*, **73**, 647.
[213] Kohls, D., Beaucage, G., Pratsinis, S. E. *et al.* (2001). In *Filled and Nanocomposite Polymer Materials* (ed. R. J. Hjelm, A. I. Nakatani, M. Gerspacher, and R. Krishnamoorti), Volume 661. Materials Research Society, Warrendale, PA.
[214] Luna-Xavier, J.-L., Bourgeat-Lami, E., and Guyot, A. (2001). *Coll. Polym. Sci.*, **279**, 947.
[215] Ho, C. C., Hill, M. J., and Odell, J. A. (1993). *Polymer*, **34**, 2019.
[216] Sharaf, M. A. and Mark, J. E. (2004). *Polymer*, **45**, 3943.
[217] Choi, J., Harcup, J., Yee, A. F. *et al.* (2001). *J. Am. Chem.*

Soc., **123**, 11420.
[218] Andrews, R. and Weisenberger, M. C. (2004). *Curr. Opinion Solid State Mater. Sci.*, **8**, 31.
[219] Sohoni, G. B. and Mark, J. E. (1987). *J. Appl. Polym. Sci.*, **34**, 2853.
[220] Mark, J. E., Abou-Hussein, R., Sen, T. Z. *et al.* (2005). *Polymer*, **46**, 8894.
[221] Mark, J. E. and Curro, J. G. (1983). *J. Chem. Phys.*, **79**, 5705.
[222] Sharaf, M. A. and Mark, J. E. (2002). *Polymer*, **43**, 643.

Polyelectrolyte Solutions and Gels

Andrey V. Dobrynin
Polymer Program, Institute of Materials Science and Department of Physics
University of Connecticut
Storrs, Connecticut, 06269-3136, USA

Introduction

Polyelectrolytes are macromolecules with ionizable groups [1–9]. In aqueous solutions charged groups dissociate, leaving charges on the chain and releasing counterions into solution. Common polyelectrolytes include polyacrylic and polymethacrylic acids and their salts, polystyrene sulfonate, DNA, RNA, and other polyacids and polybases (see Figure 9.1).

Polyelectrolytes play an important role in a diverse number of fields ranging from materials science and colloids to biophysics [8–21]. These polymers are used as rheology modifiers, adsorbents, coatings, biomedical implants, flocculants for waste-water treatment, colloidal stabilizing agents, and suspending agents for pharmaceutical delivery systems.

Polyelectrolyte solutions and gels are multicomponent systems that consist of solvent molecules, charged polymer chains, counterions, and salt ions (see Figure 9.2). The properties of these systems depend on polymer and salt concentrations, the electrostatic interactions between charges, monomer-solvent interactions controlling the solvent affinity for the polymer backbone, and the interactions between charges and solvent molecules determining solvation of the ions by a solvent. Thus, solutions and gels of charged polymers are complex systems and their exact description requires knowledge of the variety of molecular parameters. However, the universality of the system properties shows that one can simplify the problem by ignoring the atomistic details of solvent and polymers [1–9]. In this approach polyelectrolyte chains are represented by charged bead-spring chains with beads representing the groups of actual monomeric units. The solvent molecules are replaced by a continuum with the macroscopic dielectric constant ε and solvent viscosity η_0. The value of the dielectric constant determines the strength of the electrostatic interactions between charges while the value of the solvent viscosity determines the friction with the solvent. The counterions and salt ions could be considered either explicitly or as an effective charged

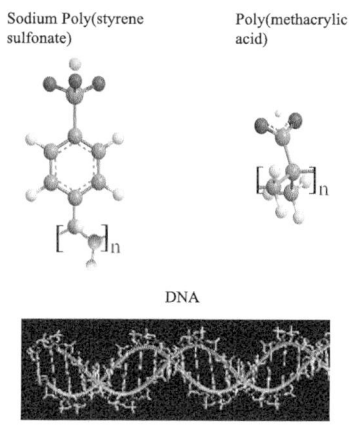

Fig. 9.1 Examples of charged polymers.

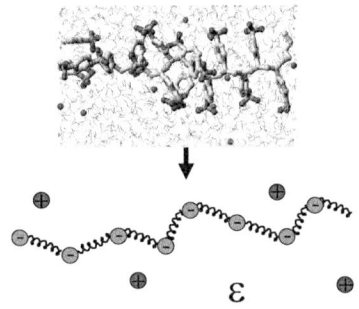

Fig. 9.2 Illustration of the mapping procedure of sodium polystyrene sulfonate (NaPSS) in water. The repeat units of NaPSS are replaced by spherical monomers ([-] circles) connected by springs and the solvent molecules are represented by a continuum with dielectric constant ε. Counterions are represented by spherical beads ([+] circles). Reproduced with permission from [8].

background resulting in screening of the electrostatic interactions between charged groups on the polymer backbone. Figure 9.2 shows an example of representation of the sodium polystyrene sulfonate chain in water by a bead-spring chain with counterions in a dielectric continuum.

In the following sections we will use this simplified model to describe properties of polyelectrolyte solutions and gels.

9.1 Dilute salt-free polyelectrolyte solutions

In dilute solutions, the distance between chains is larger than their size such that the intrachain electrostatic interactions dominate over the interchain ones. Thus, to describe chain properties one can effectively consider a single polyelectrolyte chain with counterions surrounding it in a large cell which size is on the order of the distance between chains.

Consider a polyelectrolyte chain with the degree of polymerization N, bond length b and with fraction f of the charged groups on the polymer backbone in a medium with the dielectric permittivity ε. At very low polymer concentrations almost all counterions are distributed outside the volume occupied by a chain. At these polymer concentrations the electrostatic interactions between charged groups and chain's elasticity control chain dimensions. The conformations of a polyelectrolyte chain in a dilute salt-free solution can be described by using a scaling or Flory-like approach.

Conformations of a polyelectrolyte chain

In the framework of the scaling approach it is assumed that there is a separation of the length scales [22–25], see Figure 9.3(a). At short length scales the electrostatic interactions between charged groups are weak and the chain preserves its unperturbed conformation. However, at large length scales, the electrostatic interactions dominate resulting in chain elongation. The crossover length scale between two regimes is called *electrostatic blob*. For a Θ-solvent for the uncharged polymer backbone, the relation between the blob size and the number of monomers in it is given by $D_e^0 \approx b\sqrt{g_e^0}$. The energy of the electrostatic interactions between all charged monomers inside a blob is on the order of the thermal energy $k_B T$ (where k_B is the Boltzmann constant and T is the absolute temperature).

$$\frac{l_B \left(f g_e^0\right)^2}{D_e^0} \approx l_B f^2 \left(g_e^0\right)^{3/2} / b \approx 1 \; . \tag{9.1}$$

Equation (9.1) introduces the Bjerrum length

$$l_B = \frac{e^2}{\varepsilon k_B T} \; ; \tag{9.2}$$

the distance at which the interaction between two elementary charges e in the medium with the dielectric constant ε is equal to the thermal energy k_BT. (The electrostatic system of units [26] is used in this chapter. In SI units [26], the Bjerrum length is equal to $l_B = e^2/(4\pi\varepsilon_0\varepsilon k_B T)$ where $\varepsilon_0 = 8.85 \times 10^{-12}\,\mathrm{C^2/m^2 N}$ is the dielectric permittivity of the vacuum.) Substituting relation between the number of monomers in the electrostatic blob and blob size into equation (9.1) one can obtain the following relations between the number of monomers in a blob g_e^0, its size D_e^0, and the fraction of charged monomers f:

$$g_e^0 \approx \left(uf^2\right)^{-2/3} \tag{9.3a}$$

$$D_e^0 \approx b\left(uf^2\right)^{-1/3} \tag{9.3b}$$

where u is the ratio of the Bjerrum length l_B, equation (9.2), to the bond size b:

$$u = l_B/b \ . \tag{9.4}$$

Electrostatic interactions at the length scales larger than the blob size lead to the elongation of the polyelectrolyte chain into an array of blobs. The size of the polyelectrolyte chain is estimated as the number of blobs per chain N/g_e^0 times the blob size D_e^0:

$$R_e^{\mathrm{blob}} \approx \frac{N}{g_e^0} D_e^0 \approx bN\left(uf^2\right)^{1/3} \ . \tag{9.5}$$

Note, that one can extend the presented above approach to describe conformations of a polyelectrolyte chain in good solvent conditions for the polymer backbone. This can be done by changing the relation between the blob size and the number of monomers in it to that for a swollen section of a chain, $D_e^0 \approx b(g_e^0)^{3/5}$, see [22–25] for detail.

Thus, in the framework of the scaling approach it is assumed that all electrostatic blobs have the same size independently of their location along the polymer backbone. However, the blobs in the middle of the chain interact with the rest of the chain stronger than the blobs at the chain ends [27]. This additional contribution from the larger length scales leads to the logarithmic correction to the blob size and to the chain size

$$R_e^{\mathrm{blob}} \approx \frac{N}{g_e^0} D_e^0 \left[\ln\left(N/g_e^0\right)\right]^{1/3} \approx bN\left(uf^2\right)^{1/3} \left[\ln\left(\left(uf^2\right)^{2/3} N\right)\right]^{1/3} . \tag{9.6}$$

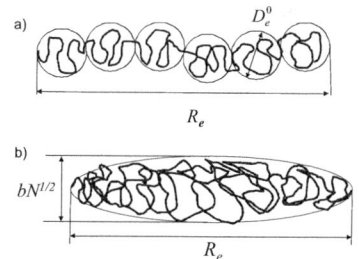

Fig. 9.3 Schematic representation of a polyelectrolyte chain as a chain of electrostatic blobs (a) and representation of a polyelectrolyte chain in an elongated conformation (b).

Equation (9.6) shows that the chain size grows faster than linear with the chain's degree of polymerization, $N(\ln(N))^{1/3}$ [27, 28]. It is important to point out that the blob model of a polyelectrolyte chain is only valid if the number of monomers per electrostatic blob is larger than unity, $g_e^0 \gg 1$. This is true if the value of the parameter uf^2 is smaller than unity. Polyelectrolytes for which $uf^2 \ll 1$ are called *weakly charged polyelectrolytes*.

Contrary to the scaling approach the Flory method looks at chain conformations at the large length scales on the order of the chain size R_e. In the framework of this approach [29] a chain size is obtained by optimizing conformational and electrostatic parts of the chain's free energy. The conformational part of free energy of a chain with the end-to-end distance R_e is equal to

$$F_{\text{conf}}(R_e) \approx k_B T \frac{R_e^2}{R_0^2} \approx k_B T \frac{R_e^2}{b^2 N} \qquad (9.7)$$

The electrostatic part of the chain's free energy is evaluated by assuming that monomers are uniformly distributed within the volume of a chain. The electrostatic interactions lead to elongation of the chain along, for example the z-axis, while its size stays unperturbed in the directions perpendicular to the elongation axis. Thus, the polyelectrolyte chain has a shape of an ellipsoid with the longitudinal size equal to R_e and with the transversal size equal to that of an ideal chain $bN^{1/2}$, see Figure 9.3(b). The electrostatic energy of a charged ellipsoid [30] with the net charge efN is proportional to

$$\frac{F_{\text{electr}}(R_e)}{k_B T} \approx \frac{l_B (fN)^2}{R_e} \ln\left(\frac{R_e}{bN^{1/2}}\right) \qquad (9.8)$$

Combining conformational and interaction parts of the chain's free energy, equations (9.7) and (9.8), one arrives at the free energy of a polyelectrolyte chain with a given value of the end-to-end distance R_e [29]

$$\frac{F_{\text{Flory}}(R_e)}{k_B T} \approx \frac{R_e^2}{b^2 N} + \frac{l_B (fN)^2}{R_e} \ln\left(\frac{R_e}{bN^{1/2}}\right). \qquad (9.9)$$

The analysis of equation (9.9) shows that the conformational part of the chain free energy – the first term on the right-hand side of equation (9.9) – increases as the value of the chain size R_e increases. This increase in the conformational part of the chain free energy is due to its entropic nature. The number of available conformations decreases with increasing end-to-end distance leading to a decrease in the chain conformational entropy. On the contrary, the electrostatic part of the chain free energy [the second term on the right-hand side of equation (9.9)] decreases with increasing a chain size R_e. Charged monomers move further apart with increasing chain size weakening the electrostatic repulsion between them. The optimal chain size corresponds to the minimum of the chain free energy as a function of the end-to-end distance R_e. Minimizing equation (9.9) with respect to the chain's end-to-end distance R_e and solving iteratively the resultant nonlinear equation, one obtains the following expression for the chain size

$$R_e^F \approx bN u^{1/3} f^{2/3} \left[\ln\left(N \left(u f^2\right)^{2/3}\right)\right]^{1/3}. \qquad (9.10)$$

The onset of elongation of a polyelectrolyte chain is at the value of its electrostatic energy, equation (9.8), on the order of the thermal energy $k_B T$:

$$\frac{l_B (fN)^2}{bN^{1/2}} \approx 1. \qquad (9.11)$$

This happens when the number of the charged monomers on the chain fN is on the order of $u^{-1/2} N^{1/4}$. At stronger electrostatic interactions, the chain size monotonically increases with increasing the fraction of charged monomers f on the polymer backbone. Note, that equation (9.11) is similar to equation (9.1) if one considers a whole chain as an electrostatic blob.

The Flory approach presented above is only valid when the chain size R_e^F is smaller than the size of the fully extended chain bN. This requirement leads to the upper bound on the chain degree of polymerization $N < u^{-2/3} f^{-4/3} \exp(u^{-1} f^{-2})$. For longer chains, quadratic dependence of the conformational part of the chain free energy on the end-to-end distance [the first term in equation (9.10)] is no longer valid and one has to take into account the nonlinear effects in chain elasticity [31,32].

Polyelectrolyte chain in poor solvent conditions

In a poor solvent for the polymer backbone there is an effective attraction between monomers which causes a neutral polymer chain without charged groups to collapse into dense spherical globule in order to minimize the number of unfavorable polymer-solvent contacts [31,32]. In the globular state close to the transition point, the interaction part of the neutral chain free energy $F_{\text{int}}(R)$ can be expanded into a power series of the number density of monomers $\rho \approx N/R_{\text{glob}}^3$ (R_{glob} is the size of the globule). This expansion is similar to the virial expansion in the theory of real gases [33]. Thus, the interaction part of the free energy can be written as [31,32]

$$\frac{F_{\text{int}}}{k_B T} \approx N\tau b^3 \rho + N b^6 \rho^2 \qquad (9.12)$$

where $\tau = 1 - \Theta/T$ is the relative deviation from the Θ-temperature, called the effective temperature. At Θ-point, the first term on the right-hand side of equation (9.12) describing contribution of the binary contacts to the chain's free energy is equal to zero. Below the Θ-temperature, the effective temperature is negative, corresponding to the effective attraction between monomers - poor solvent conditions for the polymer backbone. The equilibrium size of the globule is obtained by minimizing the interaction part of the free energy F_{int} of the globule, equation (9.12), with respect to its size R_{glob}. The optimal size of the globule is equal to [31,32]

$$R_{\text{glob}} \approx b |\tau|^{-1/3} N^{1/3} \qquad (9.13)$$

There is an important length scale - the correlation length ξ_T of the density fluctuations inside a globule (*thermal blob* size). At the length scales smaller than the thermal blob size ξ_T, the chain statistics are unperturbed by the monomer-monomer attractive interactions and are that of an ideal chain of g_T monomers ($\xi_T \approx b g_T^{1/2}$), see Figure 9.4(a). At the length scales larger than the thermal blob size, the attraction between monomers wins (the attractive interaction between two blobs is on the order of the thermal energy $k_B T$) and thermal blobs in a globule are densely packed, $\rho \approx g_T/\xi_T^3$. The polymer density ρ inside the globule scales linearly with the absolute value of the effective temperature τ,

$$\rho \approx b^{-3} |\tau| . \qquad (9.14)$$

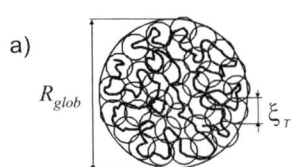

Thus, the number of monomers in a thermal blob is

$$g_T \approx |\tau|^{-2} \qquad (9.15)$$

and its size is

$$\xi_T \approx b |\tau|^{-1} . \qquad (9.16)$$

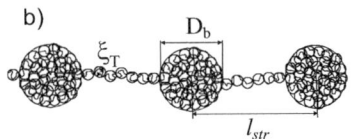

The free energy of the globule is on the order of the thermal energy $k_B T$ per thermal blob

$$\frac{F_{\text{int}}}{k_B T} \approx -\frac{N}{g_T} \approx -N\tau^2 . \qquad (9.17)$$

Fig. 9.4 (a) Schematic representation of a polymeric globule of size R_{glob} consisting of dense packing of thermal blobs of size ξ_T. (b) Schematic representation of a necklace-like polyelectrolyte chain. Reproduced with permission from [8].

In addition to this bulk contribution to the free energy of a globule, there is also a surface energy contribution. The origin of the surface energy is the difference in the number of the nearest neighbors for the thermal blobs at the surface of the globule and those in the bulk. It requires extra energy to bring a blob to the surface of the globule. The surface energy of a globule can be estimated as the number of blobs at the globule surface S/ξ_T^2 times the thermal energy $k_B T$ because any two blobs inside the globule attract each other with the energy on the order of the thermal energy $k_B T$ and blobs at the surface are missing part of this attractive energy due to fewer neighbors.

$$F_{\text{surf}} \approx k_B T S/\xi_T^2 \approx \gamma S \qquad (9.18)$$

where γ is the surface tension of the globule ($\gamma \approx k_B T/\xi_T^2$). (For exact derivation of the surface energy term see [32].)

The problem of the shape of a charged globule bears similarity with the classical problem of the instability of a charged droplet, considered by Lord Rayleigh (John William Strutt) [34]. In his classical experiments, Lord Rayleigh showed that a charged droplet is unstable and breaks into smaller droplets if its electric charge exceeds some critical value. The equilibrium state of the charged droplet with $Q_d > Q_{\text{crit}}$ is a set of the smaller droplets with charge on each of them smaller than the critical one and placed at an infinite distance from each other. The value of the critical charge is determined by the electrostatic energy, $Q_d^2/\varepsilon R_{\text{drop}}$, of the droplet of size R_{drop} with charge Q_d, and its surface energy $\gamma_w R_{\text{drop}}^2$ where γ_w is the surface tension of the air-water interface. Balancing these two energies, one finds that the critical charge Q_{crit} scales with the size of the droplet as $R_{\text{drop}}^{3/2}$. For a polyelectrolyte chain in the globular state, its surface energy $k_B T R_{\text{glob}}^2/\xi_T^2$ becomes on the order of its electrostatic energy $k_B T l_B (fN)^2/R_{\text{glob}}$ at the critical fraction of charged monomers per chain f_{crit} equal to

$$f_{\text{crit}} \approx \left(\frac{|\tau|}{uN}\right)^{1/2} \qquad (9.19)$$

Thus, upon charging, polymeric globules change their shape and size. This phenomenon was first observed by Katchalsky and Eisenberg [35] in their study of the viscosity of aqueous solutions of poly(methacrylic acid) (PMA). The viscosity of dilute PMA solution stayed almost constant at a low pH and then abruptly increased as solution pH reached some critical value, indicating a dramatic change in the chain dimensions. This dramatic change in the solution viscosity of PMA is qualitatively different from that observed in solutions of poly(acrylic acid) (PAA) for which viscosity grows smoothly with increasing neutralization, indicating smooth expansion of polymer chains. The difference in behavior of PMA and PAA is due to the fact that water is a poor solvent for PMA that has hydrophobic methyl groups while it is a good solvent for PAA. (For a historical overview of this subject see the paper by Morawetz [7].)

Rayleigh's experiments give a hint to what happens with polyelectrolytes in a poor solvent when the fraction of charged monomers f exceeds the critical value f_{crit}. Similar to a charged droplet, a polyelectrolyte chain in a poor solvent reduces its energy by splitting into a set of the smaller charged globules, called beads. Since this transformation occurs in a single chain, all these beads are connected by strings of monomers into a necklace of globules [36]. The size of the beads

$$D_b \approx b |\tau|^{-1/3} m_b^{1/3} \approx b \left(uf^2\right)^{-1/3}, \tag{9.20}$$

is determined from the stability condition when electrostatic energy of a bead with m_b monomers is on the order of its surface energy

$$k_B T \frac{l_B (fm_b)^2}{D_b} \approx k_B T \frac{D_b^2}{\xi_T^2}, \tag{9.21}$$

where the number of monomers in a bead m_b is equal to

$$m_b \approx \rho D_b^3 \approx \frac{|\tau|}{uf^2}. \tag{9.22}$$

It is interesting to point out that the size of the bead has the same scaling dependence on the fraction of the charged monomers f and interaction parameter u as the electrostatic blob size, see equation (9.3b). The length l_{str} of the strings connecting two neighboring blobs along the polymer backbone beads can be estimated by balancing the energy of the electrostatic repulsion between two closest beads with the string surface energy

$$k_B T \frac{l_B (fm_b)^2}{l_{\text{str}}} \approx k_B T \frac{l_{\text{str}} \xi_T}{\xi_T^2}. \tag{9.23}$$

Note, that the string thickness is on the order of the thermal blob size ξ_T. Solving equation (9.23) for the string length one has

$$l_{\text{str}} \approx b m_b^{1/2} \approx b \left(\frac{|\tau|}{uf^2}\right)^{1/2}. \tag{9.24}$$

The mass of a string $m_{\text{str}} \approx \rho l_{\text{str}} \xi_T^2$ is much smaller than the mass of a bead m_b

$$\frac{m_{\text{str}}}{m_b} \approx \left(\frac{uf^2}{|\tau|^3}\right)^{1/2} \ll 1 . \qquad (9.25)$$

In this case the number of beads n_b per chain is approximately equal to N/m_b. Since the length of the strings is larger than the bead size, the length of the necklace can be estimated as the number of beads n_b per chain times the distance between the centers of mass of two neighboring beads l_{str}

$$L_{\text{nec}} \approx n_b l_{\text{str}} \approx b \left(\frac{uf^2}{|\tau|}\right)^{1/2} N . \qquad (9.26)$$

In this approximation the total free energy of a polyelectrolyte chain in a poor solvent is

$$\frac{F_{\text{neck}}}{k_B T} \approx n_b \frac{l_B (fm_b)^2}{D_b} + \frac{l_B (fN)^2}{L_{\text{nec}}} - N|\tau|^2 \qquad (9.27)$$

where the first term is the electrostatic and surface energies of the beads, the second term is the electrostatic repulsion between beads which is on the order of the surface energy of the strings, and the last term is the free energy of the polymer backbone in the poor solvent. The electrostatic repulsion between beads is smaller than the electrostatic energy of the beads as long as the string length between beads is larger than the bead size. In this approximation the free energy of the necklace is equal to

$$\frac{F_{\text{neck}}}{k_B T} \approx N|\tau| \left(uf^2\right)^{1/3} - N|\tau|^2 . \qquad (9.28)$$

The number of beads n_b ($n_b \approx N/m_b$) in a necklace can only be an integer, therefore the following equation

$$f \approx \left(\frac{|\tau| n_b}{u N}\right)^{1/2} \qquad (9.29)$$

defines the set of boundaries between the states of the necklace globule with different number of beads. By changing the fraction of charged monomers f or effective temperature τ (solvent quality), the globule undergoes a cascade of transitions between states with different integer number of beads n_b per necklace.

The details of the transition and conformations of polyelectrolyte chains above the transition were described by Dobrynin et al. [36]. Performing Monte Carlo simulations of the freely jointed uniformly charged polyelectrolyte chains with fractional charge on each monomer f, they showed that the critical number of charged monomers $(fN)_{\text{crit}}$ on the chains at which charged globules becomes unstable is proportional to $\sqrt{N}$ (see Figure 9.5).

For the number of charged monomers on the chain fN above the critical value $(fN)_{\text{crit}}$, the polyelectrolyte chain first forms a dumbbell, see

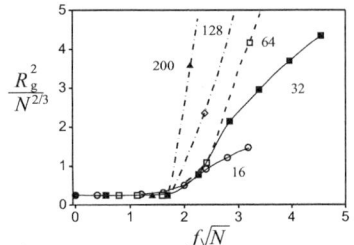

Fig. 9.5 Dependence of the reduced mean square radii of gyration of polyelectrolyte chains in a poor solvent on the reduced chain valence $fN^{1/2}$ for chains with degrees of polymerization $N = 16, 32, 64, 128,$ and 200. Results of the Monte Carlo simulations of freely jointed chains interacting via Coulomb and Lennard-Jones potential ($\varepsilon_{\text{LJ}} = 1.5 k_B T$ and $u = 2$). Reproduced with permission from [36].

Figure 9.6(b). For larger values of the net charge on the polyelectrolyte chain, the polymer forms a necklace with three beads connected by two strings, Figure 9.6(c). Thus, there is a cascade of transitions between necklaces with different number of beads as the charge on the chain increases.

It is important to point out that this abrupt conformational transition was also observed in the detailed molecular dynamics simulations of the NaPSS chains in the dielectric continuum and with explicit water [37]. Figure 9.7 shows the dependence of the reduced chain size $\langle R_g^2 \rangle^{1/2}/N^{1/3}$ on the normalized charge $Q/Q_{\text{crit}} = fN/f_{\text{crit}}N \approx f\sqrt{N}$. As in the case of the coarse-grained simulations (see Figure 9.5) the data points have collapsed into universal plot. Similarity between two plots indicates that coarse-grained simulations provide a qualitatively correct picture of the stability of the charged polymeric globule.

The effect of the solvent quality on the cascade of transitions between different pearl-necklace structures was investigated by Monte Carlo [38] and molecular dynamics [39] simulations. Chodanowski and Stoll [38] have found necklaces with up to 12 beads for a polyelectrolyte chain with degree of polymerization $N = 200$. These results of the computer simulations are in good qualitative agreement with the theoretical models [36, 40–42] of a polyelectrolyte chain in a poor solvent. Explicit solvent simulations of short polyelectrolyte chains in poor solvents showed that the necklace structure is not very stable because of the additional entropic solvent effect weakening effective monomer-monomer attraction [43]. The results of simulations with explicit solvent are in good quantitative agreement with the predictions of the solvent-accessible surface area (SASA) model for a polyelectrolyte chain in poor solvent [44].

The transition between necklaces with different number of beads can be induced by applying an external force to stretch the necklace [41, 45, 46]. Since the necklaces with different numbers of beads coexist, the chain deformation could have a saw-tooth deformation curve. However, this mean-field picture of abrupt transitions between necklaces with different numbers of beads is smeared by fluctuations in the number of monomers in a bead as well as by fluctuations of the number of beads in a necklace [46]. The smooth deformation curve of a hydrophobic polyelectrolyte chain is confirmed by computer simulations [47, 48].

For semiflexible polyelectrolyte chains the chain rigidity plays an important role in determining chain conformations in a poor solvent. A semiflexible polyelectrolyte chain adopts rings on a string conformation of collapsed toroidal globules connected by the stretched strings of monomers [49–52]. This structure optimizes the short-range monomer-monomer attractive interactions, chain's bending energy and long-range electrostatic repulsion between charged groups.

Diagram of states of a polyelectrolyte chain

This section summarizes results for single chain conformations discussed in previous sections and presents a diagram of states of a polyelectrolyte

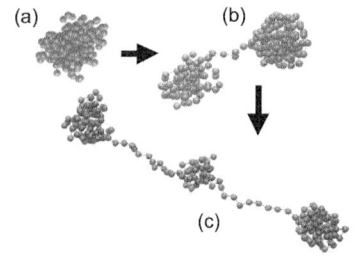

Fig. 9.6 Typical configurations of freely-jointed uniformly charged chains in poor solvent conditions for three different charge fractions: (a) a spherical globule for $f = 0$; (b) a dumbbell for $f = 0.125$ and (c) a necklace with three beads for $f = 0.15$. Reproduced with permission from [36].

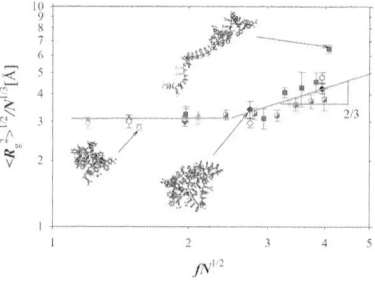

Fig. 9.7 Dependence of the reduced chain size $\langle R_g^2 \rangle^{1/2}/N^{1/3}$ of the NaPSS chain on the reduced charge $fN^{1/2}$ and different degree of sulfonation: $f = 1.0$ (circles), $f = 0.5$ (squares), $f = 0.333$ (rhombs) and $f = 0.25$ (triangles). Filled symbols represent simulations with the explicit water. Open and half-filled symbols are simulations of NaPSS chain in a medium with the dielectric permittivity, $\varepsilon = 77.73$. All simulations were performed at temperature $T = 300\,\text{K}$.

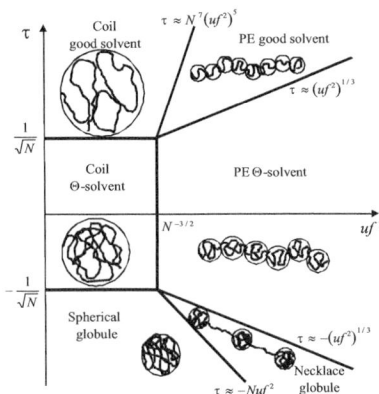

Fig. 9.8 Diagram of states of a polyelectrolyte chain.

chain in a dilute salt-free solution. As we saw in the previous sections the chain conformations depend on the solvent quality for the polymer backbone described by the value of the effective temperature τ and on the strength of the electrostatic interactions between ionized groups characterized by the value of the dimensionless parameter uf^2. Figure 9.8 shows the diagram of states of a polyelectrolyte chain in (uf^2, τ)-plane. The interval of the negative effective temperatures corresponds to the effective attraction between monomers. For small values of the interaction parameter $uf^2 \leq N^{-3/2}$ the electrostatic repulsion between charged monomers is weaker than the thermal energy $k_B T$ and chain maintains its ideal conformation. The two-body interactions between monomers are also weaker than the thermal energy $k_B T$ [see the first term in equation (9.12)] if the value of the effective temperature $|\tau| \leq N^{-1/2}$. In this interval of the effective temperatures there is less than one thermal blob per chain, $bN^{1/2} < \xi_T$.

Thus, in this regime of the diagram of states a polyelectrolyte chain has a coil-like structure with electrostatic interactions providing a minor correction to chain conformations. The effective attraction between monomers forces a polyelectrolyte chain to collapse when the value of the effective temperature τ becomes smaller than $-N^{-1/2}$. The equilibrium density inside a globule is determined by balancing the two-body attraction and three-body repulsion, equation (9.12). The charged polymeric globule remains stable as long as the electrostatic repulsion between charged monomers is weaker than the globular surface energy, equation (9.19). A polyelectrolyte chain forms a necklace globule when $uf^2 > |\tau|/N$. In the necklace globule regime the size of the necklace beads decreases with increasing the strength of the electrostatic interactions, see equation (9.20). It becomes comparable with the thickness of a string, $D_b \approx \xi_T$, at $|\tau| \approx (uf^2)^{1/3}$. The last condition determines a crossover to a polyelectrolyte chain regime with a Θ-solvent condition for the polymer backbone. In this regime the electrostatic interactions between charged monomers result in strong chain elongation, equation (9.5). The crossover to good solvent conditions for the polymer backbone takes place when the size of the electrostatic blob D_e^0 becomes on the order of the size of the thermal blob b/τ, $\tau \approx (uf^2)^{1/3}$. (Note, that in a good solvent there is also a thermal blob - the length scale at which the two-body monomer repulsion results to chain swelling [31, 32].) In the polyelectrolyte good solvent regime statistics of the chain at the length scales on the order of the electrostatic blob size is that of the self-avoiding walk of the thermal blobs [31, 32]. The electrostatic blob size in this regime can be found by using a good solvent relation between a section of the chain with g_e monomers and its size $D_e^g \approx b\tau^{1/5} g_e^{3/5}$. This leads to the electrostatic blob size to be equal to $D_e^g \approx b\tau^{2/7}(uf^2)^{-3/7}$. Finally, a polyelectrolyte chain in good solvent conditions for the polymer backbone will adopt a coil-like conformation when there is one electrostatic blob per chain or when the electrostatic repulsion between charged monomers is on the order of the thermal energy $k_B T$. This happens when the size of the electrostatic blob is on the order of the size of a neutral chain in a good solvent $R_{\text{good}} \approx b\tau^{1/5}N^{3/5}$. This condition determines the upper boundary for of the polyelectrolyte good solvent regime $\tau \approx N^7(uf^2)^5$.

Counterion condensation

The electrostatic attraction between polyelectrolyte chains and counterions increases with increasing the polymer concentration resulting in increase of counterion concentration in the vicinity of the polyelectrolyte chain. There are two groups of models that are used to describe counterion distribution in polyelectrolyte solutions as a function of the polymer concentration [2, 5, 53–62]. The first group is based on the solution of the nonlinear Poisson–Boltzmann equation describing the distribution of the electrostatic potential around a macroion [2, 53–56, 63]. This approach decouples the counterion and polymeric degrees of freedom providing the equilibrium counterion density profile for a fixed (rod-like) polymer conformation. The detailed discussion of this approach can be found in [8, 53].

The second group of models (Oosawa–Manning condensation theory) assumes a step-like counterion distribution by separating counterions into 'free' and 'condensed' [5, 57–62]. To illustrate this approach let us first consider counterion condensation on a rod-like polyelectrolyte. Let us divide counterions surrounding a rod-like polyion into two categories: counterions localized inside potential valleys along the polymer backbones (state 1) and counterions freely moving outside the region occupied by polyelectrolyte chains (state 2), see Figure 9.9 [5, 57]. Thus, the total solution volume V is divided into two regions. One region surrounds polyelectrolyte backbone with volume v where counterions are localized. (For a rod-like polyion with radius r_0 and length L the volume is equal to $v = \pi r_0^2 L$.) The total volume occupied by condensed counterions is equal to the total number of chains N_p in a solution times the localization volume v ($V_{\text{con}} = N_p v$). The outer region (state 2) is further away from polyions where free counterions are distributed. The volume of this region is $V - N_p v$. At equilibrium, the electrochemical potentials μ_{el} of counterions in both states are equal

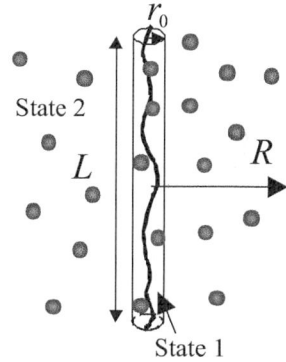

Fig. 9.9 Schematic sketch of a polyelectrolyte chain and the definition of different length scales for the two-state model. L is the length of rod-like polyion and R is the cell size ($R \ll L$). Reproduced with permission from [8].

$$\frac{\mu_{\text{el}}}{k_B T} = \ln\left(\frac{n_1}{N_p v}\right) - \frac{e\Psi_1}{k_B T} = \ln\left(\frac{n_2}{V - N_p v}\right) - \frac{e\Psi_2}{k_B T} \quad (9.30)$$

where n_1 is the number of the condensed counterions, n_2 is the number of free counterions, and Ψ_i is the electrostatic potential of counterions in the ith state. Equation (9.30) can be rewritten by introducing the fraction of condensed counterions $\beta = n_1/(n_1+n_2)$ and polymer volume fraction $\phi = N_p v/V$

$$\ln\left(\frac{\beta}{1-\beta}\right) = \ln\left(\frac{\phi}{1-\phi}\right) - \frac{e\Delta\Psi}{k_B T} \quad (9.31)$$

where $\Delta\Psi = \Psi_2 - \Psi_1$ is the difference of the electrostatic potentials between states 2 and 1. In the original Oosawa–Manning model, the polyelectrolyte chains were assumed to be rod-like with the distance $2R$ between them smaller than their length L. In this case, the typical variations of the electrostatic potential between two cylindrical regions occupied by counterions is approximately equal to that of a cylindrical capacitor [30]

$$\frac{e\Delta\Psi}{k_B T} \approx 2(1-\beta)\frac{l_B f N}{L} \ln\left(\frac{r_0}{R}\right) \approx (1-\beta)\gamma_0 \ln(\phi) \qquad (9.32)$$

where r_0 is the radius of the region 1 and γ_0 is the Manning counterion condensation parameter

$$\gamma_0 = f N l_B / L \ . \qquad (9.33)$$

Substituting equation (9.32) into equation (9.31), one can rewrite equation (9.31) in the limit of low polymer volume fractions, $\phi \ll 1$, in the following form

$$\ln\left(\frac{\beta}{1-\beta}\right) = [1 - (1-\beta)\gamma_0]\ln(\phi) \ . \qquad (9.34)$$

Figure 9.10 shows the dependence of the reduced effective linear charge density $(1-\beta)\gamma_0$ of the rod-like polyelectrolyte on its bare value γ_0 with two qualitatively different regimes. The reduced effective linear charge density $(1-\beta)\gamma_0$ increases linearly for small values of the Manning condensation parameter γ_0 then saturates as it becomes larger than unity. For the small values of the counterion condensation parameter $\gamma_0 \ll 1$, the fraction of condensed counterions β is equal to ϕ and decreases with decreasing polymer volume fraction ϕ. The fraction of condensed counterions β eventually approaches zero at infinite dilution, $\phi \to 0$. In the opposite limit of large values of the Manning counterion condensation parameter $\gamma_0 \gg 1$ the fraction of free counterions $1 - \beta$ is equal to

$$1 - \beta \approx \gamma_0^{-1}\left[1 - \ln(\gamma_0)/\ln(\phi)\right] \qquad (9.35)$$

and approaches $1/\gamma_0$ at infinite dilution, $\phi \to 0$. Thus, there is the counterion condensation phenomenon that is associated with the value of the parameter γ_0. The reduced effective linear charge density $(1-\beta)\gamma_0$ of the rod-like polyelectrolyte as 'seen' by a 'free' counterion saturates at unity. In this regime, the distance between ionized charged groups on polyion, $L/[(1-\beta)fN] \approx l_B[(1-\beta)\gamma_0]^{-1}$, is on the order of the Bjerrum length l_B.

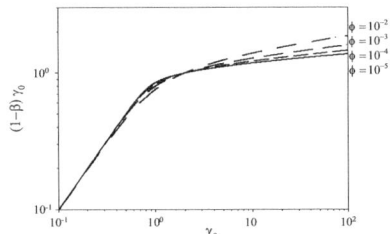

Fig. 9.10 Dependence of the reduced effective linear charge density of the rod-like polyelectrolyte on the Oosawa–Manning condensation parameter γ_0 at different polymer volume fractions ϕ. Reproduced with permission from [8].

In the model presented above a polyelectrolyte chain was approximated by a rod-like polyion and chain conformations were decoupled from the counterions condensation. In the case of flexible polyelectrolytes chain conformations are directly coupled with the intrachain electrostatic interactions between charges that are controlled by the amount of the condensed counterions [3, 8, 59–62, 64–68]. Thus, in addition to counterion configurational entropy and electrostatic interactions the chain's conformational free energy comes into play. To illustrate coupling between counterion condensation and change in chain conformation let us consider counterion condensation on the polyelectrolyte chain in a Θ-solvent for the polymer backbone. Below, I will use a scaling model to describe a chain structure and the free energy approach to describe equilibrium between free and condensed counterions. Since a fraction β of counterions is localized inside the chain volume, the fraction of the uncompensated charged groups leading to chain stretching is reduced to

$$f_* = f(1-\beta) . \tag{9.36}$$

Using a scaling model one can approximate a polyelectrolyte chain by a cylinder with length R_e and thickness on the order of the size of the electrostatic blob D_e, see Figure 9.11. All counterions that are localized inside the chain volume $v \approx ND_e^3/g_e$ will be considered as condensed counterions. The total volume occupied by condensed counterions is equal to the total number of chains N_p in solution times the localization volume v ($V_{\text{con}} = N_p v$). The free counterions distributed outside this volume occupy the volume $V - N_p v \approx V$.

The configuration part of the system free energy associated with separation of the counterions into two groups — free and condensed — is equal to

$$\frac{F_{\text{trans}}(\beta)}{k_B T} \approx f(1-\beta)NN_p \ln\left[\frac{f(1-\beta)cv_0}{e}\right] + f\beta NN_p \ln\left[\frac{\beta f \rho_e v_0}{e}\right] \tag{9.37}$$

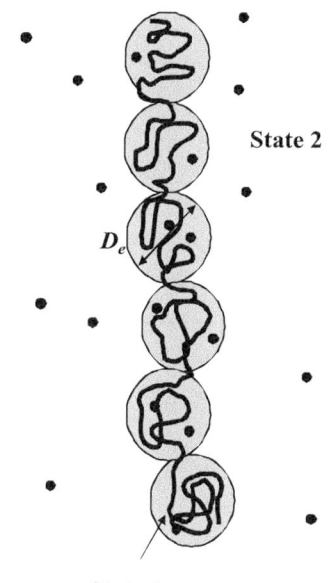

Fig. 9.11 Schematic representation of a chain of electrostatic blobs with counterions and the definition of different length scales for the two-state model of flexible polyelectrolyte chain.

where $c \approx N_p N/V$ is the polymer concentration, $\rho_e \approx g_e/D_e^3 \approx u^{1/3}f^{2/3}(1-\beta)^{2/3}/b^3$ is the monomer concentration inside the electrostatic blob, and v_0 is the monomer excluded volume, $v_0 \approx b^3$. The explicit dependence of the electrostatic blob size and the number of monomers per blob on the fraction of the condensed counterions can be obtained by substituting into equations (9.3) the fraction of the uncompensated charged groups f_* instead of the fraction of the charged groups f. The free energy of a chain due to chain stretching and electrostatic interactions between charged groups is on the order of the thermal energy $k_B T$ per electrostatic blob $F_{\text{scaling}}(\beta) \approx k_B TN/g_e$. Combining together the chain's free energy and configurational parts of the system free energy one arrives at

$$\frac{F(\beta)}{K_B TNN_p} \approx f(1-\beta) \ln\left[\frac{f(1-\beta)\phi}{e}\right]$$
$$+ f\beta \ln\left[\frac{\beta f \rho_e v_0}{e}\right] + \left(uf^2\right)^{2/3}(1-\beta)^{4/3} . \tag{9.38}$$

The equilibrium fraction of the condensed counterions is obtained by minimizing the system free energy with respect to fraction of the condensed counterions β

$$\frac{\beta}{(1-\beta)^{1/3}} \approx \frac{\phi}{u^{1/3}f^{2/3}} \exp\left(u^{2/3}f^{1/3}(1-\beta)^{1/3}\right) \tag{9.39}$$

where $\phi = cv_0$ is the polymer volume fraction. Renormalization of the chain charge will be significant when the value of the parameter $u^{2/3}f^{1/3} > 1$. In this case the average distance between charged groups $D_e^0/fg_e^0 \approx l_B/(u^{2/3}f^{1/3})$ is smaller than the Bjerrum length l_B. Analysis of equation (9.39) shows that the fraction of the condensed counterions gradually increases with increasing the polymer concentration [3, 8, 27, 59, 65, 66, 69]. The polyelectrolyte conformations are controlled by the fraction of ionized groups. Thus, counterion condensation leads to weakening of the electrostatic interactions and promotes contraction of the polyelectrolyte chains. The decrease of chain size with increasing polymer concentration is supported by the results of computer simulations, see Figure 9.12 [27, 64, 70, 71]. Computer simulations also show additional counterion accumulation at the chain ends (see for details [68, 72]). This nonuniform counterion redistribution occurs because the polyelectrolyte is stronger stretched in the middle than at the chain ends leading to the increase of the effective linear charge density toward the chain ends.

In a poor solvent a polyelectrolyte chain forms a necklace globule of dense polymeric beads connected by strings of monomers [36, 38–42, 48, 71, 73–78]. The counterion condensation can occur in avalanche-like fashion [8, 59, 65, 74]. By increasing polymer concentration or decreasing temperature one can induce a spontaneous condensation of counterions inside beads of the necklace globule. This reduces the bead's charge and results in increase of the bead mass (size), which initiate a further increase in the number of condensed counterions inside beads starting the avalanche-like counterion condensation process (see for review [8]). To illustrate this point one can use a two-state model of the counterion condensation described above. In the poor solvent conditions the main contribution to the necklace free energy comes from the electrostatic and surface energy of the beads. Thus, the free energy of the necklace can be approximated as [see equation (9.28)]:

$$F_{\text{neck}}(\beta) \approx k_B T N u^{1/3} |\tau| f^{2/3} (1-\beta)^{2/3} - k_B T N \tau^2 \tag{9.40}$$

where the fraction of charged monomers f was substituted by the effective fraction of the ionized groups f_*. Taking into account this expression for the necklace free energy and assuming the monomer density inside beads to be equal to $\rho_e \approx |\tau|/b^3$, equation (9.39) can be rewritten as follows

$$\frac{\beta}{1-\beta} \approx \frac{\phi}{|\tau|} \exp\left(\frac{u^{1/3}|\tau|}{f^{1/3}(1-\beta)^{1/3}}\right). \tag{9.41}$$

It follows from this equation that the free energy per charged monomer increases with increasing the number of bound counterions. This induces an additional influx of counterions to compensate for the growing necklace free energy and sets up an avalanche-like counterion condensation and an abrupt first order transition into a globular state.

Counterion condensation on the necklace beads was recently studied by Chepelianskii et al. [79]. Using a numerical solution of the Poisson–

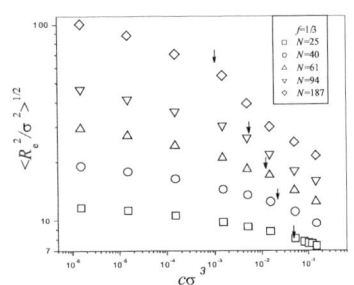

Fig. 9.12 Concentration dependence of the size of partially charged chains with $f = 1/3$. The overlap concentrations are marked by arrows. Reproduced with permission from [27].

Boltzmann equation for the spherical beads they have found that condensed counterions neutralize the bulk charge of the beads, keeping uncompensated charges only in the thin surface layer. Unfortunately, the numerical solutions of the Poisson–Boltzmann equation were obtained for individual beads. Thus, it was impossible to confirm the avalanche-like counterion condensation on the necklace globule and obtain the necklace stability region.

In addition to the reduction of the net polymeric charge weakening intrachain electrostatic repulsion the condensed counterions can also induce effective attractive interactions between charged monomers [61, 66, 67, 75, 76]. In the ion binding and counterion adsorption models condensed counterions form ionic pairs with oppositely charged ions on the polymer backbone [61, 62, 66]. The formation of the ionic pairs leads to an additional dipole-dipole and charge-dipole attractive interactions. These attractive interactions decrease the value of the effective second virial coefficient for monomer-monomer interactions shifting the position of the Θ-point. In the case of the strongly charged polyelectrolytes the shift of the Θ-temperature could be significant and change the solvent quality for the polymer backbone to poor solvent conditions as the number of condensed counterions increases. This can result in a chain collapse and completely alter scenario of the counterion condensation (see discussion about avalanche-like counterion condensation). The analysis of the effect of the counterion condensation on conformations of a polyelectrolyte chain was done by Schiessel and Pincus [66] in the framework of the scaling approach, and by Muthukumar [61] in the framework of the variational approach. These theories predict nonmonotonic dependence of the chain size on the solution dielectric constant ε and solution temperature T [solution Bjerrum length $l_B \sim 1/(\varepsilon T)$]. The chain size first increases with increasing value of the Bjerrum length, then begins to decrease as the Bjerrum length exceeds the crossover value. This nonmonotonic dependence of the chain size is the manifestation of the two-fold role of the electrostatic interactions. At low values of the Bjerrum length the intrachain electrostatic repulsion controls the chain size. These interactions become stronger with increasing the value of the Bjerrum length and forces the polyelectrolyte chain to expand. At large values of the Bjerrum length the condensed counterions reduce net polymeric charge weakening the intrachain electrostatic repulsion, which together with the dipole-dipole and charge-dipole attractive interactions, induce chain contraction.

The localization of counterions inside the chain volume can also lead to effective attractive interactions [12, 80]. These interactions are due to heterogeneous distribution of the charge density along the polymer backbone. In the case of weak electrostatic attraction the origin of these interactions is similar to the fluctuation-induced attraction in two-component plasma and is related to the local charge density fluctuations [80]. In the opposite limit of strong electrostatic interactions the effect is due to correlation-induced attraction between the counterions and the oppositely charged polymer backbone similar to the interactions in strongly

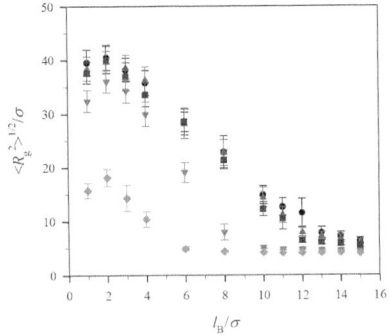

Fig. 9.13 Dependence of the chain size on the value of the Bjerrum length for polyelectrolyte chain with the degree of polymerization $N = 304$, the fraction of charged monomers $f = 1/3$ for different values of the Lennard-Jones interaction parameters $\varepsilon_{\text{LJ}} = 0.1k_BT$ (filled circles), $\varepsilon_{\text{LJ}} = 0.3k_BT$ (filled triangles), $\varepsilon_{\text{LJ}} = 0.5k_BT$ (filled squares), $\varepsilon_{\text{LJ}} = k_BT$ (inverted filled triangles), and $\varepsilon_{\text{LJ}} = 1.5k_BT$ (filled rhombs). Reproduced with permission from [76].

correlated Wigner liquids (see for review [12, 80]) or in ionic crystals such as NaCl. For example, in the case of the ionic crystal the attractive (negative) lattice energy is due to the spatial distribution (spatial correlations) of cations and anions over the lattice sites, even though the net charge of the crystal is zero. The crystal will remain stable even if it carries a small nonzero charge because of the large lattice (correlation) energy.

The effect of the fluctuation/correlation-induced attractive interactions on the conformations of a polyelectrolyte chain was studied theoretically [75, 76] and by molecular dynamics simulations [75–77]. These studies show that the fluctuation/correlation-induced-attractive interactions can cause additional chain collapse. In particular, in the case of the poor solvent conditions for the polymer backbone these studies established existence of the two different mechanisms that could lead to formation of the necklace globule [75–77]. For the values of the Bjerrum length $l_B = 1\sigma$ and 2σ, where σ is the Lennard-Jones charge diameter, the necklace structure appears as a result of competition between short-range monomer-monomer attractive interactions and electrostatic repulsion between uncompensated charges [76]. However, for the value of the Bjerrum length $l_B > 3\sigma$ the necklace structure is controlled by counterion condensation and is due to the optimization of the correlation-induced attraction between charged monomers and condensed counterions and electrostatic repulsion between uncompensated charges on the polymer backbone. Note that counterion condensation on the polymer backbone can also lead to chain collapse and necklace formation even in the good or theta solvent conditions for the uncharged polymer backbone [76]. These simulations also show nonmonotonic dependence of the chain size on the Bjerrum length confirming a two-fold role of the strength of the electrostatic interactions on the chain's properties (see Figure 9.13).

Ionization equilibrium

In the case of weak polyelectrolytes the ionic equilibrium between dissociated and undissociated ionic groups is controlled by the solution pH [4, 5]. For example, in the case of neutralization of the polyacrylic acid by sodium hydroxide the binding of the hydrogen ions is governed by the reaction

$$-\text{COOH} + \text{NaOH} \rightarrow -\text{COO}^- + \text{Na}^+ + \text{H}_2\text{O} \ . \qquad (9.42)$$

The concentration of the produced $-\text{COO}^-$ ions can be set to concentration of the added Na^+ or OH^- ions because the change in concentration of the free hydrogen ions H^+ in a solution due to this reaction is small. At equilibrium the chemical potential of the bound hydrogen ions and mobile hydrogen ions in a solution should be equal to each other. In the solution the chemical potential is controlled by a solution pH and is equal to

$$\frac{\mu_{\text{sol}}}{k_BT} = pK_0 + \ln c_{\text{H}^+} \qquad (9.43)$$

where pK_0 is the dissociation constant of the isolated $-COOH$ group equal to the difference of the chemical potentials $pK_0 = (\mu^0_{H^+} + \mu^0_{COO^-} - \mu^0_{COOH})/k_BT$. The chemical potential of the bound hydrogen ion includes entropic contribution from partitioning fN ionized groups between N monomers and contribution due to the interaction between ionized groups on the polymer backbone

$$\frac{\mu_{\text{bound}}}{k_BT} \approx \ln\left(\frac{1-f}{f}\right) - u^{2/3}f^{1/3} \ . \tag{9.44}$$

In writing equation (9.44) we have assumed for the calculation of the chain free energy with f ionized groups, that solvent is close to the Θ-solvent for the polymer backbone, $F_{\text{scaling}}(f) \approx k_BTN/g_e^0$. Equating chemical potentials one has

$$\log_{10}\left(\frac{1-f}{f}\right) \approx pK_0 - pH + u^{2/3}f^{1/3} \ . \tag{9.45}$$

This equation provides a relation between solution $pH = -\log_{10} c_{H^+}$ and the fraction of the charged monomers f on the polymer backbone for weak polyelectrolytes.

Thus, by varying solution pH one can change the fraction of charged monomers on the polymer backbone and the chain conformation [81–83]. Monte Carlo simulations of the titration of hydrophobic and hydrophilic polyelectrolytes were performed by Ulrich *et al.* [81]. These studies showed that depending on the solvent quality for polymer backbone and the $pH - pK_0$ value, a polyelectrolyte chain could be in five different conformational states: coil, collapsed spherical globule, necklace globule, sausage-like aggregate, and fully stretched chain. The effect of the charge annealing on the properties of the polyelectrolyte chain in poor solvent condition was also studied in [82]. These molecular simulations [81–83] provide a qualitative understanding of the pH induced chain collapse transition in a dilute polyelectrolyte solution [84–86].

9.2 A polyelectrolyte chain in salt solutions

In salt solutions the electrostatic interactions between charged groups on the polymer backbone are exponentially screened by the salt ions at the distances larger than the solution Debye screening length

$$\frac{U_{\text{DH}}(r_{ij})}{k_BT} = \frac{l_B q_i q_j}{r_{ij}} \exp(-r_{ij}/r_D) \tag{9.46}$$

where $r_D = (8\pi l_B c_s)^{-1/2}$ is the Debye screening length in a solution with the monovalent salt concentration c_s. The total effect of the electrostatic interactions on the chain's conformations is reduced to the local chain stiffening, which is manifested in renormalization of the chain persistence length, and to the additional chain swelling which is due to interactions

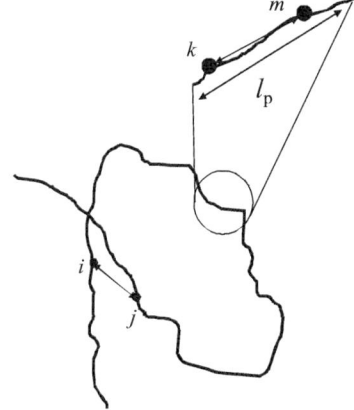

Fig. 9.14 Schematic representation of separation of the length scales for calculation of the effect of the electrostatic interactions on the electrostatic persistence length and chain swelling.

between remote along the polymer backbone charges, see Figure 9.14. The next section considers the effect of the electrostatic interactions on the local chain stiffening and introduces a concept of the electrostatic persistence length.

Electrostatic persistence length

The concept of the electrostatic persistence length was introduced by Odijk [87] and by Skolnick and Fixman [88] (collectively, OSF). They showed that electrostatic interactions between charged monomers at the distances smaller than the Debye screening length result in additional chain stiffening known as the electrostatic persistence length.

We will derive the OSF expression for electrostatic persistence length by using a discrete model of semiflexible polyelectrolyte chain is a salt solution. Consider a semiflexible chain with the number of bonds $N_b = N - 1$, bond length b, and fraction of charged monomers f. Chain conformations are described by a set of the unit vectors $\vec{n}_i$ pointing along the chain bonds. The potential energy of a semiflexible polyelectrolyte chain with the bending energy $k_B T K$ in a given conformation includes the bending energy and the electrostatic energy contributions

$$\frac{U_{\rm PE}(\{\vec{n}_i\})}{k_B T} = \frac{K}{2} \sum_{i=0}^{N_b - 2} (\vec{n}_i - \vec{n}_{i+1})^2 + \frac{l_B f^2}{2} \sum_{i \neq j}^{N} \frac{\exp(-r_{ij}/r_D)}{r_{ij}} \quad (9.47)$$

where r_{ij} is the distance between monomers i and j on the polymer backbone. The persistence length of a semiflexible chain without electrostatic interactions is equal to $l_p^0 = bK$. In writing the electrostatic part of the chain's potential energy [equation (9.47)] we have assumed that each monomer is carrying a fractional charge f. The OSF-derivation [87, 88] of the electrostatic persistence length is based on the evaluation of the change of the chain's electrostatic energy as polymer conformation deviates from a straight line. Let us consider a variation in the electrostatic energy of a polyelectrolyte chain with the bond length b by bending the chain into a circle with radius $R_c = b/[2\sin(\theta/2)]$, see Figure 9.15. The distance between two monomers separated by n bonds along the polymer backbone in such conformation is equal to

$$r(n) = 2R_c \sin(n\theta/2) = \left.\frac{b \sin(n\theta/2)}{\sin(\theta/2)}\right|_{\theta \ll 1} \approx bn\left(1 - n^2\theta^2/24\right). \quad (9.48)$$

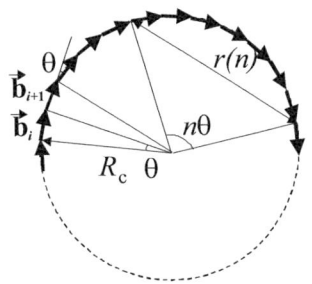

Fig. 9.15 Schematic representation of the conformation of a polyelectrolyte chain for calculation of the OSF electrostatic persistence length. Reproduced with permission from [8].

The difference between the electrostatic energy per monomer in the circular and rod-like conformations is

$$\frac{\Delta U_{\rm electr}(\theta)}{k_B T} \approx l_B f^2 \sum_{n=1}^{\infty} \left(\frac{\exp(-r(n)/r_D)}{r(n)} - \frac{\exp(-bn/r_D)}{bn}\right)$$

$$\approx \left.\frac{l_B f^2 r_D^2}{8b^3}\theta^2\right|_{b/r_D \ll 1}. \quad (9.49)$$

The final result in equation (9.49) was obtained by substituting the expression for $r(n)$, equation (9.48), into the right-hand side of equation

(9.49) and expanding it into the power series over the angle θ. A chain in the circular configuration makes a complete turn after $n_p \propto \theta^{-1}$ steps leading to the persistence length to be on the order of $b\theta^{-1}$.

In the OSF derivation of the electrostatic persistence length [87, 88], it was assumed that such bending of a chain can be induced by thermal fluctuations if the change in the electrostatic energy per persistence length $n_p \Delta U_{\text{electr}}(\theta)$ is on the order of the thermal energy $k_B T$. This leads to the typical values of the bending angle $\theta_{\text{OSF}} \approx b^2/r_D^2 u f^2$ and the OSF electrostatic persistence length equal to

$$l_p^{\text{OSF}} \approx \frac{b}{\theta_{\text{OSF}}} \approx \frac{l_B f^2 r_D^2}{4b^2}. \tag{9.50}$$

It is interesting to point out that the electrostatic persistence length is proportional to the square of the charge valence within the Debye screening length, fr_D/b.

For semiflexible polyelectrolyte chains the total persistence length is the sum of the electrostatic contribution given by equation (9.50) and of the bare persistence length $l_p^0 = bK$ [87, 88]

$$l_p = l_p^0 + l_p^{\text{OSF}} \approx l_p^0 + \frac{l_B f^2 r_D^2}{4b^2}. \tag{9.51}$$

Odijk [89] applied equation (9.50) to describe solution properties of flexible strongly charged polyelectrolytes with the electrostatic interaction parameter $uf^2 \approx 1$. In this case the electrostatic contribution to the chain persistence length l_p^{OSF} is the main factor determining the chain's bending rigidity. The additional chain stiffening of these polyelectrolytes could occur at distances substantially larger than the Debye screening length r_D. The Odijk result was extended to flexible weakly charged polyelectrolytes with $uf^2 \ll 1$ by Khokhlov and Khachaturian [24] by considering electrostatic blobs of size D_e^0 as new effective monomers

$$l_p^{KK} \approx D_e^0 + l_B \left(\frac{g_e^0 f r_D}{D_e^0}\right)^2 \approx \frac{r_D^2}{D_e^0}. \tag{9.52}$$

This expression can be easily derived from equation (9.51) by substituting for the charge valence within the Debye radius, $fg_e^0 r_D/D_e^0$, and remembering that $l_B (fg_e^0)^2/D_e^0 \approx 1$.

Since its introduction the concept of the electrostatic persistence length was one of the most controversial issues of the theory of polyelectrolyte solutions. There were numerous attempts to confirm or disprove the OSF finding for the electrostatic persistence length [3, 90–107]. This resulted in either quadratic or linear dependencies of the chain persistence length on the Debye screening length. A historical overview and critical discussion of different methods used for calculation of the electrostatic persistence length can be found in [8, 108].

One of the main reasons for the different results obtained for dependence of the electrostatic persistence length on the Debye screening length is the assumption that a polyelectrolyte chain can be approximated by a semiflexible chain with the effective chain bending constant.

Unfortunately, the real situation is more complex that was assumed in the OSF approach [87, 88]. To illustrate this, Figure 9.16 shows the bond-bond correlation function obtained from the molecular dynamics simulations of the semiflexible polyelectrolyte chain with the potential energy similar to the one given by equation (9.47) [108]. According to the results of these simulations the bond-bond correlation function describing decay of the orientational memory along the polymer backbone between unit bond vectors separated by l-bonds

$$G(l) = \frac{1}{N_b - l} \sum_{s=0}^{N_b-l-1} \langle (\vec{n}_s \cdot \vec{n}_{s+l}) \rangle \tag{9.53}$$

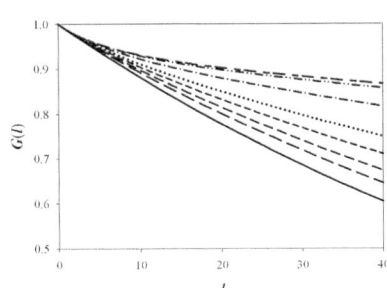

Fig. 9.16 Bond-bond correlation functions of fully charged $f = 1$ semiflexible polyelectrolyte chains with the degree of polymerization $N = 200$, the value of the Bjerrum length $l_B = 1.0\sigma$, bending constant $K = 80.0$, and different values of the Debye screening length: $r_D = 7\sigma$ (long dashed line), $r_D = 10\sigma$ (medium dashed line), $r_D = 14.25\sigma$ (short dashed line), $r_D = 20\sigma$ (dotted line), $r_D = 40\sigma$ (dashed-dotted line), $r_D = 100\sigma$ (dashed-dotted-dotted line) and $r_D = 200\sigma$ (long-short dashed line). The bond-bond correlation function of a neutral chain is shown by a solid line. Reproduced with permission from [108].

has two different functional forms. For small values of the Debye screening length $r_D < 6\sigma$ (where σ is the Lennard-Jones diameter of a monomer) the decay of the bond-bond orientational correlations can be approximated by a single exponential function

$$G(l) = \exp(-|l|/\lambda_1). \tag{9.54}$$

However, above a threshold value of the Debye screening length, $r_D > 6\sigma$, the simulation data can only be fitted by the combination of two exponential functions [108]

$$G(l) = (1-B)\exp\left(-\frac{|l|}{\lambda_1}\right) + B\exp\left(-\frac{|l|}{\lambda_2}\right). \tag{9.55}$$

This form of the bond-bond correlation function indicates that above a threshold value of the Debye screening length there are two different length scales that describe the chain orientational memory. Note, that the characteristic length scale λ_1 describes the chain's orientational correlations at large distances while parameter λ_2 is responsible for the bond-bond correlation properties at short distances along the polymer backbone. The values of the parameter λ_1 are always large such that for small separations along the polymer backbone, where the second term in the right-hand side of equation (9.55) is dominant, one can expand equation (9.55) in the power series of l/λ_1. This results in the following modification of equation (9.55)

$$G(l) \approx 1 - (1-B)\frac{|l|}{\lambda_1} + B\left(\exp\left(-\frac{|l|}{\lambda_2}\right) - 1\right) \tag{9.56}$$

Note that the last term in the right-hand side of this equation has a form characteristic of a semiflexible chain under tension [109]. It is useful to introduce an effective chain bending constant λ_e and effective force f_e by setting $\sqrt{\lambda_e/f_e} = \lambda_2$ and $\sqrt{\lambda_e f_e} = B^{-1}$. Thus, the effective short-scale chain bending rigidity is equal to $\lambda_e = \lambda_2/B$ and the effective force is $f_e = 1/\lambda_2 B$. Note, that approximation of the bond-bond correlation function in this regime by a single exponential function results in a smaller value of the effective chain bending constant and a smaller value of the chain's persistence length.

Figure 9.17 shows dependence of the chain's dimensionless bending rigidities λ_1 and λ_e on the Debye screening length [108]. As one can see from the plot the value of the chain's bending constant λ_e decreases as the value of the Debye screening length increases approaching the bare value of the chain's bending constant K at large values of the Debye screening length. However, the long-length scale bending rigidity λ_1 increases with increasing the value of the Debye radius (decreasing the salt concentration). At small values of the Debye screening length the value of the bending rigidity λ_1 approaches the chain's bare bending rigidity K. For large values of the Debye screening length far from a crossover point, a parameter λ_1 is a universal function of the Debye screening length and is almost independent of the initial value of the chain's bending energy K. It is impossible to say what exactly happens with these two bending rigidities close to a crossover point because the error in separating two length scales increases as the value of the Debye radius decreases.

A detailed analysis of the fluctuation spectrum for chain's potential energy equation (9.47) proves that the bond-bond correlation function indeed can be approximated by equation (9.55). At long length scales the spectrum of fluctuations of the semiflexible polyelectrolyte chain is that of a semiflexible chain with the effective chain bending rigidity λ_1 equal to the sum of the bare chain bending constant and the OSF contribution

$$\lambda_1 \approx K + K_{\text{OSF}} = K + \frac{uf^2 r_D^2}{4b^2} \,. \qquad (9.57)$$

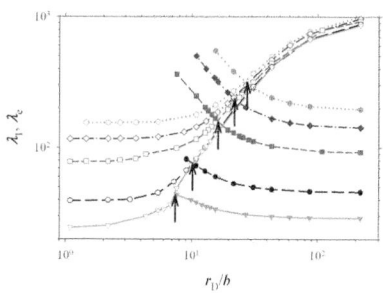

Fig. 9.17 Dependence of the bending rigidities λ_1 (open symbols) and λ_e (filled symbols) on the Debye screening length for semiflexible polyelectrolyte chains with the degree of polymerization $N = 200$, electrostatic coupling constant $A_{el} = 1.0$ and with different bending constants: $K = 25.0$ (solid line with inverted triangles), $K = 40.0$ (long-dashed line with circles), $K = 80.0$ (short-dashed line with squares), $K = 120.0$ (dashed-dotted line with rhombs) and $K = 160.0$ (grey dotted line with hexagons). Arrows show intersection points between λ_1 and λ_e. Reproduced with permission from [108].

This equation coincides with the OSF expression [87, 88] for the chain's bending rigidity. The new feature in comparison with the OSF result is the existence of the second length scale characterized by the parameter λ_2 [108]. It is worth pointing out that at low salt concentrations (large values of the Debye radius) the dependence of

$$\lambda_2 \approx \sqrt{\frac{K}{uf^2 \ln(r_D/b)}} \qquad \text{for } K < K_{\text{OSF}} \qquad (9.58)$$

on the system parameters coincides with that of a semiflexible chain with the bending rigidity K under tension by the external electrostatic force $f_e b/k_B T \approx uf^2 \ln(r_D/b)$. Thus, at short distances electrostatic interactions lead to local chain deformation which is manifested in the initial fast decay of the correlation function $G(l)$. However, for small values of the Debye screening lengths the contribution of the second exponential function is negligible at the distances along the polymer backbone larger than the Debye screening length,

$$\lambda_2 \propto r_D/b \,, \qquad \text{for } K > K_{\text{OSF}} \,. \qquad (9.59)$$

One can associate this exponential term with the high frequency modes that control chain's bending rigidity at short-length scales.

One can estimate a salt concentration at which the crossover between two different regimes in λ_2 dependence on the salt concentration may

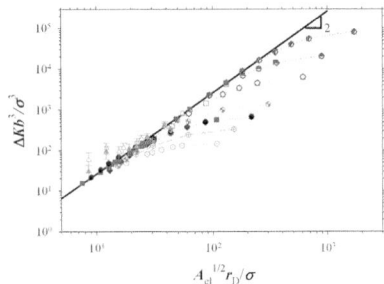

Fig. 9.18 Dependence of the parameter $\Delta K b^3/\sigma^3$ on the parameter $A_{\rm el}^{1/2} r_D/\sigma$ for semiflexible polyelectrolyte chains with the degree of polymerization $N = 200$, bending constants: $K = 40.0$ (circles), $K = 80.0$ (squares), $K = 120.0$ (rhombs) and $K = 160.0$ (hexagons) with electrostatic coupling constants: $A_{\rm el} = 0.25$ (open symbols with dot), $A_{\rm el} = 0.5$ (open symbols with cross), $A_{\rm el} = 1.0$ (filled symbols), $A_{\rm el} = 2.0$ (checkered symbols), $A_{\rm el} = 9.0$ (open symbols), $A_{\rm el} = 25.0$ (top filled symbols) and $A_{\rm el} = 100.0$ (left filled symbols). The solid line is the best fit given by the equation $\Delta K b^3/\sigma^3 = 0.264 A_{\rm el} r_D^2/\sigma^2$. The experimental data for DNA from Hagerman [110] are shown as open triangles for 587 bp and filled triangles for 434 bp. The value of the parameter $A_{\rm el} = 0.093$ for the experimental data was obtained by assuming the distance between effective charges $b = 0.17$ nm (an effective projection length between charges), degree of ionization $f_* = 0.15$ and value of the Bjerrum length $l_B = 0.7$ nm. Reproduced with permission from [110].

occur, $(r_D/b)_{\rm cr} \propto \sqrt{K/uf^2}$ [108]. Below we assume that the fraction f of charged monomers on the polymer backbone is set at the Manning–Oosawa counterion condensation value, $f \approx b/l_B$. This condition determines a low bound on the distance between ionized groups along the polymer backbone. Taking this into account one can rewrite a crossover condition in terms of the chain's persistence length and solution Bjerrum length, $(r_D)_{\rm cr} \approx \sqrt{l_p^0 l_B}$. For a DNA molecule in water at room temperature with the value of the Bjerrum length, $l_B = 0.7$ nm, and persistence length, $l_p^0 \approx 50$ nm, a crossover takes place at ionic strengths on the order of $I \approx 2.5 \times 10^{-3}$ M, which is below a biologically relevant range of the ionic strengths, $I \sim 0.1$–0.01 M. However, it is possible to observe this crossover in solutions of synthetic polyelectrolytes or single stranded DNA for which typical values of the persistence length are about 50 times smaller, $l_p^0 \approx 1$ nm. In this case a crossover value of the ionic strength is about $I \approx 10^{-1}$ M. For lower solution ionic strengths (or larger Debye screening lengths) one should expect that a short-length scale bond-bond correlation function has a form characteristic of a chain under tension. Thus, for the most relevant salt concentration range the bond-bond correlation function of the flexible polyelectrolytes at short length scales should demonstrate features characteristic of the chain under local tension. This observation is supported by the results of the computer simulations of flexible polyelectrolyte that show the bond-bond correlation function with two characteristic length scales [106].

In Figure 9.18 we verify a scaling dependence of the parameter $\Delta K = \lambda_1 - K \propto l_B f^2 r_D^2/b^3 = A_{\rm el} r_D^2 \sigma/b^3$ on the reduced value of the Debye screening length, $A_{\rm el}^{1/2} r_D/\sigma$, where the dimensionless amplitude of the strength of the electrostatic interactions $A_{\rm el} = l_B f^2/\sigma$ is introduced. Indeed, the simulation data follow expected scaling dependence. The deviation of the simulation data from the power law occurs when the value of the Debye screening length becomes comparable with the chain's contour length Nb. This is a manifestation of the finite chain length effect. In order to compare simulations with the experimental data, Figure 9.18 has the data points for salt dependence of the DNA persistence length [110].

The nonlinear effects in counterion distribution around a polyion on the electrostatic persistence length can be taken into account by solving the Poisson–Boltzmann equation [111, 112]. In the framework of this approach one can also consider the effect of the counterion condensation and the difference in the dielectric constant of the polyion and surrounding solvent on the chain's bending rigidity. Note, that in addition to the reduction of the net polymeric charge the counterion condensation can lead to a negative contribution to the chain's bending rigidity [95, 113, 114]. This contribution is due to fluctuation/correlation effects in counterion distribution along the polymer backbone.

Swelling of a polyelectrolyte chain in salt solutions

Swelling of a polyelectrolyte chain in salt solutions is one of the classical problems of polymer physics [3, 8, 29, 105, 115–119]. The first attempt to account for the effect of the electrostatic interactions on conformations of a polyelectrolyte chain was over 60 years ago by Kuhn, Kunzle, and Katchalsky [29, 115] and by Hermans and Overbeek [116]. This was done by minimizing the total chain's free energy with respect to a chain size. The total chain free energy included the chain's configurational (elastic) free energy

$$F_{\text{elast}} \approx k_B T \frac{R_e^2}{b^2 N} \qquad (9.60)$$

and the chain interaction free energy which was evaluated by taking into account two-body interactions between charged monomers

$$F_{\text{int}} \approx k_B T B_{\text{DH}} \frac{f^2 N^2}{R_e^3} , \qquad (9.61)$$

where B_{DH} is the second virial coefficient of the Debye-Hückel potential [see equation (9.46)]:

$$B_{\text{DH}} = \int d^3 r \left(1 - \exp\left(-\frac{U_{\text{DH}}(r)}{k_B T} \right) \right) \bigg|_{r_D > l_B} \approx 4\pi l_B r_D^2 . \qquad (9.62)$$

The optimization of the total chain's free energy

$$\frac{F}{k_B T} \approx \frac{R_e^2}{b^2 N} + B_{\text{DH}} f^2 \frac{N^2}{R_e^3} \qquad (9.63)$$

with respect to the chain size R_e resulted in the following expression for a chain size

$$R_e \approx \left(uf^2\right)^{1/5} b^{3/5} r_D^{2/5} N^{3/5} \propto c_s^{-1/5} N^{3/5} \qquad (9.64)$$

Thus, in salt solutions conformations of a polyelectrolyte chain are those of a self-avoiding chain with the excluded volume determined by the electrostatic repulsion between charged monomers. With decreasing salt concentration the chain size increases. It becomes comparable to the size of a polyelectrolyte chain in a salt-free solution [see equation (9.10)] at salt concentrations for which the value of the Debye screening length is on the order of $r_D \approx b(uf^2)^{1/3} N$.

This result was unchallenged until it was realized that electrostatic interactions between charged monomers can also lead to chain stiffening. The combined effect of the electrostatic interactions on the local chain stiffening and swelling was taken into account by Odijk and Houwaart [120]. Their analysis leads to a chain size,

$$R_e \propto r_D^{3/5} N^{3/5} \propto c_s^{-3/10} N^{3/5} \qquad (9.65)$$

which has a stronger salt concentration dependence than the chain size given by equation (9.64). Note, that equation (9.64) can be derived by assuming a linear scaling of a chain's persistence length l_p with the Debye screening length r_D. Thus, the discrepancy between two results is due to the different dependence of the chain persistence length on the Debye radius.

Swelling of semiflexible biological polyelectrolytes

We begin analysis of the effect of the electrostatic interactions on the chain swelling by considering the case of semiflexible (biological) polyelectrolytes for which the bare chain persistence length is large, $l_p^0 \gg b$. The chain's electrostatic energy can be represented as a sum of two terms describing electrostatic interactions between local and remote along the polymer backbone charged pairs, see Figure 9.14. The local term takes into account interactions between charged monomers that are separated by distances along the polymer backbone shorter than the chain's persistence length. These electrostatic interactions renormalize chain persistence length.

$$l_p \approx l_p^0 + C_l \frac{l_B f^2 r_D^2}{b^2} . \tag{9.66}$$

The value of the numerical coefficient C_l depends on the semiflexible chain model used to describe properties of the neutral chain [121]. The second term includes electrostatic interactions between remote particles along the polymer backbone: charged pairs located at the distances along the polymer backbone larger than the chain persistence length, see Figure 9.14. These interactions are responsible for the chain swelling. The contribution of these interactions can be taken into account by dividing a chain into segments with length on the order of the chain's persistence length and calculating the effective second virial coefficient between these segments. Since, at the length scales smaller than the chain's persistence length, conformation of a chain is close to rod-like, the energy of the electrostatic repulsion between segments can be approximated by the energy of electrostatic interaction between two charged rods. The electrostatic potential energy between two charged infinitely long rod-like segments separated by a distance d and oriented with respect to each other with an angle γ_a is [122]

$$\frac{U_{\text{int}}(d, \gamma_a)}{k_B T} \approx \frac{2\pi l_B r_D f^2 \exp(-d/r_D)}{b^2 |\sin \gamma_a|} . \tag{9.67}$$

An effective second virial coefficient per monomer of two charged segments interacting with the interaction potential equation (9.67) is equal to [122]

$$B_{\text{el}} \approx 2b^2 \int_0^{\pi/2} \sin^2 \gamma_a \, d\gamma_a \int_{d_0}^{\infty} \left(1 - \exp\left(-\frac{U_{\text{int}}(r, \gamma_a)}{k_B T}\right)\right) dr \tag{9.68a}$$

where d_0 is the chain diameter which is on the order of b. The integration over rod separation r can be performed exactly leading to

$$B_{\text{el}} \approx 2b^2 r_D \int_0^{\pi/2} \sin^2 \gamma_a \left(E_1(\zeta/\sin \gamma_a) + \ln(\zeta/\sin \gamma_a) + 0.57721\right) d\gamma_a \tag{9.68b}$$

where $\zeta = (2\pi l_B r_D f^2 / b^2) \exp(-d_0/r_D)$ is an effective interaction energy at contact. Fixman and Skolnick [122] showed that expression equation

(9.68b) has two simple asymptotic regimes

$$B_{\text{el}} \approx \begin{cases} 4\pi l_B f^2 r_D^2 & \text{for } r_D \ll b/uf^2 \\ \frac{\pi}{4} b^2 r_D \ln\left(l_B f^2 r_D/b^2\right) & \text{for } r_D \gg b/uf^2 \end{cases}. \qquad (9.69)$$

In the case of the weak electrostatic interactions, $r_D \ll b/uf^2$, the effective second virial coefficient between monomers coincides with $B_{\text{DH}}f^2$, see equation (9.62). In this range of parameters the charges on the chain can be considered as uncorrelated. However, in the case of strong electrostatic interactions, when the energy of electrostatic repulsion, $U(g) \propto k_B T(l_B g^2/r_D)$ between g charges within the Debye screening length, $g \propto fr_D/b$, is much larger than the thermal energy $U(g) \gg k_B T$ (or $uf^2 > b/r_D$), the connectivity of the charged monomers into a chain plays an important role. This is manifested in the appearance of the correlation hole with size on the order of the Debye screening length surrounding the polymer backbone [121].

The scaling analysis of the different salt concentration regimes in polyelectrolyte chain conformations can be done by using a Flory-like approach by optimizing the elastic and segment-segment interaction contributions to the chain's free energy

$$\frac{F}{k_B T} \approx \frac{R_e^2}{bl_p N} + \frac{B_{\text{el}} N^2}{R_e^3} \qquad (9.70)$$

with respect to a chain size R_e

$$R_e \approx bN^{3/5} \begin{cases} \left(uf^2 l_p r_D^2/b^3\right)^{1/5} & \text{for } r_D \ll b/uf^2 \\ \left(l_p r_D/b^2\right)^{1/5} & \text{for } r_D \gg b/uf^2 \end{cases} \qquad (9.71)$$

where l_p is a chain persistence length which includes contributions from the local electrostatic interactions, see equation (9.66). To find all possible scaling regimes of a swollen chain one has to explicitly consider the effect of renormalization of the chain persistence length by the local electrostatic interactions (see for details [121]).

The crossover expression suitable for analysis of the experimental data was derived by Dobrynin and Carrillo [121] in the framework of the Edwards–Singh variational principle [123]

$$\frac{l_p^{\text{tr}}}{l_p} - 1 - \frac{3^{1/2}}{2\pi^{3/2}} \frac{(B_{\text{el}} + B_0) N^{1/2}}{\left(l_p^{\text{tr}} b\right)^{3/2}} \left(1 - \frac{3}{2}\sqrt{\frac{l_p}{bN}} + \frac{1}{2}\left(\frac{l_p}{bN}\right)^{3/2}\right) = 0. \qquad (9.72)$$

Equation (9.72) can be considered as a self-consistent equation for the effective chain's persistence length l_p^{tr}. Note, that by using the relationship between a chain persistence length and chain size $R_e^2 = 2bl_p^{\text{tr}} N$, one can rewrite equation (9.72) in a form similar to the Flory-like expression for a chain size [124]. It is important to point out that for the case of weak electrostatic interactions their contribution to the effective monomeric virial expansion has to be supplemented by the second virial coefficient due to excluded volume (hard core) interactions, B_0.

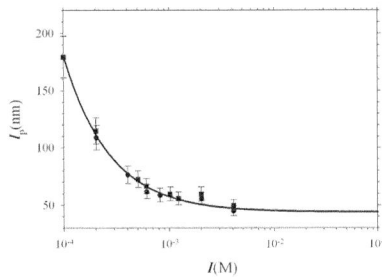

Fig. 9.19 Dependence of the DNA persistence length on the solution ionic strength I obtained by Hagerman for short fragments of DNA (587 and 434 bp) [110]. Solid line is the best fit to equation (9.72). Reproduced with permission from [121].

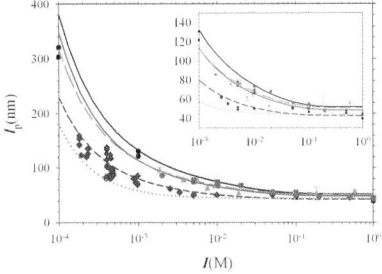

Fig. 9.20 Dependence of the DNA persistence length on the solution ionic strength I. Dark shaded rhombs show data by Maret and Weill on the lyophilized erythrocyte DNA ($M_w = 4.2 \times 10^6$ g/mol) [125]. The dark short-dashed line is the best fit to the equation (9.72) with ionization degree $f_* = 0.15$, bare persistence length $l_p^0 = 40$ nm, and degree of polymerization $N = 1.27 \times 10^4$. Light shaded triangles show data by Baumann et al. on λ-DNA ($M_w = 3.2 \times 10^7$ g/mol) [132]. The light long dashed line is the best fit with $f_* = 0.15$, $l_p^0 = 44$ nm, $N = 9.7 \times 10^4$. Dark circles [127] and squares [126] are data on T7 bacteriophage DNA ($M_w = 2.65 \times 10^7$ g/mol). The dark shaded line is the best fit with $f_* = 0.17$, $l_p^0 = 42$ nm, $N = 8.03 \times 10^4$. Black circles are data by Makita et al. on T4 DNA (165.6 Kbps) [129]. The black line is the best fit with $f_* = 0.15$, $l_p^0 = 40$ nm, $N = 3.3 \times 10^5$. The dotted line corresponds to persistence length of a short DNA without swelling effects (Figure 9.19). Reproduced with permission from [121].

Comparison with experiments

DNA is the most studied example of semiflexible polyelectrolytes [109, 110, 125–133]. The effect of salt concentration on conformational properties of DNA was studied by: light scattering [126], force-extension experiments [132], transient electric birefringence [110], magnetic birefringence [125], flow birefringence [127], and by fluorescence microscopy [129, 131]. The main goal of these research efforts was to find a salt concentration dependence of the DNA persistence length. The chain's persistence length was extracted from experimental data by assuming the following expression for a chain size

$$R_e^2 \approx 2bl_p^{\text{eff}} N \ . \tag{9.73}$$

Note, that this relation provides an actual value of the chain's persistence length only when there is no additional contribution from the chain swelling. This is usually true for relatively short chains and high salt concentrations. In this case the contribution of the pair electrostatic interactions between remote charged particles along the polymer backbone is weak. In the case of the longer chains a chain persistence length obtained from equation (9.73) includes a chain swelling effect and can not be considered as a local chain property.

Figure 9.19 presents experimental data from Hagerman for short fragments of DNA (587 and 434 bp) [110]. The data points can be fitted by a function

$$l_p^{\text{exp}} \approx 44 \, \text{nm} + \frac{0.0136}{I} \, , \tag{9.74}$$

where I is the solution ionic strength, for the monovalent salts it is equal to the solution salt concentration, $I = c_s$. Using the expression for the chain persistence length equation (9.66) and assuming that the distance between phosphate groups on the DNA $d_{\text{ph}} = 0.34$ nm (there are about 10 base pairs per DNA double helix turn with length 3.4 nm) and the Bjerrum length $l_B = 0.7$ nm, one can estimate an effective chain's ionization degree by approximating DNA by a rod with a charge density 20 charges per 3.4 nm, which gives a distance between effective charges to be equal to $b = 0.17$ nm. For the OSF model [87, 88] the numerical coefficient in front of the electrostatic correction term is $C_l = 0.25$. In this case the ionization degree is $f_{\text{OSF}}^* \approx 0.154$. The simulation data for electrostatic persistence length of semiflexible polyelectrolytes by Gubarev et al. [108] give a value of the coefficient $C_l = 0.264$, which results in a fractional charge per phosphate group to be equal to $f_* \approx 0.151$.

Figure 9.20 combines results for the effective DNA persistence length obtained by fluorescence microscopy [129], flow birefringence [127], magnetic birefringence [125], force-extension measurements [132], and light scattering [126], and fitting of the experimental data to the effective persistence length given by equation (9.72) and using the fraction of charged monomers f_* and bare chain persistence length l_p^0 as adjustable parameters. The value of the electrostatic second virial coefficient was calculated by numerical integration of equation (9.68) with a hard-core

DNA diameter $d_0 = 2\,\text{nm}$. The value of the hard-core second virial coefficient B_0 was set to $B_0 = \pi b^2 d_0/2 \approx \pi(0.17)^2 2.0/2 \approx 9.1 \times 10^{-2}\,\text{nm}^3$. The numerical constant C_l for the coefficient in front of the electrostatic correction to the chain's persistence length [see equation (9.66)] was set to $C_l = 0.264$ in agreement with the simulation results [108]. As one can see all five different data sets can be fitted by equation (9.72) with close values of the adjustable parameters $f_* = 0.15 \div 0.17$ and a bare persistence length $l_p^0 = 40 \div 44\,\text{nm}$. The agreement between theory and experiments is very good since only known parameters to describe the structure of DNA were used. It is interesting to point out that for these fractions of charged monomers, the distance between ionized groups $l_\text{ion} \approx b/f_* \approx 1 \div 1.13\,\text{nm}$ is larger than the value of the Bjerrum length $l_B \approx 0.7\,\text{nm}$. Thus, there is less than one charge per Bjerrum length.

A flexible polyelectrolyte chain in salt solutions

The results for swelling of the semiflexible polyelectrolytes can be extended to the case of flexible weakly charged polyelectrolytes by considering electrostatic blobs as new effective monomers. Thus, to modify equations obtained for semiflexible chains one has to substitute for the bond length b the electrostatic blob size D_e^0 and for the linear number charge density f/b the linear number charge density per electrostatic blob $f g_e^0 / D_e^0$. For simplicity, we will neglect counterion condensation on the polymer backbone. Using these transformations, one can estimate the electrostatic second virial coefficient per blob as

$$B_\text{el} \approx \begin{cases} l_B f^2 g_e^{0\,2} r_D^2 & \text{for } r_D \ll D_e^0 \\ D_e^{0\,2} r_D & \text{for } r_D \gg D_e^0 \end{cases}. \tag{9.75}$$

We first consider a case of weak electrostatic interactions when the Debye screening length is smaller than the electrostatic blob size D_e^0. In this range of parameters the chain persistence length is equal to the bare persistence length which for flexible chains is on the order of bond length b. In this case the Flory-like free energy of a polyelectrolyte chain in a salt solution is

$$\frac{F}{k_B T} \approx \frac{R_e^2}{b^2 N} + \frac{B_\text{el} N^2}{g_e^{0\,2} R_e^3} \approx \frac{R_e^2}{b^2 N} + \frac{l_B f^2 r_D^2 N^2}{R_e^3}. \tag{9.76}$$

Optimization of the free energy equation (9.76) with respect to the chain size R_e recovers equation (9.64)

$$R_e \approx b^{3/5} \left(u f^2\right)^{1/5} r_D^{2/5} N^{3/5}$$
$$\text{for } b \left(u f^2\right)^{-1/2} N^{-1/4} \ll r_D \ll b \left(u f^2\right)^{-1/3}. \tag{9.77}$$

Thus, in this regime the chain size scales with the salt concentration as $R_e \propto r_D^{2/5} \propto c_s^{-1/5}$. Figure 9.21 is a diagram of the state of a flexible polyelectrolyte chain in salt solutions. This regime corresponds to

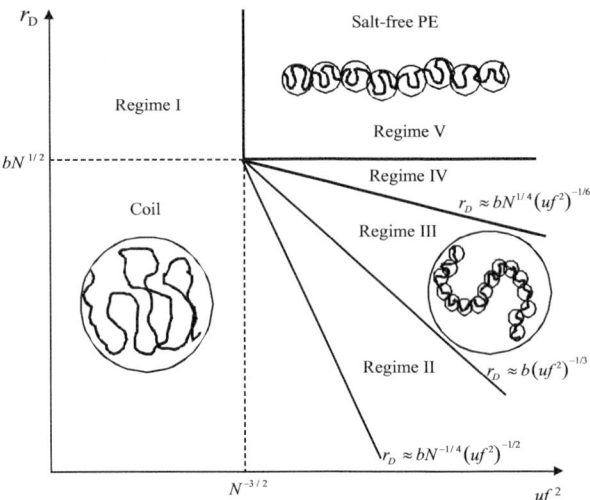

Fig. 9.21 Diagram of states of a flexible polyelectrolyte chain in a salt solution.

Regime II in the diagram of states. A polyelectrolyte chain will remain swollen with increasing salt concentration (decreasing the value of the Debye screening length) as long as its size is larger than the size of the ideal chain $bN^{1/2}$. This continues until $r_D \geq b(uf^2)^{-1/2}N^{-1/4}$. This inequality determines the low boundary for Regime II (crossover to Regime I where the polyelectrolyte chain has a coil-like conformation).

The dependence of the chain size on salt concentration in the case when the Debye radius is larger than the electrostatic blob size can be obtained by using for the chain's persistence length equation (9.52). Taking this into account one has

$$\frac{F}{k_BT} \approx \frac{R_e^2}{l_p D_e^0 (N/g_e^0)} + \frac{B_{el}N^2}{g_e^{02} R_e^3} \approx \frac{R_e^2 g_e^0}{r_D^2 N} + \frac{b^2 r_D N^2}{g_e^0 R_e^3} \ . \qquad (9.78)$$

Optimization of equation (9.78) with respect to the chain size R_e results in

$$R_e \approx b^{2/5} \left(uf^2\right)^{4/15} r_D^{3/5} N^{3/5}$$
$$\text{for } b\left(uf^2\right)^{-1/3} \leq r_D \leq b\left(uf^2\right)^{-1/6} N^{1/4} \qquad (9.79)$$

(For the upper boundary of this regime see discussion below.) In this regime (Regime III in Figure 9.21) the chain size shows slightly stronger salt concentration dependence $R_e \propto r_D^{3/5} \propto c_s^{-3/10}$.

Let us calculate the ratio between chain sizes given by the expressions equations (9.77) and (9.79). This ratio is equal to $(r_D/D_e^0)^{1/5}$. For a flexible chain with the size of the electrostatic blob $D_e^0 \approx 1.0$ nm the ratio is equal to unity at $r_D \approx D_e^0 \approx 1.0$ nm and the solution ionic strength $I \approx (0.3/1.0)^2 \approx 9 \times 10^{-2}$ M. (In obtaining the last estimate we used the following relation between the solution ionic strength I and the Debye radius: $r_D = 0.3/\sqrt{I}$ nm [134].) It becomes on the order of $(0.3/1.0\sqrt{I})^{1/5} \approx 2$ at the solution ionic strength 10^{-4} M. Thus, for the typical salt concentration range between 0.1 and 10^{-4} M the chain size changes by a factor of two indicating that there is no significant difference between the expression (9.64) and the result of equation (9.79). This makes it very difficult to observe pure scaling regimes in experiments. Computer simulations by Everaers et al. [107] show that the simulation results are closer to $R_e \propto r_D^{3/5}$ scaling dependence.

A swollen chain behavior persists until the value of the chain interaction parameter is larger than unity, $(B_{\rm el}/(l_p D_e^0)^{3/2})(N/g_e^0)^{1/2} \approx (D_e^0/r_D)^2(N/g_e^0)^{1/2} > 1$, or the value of the Debye screening length is smaller than $r_D \leq b(uf^2)^{-1/6} N^{1/4}$. For lower salt concentrations (larger values of the Debye screening length) the strength of the electrostatic interactions between remote along the polymer backbone charged pairs is too weak to swell a chain. In this salt concentration range the electrostatic interactions renormalize chain's persistence length and the chain size is equal to

$$R_e \approx (l_p D_e^0 N/g_e^0)^{1/2} \approx r_D \left(uf^2\right)^{1/3} N^{1/2} \quad (9.80)$$
$$\text{for } b\left(uf^2\right)^{-1/6} N^{1/4} \leq r_D \leq bN^{1/2}$$

The upper boundary of this regime is obtained by equating the chain's persistence length with the size of the extended chain of blobs ($R_e \approx b(uf^2)^{1/3} N$). This regime is shown as Regime IV in Figure 9.21. At the values of the Debye screening lengths, $r_D > b\sqrt{N}$, the electrostatic interactions between charged monomers result in elongation of a polyelectrolyte chain into array of blobs. The condition $r_D \approx b\sqrt{N}$ determines a crossover to a salt-free polyelectrolyte solution regime (see Regime V in Figure 9.21). Note that in a salt-free solution the electrostatic interactions between charged monomers result in chain stretching only if $uf^2 > N^{-3/2}$ (see Section 9.1).

9.3 Semidilute polyelectrolyte solutions

Overlap concentration

With increasing polymer concentration the distance between chains in a solution decreases. Polyelectrolyte chains begin to overlap when the distance between them becomes on the order of their size. This takes place at polymer concentration

$$c^* \approx \frac{N}{R_e^3}. \quad (9.81)$$

At the overlap concentration the concentration of monomers within a volume occupied by a chain is on the order of the average monomer concentration in a solution. For a strongly charged chains in salt-free solutions the chain size is close to that of a fully stretched chain $R_e \approx bN$. In this case the chain's overlap concentration scales with the chain's degree of polymerization as $c^* \approx N/R_e^3 \approx b^{-3} N^{-2}$. This dependence of the

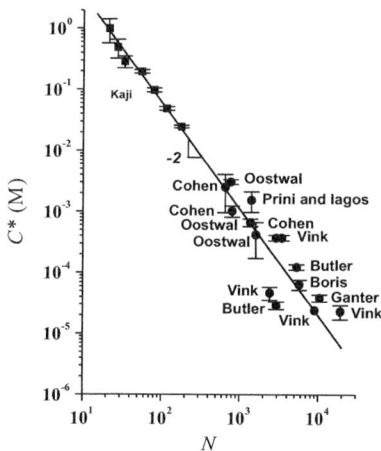

Fig. 9.22 Dependence of the overlap concentration, c^*, on chain degree of polymerization, N, in salt-free polyelectrolyte solutions of NaPSS. Data obtained from X-ray scattering (squares) and from viscosity (circles). Data assembled by D. C. Boris and R. H. Colby.

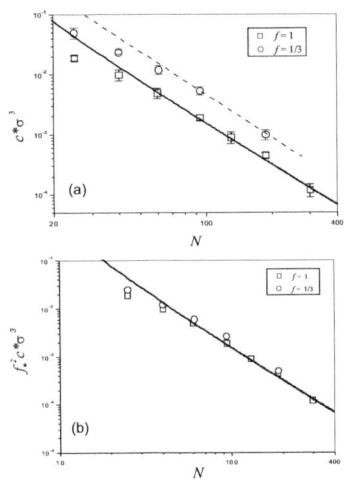

solution overlap concentration on the chain's degree of polymerization is much stronger than one observed in a solution of neutral chains for which $c^* \propto b^{-3} N^{-1/2}$. Thus, crossover to a semidilute solution regime takes place at much lower polymer concentrations in polyelectrolyte solutions than in solutions of neutral polymers. Figure 9.22 demonstrates dependence of the the overlap concentration, c^*, on the chain degree of polymerization, N, for salt-free solutions of NaPSS [135]. It is in excellent agreement with the scaling law $c^* \propto N^{-2}$. The overlap concentration for shorter polymers was obtained from X-ray scattering data (squares), while for longer polymers it was determined from viscosity data (circles).

As we pointed out in Section 9.1, weakly charged polyelectrolyte chains are non-uniformly stretched. The relation between the number of their monomers N and the chain size R_e in salt-free solutions is given by equation (9.10). In this case the overlap concentration c^* can be estimated as

$$c^* \approx \frac{N}{\left(R_e^F\right)^3} \approx b^{-3} u^{-1} f^{-2} N^{-2} \ln^{-1}\left(N/g_e\right). \qquad (9.82)$$

Fig. 9.23 (a) Dependence of the overlap concentration on the number of monomers for partially $f = 1/3$ (circles) and fully $f = 1$ (squares) charged chains. Lines are predictions of equation (9.82). (b) Universal curve for the dependence of the overlap concentration on the number of monomers. Reproduced with permission from [27].

Figure 9.23(a) shows the dependence of the overlap concentration c^* on the degree of polymerization N for partially $f = 1/3$ (circles) and fully $f = 1$ (squares) charged chains [27]. The solid and dashed lines correspond to equation (9.82) for different values of f. The dependence of the overlap concentration c^* on the degree of polymerization N follows the modified scaling predictions given by equation (9.82) for the longer chains but deviates from it for the shorter ones. This deviation from the scaling law can be explained by the finite size effect. In Figure 9.23(b), the simulation results are collapsed onto one universal curve. The numerical factor for this transformation is equal to 0.4. According to the scaling theory this factor should be inversely proportional to the square of the ratio of the effective charge densities on the chains $[f_*(1/3)/f_*(1)]^2 = 0.4$. For weakly charged chains with bare charge density $f = 1/3$, this conversion factor is equal to $[(1/3)/(1)]^2 = 0.11$. The difference between 0.11 and 0.4 can be explained by the counterion condensation. The ratio of the effective charge fractions on the two chains obtained from the osmotic coefficient is $f_*(1/3)/f_*(1) = 0.69$ (see [69]), which is close to the expected value $f_*(1/3)/f_*(1) = \sqrt{0.4} \approx 0.63$. Thus, effective charge densities f_* rather than the bare ones have to be used for the evaluation of the conversion factor. Below I will use f_* to describe the effective charge density on the polymer backbone instead of its bare value f and consider it as an adjustable parameter. Experimentally the effective fraction of charged monomers f_* can be evaluated from conductivity [136–138] or from the osmotic pressure measurements (see discussion below).

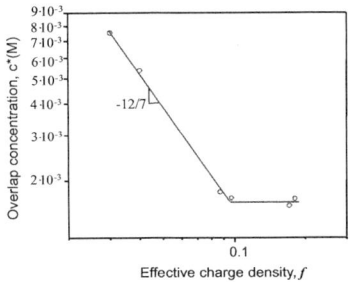

Fig. 9.24 Dependence of the overlap concentration, c^*, on the effective charge density, f_*, for 2-vinyl pyridine and N-methyl-2-vinyl pyridinium chloride random copolymer in ethylene glycol solvent at $25°C$. Data provided by Dou, S. and Colby, R. H.

The overlap concentration c^* is predicted to decrease with the fraction f of charged monomers as $c^* \propto f^{-2}$ for polyelectrolytes in a Θ-solvent for the polymer backbone [equation (9.82)] and as $c^* \propto f^{-12/7}$ for polyelectrolytes in a good solvent. Figure 9.24 confirms this scaling prediction for ethylene glycol solutions of N-methyl-2-vinyl pyridinium chloride random copolymers with fractions of quaternized monomers up to 10 mol%. At a higher fraction of quaternized monomers one reaches the Manning counterion condensation limit of one charge per Bjerrum length ($l_B = 15$ Å in ethylene glycole) and the overlap concentration c^* becomes independent of charge density, f, and saturates.

Scaling model of semidilute polyelectrolyte solutions

Correlation length

The important length scale above the overlap concentration, $c > c^*$, is the correlation length ξ — the average mesh size of the semidilute polyelectrolyte solution. The average charge of the correlation volume ξ^3 is equal to zero because the charge on the section of the chain with g_ξ monomers within the correlation length ξ is compensated by counterions. Each charged monomer experiences electrostatic repulsion from all other charged monomers within the correlation volume and electrostatic attraction to the counterion background. This corresponds to the well-known Katchalsky's cell model approximation [8, 53–55]. Electrostatic interactions within a correlation cell with radius $\xi/2$ and length ξ can be estimated by assuming that the cell has cylindrical symmetry with the polyelectrolyte chain located along the axis of the cylinder, see Figure 9.25.

Each electrostatic blob on a chain interacts with other electrostatic blobs located within the correlation length ξ. The energy of the electrostatic interactions between a blob located in the middle of the linear array of blobs of length ξ is estimated as

$$\frac{U_{\text{elec}}^{\text{blob}}}{k_B T} \approx \frac{l_B f_*^2 g_e^2}{D_e} \ln\left(\frac{\xi}{D_e}\right). \tag{9.83}$$

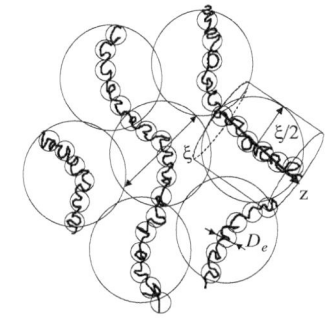

Fig. 9.25 Schematic representation of a semidilute polyelectrolyte solution. Reproduced with permission from [8].

The total electrostatic energy of a chain in a semidilute solution is equal to the electrostatic energy of a blob times the number of blobs per chain N/g_e

$$\frac{U_{\text{elec}}}{k_B T} \approx \frac{N}{g_e} \frac{l_B f_*^2 g_e^2}{D_e} \ln\left(\frac{\xi}{D_e}\right) \approx N \frac{l_B f_*^2 g_e}{D_e} \ln\left(\frac{\xi}{D_e}\right). \tag{9.84}$$

The total free energy of a chain of blobs consists of the electrostatic energy U_{elec} and the elastic free energy contribution due to stretching of the polyelectrolyte chain into array of blobs

$$F_{\text{elast}} \approx k_B T N/g_e . \tag{9.85}$$

Thus, combining the electrostatic and elastic energy contributions the chain free energy can be written as

$$\frac{F}{N k_B T} \approx \frac{1}{g_e} + \frac{l_B f_*^2 g_e}{D_e} \ln\left(\frac{\xi}{D_e}\right). \tag{9.86}$$

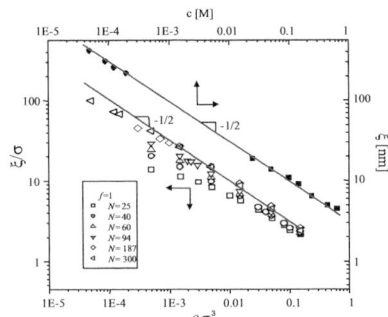

Fig. 9.26 Concentration dependence of the correlation length ξ in salt-free polyelectrolyte solutions. Filled symbols corresponds to the SANS data (circles) [139] and light scattering data (squares) [140] in solutions of NaPSS. Open symbols represent results of the molecular dynamics simulations [27]. The lines with slope $-1/2$ are shown to guide the eye. Reproduced with permission from [8].

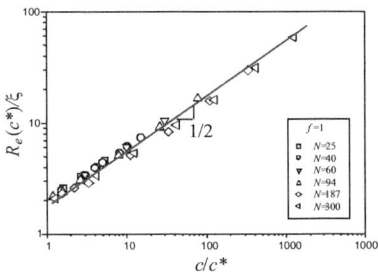

Fig. 9.27 Normalized chain size $R_e(c^*)/\xi$ as a function of reduced concentration c/c^*. The line with slope $1/2$ is shown to guide the eye. Reproduced with permission from [27].

This expression has to be minimized with respect to the electrostatic blob size D_e by taking into account the relation between the number of monomers in the electrostatic blob g_e and its size ($D_e^2 \propto b^2 g_e$). This leads to the following expression for the electrostatic blob size

$$D_e \approx D_e^* \ln^{-1/3}\left(\frac{e\xi}{D_e^*}\right) \qquad (9.87)$$

where we introduce the electrostatic blob size $D_e^* \approx b(uf_*^2)^{-1/3}$ for the effective charge fraction f_*. The correlation blobs are space-filling such that the average polymer concentration

$$c \approx N/\left(ND_e\xi^2/g_e\right) \approx g_e/D_e\xi^2 . \qquad (9.88)$$

By solving equation (9.88) for the correlation length one has

$$\xi \ln^{1/6}\left(\frac{e\xi}{D_e^*}\right) \approx \left(\frac{D_e^*}{cb^2}\right)^{1/2} . \qquad (9.89)$$

The electrostatic blob size D_e [equation (9.87)] increases logarithmically with polymer concentration. The correlation length of semidilute polyelectrolyte solution [see equation (9.89)] has only minor logarithmic corrections to the well-known scaling form $\xi \propto c^{-1/2}$ [3,8,23–25,141,142]

$$\xi \propto c^{-1/2}\ln^{-1/6}\left(\frac{ec_b}{c}\right) \propto c^{-1/2} . \qquad (9.90)$$

The concentration dependence of the number of monomers in a correlation volume ξ can be obtained by imposing the close-packing condition for chain sections of size ξ, $c \approx g_\xi/\xi^3$. This leads to the following concentration dependence of the number of monomers within a correlation volume [3,8,23–25,141,142]:

$$g_\xi \propto c\xi^3 \propto c^{-1/2}\ln^{-1/2}\left(\frac{e}{cD_e^*b^2}\right) \approx c^{-1/2}\ln^{-1/2}\left(\frac{ec_b}{c}\right) \qquad (9.91)$$

where c_b is the polymer concentration at which electrostatic blobs begin to overlap,

$$c_b \approx g_e^*/D_e^{*3} . \qquad (9.92)$$

Figure 9.26 displays the results of the molecular dynamics simulations for concentration dependence of the correlation length ξ for several chain lengths (open symbols) [27] and results of scattering experiments (filled symbols) [139,140] in the semidilute salt-free polyelectrolyte solutions. For the fixed number of monomers N, simulations show that the slope of the correlation length dependence on the polymer concentration approaches the scaling value $-1/2$ expected in the semidilute regime. However, for the same polymer concentrations, the correlation length increases with increasing number of monomers N and, finally, saturates for longer chains. This saturation of the correlation length indicates that the N-dependence of the correlation length is due to the finite size effects. For shorter chains there are not enough correlation blobs per chain to completely suppress the contributions from the chain ends. Scattering experiments in the semidilute solutions [139, 140] confirmed that the correlation length is varying reciprocally with the square root of polymer concentration, $\xi \propto c^{-1/2}$. Light scattering [140] and neutron scattering [139] data were combined to cover four orders of magnitude in concentration of salt-free solutions of polystyrene sulfonate sodium salt: $5 \times 10^{-5} < c < 5 \times 10^{-1}$ M, see filled symbols in Figure 9.26.

A plot of $R(c^*)/\xi$ as a function of c/c^* (Figure 9.27) provides an estimate (with logarithmic accuracy) of the number of correlation blobs per chain. All simulation data collapse onto a single universal curve. The y-axis of this plot is proportional (up to a logarithmic correction) to the number of correlation blobs per chain. The number of blobs per chain $R(c^*)/\xi$ in the semidilute solution increases with polymer concentration as $c^{1/2}$ in agreement with the prediction of the scaling theory. It follows from this plot that for short chains the number of correlation blobs per chain is less than 10 throughout the entire semidilute solution regime. Finite size effects dominate chain properties of such short chains. However, for longer chains far from the overlap concentration, the number of blobs approaches 100, and finite size effects are suppressed.

Persistence length and chain size

The scaling model of a polyelectrolyte chain in semidilute solutions is based on the assumption of the existence of a single length scale — the correlation length ξ. At the length scales larger than the solution correlation length ξ, other chains and counterions screen electrostatic interactions, and the statistics of the chain are those of a Gaussian chain with the effective persistence length on the order of the correlation length ξ. Thus, the polyelectrolyte chain is assumed to be flexible at the length scales on the order of the correlation length ξ. Figure 9.28 shows the results for the concentration dependence of the electrostatic persistence length $l_p(c)$ and the correlation length $\xi(c)$ in semidilute polyelectrolyte solutions. As one can see from this plot, both length scales are proportional (in fact very close) to each other. These results support the hypothesis of a single length scale in semidilute polyelectrolyte solutions.

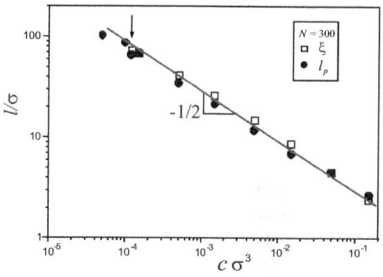

Fig. 9.28 Concentration dependence of the persistence length l_p (filled circles) and the correlation length ξ (open squares) for chains with $N = 300$ and $f = 1$. The arrow marks the location of the overlap concentration. Reproduced with permission from [27].

According to the scaling model, a chain in the semidilute salt-free polyelectrolyte solution is a random walk of correlation blobs [3, 8, 23, 25, 141, 142] with size

$$R_e \approx \xi \left(\frac{N}{g_\xi}\right)^{1/2} \propto N^{1/2} c^{-1/4} \ln^{1/12}\left(\frac{ec_b}{c}\right) \propto N^{1/2} c^{-1/4}. \quad (9.93)$$

To test the scaling hypothesis for the chain size scaling, the plot of the normalized chain size R_e/ξ as a function of the number of correlation blobs per chain N/g_ξ is shown in Figure 9.29. All points for chains with different degrees of polymerization, different fractions of charged monomers, and at different polymer concentrations collapse onto one universal line, with the slope $1/2$ as expected for Gaussian chains with

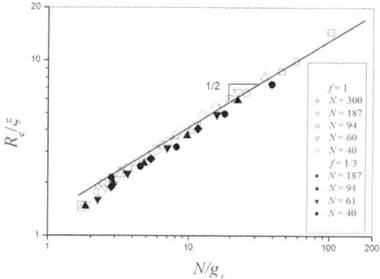

Fig. 9.29 Dependence of the reduced chain size R_e/ξ on the number of correlation blobs N/g_ξ for chains with different degrees of polymerization, fraction of charged monomers, and at different polymer concentrations. Thin solid line has slope $1/2$. Reproduced with permission from [27].

N/g_ξ correlation blobs.

Figure 9.30 shows the results of the molecular dynamics simulations for the end-to-end distance of chains as a function of reduced polymer concentration c/c^* in semidilute solutions for polyelectrolytes with different number of monomers N. The results show that the concentration dependence of the chain size can be described by the power law $R \sim c^{-\nu}$; however, the exponent ν is N-dependent. The simulation results clearly show the cross-over from the weak concentration dependence of end-to-end distance of polyelectrolyte chain, $R_e \sim c^{-0.094}$ for $N = 25$, to a stronger concentration dependence, $R_e \sim c^{-0.22}$ for $N = 300$. The concentration dependence of the chain size for the longest chains approaches the predicted value of the scaling exponent of $-1/4$.

The chain size of NaPSS in aqueous semidilute solutions with no added salt was found [143] to scale with concentration as $R_e \sim c^{-1/4}$ in agreement with equation (9.93).

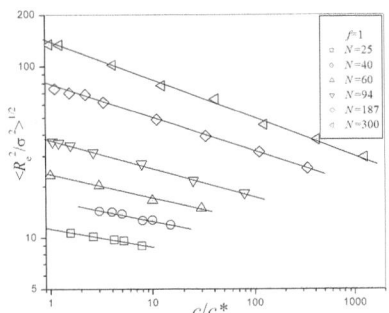

Fig. 9.30 Dependence of the root-mean-square end-to-end distance for chains with the fraction of charged monomers $f = 1$ on the reduced polymer concentration c/c^*. Reproduced with permission from [27].

Semidilute solutions with added salt

The electrostatic interactions in salt-free solutions are screened at length scales on the order of the correlation length ξ. The reason for such screening is that the Debye screening length due to counterions alone $r_D = (4\pi l_B f_* c)^{-1/2}$ is always larger than the correlation length ξ as long as the system is below the counterion condensation threshold. The counterion condensation in a Θ-solvent takes place for strongly charged polyelectrolytes for which the parameter $uf^{1/2}$ is larger than unity or when the distance between charges D_e/fg_e is smaller than the Bjerrum length l_B. In this case one has to include the sections of the chains inside the Debye screening length into the screening of electrostatic interactions. In order to calculate the contribution of sections of the chains into the Debye screening length, each chain is divided into subsections of size r_D. The charges on each of these subsections interact strongly and contribute coherently to the screening as one big charge Z

$$Z \approx \frac{f_* g_e r_D}{D_e} \approx \left(\frac{f_*}{u}\right)^{1/3} \frac{r_D}{b}. \tag{9.94}$$

The concentration of these sections is equal to

$$c_Z \approx \frac{cf_*}{Z}. \tag{9.95}$$

By assuming that each section contributes to screening independently, one can use the expression for the Debye screening length for multivalent ions to estimate the electrostatic screening length

$$r_D = \left[4\pi l_B \left(f_* c + c_Z Z^2\right)\right]^{-1/2} \approx (4\pi l_B c f_* Z)^{-1/2}. \tag{9.96}$$

Notice that for $Z \gg 1$, the sections of the chain provide the main contribution to the screening. The screening length can be determined self-consistently from equation (9.96) by substituting expression (9.94) for the section valence Z [25].

$$r_D = \left(4\pi l_B f_*^2 \frac{cg_e}{D_e}\right)^{-1/3} \approx \left(uf_*^2\right)^{-2/9} c^{-1/3}. \qquad (9.97)$$

The Debye screening length due to chain sections $r_D(Z)$ is smaller than the solution correlation length ξ. This means that there is less than one polyion section inside a Debye volume $c_Z r_D^3(Z) \ll 1$ contradicting the assumption of the Debye-Hückel theory. In semidilute solutions, the Debye screening length r_D due to counterions alone is larger than the correlation length of the solution ξ, while the screening length due to sections of the chain is smaller than ξ. Therefore, one can conclude that in this case when the counterion screening is too weak, but screening due to sections of the chains is too strong, the electrostatic screening length is on the order of the distance between chains ξ. Thus, the electrostatic screening length $r_{\rm scr}$ in the solutions of multivalent ions with valence Z can be found self-consistently by assuming that the number of counterions and salt ions inside the volume with radius on the order of the electrostatic screening length is equal to the charge valence Z.

$$c_{\rm ion} r_{\rm scr}^3 \approx (cf_* + 2c_s) r_{\rm scr}^3 \approx Z. \qquad (9.98)$$

It is important to point out that this is the minimal conjecture: it is conceivable that the electrostatic screening length expands even more, but it must expand at least up to $r_{\rm scr}$ to compensate for the charge with valence Z.

For semidilute polyelectrolyte solutions, the effective charge on a chain section inside the electrostatic screening length $r_{\rm scr}$ is equal to [see equation (9.94)]:

$$Z \approx \frac{f_* g_e r_{\rm scr}}{D_e} \approx \left(\frac{f_*}{u}\right)^{1/3} \frac{r_{\rm scr}}{b}. \qquad (9.99)$$

Substituting expression (9.99) into equation (9.98), one can find the relation between the electrostatic screening length and the ion concentration

$$r_{\rm scr} \approx \left(\frac{f_*}{u}\right)^{1/6} b^{-1/2} (cf_* + 2c_s)^{-1/2} \approx \xi_0(c) \left(1 + \frac{2c_s}{f_* c}\right)^{-1/2} \qquad (9.100)$$

where

$$\xi_0(c) \approx b\left(uf_*^2\right)^{-1/6} \left(cb^3\right)^{-1/2}$$

is the correlation length of the salt-free polyelectrolyte solution.

Using this assumption about the electrostatic screening length, one can calculate the correlation length of a semidilute polyelectrolyte solution in the presence of added salt. The calculation of the correlation length in polyelectrolyte solutions with added salt is based on the assumption that the chain conformation at length scales smaller than the screening length is that of a rod-like chain consisting of $g_{\rm scr} = g_e r_{\rm scr}/D_e$ monomers, while at length scales larger than the screening length $r_{\rm scr}$, but smaller than correlation length ξ, the sections of the chain of length r_{scr} obey self-avoiding walk statistics. On length scales longer than correlation length ξ the chain is ideal. (Note that this assumption disagrees with the OSF result but it is consistent with the result $l_p \approx r_{\text{scr}}$, see Figure 9.28. In the OSF-like calculations of the chain persistence length only salt ions are considered to screen the electrostatic interactions and effect of the chain's sections into the electrostatic screening is neglected.) This leads to the following expression for the correlation length of a polyelectrolyte solution in the presence of added salt

$$\xi(c) \approx r_{\text{scr}} \left(\frac{g_\xi}{g_{\text{scr}}}\right)^{3/5} \approx r_{\text{scr}} \left(\frac{D_e g_\xi}{r_{\text{scr}} g_e}\right)^{3/5}$$

$$\approx b \left(cb^3\right)^{-1/2} \left(uf_*^2\right)^{-1/6} \left(1 + \frac{2c_s}{cf_*}\right)^{1/4} \quad (9.101)$$

$$\approx \xi_0(c) \left(1 + \frac{2c_s}{cf_*}\right)^{1/4}$$

where g_ξ is the number of monomers in a correlation volume. The concentration dependence of the correlation length ξ in the high salt concentration regime $(c_s \gg cf_*)$ is similar to that in solutions of uncharged polymers $\xi \propto c^{-3/4}$. At low salt concentrations, equation (9.101) reproduces the salt-free result $\xi \propto c^{-1/2}$. Thus, any quantity X of the polyelectrolyte solution with salt concentration c_s can be expressed in terms of the same property X_0 in a salt-free solution as

$$X = X_0 \left(1 + \frac{2c_s}{f_* c}\right)^\alpha . \quad (9.102)$$

The results for static properties of semidilute polyelectrolyte solutions with added salt are summarized in Table 9.1.

Osmotic pressure and scattering function

In ionic systems the charge neutrality is required on both sides of the membrane across which the osmotic pressure π is measured. This condition is known as the Donnan equilibrium [5]. The electroneutrality condition leads to the partitioning of salt ions between the reservoir and the polyelectrolyte solution region

$$c_s^- = c_s^+ + f_* c \quad (9.103)$$

where c_s^+ and c_s^- are the average concentrations of positively and negatively charged salt ions in the polyelectrolyte solution. Here it is assumed that counterions are negatively charged and macroions are positively charged.

The local ion concentration $c_s^+(r)$ or $c_s^-(r)$ in polyelectrolyte solutions is related to the local value of the electrostatic potential $\Psi(r)$ by the Boltzmann distribution

$$c_s^\pm(r) = c_s^\pm \exp\left[\mp e\Psi(r)/k_B T\right] . \quad (9.104)$$

By multiplying equations (9.104) one can show that the product of the concentrations of salt ions stays constant at each point of the solution. Since salt ions can penetrate through the membrane, the chemical equilibrium on both sides of the membrane requires that this product stays constant in the reservoir as well. This leads to

$$c_s^+ c_s^- = c_s^2 \qquad (9.105)$$

where c_s is the average salt concentration in the reservoir. By solving equation (9.103) with equation (9.105) one can find the average concentrations of salt ions in the polyelectrolyte solution as functions of the average salt concentration c_s and polymer concentration c. The ionic contribution to the osmotic pressure is equal to the difference between the ideal gas pressure of salt ions in the polyelectrolyte solution and in the reservoir [5]

$$\frac{\pi_{\text{ion}}}{k_B T} = c_s^+ + c_s^- - 2c_s = \sqrt{(f_* c)^2 + 4c_s^2} - 2c_s . \qquad (9.106)$$

At low salt concentrations, $c_s \ll f_* c$, the difference is proportional to the pressure from the ideal gas of counterions. At higher salt concentrations, $c_s \gg f_* c$, the ionic part of the osmotic pressure is equal to

$$\frac{\pi_{\text{ion}}}{k_B T} \approx \frac{(f_* c)^2}{4 c_s} \approx 2\pi l_B r_D^2 (f_* c)^2 , \qquad (9.107)$$

and can be considered as the result of the effective excluded volume interactions between charged monomers on the polyelectrolyte backbone with excluded volume $l_B r_D^2$. Note, that at high salt concentrations ($c_s \gg f_* c$), the concentrations of salt ions c_s^+ and c_s^- in a polyelectrolyte solution are almost equal to those in reservoir c_s.

In addition to the ionic contribution, polyelectrolyte solutions have the polymeric contribution to their osmotic pressure. In semidilute polymer solutions, the polymeric contribution is on the order of $k_B T$ per correlation volume [31, 32]

$$\frac{\pi_{\text{pol}}}{k_B T} \approx \xi^{-3} . \qquad (9.108)$$

The total osmotic pressure of polyelectrolyte solutions can be approximated as the sum of the ionic and polymeric contributions

$$\frac{\pi}{k_B T} = \frac{\pi_{\text{ion}}}{k_B T} + \frac{\pi_{\text{pol}}}{k_B T} \approx \sqrt{(f_* c)^2 + 4c_s^2} - 2c_s + \xi^{-3} . \qquad (9.109)$$

At low salt concentrations, $c_s \ll f_* c$, the ionic contribution to the osmotic pressure $f_* c$ dominates over the polymeric contribution throughout the entire semidilute regime. At high salt concentrations, $c_s \gg f_* c$, both ionic and polymeric contributions are much smaller than those at low salt concentrations. However, for the vast majority of the systems studied so far [144–152], the ionic contribution dominates the osmotic pressure of polyelectrolyte solutions. Therefore, equation (9.106) is a good approximation of the osmotic pressure of polyelectrolyte solutions. This is also confirmed by computer simulations [69, 153].

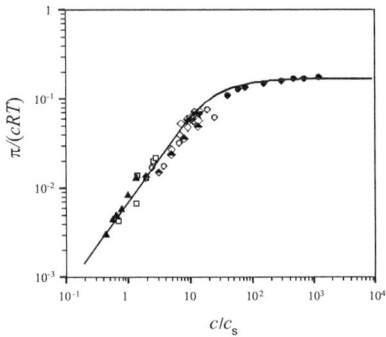

Fig. 9.31 Osmotic pressure of semidilute NaPSS solutions as a function of polymer and salt concentration. Filled circles are data of Takahashi et al. [152] for $M_w = 4.3 \times 10^5$ Da with no added salt and apparent salt concentration $c_s = 3 \times 10^{-5}$ M. Other symbols are data of Koene et al. [151] for molecular weight $M_w = 6.5 \times 10^5$ Da and salt concentrations $c_s = 5 \times 10^{-3}$ M (open circles), $c_s = 5 \times 10^{-2}$ M (open squares), $c_s = 1 \times 10^{-2}$ M (open diamonds), and $c_s = 1 \times 10^{-1}$ M (filled triangles); for $M_w = 4 \times 10^5$ Da and $c_s = 1 \times 10^{-2}$ M (bottom-filled diamonds); and for $M_w = 12 \times 10^5$ Da and $c_s = 1 \times 10^{-2}$ M (top-filled diamonds). The solid curve is equation (9.106) with $f_* = 0.16$. Reproduced with permission from [25].

The experimental data for the osmotic pressure of semidilute solutions of NaPSS obtained by Koene et al. [151] and the salt-free data of Takahashi et al. [152] are shown in Figure 9.31. While Takahashi et al. [152] did not report the salt concentration of their 'salt-free' solutions, the data clearly indicate a low level of salt (see Fig. 1 of [152]) and it was estimated from their data. Data at high salt concentrations from [152] were not used in Figure 9.31 because they correspond to a dilute solution regime. The solid curve in Figure 9.31 is the prediction of equation (9.106) with the effective fraction of free counterions f_* equal to 0.16. The agreement between the scaling theory and the experimental data is very good over a wide range of polymer and salt concentrations.

The high osmotic pressure of salt-free polyelectrolyte solutions has important consequences for the scattering function $S(q)$. The osmotic compressibility is related to the scattering at zero wavelength

$$S(0) = k_B T \frac{\partial c}{\partial \pi} . \qquad (9.110)$$

At low salt concentrations, the osmotic pressure is controlled by the concentration of free counterions. This leads to the scattering function $S(0)$ inversely proportional to the fraction of free counterions [23, 25]

$$S(0) \approx 1/f_* . \qquad (9.111)$$

The counterion pressure causes $S(0)$ to be much smaller than its value at the correlation length $S(2\pi/\xi)$. The fluctuations of polymer density on length scales larger than the electrostatic screening length are governed by the electroneutrality condition. However, these fluctuations are $f_* g_\xi$ times larger than thermal fluctuations of counterion density and would result in a prohibitively large loss of counterion entropy. Therefore the extremely high counterion pressure suppresses density fluctuations on length scales larger than the correlation length ξ. For $q > 2\pi/\xi$, the scattering function $S(q)$ decreases with q as in the case of neutral polymers. This suggests that there is a maximum in the scattering function $S(q)$ of salt-free polyelectrolyte solutions at the wave vector q on the order of $2\pi/\xi$.

The maximum in the scattering function $S(q)$ disappears at high salt concentrations. When the salt concentration c_s is on the order of $cg_\xi f_*^2$, the polymer density fluctuations on length scales larger than the correlation length ξ are no longer suppressed [$S(0) \sim g_\xi$]. In this high salt concentration regime, the scattering function is similar to that in solutions of neutral polymers [25].

Dynamics of polyelectrolyte solutions

Unentangled regime ($c^* < c < c_e$)

The hydrodynamic interactions between sections of a chain in semidilute solutions are screened on the length scales larger than the correlation length ξ [22, 31, 32, 124]. Inside the correlation blob the motion of different chain sections is strongly hydrodynamically coupled just as in dilute solutions. (This assumption for polyelectrolyte solutions was confirmed by the effective medium calculations performed by Muthukumar [154, 155] and by recent experiments [135].) The relaxation time of the section of a polyelectrolyte chain with g_ξ monomers inside the correlation blob is Zimm-like (proportional to the volume pervaded by the section) [31, 32, 124]

$$\tau_\xi \approx \eta_s \xi^3 / (k_B T) \ . \tag{9.112}$$

Each chain consists of N/g_ξ correlation blobs. The hydrodynamic interactions between these blobs are screened and therefore their motion can be described by the Rouse dynamics [31, 124]

$$\tau_{\text{Rouse}} \approx \tau_\xi (N/g_\xi)^2 \approx \frac{\eta_s b^3}{k_B T} \left(u f_*^2\right)^{1/2} \left(cb^3\right)^{-1/2} N^2, \quad c^* < c < c_e \ . \tag{9.113}$$

The chain relaxation time in this unentangled semidilute regime decreases with increasing polymer concentration as $\tau_{\text{Rouse}} \sim c^{-1/2}$. This decrease of solution relaxation time with increasing polymer concentration is unique for unentangled polyelectrolyte solutions. The relaxation time of 'normal' polymer solutions increases with increasing polymer concentration. The reason for such unusual dependence is that the chain size decreases with increasing polymer concentration while their friction coefficient (proportional to the contour length) does not change. This leads to the concentration-independent self-diffusion coefficient

$$D_{\text{self}} \approx R^2 / \tau_{\text{Rouse}} \approx \frac{k_B T}{\eta_s b} \left(u f_*^2\right)^{-1/3} N^{-1} \ , \quad c^* < c < c_e \tag{9.114}$$

that is inversely proportional to the degree of polymerization, N. The terminal modulus G of a solution in the Rouse model for unentangled polymers is $k_B T$ per chain [31, 124]. The viscosity of polyelectrolyte solutions in this regime is

$$\eta \approx G \tau_{\text{Rouse}} \approx \eta_s \left(u f_*^2\right)^{1/2} \left(cb^3\right)^{1/2} N \ , \quad c^* < c < c_e \ . \tag{9.115}$$

In the salt-free solution, the viscosity grows as the square root of concentration $\eta \sim c^{1/2}$. Thus, the scaling model of polyelectrolyte solutions recovers the well-known phenomenological Fuoss law [156].

Molecular dynamics simulations of dilute and semidilute polyelectrolyte solutions without hydrodynamic interactions were performed by Liao et al. [157] to study the Rouse chain dynamics. These simulations of the Rouse dynamics give qualitatively similar results to the experimentally observed dynamics of polyelectrolyte solutions. It was found that the chain relaxation time depends nonmonotonically on polymer concentration, see Figure 9.32. The chain relaxation time decreases with increasing polymer concentration in a dilute solution regime. This decrease in the chain relaxation time is due to chain contraction induced by counterion condensation. In the semidilute solution regime, the chain

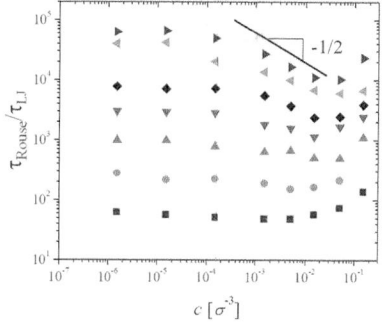

Fig. 9.32 Concentration dependence of the chain relaxation time for the system of fully charged chains. $N = 25$ (dark squares); $N = 40$ (light circles); $N = 64$ (upward triangles); $N = 94$ (downward triangles); $N = 124$ (black diamonds); $N = 187$ (left triangles); $N = 247$ (right triangles). Reproduced with permission from [157].

relaxation time decreases with polymer concentration as $c^{-1/2}$. In this concentration range, the chain relaxation time follows the usual Rouse scaling dependence on the chain degree of polymerization, $\tau_{\text{Rouse}} \propto N^2$. At high polymer concentrations, the chain relaxation time begins to increase with increasing polymer concentration. The crossover polymer concentration to the new scaling regime does not depend on the chain degree of polymerization, indicating that the increase in the chain relaxation time is due to the increase of the effective monomeric friction coefficient. The analysis of the spectrum and of the relaxation times of Rouse modes confirms the existence of the single correlation length ξ, which describes both static and dynamic properties of semidilute solutions. These simulations also show that the unentangled semidilute solution regime is very wide. The longest chains with $N = 373$ and 247 start to overlap at about $10^{-4}\sigma^{-3}$ and don't show any effect of entanglements up to the highest polymer concentration $0.15\,\sigma^{-3}$. Thus, the unentangled semidilute solution regime spans three orders of magnitude above the overlap concentration.

Entanglement criterion

The unentangled semidilute regime of neutral polymer chains exists above the chain overlap concentration c^* within the polymer concentration range $c^* < c < c_e$, where c_e is the polymer concentration corresponding to the onset of entanglements. The physical reason for the unentangled semidilute regime is that the topological constraints between polymers require significant chain overlap. This leads to the diameter of the topological tube constraining transversal motion of the chain to be larger than the correlation length. It was established experimentally that at the entanglement onset, each chain has to overlap with n other chains (see for discussion [31, 158], with $5 < n < 10$. To estimate the entanglement concentration c_e for polyelectrolytes, we assume that it is necessary to have n chains within the volume occupied by a polyelectrolyte chain to topologically constrain its motion. Thus, the monomer concentration required for chain entanglement is

$$c_e \approx \frac{nN}{R_e^3(c_e)} \approx c^* \frac{nR_e^3(c^*)}{R_e^3(c_e)} \approx n\,(c^*)^{1/4}\,c_e^{3/4} \approx n^4 c^*, \qquad (9.116)$$

where we use the following relation between the chain size at entanglement concentration c_e and that at overlap concentration c^*: $R(c_e) \approx R(c^*)(c^*/c_e)^{1/4}$.

It follows from the last equation that the unentangled semidilute regime in salt-free solutions could be three to four decades wide ($10^3 < c_e/c^* < 10^4$) [25, 142]. The physical reason for this unusually wide unentangled regime is the strong concentration dependence of the chain size ($R_e \propto c^{-1/4}$). It is only slightly weaker than the concentration dependence of the distance between the centers of mass of neighboring chains ($R_{\text{cm}} \propto c^{-1/3}$). Therefore the number of chains overlapping with a selected chain has very weak concentration dependence

$[(R_e/R_{cm})^3 \approx (c/c^*)^{1/4}]$. The viscosity at the entanglement concentration c_e is $\eta(c_e) \approx n^2 \eta_s \approx 50 \eta_s$, as in the case of neutral polymers.

Semidilute-entangled regime ($c > c_e$)

Entanglements are characterized by the tube diameter a (the mesh size of the temporary entanglement network) (see Figure 9.33). The polymer strand between entanglements is a random walk of $N_e/g_\xi \approx (a/\xi)^2$ correlation blobs, where N_e is the number of monomers in an entanglement strand. There are n such strands inside the volume a^3, so that $(a/\xi)^3 \approx nN_e/g_\xi$ leading to $n = a/\xi$. Thus, the tube diameter a is proportional to the correlation length. The longest relaxation time of a polymer chain can be calculated using the reptation theory [31, 124]. By assuming that the dynamics of the chain at the length scales smaller than a correlation blob size is Zimm-like, and the relaxation of a polymer strand between entanglements is Rouse-like, the longest relaxation time of the chain is concentration-independent

$$\tau_{\text{rep}} \approx \tau_\xi \left(\frac{N_e}{g_\xi}\right)^2 \left(\frac{N}{N_e}\right)^3 \approx \frac{\eta_s b^3}{k_B T} u f_*^2 n^{-2} N^3 . \quad (9.117)$$

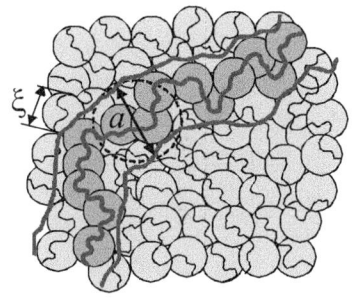

Fig. 9.33 Schematic representation of an entangled polyelectrolyte solution. Reproduced with permission from [8].

The plateau modulus in this regime is $G \approx \nu_e k_B T$, where ν_e is the number density of entanglement strands. The volume per entanglement strand is $\xi^3 N_e/g_\xi \approx \xi a^2$. The scaling model predicts the plateau modulus

$$G \approx \frac{k_B T}{\xi a^2} \approx \frac{k_B T}{b^3} n^{-2} \left(u f_*^2\right)^{1/2} \left(cb^3\right)^{3/2} \quad (9.118)$$

and the solution viscosity in the semidilute-entangled regime

$$\eta \approx \tau_{\text{rep}} G \approx \eta_s n^{-4} N^3 \left(u f_*^2\right)^{1/2} \left(cb^3\right)^{3/2} . \quad (9.119)$$

In salt-free entangled polyelectrolyte solutions, the viscosity is predicted to grow faster than linearly with polymer concentration c.

Any dynamic property x of a polyelectrolyte solution with added salt can be expressed in terms of the same property x_0 in salt-free solutions as

$$x = x_0 \left(1 + 2c_s/f_* c\right)^\alpha \quad (9.120)$$

by analogy with the similar relation for static properties, see equation (9.102). The results for dynamic properties of semidilute solutions are summarized in Table 9.1.

Comparison with experiments

Figure 9.34 shows the dependence of the specific viscosity of salt-free solutions of 2-vinyl pyridinium chloride and N-methyl-2-vinyl pyridinium chloride random copolymers on the ratio of polymer concentration c to the overlap concentration c^*. These data clearly show linear concentration dependence of specific viscosity in dilute solutions ($\eta_{\text{sp}} \approx c/c^*$

Table 9.1 Scaling relations for semidilute solutions of polyelectrolytes in a Θ-solvent for polymer backbone.

$$R \approx bN^{1/2}B_e^{-1/4}\left(cb^3\right)^{-1/4}y^{-1/8}$$
$$\xi \approx bB_e^{1/2}\left(cb^3\right)^{-1/2}y^{1/4}$$

	Unentangled	Entangled
τ	$\left(\eta_s b^3/k_B T\right) B_e^{-3/2} N^2 \left(cb^3\right)^{-1/2} y^{-3/4}$	$\left(\eta_s b^3/k_B T\right) N^3 n^{-2} B_e^{-3} y^{-3/2}$
G	$k_B T c/N$	$\left(k_B T/b^3\right) n^{-2} B_e^{-3/2} \left(cb^3\right)^{3/2} y^{-3/4}$
η	$\eta_s N B_e^{-3/2} \left(cb^3\right)^{1/2} y^{-3/4}$	$\eta_s \left(N^3/n^4\right) B_e^{-9/2} \left(cb^3\right)^{3/2} y^{-9/4}$
D_{self}	$\left(k_B T/\eta_s b\right) N^{-1} B_e y^{1/2}$	$\left(k_B T/\eta_s b\right) n^2 N^{-2} B_e^{5/2} \left(cb^3\right)^{-1/2} y^{5/4}$

where $y = (1 + 2c_s/cf_*)$, and the parameter $B_e \approx (uf_*^2)^{-1/3}$.

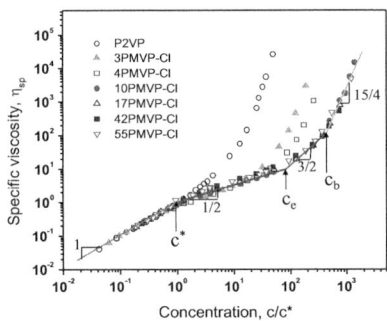

Fig. 9.34 Reduced concentration, c/c^*, dependence of the specific viscosity, $\eta_{\rm sp}$, at 25°C in ethylene glycol solvent of the random copolymer 2-vinyl pyridine and N-methyl-2-vinyl pyridinium chloride (PMVP-Cl) with various charge densities [numbers in the legend correspond to the extent (mole %) of quaternization] and the uncharged neutral parent poly (2-vinyl pyridine) (P2VP) of $M_w = 364000\,\text{Da}$ (open circles). Data provided by Dou, S. and Colby, R. H.

for $c < c^*$) as well as concentration regimes corresponding to unentangled solutions with $\eta_{\rm sp} \sim c^{1/2}$ and entangled solutions with $\eta_{\rm sp} \sim c^{3/2}$. There is indeed a wide range of polymer concentrations (two orders of magnitude for chains with higher charged fractions) where solution viscosity follows the Rouse dynamics [equation (9.115)] manifested by the Fuoss law [156], $\eta_{\rm sp} \sim c^{1/2}$. This regime of unentangled semidilute polyelectrolyte solutions has much wider concentration range than similar regime in solutions of neutral polymers (open circles in Figure 9.34). This concentration dependence of specific viscosity is general for salt-free polyelectrolyte solutions [135].

The crossover at concentration c_e to the entangled regime is also seen in the specific viscosity data (Figure 9.34). According to the Kavassalis-Noolandi conjecture [158], the crossover to the entangled regime occurs when there is a universal number n of overlapping chains to form an entanglement. Experimental data indicate that the Kavassalis-Noolandi conjecture [158] for entanglements used in the scaling model requires some modification [135]. It is possible that the properties of the confining tube for charged polymers changes with concentration leading to the concentration-dependent parameter n.

The correlation length ξ becomes smaller than electrostatic blob size D_e at polymer concentrations higher than the crossover polymer concentration c_b. Electrostatic interactions are not important in polyelectrolyte solutions with concentration c above c_b and specific viscosity is expected to have concentration dependence similar to that in solutions of entangled neutral polymers, $\eta_{\rm sp} \propto c^{15/4}$. This quasi-neutral regime of polyelectrolyte solutions at high polymer concentrations $c > c_b$ is confirmed by the data in Figure 9.34. A similar trend in concentration dependence of the solution viscosity was recently reported by Truzzolillo et al. [138].

The different concentration regimes in the solution terminal modulus are shown in Figure 9.35. The terminal modulus is $k_B T$ per chain ($G \propto k_B T c/N$) in the unentangled regime. The modulus at polymer concentrations above entanglement onset c_e increases rapidly with increasing concentration in the way expected by the scaling model $G \propto c^{3/2}$.

Semidilute polyelectrolyte solutions in a poor solvent for the polymer backbone

In a poor solvent for the polymer backbone polyelectrolyte chains form necklace-like globules. There are three different length scales that characterize the necklace globule: the string length l_{str}, the bead size D_b, and the thermal blob size ξ_T determining the length scale of density fluctuations inside beads. These three different length scales determine the properties of semidilute polyelectrolyte solutions in a poor solvent for the polymer backbone [60, 74].

The crossover from dilute to semidilute solutions takes place when the size of the necklace becomes comparable to the distance between chains

$$c^* \approx \frac{N}{L_{\text{nec}}^3} \approx b^{-3}N^{-2}\left(\frac{|\tau|}{uf_*^2}\right)^{3/2} \approx \frac{m_b^{3/2}}{b^3 N^2}. \quad (9.121)$$

In a semidilute solution ($c > c^*$), the configuration of the chain on length scales shorter than the correlation length ξ is similar to that in dilute solutions. On length scales longer than the correlation length ξ, the chain is assumed to be a random walk of correlation segments ξ. The correlation length at the overlap concentration $\xi(c^*)$ is equal to the necklace size L_{nec}. In the semidilute regime ($c > c^*$) the correlation length is independent of the degree of polymerization.

$$\xi \approx L_{\text{nec}}\left(\frac{c^*}{c}\right)^{1/2} \approx b\left(\frac{|\tau|}{uf_*^2}\right)^{1/4}(cb^3)^{-1/2} \approx bm_b^{1/4}(cb^3)^{-1/2}. \quad (9.122)$$

The correlation length is inversely proportional to the square root of polymer concentration. In the case of the necklace-like chain, the number of monomers in the correlation volume is

$$g_\xi \approx c\xi^3 \approx m_b^{3/4}(cb^3)^{-1/2}. \quad (9.123)$$

The normal dependence of the correlation length ξ on polymer concentration [see equation (9.90)] continues while ξ is larger than the length of the string l_{str} between neighboring beads. These two lengths become of the same order of magnitude ($\xi \sim l_{\text{str}}$) at polymer concentration [60, 74]

$$c_{\text{Bead}} \approx b^{-3}\left(\frac{uf_*^2}{|\tau|}\right)^{1/2} \approx b^{-3}m_b^{-1/2} \quad (9.124)$$

For higher polymer concentrations $c > c_{\text{Bead}}$, the electrostatic interaction between beads is screened, and there is one bead per every correlation volume ξ^3. The correlation volumes are space filling $c \approx g_\xi \xi^{-3}$. The number of monomers inside the correlation volume for $c > c_{\text{Bead}}$ is

$$g_\xi \approx c\xi^3 \approx m_b \quad (9.125)$$

because most of the polymer mass is in the beads ($m_b/m_{\text{str}} \gg 1$). Therefore the correlation length decreases inversely proportional to the one-third power of polymer concentration:

$$\xi \approx \left(\frac{|\tau|}{uf_*^2}\right)^{1/3} c^{-1/3} \approx \left(\frac{c}{m_b}\right)^{-1/3}. \quad (9.126)$$

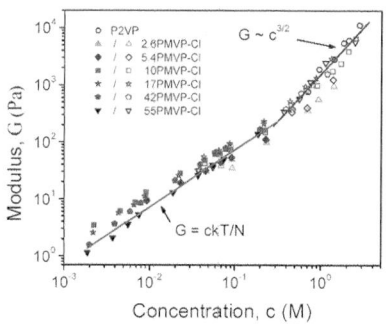

Fig. 9.35 Concentration dependence of the terminal modulus calculated from steady shear relaxation time (filled symbols) and from the oscillatory shear relaxation time (open symbols) at 25°C in ethylene glycol solvent of random copolymer 2-vinyl pyridine and N-methyl-2-vinyl pyridinium chloride (PMVP-Cl) with various charge densities [numbers in the legend correspond to the extent (mole %) of quaternization] and the uncharged neutral parent poly (2-vinyl pyridine) (P2VP) of $M_w = 364000$ Da (open circles). Data provided by S. Dou and R. H. Colby.

This concentration dependence of the correlation length ξ is unusual for semidilute solutions of polymers and is more typical for the distance between chains in dilute solutions (the system behaves as a dilute solution of beads). The main difference between the $c^{-1/3}$ concentration dependence of equation (9.126) and the similar concentration dependence of the correlation length in dilute polyelectrolyte solutions is that correlation length in semidilute solutions [equation (9.126)] is N-independent, while in dilute polyelectrolyte solutions: $\xi \approx (c/N)^{-1/3}$.

In the framework of the scaling model the electrostatic interactions are screened on the length scales longer than the correlation length ξ, the configuration of a polyelectrolyte chain is that of a random walk of size ξ

$$R_e \approx \xi \left(\frac{N}{g_\xi}\right)^{1/2} \approx bN^{1/2} \begin{cases} (c_{\text{Bead}}/c)^{1/4} & \text{for } c^* < c < c_{\text{Bead}} \\ (c_{\text{Bead}}/c)^{1/3} & \text{for } c_{\text{Bead}} < c < c_b \end{cases} \quad (9.127)$$

where c_b is the polymer concentrations at which the beads start to overlap. This crossover concentration is on the order of the concentration inside the beads $c_b \approx b^{-3}|\tau|$. It is important to point out that at the crossover concentration between string-controlled and bead-controlled regimes, the chain size is approximately equal to the Gaussian chain size $bN^{1/2}$, but its conformation is far from that of an ideal chain. In the bead-controlled regime, the chain size is smaller than the Gaussian chain size. However, above the bead overlap concentration, the chain size is once again on the order of the Gaussian chain size. But now the chain configuration is that of an ideal chain because at these high polymer concentrations the electrostatic interactions are almost completely screened. Thus, the existence of three different length scales is manifested in non-monotonic dependence of the chain size on polymer concentration [60]. This chain size dependence on polymer concentration is reflected in the dynamic properties of polyelectrolyte solutions and is expected to lead to a dramatic increase in the solution viscosity close to bead overlap concentration. The scaling predictions of the dynamic properties of polyelectrolyte solutions in poor solvent for polymer backbone are summarized in Table 9.2.

The existence of different concentration regimes in semidilute polyelectrolyte solutions was confirmed by the experiments of Essafi et al. [159] on partially sulfonated poly(styrene sulfonate) in water. It was reported that the exponent of the concentration dependence of the correlation length $\xi(c)$ depends on the degree of sulfonation. For example, this exponent is -0.38 for 40% sulfonation throughout the whole semidilute regime while it is close to its classical value -0.5 for the fully sulfonated samples. Similar observations were reported by Heitz et al. [160] for poly(methacrylic acid) in water as a function of its neutralization. The exponent of concentration dependence of the correlation length $\xi(c)$ was found to change from -0.43 to -0.31 as the neutralization decreased from 0.95 to 0.09. Evidence of the intrachain correlation peak associated with the presence of beads was observed in the Kratky plot of

Table 9.2 Scaling relations for semidilute solutions of polyelectrolytes in poor solvent for the polymer backbone.

$$\text{String-controlled regime}$$
$$R \approx bN^{1/2}m_b^{-1/8}\left(cb^3\right)^{-1/4}y^{-1/8}$$
$$\xi \approx bm_b^{1/4}\left(cb^3\right)^{-1/2}y^{1/4}$$

	Unentangled	Entangled
τ	$(\eta_s b^3/k_B T)\,N^2 m_b^{-3/4}\left(cb^3\right)^{-1/2}y^{-3/4}$	$(\eta_s b^3/k_B T)\,N^3 n^{-2} m_b^{-3/2} y^{-3/2}$
G	$k_B T c/N$	$(k_B T/b^3)\,n^{-2} m_b^{-3/4}\left(cb^3\right)^{3/2} y^{-3/4}$
η	$\eta_s N m_b^{-3/4}\left(cb^3\right)^{1/2} y^{-3/4}$	$\eta_s\left(N^3/n^4\right) m_b^{-9/4}\left(cb^3\right)^{3/2} y^{-9/4}$

$$\text{Bead-controlled regime}$$
$$R \approx bN^{1/2}m_b^{-1/6}\left(cb^3\right)^{-1/3}y^{-5/24}$$
$$\xi \approx bm_b^{1/3}\left(cb^3\right)^{-1/3}y^{5/12}$$

	Unentangled	Entangled
τ	$(\eta_s/k_B T)\,N^2 m_b^{-1} c^{-1} y^{-5/4}$	$(\eta_s b^3/k_B T)\,N^3 n^{-2} m_b^{-2} y^{-5/2} c^{-1}$
G	$k_B T c/N$	$(k_B T/n^2)\,m_b^{-1} c y^{-5/4}$
η	$\eta_s N m_b^{-1} y^{-5/4}$	$\eta_s\left(N^3/n^4\right) m_b^{-3} y^{-15/4}$

where $y = (1 + 2c_s/cf_*)$, and $m_b \approx |\tau|/uf_*^2$ is the number of monomers in a bead.

the data obtained in the small angle neutron scattering experiments by Essafi et al. [159]. The evolution of the polyelectrolyte solution with increasing polymer concentration in solvophobic polyelectrolytes in a series of polar organic solvents [161] shows that the scaling exponent of $\xi(c)$ changes from -0.45 to -0.13. Similar behavior was reported for partially sulfonated polystyrene by the Williams group [162, 163]. This concentration dependence of the correlation length is in qualitative agreement with the predictions of the necklace model for the crossover to the bead-controlled regime.

The small angle neutron scattering (SANS) spectra [164] measured in dilute solutions of water/acetone mixtures of poly(methacryloylethyl trimethylammonium methyl sulfate) can be analyzed using the necklace model of polyelectrolyte chains. According to the results of these experiments, each polyelectrolyte chain consists of a sequence of dense beads connected by regions of loose polymer. The radius of these dense beads is about 28 nm. Each molecule has about 3–4 dense beads with the volume fraction of polymer inside these globular sections close to 8%.

Recent nuclear magnetic resonance (NMR) studies [165] of semidilute solutions of poly(styrene sulfonic acid) PSS, poly(methacrylic acid) PMA, and poly(acrylic acid) PAA in water/methanol mixtures are consistent with the necklace-like structure of polyelectrolytes in poor solvent conditions. These observations indicate that, while parts of polyelectrolyte chains have compact globule-like conformation, the segments of

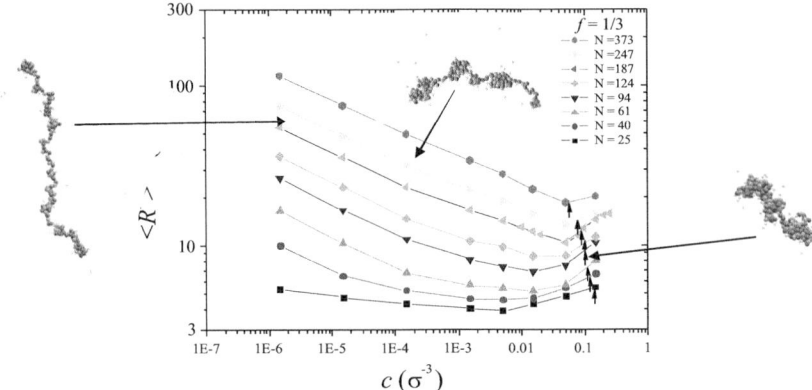

Fig. 9.36 Concentration dependence of the chain size for partially charged chains with fraction of charged monomers $f = 1/3$ in poor solvent with interaction parameters $l_B = 3\sigma$, $\varepsilon_{LJ} = 1.5 k_B T$. Different color symbols and lines correspond to the different number of monomers per chain, as indicated in the figure. The positions of the overlap concentration c^* are marked by the arrows. Inserts show typical chain conformations. Reproduced with permission from [75].

the chains connecting these compact globules retain flexibility similar to those observed in Θ and good solvents. The fraction of the mass of polyelectrolytes in compact globules varies from 27% to 32% depending on the polymeric system.

Atomic force microscopy (AFM) images of necklace globules of poly(2-vinylpyridine) and poly(methacryloyloxyethyl dimethylbenzylammonium chloride) adsorbed at mica surface were reported by Kiriy et al. [166] and by Minko et al. [167]. These images clearly show an abrupt conformational transition from elongated chains to compact globules through the intermediate necklace-like globule conformations with increasing ionic strength of the solution.

Molecular dynamics simulations of partially charged polyelectrolytes in poor solvent conditions were performed by the Mainz group [48,71,73] and by Liao et al. [75]. These simulations have confirmed that polyelectrolyte chains at low polymer concentrations form necklaces of beads connected by strings, see Figure 9.36. As the polymer concentration increases, the fraction of condensed counterions on the chain increases and chains shrink by decreasing the length of strings and the number of beads per chain. At higher polymer concentrations, polymer chains interpenetrate leading to a concentrated polyelectrolyte solution. In this range of polymer concentrations the chain size is observed to increase toward its Gaussian value. The non-monotonic dependence of the chain size on polymer concentration shown in Figure 9.36 is in a qualitative agreement with the scaling model described above [74].

9.4 Phase separation in polyelectrolyte solutions

Mean-field approach

In the poor solvent conditions for the polymer backbone (below Θ-temperature) the solutions of uncharged polymers are unstable with respect to phase separation into concentrated and dilute polymeric phases [31, 32]. This phenomenon is driven by the minimization of the number of unfavorable polymer-solvent contacts and the relatively small entropic penalty for the chain partitioning between concentrated and dilute polymeric phases. For polymers, the entropic penalty is on the order of the thermal energy $k_B T$ per chain that is much smaller than that in the mixture of similar low molecular compounds at the same concentrations. Charging polymers can significantly improve solution solubility. In the salt-free solutions, the Donnan equilibrium requires that the electroneutrality condition should be satisfied in both concentrated and dilute phases in order to avoid huge electrostatic energy penalty caused by charge inhomogeneities. Thus, separating polyelectrolyte chains into concentrated and dilute phases will result in additional entropic penalty due to the redistribution of the counterions. In order to show this, consider Flory-Huggins model of polyelectrolyte solution [8, 168, 169] with polymer volume fraction $\phi = cb^3$

$$\frac{F(\phi)}{k_B T} = \frac{V}{b^3}\left(\frac{\phi}{N}\ln(\phi) + f\phi\ln(\phi) + \frac{\tau}{2}\phi^2 + \frac{\phi^3}{6}\right) + \frac{\Delta F_{\text{ionic}}(\phi)}{k_B T} \quad (9.128)$$

where $\tau = 1 - \Theta/T$ is the effective temperature. The first two terms in equation (9.128) are contributions of chains and counterions to the entropy of mixing, while the third and fourth terms correspond to short-range interactions between polymers and solvent, respectively. Finally, the last term describes the contribution to the solution free energy from the density fluctuations. The form of this term depends on the assumptions made about the effect of polymeric degrees of freedom on density and charge fluctuations in the system. If the polymeric effects are completely ignored, the fluctuation term is the well-known Debye-Hückel correction due to charge density fluctuations (see for details [33]). In the limit $r_D/b \gg 1$ it reduces to the Debye-Hückel law, $-k_B T V/r_D^3$. The fluctuation correction has a more complicated form if the effect of the connectivity of charged monomers into polymeric chains is taken into account (see for details [170–172]).

The fluctuation corrections to the free energy of the solution of weakly charged polyelectrolytes with $f \ll 1$ can be neglected and the terms in the brackets on the right-hand side of equation (9.128) are sufficient to determine the stability region of the polyelectrolyte solution. The spinodal of the polyelectrolyte solution is given by the following equation

$$\frac{\partial^2 F(\phi)}{\partial \phi^2} = \left(f + \frac{1}{N}\right)\phi^{-1} + \tau + \phi = 0 \ . \quad (9.129)$$

If there is more than one ionized group per chain, $fN \gg 1$, the $1/N$ term can be neglected and the location of the stability region is determined by the counterion entropy and by the strength of polymer solvent interactions. In this case the critical point is located at

$$\phi_{\text{cr}} = \sqrt{f} \quad \text{and} \quad \tau_{\text{cr}} = -2\sqrt{f} \ . \tag{9.130}$$

At effective temperature τ below τ_{cr} the polyelectrolyte solutions are phase separated over some concentration range. The dependence of the spinodal line of the salt-free polyelectrolyte solution on the fraction of charged monomers f on the polymer backbone is shown in Figure 9.37. With increasing fraction of charged monomers f the two-phase region moves toward lower effective temperatures and is located at lower temperatures than in solutions of neutral polymers (for neutral polymers $\tau_{\text{cr}} \approx -2/\sqrt{N}$) making polyelectrolyte solutions more stable. Monte Carlo simulations of the phase separation in polyelectrolyte solutions have confirmed that ϕ_{cr} is independent of the degree of polymerization N [173].

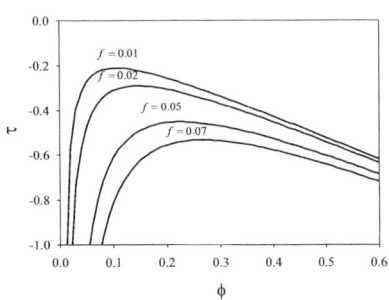

Fig. 9.37 Spinodal line given by equation (9.137) for weakly charged polyelectrolytes with the degree of polymerization $N = 1000$. Reproduced with permission from [8].

The addition of salt to polyelectrolyte solutions lowers the penalty due to counterion redistribution thus promoting phase separation. At high salt concentrations, the electrostatic interactions between charged monomers are exponentially screened leading to the renormalization of the second virial coefficient between monomers. The Flory-Huggins free energy of polyelectrolyte solutions at high salt concentrations is

$$\frac{F(\phi)}{k_B T} = \frac{V}{b^3}\left[\frac{\phi}{N}\ln(\phi) + \left(\tau + \frac{f^2}{2c_s b^3}\right)\frac{\phi^2}{2} + \frac{\phi^3}{6}\right] + \frac{\Delta F_{\text{ionic}}(\phi)}{k_B T}. \tag{9.131}$$

Taking the second derivative of equation (9.131) with respect to polymer volume fraction ϕ, one obtains the following equation for the spinodal line of the phase diagram:

$$\frac{1}{N\phi} + \tau + \frac{f^2}{2c_s b^3} + \phi = 0 \tag{9.132a}$$

with the critical point located at volume fraction

$$\phi_{\text{cr}} = \frac{1}{\sqrt{N}} \tag{9.132b}$$

and effective temperature

$$\tau_{\text{cr}} = -\frac{2}{\sqrt{N}} - \frac{f^2}{2c_s b^3} \ . \tag{9.132c}$$

Thus, addition of salt leads to the increase of the critical temperature. The shift of the critical point is inversely proportional to the salt concentration c_s at high salt concentrations. For a recent review of experimental results of the phase separation in polyelectrolyte solutions see the paper by Volk et al. [174]. Modification of the Flory approach that takes into account the effect of counterion condensation and difference in the dielectric constant in dilute and concentrated phases was recently presented by the Muthukumar's group [175].

Finally, it is important to point out that the Flory-Huggins lattice consideration of the polyelectrolyte solutions presented above incorrectly describes dilute polyelectrolyte solutions. In the Flory-Huggins approach the monomers are uniformly distributed over the whole volume of the system leading to an underestimation of the effect of the short-range monomer-monomer interactions and of the intrachain electrostatic interactions. A similar problem appears in the Flory-Huggins theory of the phase separation of polymer solutions (see for discussion [31, 32]). This leads to the incorrect expression for the low polymer density branch of the phase diagram.

Microphase separation

Phase separation into dilute and concentrated phases is not the only option for polyelectrolytes in a poor solvent for the polymer backbone. It was shown that weakly charged polyelectrolytes could locally violate electroneutrality condition thus minimizing the entropy loss due to the redistribution of counterions. This local violation of electroneutrality leads to the formation of mesophases — alternating regions with high and low polymer densities. These mesophases appear as a result of optimization of the electrostatic and short-range interactions. The possibility of this type of instability of the homogeneous phase in solutions of weakly charged polyelectrolytes was first discovered by Borue and Erukhimovich [170] and then studied in a series of papers by Joanny and Leibler [176], and Khokhlov and Nyrkova [168]. The phase separation and mesophase formation in solutions of weakly charged polyelectrolytes in poor solvent for the polymer backbone was investigated by Dormidontova et al. [177].

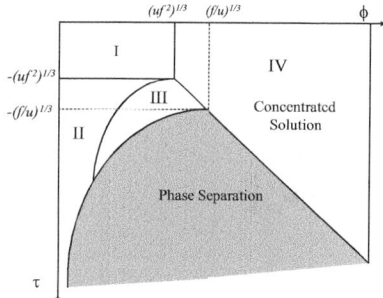

Fig. 9.38 Phase diagram of polyelectrolytes in a poor solvent. Logarithmic scales. Reproduced with permission from [60].

Necklace model of phase separation

A different approach to the phase separation in polyelectrolyte solutions was proposed by Dobrynin and Rubinstein [60]. They used a two-state model to describe counterion condensation inside beads of the necklace-like globules in dilute solutions. Within this approximation, the free energy of the dilute solution of necklaces is

$$\frac{F_{\text{neck}}(\phi)}{k_B T} \approx \frac{V}{b^3}\left(\frac{\phi}{N}\ln(\phi) + \phi f \ln(f\beta|\tau|) + \phi u^{1/3}(1-\beta)^{2/3} f^{2/3}|\tau| - \phi\tau^2\right), \quad (9.133)$$

where β is the fraction of the condensed counterions. The first term on the right-hand side of equation (9.133) describes polymer contribution to the entropy of mixing, the second term is the entropy of counterions localized inside beads, the third term is the free energy of the beads, and, finally, the last term is the contribution to the necklace free energy due to the short-range monomer-monomer interactions inside beads. The dense polymeric phase was described by adopting the virial expansion of the Flory-Huggins free energy, see equation (9.128). The phase diagram was calculated by comparing the chemical potential of chains and the osmotic pressure in dilute and concentrated phases. Figure 9.38 shows the resultant phase diagram of salt-free polyelectrolyte solutions at poor solvent conditions for the polymer backbone. The shaded area corresponds to the two-phase region. Above the critical temperature (for the values of the parameter $|\tau| < (f/u)^{1/3}$), the polyelectrolyte solution is stable with respect to phase separation. In regime I of the phase diagram (Figure 9.38), the thermal blob size $\xi_T \approx b/|\tau|$ is larger than the bead size $D_b \approx b(uf^2)^{-1/3}$. For the range of the effective temperatures $|\tau| < (uf^2)^{1/3}$ the polymer-solvent interactions are not strong enough to collapse chains, and they adopt the conformations similar to the ones for polyelectrolytes in a Θ-solvent (regime I). In regime II the polymer-solvent interactions dominate at small length scales and polyelectrolytes form necklace globules. The chains can either be in a dilute regime for polymer concentration below the overlap concentration c^* [see equation (9.121)] or in a semidilute regime for $c > c^*$. In the semidilute regime, the correlation length ξ decreases with increasing polymer concentration c, equation (9.122). The correlation length ξ becomes on the order of the length of a string l_{str} connecting two neighboring beads at polymer concentration $c \approx c_{\text{Bead}}$ (see equation (9.124)). At this polymer concentration the system crosses over into the so-called bead-controlled regime in which there is one bead per every correlation volume ξ^3 and polyelectrolyte solution can be viewed as a strongly correlated charged colloidal liquid of beads (regime III). The structure of the polyelectrolyte solution in the bead-controlled regime resembles that of microphase separated one with beads forming polymer-rich spherical domains that are surrounded by a solution of strings with the excess of the solvent.

For the effective temperature $|\tau| \approx (f/u)^{1/3}$, the correlation length ξ becomes on the order of the size of a bead D_b at polymer concentration $cb^3 \approx |\tau|$. At higher polymer concentrations, the system crosses over into the concentrated polyelectrolyte solution (regime IV). However, if the value of the parameter $|\tau|$ is larger than $(f/u)^{1/3}$, the system will phase separate into a concentrated polymer solution and a solution of necklaces. The left boundary of the two-phase region is

$$\phi_{\text{dil}} \approx |\tau| \exp\left(-\left(\frac{u}{f}\right)^{1/3}|\tau|\right) \tag{9.134}$$

while the right boundary of the two-phase region is

$$\phi_{\text{con}} \approx |\tau| . \tag{9.135}$$

Thus, the polymer volume fraction in the concentrated phase at the two-phase boundary is the same as inside the beads of the necklace globule in the dilute phase. Two lines given by equations (9.134) and (9.135) intersect at $\phi_{\text{cr}} \approx (f/u)^{1/3}$ and $\tau_{\text{cr}} \approx -(f/u)^{1/3}$.

The phase separation in solutions of hydrophobic polyelectrolyte was observed in MD simulations of Chang and Yethiraj [178]. They have found that solutions of necklaces phase separate with increasing polymer concentration. Polyelectrolytes in the dense phase form spherical, cylindrical, and lamellar structures depending on polymer concentration.

9.5 Polyelectrolyte gels

Polyelectrolyte gels consist of cross-linked polyelectrolyte chains. Networks made from polyelectrolyte chains are quite common in both nature and industry [179–185]. These materials have the ability to absorb up to thousand times their dry weight of water. This swelling ability of polyelectrolyte gels results from the high osmotic pressure created by dissociated counterions. In salt solutions the excess of salt ions reduces the osmotic imbalance and decreases the gel swelling. At high salt concentrations the properties of polyelectrolyte gels are similar to those of neutral gels. The unique swelling ability of polyelectrolyte gels and their response to the ionic environment have lead to utilization of polyelectrolyte gels as the superabsorbent materials in medical, chemical, and agricultural applications, as ion-exchange resins, as the carrier for drug delivery targeting specific organs [184, 185], and as sensors and actuators [186]. We begin with a discussion of properties of polyelectrolyte gels by considering gel swelling in salt-free solutions.

Swelling of polyelectrolyte gels in salt-free solutions

Consider a polyelectrolyte gel made of N_s charged strands with the number of monomers N between the cross-links (see Figure 9.39). The fraction of charged monomers on these strands is equal to f. For simplicity, we will assume that gel swells up to volume V from its dry state with volume V_{dry}. The polymer volume fraction in the swollen state is equal to [31, 187]

$$\phi = \frac{V_{\text{dry}}}{V} . \qquad (9.136)$$

Let us assume that the gel was cross-linked at a polymer volume fraction ϕ_0 and the network occupied volume V_0. When gel swells in a solvent it undergoes uniform swelling by the same amount in all directions. The linear deformation ratio λ in each direction is related to the volume ratio as [31, 187]

$$\lambda = \left(\frac{V}{V_0}\right)^{1/3} = \left(\frac{\phi_0}{\phi}\right)^{1/3} . \qquad (9.137)$$

In a swollen state each network strand is stretched because cross-links move further apart. The elastic part of the strand free energy with the end-to-end distance R_e is

$$F_{\text{conf}}(R_e) \approx k_B T \frac{R_e^2}{Nb^2} . \qquad (9.138)$$

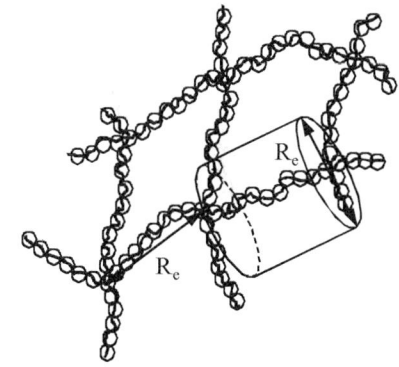

Fig. 9.39 Schematic representation of a polyelectrolyte gel as chains of tension blobs (see text for details).

Assuming the affine deformation of the network strands, the end-to-end distance of the gel strands in the swollen state is related to their size in the preparation state as follows $R_e = \lambda R^0$ where R^0 is the size of the polymeric strand in a preparation state. The total elastic free contribution to the gel free energy due to strand deformation is equal to the contribution from the individual strands times the number of strands in a network N_s [31, 187]

$$F_{\text{elast}}(\phi) \approx N_s F_{\text{conf}}(\lambda R_0) \approx k_B T N_s \frac{\lambda^2 \langle R^{0^2} \rangle}{Nb^2} \approx \frac{k_B T V}{Nb^3} \phi_0^{2/3} \phi^{1/3} \tag{9.139}$$

In simplifying the last expression we have assumed that the gel was prepared in a Θ-solvent condition for the polymer backbone, $\langle R^{0^2} \rangle \approx b^2 N$, and taken the number of polymeric strands in a gel: $N_s \approx \phi/b^3 N$.

The dissociated counterions are localized inside a gel maintaining the gel electroneutrality. The configurational entropy contribution due to counterion localization inside gel volume is equal to

$$F_{\text{count}}(\phi) \approx \frac{k_B T V}{b^3} f \phi \ln(\phi). \tag{9.140}$$

The last contribution to the gel free energy comes from the energy of electrostatic interactions between counterions, charged monomers on the polymer backbone, and between counterions and charged monomers. This contribution to the system free energy can be evaluated by taking into account the electroneutrality of the volume per gel strand which contains fN charged monomers and the same amount of counterions. Note that one can envision a polyelectrolyte gel as a polyelectrolyte solution at overlap concentration c^*. Using the same cell model approximation as in the case of semidilute polyelectrolyte solutions, one can approximate a polyelectrolyte strand between cross-links with its counterions by a coaxial cylindrical capacitor of length R_e having the charge efN and with the outer diameter on the order of R_e. The total contribution of these interactions to the gel free energy is estimated as follows [30]

$$F_{\text{int}}(\phi) \approx k_B T N_s \frac{l_B f^2 N^2}{R_e} \ln\left(\frac{R_e}{d}\right)$$

$$\approx k_B T \frac{V}{b^3} N^{1/2} u f^2 \phi^{4/3} \phi_0^{-1/3} \ln\left(\frac{bN^{1/2}\phi_0^{1/3}}{d\phi^{1/3}}\right) \tag{9.141}$$

where d is the strand thickness. The total gel free energy is equal to the sum of the elastic, configurational, and interaction free energy contributions

$$F(\phi) \approx F_{\text{elast}}(\phi) + F_{\text{count}}(\phi) + F_{\text{int}}(\phi). \tag{9.142}$$

The equilibrium gel volume is obtained by minimizing the gel free energy with respect to the gel volume V

$$\frac{\partial F(\phi)}{\partial V} \approx \frac{k_B T}{Nb^3} \phi_0^{2/3} \phi^{1/3} - \frac{k_B T}{b^3} f\phi \approx G(\phi) - \pi(\phi) = 0. \tag{9.143}$$

In writing equation (9.143) we have neglected the derivative from the interaction term. It will be demonstrated below that for polyelectrolyte gels in salt-free solutions, this contribution is small. Thus, the equilibrium swelling of the polyelectrolyte gels is determined by balancing the gel shear modulus $G(\phi)$ with the osmotic pressure of counterions $\pi(\phi)$. At equilibrium the polymer volume fraction inside swollen gel is equal to

$$\phi \approx \phi_0 (fN)^{-3/2} . \tag{9.144}$$

The equilibrium swelling ratio Q is the ratio of the gel volume in the swollen state to the volume in the dry state. For a polyelectrolyte gel in a salt-free solution this ratio is

$$Q \approx \phi_0^{-1} (fN)^{3/2} . \tag{9.145}$$

If the network was prepared in the dry state $\phi_0 = 1$, the swelling ratio depends only on the net charge of the polymeric strands between cross-links. For example, if the number of monomers between cross-links $N = 100$ and only 25% of these monomers is charged, the volume of the polyelectrolyte gel will increase 125 times upon swelling. The value of the gel shear modulus at the equilibrium swelling is equal to

$$G \approx \frac{k_B T}{b^3} f\phi \approx \frac{k_B T f}{b^3 Q} . \tag{9.146}$$

At equilibrium each polymeric strand between cross-links has end-to-end distance on the order of

$$R_e = \lambda R^0 \approx b f^{1/2} N . \tag{9.147}$$

Thus, each strand is strongly stretched. We can use a tension blob picture to understand structure of these strands, see Figure 9.39. Let us assume that at length scales smaller than the tension blob size ξ_s the statistics of a section of a strand with g_s monomers is that of an ideal chain $\xi_s \approx b g_s^{1/2}$. However on length scales larger than the blob size, a strand is strongly stretched forming an array of tension blobs

$$R_e \approx \xi_s N/g_s \approx bN/g_s^{1/2} . \tag{9.148}$$

Comparing equations (9.148) and (9.147) one can conclude that each tension blob has $1/f$ monomers or there is one charged group per each tension blob.

Now, let us show that the contribution of the electrostatic interactions [see equation (9.141)] into the system free energy is smaller than those from elastic and counterion osmotic terms. Due to strong stretching of the network strands each strand contributes on the order of thermal energy $k_B T$ per each tension blob to the gel elastic energy, $F_{\text{conf}} \approx k_B T f N$. The contribution of the electrostatic interactions per strand is estimated as $k_B T u f^{3/2} N$ with logarithmic accuracy. Thus, the energy of electrostatic interactions is weaker than the elastic energy due to osmotic gel swelling if $u f^{1/2} < 1$ which is true for weakly charged strands above

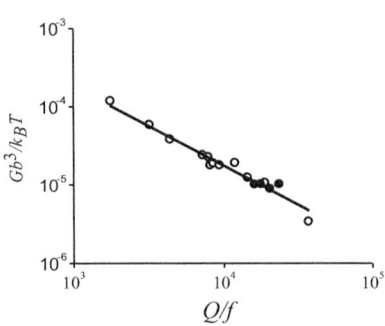

Fig. 9.40 Dependence of the modulus of swollen gel on the equilibrium swelling ratio for weakly charged gels made by copolymerization of acrylamide, bisacrylamide and sodium methacrylate (data from Ilavsky and Hrouz). Open symbols have $f = 0.012$ and filled symbols have variable f between $0.004 \leq f \leq 0.024$. The solid line has slope -1. Reproduced with permission from [194].

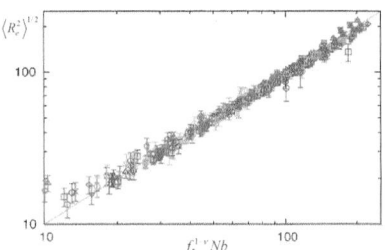

Fig. 9.41 The equilibrium swelling behavior of salt-free polyelectrolyte gels in the good solvent and close to the Θ-solvent conditions for the polymer backbone. The data points follow the scaling expression $R_e \approx b f_*^{1-\nu} N$, where the exponent ν is equal to $1/2$ in Θ-solvent and 0.6 in good solvent. f_* is the effective fraction of the charged groups on the polymeric strand with N monomers. Reproduced with permission from [191]

the counterion condensation threshold. If the opposite inequality holds, the counterion condensation reduces strand charge to the critical value and only osmotically active ('free' counterions) contribute to the gel swelling.

The dominant role of the osmotically active counterions in gel swelling is confirmed experimentally [188, 189] and by computer simulations of the polyelectrolyte networks [190–192]. Ilavsky and Hrouz [188,193] have studied polyelectrolyte gels made by copolymerization of acrylamide, bisacrylamide, and sodium methacrylate and swollen in water (with no added salt). These gels were made by varying N and f in the range $72 \leq N \leq 11000$ and $0.004 \leq f \leq 0.024$. The data are shown in Figure 9.40. All data obey the prediction of equation (9.146), and the solid line has the expected slope -1, confirming that $G \propto (Q/f)^{-1}$.

In a series of papers the Linse's group has used Monte Carlo simulations to model salt-free regular polyelectrolyte networks with diamond-like topology and explicit counterions [192]. These simulations have shown that the gel volume increases with increasing the fraction of charged monomer on the polymeric strands forming a gel. It also increases with decreasing the cross-linking density and increasing chain stiffness. This dependence of the gel swelling on the network parameters is in line with the notion that polyelectrolyte gel swelling is controlled by the osmotic pressure of counterions and by the elasticity of the polymeric strands forming a gel. The increase in the fraction of the charged monomers on polymer backbones leads to increase of the osmotic pressure of the counterions which forces the gel to increase its volume. The decrease in the number of cross-links (increase in the degree of polymerization of the polymeric stands between cross-links) and increase of the chain stiffness (increase of the chain's Kuhn length) both result in the decrease of the gel shear modulus, which also promotes network swelling. The effect of the counterion osmotic pressure and chain's elasticity on the gel swelling has also been studied by the Mainz group in molecular dynamics simulations of the salt-free polyelectrolyte networks [191, 195, 196]. These simulations established that only 'free' (osmotically active) counterions contribute to the gel swelling, see Figure 9.41. This was verified for the polyelectrolyte networks in the good and almost theta solvent conditions for the polymer backbone. It is important to point out that the swollen polyelectrolyte gels deform affinely.

The simulations of polyelectrolyte gels with explicit counterions and explicit solvent particles were reported by Lu and Hentschke [197]. Using a 'two-box-particle' transfer molecular dynamics simulation method they modeled an extremely highly crosslinked network in equilibrium with a dipolar Stockmayer solvent. In these simulations each bead of the polyelectrolyte strand forming a network was carrying a partial charge eq and each corresponding neutralizing counterion had partial charge $-eq$. The swelling ratio of the polyelectrolyte network has a maximum as a function of the partial charge eq. This nonmonotonic dependence of the swelling ratio was attributed to a competition between electrostatic repulsion and the network conformational entropy. The maximum becomes less pronounced with increasing the dipole moment of the Stockmayer fluid.

Edgecombe and Linse have studied the role of the chain's polydispersity and topological network defects on gel swelling [190]. They found that polydisperse networks swell less than regular gels. In polydisperse gels the short chains between cross-links are more stretched and the long ones are less stretched as compared with strands of polyelectrolyte networks made of monodisperse chains. Thus, the deformation of the short polyelectrolyte chains controls the polydisperse gel swelling. The normal stress in uniaxial stretching simulations for neutral and polyelectrolyte gels follows roughly exponential dependence on the deformation ratio λ. The agreement between simulation data and theoretical models of uniaxial gel deformation was better for the non-Gaussian network theory than the Gaussian theory. However, both models show significant deviation from the simulation data in the limit of the large network deformations.

The strong nonlinear effect in strand deformations was observed in experiments by Okay's group [189]. They found that the equilibrium swelling ratio Q for series of PNIPA gels deviates from the expected scaling law $Q \propto (fN)^{3/2}$ showing a weaker dependence on the number of charged monomers, $Q \propto (fN)^{0.75}$. This discrepancy between model and experimental results was explained by nonlinear chain elasticity in the limit of the strong stretching of the polymeric strands between cross-links.

Effect of added salt on gel swelling

In salt solutions the Donnan equilibrium controls partitioning of salt ions between gel and outside solution. Counterions and salt ions are distributed in such a way to maintain electroneutrality of a gel

$$c_s^- = c_s^+ + fc \tag{9.149}$$

where c_s^+ and c_s^- are the average concentrations of positively and negatively charged salt ions, respectively, and $c = \phi/b^3$ is the polymer concentration inside the polyelectrolyte gel. Here we have assumed that counterions are negatively charged and macroions are positively charged. Because salt ions can leave and enter the gel, the chemical equilibrium between salt ions inside and outside the gel requires that

$$c_s^+ c_s^- = c_s^2 \tag{9.150}$$

where c_s is the average salt concentration outside the gel.

The total free energy of the system consisting of a polyelectrolyte gel and surrounding gel salt solution (reservoir) has contributions from the gel elastic free energy [see equation (9.139)] and entropic contribution due to partitioning salt ions and counterions between gel and reservoir:

$$\frac{F_{\text{tot}}(\phi)}{k_B T} \approx \frac{V}{b^3 N}\phi^{1/3}\phi_0^{2/3} + Vc_s^+ \ln\left(c_s^+ b^3\right) + Vc_s^- \ln\left(c_s^- b^3\right) + 2V_r c_s \ln\left(c_s b^3\right) \tag{9.151}$$

where V_r is the volume of the reservoir which is equal to the system volume V_s minus the gel volume V, $V_r = V_s - V$. The equilibrium gel swelling is obtained by optimizing the system free energy equation (9.151) with respect to the gel volume V. This procedure results in

$$\frac{\partial F_{\text{tot}}(\phi)}{\partial V} \approx \frac{k_B T}{b^3 N} \phi^{1/3} \phi_0^{2/3} - k_B T \left(c_s^+ + c_s^- - 2c_s \right)$$
$$= \frac{k_B T}{b^3 N} \phi^{1/3} \phi_0^{2/3} - \pi_{\text{ion}} = 0 \,. \qquad (9.152)$$

The ionic contribution to the osmotic pressure in the case of the Donnan equilibrium has already been evaluated in equation (9.106).

$$\frac{\pi_{\text{ion}}}{k_B T} = c_s^+ + c_s^- - 2c_s = \sqrt{(fc)^2 + 4c_s^2} - 2c_s \,. \qquad (9.153)$$

At low salt concentrations, $c_s \ll fc$, the ionic part of the osmotic pressure is equal to the pressure from the ideal gas of localized inside gel free counterions. At higher salt concentrations, $c_s \gg fc$, the ionic part of the osmotic pressure is equal to

$$\frac{\pi_{\text{ion}}}{k_B T} \approx \frac{(fc)^2}{4c_s} \approx 2\pi l_B r_D^2 (fc)^2 \qquad (9.154)$$

and can be considered as the result of the effective excluded volume interactions between charged monomers on the polyelectrolyte backbone with excluded volume $l_B r_D^2$. At this high salt concentration, the concentrations of the salt ions c_s^+ and c_s^- inside a polyelectrolyte gel are almost equal to those in reservoir c_s.

Solving equation (9.152) for the polymer volume fraction ϕ in the high salt concentration limit one obtains

$$\phi \approx \frac{\phi_0^{2/5} \phi_s^{3/5}}{N^{3/5} f^{6/5}} \,. \qquad (9.155)$$

Thus, the polymer volume fraction inside a gel increases with increasing the salt concentration. The polyelectrolyte gels occupy smaller volumes in salt solutions in comparison with that in salt-free solutions. This is due to decrease of the ionic osmotic pressure with increasing the salt concentration. The gel shear modulus at equilibrium swelling

$$\frac{Gb^3}{k_B T} \approx \frac{\phi_0^{4/5} \phi_s^{1/5}}{N^{6/5} f^{2/5}} \qquad (9.156)$$

increases with increasing the salt concentration.

The Flory-Rehner-like calculations [31, 198, 199] of the properties of polyelectrolyte gels presented above provide a qualitative understanding of the polyelectrolyte gel properties. The modern approach to the gel elasticity considers a chain in a solution as a reference state for gel swelling [31]. In this approach the elastic part of the gel free energy is estimated as follows

$$F_{\text{elast}}(c) \approx k_B T \frac{Vc}{N} \left(\frac{\lambda R^0}{R_{\text{ref}}} \right)^2 \qquad (9.157)$$

where $R_{\text{ref}} \approx \xi(N/g_\xi)^{1/2}$ is the size the strand would have if it were a free (unconnected) chain of N monomers in a polyelectrolyte solution at the same polymer concentration c. In this approach one can also consider the case when a gel is prepared by cross-linking chains in a semidilute polyelectrolyte solution. In this case the strand size R^0 in the preparation state is unperturbed by cross-links and thus identical to a free polyelectrolyte chain of N monomers at a polyelectrolyte concentration c_0 and salt concentration c_s^0, see Table 9.1. Using the expression for the chain size at high salt concentrations one can calculate the expression for the gel elastic energy as follows [194]:

$$F_{\text{elast}} \approx k_B T \frac{V}{Nb^3} \phi^{7/12} \phi_0^{5/12} \left(\frac{\phi_s}{\phi_s^0}\right)^{1/4}. \qquad (9.158)$$

In writing expression (9.158) we have assumed that the solvent and the fraction of the ionized groups do not change between preparation and testing states. Note, that equation (9.158) has a different scaling form the expression for the elastic free energy (9.139), which was derived in the framework of the Flory-Rehner approach [31, 198, 199]. The equilibrium gel swelling is obtained by balancing the ionic osmotic pressure with the derivative of gel elastic free energy with respect to a gel volume. This leads to

$$\phi \approx \frac{\phi_0^{5/17} \phi_s^{15/17}}{N^{12/17} f^{24/17} (\phi_s^0)^{3/17}}. \qquad (9.159)$$

This equation demonstrates much stronger salt concentration dependence than equation (9.155). A detailed description of a scaling model of swelling of the polyelectrolyte gels and comparison model predictions with experiments [200] can be found in [194].

Collapse of polyelectrolyte gels

In Section 9.4 it was shown that with decreasing temperature a polyelectrolyte solution phase separates into dilute and concentrated polymer phases. In the case of the gel the phase separation phenomena is manifested in expulsion of the solvent from the gel interior and significant reduction in the gel volume. The theoretical model of the gel collapse as a function of the solvent quality for the polymer backbone and the gel ionization degree was developed by Dusek and Patterson in 1968 [201]. The first experimental study of the collapse of the polyelectrolyte gels was reported by the Tanaka's group in 1980 [202, 203]. They observed an abrupt volume change in hydrolyzed acrylamide gel by varying the solution pH. Now, it is well understood that the gel collapse is a result of optimization of the gel elastic energy, counterion entropy, and polymer-solvent interactions (see for review [198, 203]). In the simplest form the gel free energy includes the gel elastic free energy, the configurational entropy of localized inside gel counterions, and the energy of the polymer-solvent interactions

$$F_{\text{gel}}(\phi) = \frac{3G_0 V}{2} \phi^{1/3} + \frac{k_B T V}{b^3} \left(f\phi \ln(\phi) + \frac{\tau}{2}\phi^2 + \frac{\phi^3}{6} \right) \qquad (9.160)$$

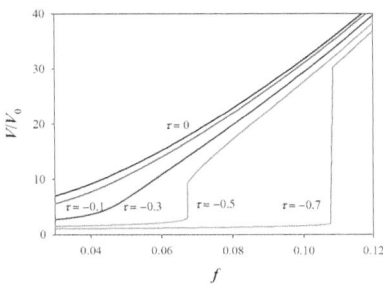

Fig. 9.42 Dependence of the reduced gel volume V/V_0 on the fraction of the ionized groups on the polymer backbone f for polyelectrolyte gels with the value of the gel shear modulus in the dry state $G_0 b^3 / k_B T = 10^{-2}$.

where $G_0 \approx k_B T / b^3 N$ is the gel shear modulus in the preparation state which for simplicity is assumed to be a dry state, $\phi_0 = 1$. The equilibrium polymer volume fraction inside a gel is found by minimizing the gel free energy with respect to a gel volume V:

$$0 = \frac{G_0 b^3}{k_B T} \phi^{1/3} - f\phi - \frac{\tau \phi^2}{2} - \frac{\phi^3}{3} \qquad (9.161)$$

The solution of equation (9.161) provides a relation between gel swelling, the solvent quality for the polymer backbone, and the gel degree of ionization. For some values of the effective temperature τ and fraction of charged monomers f there are three nonzero solutions which satisfy equation (9.161). These solutions correspond to two minima and one maximum of the gel free energy per strand as a function of the gel volume fraction ϕ. The value of the gel volume fraction ϕ corresponding to the lower minimum of the gel free energy per strand [see equation (9.160)] represents an equilibrium gel state. Coexistence of two different gel states leads to an abrupt transition between the swollen and the collapsed states with decreasing the fraction of the charged monomer on the polymer backbone. Note that for weak polyelectrolytes the variations in solution pH lead to a change in the chain's degree of ionization. Figure 9.42 displays the dependence of the gel volume on the fraction of the charged monomers f (gel degree of ionization). For polyelectrolyte gels at effective temperatures $\tau = -0.5$ and -0.7 there is an abrupt transition between swollen and collapsed gel states. For other effective temperatures the gel volume decreases monotonically with decreasing the fraction of the ionized groups on the polymeric strands forming a gel.

At high salt concentrations the jump in the gel volume upon gel collapse decreases. This is due to the decrease of the osmotic pressure of the salt ion which drives the gel swelling. In this high salt concentration regime the effect of the salt ion is effectively reduced to renormalization of the second virial coefficient due to electrostatic interactions between charged groups [see equations (9.131) and (9.154)]. Taking this into account, equation (9.161) can be rewritten as follows

$$0 = \frac{G_0 b^3}{k_B T} \phi^{1/3} - \left(\tau + \frac{f^2}{2\phi_s}\right) \frac{\phi^2}{2} - \frac{\phi^3}{3} . \qquad (9.162)$$

In addition to variations in the gel ionization degree, a gel collapse can be induced by adding the surfactant molecules [204–206], multivalent ions [207, 208], and linear charged chains [209, 210]. In all these cases the main reason for the gel collapse is a decrease of the osmotic contribution due to counterion localization inside a gel.

The classical model of the gel collapse described above does not take into account the fact that in poor solvent conditions for the polymer backbone, polymeric strands connecting the nodes of the polyelectrolyte network can have necklace-like conformations similar to those observed in polyelectrolyte solutions. In the case of the necklace globule, the surface energy of strings connecting the necklace beads provides the main contribution to the gel elasticity. The detailed study of the network structure in the poor solvent conditions for the polymer backbone as a function of the fraction of charged monomers f and the value of the Bjerrum length was performed by Mann et al. [191]. They observed the formation of the following chain conformations within polyelectrolyte network: collapsed conformations for small charge fractions or very strong electrostatic interactions (large values of the Bjerrum length); necklace structure for moderate to high charge fractions and not very strong electrostatic coupling; stretched structures for large fractions of charged monomers f and moderate values of the Bjerrum lengths l_B; and finally the 'sausage'-like conformation for large Bjerrum lengths. Note that the diagram of states for the polyelectrolyte networks has regimes similar to those observed at the diagram of state for the single polyelectrolyte chain. However, the boundaries between different conformational regimes are shifted due to the additional effect of the chain connectivity into the network on the backbone conformations.

A scaling model of the polyelectrolyte gel in a poor solvent for the polymer backbone, which takes into account the necklace structure of the network strands, was developed by Johner et al. [211]. Using a model of a chain under tension [46] they have calculated the dependence of the equilibrium shear modulus of the polyelectrolyte gel as a function of the fraction of charged monomers f and effective temperature τ characterizing the solvent quality for the polymer backbone. This model predicts an abrupt gel collapse transition in some regions of the gel diagram of states. Unfortunately, the predictions of this model are still awaiting experimental verification.

Note that similar to polyelectrolyte solutions, in poor solvent conditions for the polymer backbone, polyelectrolyte gels can also undergo a microphase separation transition which is manifested by the appearance of heterogeneous polymer density distribution inside a gel [212–215].

Conclusions

In this chapter I have presented an overview of properties of polyelectrolyte solutions and gels. Electrostatic interactions between charges lead to the rich behavior of these polymeric systems. For example, a polyelectrolyte chain in a poor solvent for the polymer backbone forms an unusual necklace-like structure of dense polymeric beads connected by strings of monomers. In salt-free polyelectrolyte solutions the electrostatic interactions between charged groups on the polymer backbone result in strong chain elongation with chain size scaling almost linearly with the chain degree of polymerization. Due to this strong dependence of the chain size on the chain degree of polymerization, the crossover to the semidilute polyelectrolyte solution regime occurs at much low polymer concentrations than in solutions of neutral polymers. The main contribution to the osmotic pressure in polyelectrolyte solutions comes from the ionic component. In polyelectrolyte gels this ionic osmotic pressure leads to a significant gel swelling and absorption of up to a thousand times of gel dry weight of water.

I hope that the topics discussed in this chapter will be useful in future studies of more complicated ionic systems such as polyelectrolyte brushes [216], polyelectrolyte-protein complexes [217,218], ionomers [219], charged polymers at surfaces and interfaces [8,220], and multilayer formation by charged molecules [14].

Acknowledgments

I would like to acknowledge the support of the donors of the Petroleum Research Fund, administered by the American Chemical Society under grants PRF#44861-AC7 and PRF#49866-ND7 and the National Science Foundation under grant DMR#1004576. The author thanks Michael Rubinstein, Ralph Colby, Jan-Michael Carrillo, Alexander Grosberg, Boris Shklovskii, Jean-Francois Joanny, Ludwik Leibler, Monica Olivera de la Cruz, Sanat Kumar, Robert Weiss, Thomas Seery, and Paul Dubin for numerous discussions on the topics of charged polymers.

References

[1] Masanori, H. (ed.) (1993). *Polyelectrolytes*. Marcel Dekker, New York.
[2] Forster, S. and Schmidt, M. (1995). *Adv. Polym. Sci.*, **120**, 51–133.
[3] Barrat, J. L. and Joanny, J.-F. (1996). *Adv. Chem. Phys.*, **94**, 1–66.
[4] Tanford, C. (1961). *Physical Chemistry of Macromolecules*. Wiley, New York.
[5] Oosawa, F. (1971). *Polyelectrolytes*. Marcel Dekker, New York.
[6] Radeva, T. (2001). *Physical Chemistry of Polyelectrolytes*. Marcel Dekker, New York.
[7] Morawetz, H. (2002). *J. Polym. Sci. Part B: Polym. Phys.*, **40**(11), 1080–1086.
[8] Dobrynin, A. V. and Rubinstein, M. (2005). *Prog. Polym. Sci.*, **30**(11), 1049–1118.
[9] Dobrynin, A. V., Colby, R. H., and Rubinstein, M. (2004). *J. Polym. Sci. Part B: Polym. Phys.*, **42**(19), 3513–3538.
[10] Dobrynin, A. V. (2008). *Curr. Opin. Coll. & Interface Sci.*, **13**(6), 376–388.
[11] Khokhlov, A. R. and Khalatur, P. G. (2005). *Curr. Opin. Coll. & Interface Sci.*, **10**(1–2), 22–29.
[12] Grosberg, A. Y., Nguyen, T. T., and Shklovskii, B. I. (2002). *Rev. Mod. Phys.*, **74**(2), 329–345.
[13] Cooper, C. L., Dubin, P. L., Kayitmazer, A. B. *et al.* (2005). *Curr. Opin. Coll. & Interface Sci.*, **10**(1–2), 52–78.
[14] Decher, G. and Schlenoff, J. B. (2003). *Multilayer Thin Films*. Wiley-VCH, New York.
[15] Boroudjerdi, H., Kim, Y. W., Naji, A. *et al.* (2005). *Phys. Rep.*, **416**(3–4), 129–199.

[16] Belyi, V. A. and Muthukumar, M. (2006). *Proc. Natl. Acad. Sci. USA*, **103**(46), 17174–17178.
[17] Koltover, I. (2004). *Nature Mater.*, **3**(9), 584–586.
[18] Wong, G. C. L. (2006). *Curr. Opin. Coll. & Interface Sci.*, **11**(6), 310–315.
[19] Schiessel, H. (2003). *J. Phys.-Condensed Matter*, **15**(19), R699–R774.
[20] Bloomfield, V. A. (1997). *Biopolymers*, **44**(3), 269–282.
[21] Kabanov, A. V. (2006). *Adv.d Drug Del. Rev.*, **58**(15), 1597–1621.
[22] de Gennes, P. G. (1979). *Scaling Concepts in Polymer Physics*. Cornell University Press, Intaca, NY.
[23] de Gennes, P. G., Pincus, P., Brochard, F. *et al.* (1976). *J. Phys. (France)*, **37**, 1461–1476.
[24] Khokhlov, A. R. and Khachaturian, K. A. (1982). *Polymer*, **23**, 1742–1750.
[25] Dobrynin, A. V., Colby, R. H., and Rubinstein, M. (1995). *Macromolecules*, **28**(6), 1859–1871.
[26] Jackson, J. D. (1998). *Classical Electrodynamics*. Wiley, New York.
[27] Liao, Q., Dobrynin, A. V., and Rubinstein, M. (2003). *Macromolecules*, **36**(9), 3386–3398.
[28] Migliorini, G., Rostiashvili, V., and Vilgis, T. A. (2001). *Eur. Phys. J. E*, **4**, 475–487.
[29] Kuhn, W., Kunzle, O., and Katchalsky, A. (1948). *Helvetica Chimica Acta*, **31**, 1994–2037.
[30] Landau, L. D. and Lifshitz, E. M. (1984). *Electrodynamics of Continuous Media*. Addison-Wesley, Reading, MA.
[31] Rubinstein, M. and Colby, R. H. (2003). *Polymer Physics*. Oxford University Press, New York.
[32] Grosberg, A. Y. and Khokhlov, A. R. (1994). *Statistical Physics of Macromolecules*. AIP Press, New Yourk.
[33] McQuarrie, D. A. (1976). *Statistical Mechanics*. Harper & Row, New York.
[34] Rayleigh, L. (1882). *Philos. Mag.*, **14**, 184–186.
[35] Katchalsky, A. and Eisenberg, H. (1951). *J. Polym. Sci.*, **6**, 145–154.
[36] Dobrynin, A. V., Rubinstein, M., and Obukhov, S. P. (1996). *Macromolecules*, **29**, 2974–2979.
[37] Carrillo, J. M. Y. and Dobrynin, A. V. (2010). *J. Phys. Chem. B*, **114**, 9391–9399.
[38] Chodanowski, P. and Stoll, S. (1999). *J. Chem. Phys.*, **111**(13), 6069–6081.
[39] Lyulin, A. V., Dunweg, B., Borisov, O. V. *et al.* (1999). *Macromolecules*, **32**(10), 3264–3278.
[40] Solis, F. J. and Olvera de La Cruz, M. (1998). *Macromolecules*, **31**, 5502–5506.
[41] Balazs, A. C., Singh, C., Zhulina, E. *et al.* (1997). *Progress in Surface Science*, **55**(3), 181–269.

[42] Migliorini, G., Lee, N., Rostiashvili, V. *et al.* (2001). *Eur. Phys. J. E*, **6**(3), 259–270.

[43] Chang, R. W. and Yethiraj, A. (2006). *Macromolecules*, **39**(2), 821–828.

[44] Reddy, G. and Yethiraj, A. (2006). *Macromolecules*, **39**(24), 8536–8542.

[45] Tamashiro, M. N. and Schiessel, H. (2000). *Macromolecules*, **33**(14), 5263–5272.

[46] Vilgis, T. A., Johner, A., and Joanny, J.-F. (2000). *Eur. Phys. J. E*, **2**(3), 289–300.

[47] Holm, C., Joanny, J.-F., Kremer, K. *et al.* (2004). *Adv. Polym. Sci.*, **166**, 67–111.

[48] Limbach, H. J., Holm, C., and Kremer, K. (2002). *Europhys. Lett.*, **60**(4), 566–572.

[49] Iwaki, T. (2006). *J. Chem. Phys.*, **125**(22), 224901.

[50] Iwaki, T. and Yoshikawa, K. (2004). *Europhys. Lett.*, **68**(1), 113–119.

[51] Sakaue, T. (2004). *J. Chem. Phys.*, **120**(13), 6299–6305.

[52] Sakaue, T. and Yoshikawa, K. (2006). *J. Chem. Phys.*, **125**(7), 074904.

[53] Schmitz, K. S. (1993). *Macroions in Solution and Colloidal Suspension*. Wiley-VCH, Weinheim.

[54] Alfrey, T., Berg, P. V., and Morawetz, H. (1951). *J. Polym. Sci.*, **7**, 543–551.

[55] Fuoss, R. M., Katchalsky, A., and Lifson, S. (1951). *Proc. Natl. Acad. Sci. USA*, **37**, 579–586.

[56] Alexandrowicz, Z. and Katchalsky, A. (1963). *J. Polym. Sci.: Part A*, **1**, 3231–3260.

[57] Manning, G. S. (1969). *J. Chem. Phys.*, **51**(3), 924–933.

[58] Manning, G. S. and Ray, J. (1998). *J. Biomolec. Str. & Dyn.*, **16**(2), 461–476.

[59] Raphael, E. and Joanny, J.-F. (1990). *Europhys. Lett.*, **13**, 623–628.

[60] Dobrynin, A. V. and Rubinstein, M. (2001). *Macromolecules*, **34**(6), 1964–1972.

[61] Muthukumar, M. (2004). *J. Chem. Phys.*, **120**(19), 9343–9350.

[62] Kramarenko, E. Y., Khokhlov, A. R., and Yoshikawa, K. (2000). *Macromol. Theor. Sim.*, **9**(5), 249–256.

[63] Deshkovski, A., Obukhov, S., and Rubinstein, M. (2001). *Phys. Rev. Lett.*, **86**(11), 2341–2344.

[64] Dobrynin, A. V. (2004). In *Simulation Methods for Polymers* (ed. M. Kotelyanskii and D. N. Theodorou), pp. 259–312. Marcel Dekker, New York.

[65] Khokhlov, A. R. (1980). *J. Phys. A*, **13**, 979–987.

[66] Schiessel, H. and Pincus, P. (1998). *Macromolecules*, **31**(22), 7953–7959.

[67] Winkler, R. G., Gold, M., and Reineker, P. (1998). *Phys. Rev. Lett.*, **80**(17), 3731–3734.

[68] Castelnovo, M., Sens, P., and Joanny, J.-F. (2000). *Eur. Phys. J. E*, **1**(2–3), 115–125.
[69] Liao, Q., Dobrynin, A. V., and Rubinstein, M. (2003). *Macromolecules*, **36**(9), 3399–3410.
[70] Stevens, M. J. and Kremer, K. (1995). *J. Chem. Phys.*, **103**(4), 1669–1690.
[71] Micka, U., Holm, C., and Kremer, K. (1999). *Langmuir*, **15**(12), 4033–4044.
[72] Limbach, H. J. and Holm, C. (2001). *J. Chem. Phys.*, **114**(21), 9674–9682.
[73] Micka, U. and Kremer, K. (2000). *Europhys. Lett.*, **49**(2), 189–195.
[74] Dobrynin, A. V. and Rubinstein, M. (1999). *Macromolecules*, **32**(3), 915–922.
[75] Liao, Q., Dobrynin, A. V., and Rubinstein, M. (2006). *Macromolecules*, **39**(5), 1920–1938.
[76] Jeon, J. and Dobrynin, A. V. (2007). *Macromolecules*, **40**, 7695–7706.
[77] Limbach, H. J. and Holm, C. (2002). *Comp. Phys. Commun.*, **147**(1–2), 321–324.
[78] Limbach, H. J., Holm, C., and Kremer, K. (2004). *Macromol. Symp.*, **211**, 43–53.
[79] Chepelianskii, A., Mohammad-Rafiee, F., Trizac, E. *et al.* (2009). *J. Phys. Chem. B*, **113**(12), 3743–3749.
[80] Levin, Y. (2002). *Rep. Prog. Phys.*, **65**(11), 1577–1632.
[81] Ulrich, S., Laguecir, A., and Stoll, S. (2005). *J. Chem. Phys.*, **122**, 094911-1–9.
[82] Uyaver, S. and Seidel, C. (2004). *J. Phys. Chem. B*, **108**(49), 18804–18814.
[83] Uyaver, S. and Seidel, C. (2009). *Macromolecules*, **42**(4), 1352–1361.
[84] Wang, S. Q., Granick, S., and Zhao, J. (2008). *J. Chem. Phys.*, **129**(24), 241102.
[85] Vallat, P., Catala, J. M., Rawiso, M. *et al.* (2008). *Europhys. Lett.*, **82**(2), 28009.
[86] Kirwan, L. J., Papastavrou, G., Borkovec, M. *et al.* (2004). *Nano Lett.*, **4**(1), 149–152.
[87] Odijk, T. (1977). *J. Polym. Phys. Part B: Polym. Phys.*, **15**, 477–483.
[88] Skolnick, J. and Fixman, M. (1977). *Macromolecules*, **10**(5), 944–948.
[89] Odijk, T. (1979). *Macromolecules*, **12**, 688–693.
[90] Netz, R. R. and Orland, H. (1999). *Eur. Phys. J. B*, **8**(1), 81–98.
[91] Ha, B. Y. and Thirumalai, D. (1999). *J. Chem. Phys.*, **110**(15), 7533–7541.
[92] Li, H. and Witten, T. A. (1995). *Macromolecules*, **28**(17), 5921–5927.
[93] Schmidt, M. (1991). *Macromolecules*, **24**, 5361–5364.
[94] Barrat, J. L. and Joanny, J.-F. (1993). *Europhys. Lett.*, **24**, 333–

338.

[95] Dobrynin, A. V. (2006). *Macromolecules*, **39**, 9519–9527.
[96] Schafer, H. and Seidel, C. (1997). *Macromolecules*, **30**, 6658–6661.
[97] Ullner, M. and Woodward, C. E. (2002). *Macromolecules*, **35**(4), 1437–1445.
[98] Micka, U. and Kremer, K. (1996). *Phys. Rev. E*, **54**(3), 2653–2662.
[99] Micka, U. and Kremer, K. (1997). *Europhys. Lett.*, **38**(4), 279–284.
[100] Tricot, M. (1984). *Macromolecules*, **17**, 1698–1704.
[101] Forster, S., Schmidt, M., and Antonietti, M. (1992). *J. Phys. Chem.*, **96**, 4008–4014.
[102] de Nooy, A. E. J., Besemer, A. C., van Bekkum, H. *et al.* (1996). *Macromolecules*, **29**, 6541–6547.
[103] Nishida, K., Urakawa, H., Kaji, K. *et al.* (1997). *Polymer*, **38**(24), 6083–6085.
[104] Tanahatoe, J. J. and Kuil, M. E. (1997). *J. Phys. Chem. B*, **101**(45), 9233–9239.
[105] Beer, M., Schmidt, M., and Muthukumar, M. (1997). *Macromolecules*, **30**, 8375–8385.
[106] Nguyen, T. T. and Shklovskii, B. I. (2002). *Phys. Rev. E*, **66**(2), 021801-1–7.
[107] Everaers, R., Milchev, A., and Yamakov, V. (2002). *Eur. Phys. J. E*, **8**(1), 3–14.
[108] Gubarev, A., Carrillo, J. M. Y., and Dobrynin, A. V. (2009). *Macromolecules*, **42**(15), 5851–5860.
[109] Marko, J. F. and Siggia, E. D. (1995). *Macromolecules*, **28**(26), 8759–8770.
[110] Hagerman, P. J. (1981). *Biopolymers*, **20**, 1503–1535.
[111] Le Bret, M. (1982). *J. Chem. Phys.*, **76**, 6243–6255.
[112] Fixman, M. (1982). *J. Chem. Phys.*, **76**, 6346–6353.
[113] Golestanian, R., Kardar, M., and Liverpool, T. B. (1999). *Phys. Rev. Lett.*, **82**(22), 4456–4459.
[114] Nguyen, T. T., Rouzina, I., and Shklovskii, B. I. (1999). *Phys. Rev. E*, **60**(6), 7032–7039.
[115] Katchalsky, A., Kunzle, O., and Kuhn, W. (1950). *J. Polym. Sci.*, **5**, 283–300.
[116] Hermans, J. J. and Overbeek, J. T. G. (1948). *Rec. Trav. Chim.*, **67**, 761–776.
[117] Muthukumar, M. (1987). *J. Chem. Phys.*, **86**(12), 7230–7235.
[118] Muthukumar, M. (1996). *J. Chem. Phys.*, **105**(12), 5183–5199.
[119] Ghosh, K., Carri, G. A., and Muthukumar, M. (2001). *J. Chem. Phys.*, **115**, 4367–4375.
[120] Odijk, T. and Houwaart, A. C. (1978). *J. Polym. Sci. Polym. Phys. Ed.*, **16**, 627–639.
[121] Dobrynin, A. V. and Carrillo, J. M. Y. (2009). *J. Phys. - Cond. Mat.*, **21**(42), 424112.
[122] Fixman, M. and Skolnick, J. (1978). *Macromolecules*, **11**, 863–867.
[123] Edwards, S. F. and Singh, P. (1979). *J. Chem. Soc. - Faraday Trans. II*, **75**, 1001–1019.

[124] Doi, M. and Edwards, S. F. (ed.) (1986). *The Theory of Polymer Dynamics*. Oxford University Press, New York.

[125] Maret, G. and Weill, G. (1983). *Biopolymers*, **22**, 2727–2744.

[126] Sobel, E. S. and Harpst, J. A. (1991). *Biopolymers*, **31**, 1559–1564.

[127] Cairney, K. L. and Harrington, R. E. (1982). *Biopolymers*, **81**, 923–934.

[128] Saleh, O. A., McIntosh, D. B., Pincus, P. et al. (2009). *Phys. Rev. Lett.*, **102**(6), 068301.

[129] Makita, N., Ullner, M., and Yoshikawa, K. (2006). *Macromolecules*, **39**(18), 6200–6206.

[130] Hugel, T., Grosholz, M., Clausen-Schaumann, H. et al. (2001). *Macromolecules*, **34**(4), 1039–1047.

[131] Hsieh, C. C., Balducci, A., and Doyle, P. S. (2008). *Nano Lett.*, **8**(6), 1683–1688.

[132] Baumann, C. G., Smith, S. B., Bloomfield, V. A. et al. (1997). *Proc. Natl. Acad. Sci. USA*, **94**(12), 6185–6190.

[133] Abels, J. A., Moreno-Herrero, F., van der Heijden, T. et al. (2005). *Biophys. J.*, **88**(4), 2737–2744.

[134] Israelachvili, J. (ed.) (1992). *Intermolecular and Surface Forces*. Academic Press, London.

[135] Boris, D. C. and Colby, R. H. (1998). *Macromolecules*, **31**(17), 5746–5755.

[136] Bordi, F., Cametti, C., Tan, J. S. et al. (2002). *Macromolecules*, **35**(18), 7031–7038.

[137] Bordi, F., Colby, R. H., Cametti, C. et al. (2002). *J. Phys. Chem. B*, **106**(27), 6887–6893.

[138] Truzzolillo, D., Bordi, F., Cametti, C. et al. (2009). *Phys. Rev. E*, **79**(1), 011804-1–9.

[139] Nierlich, M., Williams, C. E., Boue, F. et al. (1979). *J. Phys. (Paris)*, **40**, 701–704.

[140] Drifford, M. and Dalbiez, J. P. (1984). *J. Phys. Chem.*, **88**, 5368.

[141] Pfeuty, P. (1978). *J. Phys. France Coll.*, **39**, C2–149.

[142] Rubinstein, M., Colby, R. H., and Dobrynin, A. V. (1994). *Phys. Rev. Lett.*, **73**(20), 2776–2779.

[143] Nierlich, M., Boue, F., and Lapp, A. (1985). *J. Phys. (Paris)*, **46**, 649–655.

[144] Kakehashi, R. and Maeda, H. (1996). *J. Chem. Soc. - Faraday Trans.*, **92**(17), 3117–3121.

[145] Kakehashi, R. and Maeda, H. (1996). *J. Chem. Soc. - Faraday Trans.*, **92**(22), 4441–4444.

[146] Kakehashi, R., Yamazoe, H., and Maeda, H. (1998). *Coll. Polym. Sci.*, **276**(1), 28–33.

[147] Raspaud, E., de Conceicao, M., and Livolant, F. (2000). *Phys. Rev. Lett.*, **84**, 2533–2536.

[148] Hansen, P. L., Podgornik, R., and Parsegian, A. V. (1997). *Phys. Rev. E*, **64**, 021907-1–4.

[149] Reddy, M. and Marinsky, J. A. (1970). *J. Phys. Chem.*, **74**, 3884–3891.

[150] Wang, L. and Bloomfield, V. A. (1990). *Macromolecules*, **23**, 804–809.
[151] Koene, R. S., Nicolai, T., and Mandel, M. (1983). *Macromolecules*, **16**, 231–236.
[152] Takahashi, A., Kato, N., and Nagasawa, M. (1970). *J. Phys. Chem.*, **74**, 944–946.
[153] Chang, R. and Yethiraj, A. (2005). *Macromolecules*, **38**(2), 607–616.
[154] Muthukumar, M. (1997). *J. Chem. Phys.*, **107**(7), 2619–2635.
[155] Muthukumar, M. (2001). *Polymer*, **42**(13), 5921–5923.
[156] Fuoss, R. M. (1948). *J. Polym. Sci.*, **3**, 603–604.
[157] Liao, Q., Carrillo, J. M. Y., Dobrynin, A. V. *et al.* (2007). *Macromolecules*, **40**(21), 7671–7679.
[158] Kavassalis, T. A. and Noolandi, J. (1988). *Macromolecules*, **21**, 2869–2879.
[159] Essafi, W., Spiteri, M. N., Williams, C. *et al.* (2009). *Macromolecules*, **42**(24), 9568–9580.
[160] Heitz, C., Rawiso, M., and Francois, J. (1999). *Polymer*, **40**(7), 1637–1650.
[161] Waigh, T. A., Ober, R., Williams, C. E. *et al.* (2001). *Macromolecules*, **34**(6), 1973–1980.
[162] Qu, D., Baigl, D., Williams, C. E. *et al.* (2003). *Macromolecules*, **36**(18), 6878–6883.
[163] Baigl, D., Sferrazza, M., and Williams, C. E. (2003). *Europhys. Lett.*, **62**(1), 110–116.
[164] Aseyev, V. O., Klenin, S. I., Tenhu, H. *et al.* (2001). *Macromolecules*, **34**(11), 3706–3709.
[165] Lee, M. J., Green, M. M., Mikes, F. *et al.* (2002). *Macromolecules*, **35**(10), 4216–4217.
[166] Kiriy, A., Gorodyska, G., Minko, S. *et al.* (2002). *J. Amer. Chem. Soc.*, **124**(45), 13454–13462.
[167] Minko, S., Kiriy, A., Gorodyska, G. *et al.* (2002). *J. Amer. Chem. Soc.*, **124**(13), 3218–3219.
[168] Khokhlov, A. R. and Nyrkova, I. A. (1992). *Macromolecules*, **25**, 1493–1502.
[169] Muthukumar, M. (2002). *Macromolecules*, **35**(24), 9142–9145.
[170] Borue, V. Y. and Erukhimovich, I. Y. (1988). *Macromolecules*, **21**, 3240–3249.
[171] Vilgis, T. A. and Borsali, R. (1991). *Phys. Rev. A*, **43**, 6857–6873.
[172] Ermoshkin, A. V. and de la Cruz, M. O. (2003). *Phys. Rev. Lett.*, **90**(12), 125504–1–4.
[173] Orkoulas, G., Kumar, S. K., and Panagiotopoulos, A. Z. (2003). *Phys. Rev. Lett.*, **90**, 048303.
[174] Volk, N., Vollmer, D., Schmidt, M. *et al.* (2004). *Adv. Polym. Sci.*, **166**, 29–65.
[175] Lee, C. L. and Muthukumar, M. (2009). *J. Chem. Phys.*, **130**(2), 024904.
[176] Joanny, J.-F. and Leibler, L. (1990). *J. Phys.*, **51**, 545–557.

[177] Dormidontova, E. E., Erukhimovich, I. Y., and Khokhlov, A. R. (1994). *Macromol. Theor. Sim.*, **3**(4), 661–675.

[178] Chang, R. W. and Yethiraj, A. (2003). *J. Chem. Phys.*, **118**(14), 6634–6647.

[179] Kwon, H. J., Osada, Y., and Gong, J. P. (2006). *Polymer Journal*, **38**, 1211–1219.

[180] Osada, Y. and Khokhlov, A. R. (ed.) (2001). *Polymer Gels and Networks*. Marcel Dekker.

[181] Harland, R. S. and Prudhomme, R. K. (ed.) (1998). *Polyelectrolyte gels: Properties, preparation, and applications*. American Chemical Society, Washington, DC.

[182] Gong, J. P. and Osada, Y. (2002). *Progress in Polymer Science*, **27**(1), 3–38.

[183] Huck, W. T. S. (2008). *Materials Today*, **11**(7–8), 24–32.

[184] Malmsten, M. (2006). *Soft Matter*, **2**(9), 760–769.

[185] Peppas, N. A., Bures, P., Leobandung, W. *et al.* (2000). *Eur. J. Pharmaceutics and Biopharmaceutics*, **50**(1), 27–46.

[186] Kaneko, D., Gong, J. P., and Osada, Y. (2002). *J. Mater. Chem.*, **12**(8), 2169.

[187] Treloar, L. R. G. (ed.) (2005). *The Physics of Rubber Elasticity*. Clarendon Press, Oxford.

[188] Ilavsky, M. (1982). *Macromolecules*, **15**, 782–788.

[189] Gundogan, N., Melekaslan, D., and Okay, O. (2002). *Macromolecules*, **35**(14), 5616–5622.

[190] Edgecombe, S. and Linse, P. (2007). *Macromolecules*, **40**(10), 3868–3875.

[191] Mann, B. A., Kremer, K., and Holm, C. (2006). *Macromol. Symp.*, **237**, 90–107.

[192] Schneider, S. and Linse, P. (2003). *J. Phys. Chem. B*, **107**(32), 8030–8040.

[193] Ilavsky, M. and Hrouz, J. (1983). *Polym. Bull.*, **9**, 159–166.

[194] Rubinstein, M., Colby, R. H., Dobrynin, A. V. *et al.* (1996). *Macromolecules*, **29**(1), 398–406.

[195] Mann, B. A., Everaers, R., Holm, C. *et al.* (2004). *Europhys. Lett.*, **67**(5), 786–792.

[196] Mann, B. A., Holm, C., and Kremer, K. (2005). *J. Chem. Phys.*, **122**(15), 154903.

[197] Lu, Z. Y. and Hentschke, R. (2003). *Phys. Rev. E*, **67**(6), 061807.

[198] Khokhlov, A. R., Starodubtsev, S. G., and Vasilevskaya, V. V. (1993). *Adv. Polym. Sci.*, **109**, 123–171.

[199] Flory, P. J. (ed.) (1953). *Principles of Polymer Chemistry*. Cornell University Press, Ithaca, NY.

[200] Skouri, R., Schosseler, F., Munch, J. P. *et al.* (1995). *Macromolecules*, **28**(1), 197–210.

[201] Dusek, K. and Patterson, D. (1968). *J. Polym. Sci., Part A-2*, **6**, 1209–1216.

[202] Tanaka, T., Fillmore, D., Sun, S.-T. *et al.* (1980). *Phys. Rev. Lett.*, **45**(20), 1636–1639.

[203] Shibayama, M. and Tanaka, T. (1993). *Adv. Polym. Sci.*, **109**, 1–62.

[204] Zhang, Y.-Q., Tanaka, T., and Shibayama, M. (1992). *Nature*, **360**, 142–144.

[205] Lynch, I., Sjostrom, J., and Piculell, L. (2005). *J. Phys. Chem. B*, **109**(9), 4258–4262.

[206] Hansson, P. (2006). *Curr. Opin. Coll. & Interface Sci.*, **11**(6), 351–362.

[207] Yin, D. W., de la Cruz, M. O., and de Pablo, J. J. (2009). *J. Chem. Phys.*, **131**(19), 194907.

[208] Horkay, F., Tasaki, I., and Basser, P. J. (2000). *Biomacromolecules*, **1**(1), 84–90.

[209] Edgecombe, S. and Linse, P. (2006). *Langmuir*, **22**(8), 3836–3843.

[210] Bysell, H., Schmidtchen, A., and Malmsten, M. (2009). *Biomacromolecules*, **10**(8), 2162–2168.

[211] Johner, A., Vilgis, T. A., and Joanny, J.-F. (1999). *Macromol. Symp.*, **146**, 223–226.

[212] Shibayama, M., Ikkai, F., Shiwa, Y. et al. (1997). *J. Chem. Phys.*, **107**(13), 5227–5235.

[213] Ikkai, F., Shibayama, M., and Han, C. C. (1998). *Macromolecules*, **31**(10), 3275–3281.

[214] Ikkai, F. and Shibayama, M. (2007). *Polymer*, **48**(8), 2387–2394.

[215] Zeldovich, K. B., Dormidontova, E. E., Khokhlov, A. R. et al. (1997). *J. Phys. II*, **7**(4), 627–635.

[216] Ruhe, J., Ballauff, M., Biesalski, M. et al. (2004). *Adv. Polym. Sci.*, **165**, 79–150.

[217] Dubin, P., Bock, J., Davis, R. et al. (ed.) (1994). *Macromolecular Complexes in Chemistry and Biology*. Springer, New York.

[218] Dubin, P. L. and Farinato, R. S. (ed.) (1999). *Colloid-Polymer Interactions: From Fundamentals to Practice*. Wiley-Interscience, New York.

[219] Eisenberg, A. and Kim, J.-S. (ed.) (1998). *Introduction to Ionomers*. Wiley, New York.

[220] Netz, R. R. and Andelman, D. (2003). *Phys. Rep.*, **380**(1–2), 1–95.

Fluid Transport in Gels

Masao Doi

Department of Applied Physics, University of Tokyo
Tokyo 1138656, Japan

Introduction

A gel is an elastic continuum which contains fluid; it consists of two parts, the elastic part and the fluid part. The elastic part is usually made of polymers which form a three-dimensional network and give elasticity to the gel (see Figure 10.1). The fluid part is made of small molecules which can move around in the network. In this chapter, the words "polymer" and "solvent" are often used to represent the network part and the fluid part.

This state of matter (elastic materials containing fluids) is seen in everyday life, e.g. in foods, cosmetics, and living materials. Gels are also important in technology. Many industrial processes in printing, coating, and device making use this state of matter. Gels are also used in agricultural, biological, and environmental applications.

The amount of fluids contained in gels can be controlled by various ways. It can be controlled by changing thermodynamic parameters such as temperature, solvent composition, and pH, etc; for example, hydro gels shrink when salt is added. The amount of solvent in the gel can be changed by mechanical forces; fluid in a gel can be squeezed out by compressional force. Finally the fluid in ionic gels can be controlled by electric field; solvent flow is induced by an electric field as we shall see later.

The purpose of this chapter is to discuss the transport of fluids in gels and their effect on the deformation of gels. It is important to note that the transport of fluids is coupled with the deformation of gels. For example when a gel slab is bent, fluid is squeezed out from the bent side. Conversely, when a solvent is transported forcefully by an electric field, the gel slab is bent. We shall develop a continuum model which describes such coupled phenomena.

The construction of this chapter is as follows. In Section 10.1, the equilibrium state of gels under the action of mechanical forces is discussed. In Section 10.2, the dynamics of gels, especially the relaxation of mechanical responses caused by solvent flow, is discussed. In Section 10.3, the effect of electric field on solvent flow and the gel deformation is discussed.

The discussion in Sections 10.1 and 10.2 can be applied both for ionic and non-ionic gels as long as no electric field is applied and solvent com-

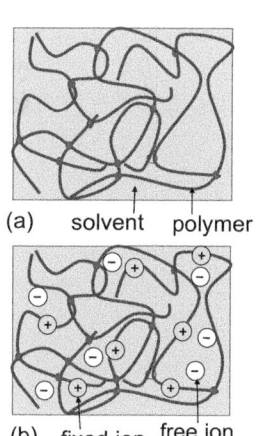

Fig. 10.1 Examples of gels: (a) non-ionic polymer gel, (b) ionic polymer gel.

position remains unchanged. This subject (i.e. the coupling of fluid transport and gel deformation) has recently been reviewed in more general situations [1]. Hence we shall restrict the discussion in these Sections to the minimum level needed for the discussion in Section 10.3. Section 10.3 is the main part of this chapter, and it applies specifically for ionic gels.

10.1 Equilibrium state of non-ionic gels

Description of deformation

We first consider the equilibrium state of non-ionic gels under mechanical forces. To describe the state of an elastic material, we need a certain state for reference. Let us consider a state of the gel in equilibrium when all external forces are removed, and call this the reference state. Each point of the polymer network can be labeled by the position vector $\boldsymbol{x}$ at the reference state, then the deformation of the gel is described by the function $\tilde{\boldsymbol{x}}(\boldsymbol{x})$ which denotes the position of the material point $\boldsymbol{x}$ after deformation. The mapping $\boldsymbol{x} \to \tilde{\boldsymbol{x}}(\boldsymbol{x})$ is usually a non-linear mapping since gel can undergo a very large deformation. However, the local deformation is always represented by a linear mapping. Consider a vector $d\boldsymbol{x}$ joining two nearby material points in the reference state. After deformation, the vector is transformed to

$$d\tilde{\boldsymbol{x}} = d\boldsymbol{x} \cdot \frac{\partial}{\partial \boldsymbol{x}} \tilde{\boldsymbol{x}}(\boldsymbol{x}) \tag{10.1}$$

or in terms of components,

$$d\tilde{x}_\alpha = F_{\alpha\beta} dx_\beta, \tag{10.2}$$

where α and β are the x, y, z components of the vector and $F_{\alpha\beta}$ is defined by

$$F_{\alpha\beta}(\boldsymbol{x}) = \frac{\partial \tilde{x}_\alpha}{\partial x_\beta}. \tag{10.3}$$

(In this chapter, we use the standard summation convention in which summation is taken over the repeated indices.)

A volume element dV in the gel at the reference state will occupy the volume $det(\boldsymbol{F})dV$ after deformation. Notice that when the volume of gel changes, it has to take in solvent from the surrounding, or expel it out. Under usual mechanical stresses, which is of the order of MPa, the change of the specific volume of solvent and polymer is very small, and we may assume that the increase of the gel volume is equal to the volume of solvent that the gel has taken in from the surrounding. This assumption is called incompressible condition.

If the volume fraction of polymer in the reference state is ϕ_0, the volume fraction ϕ after the deformation is given by

$$\phi(\boldsymbol{x}) = \frac{\phi_0}{det(\boldsymbol{F})}. \tag{10.4}$$

Free energy

Let $\mathcal{A}[\tilde{\boldsymbol{x}}]$ be the total free energy of the gel in the deformed state characterized by the function $\tilde{\boldsymbol{x}}(\boldsymbol{x})$. The free energy $\mathcal{A}[\tilde{\boldsymbol{x}}]$ is an integral of the local free energy and can be written as

$$\mathcal{A} = \int d\boldsymbol{x}\, A(F_{\alpha\beta}(\boldsymbol{x})). \tag{10.5}$$

Here $A(F_{\alpha\beta})$ is the free energy density, i.e. the free energy of deformation per unit volume in the reference state.

The form of the free energy $A(F_{\alpha\beta})$ is discussed in many literature sources [2,3]. An example is

$$A(F_{\alpha\beta}) = \frac{G}{2}(F_{\alpha\beta}^2 - 3) + A_{mix}(\phi), \tag{10.6}$$

where G is the shear modulus (which can be related to the cross-link density of the polymer network) and A_{mix} stands for the free energy of mixing of polymer and solvent.

Free energy for small deformation

Polymeric gels can bear a very large strain (of several hundreds %) without being broken. To describe such large deformation, one needs the general expression of free energy such as equation (10.6) [1]. However, the mathematics becomes rather complicated. Here, we restrict ourselves to simple cases where deformation is small and the linear theory of elasticity can be used.

Let $\boldsymbol{u}(\boldsymbol{x})$ be the displacement vector of the material point $\boldsymbol{x}$ [i.e., $\boldsymbol{u}(\boldsymbol{x}) = \tilde{\boldsymbol{x}}(\boldsymbol{x}) - \boldsymbol{x}$]. We assume that $\partial u_\alpha/\partial x_\beta$ is small and consider only the lowest order terms with respect to $\partial u_\alpha/\partial x_\beta$. For an isotropic material, the free energy density is written as [4]

$$A = \frac{1}{2}Kw^2 + \frac{1}{4}G\left(\frac{\partial u_\alpha}{\partial x_\beta} + \frac{\partial u_\beta}{\partial x_\alpha} - \frac{2}{3}w\delta_{\alpha\beta}\right)^2, \tag{10.7}$$

where

$$w = \nabla \cdot \boldsymbol{u} = \frac{\partial u_\alpha}{\partial x_\alpha} \tag{10.8}$$

is the volume strain (the ratio between the volume change ΔV and the original volume V), and K and G are material constants called the osmotic modulus and the shear modulus, respectively, K stands for the restoring force when the volume of the gel is changed, and G stands for the restoring force for a deformation that preserve the volume of the gel (such as a shear).

The constant K and G can be derived from $A(F_{\alpha\beta})$. If $A(F_{\alpha\beta})$ is given by equation (10.6), G is given by the same G in equation (10.6), and K is given by

$$K = \phi_0^2 \frac{\partial^2}{\partial \phi^2}\left(3G\left(\frac{\phi_0}{\phi}\right)^{2/3} - 3G + \tilde{A}_{mix}(\phi)\right) \quad \text{at} \quad \phi = \phi_0. \tag{10.9}$$

Equilibrium state

Let us first consider the equilibrium state of a gel under the action of mechanical forces. Suppose that a weight is placed on top of a gel, then the gel will deform and eventually takes an equilibrium state (see Figure 10.2). Given the form of the free energy, the deformation of the gel at equilibrium can be calculated by the condition that the free energy be minimum at equilibrium, i.e. the variation of $\mathcal{A}$ with respect to hypothetical variation of u_α be zero:

$$\frac{\delta \mathcal{A}}{\delta u_\alpha} = 0 \tag{10.10}$$

For the free energy expression of equations (10.5) and (10.7), equation (10.10) can be written as

$$\delta \mathcal{A} = \int d\boldsymbol{x} \sigma_{\alpha\beta} \frac{\partial \delta u_\alpha}{\partial x_\beta} \tag{10.11}$$

where $\sigma_{\alpha\beta}$ is given by

$$\sigma_{\alpha\beta} = Kw\delta_{\alpha\beta} + G\left(\frac{\partial u_\alpha}{\partial x_\beta} + \frac{\partial u_\beta}{\partial x_\alpha} - \frac{2}{3}w\delta_{\alpha\beta}\right). \tag{10.12}$$

which is precisely the same as the expression of the stress tensor in the theory of elasticity.

Using the stress tensor, one can show that the equilibrium condition (10.10) gives the force balance equation in the bulk

$$\frac{\partial \sigma_{\alpha\beta}}{\partial x_\beta} = 0 \tag{10.13}$$

and the boundary condition

$$\sigma_{\alpha\beta} n_\beta = f_\alpha , \tag{10.14}$$

where $\boldsymbol{n}$ is the unit vector normal to the surface and $\boldsymbol{f}$ is the force acting on the unit area of the surface of the gel.

Effect of solvent

The above set of equations is precisely the same as those for the theory of elasticity and there is no term which represents the effect of solvent. In gels, however, solvent plays an important role in equilibrium and in non-equilibrium state.

Consider the case shown in Figure 10.2. When the weight is placed, the gel is deformed and mechanical equilibrium is attained. This state, however, is not the true equilibrium state since the solvent is not in equilibrium. The weight tends to squeeze out the solvent from the gel, but this takes a very long time; the equilibration of solvent is attained by the diffusion of solvent and the time can be hours or days. Immediately after the weight is placed, the volume of the gel is the same as that before

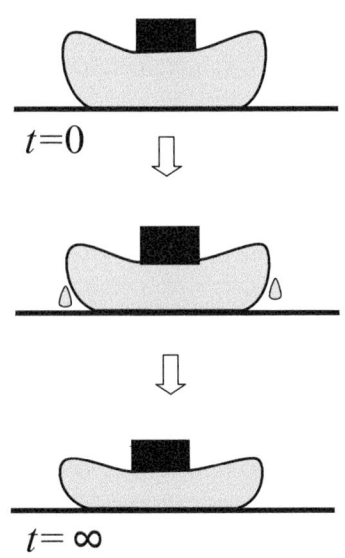

Fig. 10.2 Deformation of a gel when a weight is placed. Initially the gel deforms keeping its volume constant. This gives a pseudo-equilibrium state. As time goes on, solvent bleeds out and the gel eventually reaches the true equilibrium state.

the deformation. Therefore the pseudo-equilibrium state is determined by the condition that the free energy $\mathcal{A}$ be minimum under the constraint

$$\frac{\partial u_\alpha}{\partial x_\alpha} = 0 \tag{10.15}$$

This constraint has to be satisfied for all points x in the gel. If the constraint is taken into account, the functional to be minimized is

$$\mathcal{A}' = \mathcal{A} - \int d\boldsymbol{x} p(\boldsymbol{x}) \frac{\partial u_\alpha}{\partial x_\alpha} \tag{10.16}$$

where $p(\boldsymbol{x})$ is the Lagrangian multiplier. Minimizing equation (10.16) with respect to $\boldsymbol{u}$, one can show that the pseudo-equilibrium state is determined by

$$\frac{\partial}{\partial x_\beta}(\sigma_{\alpha\beta} - p\delta_{\alpha\beta}) = 0 \tag{10.17}$$

and the boundary condition

$$(\sigma_{\alpha\beta} - p\delta_{\alpha\beta})n_\beta = f_\alpha . \tag{10.18}$$

Equations (10.17) and (10.18) indicate that the mechanical stress is given by $\sigma_{\alpha\beta} - p\delta_{\alpha\beta}$. The extra term $p\delta_{\alpha\beta}$ arises from the constraint that the local volume of the gel is unchanged, i.e. for the same reason as the pressure term in the hydrodynamics of incompressible fluids. We can interpret p as the pressure of the fluid. The pressure is zero (or constant) in the true equilibrium state, but is non-zero in the pseudo-equilibrium state where only the mechanical equilibrium is attained.

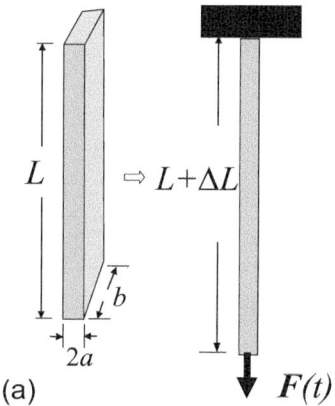

Examples

Stretched gel sheet

We now give examples of analysis for deformation of gels under mechanical forces. Consider a sheet of a gel of thickness $2a$, width b, and length L with $L \gg b \gg a$ [see Figure 10.3(a)]. Suppose that the gel sheet is stretched along the longitudinal of the sheet and the length is changed from L to $L + \Delta L$. We shall calculate the force $F(t)$ needed to keep the elongation ΔL constant.

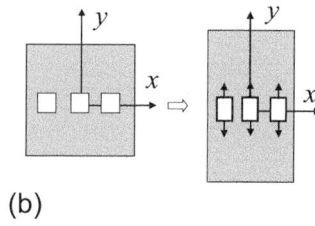

We take the coordinate shown in Figure 10.3(b). The origin is taken at the center of the sheet, and x and y axes are normal and tangent to the sheet respectively. In the present situation, the strain tensor is diagonal. Let $\epsilon_x, \epsilon_y, \epsilon_z$ be the diagonal components of the strain tensor. ϵ_y is given by the change of the longitudinal length

Fig. 10.3 (a) Overall view of the stretched sheet. (b) Local deformation.

$$\epsilon_y = \frac{\Delta L}{L} . \tag{10.19}$$

The other components, ϵ_x, ϵ_z, are unknowns and must be determined by the mechanical equilibrium condition.

Let us assume that the sheet is uniformly stretched, i.e. ϵ_x, ϵ_y, and ϵ_z are constant independent of position. (As we shall see in Section 10.2, this condition is not true in general, but it is true in the initial and in the final state of relaxation.) Then the free energy is written as

$$\mathcal{A} = V\left[\frac{1}{2}K(\epsilon_x + \epsilon_y + \epsilon_z)^2 + \frac{1}{3}G\left((\epsilon_x - \epsilon_y)^2 + (\epsilon_y - \epsilon_z)^2 + (\epsilon_z - \epsilon_x)^2\right)\right]. \tag{10.20}$$

where $V = 2abL$ is the volume of the gel.

Immediately after the deformation, the volume of the gel cannot change, so $\epsilon_x + \epsilon_y + \epsilon_z = 0$. Minimization of equation (10.20) under this constraint gives

$$\epsilon_x = \epsilon_z = -\frac{1}{2}\epsilon_y \tag{10.21}$$

and the free energy

$$\mathcal{A} = V\frac{3}{2}G\epsilon_y^2. \tag{10.22}$$

Let $f(t)$ be the force acting on the unit area of the cross section of the sheet. The force is given by

$$f(t) = \frac{F(t)}{2ab} = \frac{1}{2ab}\frac{\partial \mathcal{A}}{\partial \Delta L} = \frac{1}{V}\frac{\partial \mathcal{A}}{\partial \epsilon_y}. \tag{10.23}$$

Equations (10.22) and (10.23) give the initial force

$$f(0) = 3G\epsilon_y. \tag{10.24}$$

Therefore the initial Young's modulus is $3G$.

The true equilibrium state is obtained by minimizing equation (10.20) with respect to ϵ_x and ϵ_z without any constraint. This gives

$$\epsilon_x = \epsilon_z = -\frac{1}{2}\frac{K - \frac{2}{3}G}{K + \frac{1}{3}G}\epsilon_y \tag{10.25}$$

This is larger than the initial value, equation (10.21). Hence the volume of the gel increases. (Here we are assuming that the gel is placed in the bath of solvent.) The volume strain at equilibrium is

$$w_{eq} = \frac{G}{K + \frac{1}{3}G}\epsilon_y \tag{10.26}$$

Equations (10.20) and (10.25) give the force at equilibrium

$$f_{eq} = 3G\frac{K}{K + \frac{1}{3}G}\epsilon_y \tag{10.27}$$

Thus, as time goes on, the force becomes smaller. Explicit time dependence will be discussed later.

Bent gel sheet

Next, consider that the same sheet is held at one end, and is bent by the force $F(t)$ acting at the other end as shown in Figure 10.4. The

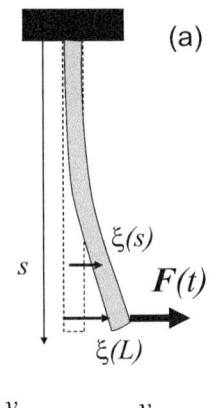
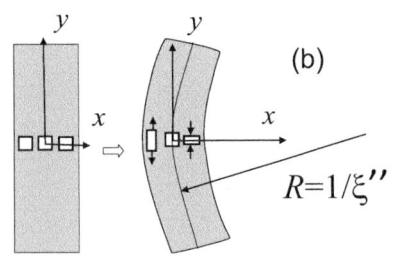

Fig. 10.4 (a) Overall view of the stretched sheet. (b) Local deformation.

mechanical equilibrium for a bent plate is discussed in the textbooks of elasto-mechanics, e.g. [4].

We take the coordinate s along the longitudinal of the sheet (the origin being at the fixed point) as it is shown in Figure 10.4(a) and define the bending moment $M(s)$ which acts on the cross-section of the sheet at s. In the situation shown in Figure 10.4(a), $M(s)$ is given by

$$M(s) = F(L-s). \tag{10.28}$$

Let $\xi(s)$ be the displacement of the sheet at point s. The bending moment $M(s)$ is proportional to the curvature $1/R = d^2\xi/ds^2$, and can be written as

$$M(s) = B_b \frac{d^2\xi(s)}{ds^2} = B_b \xi'', \tag{10.29}$$

where B_b is the bending modulus. If B_b is constant, equations (10.28) and (10.29) are easily solved to give the following relation between the applied force and the displacement of the gel at the bottom $(s = L)$

$$\xi(L) = \frac{1}{6B_b} L^3 F. \tag{10.30}$$

The bending modulus B_b can be expressed by the elastic constant of the gel and the geometry of the sheet. Consider the local deformation of a bent sheet. When the sheet is bent, the outer part is extended, while the inner part is compressed as it is shown in Figure 10.4(b). We take a local coordinate: the x axis is perpendicular to the sheet, and the y axis is along the longitudinal of the sheet. The origin is taken at the middle of the sheet. It can be shown that ϵ_z is always equal to zero since the overall volume strain $\int_{-a}^{a} dx w(x)$ remains zero in bending.

For a thin sheet, the displacement of the point (x,y) near the origin is written as

$$u_x = u(x,t) \tag{10.31}$$
$$u_y = -xy\xi''. \tag{10.32}$$

Therefore, the strain tensor is given by

$$\epsilon_{xx} = \frac{\partial u}{\partial x}, \quad \epsilon_{yy} = -\xi'' x \tag{10.33}$$

and the other components are zero. The volume strain w is given by

$$w = \epsilon_{xx} + \epsilon_{yy} = \frac{\partial u}{\partial x} - \xi'' x. \tag{10.34}$$

The components of the stress tensor are expressed in terms of w and ξ'' as follows

$$\sigma_{xx} = \left(K + \frac{4}{3}G\right)w + 2G\xi'' x \tag{10.35}$$

$$\sigma_{yy} = \left(K - \frac{2}{3}G\right)w - 2G\xi'' x \tag{10.36}$$

$$\sigma_{xy} = 0 \tag{10.37}$$

The mechanical equilibrium condition (10.17) is written as

$$\frac{\partial}{\partial x}(\sigma_{xx} - p) = 0. \tag{10.38}$$

Therefore $\sigma_{xx} - p$ is constant, which is equal to zero since there is no force acting on the surface at $x = a$. Thus the pressure is given by

$$p = \left(K + \frac{4}{3}G\right)w + 2G\xi''x \tag{10.39}$$

The bending moment is given by [4]

$$M = -b\int_{-a}^{a} dx(\sigma_{yy} - p)x \tag{10.40}$$

Using equations (10.36) and (10.39), this can be written as

$$M = b\int_{-a}^{a} dx\,(2Gw + 4G\xi''x)\,x \tag{10.41}$$

Immediately after the gel is bent, the volume of the gel is unchanged, so $w = 0$. In this case, equation (10.41) gives

$$M(0) = \frac{8}{3}Gba^3\xi'' \tag{10.42}$$

On the other hand, at equilibrium, p becomes equal to zero. Therefore from equation (10.39), the equilibrium swelling is given by

$$w_{eq}(x) = -\frac{2G}{K + \frac{4}{3}G}\xi''x \tag{10.43}$$

which indicates that the outer part $(x < 0)$ is swollen while the inner part $(x > 0)$ is de-swollen.

The bending moment at equilibrium is obtained by equations (10.41) and (10.43)

$$M_{eq} = \frac{8}{3}Gba^3\frac{K + \frac{1}{3}G}{K + \frac{4}{3}G}\xi'' \tag{10.44}$$

This is also smaller than the initial value (10.42). Therefore the force $F(t)$ relaxes in time.

10.2 Dynamics of non-ionic gels

Conservation of solvent volume

Having discussed the initial pseudo-equilibrium state and the final equilibrium state, we now consider the transient state, i.e. how the forces $F(t)$ defined in Figures 10.3 and 10.4 relax from the initial value to the final equilibrium value.

The slow relaxation process of gel is essentially determined by the permeation of solvent. As we have seen, when the gel is deformed, the mechanical equilibrium is attained very quickly, (within the damping time of the sound wave), while the permeation of water takes place very slowly. Therefore we may assume that in the permeation process of solvent, the gel is in mechanical equilibrium for a given value of volume strain $w(\boldsymbol{x},t)$, i.e. the deformation of the gel is obtained by minimizing the free energy $\mathcal{A}$ under the constraint that $w(\boldsymbol{x},t)$ is fixed. This constraint gives the term of the fluid pressure p in the mechanical equilibrium equation (10.17).

The time evolution of w is determined by the permeation of the solvent. Let $\boldsymbol{j}_s$ be the flux of solvent volume, then the conservation of the solvent volume (or the incompressible condition for the gel) is written as

$$\frac{\partial w}{\partial t} = -\nabla \cdot \boldsymbol{j}_s . \qquad (10.45)$$

Darcy's law

The solvent flux $\boldsymbol{j}_s$ is driven by the gradient of the pressure $p(\boldsymbol{x})$; the solvent will flow from the region of high pressure to the region of low pressure. It is natural to assume that the fluid flux is proportional to the gradient of the pressure:

$$\boldsymbol{j}_s = -\kappa \nabla p . \qquad (10.46)$$

This empirical law is called Darcy's law. The constant κ is called the Darcy coefficient. For isotropic material, κ is a scalar and depends on the local polymer concentration. Since we are considering the case of small deformation, κ is assumed to be constant.

Stress-diffusion coupling

The mechanical equilibrium condition (10.17), the conservation equation for the solvent volume (10.45), and the Darcy's law (10.46) give a closed set of equations which determine the permeation of solvent, and the deformation of gel. From equations (10.12) and (10.17) it follows

$$\left(K + \frac{G}{3}\right)\frac{\partial w}{\partial x_\alpha} + G\nabla^2 \frac{\partial^2 u_\alpha}{\partial x_\beta^2} = \frac{\partial p}{\partial x_\alpha}. \qquad (10.47)$$

Equations (10.45) and (10.46) give

$$\frac{\partial w}{\partial t} = \kappa \frac{\partial^2 p}{\partial x_\alpha^2} . \qquad (10.48)$$

Equations (10.8), (10.47), and (10.48) are the basic set of equations which describes the coupling of gel deformation and the solvent permeation.

Boundary conditions

In order to solve the partial differential equations given above, boundary conditions are needed. There are two types of boundary conditions.

One is for the force balance at the boundary, which is given by equation (10.18). The other is for the permeation of solvent. If the solvent can pass through the surface freely, the pressure has to be continuous at the boundary. If we take the pressure at the outer solution zero, the condition at free boundary is written as

$$p = 0, \quad \text{free boundary} \tag{10.49}$$

On the other hand, if the permeation of the solvent is blocked at the boundary, the normal component of the fluid flux has to be zero. Hence

$$\boldsymbol{j}_s \cdot \boldsymbol{n} = 0, \quad \text{blocked boundary}. \tag{10.50}$$

Examples

Relaxation of a stretched gel sheet

Let us consider the relaxation of the stretched sheet shown in Figure 10.3. For a thin sheet, the strain must be uniform in the y and z directions, but the strain in the x direction may vary with position, i.e. the displacement vector $\boldsymbol{u}$ can be written as

$$u_x = u(x,t), \quad u_y = \epsilon_y y, \quad u_z = \epsilon_z(t) z, \tag{10.51}$$

where ϵ_y is a given quantity, while $u(x,t)$ and $\epsilon_z(t)$ are the quantities to be determined. Using the volume strain

$$w(x,t) = \frac{\partial u}{\partial x} + \epsilon_y + \epsilon_z(t); \tag{10.52}$$

the stress tensor is written as

$$\sigma_{xx} = \left(K + \frac{4}{3}G\right)w - 2G(\epsilon_y + \epsilon_z) \tag{10.53}$$

$$\sigma_{yy} = \left(K - \frac{2}{3}G\right)w + 2G\epsilon_y \tag{10.54}$$

$$\sigma_{zz} = \left(K - \frac{2}{3}G\right)w + 2G\epsilon_z. \tag{10.55}$$

The force balance equation $\partial(\sigma_{xx} - p)/\partial x = 0$ and the boundary condition at $x = \pm a$ give the following expression for the pressure:

$$p = \sigma_{xx} = \left(K + \frac{4}{3}G\right)w - 2G(\epsilon_y + \epsilon_z). \tag{10.56}$$

Equations (10.48) and (10.56) give

$$\frac{\partial w}{\partial t} = D_s \frac{\partial^2 w}{\partial x^2} \tag{10.57}$$

where

$$D_s = \kappa\left(K + \frac{4}{3}G\right). \tag{10.58}$$

This stands for the diffusion coefficient of solvent in the gel.

To determine $\epsilon_z(t)$, we use the condition that the total force acting on the cross-section normal to the z axis is zero:

$$\int_{-a}^{a} dx(\sigma_{zz} - p) = 0 \tag{10.59}$$

Equations (10.55), (10.56), and (10.59) give

$$\epsilon_z = \frac{1}{2}(\bar{w} - \epsilon_y) ; \tag{10.60}$$

where $\bar{w}$ stands for the average of w for x:

$$\bar{w}(t) = \frac{1}{2a} \int_{-a}^{a} dx w(x,t) . \tag{10.61}$$

At $x = \pm a$, solvent can pass freely. Hence equations (10.49) and (10.56) give

$$\left(K + \frac{4}{3}G\right) w = 2G(\epsilon_y + \epsilon_z(t)) = G(\bar{w}(t) + \epsilon_y), \quad \text{at } x = \pm a . \tag{10.62}$$

Notice that the right hand side of equation (10.62) is not constant but changes with time as $\bar{w}(t)$ varies with time. This type of boundary value problem appears frequently in gel dynamics [1]. To determine the long-time behavior, we assume the following form for the solution:

$$w(x,t) = w_{eq} + g(x)e^{-t/\tau} \tag{10.63}$$

where w_{eq} is the equilibrium value given by equation (10.26), and $g(x)$ and τ are the solutions of the following equations:

$$-\frac{g}{\tau} = D_s \frac{\partial^2 g}{\partial x^2} \tag{10.64}$$

and

$$\left(K + \frac{4}{3}G\right) g = G\bar{g}, \quad \text{at } x = \pm a \tag{10.65}$$

where $\bar{g}$ stands for the average of g for x:

$$\bar{g} = \frac{1}{2a} \int_{-a}^{a} dx g(x) . \tag{10.66}$$

Equations (10.64)-(10.66) set an eigenvalue problem: the task is to find the value of τ which gives the non-zero function $g(x)$ that satisfies both equations (10.64) and (10.65). The solution of this eigenvalue problem is given by

$$\tau_s = \frac{a^2}{D_s \chi^2} \tag{10.67}$$

where χ is the solution of the equation

$$\frac{1}{\chi} \tan(\chi) = \frac{K + \frac{4}{3}G}{G} . \tag{10.68}$$

The stress relaxes from the initial value [equation (10.24)] to the final equilibrium value [equation (10.27)] with the relaxation time τ_s. Although the expression for τ_s looks complicated, τ_s is essentially determined by the diffusion time a^2/D_s since the value of χ is of the order of unity.

Relaxation of a bent gel sheet

Relaxation of the bent sheet can be analyzed in a similar way as the stretched sheet. Since mechanical equilibrium is always maintained, equations (10.31)-(10.41) are valid in the relaxation process. From equations (10.39) and (10.48) it follows

$$\frac{\partial w}{\partial t} = D_s \frac{\partial^2 w}{\partial x^2} \qquad (10.69)$$

where D_s is given by equation (10.58). At the boundary $x = \pm a$, pressure p has to be zero. This condition and equation (10.39) give the following boundary condition for w

$$\left(K + \frac{4}{3}G\right) w = \mp 2G\xi'' a \qquad \text{at} \qquad x = \pm a \qquad (10.70)$$

The solution of equation (10.69) with the boundary condition (10.70) and the initial condition $w(x,0) = 0$ is sketched in Figure 10.5(a). The solution approaches to the equilibrium value

$$w_{eq} = -\frac{2G}{\left(K + \frac{4}{3}G\right)}\xi'' x \qquad (10.71)$$

with the relaxation time

$$\tau_s = \frac{4a^2}{\pi^2 D_s}. \qquad (10.72)$$

Accordingly, the force $F(t)$ relaxes as it is shown in Figure 10.5(b).

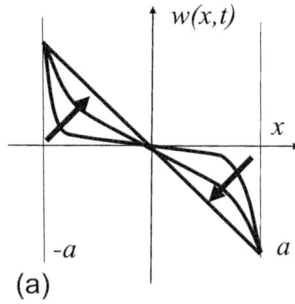

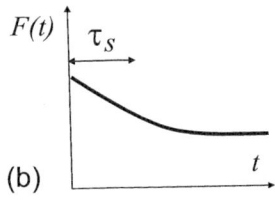

Fig. 10.5 (a) Time variation of the volume strain $w(x,t)$ of a bent gel sheet shown in Figure 10.4. (b) Time variation of the force $F(t)$ that is needed to keep the curvature of the gel sheet at a constant value.

10.3 Dynamics of ionic gels

Electro-mechanical coupling in ionic gels

Ionic gels are the gels containing ionic solvents. Typical examples of ionic gels are the gels made of polyelectrolytes [see Figure 10.1(b)]. Dissociation of ionic groups leaves ions fixed on the polymer network and free counter ions which can move around in the solvent.

If electric field is applied to such a gel, free ions start to move and drag surrounding solvent. This causes a macroscopic flow of solvent through the polymer network. Conversely, if the pressure gradient is applied to the gel, free ions are carried by solvent and there appears an electric current. These effects are observed in the experimental setup shown in Figure 10.6 [5]. In Figure 10.6(a), an ionic gel is sandwiched between two porous walls, and an electric field is applied across the gel. It is observed that the solvent flows from one chamber to the other. This phenomena is called electro-osmosis. In Figure 10.6(b), pressure drop Δp is applied in one chamber to create solvent flux across the gel. It is observed that the electric potential difference $\Delta \psi$ appears between the two chambers. The potential difference is called streaming potential.

If the ionic gel is deformable, the electric field can induce deformation of the gel. An example is shown in Figure 10.7. If an electric potential

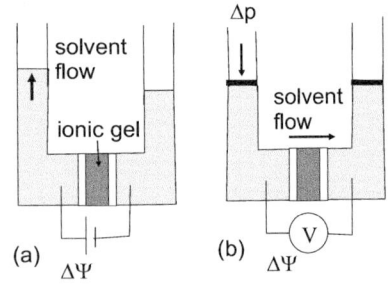

Fig. 10.6 (a) Effect of electro-osmosis. (b) Effect of streaming potential.

ψ_{ext} is applied to the electrode sandwiching a sheet of an ionic gel, the gel deforms as is shown in Figure 10.7(a). Conversely, if the sheet is deformed by a force applied at the end of the sheet, an electric current appears in the outer circuit, Figure 10.7(b). Thus the ionic gels can be used as electro-mechanical transducers.

These phenomena indicate that, in ionic gels, the electric field or electric current is also involved in the diffusio-mechanical coupling phenomena discussed in the previous Sections. In this Section, we shall extend the previous theory and set up equations which describes the electro-diffusio-mechanical coupling, and analyze the phenomena shown in Figure 10.7.

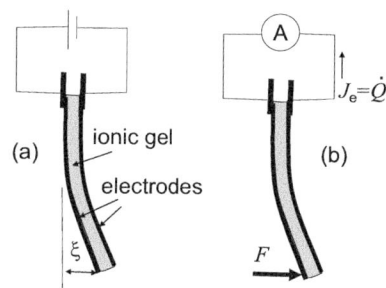

Fig. 10.7 Electro-mechanical tranducer made of an ionic gel. (a) Electric potential ψ_{ext} applied accross the gel induces the bending of the gel. (b) Bending of the gel induces electric current in the outer circuit.

Transport equations in ionic gels

Modeling the dynamics of ionic gels is much more complex than that of non-ionic gels since there are many variables to be determined such as the electric potential, concentration of ions, pH, etc. in addition to the displacement vector $\boldsymbol{u}(\boldsymbol{x},t)$ and pressure $p(\boldsymbol{x},t)$. Although it is possible to set up equations taking all these fields into account [6, 7], analysis of such equations becomes quite complicated. Here we shall discuss the simplest model regarding the gel as a homogeneous conductor with constant electrical conductivity σ_e. This model, originally proposed by de Gennes et al. [8, 9] ignores changes of ion distribution and electric double layer, but the model describes the essential feature of the electro-mechanical coupling.

Let $\psi(\boldsymbol{x},t)$ be the electric potential at position $\boldsymbol{x}$ and time t. In the case of non-ionic gels, solvent flux $\boldsymbol{j}_s$ is caused by the pressure gradient ∇p only. In ionic gels, solvent flux is also caused by the gradient of electric potential $\nabla \psi$, see Figure 10.8. A simple model for the electro-mechanical coupling is therefore

$$\boldsymbol{j}_s = -\kappa \nabla p - \lambda \nabla \psi \qquad (10.73)$$

where λ is a constant representing the electro-mechanical coupling.

Conversely, electric current $\boldsymbol{j}_e$ is induced not only by the gradient of electric potential, but also by the pressure gradient, and the transport equation is written as

$$\boldsymbol{j}_e = -\sigma_e \nabla \psi - \lambda' \nabla p \qquad (10.74)$$

where λ' is another constant representing the electro-mechanical coupling. By the reciprocal relation of Onsager, λ' is equal to λ [8]:

$$\lambda = \lambda'. \qquad (10.75)$$

Since the total charge density is zero everywhere in the ionic gel, the current has to satisfy the condition

$$\nabla \cdot \boldsymbol{j}_e = 0. \qquad (10.76)$$

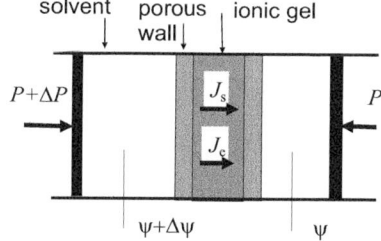

Fig. 10.8 Dynamic electro-mechanical coupling. The gradient of pressure p and the electric potential ψ induces the flux of solvent j_s and the electric current j_e across the ionic gel.

Equations (10.73)-(10.76) together with the force balance equation (10.47) and the volume conservation equation (10.45) give a complete set of equations which describe the electro-mechanical coupling in ionic gels.

464 *Fluid Transport in Gels*

Gel-electrode composite

The above set of equations describe the dynamics of the bulk phase of ionic gels. In order to discuss the phenomena shown in Figure 10.7, it is necessary to consider the boundary condition at the interface between the ionic gel and the electrode. It is important to note that this interface is usually not a flat two-dimensional surface, but a surface having a very complex fractal shape and occupying a three dimensional bulk region. An example is shown in Figure 10.9(a), where the electrodes are made of conductive particles (of carbons or metals) embedded in ionic gel. The particles are electrically connected to each other and form a network of conductors. We call such material gel-electrode composite.

The gel-electrode composite consists of two continuous phases. Both are electrically conductive, but the carriers are different. The carriers in the electrode are electrons, while the carriers in the gel are ions. We assume that the current does not flow across the interface, and therefore an electric double layer is formed at the interface. Figure 10.9(b) shows the model of the electrical property of the gel-electrode composite. Let $q(\boldsymbol{x})$ be the ionic charge density stored at position $\boldsymbol{x}$ of the gel-electrode composite, and c be the capacitance of the double layer per unit volume, then

$$q(\boldsymbol{x}) = c(\psi(\boldsymbol{x}) - \psi_e(\boldsymbol{x})) \qquad (10.77)$$

where $\psi(\boldsymbol{x})$ and $\psi_e(\boldsymbol{x})$ are the electric potentials of the gel and the electrode, respectively. We shall assume that the electrical conductivity in the electrode is very high and that the potential $\psi_e(\boldsymbol{x})$ is constant and is equal to the potential at the outer circuit.

The conservation equation for the ionic charge in the gel-electrode composite is now written as

$$\frac{\partial q}{\partial t} = -\nabla \cdot \boldsymbol{j}_e \qquad (10.78)$$

Note that the electro-neutrality condition is satisfied everywhere in the gel-electrode composite since the ionic charge density is always equal to the electronic charge density in the capacitor.

Equations (10.77) and (10.78) together with the previous equations for ionic gels [equations (10.73),(10.74), (10.47), (10.45)] give a complete set of equations for the dynamics in the gel-electrode composite.

Electro-mechanical coupling in equilibrium

Although the above model gives a complete set of equations for the gel-electrode composite, it has been shown that an additional effect needs to be considered to explain experimental results [7,10]. According to the above model, the equilibrium conformation of the gel-electrode composite is not affected by electric potential. This is seen as follows. At equilibrium, there should be no flux ($\boldsymbol{j}_s = \boldsymbol{j}_e = 0$), and therefore, according to equations (10.73) and (10.74), p and ψ must be constant

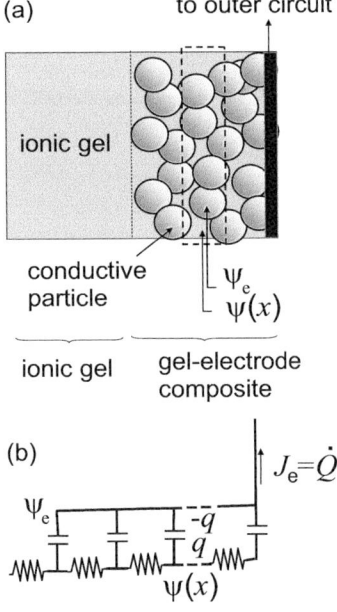

Fig. 10.9 (a) Schematic picture of gel-electrode composite. The conductive particles are embedded in an ionic gel and form a three-dimensional continuous phase of high electrical conductivity. When electric potential is applied, an electric double layer is formed at the surface of conductive particles. (b) Equivalent circuit representing the electrical property of the gel-electrode composite.

throughout the gel. Therefore, the equilibrium conformation of the gel-electrode composite is not affected by the electric potential. This contradicts the experimental observations that gels remain to be bent as far as the electric field is applied [7, 10]. To resolve this discrepancy, it is necessary to consider another electro-mechanical coupling: the static electro-mechanical coupling working in the gel-electrode composite at equilibrium.

When electric field is applied, ions move into (or out of) the gel-electrode composite to charge the electric double layer. Therefore the equilibrium volume of the composite should change depending on the charge density q. The simplest way of considering this effect is to add the following term to the free energy density

$$A_{em} = \frac{1}{2c}q^2 - q(\psi - \psi_e) - \mu w q . \tag{10.79}$$

The first two terms stand for the usual electric energy of the capacitor, and the last term stands for the electro-mechanical coupling.

The total free energy density is then written as

$$A = \frac{1}{2}Kw^2 + \frac{1}{4}G\left(\frac{\partial u_\alpha}{\partial x_\beta} + \frac{\partial u_\beta}{\partial x_\alpha} - \frac{2}{3}w\delta_{\alpha\beta}\right)^2 + \frac{1}{2c}q^2 - q(\psi - \psi_e) - \mu w q \tag{10.80}$$

If the swelling takes place isotropically, the equilibrium condition ($\partial A/\partial w = 0$ and $\partial A/\partial q = 0$) give

$$w = \frac{\mu}{K}q \tag{10.81}$$

$$q = c(\psi - \psi_e) + c\mu w . \tag{10.82}$$

Equation (10.81) indicates that charging the electric double layer in the gel-electrode composite induces the volume change of the material. Equation (10.82) indicates the reverse effect, i.e. volume change of the gel-electrode composite induces the charging of the electric double layer. We shall call such effect static electro-mechanical coupling to distinguish it from the dynamic electro-mechanical coupling represented by equations (10.73) and (10.74). The static electro-mechanical coupling has the result that ions have to move in or out of the gel-electrode composite in order to charge the capacitor. The mechanism has been proposed by several authors.

If the static electro-mechanical coupling effect is taken into account, equation (10.77) is replaced by

$$q(\boldsymbol{x}, t) = c(\psi(\boldsymbol{x}, t) - \psi_e(t)) + c\mu w(\boldsymbol{x}, t) \tag{10.83}$$

and the expression for the stress tensor (10.12) is replaced by

$$\sigma_{\alpha\beta} = Kw\delta_{\alpha\beta} + G\left(\frac{\partial u_\alpha}{\partial x_\beta} + \frac{\partial u_\beta}{\partial x_\alpha} - \frac{2}{3}w\delta_{\alpha\beta}\right) - \mu q \delta_{\alpha\beta} \tag{10.84}$$

The last terms in equations (10.83) and (10.84) stand for the static electro-mechanical coupling.

Examples

Electro-mechanical transducer

Having formulated the basic equations, we now apply the above model to the situation shown in Figure 10.7.

Since the essential effect of the electro-mechanical coupling is taking place in the gel-electrode composite, we consider the device shown in Figure 10.10. Two layers of gel-electrode composite are connected by a very thin layer of ionic gel. We assume that the system has a mirror symmetry with respect to the mid-plane of the device, and therefore the profile of pressure $p(x,t)$, and electric potential $\psi(x,t)$ are odd functions of x, i.e. $p(-x,t) = -p(x,t)$ and $\psi(-x,t) = -\psi(x,t)$. Therefore, we shall limit our analysis in the region of $0 < x < a$, where a is the thickness of the gel-electrode composite.

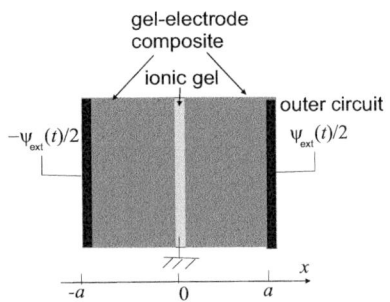

Fig. 10.10 Simple electro-mechanical transducer. Two gel-electrode composites sandwich a very thin layer of ionic gel. The two electrode layers are assumed to be equivalent. The electric potential at the center is taken to be zero.

Summary of equations

For the system shown in Figure 10.10, the basic set of equations become as follows. The conservation equations for solvent volume and ionic charge are written as

$$\frac{\partial w}{\partial t} = -\frac{\partial j_s}{\partial x} \tag{10.85}$$

$$\frac{\partial q}{\partial t} = -\frac{\partial j_e}{\partial x} \ . \tag{10.86}$$

The solvent flux j_s and electric current j_e are given by the coupled transport equations (10.73) and (10.74)

$$j_s = -\kappa \frac{\partial p}{\partial x} - \lambda \frac{\partial \psi}{\partial x} \tag{10.87}$$

$$j_e = -\sigma_e \frac{\partial \psi}{\partial x} - \lambda \frac{\partial p}{\partial x} \ . \tag{10.88}$$

The pressure p is equal to σ_{xx} and is given by equations (10.39) and (10.84):

$$p = \left(K + \frac{4}{3}G\right) w + 2G\xi'' x - \mu q \ . \tag{10.89}$$

The electric potential ψ is given by equation (10.83):

$$\psi = \psi_e + \frac{q}{c} - \mu w \ . \tag{10.90}$$

The boundary conditions for these equations are as follows. Since p and ψ are odd functions of x, they must vanish at $x = 0$, i.e.

$$p = \psi = 0 \quad \text{at} \quad x = 0 \ . \tag{10.91}$$

At $x = a$, the flux must vanish:

$$j_s = j_e = 0 \quad \text{at} \quad x = a \ . \tag{10.92}$$

The electric current J_{ext} flowing in the outer circuit (see Figure 10.9) is given by

$$J_{ext} = \frac{dQ}{dt} \qquad (10.93)$$

where Q is the total ionic charge in the gel-electrode composite

$$Q(t) = \int_0^a dx\, q(x,t) \, . \qquad (10.94)$$

The bending moment of the sheet is given by equation (10.41):

$$M(t) = \frac{8G}{3} a^3 b \xi'' + 4Gb \int_0^a dx w(x,t) x \, . \qquad (10.95)$$

To analyze the situation of Figure 10.7(a), we solve the above set of equations for given $\psi_e = \psi_{ext}/2$ and calculate ξ'' in the force free condition of $M = 0$. On the other hand, to analyze the situation of Figure 10.7(b), we solve the above set of equations for given ξ'' and $\psi_e = 0$ (closed circuit) and calculate $J_{ext} = \dot{Q}(t)$.

Strategy to solve the equations

Solving the above set of equations is still complicated. To simplify the analysis, we make further assumptions.

(a) **Weak coupling:** Since the electro-mechanical coupling is usually weak, it is enough to calculate the first-order effect in the electro-mechanical coupling constants λ and μ. We can therefore use the perturbation method. For example, to calculate the mechanical response to electrical stimuli, Figure 10.7(a), we first solve the electrical response ignoring the electro-mechanical coupling, and then calculate the mechanical response as the first-order perturbation with respect to the coupling constants λ and μ.

If the electro-mechanical coupling effect is ignored, equations (10.85)-(10.90) give the following two diffusion equations for w and q.

$$\frac{\partial w}{\partial t} = D_s \frac{\partial^2 w}{\partial x^2} \qquad (10.96)$$

$$\frac{\partial q}{\partial t} = D_e \frac{\partial^2 q}{\partial x^2} \qquad (10.97)$$

where D_s and D_e are given by

$$D_s = \kappa \left(K + \frac{4}{3} G \right) \qquad (10.98)$$

$$D_e = \frac{\sigma_e}{c} \, . \qquad (10.99)$$

(b) **Different time scales:** Equations (10.96) and (10.97) indicate that the mechanical response time is of the order of $\tau_s \simeq a^2/D_s$, while the electrical response time is of the order of $\tau_e \simeq a^2/D_e$. Usually, τ_e is much shorter than τ_s. We shall make use of this fact in the following analysis.

Mechanical response to electric stimuli

We now analyze the situation shown in Figure 10.7(a). At time $t = 0$, an electric potential $\psi_{ext} = 2\psi_e$ is applied to the outer circuit, then the sheet will bend. We shall calculate the time evolution of the bending of the sheet. The bending is expressed by the curvature of the sheet $\xi'' = d^2\xi/ds^2$. If the sheet is homogeneous along s, ξ'' is independent of s, and the displacement at the bottom of the sheet is given by $\xi(L) = (1/2)\xi''L^2$.

The zero-th order solution, i.e. the time evolution of charge distribution $q(x,t)$ is obtained by solving equation (10.97). Since $q = c(\psi - \psi_e)$ in the zero-th order, the boundary conditions (10.91) and (10.92) give the following

$$q = -c\psi_e \quad \text{at} \quad x = 0 \qquad (10.100)$$

and

$$\frac{\partial q}{\partial x} = 0 \quad \text{at} \quad x = a. \qquad (10.101)$$

The solution is sketched at the top of Figure 10.11(b,c): $q(x,t)$ approaches to the equilibrium solution $q_{eq} = -c\psi_e$ with relaxation time τ_e.

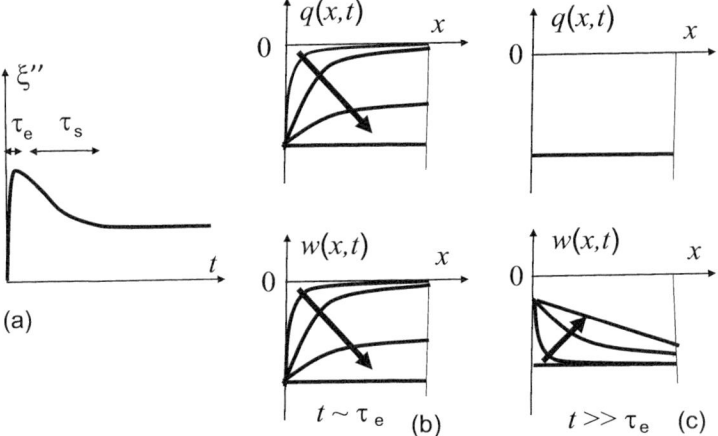

Fig. 10.11 (a) Bending of the sheet when a constant electric potential Ψ_{ext} is applied. (b) Short-time behavior of ionic charge density $q(x,t)$ and volume strain $w(x.t)$. (c) Long-time behavior of ionic charge density $q(x,t)$ and volume strain $w(x.t)$.

The mechanical response to this solution is obtained by calculating the first-order perturbation to this solution. By ignoring the higher order terms with respect to the coupling constants λ and μ, we have the following equations for $w(x,t)$:

$$\frac{\partial w}{\partial t} = D_s \frac{\partial^2 w}{\partial x^2} + \left(\frac{\lambda}{c} - \kappa\mu\right)\frac{\partial^2 q}{\partial x^2}. \qquad (10.102)$$

This equation can be rewritten by equation (10.97) as

$$\frac{\partial w}{\partial t} = D_s \frac{\partial^2 w}{\partial x^2} + \frac{\tilde{\lambda}}{\sigma_e}\frac{\partial q}{\partial t} \qquad (10.103)$$

where

$$\tilde{\lambda} = \lambda - \mu\kappa c. \qquad (10.104)$$

For a given $q(x,t)$, equation (10.103) is easily solved for $w(x,t)$. The solution is simplified by the fact that $\tau_e \ll \tau_s$.

Short time: For $t \simeq \tau_e$, the first term on the right-hand side of equation (10.103) is negligibly small, and equation (10.103) can be approximated as

$$\frac{\partial w}{\partial t} = \frac{\tilde{\lambda}}{\sigma_e} \frac{\partial q}{\partial t} . \tag{10.105}$$

This is solved as

$$w(x,t) = \frac{\tilde{\lambda}}{\sigma_e} q(x,t) . \tag{10.106}$$

Therefore $w(x,t)$ increases from the initial profile $w(x,0) = 0$ to the final profile

$$w(x,t)|_{t=\tau_e} = -\frac{\tilde{\lambda}c}{\sigma_e}\psi_e \tag{10.107}$$

with the relaxation time τ_e as shown in the bottom of Figure 10.11(b).

For given $w(x,t)$, the bending of the sheet ξ'' is obtained by the condition that the bending moment $M(t)$ is zero. By equation (10.95), this condition gives

$$\xi'' = -\frac{3}{2a^3} \int_0^a dx w(x,t) x . \tag{10.108}$$

For equation (10.106), the bending is given by

$$\xi''|_{t=\tau_e} = \frac{3}{4a} \frac{c\tilde{\lambda}}{\sigma_e}\psi_e . \tag{10.109}$$

The quick increase of $w(x,t)$ in the short time is a result of the kinetic coupling constant λ: the solvent flux induced by the electric field causes a quick change of the gel volume and gives a quick bending of the gel sheet.

Long time: For $t \gg \tau_e$, $q(x,t)$ does not change anymore. Hence equation (10.103) reduces to

$$\frac{\partial w}{\partial t} = D_s \frac{\partial^2 w}{\partial x^2} . \tag{10.110}$$

This can be solved for the boundary conditions $p = 0$ at $x = 0$ and $j_e = 0$ at $x = a$, which are written as

$$w = -\frac{\mu c}{K + (4/3)G}\psi_e \quad \text{at} \quad x = 0 \tag{10.111}$$

$$\frac{\partial w}{\partial x} = -\frac{2G}{K + (4/3)G}\xi'' \quad \text{at} \quad x = a \tag{10.112}$$

where ξ'' is given by equation (10.107). As in the problem in Section 10.2, the boundary condition includes an unknown ξ'' which depends on the function to be determined. Such a boundary-value problem can be solved in the same way as in Section 10.2, and the solution is illustrated at the bottom of Figure 10.11(c): $w(x,t)$ relaxes from the initial profile at $t =$

τ_e, to the equilibrium profile with relaxation time τ_s. The equilibrium profile is given by

$$w_{eq}(x) = -\frac{\mu c \psi_e + 2G\xi''_{eq}x}{K + (4/3)G} \qquad (10.113)$$

where ξ''_{eq} is determined by equation (10.107).

$$\xi''_{eq} = \frac{3}{4a}\frac{\mu}{K+(1/3)G}\psi_e. \qquad (10.114)$$

The characteristic behavior of the bending is shown in Figure 10.11(a). When the external potential ψ_e is applied, the sheet bends quickly with relaxation time τ_e, then it slowly relaxes to the equilibrium value with relaxation time τ_s. Usual observation is that the bending shows an overshoot, i.e. $\xi''|_{t=\tau_e} > \xi''_{eq}$, but there is no reason that this should be so. The equilibrium bending ξ''_{eq} can be larger than the first bending $\xi''|_{t=\tau_e}$, or can change the sign.

Electric response to mechanical stimuli

Next we analyze the situation shown in Figure 10.7(b). At time $t = 0$, the sheet is bent (i.e. ξ'' is given), then electric current J_{ext} is induced in the outer circuit. If the resistance of the outer circuit is negligibly small, the external potential is equal to zero. Hence

$$\psi_e = 0. \qquad (10.115)$$

In the following, we focus on the total amount of charge carried by this current

$$Q(t) = \int_0^t dt' J_{ext}(t'). \qquad (10.116)$$

The response of $w(x,t)$ for a given bending ξ'' can be analyzed in the same way as in Section 10.2. The boundary conditions in the present problem are $p = 0$ at $x = 0$ and $j_s = 0$ at $x = a$, which gives the following conditions for w

$$w(x,t) = 0 \quad \text{at} \quad x = 0 \qquad (10.117)$$

$$\frac{\partial w}{\partial x} = -\frac{2G\xi''}{K+\frac{4}{3}G} \quad \text{at} \quad x = a. \qquad (10.118)$$

The solution is shown at the bottom of Figure 10.12(b,c). For given $w(x,t)$, the electric current is given by equations (10.88), (10.89), and (10.90):

$$j_e = -D_e \frac{\partial q}{\partial x} - \lambda \left[\left(K + \frac{4}{3}G\right)\frac{\partial w}{\partial x} + 2G\xi''\right]. \qquad (10.119)$$

Equations (10.86) and (10.119) give an equation for $q(x,t)$. Again, the equation is solved as follows.

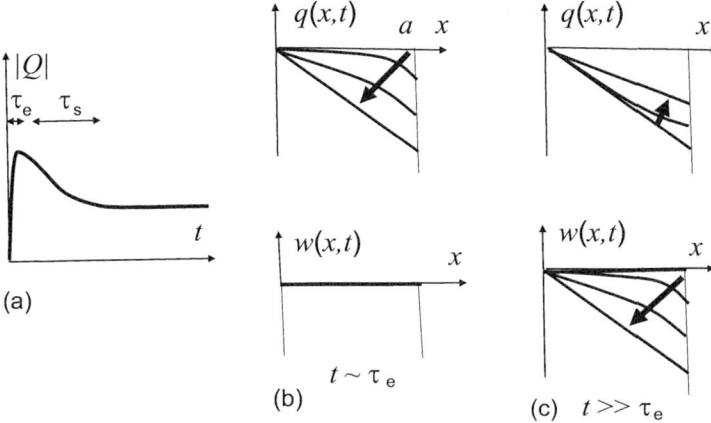

Fig. 10.12 (a) The total amount of charge transported by the induced electric current J_{ext} when the sheet is bent at time $t = 0$. (b) Short time behavior of ionic charge density $q(x,t)$ and volume strain $w(x.t)$, (c) Long time behavior of ionic charge density $q(x,t)$ and volume strain $w(x.t)$.

Short time: In the short time of $t \simeq \tau_e$, $w(x,t)$ is regarded to be zero. Then equations (10.86) and (10.119) give

$$\frac{\partial q}{\partial t} = D_e \frac{\partial^2 q}{\partial x^2} . \tag{10.120}$$

Using equations (10.90) and (10.91), the boundary condition $\psi = 0$ at $x = 0$ now gives

$$q(x, t) = 0 \quad \text{at} \quad x = 0. \tag{10.121}$$

On the other hand, at $x = a$, the boundary condition $j_e = 0$ and equation (10.119) now gives:

$$\frac{\partial q}{\partial x} = -\frac{2G}{D_e} \lambda \xi'' \quad \text{at} \quad x = a \tag{10.122}$$

The solution of equation (10.120) for these boundary conditions and the initial condition $q(x.0) = 0$ gives the solution illustrated at the top of Figure 10.12(b), i.e. $q(x.t)$ approaches to the steady-state solution

$$q(x, \tau_e) = -\frac{2G}{D_e} \lambda \xi'' x \tag{10.123}$$

with relaxation time τ_e. The displaced charge is then given by

$$Q(\tau_e) = \int_0^a dx q(x, \tau_e) = -\frac{a^2}{D_e} \lambda G \xi'' . \tag{10.124}$$

When the sheet is bent, the pressure gradient $\partial p / \partial x \simeq G \xi''$ appears across the gel, and this induces an electric current $\lambda G \xi''$. The total amount of charge carried by this current in time τ_e is $\lambda G \xi'' \tau_e$, which is equal to equation (10.123).

Long time: After a long time $w(x,t)$ changes as it is shown in the bottom of Figure 10.12(c). Since the response of $q(x,t)$ is fast, we may assume that $q(x,t)$ is in a steady state for given $w(x,t)$. The steady state condition $\partial q / \partial t = 0$ under the boundary condition (10.86) gives

$$j_e(x, t) = 0. \tag{10.125}$$

This expression is written by use of equations (10.86) and (10.90) as

$$-D_e \frac{\partial q}{\partial x} + \sigma_e \mu \frac{\partial w}{\partial x} - \lambda \left[\left(K + \frac{4}{3}G \right) \frac{\partial w}{\partial x} + 2G\xi'' \right] = 0. \quad (10.126)$$

Thus $q(x,t)$ is obtained as

$$q(x,t) = c \left[\mu - \frac{\lambda}{\sigma_e} \left(K + \frac{4}{3}G \right) \right] w(x,t) + \frac{2G\lambda}{D_e} \xi'' x . \quad (10.127)$$

The time variation of $q(x,t)$ is illustrated in the top of Figure 10.12(c). At equilibrium, $q_{eq}(x) = c\mu w_{eq}(x)$. This gives

$$q_{eq}(x) = \mu c \frac{2G}{\left(K + \frac{4}{3}G \right)} \xi'' x . \quad (10.128)$$

Hence

$$Q_{eq} = \mu c \frac{G}{\left(K + \frac{4}{3}G \right)} \xi'' a^2 . \quad (10.129)$$

Figure 10.12(a) shows the time dependence of the displaced charge. The behavior of $Q(t)$ is characterized by two time scales. After the initial quick response with relaxation time τ_e, $Q(t)$ approaches to the equilibrium value Q_{eq} with relaxation time τ_s. Again Q_{eq} can be larger or smaller than the initial response $Q(\tau_e)$.

Summary

We have developed a continuum model which describes the coupled phenomena of electric current (j_e), solvent flux (j_s), and deformation of gel network (u). The model is a generalization of the diffusio-mechanical coupling model of non-ionic gels.

Two effects are important in the electro-diffusio-mechanical coupling. One is the dynamic coupling which appears in the transport equation for the solvent flux and electric current. The effect is represented by the parameter λ. The other is the static coupling which is important in the gel-electrode composite, where the equilibrium volume changes with the charge stored in the electric double layer. The effect is represented by the parameter μ.

The model has been applied to a simple system of an electro-mechanical transducer. It is shown that there are essentially two relaxation times, the electrical relaxation time τ_e, and the mechanical relaxation time τ_s, but various types of response behaviors are possible depending on the sign and the magnitude of the coupling constants λ and μ.

In this chapter, we have assumed that there is no chemical reaction taking place at the interface between the gel and the electrode. Modeling the effects of chemical reaction is important in various applications in batteries and fuel cells, but this will be a future work.

Acknowledgment

This work emerged from collaboration with many people. I thank Dr. T. Yamaue and Dr. K. Asaka in formulating the dynamic electro-mechanical coupling. I also thank my former students, Mr. K. Nakamura, Mr. K. Takahashi, and Mr. T. Yonemoto whose work was essential in setting up the static electro-mechanical coupling.

References

[1] M. Doi, *J. Phys. Soc. Jpn.* **78** (2009) 052001
[2] M. Doi, *Introduction to Polymer Physics*, Oxford Univ. Press, (1996).
[3] M. Rubinstein and R. Colby, *Polymer Physics*, Oxford Univ. Press, (2003).
[4] L. D. Landau, L. P. Pitaevskii, A. M. Kosevich and E.M. Lifshitz, *Theory of Elasticity*, Butterworth-Heinemann, (1986).
[5] J.H. Masliyah and S. Bhattacharjee, *Electrokinetic and Colloid Transport Phenomena*, Wiley (2006).
[6] M. Doi, M. Matsumoto and Y. Hirose, *Macromolecules*, **25** (1992) 5504
[7] T. Yamaue, H. Mukai, K. Asaka and M. Doi, *Macromolecules*, **38** (2005) 1349
[8] P.-G. de Gennes, K. Okumura, M. Shahinpoor, J. Kim, *Europhys. Lett.* **50** (2000) 513.
[9] K. Asaka, K. Oguro, Y. Nishimura, M. Mizuhata, H. Takenaka, *Polym. J.* **27** (1995) 436
[10] M. Doi, K. Takahashi, T.Yonemoto, T. Yamaue, *React. Func. Polym.* **73** (2013) 891.

Order and Disorder in the Extracellular Matrix

Astrid van der Horst[1,2] and Gijs J. L. Wuite[1]

[1]Faculty of Sciences/Division of Physics and Astronomy
VU University, 1081 HV, Amsterdam, The Netherlands
[2]Currently at the Department of Radiation Oncology
Academic Medical Center, University of Amsterdam, The Netherlands

The mechanical and structural properties of the tissues in our bodies are in large part determined by the extra-cellular matrix (ECM), the fibrillar network of proteins surrounding the cells. The main component of the ECM is collagen, a scleroprotein or fibrous protein. *In vivo*, collagen self-assembles into fibrils and fibers that supply rigidity and tensile strength to the ECM. The protein is often found together with elastin, another self-assembling ECM scleroprotein, whose properties of elasticity, extensibility, and elastic recoil supplement the mechanical properties of collagen.

Both elastin and collagen self-organize into highly hierarchical structures: molecules form fibrils, which associate into larger fibrils, fibers, and a wide variety of fibrillar networks. Also *in vitro*, in the absence of cells and other proteins, these two proteins each will self-assemble into fibrillar matrices, very similar to their *in vivo* counterparts. It is the architectures of such matrices, more so than the mechanical properties of individual molecules, that determine local mechanical properties, thereby meeting the demands of specific tissues.[1]

The interest from the field of physics in collagen and elastin does not only stem from their unique structural characteristics and mechanical function in the body. In addition, this interest is raised by their intrinsic capacity for hierarchical self-organization, which makes collagen and elastin extremely fitting to function as model systems for the development of biomimetic synthetic biopolymers. In the search for such engineered materials, used for example for tissue engineering, understanding the physical basis of the elastomeric and mechanical properties of these two proteins and their ability to self-organize is crucial. Thus far, however, the enormous potential of these extraordinary molecules has been largely unexplored, due to a lack of understanding of the link between molecular structure and macroscale function.

In the effort to relate peptide sequence to the mechanical properties of materials and their propensity to self-assemble, an increasing and paramount role is being played by nano- and micrometer-scale me-

[1]Certain tissues require very different mechanical properties than others, for example arteries versus intervertebral discs. Differences in the local architectures of elastin and collagen supply these distinctive mechanical qualities.

chanical probing techniques, such as atomic force microscopy (AFM), micro-electro-mechanical systems (MEMS), and optical tweezers (OTs). These techniques enable direct measurement of structural properties at multiple levels in the hierarchical organization, ranging from the single-molecule to the fibril and matrix level. Such measurements, together with computer simulations, allow relating mechanical properties of single molecules to those of fibrils and matrices. This has supplied an immense contribution to obtaining a better and more general comprehension of mechanisms of self-assembly and elastomeric properties of both engineered and natural materials, including spider silks.[2]

[2] The propensity of molecules to self-assemble into fibers and matrices is an extremely useful asset in the engineering of new materials. This makes the proteins collagen and elastin, which possess this quality, particularly suitable as model systems for synthetic polypeptides.

In the structure-function relations of collagen and elastin, the interplay between order and disorder and the influence of this on their structural configurations and properties, is a re-occurring theme. The difference between the functions of these two scleroproteins is reflected in the differences in their structural order. As we will see in this chapter, collagen is highly ordered at all levels of its hierarchical organization and this order plays a key role in its mechanical properties. Elastin, on the other hand, is an intrinsically disordered protein, and it is this lack of order that endows this protein with its unique elastomeric properties.

Due to the complexity of biological systems, the descriptions in this chapter are not inclusive of all detail, but aim at giving a relevant view of the highly regulated self-assembling processes and resulting structural architectures of ECM scleroproteins, thereby highlighting the importance of order and disorder in the functioning of these proteins.

11.1 The building blocks: Protein synthesis and structure

Proteins are comprised of chains of amino-acids that all have the same molecular backbone structure with a length of ~ 0.3 nm per amino-acid residue. However, each of the 21 amino-acids (denoted by either a three-letter or single-letter abbreviation) has a unique side-chain that bestows it with different properties of polarity, charge, size, and configurational flexibility.

The sequence of amino-acids residues is called the *primary structure* of a protein. The amino-acid chains, or polypeptides, will contain local structural motifs, the so-called *secondary structure* of a protein. These motifs are stabilized by hydrogen bonding and include, for example, α-helices, for which the backbone of the polypeptide forms a helix with hydrogen bonds between residues i and $i+4$, β-turns (a single turn with an H-bond between residues i and $i+3$), and β-sheets, which are formed by the association of multiple β-strands (peptide strands in almost fully extended form) into a hydrogen-bonded sheet-like structure.

After synthesis, most proteins will fold into a globular 3-dimensional configuration, the *tertiary structure*. The folding process of these globular proteins is guided and stabilized by hydrophobic interactions and hydrogen bonding and the resulting unique 3D configuration, with local secondary structure, is determined by the peptide sequence of the protein. Scleroproteins, however, such as collagen and elastin, are characterized by a lack of such a predetermined three-dimensional structure and instead maintain their linear, string-like nature.

Before being cross-linked into polymers, collagen and elastin networks self-assemble from their respective soluble monomeric precursors, *tropocollagen* and *tropoelastin*. In this section, we describe the synthesis and structure of these molecular building blocks.

Tropocollagen

The distinguishing feature of a collagen is the presence of a triple helix domain consisting of a repetitive triplet amino-acid motif that includes glycine (Gly). Collagens are found in all vertebrates, and so far over 25 human collagen types have been identified [1]. Based on the extent of the triple helix domain, the supramolecular structure, and its location and function, the collagen superfamily is divided into several subfamilies, including the *classical fibrillar collagens*. The collagens in this subfamily have a large central triple helix region and have the ability to self-assemble into fibrils. A clear overview on the structure of collagens is given in [2]. In view of the objective of this chapter, however, we limit our discussion here to *classical fibrillar collagens*, i.e. types I, II, III, V, and XI.

Triple helix

Collagens consist of polypeptide strands called α-*chains*, which have a Gly-Xxx-Yyy sequence motif. Xxx and Yyy can be any amino-acid, but are often proline. Proline (Pro) at the Yyy position is frequently hydroxilated to form hydroxyproline (Hyp), making Gly-Pro-Hyp the most common triplet present [3]. The chains are left-handed coils and the abundant presence of proline and hydroxyproline stabilizes this helical nature due to steric hindrance of their side-chains [4]. The *quaternary structure* of collagen is a trimer: three α-chains can coil around each other, with a one-residue stagger, into a 300-nm-long right-handed triple helix held together by interchain hydrogen bonds. Glycine is the smallest amino-acid (with a mere hydrogen atom as its side-chain) and the occurrence of this amino-acid residue at every third position permits tight packing of the stands into the 1.6-nm-wide triple helix. [3]

The chains of the classical fibrillar collagens have an uninterrupted $(Gly-Xxx-Yyy)_n$ sequence, with, depending on the type, $n = 337 - 343$. Types II and III are *homotrimeric* collagens: with only one type of α-chain, the molecules consist of three identical chains. The other three classical fibrillar collagens are *heterotrimeric*. Collagen I always has two $\alpha 1$-chains and one $\alpha 2$-chain. Collagen V and XI each have three different α-chains, whereas type V can have a different *chain stoichiometry*: it either consists of two $\alpha 1$-chains and one $\alpha 2$-chain, or of three different chains.

---

[3] Without glycine at every third position in each α-chain (Gly-Xxx-Yyy), three procollagen chains would not be able to properly pack together into a triple helix.

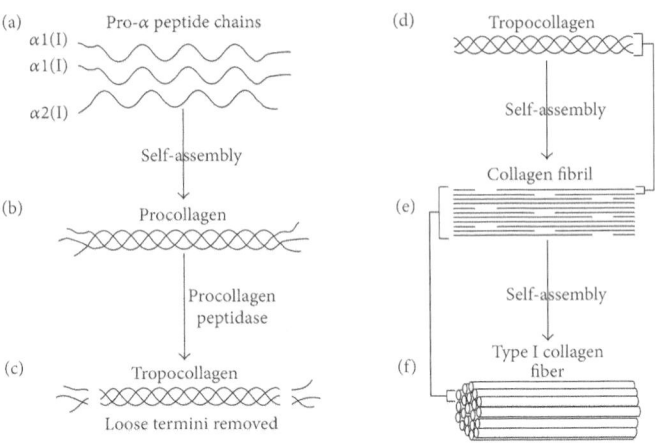

Fig. 11.1 The process of collagen synthesis. (a) For collagen type I, two identical $\alpha 1(I)$ and one $\alpha 2(I)$ peptide chains self-assemble to form procollagen (b). (c) Removal of the propeptides by procollagen peptidase creates a type I tropocollagen molecule (d). Tropocollagen molecules self-assemble to form a growing collagen fibril (e). Self-assembly of collagen fibrils forms a type I collagen fiber (f). Figure taken from [5].

Synthesis

The collagen α-chains are synthesized inside cells. At this stage, the chains contain a globular *propeptide* on either end. These non-helical peptide domains, named C- and N-propeptide, respectively, play a role in chain association. After enzymatic post-translational modification (including hydroxylation of proline and cis-trans isomerization of peptide bonds to ensure the left-handed helical configuration of the chain), three α-chains will associate at the C-propeptide and wind around each other to form a *procollagen*, Figure 11.1(a) and (b). Even though multiple types of collagen might be produced simultaneously by a single cell, each procollagen molecule will be formed of strands of its own type; chain recognition at the C-propeptide ensures correct chain stoichiometry and thus the correct homo- or heterotrimeric combinations.

After chain association, the procollagen molecules are excreted from the cell. The propeptides are cleaved off by enzymes, Figure 11.1(c), resulting in *tropocollagen*, consisting almost entirely of a central triple helix domain, which is, in the case of types I, II, and III, flanked by *telopeptides*, small, non-helical end-domains that are involved in cross-linking. For collagen types V and XI, cleaving of the N-propeptides can be partial or absent. When incorporated in heterotypic fibrils (such as types I/V fibrils in the cornea), the triple helix can be buried inside the fibril, but these large extensions will be located at the surface, thereby obstructing fibril growth and limiting fibril diameter [6].

Thermal stability

The temperature at which human tropocollagen is thermally stable lies several degrees below body temperature [7] and depends on the ratio of proline hydroxylation, a feature exploited in tuning the thermal stability to the different body temperatures of the various species [3, 8]. The importance of helix stability is demonstrated by the severe consequences of a lack of ascorbic acid (vitamin C), which is required for

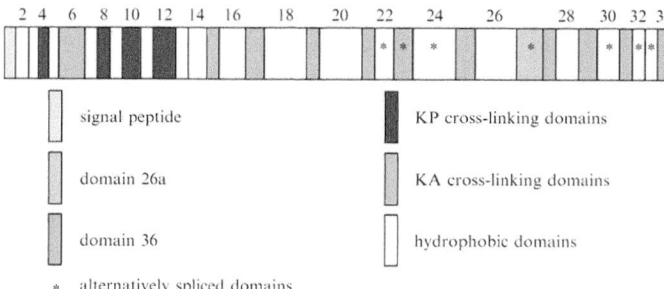

Fig. 11.2 Schematic view of the 34 exon domains of human tropoelastin, showing the typical alternating hydrophobic/cross-linking domain structure of the protein [11].

post-translational hydroxylation; a prolonged absence of this vitamin in the body inhibits collagen synthesis, causing *scurvy*, with symptoms of loss of teeth and open skin wounds [9].

Tropoelastin

Human tropoelastin (hTE) most commonly (depending on splicing) consists of 786 amino-acids, encoded by 34 exons (the regions of a gene that are translated into a peptide sequence) [10]. One of the most prominent and defining features of this protein is a structure of alternating *hydrophobic* and *cross-linking* domains, see Figure 11.2 [11–13].

Within the hydrophobic domains, which are rich in glycine (G), valine (V), proline (P), and alanine (A), highly repetitive sequences can be located, such as the 7-fold PGVGVA repeat in the human exon 24 region and the PGVGV repeat (10× in chicken [14], and 11× in bovine and porcine tropoelastin [15]).

The covalent cross-linking of tropoelastin into polymeric elastin is established through cross-linking of lysine (K) residues. These are found in combination with either alanine (KA domains) or proline (KP domains). Most common in human tropoelastin is the KA cross-linking domain. In such a domain, two lysine residues are located within a longer stretch of alanine residues, separated by two or three A residues (e.g. AAAAAAKAAAKAA in hTE exon 29). The alanine residues are thought to adopt a helical configuration and separation by two or three residues places the two lysines on the same side of this helix, providing the opportunity to form zero-length cross-links of four lysine residues by desmosine bridges [16, 17].

Synthesis

Tropoelastin is synthesized inside cells, where binding of a chaperone protein prevents intracellular aggregation [18]. After being excreted from the cell, the protein, with a contour length of ∼ 180 nm, forms globules half a micrometer in diameter on the cell surface [19], similar to coacervate droplets seen *in vitro*, see Section 11.3. *In vivo*, these droplets are subsequently deposited onto fibrillin microfibrils to form *elastic fibers*, a process mediated by multiple other proteins [20][4]. However, also *in*

[4]Cross-linked elastin is an extremely durable material. In the body, the protein is laid down during developement, lasting a lifetime.

vitro elastin will self-assemble into fibrillar structures, without fibrillin microfibrils or other proteins present.

Elastin is found in nearly all vertebrates and in almost all species a single gene encodes for tropoelastin. Differences in the peptide sequence, however, can be introduced by alternative splicing (meaning that the amino-acid sequence of the protein can differ due to a variation in DNA transcription), which can be tissue dependent and dependent on the stage of development, linking it to function [21].

Recombinantly-made polypeptides

Both for collagen and for elastin it is problematic to obtain their monomeric precursors from tissue, due to the extensive cross-linking of the proteins into insoluble polymers *in vivo*. The introduction of recombinant techniques, however, has opened up the opportunity to produce tropocollagen and tropoelastin using bacteria or cell lines [22].

A large advantage of these recombinant techniques is the possibility to not only obtain monomeric proteins as they are found *in vivo*, but in addition change their amino-acid sequence at will. By mutating the sequence, the effect of certain amino-acids or sequence domains can be investigated in comparative studies *in vitro*. Examples are collagen-like polypeptides containing a $(Gly-Xxx-Yyy)_n$ sequence [23], and elastin-like polypeptides (ELPs) based on certain amino-acid repeats, such as $(GVGVP)_{251}$ [24] or on domains of elastin preserving the alternating domain structure of this protein [25]. Moreover, new polypeptides can be designed, e.g. to be used in tissue engineering [26, 27]. By changing their sequence, the properties of these polypeptides can in principle be tailored to local requirements such as, for example, the elasticity of synthetic skin.

11.2 Hierarchical supramolecular organization

Extracellular matrix scleroprotein structures are built up in a highly hierarchical manner. *In vivo*, their organization is mediated by, amongst other factors, cell configuration, enzymes, and proteins. Also *in vitro* though, supramolecular organization of pure elastin and collagen can take place. In this section, we describe the collagen and elastin protein structures at different levels of their organization.

Collagen organization

Collagen's hierarchical organization ranges from single molecules, with a diameter of ~ 1.6 nm, to 4-nm-wide *microfibrils*, fibrils (20–500 nm), fibers (several μm), Figure 11.1(d-f), and a wide variety of fibrillar structures. Collagen types I (found abundantly in skin, lung, and bone) and II (in cartilage) are the major fibrillar collagens and types III, V, and XI are the minor fibrillar collagens, which are often found together in heterotypical fibers with type I or II [2]. There are differences between the various fibrillar collagen types, but the following description of supramolecular organization of collagen type I can in many ways be regarded as generic for all classical fibrillar collagens.

D-banding

The most outstanding feature of collagen fibrils is their *banded pattern*, with a period D of 67 nm, which is easily visible in electron microscopy (EM) imaging, Figure 11.3 [28]. The association of tropocollagen during fibrillogenesis is an entropy-driven process whereby a pseudo-periodicity of 234 amino-acids in the peptide sequence dictates the preferred 67-nm staggered assembly of adjacent collagen molecules [29].

Since the length of the molecule (~ 300 nm) is equal to $\sim 4.47D$, a gap region of $\sim 0.53D$ is present between longitudinal neighbours. Within a fibril, this creates regions of lower electron density (where a gap is located) and higher electron density (the overlap region), visible in EM as a striped pattern of lighter and darker regions [29]. The dimension of the D-period has also been obtained by X-ray diffraction [32] and AFM [33].

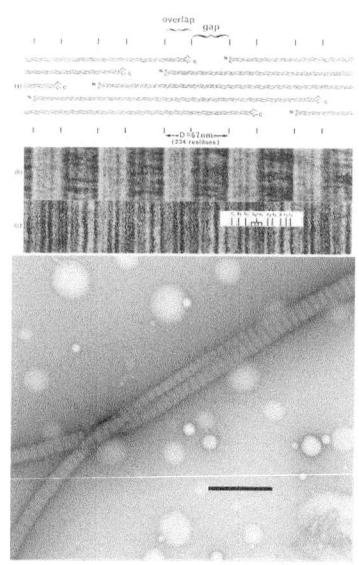

Fig. 11.3 Schematic (top) denoting the gap and overlap regions that produce the D-banding of collagen fibrils visible in the positively and negatively stained EM images (middle) [30]. Bottom: EM image of a collagen fibril displaying D-banding. Scale bar=500 nm [31].

Microfibril

As hypothesized by Smith [28], the smallest three-dimensional configuration that would ensure a $1D$ stagger between a molecule and its neighbours is a hollow cylinder with five molecules in its cross-section. Based on the diameter of a molecule, such a cylinder is expected to measure 4 to 5 nm in width.

The existence of these so-called *microfibrils* as intermediate structures has since been supported by experimental results from X-ray diffraction [34], EM imaging [35], and AFM imaging [33]. Direct observation of fibril growth using AFM showed 3.7 ± 1.3 nm lateral size steps [36], which compares well with the diameter of a single microfibril. X-ray diffraction studies by Wess *et al.* indicated a left-handed stagger of molecules within microfibrils [37]. Isolation of individual microfibrils, however, for example for mechanical testing, has proven difficult, possibly due to cross-linking [38].

Cross-links between adjacent molecules are established in the overlap region, between the C-telopeptide of one molecule and a specific sequence in the triple helix domain of a neighbouring molecule [39]. The 67-nm staggering ensures lining up of these two sites, thereby granting the opportunity for cross-linking, a requirement for mechanical integrity [40].

Lateral organization

From the fact that the D-banding is visible in fibrils over a lateral range larger than the width of a microfibril, it can be concluded that neighbouring microfibrils align axially such that their D-staggers are in register with each other. Details on the cross-sectional organization of molecules within a fibril, however, have long been subject to discussion.

This lateral organization of fibrils is tissue dependent. Especially in tissues such as tendon, the organization is very unidirectional and X-ray diffraction shows that part of the molecules within the fibers is organized in a well-defined three-dimensional (3D) crystal structure [41, 42]. Combining this lateral crystallinity with the deemed existence of the 5-stranded microfibril led to the hypothesis that the lateral organization in such structures is quasi-hexagonal, with the microfibrils compressed to accommodate this quasi-hexagonal lattice [42]. The X-ray diffraction data, however, also indicate that most of the molecules are in a state of liquid-like disorder [41–43].

Looking at the orientational organization of the collagen with fibrils, EM and AFM imaging showed that the microfibrils are tilted with respect to the main axis of the fibril [34, 44] and two models were proposed [44]: a constant-pitch model (pitch independent of radial distance and the angle variable with radius) and a constant-angle model (angle independent of radial distance). 3D tomography revealed that the same tilt angle can be found throughout the fibril [45], confirming the constant-angle model. Depending on the tissue and on collagen types within fibrils, the angle of this tilt is either 5° or 17° [46]. This tilt of molecules within a fibril would also explain the shorter D-period (~ 64 nm) observed in certain tissues; due to the helical configuration, the D-period of 67 nm will appear as $D \times \cos(17°) = 64$ nm [44]. These findings contributed to a proposed configuration in which the left-handed staggered microfibrils are associated into a right-handed helical manner to form quasi-hexagonally packed fibrils; a model consistent with X-ray diffraction patterns [47].

Elastin organization

Like collagen, elastin is characterized by a hierarchical organization. Even though *in vivo* elastin is associated with fibrillin microfibrils that act as a scaffold, tropoelastin will also self-assemble into fibrillar structures *in vitro*. The first step in this process of organization, both *in vivo* and *in vitro*, is coacervation [48, 49].

Coacervation

The term coacervation, introduced by Bungenberg de Jong and Kruyt [50, 51], denotes a temperature-induced liquid-liquid phase separation. Tropoelastin in solution will, upon raising the temperature above its so-called *coacervation temperature* T_c, come out of solution and form elastin-rich droplets, Figure 11.4. This process is reversible; upon lowering the temperature, the tropoelastin will go back into solution.

As mentioned earlier, after excretion from the cell, elastin forms globules at the cell surface. The coacervation temperature T_c depends on concentration, salt concentration, and pH and is 37°C for human tropoe-

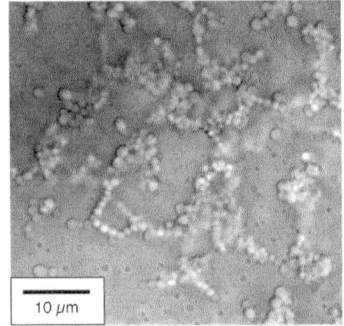

Fig. 11.4 Elastin polypeptide coacervate droplets (with fibrillin) [52].

lastin at physiological salt concentration and pH [53].

The mechanism behind the process of elastin's coacervation is thought to be based on hydrophobic hydration. The ordered, clathrate-like, structures of water molecules surrounding the hydrophobic domains will be disrupted when a certain temperature is reached, allowing these domains to associate. The driving force is the entropic gain from disruption of the ordered water molecules, which outweighs the entropic loss from tropoelastin association [49].

In vitro, coacervation can be observed by the naked eye as an instant transformation of a clear tropoelastin solution to a turbid solution. Dynamic light scattering shows the rapid change from 15 nm diameter single tropoelastin molecules to 4–6 μm diameter droplets [54]. During this process, touching droplets will coalesce, increasing their average diameter. With passing time, however, the droplets will cease to do so and, in addition, will not go back into solution upon lowering the temperature [55]. This stabilization is thought to be due to a maturation process during which the tropoelastin at the surface of the droplet becomes partially organized, as evidenced by CD measurements on ELPs mimicking the alternating domain structure of tropoelastin [52, 56].

Hierarchical supramolecular structures

In vivo, elastin is found in association with 10-12 nm diameter fibrillin *microfibrils* (not to be confused with the aforementioned collagen microfibrils), forming *elastic fibers*, whereby the elastin is located at the core of these fibers and the fibrillin microfibrils lie parallel along their long axis.

Electron microscopy of *in vivo* elastin structures shows a fibrillar substructure of aligned elastin *filaments* of 4 to 5 nm in diameter [57, 58]. In addition, a 4-nm periodicity was seen along these filaments [57]. Filaments have also been observed *in vitro*; a few minutes after coacervation, EM images showed filaments of homogeneous 5 nm width and varying lengths (50–1000 nm) and fibrillar aggregates formed by lateral apposition of such filaments [55]. After several hours above the coacervation temperature, the fibrillar aggregates were the predominant structure. An incubation period of 10–24 hours resulted in larger branched fibers with an amorphous appearance, identical to those seen in normal tissue [57]. However, after short sonication of these larger structures, their fibrillar substructure was visible.

While the debate on the precise mechanism of association and the organization of individual proteins within the elastin structures is ongoing, the forming of 5-nm wide tropoelastin filaments does point in the direction of preferred association of the protein in an end-to-end fashion [55]. In recent work, Baldock *et al.* have used small-angle X-ray scattering and neutron scattering to determine the shape of tropoelastin in solution [59]. Combining this deemed shape with the prevalent connection between exon domains 10, 19, and 25, found in elastin, the authors proposed a head-to-tail model for tropoelastin association [59].

At present, however, it remains unclear how the development from coacervate droplets into tropoelastin filaments takes place.

11.3 Linking structure and function

The mechanical function of the extra-cellular matrix is to supply support and structural integrity under external forces. In the soft ECM, particularly the behavior under *tensile* force is of importance.

In the various tissues in the body, collagen and elastin matrices are often found together. These two proteins differ greatly in their Young's modulus (> 3 orders of magnitude) and complement each other in their function; at low strain, elastin supplies the tissue with remarkable elastic recoil, while at higher strain, the material is given tensile strength by collagen, preventing failure.[5]

The differences in the functions of elastin and collagen are identifiable by the differences in their organization. Both proteins self-assemble into hierarchically organized fibrillar structures. However, while collagen displays utmost order at all levels of this hierarchy, elastin is inherently disordered. In this section, we will describe the relation between structure and function at the lower levels of organization and the role that order and disorder played in both proteins.

The architecture of the networks that are built up from these proteins, including structural parameters such as fibril diameter, primary directionality, and larger-scale 3D configuration, is an important determinant of their function as well. As an example, we start here with a description of collagen's organization in the cornea, to illustrate structure-function relations at the matrix level.

The cornea: Extreme collagen organization

One of the most striking examples of collagen organization can be found in the eye. While being subject to internal pressure and external forces, it is of vital importance to proper vision that the eyeball maintains its shape. Throughout the *sclera*, the white of the eye, tensile strength and resilience is supplied by lamellae of compact collagen fibers. These lamellae branch and interlace extensively and the preferred direction, density, and diameter of the fibers within vary with position, depending on the local mechanical requirements [60].

Part of the outer surface, however, the *cornea*, holds the additional constraint that it must be transparent for visible light. In addition, the *lens shape* of the cornea ought to be very well sustained, as most of the focusing of the incoming light happens here and not at the lens of the eye. Supplying structural integrity while transmitting over 90% of the visible light is achieved by an *exceptionally regulated* organization of the collagen structures present in the cornea [61].

The bulk of the cornea consists of the *corneal stroma*, which is in its ~ 8 mm diameter central region comprised of over 300 layers of parallel-oriented collagen fibrils, consisting mostly of collagen types I and V [62].

---

[5] When a tensile force F is applied to a material, the imposed *tensile stress* σ is the force per unit area and the resulting relative elongation $\Delta L/L_0$ is the dimensionless *strain* ε. The tensile elasticity of a material can be expressed in the tensile elastic modulus, or Young's modulus E, which is the ratio between stress and strain: $E = \sigma/\varepsilon$.

The direction within neighbouring layers alternates between horizontal and vertical, perpendicular to incoming light, Figure 11.5. The fibrils display D-banding and are built up from microfibrils that are tilted over 17° with respect to the fibril axis, Figure 11.6 [63].

A prominent feature is the exceptional monodispersity of the diameter of the fibrils, clearly visible in EM images. In addition, the centre-to-centre distance between fibers is extremely regular. Both diameter and distance are species and age dependent. For the human eye, the diameter has been determined, using X-ray diffraction, to be 31 nm [64], while the centre-to-centre distance was found to be 62 nm [65]. These dimensions, however, are dependent on hydration of the tissue.

The refractive index of the fibrils does not match that of the surrounding medium, so light will be scattered. It is understood that it is the uniform size of the parallel fibrils and the uniform spacing between them that ensure transparency of the corneal stroma through *destructive interference* of scattered light in all but the forward direction [66]. Hart and Farrell showed theoretically that transparency could not only be achieved by perfect crystalline lattices, but also by the quasi-ordered configurations as seen in corneal fibril organization *in vivo* [67].

Thus, by employing different levels of collagen's ordered hierarchical architecture, the cornea has been optimized to transmit visible light. It is of great interest to understand how the monodispersity of the fibril diameter is achieved and how the distance between fibrils and their direction are regulated during fibrillogenesis. This interest stems not in the least from the desire to be able to construct synthetic biomimetic cornea for tissue replacement.

The various techniques used to align collagen fibrils *in vitro* include electrospinning, hydrodynamic flow, and magnetic fields [68]. In the latter method, Torbet *et al.* exploited the inherent diamagnetic anisotropy of collagen to align self-assembled collagen fibrils in a strong (7 Tesla) magnetic field [68]. By rotating the sample over 90° with respect to the field before incubating each subsequent layer, the authors produced a matrix of orthogonally oriented layers of collagen I fibrils displaying D-banding.

In vivo, the presence of collagen type V plays a role in limiting the diameter of the fibrils, but also proteoglycans such as decorin and lumican have been shown to reduce the fibril diameter, both *in vitro* and *in vivo* [61]. In the work of Torbet *et al.*, the addition of decorin and lumican prior to fibrillogenesis indeed reduced the fibril average diameter and diameter polydispersity, thereby improving the transparency of the synthetic matrix [68].

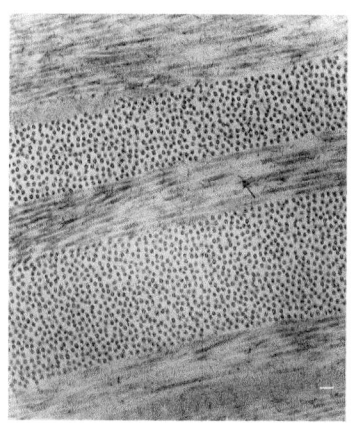

Fig. 11.5 Plywood configuration of collagen in the cornea [61]. In this cross-section, five of the over 300 layers with alternating orientation of the collagen fibrils are visible. The monodispersit and regular spacing of the fibrils allow for light transmission. Scale bar is 100 nm.

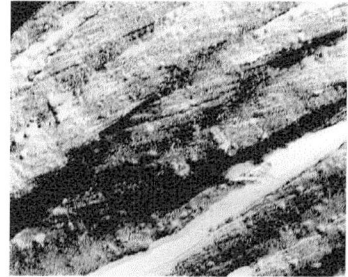

Fig. 11.6 Twisted-rope structure of bovine cornea collagen [63]. Visible is the 17° tilt angle of the microfibrils with the fibril axis.

Collagen structure-function relations

The major function of collagen is to provide rigidity and tensile strength to the ECM. For proper functioning, however, the tissues in our body cannot be completely rigid; when forces are applied, e.g. externally or during movement, tissues must comply to some extent, making bending

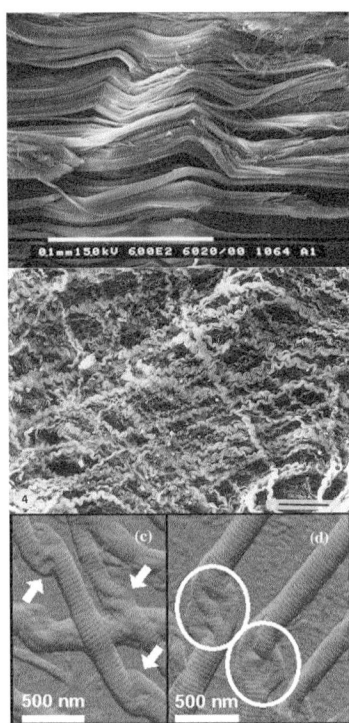

Fig. 11.7 Twisted-rope configuration of collagen. (a) SEM image showing the macroscopic crimp in the Achilles tendon of a rat. Scale bar is 100 μm [70]. (b) Collagen bundles in the small intestine. Scale bar is 50 μm [46]. (c) AFM image of digital tendon fibrils displaying unwinding of the twisted ropes [71].

flexibility and elasticity essential properties. These properties are met by the architecture of the collagen structures, both at the level of molecules organized into microfibrils, as well as in the organization throughout the different hierarchical levels.

Alternating handedness in collagen ropes

In many tissues, collagen fibers are present in a not fully extended form. Instead, they display buckling on a 10–100 μm scale, which, as visible in light microscopy, is stretched out when low forces are applied, Figure 11.7(a) [69].

This macroscopic crimp has been recognized as part of a larger pattern, namely that of a wire rope-like build-up of collagen structures with alternating helical handedness, spanning five successive hierarchical levels [63, 70, 71]. The α-chains are *left*-handed helices that twist together into a *right*-handed tropocollagen triple helix. Five of such tropocollagen molecules assemble together into a *left*-handed staggered microfibril, which in turn is arranged into *right*-handed fibrils. At the next level, multiple fibrils twist around each other to form larger *left*-handed fibrils, which can be seen in collagen structures in, e.g. tendons and ligaments, Figure 11.7.

Such twisted-rope construction elements are known to maximize axial stiffness while minimizing bending stiffness. The alternating handedness supplies torsional stability to the twisted rope and a tensile force will compact the strands, increasing their lateral interaction. In addition, a rupture in one of the strands will be stopped at the strand boundary and not propagate throughout the entire structure. This combination of mechanical properties is evidently what made twisted-rope configuration so extensively used also in human-made products, ranging from rope and yarn to the metal cables of suspension bridges.

Microscopic crimp

After the collagen fibers are straightened out, further stretching of tissue will affect the collagen organization at lower levels in its hierarchy, starting with the collagen molecules within the microfibers.

Although different methods to determine tropocollagen's stiffness have given varying results, direct measurement of force-extension curves for single molecules have shown tropocollagen to be very flexible [72, 73]. These measurements have been conducted using optical tweezers, which enable the recording of applied picoNewton-scale forces while manipulating the end-to-end distance of a single molecule [74]. By fitting a worm-like chain (WLC) model to force-extension curves obtained with optical tweezers, the persistence lengths L_P (a measure for stiffness) for collagen type I and type II have been determined to be 14.5 ± 7.3 nm [72] and 11.2 ± 8.4 nm [73], respectively. Molecular dynamics modelling of the stretching of tropocollagen has predicted an L_P of 16 nm, corroborating these experimental values and indicating tropocollagen to be a flexible string with considerable entropic elasticity [75], i.e. upon stretching, the material loses entropy resulting in a restoring force.

Associated and cross-linked into microfibrils, however, the molecules are packed in an extended configuration, increasing order and limiting this entropic potential [75]. Yet a certain amount of disorder remains present within the microfibrils, as indicated by diffuse scatter seen in X-ray diffraction measurements [37]. This disorder is thought to mostly occur in the gap region, where the packing density and the content of proline and hydroxyproline are lower relative to the overlap region [76]. Under mechanical load, this molecular crimp is straightened out. Thus, the entropic disorder in the gap region is reduced, as shown by time-resolved small-angle X-ray diffraction measurements on tendon during the application of force [77,78]. As a consequence, collagen fibrils display entropic elasticity under moderate strain.

After straightening of both fibrillar and molecular crimp, no additional entropic extension is possible. Further increase in strain will lead to stretching of the collagen triple helices and sliding of the molecules within the fibrils, implying stretching of the cross-links, and finally breaking of the structure [78].

Elastin structure-function relations: Maintaining disorder

The elasticity provided by elastin to the ECM is widely accepted to be entropy-based, although the exact molecular basis of this elasticity has not yet been uncovered [11]. The importance of *configurational entropy* of the peptide chain for elastin's functioning, however, is evident [59,79,80], establishing the presence of extensive configurational disorder as a prerequisite for the performance of the protein.

Early models of elastin's organization have, therefore, described elastin as amorphous or rubber-like; random chains are cross-linked into a disordered material. This, however, is at odds with both the *in vivo* and *in vitro* observed substructure of 5-nm filaments within elastin fibers [55,57]. For the self-organization of tropoelastin into coacervates and subsequent fibrillar structures, a certain level of order is required; the proteins must associate and be presented with opportunity for cross-linking [48]. The need for such order is reflected in several other models describing tropoelastin association, including tiled structures [81] and head-to-tail stacking of monomers [59].

Although thus far none of the proposed models has been widely accepted, combined experimental and computational research has provided clues as to how these apparent contradicting demands of order and disorder are balanced and met by the primary structure of tropoelastin.

Secondary structure in tropoelastin

As mentioned earlier, most proteins possess a large amount of secondary order motifs that, upon hydrophobic collapse, will fold into a well-defined 3D structure. Tropoelastin, however, will not undergo this kind of folding and, in contrast, will continue to display a high level of backbone flexibility and low amounts of secondary order. The order present is short-ranged and rich in poly-proline-II structure (a helical configuration lacking H-bonds) and transient β-turns [80, 82]. Also when cross-linked into an elastin matrix, the protein is highly flexible, as shown by solid-state NMR, which indicates an absence of α-helices and β-sheets in elastin [83].

The lack of long-range order, combined with elastin's insolubility, inhibits the use of structural determination techniques such as X-ray crystallography and solution NMR. The use of computer simulation techniques, however, has been valuable in providing additional pointers regarding the basis of elastin's disorder [80].

Maintaining disorder

Using molecular dynamics simulations, Rauscher *et al.* have looked at the behavior of ELPs that consist of different repetitive hydrophobic domain motifs found in tropoelastin, such as PGV and GVA [80]. By comparing different combinations of sequences, e.g. $(PGVPGV)_6$, $(GVAPGV)_6$, and $(GVAGVA)_6$, the authors investigated the influence of certain amino-acids on the propensity of the polypeptide to form elastomeric or amyloid structures, the latter of which are stiff structures containing cross-β-sheets [84]. Their results show that elastomer-forming polypeptides are characterized by higher levels of backbone hydration and conformational disorder and that these properties are determined by glycine (G) and proline (P) content. The effect of these two amino-acids can be explained by their conformational flexibility.

The most important determinant is proline, which displays exceptional conformational *rigidity*, it being the only amino-acid whose side-chain (a cyclic structure) is attached to its backbone structure. As a consequence of this steric constraint, proline functions as a breaker of secondary structure including β-sheets; its fixed backbone configuration is incompatible with the extended backbone configuration of β-strands. The reduced ability to form hydrogen-bonded turns and β-sheets increases backbone hydration.

Glycine, on the other hand, is highly flexible as a result of its extremely small side-chain. In an aqueous environment, this flexibility prevents the forming of secondary structures due to the high entropic penalty of constraining the backbone within hydrogen-bonded turns.

The findings from simulations are in agreement with experimental data for several of the investigated recombinant polypeptides [25, 80]. A comparison with other elastomeric and amyloidogenic structures found in nature, including resilin (an elastomer found in, e.g. insect wing hinges) and several spider silks, points in the direction of a combined glycine/proline content threshold $[(2P+G) > 0.6]$ above which a protein will display elastomeric assembly [80][6]. In hTE, the residues proline and glycine make up 14% and 35% of the hydrophobic domain sequences, respectively [85].

---

[6]The proline and glycine content of different spider silks differs as each type requires its own specialized set of mechanical properties. The structural treads of the web demand toughness, whereas the captive thread requires more elasticity, not to break when a prey enters the web.

Tropoelastin aggregation

For tropoelastin to associate into a fibrillar matrix through the process of coacervation, some ordering of the protein must take place; the cross-linking domains have to be lined up to facilitate cross-linking [86].

The proline residues in the hydrophobic domains of tropoelastin, important for maintaining disorder, are not equally distributed throughout the polypeptide. Near the C-terminus, the average proline spacing is significantly higher than the 4 to 8 residue average in the rest of the protein, especially in exon domain 30, with a proline spacing of 16 residues in human and 30 in rat tropoelastin [85].

By selective deletion of exon regions of the protein, Kozel *et al.* identified exon 30 as a major determinant in tropoelastin assembly [87]. Protein lacking exon 30 showed a large decrease in fiber assembly *in vitro*. The mechanism behind this is thought to rely on the forming of β-sheet structures and by itself exon 30 indeed forms amyloidogenic structures with high β-structure content.

Using ELPs, Muiznieks and Keeley have experimentally investigated the effect of proline content and proline spacing on the propensity of polypeptides to coacervate and self-assemble into aggregates [85]. The authors found that an increase in proline spacing decreased the ability of polypeptides to reversibly self-associate. The coacervate droplets that formed were smaller, clustered into dense networks, and were enriched in β-structure. These results are in agreement with the effects seen for proline to glycine point mutations in ELPs, which were shown to promote the formation of amyloid-like fibers [25].

Collagen and elastin in concert: Complementary mechanical properties

In many tissues, elastin and collagen are found together, each fulfilling its own specific role in the overall mechanical properties of these tissues. Their complementary roles are well illustrated by the measured force-extension response of the aortic wall, which consists mostly of elastin, collagen and smooth muscle cells.

With every heart beat, the aortic wall, in which collagen fibrils and elastin lamellae are arranged in a wavy concentric manner, expands and contracts. To investigate the basis of the stress-strain behavior of the aorta, Sokolis *et al.* combined macroscopic tensile experiments with light microscopy to image the changes in arrangement and orientation of the collagen and elastin structures during these experiments [88].

The authors found that the measured elastic modulus (tangent of the σ/ε curve) increased non-linearly with strain and three distinct sections could be recognized. At low stress ($\sigma < 40\,\mathrm{kPa}$), the elastic modulus increased non-linearly up to $\sim 175\,\mathrm{kPa}$ at $\varepsilon = \sim 0.3$. From the concurrently obtained microscopy images it could be seen that in this regime the wavy, crumpled elastic fibers were stretched out until taut, while the collagen fibers reoriented in the direction of stress but remained

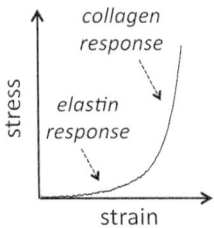

Fig. 11.8 Stress-strain curve representative of tissues containing both collagen and elastin. Elastin is responsible for the initial softer response and collagen for the latter stiffer response. Figure adapted from Muiznieks and Keeley [89].

corrugated, i.e. showed macroscopic crimp.

In the second regime ($40\,\mathrm{kPa} < \sigma < 120\,\mathrm{kPa}$), the collagen started to reorient and straighten more, while the modulus increased linearly with strain up to 300 kPa at $\varepsilon = 0.7$. In the third section ($\sigma > 120\,\mathrm{kPa}$), the linear increase with strain was much steeper, up to a modulus of 5 MPa at a strain of ~ 1, with the collagen appearing straight, densely packed and uncoiling.

From this it was concluded that for the aorta, the elastic modulus at low and high stresses is determined mostly by elastin and collagen, respectively, while in the intermediate (physiological) regime both proteins contribute, Figure 11.8 [88].

Relating such macroscopic results, from tissues or isolated protein matrices, to the mechanical properties of the fibrils and molecules they are built up from, will help uncover the basis of the overall structural properties. This, however, is not an easy and straightforward task.

Direct mechanical probing of proteins at the molecular and fibrillar level has been made possible with the development of nano-manipulation techniques such as AFM [90], OTs [74], and MEMS. These techniques, each with different experimental properties of force (and distance) range and resolution making them suitable for a variety of experimental configurations, can be used to obtain properties such as elastic moduli, yield strain (the strain at which permanent deformation of the material begins and the strain ceases to be completely elastic), and fracture stress (the maximum stress the material can withstand).

AFM has been used for single-molecule tensile experiments on collagen [91], ELPs based on elastin's hydrophobic domains [92], and tropoelastin [59]. Fitting a WLC model to the force-extension curves of tropoelastin yielded an L_P of 0.36 ± 0.14 nm [59]. This corresponds to a Young's modulus E of 2.9 kPa, assuming a diameter of 10 nm for the molecule. With the modulus depending strongly on the radius R of the molecule ($E \sim R^{-4}$), uncertainty in R makes it difficult to obtain an accurate Young's modulus from such single-molecule results. From the force-extension curves measured with OTs for collagen type I, a Young's modulus of 0.35 to 12.2 GPa, depending on the assumed molecular radius, was found [72].

AFM tensile and nano-indentation measurements have been performed on collagen fibrils [93–96], while Koenders et al. performed nano-indentation on native elastic fibers of which the fibrillin had been removed (leaving the elastin), and found the Young's modulus of 0.3–1.5 MPa [97].

For collagen, AFM indentation yielded a modulus of 1 GPa [95] and AFM pulling experiments on fibrils yielded elastic moduli of 32 MPa (in air) [93] and 0.2 GPa (low strain) to 0.5 GPa (4% strain) [99]. In addition, tensile measurements on collagen fibrils have been performed using MEMS, Figure 11.9 [98, 100, 101], resulting in an elastic modulus of 0.47 ± 0.41 GPa (low strain) in vitro [98].

The results mentioned here clearly show the orders of magnitude difference between the tensile elastic modulus of elastin and that of collagen. It is also clear, however, that different measurements can yield

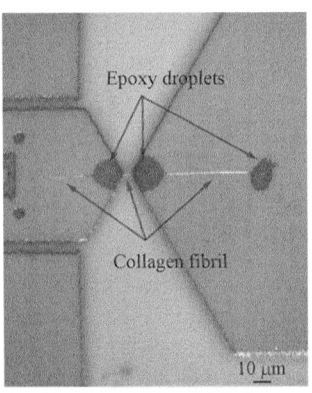

Fig. 11.9 Micro-electro-mechanical system (MEMS) used for an in vitro fracture test of a collagen fibril. The fibril is fixed to the stationary (right) and moveable (left) pads [98].

greatly differing results. Aside from different properties being measured (indentation versus tensile) and different experimental circumstances affecting the results (e.g. hydration or cross-link density), the strain at which the experiments are performed is of great importance, as the elastic modulus is strain dependent [40].

When comparing results between the various levels of organization, a decrease in stiffness can be seen going from single molecule to collagen fibril and subsequently tendon [102]. Softening of fibrils can be explained in part by the weak intermolecular interactions [40], while for fibers and tissues (tendon) the interaction between fibrils, with the viscoelastic behavior of proteoglycans, should be taken into consideration.

Conclusion

The proteins collagen and elastin possess many similarities: both are extracellular matrix scleroproteins that self-assemble into hierarchical fibrillar matrices, *in vivo* and *in vitro*. Their structural function in the body, however, differs greatly. While collagen provides tensile strength and resilience, elastin supplies elasticity and elastic recoil. This distinct difference in function is reflected in the organization of the monomeric precursors within the polymeric matrices.

Collagen displays an exceptionally high degree of order at all levels of its organization, ranging from the rigorous $(Gly\text{-}Xxx\text{-}Yyy)_n$ structure of the α-strands, to the D-banded microfibrils packed into semi-hexagonal liquid-crystalline fibrils, all with alternating handedness of twisted-rope configurations at subsequent levels of organization. This strict configurational ordering is necessary for the functioning of the collagen matrix, supplying stiffness and ensuring its structural integrity. These properties are balanced by the low-strain elasticity bestowed on the material by the molecular disorder within the collagen microfibrils.

In elastin, on the other hand, the imposed local propensity for secondary order appears to be a requirement for assembly, through the process of coacervation, into an elastomeric matrix. This presence of order is combined with a high level of inherent disorder of the protein, a requirement for elastin to function as an entropic elastomer.

Interestingly, in both proteins, the amino-acids proline and glycine play a key role in the balance between order and disorder due to their distinct configurational properties. In collagen, the small size of glycine, located at every third position in the primary structure, allows the packing into the molecular triple helix configuration. As a consequence of the constricted backbone configuration of proline and even more so hydroxyproline, the thermal stability of this triple helix depends strongly on its proline content and proline hydroxylation ratio. The lower proline and hydroxyproline content in the gap regions of the tropocollagen makes the molecules locally more flexible, allowing for the small amount of localized disorder.

In elastin, the conformational constraints imposed by proline inhibit extensive secondary structure, ensuring enough disorder for entropic elasticity. A similar effect on order is endowed by glycine, whose large flexibility frustrates secondary order.

Evidently, larger-scale mechanical properties are determined by the manner in which the molecules are assembled into fibril and fibers. This, in turn, is determined by the primary structure of the proteins. The sequence of amino-acids, their number and specific location within the protein, determine the flexibility, (lack of) secondary structure, alignment, and with that the ability and propensity of the protein to organize into cross-linked matrices with very well-regulated amounts of order and disorder.

Computer simulation techniques have greatly advanced the investigation into the link between the structure of the proteins at the molecular level and their function at the matrix level. Physical measurements to verify the outcome of such calculations, however, remain crucial. In this, the application of micro- and nano-manipulation techniques has played an increasingly important role. The ability to mechanically probe the protein assemblies at the intermediate levels of organization—from single molecules using OTs and AFM, to fibrils with AFM and MEMS—has supplied otherwise unattainable details on the structural and elastomeric properties of elastin and collagen microfibrils and fibers. This information has furthered the establishment of the relation between amino-acid sequence and structural properties at higher hierarchical levels.

An increased knowledge of the relation between the sequence of collagen and elastin and their structure and function at higher levels of organization, will allow directed designing of materials based on their sequences.[7] Supplementing the wide variety of available measurements techniques, including spectroscopy, micromanipulation, and NMR, with recombinant cloning techniques and computer simulations, has proven to be vital in these efforts to establish structure-function relations for elastin and collagen.

Acknowledgment

This work is part of the research programme of the Foundation for Fundamental Research on Matter (FOM), which is part of the Netherlands Organisation for Scientific Research (NWO).

References

[1] Ricard-Blum, S. and Ruggiero, F. (2005). *Pathologie Biologie*, **53**, 430–442.

[2] Hulmes, D. J. S. (2008). In *Collagen: Structure and Mechanics* (ed. P. Fratzl), pp. 15–47. Springer, New York.

[3] Ramshaw, J. A. M., Shah, N. K., and Brodsky, B. (1998). *J. Struct. Biol.*, **122**, 86–91.

[4] Schimmel, P. R. and Flory, P. J. (1968). *J. Molec. Biol.*, **34**, 105–120.

---

[7] *De novo* materials, based on elastin and/or collagen sequences, are being developed for tissue engineering. Such synthetic materials can assist or replace existing tissues in the body, e.g. blood vessels or skin.

[5] Kruger, T.E., Miller, A. H., and Wang, J. (2013). *Sci. World J.*, **2013**, 812718.
[6] Birk, D. E., Fitch, J. M., and Linsenmayer, T. F. (1986). *Invest. Ophthalmol. Vis. Sci.*, **27**, 1470–1477.
[7] Leikina, E., Mertts, M. V., Kuznetsova, N. *et al.* (2002). *Proc. Natl. Avad. Sci. USA*, **99**, 1314–1318.
[8] Privalov, P. L., Tiktopulo, E. I., and Tischenko, V. M. (1979). *J. Molec. Biol.*, **127**, 203–216.
[9] van B. Robertson, W. (1964). *Biophys. J.*, **4**, 93–106.
[10] Muiznieks, L. D., Weiss, A. S., and Keeley, F. W. (2010). *Biochem. Cell Biol.*, **88**, 239–250.
[11] Mithieux, S. M. and Weiss, A. S. (2005). *Adv. Protein Chem.*, **70**, 437–461.
[12] Mithieux, S. M., Tu, Y., Korkmaz, E. *et al.* (2009). *Biomaterials*, **30**, 431–435.
[13] Wise, S. G. and Weiss, A. S. (2009). *Intl. J. Biochem. Cell Biol.*, **41**, 494–497.
[14] Bressan, G. M., Argos, P., and Stanley, K. K. (1987). *Biochemistry*, **26**, 1497–1503.
[15] Raju, K. and Anwar, R. A. (1987). *J. Biol. Chem.*, **262**, 5755–5762.
[16] Foster, J. A., Bruenger, E., Rubin, L. *et al.* (1976). *Biopolymers*, **15**, 833–841.
[17] Tamburro, A. M., Bochicchio, B., and Pepe, A. (2003). *Biochemistry*, **42**, 13347–13362.
[18] Hinek, A. and Rabinovitch, M. (1994). *J. Cell Biol.*, **126**, 563–574.
[19] Kozel, B. A., Rongish, B. J., Czirok, A. *et al.* (2006). *J. Cell. Physiol.*, **207**, 87–96.
[20] Wagenseil, J. E. and Mecham, R. P. (2007). *Birth Defects Res. Part C*, **81**, 229–240.
[21] Heim, R. A., Pierce, R. A., Deak, S. B. *et al.* (1991). *Matrix*, **11**, 359–366.
[22] Martin, S. L., Vrhovski, B., and Weiss, A. S. (1995). *Gene*, **154**, 159–166.
[23] Olsen, D., Yang, C., Bodo, M. *et al.* (2003). *Adv. Drug Delivery Rev.*, **55**, 1547–1567.
[24] Urry, D. W., Hugel, T., Seitz, M. *et al.* (2002). *Phil. Trans. Royal Soc. B*, **357**, 169–184.
[25] Miao, M., Bellingham, C. M., Stahl, R. J. *et al.* (2003). *J. Biol. Chem.*, **278**, 48553–48562.
[26] Daamen, W. F., Veerkamp, J. H., van Hest, J. C. M. *et al.* (2007). *Biomaterials*, **28**, 4378–4398.
[27] Kyle, S., Aggeli, A., Ingham, E. *et al.* (2009). *Trends Biotech.*, **27**, 423–433.
[28] Smith, J. W. (1968). *Nature*, **219**, 7307–7312.
[29] Hodge, A. J. and Petruska, J. A. (1963). In *Aspects of Protein Structure* (ed. G. N. Ramachandran), pp. 289–300. Academic Press, New York.

[30] Kadler, K. E., Holmes, D. F., Trotter, J. A. *et al.* (1996). *Biochem. J.*, **316**, 1–11.
[31] Kadler, K. E., Holmes, D. F., Graham, H. *et al.* (2000). *Matrix Biol.*, **19**, 359–365.
[32] Fraser, R. D. B., MacRae, T. P., Miller, A. *et al.* (1983). *J. Molec. Biol.*, **167**, 497–521.
[33] Baselt, D. R., Revel, J.-P., and Baldeschwieler, J. D. (1993). *Biophys. J.*, **65**, 2644–2655.
[34] Hulmes, D. J. S., Jesior, J.-C., Miller, A. *et al.* (1981). *Proc. Natl. Acad. Sci. USA*, **78**, 3567–3571.
[35] Veis, A., Miller, A., Leibovich, S. J. *et al.* (1979). *Biochim. Biophys. Acta*, **576**, 88–98.
[36] Cisneros, D. A., Hung, C., Franz, C. M. *et al.* (2006). *J. Struct. Biol.*, **154**, 232–245.
[37] Wess, T. J., Hammersley, A. P., Wess, L. *et al.* (1998). *J. Struct. Biol.*, **122**, 92–100.
[38] Fratzl, P. (ed.) (2008). *Collagen: Structure and Mechanics.* Springer, New York.
[39] Eyre, D. R., Paz, M. A., and Gallop, P. M. (1984). *Ann. Rev. Biochem.*, **53**, 717–748.
[40] Buehler, M. J. (2008). *J. Mech. Behavior Biomed. Mater.*, **1**, 59–67.
[41] Fratzl, P., Fratzl-Zelman, N., and Klaushofer, K. (1993). *Biophys. J.*, **64**, 260–266.
[42] Hulmes, D. J. S., Wess, T. J., Prockop, D. J. *et al.* (1995). *Biophys. J.*, **68**, 1661–1670.
[43] Prockop, D. J. and Fertala, A. (1998). *J. Struct. Biol.*, **122**, 111–118.
[44] Raspanti, M., Ottani, V., and Ruggeri, A. (1989). *Intl. J. Biol. Macromol.*, **11**, 367–371.
[45] Holmes, D. F., Gilpin, C. J., Baldock, C. *et al.* (2001). *Proc. Natl. Avad. Sci. USA*, **98**, 7307–7312.
[46] Ottani, V., Raspanti, M., and Ruggeri, A. (2001). *Micron*, **32**, 251–260.
[47] Orgel, J. P. R. O., Miller, A., Irving, T. C. *et al.* (2001). *Structure*, **9**, 1061–1069.
[48] Vrhovski, B. and Weiss, A. S. (1998). *Eur. J. Biochemistry*, **258**, 1–18.
[49] Yeo, G. C., Keeley, F. W., and Weiss, A. S. (2010). *Adv. Coll. & Interface Sci.*, **167**, 94–103.
[50] Bungenberg de Jong, H. G. and Kruyt, H. R. (1929). *Proc. Koninklijke Nederlandse Akad. Wetenschap.*, **32**, 849–856.
[51] Bungenberg de Jong, H. G. (1949). In *Colloid Science* (ed. H. R. Kruyt), Volume II. Elsevier, Amsterdam.
[52] Cirulis, J. T., Bellingham, C. M., Davis, E. C. *et al.* (2008). *Biochemistry*, **47**, 12601–12613.
[53] Vrhovski, B., Jensen, S., and Weiss, A. S. (1997). *Eur. J. Biochem.*, **250**, 92–98.

[54] Clarke, A. W., Arnspang, E. C., Mithieux, S. M. et al. (2006). Biochemistry, **45**, 9989–9996.
[55] Bressan, G. M., Pasquali-Ronchetti, I., Fornier, C. et al. (1986). J. Ultrastruct. Molec. Struct. Res., **94**, 209–216.
[56] Urry, D. W., Starcher, B., and Partridge, S. M. (1969). Nature, **222**, 795–796.
[57] Gotte, L., Giro, M. G., Volpin, D. et al. (1974). J. Ultrastruct. Res., **46**, 23–33.
[58] Pasquali Ronchetti, I., Alessandrini, A., Baccarani Contri, M. et al. (1998). Matrix Biol., **17**, 75–83.
[59] Baldock, C., Oberhauser, A. F., Ma, L. et al. (2011). Proc. Natl. Avad. Sci. USA, **108**, 4322–4327.
[60] Watson, P. G. and Young, R. D. (2004). Exp. Eye Res., **78**, 609–623.
[61] Meek, K. M. (2008). In Collagen, Structure and Mechanics (ed. P. Fratzl), pp. 359–396. Springer, New York.
[62] Meek, K. M. and Boote, C. (2004). Exp. Eye Res., **78**, 503–512.
[63] Ottani, V., Martini, D., Franchi, M. et al. (2002). Micron, **33**, 587–596.
[64] Meek, K. M. and Leonard, D. W. (1993). Biophys. J., **64**, 273–280.
[65] Gyi, T. J., Meek, K. M., and Elliott, G. F. (1988). Intl. J. Biol. Macromol., **10**, 265–269.
[66] Maurice, D. M. (1957). J. Physiol., **136**, 263–286.
[67] Hart, R. W. and Farrell, R. A. (1969). J. Opt. Soc. Amer., **59**, 766–774.
[68] Torbet, J., Malbouyres, M., Builles, N. et al. (2007). Biomaterials, **28**, 4268–4276.
[69] Diamant, J., Keller, A., Baer, E. et al. (1972). Phil. Trans. Royal Soc. B, **180**, 293–315.
[70] Franchi, M., Ottani, V., Stagni, R. et al. (2010). J. Anatomy, **216**, 301–309.
[71] Bozec, L., van der Heijden, G., and Horton, M. (2007). Biophys. J., **92**, 70–75.
[72] Sun, Y.-L., Luo, Z.-P., Fertala, A. et al. (2002). Biochem. Biophys. Res. Comm., **295**, 382–386.
[73] Luo, Z.-P., Sun, Y.-L., Fujii, T. et al. (2004). Biorheology, **41**, 247–254.
[74] Neuman, K. C. and Block, S. M. (2004). Rev. Sci. Instr., **75**, 2787–2809.
[75] Buehler, M. J. and Wong, S. Y. (2007). Biophys. J., **93**, 37–43.
[76] Fraser, R. D. B. and Trus, B. L. (1986). Biosci.Rep., **6**, 221–226.
[77] Misof, K., Rapp, G., and Fratzl, P. (1997). Biophys. J., **72**, 1376–1381.
[78] Fratzl, P., Misof, K., and Zizak, I. (1997). J. Struct. Biol., **122**, 119–122.
[79] Hoeve, C. A. J. and Flory, P. J. (1974). Biopolymers, **13**, 677–686.
[80] Rauscher, S., Baud, S., Miao, M. et al. (2006). Structure, **14**, 1667–1676.

[81] Gray, W. R., Sandberg, L. B., and Foster, J. A. (1973). *Nature*, **246**, 461–466.

[82] Debelle, L. and Tamburro, A. M. (1999). *Intl. J. Biochem. Cell Biol.*, **31**, 261–272.

[83] Pometun, M. S., Chekmenev, E. Y., and Wittebort, R. J. (2004). *J. Biol. Chem.*, **279**, 7982–7987.

[84] Petkova, A. T., Ishii, Y., Balbach, J. J. *et al.* (2002). *Proc. Natl. Avad. Sci. USA*, **99**, 16742–16747.

[85] Muiznieks, L. D. and Keeley, F. W. (2010). *J. Biol. Chem.*, **285**, 39779–39789.

[86] Wise, S. G., Mithieux, S. M., Raftery, M. J. *et al.* (2005). *J. Struct. Biol.*, **149**, 273–281.

[87] Kozel, B. A., Wachi, H., Davis, E. C. *et al.* (2003). *J. Biol. Chem.*, **278**, 18491–18498.

[88] Sokolis, D. P., Kefaloyannis, E. M., Kouloukoussa, M. *et al.* (2006). *J. Biomech.*, **39**, 1651–1662.

[89] Muiznieks, L. D. and Keeley, F. W. (2013). *Biochim. Biophys. Acta*, **1832**, 866–875.

[90] Binnig, G., Quate, C. F., and Gerber, Ch. (1986). *Phys. Rev. Lett.*, **56**, 930–933.

[91] Bozec, L. and Horton, M. (2005). *Biophys. J.*, **88**, 4223–4231.

[92] Valiaev, A., Lim, D. W., Schmidler, S. *et al.* (2008). *J. Amer. Chem. Soc.*, **130**, 10939–10946.

[93] Graham (2004). *Experimental Cell Research*, **299**, 335–342.

[94] Gutsmann, T., Fantner, G. E., Kindt, J. H. *et al.* (2004). *Biophys. J.*, **86**, 3186–3193.

[95] Strasser, S., Zink, A., Janko, M. *et al.* (2007). *Biochem. Biophys. Res. Comm.*, **354**, 27–32.

[96] Wenger, M. P. E., Bozec, L., Horton, M. A. *et al.* (2007). *Biophys. J.*, **93**, 1255–1263.

[97] Koenders, M. M. J. F., Yang, L., Wisman, R. G. *et al.* (2009). *Biomaterials*, **30**, 2425–2432.

[98] Shen, Z. L., Dodge, M. R., Kahn, H. *et al.* (2010). *Biophys. J.*, **99**, 1986–1995.

[99] van der Rijt, J. A. J., van der Werf, K. O., Bennink, M. L. *et al.* (2006). *Macromolecular Bioscience*, **6**, 697–702.

[100] Eppell, S. J., Smith, B. N., Kahn, H. *et al.* (2006). *J. Royal Soc. Int.*, **3**, 117–121.

[101] Shen, Z. L., Dodge, M. R., Kahn, H. *et al.* (2008). *Biophys. J.*, **95**, 3956–3963.

[102] Sasaki, N. and Odajima, S. (1996). *J. Biomech.*, **29**, 1131–1136.

Cell Cytoskeleton

Matthieu Piel and Raphael Voituriez
[a] UMR 144 Institut Curie/CNRS, 26 rue d'Ulm, 75248 Paris Cedex 05, France
[b] Laboratoire de Physique Théorique de la Matière Condensée, UMR 7600 CNRS/UPMC, F-75005 Paris, France
[c] Laboratoire Jean Perrin, FRE 3231 CNRS/UPMC, F-75005 Paris, France

The nature of cell components – polymers, lipids, ions, and other solutes – make the cell, as a material, perfectly suited for soft matter physics. Its unique out-of-equilibrium properties render it all the more exciting, asking for new developments. In this chapter, we present theoretical tools derived from a soft matter physics coarse-grained approach, leading to a continuous description of the "active" part of the cell cytoskeleton – which mostly corresponds to actin and tubulin polymers and associated molecular motors. This recent theoretical framework describing what is often called an "active gel" is now used to model many different biological phenomena, even some unrelated to the cytoskeleton itself, which initially inspired this active gel theory. We then give two examples of models for important cellular functions, cell migration and cell polarity, based on this type of description of the cytoskeleton (in this case mostly the actin cytoskeleton). Each section dealing with theoretical aspects is preceded by a concise description of the biological background and questions. These sections are meant to set the stage for the development of the theoretical arguments and to point the reader to the relevant biological literature. This chapter does not aim at describing exhaustively the very broad contribution of soft matter physics to biology, but hopefully gives a rather precise account of how coarse-grained active gel theory was built and how it contributed to model some properties of the cell cytoskeleton and its role at the cell level.

Introduction

Biophysics has grown into a major field of biology, if not yet of physics. From the biological perspective, it is often viewed as an integrative effort to describe how complex biological functions emerge from basic physical laws, but it also brings incentives and tools to get biology further on the quantitative side. One historical trend is more linked to "-omics" approaches of the last two decades and the consecutive need to handle and analyze the wealth of data produced (Figure 12.1), while another approach relies more on the possibility to obtain quantitative measures on biological material, in particular thanks to the development of mi-

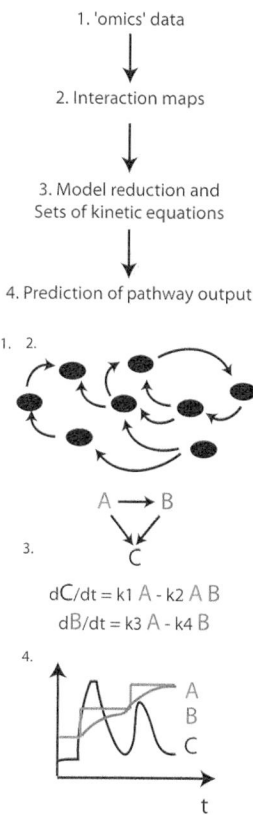

Fig. 12.1 "omics" systems biology. From 1 to 4 are the successive steps followed when studying and modeling biological "networks". Black dots are biological entities (enzymes, transcription factors, genes) which interact with each other (arrows). The arrows can represent very different types of interactions (protein/protein interaction, catalytic activity, induction of gene transcription) depending on the study. A simple example of a minimal network is shown (the feed-forward loop) with the typical set of kinetic equations for such a network, and the dynamic behavior of the network output (quantity of species C, in black) when A (input signal) follows step increments.

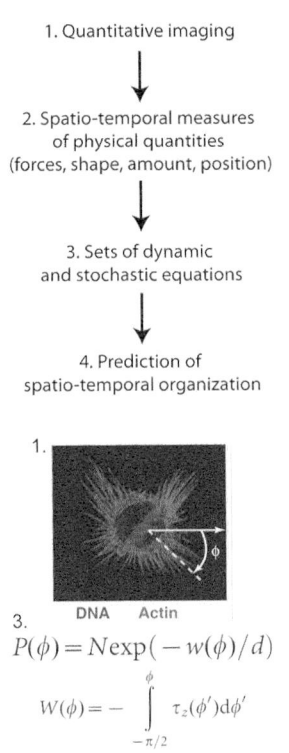

Fig. 12.2 Physics of the cell. From 1 to 4 are successive steps followed when studying and modeling physical aspects of cell behavior. The image shows a mitotic cell attached to an adhesive micro-pattern via retraction fibers (actin in light shading). The chromosome metaphase plate (light shade across the dark nucleus) shows the orientation of the mitotic spindle which also dictates the cell division plane, whose angle relative to the adhesive micropattern is measured by Φ. The equations depict the probability density as a function of Φ predicted from the model (see [3]). The bottom graph shows in dark shade the measured distribution of angles and in light shade the model prediction.

croscopy technologies (see Figure 12.2). The types of data produced are of a very different kind and as a consequence call for different areas of expertise to produce and analyze them. The first approach relies on data produced by automated screening platforms and molecular chips (DNA, RNA, or protein chips). Related models are often based on sets of kinetic equations describing biological interaction networks (e.g. [1], see Figure 12.1), which take stochastic effects into account, but without consideration to spatial aspects. On the other hand, analysis of microscopy data requires a different physics, often related to soft matter physics (membranes, polymers, liquid crystals), which deals with the materials the cell is made of and the forces involved. It aims at describing the spatio-temporal dynamics of biological objects (e.g. [2], see Figure 12.2), but is often still unable to account for the full complexity of biological phenomena, in particular their multiple regulations by signaling pathways. In this chapter we will limit ourselves to the second aspect of biophysics, often called physics of the cell, and to single cells as opposed to tissues. We will first present biological background on the cytoskeleton and then models coming from soft matter physics. In a second part we will review how this particular type of model contributed to the understanding of cytoskeleton-based cell functions like cell motility and cell polarity.

It is now possible, thanks to a number of micro-tools, ranging from micro-pipettes to AFM and micro-pillars, as well as optical tools, to apply and measure precise forces on cells (e.g. [4]) and single molecules (e.g. [5]). This has enabled rapid progress of the field of cell mechanics, as precise measurements have called for new physical models. Some models choose to ignore the molecular details and to focus on the actual behavior of the whole cell as a material, at the temporal and spatial scales studied, but other more-detailed models have tried to account for the wealth of literature accumulated on the main molecular players of cell mechanics, usually commonly referred to as the cell cytoskeleton. In parallel, *in vitro* reconstituted systems from purified components have allowed a better characterization of physical properties of cytoskeletal elements and the complex supra-molecular assemblies they can form, leading to the concept of self-organization [6, 7] and associated models and simulations [8].

It was early discovered (end of 19th century) that living cells possess an internal complex architecture, made of filaments and organelles. Purification techniques allowed the isolation of contractile elements (composed of actin filaments and myosin motors) about 60 years ago (A. Szent-Gyorgyi and colleagues) and, since, the genetic and biochemical characterization of a growing number of elements forming the cytoskeleton was performed. Almost 30 years ago, a major paradigm emerged: some of these elements are highly dynamic and constantly being remodelled. In particular, microtubules, which were considered rigid static rods, are in fact constantly polymerizing and depolymerizing [9], an out-of-equilibrium phenomenon called dynamic instability, and treadmilling of actin filaments was described [10]. Since then, many characteristics of these out-of-equilibrium assemblies of proteins and membranes

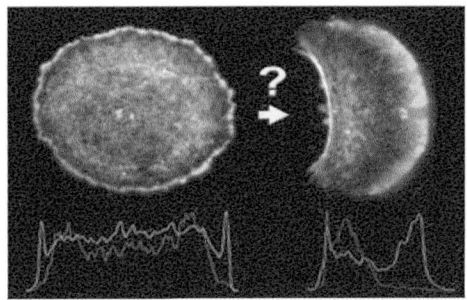

Fig. 12.3 Symmetry breaking of actin cytoskeleton in motile cell fragments (from [11]). In the spherically symmetric static state (left), actin and myosin are homogenously distributed. The transition to the motile non-symmetric (polarized) state (right), associated to a phase transition of the actomyosin cytoskeleton is not yet fully understood.

were studied. In parallel, their functions in cells were worked out. Cell motility, cell morphogenesis, membrane trafficking, cell polarity, and cell division are all examples of the spatial and dynamic organization of the cell that rely for a large part on the cytoskeleton. Nevertheless, how cells exploit this cytoskeletal tool box is highly versatile: for example, some cells rely mainly on actin filaments for intracellular transport (yeasts), while others also rely on microtubules (vertebrates).

Part of the effort has been to assemble a sort of general "phase diagram" of the cytoskeleton, both theoretically and experimentally. However, the complexity of this molecular architecture is such that a global comprehensive description of cell behavior relying on this microscopic knowledge seems out of reach. An alternative approach has consisted of building a phenomenological description of the cell cytoskeleton at a larger scale. Indeed, the cytoskeleton can be seen as a physical material showing a wide variety of phases, with specific properties of density and organization of its filaments. It presents two coupled order parameters (the macroscopic polarization of the filament network and the density), associated with its liquid crystal, and viscoelastic gel properties. Its main specificity, which makes it a material of a completely novel type, is its active character; it is driven out of equilibrium by an internal chemical energy source (ATP), which allows motor proteins to exert stresses that deform the polymer network. This is the kind of approach mainly reviewed in Section 12.2. How this description of the cell cytoskeleton can contribute to understanding cellular functions like cell migration and cell polarity will be reviewed in Sections 12.3 and 12.4.

12.1 Cytoskeleton components

In this Section we briefly review the main players of cell mechanics, to set the stage for a more detailed review of active gel models later in the text. For more details on the cytoskeleton we refer to the numerous reviews already published on each specific aspect.

Actin filaments and gels

Actin filaments are often short (a few μm), flexible (persistence length of approximately 15 μm) and highly dynamic (typical half-life in the

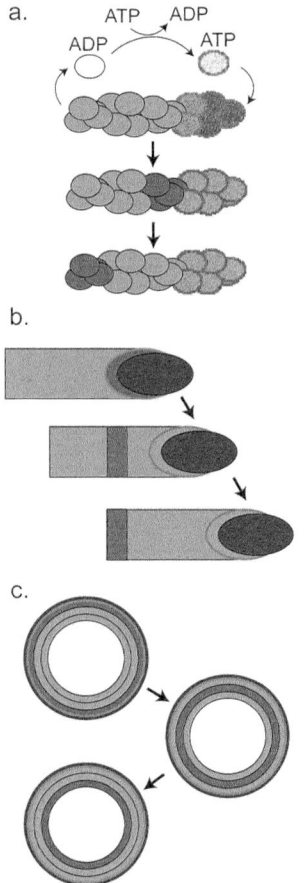

Fig. 12.4 Actin treadmilling: a. Treadmilling actin filament. ATP bound monomers (dark shading) turn to ADP bound (light shading). A subset of monomers are labeled in darker shading to highlight their journey from front to back as the filament treadmills. b. Treadmilling actin comet tail. Bacteria, vesicles or beads can nucleate actin (light-shaded tail) on their surface. ATP bound monomers are added at the surface of the object (dark outline). Due to friction on the comet, new addition of monomers propels the object. The dark stripe shows a subset of monomers incorporated in the top picture to highlight treadmilling of the comet. c. Treadmilling actin cortex. In animal cells, a mesh of actin filaments (light shading) forms at the inner surface of the plasma membrane. Monomers are added at the level of the membrane and removed further inside.

cell cortex of a few seconds). They spontaneously assemble *in vitro* above a critical concentration of monomers and remain dynamic in the presence of ATP, their most basic behavior being treadmilling (see Figure 12.4): net polymerization at one end and depolymerization at the other end [10]. In cells, they are associated to a large number of proteins, which fall in various classes: nucleation promoting factors (initiate filament assembly at low monomer concentration), capping (stabilize filaments ends), severing/destabililzing enzymes, crosslinkers, motors (for a general review of actin-binding proteins see [12]). These proteins regulate the amount of filaments, their localization in cells, their dynamics and their assembly into larger structures. *In vitro* reconstituted systems with a minimal set of purified components recapitulating actin dynamics have been developed since the late 1990's [13], one of the best known being the actin comet tail ensuring the motility of the lysteria pathogen using proteins from its host cell [13]. This system has given rise to very detailed modeling and simulations [14], although going from the molecular to the macroscopic scale is still under debate. The initiation of the movement has been of particular interest for physicists and can be pinpointed as the starting point of the spreading of the idea of symmetry breaking in the cell biology community [15]. More recently micro-patterning of actin nucleators on surfaces has opened the way to the study of complex actin based architectures *in vitro* [16].

Another level of complexity is added when there are molecular motors. These molecules, called myosins for actin-specific motors, can move filaments relative to each other, using ATP as an energy source, leading to contractility of actin gels, and spontaneous oscillations [17]. Actomyosin systems indeed drive a large part of the force production at the cell level, and can form higher-order structures like sarcomeres, at the origin of force production in animals. An experimental tour de force recently allowed reconstitution of active actin and acto-myosin gels anchored to the inside of liposome membranes [18], adding a third essential component to the repertoire of actin-based reconstituted systems now open for investigation by the biophysics community. In cells, actomyosin networks are indeed often associated to membrane organelles and to the cell plasma membrane forming the cell cortex [19].

Microtubules and specialized microtubule-based organelles

Microtubules are tubular polymers of 25 nm in diameter. They can be very long (several tens of μm) and are very rigid (persistence length of several mm). Like actin filaments, they can spontaneously assemble *in vitro* and are dynamic in the presence of GTP. They are also polar, due to the polarized assembly of the monomers (tubulin dimers). Their basic behavior is called dynamic instability (see Figure 12.5) and consists of repeated phases of growth and disassembly of one end of the microtubule (called plus end) while the other end is less dynamic [9]. Associated proteins fall in similar classes as for actin and associated motors are called kinesins (moving towards plus ends) and dynein (moving towards minus end). Minimal reconstituted systems, models, and simulations have also been achieved during the last 15 years [8, 20, 21]. Microtubules, like actin microfilaments in sarcomeres, are the basic blocks of highly complex motile organelles called axonems [22]. Inside cells, microtubules also often display ordered structures, with important function, in particular for cell division (recent models have been developed for mitotic spindle assembly [23]).

Molecular motors

Molecular motors are proteins associated to both actin and microtubule filaments which can convert ATP into physical work. They are the main source of force generation in the cytoskeleton. Purification of functional motors and single-molecule microscopy have allowed in-depth investigation of the physical principles underlying the properties of these amazing proteins. The duty cycle of these motors has been described in details thanks to a combination of genetics, ultrastructure, and single-molecule microscopy [24]. A physical model based on a biased thermal ratchet, has been proposed early on (see Figure 12.6) and experiments have since confirmed that it is a good generic description of molecular motors, accounting for most of their properties and can thus be used as a component of higher-level models [25]. Recently, the first AFM imaging of a live myosin motor walking on a filament [26] showed that the description of the motor molecule at such a level of detail calls for more complex models which could account, for example, for the hand-over-hand walking behavior.

Intermediate filaments

Intermediate filaments have generated less attention in the biophysics community, as they do not display the diversity of associated proteins (no motor is associated to these filaments), are much less dynamic, and do not form higher-order structures. They are assembled in large meshes, which can represent a large amount of the cell mass (e.g. keratinocytes), but other cells are almost completely devoid of such filaments (e.g. neutrophiles). They are often described as passive elements contributing to the mechanical stability and resistance of particular cells and tissues. They are not well-conserved amongst eukaryotes, unlike actin filaments and microtubules. One of the most studied type of intermediary filaments are lamins, which are essential elements of the nuclear cytoskeleton, forming a thick mesh under the nuclear membrane. These filaments have an important mechanical function, accounting for a large part of the stiffness of the nucleus and thus protecting chromatin from external constraints [27]. They are also the base for the anchoring of the nucleus to the cytoskeleton through the LINC complex [28]. Importantly, they also anchor specific regions of chromatin called LADs (for lamin-associated domains), thus contributing to the regulation of gene expression [29]. There are several hundreds of pathological mutations reported in the single Lamin A/C gene, defining a family of diseases called

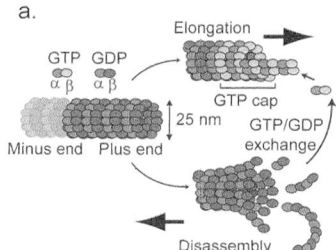

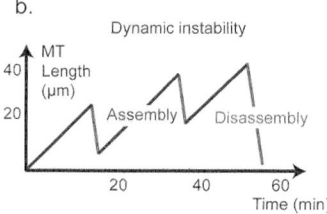

Fig. 12.5 Microtubule dynamic instability. a. Microtubules usually have a stabilized minus end, anchored through the nucleation complex, and a free, dynamic end (plus end), which elongates by addition of GTP bound dimers of α and β tubulin, forming a "GTP cap". Dimers can interact both end-on and laterally, forming a tubular polymer. Plus end can stochastically switch to a fast disassembly phase, an event called a "catastrophe" b. The alternating assembly and disassembly phases result in a fluctuating microtubule length, which is termed "dynamic instability". Assembly and disassembly rates, together with the statistics of catastrophe and rescue (transition from disassembly to assembly) are the four parameters which describe this behavior and set the distribution of microtubule length.

laminopathies, affecting specific tissues, like muscular distrophies.

The next three elements presented below are not included in the classical cytoskeleton elements, but they have important contributions to the physics of the cell and are thus introduced here, even if they are not specifically taken into account in the models presented in this chapter.

The plasma membrane and glycocalix

The plasma membrane has both mechanical properties as a lipid bilayer, and is a barrier to diffusion of molecules, which can generate osmotic pressure. This seemingly simple element of the cell is at the core of a highly diverse and complex system regulating the cell "boundary": it contains a large variety of lipids, binding proteins, transmembrane proteins, and channels (in particular ion channels and transporters which can regulate osmolarity and membrane potential). Some lipids, as well as some transmembrane proteins, bear sugar chains, which form an outer coat called the glycocalix. The glycocalix acts as a protective barrier and also regulates binding to other cells (including parasites), surfaces, and the extra-cellular matrix, as well as recognition by the immune system.

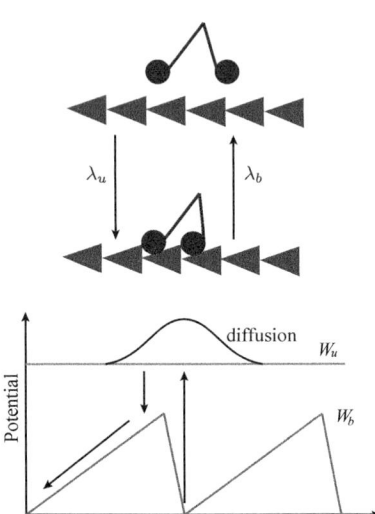

Fig. 12.6 The 2-state model of molecular motors in its simplest form assumes that the hydrolysis of an ATP-bound molecule could induce a conformational switch of the motor protein from a state tightly bound to the filament, which experiences an asymmetric potential $W_b(x)$, to a state weakly bound, which experiences an almost flat potential $W_u(x)$. In the asymmetric potential, the motor will fall in the nearest accessible minimum, while it can freely diffuse with diffusion coefficient D in the unbound state. If the time $1/\lambda_u$ spent in the unbound state is such that $a < \sqrt{D/\lambda_u} < b$ – note that in this regime energy is dissipated by the motor – the motion is on average biased to the left, as in the example of a typical trajectory indicated by arrows.

The simplest *in vitro* system to mimic the plasma membrane is the liposome. Lipid composition in liposomes can be controlled and different lipids can be mixed, which can already generate interesting behaviors, like curvature-dependant segregation [30]. Such liposomes constitute a platform for biophysical studies of membrane-associated proteins. Recent examples include reconstitution of membrane invagination and even membrane scission (with an actin cortex [31]).

The plasma membrane contains channels which can be activated by stress. They can constitute the first level for mecanotransduction mechanisms [32] (see note 1 below). Plasma membrane surface area and tension are also regulated (for example through membrane trafficking), but there is currently little biological knowledge and there is no physical model explaining cell volume homeostasis, in particular during cell proliferation.

The plasma membrane also has many physical and biochemical links to the cytoskeleton: proteins such as ERMS (Ezrin Radixin Moesin) physically anchor the actin cortex to the plasma membrane and/or to other membrane-associated proteins, and conversely proton and ion fluxes close to the plasma membrane have a direct effect on cytoskeletal polymers or associated proteins and regulatory pathways (e.g. [33]). Modification of lipids plays an important role in regulating interactions of cytoskeletal elements with the plasma membrane, the most studied aspect being the addition and removal of phosphates on phosphoinositides. Despite the wealth of biological literature on the subject, this interplay between the cytoskeleton and the plasma membrane (including its associated proteins) is rarely included in physical models, as are cell volume regulation and water flows potentially generated inside the cells when it deforms. One aspect has been recently under the spotlight: the detachment of the plasma membrane from the underlying acto-myosin cortex, leading to formation of blebs [34]. These observations also led to the proposal of a new type of model for the cytoskeleton as a poro-elastic material [19, 35].

The cell wall

Many cell types have thick cell walls composed of cross-linked sugars, which accounts for their passive mechanical properties. It is the case for bacteria, yeasts, and plants. Such cells are usually strongly hyperosmotic, their plasma membrane is pushed against the cell wall, which generates the turgor pressure responsible for their growth [36]. Cell wall dynamics and morphogenesis are subjects of intense study, including physical modeling [37]. One interesting aspect is the mechanism that allows anisotropic growth of walled cells, which involves understanding how a new cell wall is added upon cell growth [38]. The interplay between the cell wall (imposing cell shape) and the cell cytoskeleton (responsible for transport of components and thus directing growth), is thought to be at the origin of morphogenetic rules in walled cells and organizms [39].

Extracellular matrix

For cells living in tissues, the extra-cellular matrix (ECM) plays the role of a soft and dynamic cell wall. ECM has a major impact on cell mechanical behavior and even on the gene expression profile and differentiation into specific cell types [40]. Cells adapt the force they generate to the stiffness of the matrix, which is one of the important parameters monitored by cells in yet unclear ways: this is one of the most studied aspects of the function of mechanotransduction pathways[1] (see [41]). ECM also has signaling functions: several transmembrane proteins act both as physical anchors to the extracellular matrix (integrins) or other cells (cadherins) and as signal transduction receptors. The physical properties of ECM have been studied in depth in the context of tissue engineering, as well as cell adhesion to ECM and mechanical coupling between the cell and the surrounding ECM. Nevertheless, the *in vivo* heterogeneity and dynamic nature of these meshes (due to remodeling by cells) still make it difficult to account for in physical terms.

12.2 Coarse-grained models of the cytoskeleton

In this section we introduce the basic concepts that can be used to model cytoskeleton dynamics. The central idea is to coarse–grain the molecular complexity of the cytoskeleton and derive phenomenological dynamical equations for macroscopic continuous variables such as the density, velocity field, or orientation field of cytoskeletal filaments. Such an approach is complementary to many theoretical works that start from

---

[1] The cell cytoskeleton is a major player of an important phenomenon called mechanotransduction by which forces can be converted into biochemical signaling, leading to cell reorganization and/or to modification in the gene expression profile. Three major types of mechanotransduction have been studied, all involving cytoskeletal elements: 1) The regulation of ion channels (stretch activated channels); 2) The modification of protein conformation upon strain, which can reveal new enzymatic or interaction sites (e.g. the so called "molecular clutch" occurring at the level of focal adhesions); 3) The deformation of the nucleus itself by external or internal stresses. On the biochemical signaling side, the YAP/TAZ transcription factors have recently been under the spotlights as essential players in mechanotransduction pathways.

microscopic considerations at the molecular level and show how the interactions between cytoskeletal filaments and molecular motors can lead to active behaviors at larger scales. Microscopic theories, which will not be discussed here, require a detailed description at the molecular level of both cytoskeletal filaments and molecular motors. Kruse and Jülicher [42] and Marchetti and Liverpool [43] have developed such microscopic approaches on the analytical side, which yield, after a proper coarse-graining is performed, to dynamical equations for macroscopic variables. On the numerical side, detailed microscopic models have now been designed (see [44] for example). In contrast to such microscopic approaches, the hydrodynamic theory which we present here starts directly on macroscopic scales, and it has now been proven that both approaches are compatible. It should be noted that coarse-grained theories by definition require the introduction of hydrodynamic variables and imply that a local averaging of microscopic variables (such as the orientation of cytoskeletal filaments) can be performed. Such descriptions are therefore valid only in the limit of a large number of microscopic agents (filaments or motors), and are *a priori* more suited to model the actin cytoskeleton rather than microtubules, for which microscopic models might be more relevant.

We will focus here on the description of the two main mechanisms that can induce spontaneous motion, namely self-polymerization and actin/myosin contractility, which both rely on the out-of-equilibrium nature of the cell. The resulting generalized hydrodynamics is often called active gel or active matter theory because of its out-of-equilibrium nature. This theory, based essentially on symmetry arguments, is very general and actually applies to various out-of-equilibrium systems such as bacterial colonies, bird flocks or fish schools [45]. Today it constitutes a very intense field of research and we will focus here only on its application to cytoskeleton dynamics.

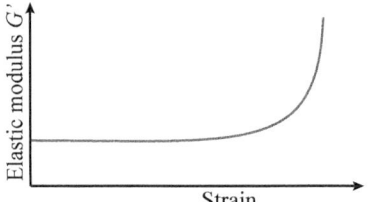

Fig. 12.7 To measure the rheological properties of biological gels or even cells, one applies a force and measures the resultant deformation; the elastic modulus, G', is defined as the ratio of the stress, σ, to the strain, γ; thus, $G' = \sigma/\gamma$. The gel can also have a viscous or dissipative response, and the resultant stress depends on the strain rate, $\dot{\gamma}$, defining the loss modulus, $G'' = \sigma/\dot{\gamma}$. It is found generally in crosslinked gels of biopolymers that the elastic modulus strongly depends on strain at large enough strains, where the gel dramatically stiffens.

The cytoskeleton as a viscoelastic gel

From a physical point of view, the cytoskeleton is a network of long semiflexible filaments made up of protein subunits, interacting with other proteins that can act as cross-linkers and therefore dynamically tune its rheological properties. Importantly, the cross-links are established by physical interactions such as dipolar interactions or ionic bonds, and therefore have a short lifetime (typically minutes or shorter) as opposed to covalent bonds. This implies that the cytoskeleton is viscoelastic, namely that at short time scales it behaves as a solid, typically characterized by an elastic modulus, while it flows as a viscous liquid at long times. Such systems are generically referred to as physical gels, as opposed to chemical gels for which cross-links are formed by covalent bonds, and which therefore behave as solids even at long times. In practice, the observed rheological properties of the cytoskeleton are quite complex and in some regimes can display strong non-linear effects that have been evidenced by *in vitro* experiments on biofilament gels [46, 47] (see Figure 12.7). The rheology of single cells has also been studied intensively and reveals unusual dynamical effects such as transient fluidization in response to stress [48]; we will focus here only on the basic features and the simplest model of a continuous viscoelastic medium, which is the linear Maxwell model (see Figure 12.8).

We denote the components of the velocity field of the gel by v_i, the strain rate by $u_{ij} = (\partial_i v_j + \partial_j v_i)/2$, and the vorticity tensor by $\omega_{ij} = (\partial_i v_j - \partial_j v_i)/2$. The constitutive equation of the Maxwell model relating the strain rate to the deviatory stress tensor σ_{ij} is written as:

$$2\eta u_{ij} = \left(1 + \tau \frac{D}{Dt}\right)\sigma_{ij}, \tag{12.1}$$

where η is the shear viscosity and τ is the typical relaxation time of the drag related to the elastic modulus $E = \eta/\tau$. To ensure invariance with respect to translation and rotation, we use the co-rotational derivative [49]

$$\frac{D\sigma_{ij}}{Dt} = \frac{\partial \sigma_{ij}}{\partial t} + v_k \partial_k \sigma_{ij} + \omega_{ik}\sigma_{kj} + \omega_{jk}\sigma_{ki}, \tag{12.2}$$

instead of the usual time derivative. For very short relaxation times τ, equation (12.1) describes a viscous fluid and for long relaxation times, it describes an elastic material. For polymer gels in practice τ widely varies with parameters such as polymer concentration, pressure, or the cross-linker concentration. For dilute polymer gels, τ is very small and the gel behaves as a viscous fluid. At higher concentrations, or for more cross-linked gels, τ becomes very large and the gel behaves as an elastic medium. In practice each of these limits is considered separately; in particular for modeling purposes the long time behavior is often more relevant and only the viscous limit is considered.

The constitutive equation has to be completed by a continuity equation which ensures the mass conservation. If we model the cytoskeleton as a single component medium, composed of polymers plus the solvent, it can be considered as incompressible in a first approximation, such that:

$$\nabla \cdot \mathbf{v} = \partial_i v_i = 0. \tag{12.3}$$

Finally, on general grounds the dynamics of the gel is governed by the Navier-Stokes equation. At the scale of the cytoskeleton however, typical Reynolds numbers are very small and viscous forces dominate over inertial forces. The Navier-Stokes equation then simplifies to the balance of forces:

$$\partial_i(\sigma_{ij} - P\delta_{ij}) = 0. \tag{12.4}$$

Equations (12.1),(12.3), and (12.4), completed by adequate boundary conditions, fully describe the dynamic of a generic viscoelastic gel. We will see next how they should be modified to incorporate the active processes at work in the cytoskeleton.

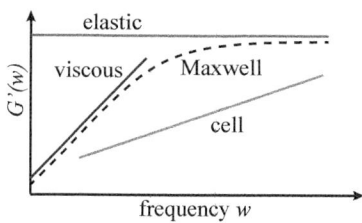

Fig. 12.8 More generally, the elastic and loss moduli $G'(\omega)$ and $G''(\omega)$ depend on the frequency and are measured by applying a small oscillatory stress at a frequency ω. For an elastic medium one has $G'(\omega) = C$ constant, and $G'(\omega) \propto \omega$ for a viscous fluid. It is found generally for single cells that $G'(\omega) \propto \omega^\alpha$, while the Maxwell model interpolates between elastic and viscous behaviors.

Self-polymerization: Treadmilling

The first peculiarity of the cytoskeleton, viewed as a physical gel, is that its constitutive polymers are not at equilibrium but continuously undergo a polymerization/depolymerization process. This feature affects the mass conservation of the system and has therefore to be taken into account in the continuity equation (12.3) above. Actin filaments and microtubules even if structurally very different, share basic common features in their growth dynamics, and we will focus in this section on actin, which is more directly involved in the mechanical properties of the cytoskeleton.

Filamentous (F)-actin is asymmetric and the two extremities of a filament retain different kinetic characteristics. Actin monomers assemble much more rapidly at the 'barbed end' compared to the 'pointed end' (these names correspond to the arrowhead appearance of myosin heads bound to actin filaments). The critical concentration of the pointed end is higher than that of the barbed end. When F-actin and globular G-actin monomers are at steady state, the global critical concentration is intermediate between those of the two ends. So, at this stage, there is a net loss of molecules at the pointed end and a net addition at the barbed end. The two rates balance, which leads to treadmilling – a net flow of actin subunits through the filament. Such a difference of the kinetic constants at both ends is made possible by ATP hydrolysis, and therefore directly results from the out-of-equilibrium nature of the cytoskeleton. Monomeric actin binds either ATP or ADP. ATP monomers assemble at a far higher rate than ADP ones. Following assembly on a treadmilling filament, ATP is hydrolysed to ADP and this induces a change in the filament conformation, resulting in a less stable form at the pointed end, which depolymerizes. A treadmilling filament therefore contains ATP-bound subunits at the barbed end, whereas the ones at the pointed end are ADP-bound. This mechanism can intrinsically generate transport of molecules, and therefore motion.

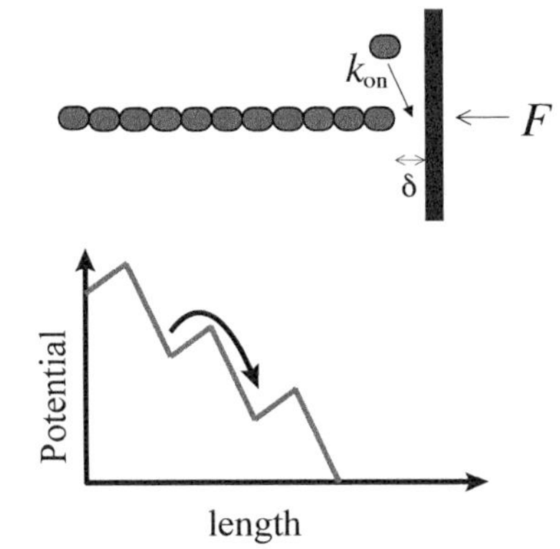

The polymerization of a filament against a barrier can be described using a standard ratchet model [50]. In the simplest ratchet model the polymerization speed of a filament at the cell membrane is given by the bare attachment rate k_{on} in the absence of any force, multiplied by the probability of thermal fluctuations producing a gap large enough for a new subunit to attach (see Figure 12.9). The energy required for a gap the size of a subunit δ is given by $F\delta$, where F is the load. Therefore the polymerization speed is given by:

$$\mathbf{v}_p = k_{on} \delta e^{-F\delta/k_B T} \mathbf{n}, \qquad (12.5)$$

where $\mathbf{n}$ denotes the normal to the surface. In this minimal model outlined here we are assuming the filament consists of a single protofilament (in reality actin has two filaments) and we neglect the disassociation rate k_{off} from the polymerizing barbed end. In other words we are assuming that all the depolymerization only occurs at the pointed end of the actin filament. Implicit in the model outlined here is also the assumption that

Fig. 12.9 Cartoon of a ratchet model of filament polymerization. Bound monomers have a lower energy than unbound monomers, what favors polymerization. The binding of each monomer, however, requires it to overcome an energy barrier, which corresponds to an elastic deformation of amplitude δ of the filament. This barrier depends on the applied force F.

the diffusion over the distance δ is fast compared to the rate of addition of subunits k_{on} and therefore does not influence the polymerization speed.

Generically, polymerization/depolymerization modifies the continuity equation (12.3) by adding a source and a decay term $k'_{on} c_m - k'_{off}$ on the right-hand side, where c_m denotes the free monomer concentration and the macroscopic polymerization rate k'_{on} is proportional to the polymerization speed v_p. Actin polymerization and therefore k'_{on} is mediated by specific nucleators such as the Arp2/3 complex, which is not homogeneous in the cell. Hence k'_{on} (as well as c_m) depends on the position. In practice nucleators are mainly localized close to the cell membrane. In this case polymerization enters only as a boundary term by setting $v_i n_i = v_p$ at the membrane, where $\mathbf{n}$ denotes the normal vector and one can take $k'_{on} = 0$ in the bulk.

Actin-myosin contractility

So far the only "active" (out-of-equilibrium) property we have considered is the self-polymerization. In this section we introduce the active stress produced by the activity of molecular motors such as myosin. The extension of standard hydrodynamic equations for a passive gel to include such an active stress is known as the theory of active gels. It has been developed over recent years [43, 51–54] and is proving to be successful in describing a wide variety of biological phenomena beyond the modelling of cytoskeleton dynamics. Here we provide an outline of the main equations.

Polar gel – Before we consider the active terms we first extend the equations for a viscoelastic gel given on page 504 to the case of a polar viscoelastic gel. For the sake of simplicity we consider the case of a two-dimensional layer of active gel. We assume the directions of the actin filaments are described by the normalized polarization field $\mathbf{p} = (\cos\theta, \sin\theta)$ where θ is the angle with the x axis. Here $\mathbf{p}$ is defined as the local (i–e over a mesoscale) average of the orientation of individual filaments. Including polarization, the constitutive equation (12.1) in the viscous fluid limit ($\tau \to 0$) is analogous to that of a polar liquid crystal

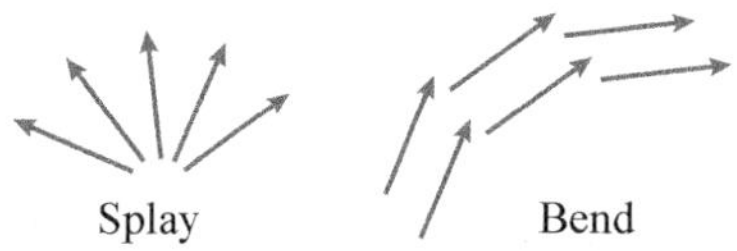

$$2\eta u_{ij} = \sigma_{ij} - \frac{\nu}{2}(p_i h_j + p_j h_i) + \frac{1}{2}(p_i h_j - p_j h_i), \quad (12.6)$$

where

$$h_i = -\frac{\delta F}{\delta p_i} + \lambda' p_i \quad (12.7)$$

is the field conjugate to the polarization, called the molecular field, and where F is the free energy. The molecular field is only defined within the transformation $\mathbf{h} \to \mathbf{h} + \lambda'\mathbf{p}$, where λ' is an arbitrary function of time and space, since this transformation leaves the torque $\mathbf{\Gamma} = \mathbf{p} \times \mathbf{h}$ unchanged. The standard free energy for a polar liquid crystal is given

Fig. 12.10 Splay and bend configurations in polar systems have an energy cost, which is taken into account in the Frank elastic energy.

by the Frank elastic energy [55]

$$F = \int \mathrm{d}x \mathrm{d}z \left[\frac{K_1}{2}(\nabla \cdot \mathbf{p})^2 + \frac{K_3}{2}(\nabla \times \mathbf{p})^2 \right], \quad (12.8)$$

where $K_{1,3}$ are the splay and bend elastic moduli (see Figure 12.10). In two dimensions the twist term is zero. In the one-constant approximation $K_1 = K_3 = K$ which leads to

$$h_i = K\nabla^2 p_i + \lambda' p_i.$$

The molecular field is conveniently written in terms of its components parallel and perpendicular to the polarization:

$$h_\parallel = h_x p_x + h_z p_z = -K((\partial_x \theta)^2 + (\partial_z \theta)^2) + \lambda'$$
$$h_\perp = h_x p_z - h_z p_x = K\nabla^2 \theta.$$

The second term of the right-hand side of equation (12.6) is the coupling between the mechanical stress and the polarization field, characterized by the coefficient ν as is standard for liquid-crystal hydrodynamics. The third term on the right-hand side of equation (12.6) is the antisymmetric part of the stress tensor describing the torque acting on each volume element.

The equation for the polarization flux is given by

$$\frac{\mathrm{D} p_i}{\mathrm{D} t} = \frac{1}{\gamma} h_i - \nu u_{ij} p_j \quad (12.9)$$

where γ is the rotational viscosity. The derivative is the co-rotational derivative

$$\frac{\mathrm{D}\phi_i}{\mathrm{D}t} = \frac{\partial \phi_i}{\partial t} + (v_k \partial_k)\phi_i + \omega_{ij}\phi_j \quad (12.10)$$

to ensure invariance with respect to rotation as well as translational flow.

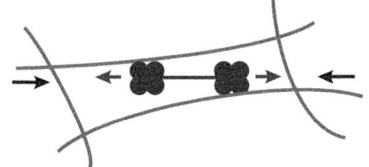

Fig. 12.11 Myosin II motors form double-headed mini-filaments, with each head moving along actin filaments in the opposite direction. This results in a contractile force dipole acting on the actin network (black arrows).

The active contribution – We are finally ready to add the active contribution to the polar gel. A systematic derivation of this contribution relies on Onsager linear response theory and can be found in [51–53]. More simply, symmetry arguments imply that to linear order the active stress is given by $\sigma_{ij}^{\text{active}} = \tilde{\zeta}\Delta\mu\, p_i p_j$ where $\Delta\mu$ is the chemical potential of ATP hydrolysis which provides the energy to the molecular motors driving the system out of equilibrium. The phenomenological coefficient $\tilde{\zeta}$ determines the nature and strength of the activity and it is to linear order proportional to the myosin concentration. In general the myosin concentration and therefore $\tilde{\zeta}$ may not be homogeneous and we therefore allow it to vary with position. Acto-myosin systems are thought to be contractile which corresponds to $\tilde{\zeta} < 0$ (see Figure 12.11). With this active stress the constitutive equation (12.6) becomes:

$$2\eta u_{ij} = \sigma_{ij} - \frac{\nu}{2}(p_i h_j + p_j h_i) + \frac{1}{2}(p_i h_j - p_j h_i) + \tilde{\zeta}\Delta\mu\, p_i p_j. \quad (12.11)$$

Locally, there are two forces acting on the gel: the deviatory stress tensor σ_{ij} and the pressure P that ensures the incompressibility of the gel. In general the equation for the polarization flux, equation (12.9), also gains an extra term due to the activity. This term, $\lambda \Delta \mu\, p_i$, is characterized by the coefficient λ, which if positive, tends to align the filaments enhancing the local polarization. However if the polarization has a fixed modulus, as in the case we consider here, this term simply modifies the parallel component of the molecular field $h_\parallel$ (modifying λ') which is itself determined by the modulus of the polarization. Using this modified molecular field $\tilde{h}_i = h_i + \gamma \lambda p_i$ in equation (12.11) simply changes the coefficient ζ to $\tilde{\zeta} = \zeta + \lambda \nu \gamma$. For simplicity we therefore directly wrote $\tilde{\zeta}$ in equation (12.11) and leave the polarization flux as in equation (12.9).

The constitutive equation (12.11), together with the dynamic equation (12.9), the definition of the molecular field (12.7), the continuity equation (12.3), and the force balance equation (12.4) define the active gel theory in its simplest form. In the last few years several developments have been accomplished. In particular, we have considered here only a one component gel, whereas the cytoskeleton is clearly multi-component. A systematic derivation of a linear theory for a multicomponent active gel can be performed [56], but it requires the introduction of a large number of coupling coefficients and goes beyond the scope of this chapter. The most important effect which is missed in the one-component theory is solvent permeation, which can be an important source of dissipation. This effect is discussed in [56]. A second restriction lies in the linear nature of the equations, which implicitly requires the system to be close to equilibrium. The linear theory should therefore be considered only as a first step in the modelling of the cytoskeleton. In principle, non-linear terms could be constructed systematically, but due to the rather large number of physical variables, this procedure is in practice unrealistic, and only non-linearities with clear physical origin should be first introduced. Finally, the above theory involves only average values of physical variables and ignores their fluctuations, which can be either of thermal or active origin. Although the effect of thermal noise has been discussed in [56], the treatment of the active noise remains a challenge because of the lack of systematic approaches.

Spontaneous flows in active gels

We now turn to examples of applications of the above active-gel theory. For the sake of simplicity we will again focus on two-dimensional films of active gel. We will in particular emphasize the spectacular consequences of the active term on the gel dynamics and discuss two examples where a spontaneous flow of gel is generated in the absence of any external field.

Motion induced by polarization gradients – Here we consider a film of thickness b, lying on the (x, y) plane, and confined between the planes $z=0$ and $z=b$. We assume invariance along the y direction, which makes the problem two-dimensional, Figure 12.12. For this geometry, both the polarization angle θ and the velocity v_x depend only on z. In the case of an incompressible film, the velocity v_z in the direction perpendicular to the film vanishes. The constitutive equation for the stress, (12.11), and the dynamic equation (12.9) can then be recast into a single equation for θ:

$$\partial_z^2 \theta = -\frac{\tilde{\zeta}\Delta\mu \sin 2\theta (1+\nu \cos 2\theta)}{K\left[4\eta/\gamma + \nu^2 + 1 + 2\nu \cos 2\theta\right]}, \quad (12.12)$$

and the velocity gradient $u = \frac{1}{2}\frac{dv_x}{dz}$ reads

$$u = \frac{\tilde{\zeta}\Delta\mu \sin 2\theta}{\gamma\left[4\eta/\gamma + \nu^2 + 1 + 2\nu \cos 2\theta\right]}. \quad (12.13)$$

In the case of a contractile stress ($\tilde{\zeta}\Delta\mu < 0$), equation (12.12) defines the characteristic thickness $l_c = (K/\tilde{\zeta}\Delta\mu)^{1/2}$. For $b \ll l_c$, the polarization angle θ is found to vary linearly with z, and one can check that unless the boundary condition at both $z = 0$ and $z = b$ imposes either $\theta = 0$ or $\theta = \pi/2$, the velocity gradient, and therefore the flow, is non-vanishing. For example, for $\theta(z=0) = \pi/2$ and $\theta(z=b) = \pi$, one finds

$$v_x = -\frac{\tilde{\zeta}\Delta\mu}{\pi\gamma\nu} \ln\left(4\eta/\gamma + \nu^2 + 1 - 2\nu \cos(z\pi/b)\right). \quad (12.14)$$

This clearly shows that for a thin film of active gel, the contractile active stress $\tilde{\zeta}\Delta\mu$ generates a spontaneous flow if a polarization gradient is imposed by the boundary conditions. This mechanism can be seen as the first possible mechanism of spontaneous motion induced by the actin/myosin contractility in the cytoskeleton.

Spontaneous flow transition – The mechanism which triggers the gel motion in the previous section relies on a polarization gradient imposed by the anchoring of the polarization field $\mathbf{p}$ at the boundary. In the case of planar anchoring ($\theta = 0$) at both boundaries, a static stationary state with homogeneous polarization and $v = 0$ exists. Actually, we will show that this static state is destabilized either for a large enough active stress or a large enough gel thickness, and eventually reaches a dynamic state characterized by a spontaneous flow.

Following [57], we consider a perturbation of the static state, where the polarization angle θ is slightly tilted and only weakly deviates from 0. Expanding equation (12.12), one obtains

$$\partial_z^2 \theta + \frac{\theta}{L_c^2} = 0 \quad (12.15)$$

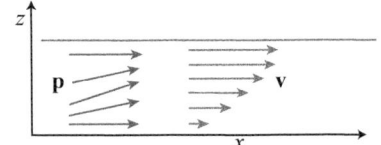

Fig. 12.12 Spontaneous flow in a layer of active gel of thickness $b > L_c$ with no slip condition at the surface ($z = 0$) and free boundary condition at $z = b$.

where $L_c = l_c\sqrt{\frac{4\eta/\gamma + (\nu+1)^2}{2(\nu+1)}}$. If the thickness b of the active film is smaller than L_c, then the only solution to equation (12.15) which satisfies the boundary conditions is $\theta = 0$ for all z. The polarization thus remains parallel to the x-direction and there is no flow in the film. At the critical value of the thickness $b = L_c$, a tilt of the polarization pops up, whose maximum amplitude can be shown to scale as $\theta_m \simeq \sqrt{b - L_c}$. This continuous transition is similar to the classical Frederiks transition of nematic liquid crystals, where a tilt of the director from the orientation imposed by the confining surfaces is due to an external magnetic field [55]. Quite remarkably, in the case of an active gel there is no external field and the tilt results from the active stress. The interesting feature is that above the transition, the polarization gradient gives rise to a non-vanishing velocity field and therefore a spontaneous global motion of the gel, see Figure 12.12.. Assuming no slip conditions at $z = 0$ and free slip conditions at $z = b$, the gel flux reads

$$Q = \int_0^b v_x dz = -\frac{4b\tilde{\zeta}\Delta\mu}{\pi[4\eta + \gamma(\nu+1)^2]} \frac{\theta_m}{}. \tag{12.16}$$

Note that the flux can occur either in the positive or in the negative direction depending on the sign of the polarization tilt even though the film is polar. This flux exists only for different hydrodynamic boundary conditions at $z = 0$ and $z = b$. For example, for a gel confined between two solid surfaces, the velocity vanishes on both surfaces of the film. The transition between a static and a flowing state occurs at a thickness $L = 2L_c$ and there is no net flux of gel but only a shear flow.

Generic phase diagram – Beyond this dynamic transition in quasi 1-dimensional geometry, it is legitimate to search for a full phase diagram of active gels [58]. In this section we follow the work of [58] and consider an infinite film of active gel. We average all properties over the thickness and therefore discuss the film properties in two dimensions. Although the film itself is in general incompressible, the two-dimensional density $c(r)$ is not constant due to variations in the film thickness. We thus consider the film as a two-dimensional compressible system. Here, we only study small density fluctuations of the film around the average density c_0, and write the local density as $c = c_0 + \rho$. The total free energy of the system F_T is the sum of the polarization free energy given by equation (12.8), and a contribution of the density F_c. For simplicity, we consider again the case of isotropic Franck constants and set $K_1 = K_3 = K$. To lowest order, the contribution of the density fluctuations couples to the polarization and can be written

$$F_c = \int dx dy \left[w\rho \nabla \cdot \mathbf{p} + \frac{\beta}{2}(\nabla \rho)^2 + \frac{\alpha}{2}\rho^2 \right]. \tag{12.17}$$

The compressibility α and the phenomenological coefficient β are related to the density fluctuations correlation length $\lambda_\rho = \sqrt{\beta/\alpha}$, while the coefficient w characterizes the coupling between density fluctuations and splay. Finally, the pressure in the film is $\Pi = c_0 \delta F_c / \delta \rho$. The equations of motion have two types of homogeneous steady states: a static

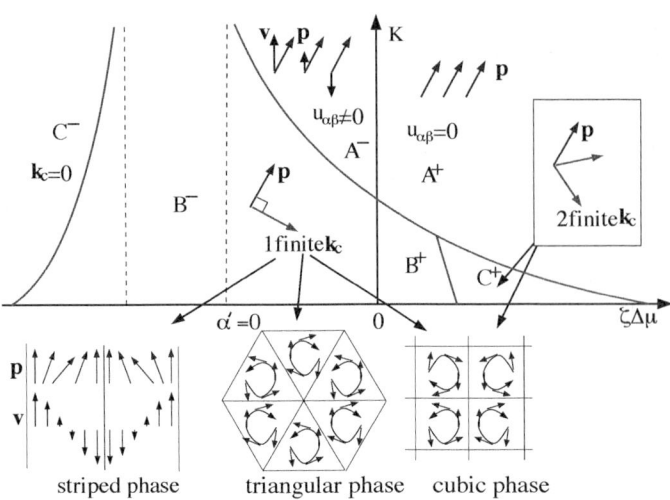

Fig. 12.13 Phase diagram of active polar films (from [58]). In region A either the static or the dynamic homogeneous steady state is macroscopically stable. In region A^-, the film spontaneously flows. In region C^-, both the static and the dynamic steady states are unstable with respect to a macroscopic perturbation (with a wave vector $k_c = 0$). In regions B and C^+, at least one of the steady states is unstable with respect to periodic perturbations of finite wave vectors k_c.

state where the velocity gradient vanishes and the polarization is uniform, and a flowing state with a finite velocity gradient and a uniform polarization, such that the angle θ_0 between the polarization and the velocity satisfies $\cos 2\theta_0 = 1/\nu$.

By studying the linear stability of these two steady states, a generic phase diagram of compressible active polar films can be obtained [58], and it is shown in Figure 12.13. Three regions can be distinguished in this diagram. In region A either the static or the dynamic homogenous steady state is macroscopically stable. In region A^-, the film spontaneously flows, quite similar to what was discussed in the previous section. In region C^-, both the static and the dynamic steady states are unstable with respect to a macroscopic perturbation (with a wave vector $k_c = 0$). In this range of parameters the activity coefficient $\tilde{\zeta}\Delta\mu$ is negative and large, which corresponds to a strong contractility. The effective compressibility is then negative and one expects a macroscopic phase separation. This could correspond to the super-precipitation observed in actin-myosin systems [59, 60] or to the formation of bundles by microtubules when the density of kinesin motors is increased. In regions B and C^+, at least one of the steady states is unstable with respect to periodic perturbations of finite wave vectors k_c. This instability is very similar to the instability described in [61] for passive nematics. In that case, the instability corresponds to the appearance of a periodic lattice of defects, which favors the apparition of splay. For the active system, this suggests a similar lattice of defects. Because of the existence of polarization gradients in this phase, the velocity gradient cannot vanish and the system flows locally. One could expect either a lamellar phase or a hexagonal phase. The hexagonal phase, for example, corresponds to an ordered array of rotating spirals (such point-like defects have been described in [52, 53]). One should however keep in mind that a linear instability at a finite wave vector does not show the existence of a stable periodic phase, and a full non-linear treatment would be necessary to assess the symmetry of the stationary phases. Note that a disordered or glassy vortex phase could also exist as a metastable state and be stabilized by the intrinsic noise of the system.

So far no systematic experimental studies of the dynamic phase diagram of actin-myosin films have been conducted. However, simulations and some experiments show the existence of phase separation and of vortex phases with flow for both microtubule and actin systems [21, 62, 63], in agreement with theoretical predictions.

12.3 Cell migration

The cytoskeleton contributes to almost all cellular functions. Nevertheless, three particular cellular functions have been the center of most studies involving biophysics of the cytoskeleton: cell migration, cell division, and cell polarity (which could all be more generally grouped under the term of cell morphogenesis). In this, and in the following Section, we will only review cell migration and cell polarity, and mostly the contribution of the acto-myosin cytoskeleton. Actin systems have been the main focus of soft-matter coarse-grained models, because actin filaments are small and dense in the cell, while microtubules have almost the cell size and tend to form more discrete assemblies, thus less suited for continuous models. Microtubule assemblies, and in particular the mitotic spindle, are more often modeled using realistic simulations, the best known example being the open source system cytosim [44] (see Figure 12.14), but simpler ones also provided useful insights (for example in the field of mitotic spindle orientation [64]).

The current paradigm is that, through evolution, as well as along developmental processes involving cell growth, migration, and division, or during specific cell responses, cytoskeletal elements are modulated, both in time and space, to give rise to the various dynamical structures found inside cells, which are a part of a highly complex n-dimensional phase diagram. Molecular studies, and then studies of cell-free and minimal *in vitro* reconstituted systems have identified parts and aspects described above and began to draw the outline of such a phase diagram. But what happens inside cells? The advance of quantitative live-cell imaging techniques, from simple GFP tagging of proteins to more sophisticated super-resolution methods including single-molecule imaging as well as speckle microscopy, but also techniques allowing measure of the molecular dynamics inside cells like FRAP, FRET, FLIM, or the recent sptPALM [65], have, for the past 15 years, provided a wealth of information on cytoskeletal dynamics inside living cells. A large part of such data relate to molecular events, and could thus be directly compared to data obtained with reconstituted systems.

One challenge has been the question of integration at the cell level. Indeed, understanding the dynamics of cytoskeletal elements at the cell scale involves taking into account the physical boundary of the cell.

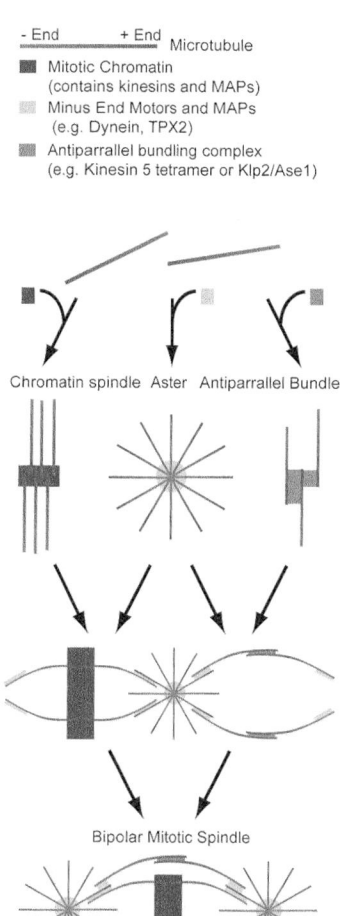

Fig. 12.14 Self-assembly of microtubules and motors/MAPs (microtubule associated proteins). Basic building blocks are shown on the top. Association of microtubules with each type of bundlers leads to a specific spatial organization. Combined two by two, these structures can associate and all together they form a bipolar mitotic spindle (bottom). How this self-assembly process depends on precise characteristics of microtubules, MAPs, and their interactions can be simulated with a software such as Cytosim. Cytosim is a simulation of the cytoskeleton. It can handle large systems of fibers with associated proteins such as molecular motors. The fibers are flexible, and the simulation engine was optimized for the inertialess equations of motion. The Brownian dynamics approach is described at http://www.cytosim.org/cytosim/index.html [44].

As was stated above, the nature of this boundary, the plasma membrane, and its associated proteins, make it a very complex one, which is thus never fully taken into account in models. A first step is to take into account the geometry. On the experimental side, micro-fabrication techniques, also developed in the past 15 years for applications to cell biology [66], have allowed such a control of cell boundaries, in terms of geometry [67], but also allowed the control of spatial and temporal signaling by fixed or diffusive ligands, as well as control of mechanical properties of the culture substrate [68]. On the theory side, the geometry of the cell has also been difficult to take into account. In the following section, we will review what soft matter models brought to the understanding of crawling migration of cells, one of the fundamental cell behaviors resulting from the dynamics of the cytoskeleton. Crawling is a spontaneous behavior of most individual non-walled cells, even of cells that might not be moving much when bound to other cells in a tissue. Most of the quantitative and modeling work has been performed on adhesive cells crawling on flat surfaces (often called 2D migration), and recent work addressed more complex and physiological environments, with a particular interest in reconstituted or artificial extra-cellular matrices and gels (often called 3D migration [69]). But no study so far addressed precisely the respective contribution of the main environmental parameters of the migration substrate (adhesion, confinement, geometry, stiffness, maybe even activity if a cell is crawling between other cells which might react to its movements). Some of these aspects will be discussed below on the theory side. Indeed one of the first questions to ask when addressing cell migration is the following: under what boundary condition a spontaneous migration behavior can be observed? If we now consider the active gel generating this cell behavior, the question becomes: in what conditions can a spontaneous flow be observed in an active gel? We mostly restrict ourselves to that latter question.

Lamellipodial motility

Taking the active gel theory as a paradigm of cytoskeleton dynamics, the analysis of Section 12.2 shows in idealized geometries (mainly infinite) that such material presents spontaneous flows. In order to model cell motility and more generally cell mechanics, one has to take into account realistic cellular geometries with proper boundary conditions, such has the interaction with a membrane or a substrate. We show below that for simplified geometries, the active gel theory actually enables us to propose several simple mechanisms of cell motility, which can vary depending on the geometry of the environment and the adhesion properties of the cell/substrate interface. Of course, such coarse-grained theories, which are the focus of this chapter, are not the only possible theoretical approaches to cell motility. In particular microscopic approaches based on the molecular description of the cytoskeleton, which most of the time involve numerical simulations, provide very useful and complementary results (see [44] for example).

Until now, the vast majority of *in-vitro* experiments of cell motility have been performed on flat substrates. In such conditions, whose relevance for *in-vivo* motility can be questioned, a prototypical crawling cell develops a thin protrusion called a lamellipodium, where the actin flow is mostly retrograde, that is, in the direction opposite to the global cell motion. At the rear of the cell however, where qualitatively the cell contracts, the actin flow is forward.

Lamellipodial motility can be discussed in the framework of active gel theory, as discussed in [51, 70]. In its simplest form, the model is as follows (see Figure 12.15). We assume that motion (whose velocity is denoted by u) is along the x axis, and that the lamellipodium thickness is measured along z. We consider that at least locally the problem is invariant along y and therefore treat it as effectively two-dimensional in the (x, z) plane. As discussed previously, mainly two microscopic mechanisms can be at play to generate motion, acto-myosin contractility and actin polymerization. As justified in previous sections, acto-myosin contractility is taken into account by modelling the cytoskeleton as a layer of thickness $h(x)$ (measured in the moving frame of the cell) of an incompressible active gel. At the tip of the lamellipodium ($x = 0$) one has $h(0) = 0$, while we assume that the cell body imposes a thickness $h(L) = h_0$ at the point $x = L$, where a drag force f_b is applied. In turn, we assume here that polymerization occurs only at the tip of the lamellipodium, where nucleators (such as the Arp2/3 complex) of density ρ_n are localized. Following equation (12.5), one then writes

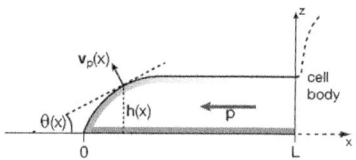

Fig. 12.15 Model of lamellipodial motility (from [51]).

$$\mathbf{v}_p = k_p \rho_n \mathbf{n}, \tag{12.18}$$

where $\mathbf{n}$ is the inward normal to the surface. Finally, in order to conserve the overall mass of actin, we assume that depolymerization takes place with speed v_d at the back of the cell ($x = L$). By conservation of the gel flux one then has $v_d = u + v(L)$.

The interaction of the lamellipodium with the substrate is mediated by complex molecular structures such as focal adhesion. In a first approximation, adhesion effects can be described by an effective viscous friction law, which linearly relates the transverse stress to the sliding velocity according to $\sigma_{zx} = \xi v_x$, where ξ characterizes the friction. The cell surface $z = h(x)$ is a free surface and both the normal and tangential stresses vanish : $\sigma_{nt} = \sigma_{nn} - P = 0$. With this set of boundary conditions, the constitutive equation (12.11), together with the continuity equation (12.3) and the force-balance equation (12.4) can be solved, and yield the following expression of the lamellipodium velocity as a function of the physical parameters (see [51, 70] for details):

$$u \simeq v_d - \left(\frac{h_0}{4\eta\xi}\right)^{1/2} (\zeta \Delta \mu + f_b/h_0). \tag{12.19}$$

Importantly, this model shows that the major contribution to the cell velocity is the depolymerization velocity, which indicates that actin treadmilling plays a crucial role in lamellipodial motility. In turn the effect

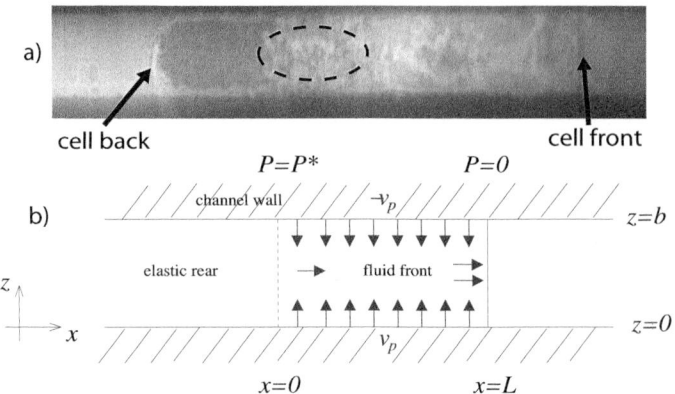

Fig. 12.16 a) RICM image of a dentritic cell moving (here to the right) in a channel of $5\mu m$ width. The dark zone at the back of the cell (left) indicates a large contact of the membrane with the channel wall (independent of the nucleus, dotted line), and therefore a high normal constraint, compared to the front (right). The typical observed velocity reaches 12-$15\mu m$/min in channels and 4-$6\mu m$/min on a flat surface. b) Channel geometry and model. Figure from [75].

of contractility indeed favors motion by increasing the speed (recall that $\zeta\Delta\mu < 0$), but can be shown to actually contribute only to 10% of the total velocity. Last, the velocity field of the gel (with respect to the substrate) can be calculated and proves to be backward in the front part of the lamellipodium, and forward at the back. Hence, the cell velocity is necessarily smaller than the polymerization velocity at the leading edge. Such a reasonably simple model therefore captures the main features of lamellipodial motility as observed in experiments on flat substrates.

Motility in confinement

In a recent paper [71] (see also [72] for another cell type), it has been observed both *in-vivo* and *in-vitro* that mutant dentritic cells (DC) that are unable to produce active integrin complexes (adhesion proteins) display sustained motility in confined environments (tissues or synthetic polymeric gels), whereas they fail to move on flat two-dimensional substrates due to their reduced adhesion ability. These observations suggest the existence of an alternative mechanism of motility to the adhesion dependent picture outlined above. Active gel theory enables one to propose an alternative mechanism of motility that accounts for these observations. This mechanism is mainly powered by actin polymerization at the cell membrane and strongly relies on geometric confinement. Interestingly, it does not necessitate strong specific adhesion, and yields velocities potentially larger than the polymerization velocity, as opposed to the case of lamellipodial motility. Such confinement-induced motility mechanism, backed by *in-vitro* experiments of DC motility in microfabricated channels [73, 74] (see Figure 12.16), could aid in understanding how cells migrate in complex crowded environments such as living tissues.

The model, which relies on the hydrodynamic theory of active gels introduced above, was presented in [75] and reads as follows. Let us consider an incompressible viscoelastic film confined in a bidimensional channel of width b. This bidimensional geometry mimics experimental conditions of micro fabricated channels of rectangular section, Figure

12.16. The axes are defined with x along and z across the channel. The confining walls are placed at $z = 0$ and $z = b$. The gel is described as previously by a linear Maxwell model of visco-elasticity, see equation (12.1). In what follows it is assumed that the gel is either in the elastic or viscous regime. For simplicity the gel is assumed to be incompressible, which means that the characteristic time τ depends on pressure P only. This defines a critical pressure such that $\tau(P < P^*) = 0$ (viscous regime) and $\tau(P > P^*) = \infty$ (elastic regime). Next it is supposed that the gel is polymerized at the gel/substrate interface with velocity $v_p \equiv v_z(x, z = 0) = -v_z(x, z = b)$ in the viscous regime, as depicted in Figure 12.16. This is justified by the common observation of actin polymerization activators such as WASP proteins preferentially located along the cell membrane [76]. In the case of DCs, actin filaments can anchor perpendicularly to the cell membrane, forming structures called podosomes where polymerization takes place, therefore inducing an inward flow of actin as it is assumed in the model. Finally viscous friction is assumed at the channel walls $z = 0$ and $z = b$, which writes $\sigma_{xz} = \xi v_x$ where ξ characterizes the friction.

We now derive the dynamical equations of the system in the lubrication approximation ($b \ll L$ where L is the typical length of the system). In this limit the Reynolds number is small and the velocity field $v_i(x, z)$ can be obtained from the force balance $\partial_x P = \eta \partial_z^2 v_x$ and the condition of incompressibility $u_{xx} + u_{zz} = 0$. Defining the average velocity along the channel $v(x) = (1/b) \int_0^b v_x(x, z) dz$, one obtains the following Darcy's law:

$$v(x) = -\frac{b^2}{12\eta}\left(1 + \frac{6\eta}{b}\xi^{-1}\right)\frac{dP}{dx}. \qquad (12.20)$$

Including the depolymerization of the gel k_d, mass conservation of the gel reads $\frac{dv}{dx} = 2v_p/b - k_d$, which gives in turn:

$$\frac{d}{dx}\left[(1 + \tilde{\xi}^{-1})\frac{dP}{dx}\right] = \frac{12\eta(2v_p - bk_d)}{b^3}, \qquad (12.21)$$

where v_p and the nondimensional friction $\tilde{\xi} \equiv \xi b/6\eta$ can be a priori functions of P and x.

Two boundary conditions are needed to determine the pressure profile $P(x)$. We neglect the friction with the surrounding fluid in the channel and set the pressure at the leading edge, which is assumed to coincide with the point $x = L$, as $P(L) = 0$, which gives the first boundary condition. We look for stationary states with broken symmetry and positive velocity and therefore the pressure is a decreasing function of x. One can then argue that if the system is large enough, there exists a travelling front of gel of length L in the fluid phase, travelling at velocity V. The back boundary of this front fluid part coincides with the point $x = 0$ where the pressure reaches the threshold P^*, behind which is an elastic part. Such a denser elastic region at the back of DCs, called the uropod, is indeed well reported [77], and is characterized by a higher concentration of cross-linkers. Assuming that the friction of the uropod with the channel walls is very large, the velocity of the elastic part with

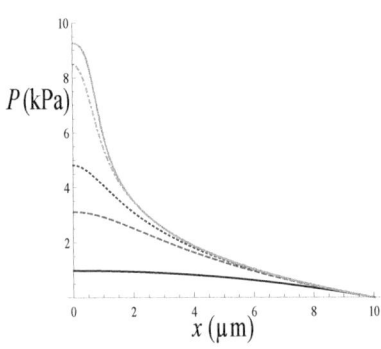

Fig. 12.17 Pressure profile. Curves from the lowest up: numerical value P^0 for $\beta = 0$, $\alpha = 0.01\text{kPa}^{-1}$; $\beta = 1$ and $\alpha = 0.01\text{kPa}^{-1}$; approximation scheme developed in [75]) for $\beta = 1$, $\alpha = 0$; numerical value P^1 for $\beta = 1$, $\alpha = 0$. Other parameters (estimates from [78, 79]): $L = 10\mu m$, $b = 1\mu m$, $\eta = 10\text{kPas}$, $k_d = 0.1\text{s}^{-1}$, $v_p^0 = 0.1\mu m s^{-1}$, $\xi_0 = 0.1\text{kPa s } \mu m^{-1}$. v_p^0 is taken as the speed of DCs on a surface which is expected to be the actin polymerization speed [80]. ξ_0 is taken as very small (lowest estimate in [78] which is 100-fold smaller than in keratocytes [51,70]) to mimic the low adhesion of integrin knockout DCs).

respect to the substrate should be set to $v(0) = 0 = \frac{dP}{dx}|_{x=0}$, giving the second boundary condition which allows the explicit calculation of the pressure field. To conserve the total cell mass, we further assume that in the uropod the gel depolymerizes at the speed of the leading edge $V = v(L) + v_p(L)$ (which defines the overall speed of the cell). A high depolymerization rate in the uropod can be justified by the high pressure and a depletion of free actin monomers due to the forward flow. With these hypotheses, the model presented above mimics a cell moving in a channel with velocity V, and shows that the confinement induced motility mechanism can indeed be used by cells. The self-consistent condition $P(0) = P^*$ gives an equation enabling the calculation of the length L of the moving cell. We stress that the flow velocity is *forward*, i.e. in the same direction as the moving leading edge.

Qualitatively, the value of the length L is dictated by the steepness of the pressure gradient, and therefore by the friction $\tilde{\xi}$. If $\tilde{\xi}$ is small, then only very long cells can move according to this mechanism. The coupling of $\tilde{\xi}$ with the pressure field actually enables small cells to move even with a low bare friction. It is shown [78] that the friction coefficient $\tilde{\xi}$ depends on the normal constraint in the case of a polymeric gel. Qualitatively a high normal constraint increases the attachment rate of polymers onto the channel walls by lowering the entropic barrier, and decreases the detachment rate. In the regime of moderate tangential speed, one has $\tilde{\xi} = \tilde{\xi}_0 e^{\beta(P-\sigma_{nn})}$ [78], where in our geometry the normal stress $\sigma_{nn} = \sigma_{zz}$ for both walls at $z = 0$ and $z = b$. Next, following equation (12.5), the polymerization speed at the cell membrane is assumed to depend on the normal constraint in the gel according to $v_p = v_p^0 e^{-\alpha(P-\sigma_{zz})}$. These assumptions make equation (12.21) an autonomous equation for P, which is completed by the two boundary conditions discussed above. Such an equation cannot be solved analytically in the general case. Approximate schemes and numerical integration can be developed [75] and yield the pressure profile given in Figure 12.17.

Interestingly, it is shown in Figure 12.17 that the critical pressure P^* in the gel can be obtained even for a very low bare-friction coefficient ξ_0, since the exponential dependence of the friction on the pressure field yields very steep pressure gradients and permits the effective friction to reach large values for rather small L. This mechanism of spontaneous motion relies on a pressure build up to P^* in the gel which is here induced by confinement. Indeed, it can be shown that the minimal size L of the cell for this mechanism to work increases faster than linearly with b [75], indicating that this mechanism would not be significant in the case of a gel on a flat open substrate. In this case, which depicts the lamellipodium of a cell lying on a flat substrate [51,70], the typical confining length b is large (of the order of the cell size), and the typical length L necessary to build a strong pressure gradient is very large ($L >$ cell size). As the pressure gradient in the cell is then much weaker than in the confined case, the friction with the substrate remains close to its bare value, yielding a much smaller momentum transfer with the substrate. Additionally, we then expect that the pressure remains below

P^*, and that no elastic phase is formed, thus preventing forward flow. The flow direction, and therefore the direction of the pressure gradient in the gel, constitutes the main difference between the confinement induced mechanism of motility and the standard picture of cells lying on flat substrates. Experimentally the pressure field can be quantified indirectly by measuring the effective contact area of the cell membrane using reflection interference contrast microscopy (RICM). The comparison between Figures 12.18 and 12.19 shows in the confined case a larger contact area at the back of the cell, indicating a backward pressure gradient in qualitative agreement with the theoretical prediction. Interestingly, for the parameter values used in Figure 12.17 the front velocity is calculated to be $\sim 10 \mu m/\text{min}$ which is close to the velocity that can be reached by DCs in collagen matrices [71] and in channels (up to 12-15μm/min) which is significantly larger than the polymerization velocity taken as the speed on a flat surface, 4-6 μm/min [73, 81]. Last, it is shown in [75] that the effect of contractility can also be taken into account and increases the actin flow as expected, even if it is not necessary to generate motility in this case.

Finally, the motility mechanism of DCs in confined environments is strikingly different from the standard picture of cell motility on open flat substrates, and is well captured by the model of confinement induced motility presented in this section (see Figure 12.19). Importantly, this mechanism is widely independent of adhesion properties with the substrate, since the mechanism relies on an enhancement of friction due to a pressure build-up, and does not require specific adhesion proteins. In particular, this result is compatible with the experiments of [71, 74], where it is found that integrin knocked-out DCs are motile only in confined environments. In this model the pressure build-up relies on actin polymerization. The contraction of the actin cortex at the back of the cell, not described in this approach, could also help increase the pressure and therefore friction.

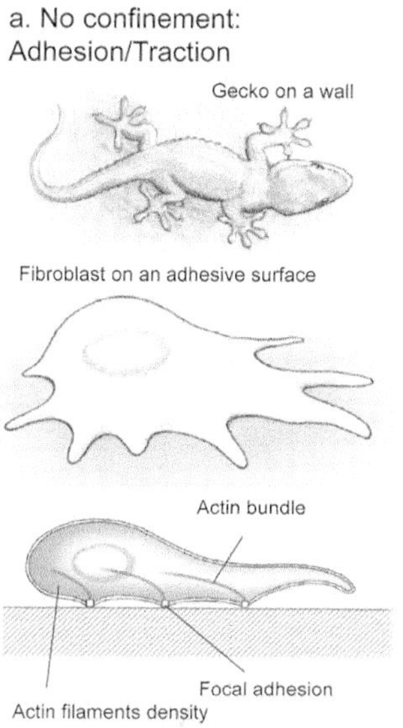

Fig. 12.18 Migration without confinement, the only way to transfer momentum to a surface is to stick to it, like a gecko does on a wall. Typical lamellipodial migration, for example performed by fibroblasts spread on an adhesive surface, involves protrusion at the front, formation of new adhesions, and retraction/de-adhesion at the back. Such cells often form contractile actin bundles bound to focal adhesions, called stress fibers.

12.4 Cell polarity

Almost independently of their mechanical properties, most cells display a stereotypical internal organization often referred to as cell polarity (see Figure 12.20). This property also relies for a large part on the cytoskeleton, but alternative reaction/diffusion mechanisms have also been proposed. This question is particularly fascinating in a physical perspective, as it relates to the question of symmetry breaking. It is also particularly complex as it is at the crossroad of the cytoskeleton and the cell signaling fields. A cell is polarized when it has developed a main axis of organization. Polarization can occur spontaneously or be triggered by external signals, like gradients of signaling molecules, light, forces, or even electrical fields. It involves a reorganization of the cell cytoskeleton and intracellular organelles and eventually results in a specific cell response like directional migration or growth, oriented division,

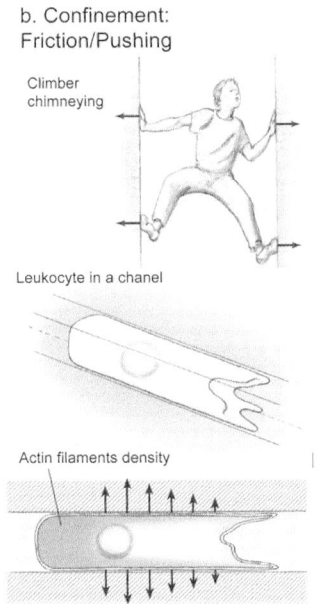

Fig. 12.19 Migration under confinement, adhesion can be replaced by friction. Friction requires a pushing force made possible by confinement, like a chimney climber. Fast migrating cells like leukocytes are poorly adhesive and are thought to perform an equivalent to chimneying. This could be made possible by the pressure generated by the acto-myosin cytoskeleton as it polymerizes and contracts.

and oriented secretion. This fundamental feature of an individual cell is an important property for the organization of tissues in multicellular organizms [82]. The role of the cytoskeleton in cell polarity has been studied in detail in model organizms, in particular in budding yeast, an organizm that often serves as a model system for cell polarity, in the amoeba Dictyostelium discoideum for actin-based polarity [83, 84], and in fission yeast for the role of microtubules [85]. Both systems have lead to a large modeling literature. In both cases two main types of models are proposed: cytoskeleton-independent reaction-diffusion models (Turing type, see [86, 87]) and cytoskeleton dependent models, the cytoskeleton having usually a function in directional transport of polarity factors [88, 89] or combination of transport and reaction diffusion [90] (in this case polarity factors are not directly transported by cytoskeleton polymers but by advection due to an acto-myosin cortical flow). These two types of models have been recently reviewed [91]. In the case of budding yeast, it seems that a cytoskeleton-independent mechanism dominates for the budding process [92], while chemotropism, which involves polarization in an external gradient of pheromone, relies strictly on the actin cytoskeleton. This shows that even within a single organizm, polarization can rely on different mechanisms depending on the physiological context.

Going back to the elementary components of the cell, one interesting question is: at which level can polarization take place? For example, what can a bare lipid bilayer do on its own? As soon as it is composed of more than one type of lipid, segregation phenomena can occur, that can also be influenced by membrane curvature, to give rise to a macroscopic spatial organization [30, 93]. Such phase separation phenomena, well described by physicists, could be viewed as a very basic intrinsic capacity of the cell membrane to polarize. When vesicles interact with surfaces (e.g. electrostatic interactions), they can transiently undergo directional motion over large distances [94]. If membrane proteins are added, the dynamics of the system becomes complex enough to recapitulate such events as membrane invagination resembling endocytosis. It is also important to note that some membrane-binding proteins are able to set diffusion barriers defining sub-regions of the cell cortex [95]. So the cell membrane on its own (with its associated proteins) is already a layer with intrinsic capacity to organize at a macroscopic scale and thus define a polarity at the cell level. But the most discussed player for animal or yeast cell polarity is the actin-based cortical layer.

In short, what are the basic concepts underlying actin systems self-assembly properties and their role in cell polarity? Nucleation of a large number of actin filaments on a surface, with an homogeneous repartition of nucleating complexes, can result in a gel showing spontaneous symmetry breaking, leading to directional movement of cells (the basis of Lysteria monocytogen motility) or beads [15]. The large-scale organization of the actin cortex in cells is governed by a variety of parameters, amongst which two have been well studied in the recent years, and are both under the control of small GTPases: actin filament nucleation (regulated by

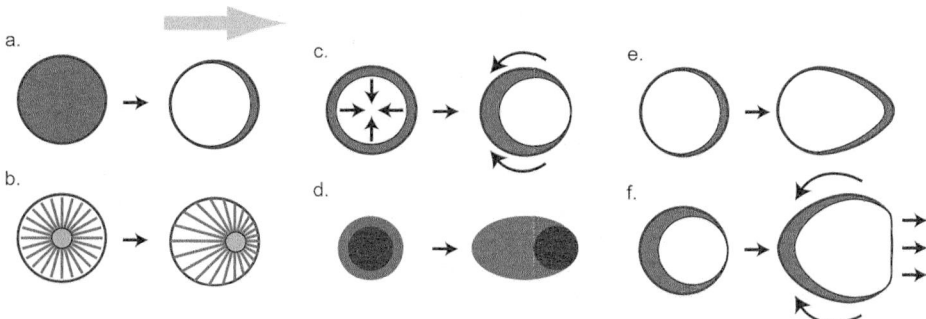

Fig. 12.20 Polarization depicts a phenomenon in which an isotropic object turns into one with a main axis of organization. This axis is usually oriented (grey arrow) a. Case of a diffusible molecule (lighter shading) that switches from a uniform distribution to an anisotropic accumulation on the plasma membrane (outer dark circle). This is the most basic polarization phenomenon. b. Case of an intracellular organelle (the centrosome, with astral microtubules), which moves from a central location to the cell periphery. This phenomenon is often associated with polarized secretion from the cell. c. Polarization can also involve mechanics and be dynamic. Here is shown the case of an actin cortex switching from an isotropic centripetal flow to a "retrograde" polarized flow. d. An actin meshwork polymerizing on an outer surface (bacteria, bead, vesicle, in darker shade) can spontaneously switch to an anisotropic growth leading to propulsion of the object. e. Polarization of a diffusible molecule (e.g. CDC 42) can lead to polarized growth (e.g. yeast cells) f. Polarization of acto-myosin dynamics can lead to shape changes and to cell migration when coupled to friction or adhesion.

the Rac pathway) and actin cortex active contractility driven by myosin (regulated by the Rho pathway), two pathways which often interact with each other, mostly negatively (one inhibiting the other). How various feedbacks can produce spontaneous and stable polarization of the actin cytoskeleton has been well documented in yeast [88, 96]. Feedbacks can simply be biochemical loops (eventually producing spatial organization), or can involve active transport of regulatory proteins on actin filaments (see discussion in the next Section). The effect of local control of contractility on global cell polarity is also well documented [90, 97], with a recent emphasis on the blebbing phenomenon, through which the membrane can locally detach from the actin cortex [98] and on acto-myosin driven cortical flows [90, 99, 100]. Coupling between the cell membrane and the cell cortex indeed constitutes another layer in cell polarization processes. The proteins involved are regulated by coupling to particular lipids and by the action of various kinases and phosphastases [101]. Finally, transmembrane receptors, some of which are involved in cell/cell and cell/matrix adhesion (like integrins and cadherins) link cell cortex polarity with external cues.

The microtubule network, first observed with the help of the early staining techniques in the late 19th century (and then called an aster) has immediately been proposed to play a major role in cell polarity due to its radial organization and position close to the cell center. The two major types of microtubule assemblies found in higher eukaryotes are microtubule bundles and asters. They both provide cell scale polarity due to the intrinsic polarity of microtubules and the capacity of motor proteins and MAPS to distinguish between the two ends when they bind on a microtubule, thus giving to microtubule assemblies the property to sort proteins and organelles inside the cell. These arrays of long dynamic rods can develop some angular anisotropy in an anisotropic environment (gradients of molecules affecting the dynamic instability [102], or the classical search and capture mechanism, when interacting with other cellular organelles, either chromosomes or a polarized cell cortex [103]). They can also get closer to the plasma membrane during various polarization processes, inducing a repositioning of bound and transported organelles, like the golgi apparatus, leading to polarized secretion and eventually to a local cortical modification (e.g. immune synapse [104]). Microtubule networks can vary a lot, from species to species and from cells to cells in a given organism, forming highly specialized structures like axonems (used in cell motility or as sensing organs), rings (used for fission in plants), polarized bundles found in epithelial cells and in neurons (used for transport and for their mechanical properties), and spindles assembled around chromosomes for their segregation during mitosis. In the following Section we will discuss the contribution of active transport on cytoskeletal elements to the polarization processes.

Modeling cell polarization

The discussion of Section 12.2 has shown that active gels can be characterized by a rich phase diagram. In particular it was demonstrated that patterned phases can emerge due to the destabilization of homogeneous states by active stresses. Such instabilities can be seen as simple potential mechanisms of pattern formation and therefore polarization in cells. To support this idea, one needs to consider more realistic cellular geometries and to take into account the coupling of the cytoskeleton to the membrane and outside environment. Recent works [79, 105] have shown that such coupling leads to shape instabilities that could be seen as purely "mechanical" mechanisms of cell polarization.

Besides the clear importance of such polarization mechanisms based on mechanical instabilities, it is well known that the early stage of cell polarization is in many cases characterized by the emergence of non-homogeneous distributions of marker molecules. In such cases, the microscopic dynamics of the markers needs to be taken into account and coarse-grained descriptions of the cytoskeleton alone are insufficient. We will now focus on the modelling of the dynamics of polarity markers. From the physical point of view, the spontaneous occurrence of polarized states even in the absence of external fields implies that a non-equilibrium process that governs the markers dynamics is at work. As discussed in section 12.2 the cell is indeed a system driven out of equilibrium and several mechanisms that result in non-homogeneous distributions of marker molecules can be proposed.

First, reaction-diffusion systems, the simplest of which is the Turing system, are known to yield patterned phases. In this example the homogeneous state is destabilized by the turnover of a slow (or local) activator and a fast (or lateral) inhibitor. Even if stationary states can be reached, such systems are intrinsically out of equilibrium due to the non-conservation of particles of each species, which makes pattern formation possible. While the Turing system is not realistic in its simplest form, possible realizations of similar reaction-diffusion mechanisms involving only a few species can however be found at the cell scale [92,106], a representative example being the Min system in bacteria [107].

Second, and as discussed at length in previous sections, molecular motors are systems out of equilibrium. Their action impacts on mechanical properties but also on transport properties of markers in the cell, whose transport cannot be simply described as a standard thermal Brownian motion. Alternatively to the Turing mechanism, the idea has emerged that cytoskeleton proteins are involved in the polarization process as regulatory factors, in particular for the example of budding yeast [88,89]. In such models it is generally argued that the active transport of markers along cytoskeleton filaments induces a positive feedback loop which favors the emergence of polarized states. Such coupling of the dynamics of molecular markers to the active transport of molecular motors constitutes the topic of this section.

More precisely, we discuss hereafter the dynamics of polarization, i.e. the mechanisms that enable a cell to switch from a symmetric homogeneous state to a polarized state characterized by a non-symmetric distribution of markers. Such symmetry breaking can be either spontaneous – then resulting from a dynamical instability – or driven by an external field depending on the biological function it should fulfill. We will focus here on polarization mechanisms involving active transport processes. Importantly, we will discuss only early stages of polarization and focus on the build-up of non-homogeneous distributions of polarity markers before any mechanical deformation of the cell has occurred. This implicitly assumes that the time scale of the mechanical deformation is much longer than the time scale of polarity marker reorganization. While this seems reasonable for instance for yeast, the coupling of markers dynamics to the cell shape dynamics could play in some cases a key role, as discussed in [91,99].

The molecular basis of spontaneous and driven cell polarization has been much debated over the past decade, and is likely to involve several pathways including historical marks (such as that left by the previous budding event in yeast [76]). However, it is widely recognized that the cytoskeleton plays a crucial role in cell polarization. For example, the efficiency of formation of polar caps in yeast is reduced when actin transport is disrupted using LatA or myosin mutants, and the polar caps formed are unstable [88, 108]. In the case of neurons, it has been shown that the polarization of the growth cone is suppressed when microtubules are depolymerized with nocodazole [109]. To account for these observations, it is generally argued that the cytoskeleton filaments mediate an effective positive feedback in the dynamics of polarization markers [88,89]. This arises from the molecular markers not only diffusing in the cell cytoplasm, but also being actively transported by molecular motors along cytoskeleton filaments. In turn, the markers enhance the polymerization of cytoskeletal filaments which carry more markers.

Actually, the geometry of cytoskeleton filaments plays a crucial role in this mechanism, and it can be shown that despite such positive feedback loop, homogeneous states can be stable and therefore unable to polarize spontaneously, as we discuss below. We stress that here we, as many authors, interpret spontaneous polarization as a dynamical instability of the system, be it in the context of reaction diffusion or active transport models. Alternatively, in Altshuler *et al.* [110] suggest that long-lived fluctuations of markers, which can occur in the case of small numbers of markers, could also be interpreted as spontaneously polarized states. This mechanism, however, implies a fine tuning of the parameters and seems less robust.

Active transport induced polarization. – In this section, we present a minimal model of cell polarization that explicitly takes into account the active transport of markers along cytoskeleton filaments, and show that the geometry of the cytosketon plays a crucial role in the emergence -or not- of spontaneous polarization [111]. Following [111], we assume that polarization markers can be either on the membrane or in the cytoplasm of the cell. For simplicity we assume that the cell is essentially two-dimensional and we neglect curvature effects. The membrane boundary is then taken as a 1D-line along the x-axis and the cytoplasm is parametrized by (x, z). This geometry is sketched in Figure 12.21. The dynamical equations for the concentration of markers on the membrane, $\mu(x, t)$, and in the cytoplasm, $c(x, z, t)$, are given by:

$$\partial_t \mu(x,t) = D_m \partial_x^2 \mu(x,t) + j_m \tag{12.22}$$
$$\partial_t c(x,z,t) = - \nabla \cdot \mathbf{j_b} \tag{12.23}$$

where $j_{m,b}$ are the flux of molecules arriving at the membrane and the flux in the bulk, respectively:

$$j_m = k^{\text{on}} c(x,0,t) - k^{\text{off}} \mu(x,t) \tag{12.24}$$
$$\mathbf{j_b} = - D_b \nabla c(x,z,t) + \mathbf{v} c(x,z,t). \tag{12.25}$$

$D_{m,b}$ are the diffusion constants in the membrane and the bulk of the cell, respectively, and $k^{\text{on,off}}$ are the rates of attachment and detachment to the membrane. In the bulk we assume that the rate of molecules switching between filament-bound and -free is high such that the flux of particles in the cell bulk, $\mathbf{j_b}$, is effectively directed diffusion where $\mathbf{v}$ is the velocity of directed transport of markers along cytoskeleton filaments [112].

The directed transport $\mathbf{v}$, which drives the system out of equilibrium, depends on the geometry of filaments, which we assume are determined by the concentration of markers on the membrane $\mu(x,t)$. This makes the term involving $\mathbf{v}$ in equation (12.23) a nonlinear coupling term. We consider two types of idealized geometries for active transport motivated by two general biological classes of (a) microtubule cytoskeleton and (b) cortical actin systems:

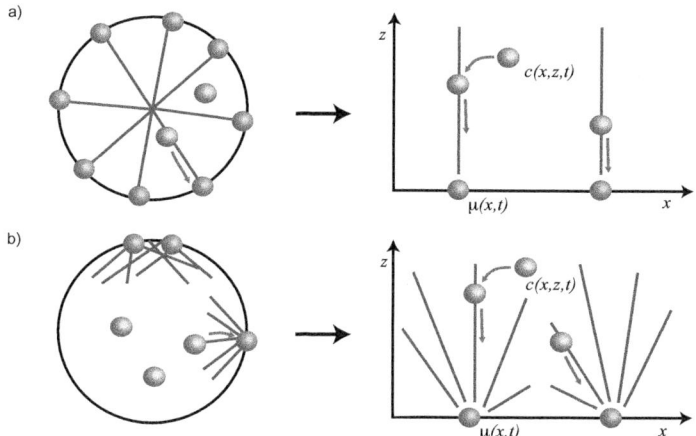

Fig. 12.21 Model geometry. Filaments nucleated at: a) the cell center, and b) the membrane. Arrows indicate the velocity of directed transport. Concentration of molecular markers μ on the membrane and c in the bulk.

a) Filaments that grow from a nucleating center in the cytoplasm towards the membrane can be taken in a first approximation to be perpendicular to the membrane as sketched in Figure 12.21(a). Assuming a "search and capture model" [113], their orientation is regulated by the polarity markers due to preferential stabilization and growth in regions of high marker concentration. With this geometry, in our model, the velocity of directed transport can be written simply as $\mathbf{v} = -\alpha\mu(x,t)\mathbf{u}_z$ where α is a parameter for the strength of coupling. In general microtubules are found with such geometry, e.g. in neuron growth cones, for which observations suggest the polarity markers, GABA receptors, are associated with and regulate the growing microtubule ends [109].

b) Filaments that are polymerized from nucleators localized at the membrane can be approximated by a superposition of asters centered on the membrane as sketched in Figure 12.21(b). We model this as a field at position $\mathbf{r} = (x,z)$ proportional to the filament concentration which decreases with distance from the nucleation point $\mathbf{r}' = (x',z')$ on the membrane such that:

$$\mathbf{v} = -\alpha \int_{-L/2}^{L/2} \frac{\mathbf{r}-\mathbf{r}'}{|\mathbf{r}-\mathbf{r}'|^2} \mu(x',t) \mathrm{d}x' , \qquad (12.26)$$

where α is the coupling parameter and L is the cell perimeter. Note that we expect that other decaying functions of $|\mathbf{r}-\mathbf{r}'|$ would not qualitatively change the results. This geometry mimics the organization of cortical actin (see also [105]), for instance in budding yeast [88, 114] where molecular markers (e.g. Cdc42) are transported along the filaments towards the membrane by myosin V molecular motors [88]. When active, these molecules in turn induce actin nucleation at the membrane thereby creating a positive feedback.

Finally, the above equations are completed by the condition for the

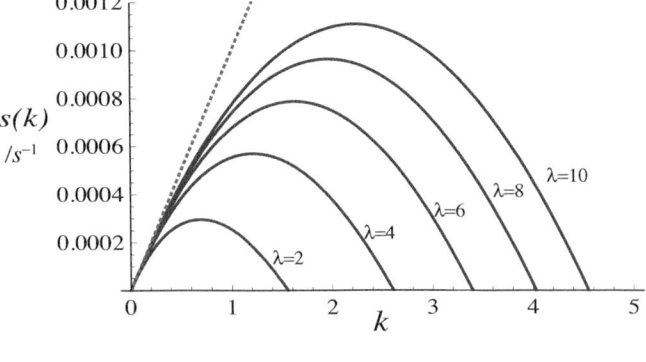

Fig. 12.22 Dispersion relation $s(k)$. k and λ are plotted in dimensionless units normalized by the inverse of the cell radius, $R^{-1} = 2\pi/L$. Parameter values used are: $k^{\text{on}} = 1\mu\text{ms}^{-1}$, $k^{\text{off}} = 0.1\text{s}^{-1}$, $D_b = 0.1\mu\text{m}^2\text{s}^{-1}$, $D_m = 0.01\mu\text{m}^2\text{s}^{-1}$, and $R = 10\mu\text{m}$. The dotted line shows the small k asymptote $s = (D_b k^{\text{off}}/k^{\text{on}})k$.

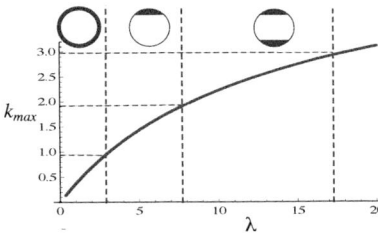

Fig. 12.23 The most unstable mode $k_{\max}$ as a function of $\lambda = \alpha\mu^0\pi/D_b$ (both normalized by R^{-1} as in Figure 12.22). Cartoons of the distribution of markers are shown above the curve.

conservation of total number of polarity markers: $M = \int_{-L/2}^{L/2} dx\, \mu(x,t) + \int_{-L/2}^{L/2} dx \int_0^\infty dz\, c(x,z,t)$. Note that we can assume that the system is not bounded in the z direction since we will only consider concentration profiles exponentially decaying with z.

To determine whether spontaneous polarization occurs in such systems we perform a linear stability analysis of equations (12.22) and (12.23) around the homogeneous out of equilibrium steady state solution which reads $\mu^0 = (k^{\text{on}}/k^{\text{off}})c^0(0)$, $c^0(z) = c^0(0)e^{-\lambda z}$. Here $\lambda = \alpha\mu^0/D_b$ for case (a) and $\lambda = \alpha\mu^0\pi/D_b$ for case (b). The conservation of markers fixes the value of $c^0(0)$ in terms of M. We perturb the system in the infinite size L limit about the homogeneous steady state such that $\mu(x,t) = \mu^0 + \mu_k e^{ikx+st}$, $c(x,z,t) = c^0(z) + c_k(z)e^{ikx+st}$.

In the earlier case (a) of a schematic microtubule system, the analysis of the resulting dispersion relation of the problem shows that the real part of $s(k)$ is always negative, which indicates that the homogeneous state is linearly stable and there is no spontaneous polarization. This actually compares well to experiments in neuron growth cones [109]. When the gradient was removed, the distribution of receptors returned to a symmetric distribution indicating the stability of the homogeneous state and the lack of spontaneous polarization, as expected from our results.

In the case (b) of a schematic cortical actin system modeled by equation (12.26), when the coupling to the cytoskeleton is switched on ($\lambda > 0$), we find that to linear order in k, the dispersion relation gives the positive solution $s = (D_b k^{\text{off}}/k^{\text{on}})|k|$, which indicates that the homogeneous solution is unstable and spontaneous polarization occurs. The full solution $s(k)$ is plotted in Figure 12.22, revealing a maximum at finite $k = k_{\max}$ which corresponds to an instability of finite characteristic length $2\pi/k_{\max}$. Figure 12.23 shows that the most unstable mode $k_{\max}$ increases with λ, and therefore with the strength of the coupling α and the number of markers M. In the case of a cell of perimeter L, we conclude that for a weak coupling $2\pi/k_{\max} > L$ and no patches will be seen. Increasing the coupling α (or M), λ reaches a series of thresholds defined by $2\pi/k_{\max} = L/n_p$ and above which n_p patches on the cell will grow. This generic behavior predicted in the case of cortical

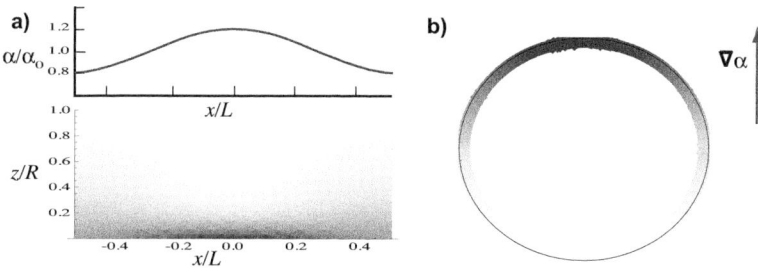

Fig. 12.24 Density of polarity markers $c(x, z)$ for driven polarization with $\alpha(x)$: (a) plotted in the $\{x, z\}$ model plane with parameters as in Figure 12.22 and $\lambda = 6$, $k = 1$, $M = 1000$, $\alpha_k/\alpha_0 = 0.2$, and (b) mapped onto a cartoon circular cell. Darker shading represents higher density.

actin-mediated transport agrees qualitatively with available experimental observations of spontaneous polarization in yeast. Indeed, it is found that spontaneous polarization occurs only above a threshold value of uniform concentration of pheromone (1nM α-factor, which is assumed to drive the coupling α of the model), while it is shown [88] that the number of patches indeed grows with the quantity of markers (Cdc42). Note that the model also predicts that increasing the size of the cell would increase the number of patches, which could be tested experimentally. Finally, the variation of the timescale for the polarization of a cell, $\tau = 1/s$, with the parameter λ corresponds qualitatively well to the data showing decreasing time for spontaneous yeast polarization with increasing concentrations of pheromone.

The model also accounts quantitatively for driven polarization, for instance in the presence of an activator gradient. We assume that such a gradient results in an asymmetric activation of the markers on the membrane and therefore a spatial inhomogeneity in the coupling parameter α. We consider a perturbation $\alpha = \alpha_0 + \alpha_k e^{ikx}$ (where $k = 2\pi/L$ mimics a cell in a constant gradient) and therefore look for solutions at linear order $\mu(x) = \mu^0 + \mu_k e^{ikx}$, and $c(x, z) = c^0(z) + c_k(z)e^{ikx}$.

In the case (a) of a microtubule system, equation (12.23) with this perturbation is solved by $c_k(z) = A_d e^{-\lambda_d z} + (c_k(0) - A_d)e^{-\rho_d z}$ with $\rho_d = \lambda_d/2 + (\lambda_d^2 + 4k^2)^{1/2}/2$, where $\lambda_d = \alpha_0 \mu^0/D_b$, and A_d is fixed by conservation of flux at the membrane. This solution gives the polarized out of equilibrium steady state of the cell in response to an external gradient (see Figure 12.24), and qualitatively reproduces experiments on neuron growth cones [109]. Note that such a response is linear in the perturbation: it therefore occurs for arbitrary small activator gradients and no threshold is involved. Furthermore, the response is linearly stable, which implies that fluctuations in the external activator concentration will be damped out in the response, and the cell will polarize on average only along the activator gradient.

With the same method we can find the stationary state also in the case (b) of a cortical actin system. This analysis however only applies in the regime of small coupling α_0 where the steady-state solution is stable as discussed earlier. In the unstable regime in an activator gradient, the cell is likely to polarize spontaneously in the 'wrong' direction due to concentration fluctuations, with a probability that increases for a short polarization time τ. There is therefore a trade off for the biological system between the ability to polarize fast and the ability to robustly polarize in the direction of a weak gradient. Sensitivity is compromised in cases such as mating yeast where it is important to polarize fast to gain a mating partner. We expect this to be the case for many short range cell-cell interactions where local gradients are strong, masking the effects of fluctuations thereby decreasing the chance of spontaneous polarization in the wrong direction. On the other hand, in cases such as neuron growth cones, the speed of polarization is sacrificed in favor of accurate response to weak gradients.

Finally, the positive feedback triggered by the coupled dynamics of molecular polarity markers and the cytoskeleton is not sufficient to produce spontaneous polarization. Whether the system can polarize spontaneously or not crucially depends on the geometry of cytoskeletal filaments. Active transport of markers along filaments oriented in a centered aster, such as microtubules, leads to a stable homogeneous state, and polarized states occur only in response to a spatial gradient of activator concentration. This result is compatible with [110], where it was argued on the basis of a generic one-dimensional model with only local positive feedback, that homogeneous states are indeed stable and that polarized states occur only transiently due to stochastic fluctuations. On the contrary, in the case of filament asters centered on the membrane, such as cortical actin, active transport induces a non-local coupling which destabilizes the homogeneous states and therefore leads to spontaneous polarization above a threshold uniform concentration of activator. Since the robustness of polarization along a weak gradient is compromised if the cell is able to polarize spontaneously, this is of great importance in biological situations where there is either a robust response to a gradient or fast spontaneous polarization is (dis)advantageous.

Conclusions

In this chapter we briefly presented the biological background on the cell cytoskeleton, pointing to the relevant literature and recent reviews. We also underlined important questions related to this force-generation part of the cell. We then described how this biological knowledge led to a proposal of a new kind of soft matter model giving a coarse-grained description of cytoskeletal polymers and associated molecular motors. This type of model is useful in describing many biological out-of-equilibrium phenomena. Here we presented two developments of such models to describe cell migration and cell polarization, two fundamental properties of single cells. These two properties of cells, together with cell growth and division, have been under detailed investigation for decades, but a major evolution during the past 10-15 years has been the ever-growing contribution of biophysical approaches in major discoveries. This led to a growing body of quantitative methods and data, but also to a strong development of physical modeling.

In many cases, several models have been proposed to account for a given biological process. Experimental evidence is unfortunately often not strong enough to distinguish between proposed models, and the list of models increases, with the risk of also increasing the confusion of the biological community regarding the function and relevance of models in the understanding of biological phenomena. A stereotypical example is the field of budding yeast polarization. Biological data point to a requirement for the actin cytoskeleton in the case of polarization in response to an external stimulation by sexual pheromones (formation of the mating projection or "shmoo"), and also to a role of actin in polarization during the budding process. But in the latter case, actin is not essential at least for part of the process (formation of a "polar cap", which will determine the site of emergence of the bud). Another actin-independent process exists (based on a Bem1 dependent aggregation of active Cdc42). This complex biological situation led to the proposal of two different types of models: one of "Turing" diffusion-reaction type (reviewed in [92]) and the other introducing transport on actin filaments as an important ingredient.

It is interesting to note that there are two largely independent issues here. One is the relevance of each of these models to the biological phenomenon they claim to describe. The other is the interest of the models *per se*. Indeed, a good model is good independently of its relevance for a given phenomenon, and a good experiment is good independently of its accounting by an existing model. The only bad thing in biophysics, which is unfortunately often the case, is vaguely coupling a cheap model to unreliable experimental data. Answering the first issue requires experiments to clearly establish the contribution of each process modeled and the relevance of the proposed models. The second issue is very different, and the interest of the model could be revealed by its relevance for other independent processes. It is indeed clear that models accounting for transport on cytoskeletal polymers have a very wide range of applications. Several recent experimental works point to the importance of transport for cell polarity and morphogenesis [90, 115], and theoretical work points to the new perspectives opened by such phenomena [99, 112]. An interesting recent perspective article [91] even called this type of reaction/diffusion plus transport models "Turing's next step", pointing to the morphogenetic potential of such processes. At this point in the development of both experiments and models, critical reviews that challenge various models with existing data, integrating the subtleties of the various biological processes modeled, would be very important for the progress of such fields as cell migration and cell polarity. Recent reviews have undertaken such approaches [116] but they are still too few.

In conclusion, while the self-organization and dynamic properties of simple sets of cytoskeletal elements can now be explained by physical models, with a rather large consensus, it is not yet clear how, in a real cell, these elements are integrated to give rise to mechanical properties of cells, or to such basic phenomena as cell migration, cell polarity, and division. It is becoming clear that modelling the cytoskeleton is not enough to account for these phenomena. At least one other element has to be considered: the plasma membrane and its associated proteins, in particular ion channels and transporters, but also its association with the extracellular matrix, or with other cells in multi-cellular organisms, or with the cell wall for walled cells like plant cells, yeast and bacteria. These elements combine diffusion barriers and regulation of flows of osmolites, and can thus regulate osmotic pressure, a source of forces much larger than the cytoskeleton can produce.

Acknowledgement

We thank Jean-François Joanny, Jacques Prost, Olivier Collin, and Ana-Maria Lennon-Duménil for helpful discussions.

References

[1] Alon, U. (2007). *Nature Rev. Genet.*, **8**, 450–61.
[2] Kruse, K. and Jülicher, F. (2005). *Curr. Opin. Cell Biol.*, **17**, 20–26.
[3] Théry, M., Jiménez-Dalmaroni, A., Racine, V. et al. (2007). *Nature*, **447**, 493–6.
[4] Yang, M. T., Fu, J., Wang, Y.-K. et al. (2011). *Nature Protoc.*, **6**, 187–213.
[5] Neuman, K. C. and Nagy, A. (2008). *Nature Methods*, **5**, 491–505.
[6] Surrey, T., Nedelec, F., Leibler, S. et al. (2001). *Science*, **292**, 1167–1171.
[7] Nédélec, F., Surrey, T., and Karsenti, E. (2003). *Curr. Opin. Cell Biol.*, **15**, 118–24.
[8] Karsenti, E., Nédélec, F., and Surrey, T. (2006). *Nature Cell Biol.*, **8**, 1204–11.
[9] Mitchison, T. and Kirschner, M. (1984). *Nature*, **312**, 237–242.
[10] Wegner, A. (1976). *J. Mol. Biol.*, **108**, 139–50.
[11] Verkhovsky, A. B., Svitkina, T. M., and Borisy, G. G. (1999). *Curr. Biol.*, **9**(1), 11–20.
[12] dos Remedios, C. G., Chhabra, D., Kekic, M. et al. (2003). *Physiol. Rev.*, **83**, 433–73.
[13] Loisel, T. P., Boujemaa, R., Pantaloni, D. et al. (1999). *Nature*, **401**, 613.
[14] Burroughs, N. J. and Marenduzzo, D. (2007). *Phys. Rev. Lett.*, **98**, 238302.
[15] van Oudenaarden, A. and Theriot, J. A. (1999). *Nature Cell Biol.*, **1**, 493–9.
[16] Reymann, A.-C., Boujemaa-Paterski, R., Martiel, J.-L. et al. (2012). *Science*, **336**(6086), 1310–1314.
[17] Plaçais, P. Y., Balland, M., Guérin, T. et al. (2009). *Phys. Rev. Lett.*, **103**, 158102–.
[18] Pontani, L., van der Gucht, J., Salbreux, G. et al. (2009). *Biophys. J.*, **96**, 192–8.
[19] Salbreux, G., Charras, G., and Paluch, E. (2012). *Trends Cell Biol.*, **22**, 536–545.

[20] Holy, T. E., Dogterom, M., Yurke, B. *et al.* (1997). *Proc. Natl. Acad. Sci. USA*, **94**, 6228–31.
[21] Nedelec, F. J., Surrey, T., Maggs, A. C. *et al.* (1997). *Nature*, **389**, 305–308.
[22] Satir, P., Mitchell, D. R., and Jékely, G. (2008). *Curr. Top. Dev. Biol.*, **85**, 63–82.
[23] Loughlin, R., Wilbur, J. D., McNally, F. J. *et al.* (2011). *Cell*, **147**, 1397–407.
[24] Veigel, C. and Schmidt, C. F. (2011). *Nature Rev. Mol. Cell Biol.*, **12**, 163–76.
[25] Jülicher, F. and Prost, J. (1995). *Phys. Rev. Lett.*, **75**, 2618–2621.
[26] Kodera, N., Yamamoto, D., Ishikawa, R. *et al.* (2010). *Nature*, **468**, 72–6.
[27] Zwerger, M., Ho, C. Y., and Lammerding, J. (2011). *Annu. Rev. Biomed. Eng.*, **13**, 397–428.
[28] Mellad, J. A., Warren, D. T., and Shanahan, C. M. (2011). *Curr. Opin. Cell Biol.*, **23**, 47–54.
[29] Kind, J. and van Steensel, B. (2010). *Curr. Opin. Cell Biol.*, **22**, 320–5.
[30] Roux, A., Cuvelier, D., Nassoy, P. *et al.* (2005). *EMBO J.*, **24**, 1537–45.
[31] Römer, W., Pontani, L.-L., Sorre, B. *et al.* (2010). *Cell*, **140**, 540–53.
[32] Vogel, V. and Sheetz, M. (2006). *Nature Rev. Mol. Cell Biol.*, **7**, 265–75.
[33] Kobayashi, T. and Sokabe, M. (2010). *Curr. Opin. Cell Biol.*, **22**, 669–76.
[34] Charras, G. and Paluch, E. (2008). *Nature Rev. Mol. Cell Biol.*, **9**, 730–6.
[35] Charras, G. T., Mitchison, T. J., and Mahadevan, L. (2009). *J. Cell Sci.*, **122**, 3233–41.
[36] Minc, N., Boudaoud, A., and Chang, F. (2009). *Curr. Biol.*, **19**, 1096–101.
[37] Fayant, P., Girlanda, O., Chebli, Y. *et al.* (2010). *Plant Cell*, **22**, 2579–93.
[38] Campàs, O. and Mahadevan, L. (2009). *Curr. Biol.*, **19**, 2102–7.
[39] Hamant, O., Heisler, M. G., Jönsson, H. *et al.* (2008). *Science*, **322**(5908), 1650–5.
[40] Brown, A. E. X. and Discher, D. E. (2009). *Curr. Biol.*, **19**, R781–9.
[41] Discher, D. E., Mooney, D. J., and Zandstra, P. W. (2009). *Science*, **324**, 1673–1677.
[42] Kruse, K. and Jülicher, F. (2000). *Phys. Rev. Lett.*, **85**, 1778–1781.
[43] Liverpool, T. B. and Marchetti, M. C. (2003). *Phys. Rev. Lett.*, **90**, 138102.
[44] Nedelec, F. and Foethke, D. (2007). *New J. Phys.*, **9**(11), 427.
[45] Toner, J., Tu, Y., and Ramaswamy, S. (2005). *Ann. Phys.*, **318**, 170–244.

[46] Gardel, M. L., Shin, J. H., MacKintosh, F. C. *et al.* (2004). *Science*, **304**(5675), 1301–1305.

[47] Storm, C., Pastore, J. J., MacKintosh, F. C. *et al.* (2005). *Nature*, **435**, 191–194.

[48] Trepat, X., Deng, L., An, S. S. *et al.* (2007). *Nature*, **447**(7144), 592–595.

[49] Larson, R. (1998). *Constitutive Equations for Polymer Melts and Solutions.* Butterworth-Heinemann, Waltham, MA.

[50] Dogterom, M., Janson, M.E., Faivre-Moskalenko, C. *et al.* (2002). *Appl. Phys. A: Mater. Sci. Proc.*, **75**, 331–336.

[51] Jülicher, F., Kruse, K., Prost, J. *et al.* (2007). *Phys. Rep.*, **449**, 1–76.

[52] Kruse, K., Joanny, J.-F., Jülicher, F. *et al.* (2004). *Phys. Rev. Lett.*, **92**, 078101.

[53] Kruse, K., Joanny, J.-F., Jülicher, F. *et al.* (2005). *Eur. Phys. J. E*, **16**, 5–16.

[54] Simha, R. A. and Ramaswamy, S. (2002). *Phys. Rev. Lett.*, **89**, 058101.

[55] de Gennes, P. G. and Prost, J. (1993). *The Physics of Liquid Crystals.* Oxford. Univ. Press, Oxford.

[56] Joanny, J.-F., Jülicher, F., Kruse, K. *et al.* (2007). *New J. Phys.*, **9**, 422.

[57] Voituriez, R., Joanny, J.-F., and Prost, J. (2005). *Europhys. Lett.*, **70**, 404–410.

[58] Voituriez, R., Joanny, J.-F., and Prost, J. (2006). *Phys. Rev. Lett.*, **96**, 028102.

[59] Takiguchi, K. (1991). *J. Biochem.*, **109**, 520–527.

[60] Tanaka-Takiguchi, Y., Kakei, T., Tanimura, A. *et al.* (2004). *J. Mol. Biol.*, **341**, 467–476.

[61] Blankschtein, D. and Hornreich, R. M. (1985). *Phys. Rev. B*, **32**, 3214–3228.

[62] Schaller, V., Weber, C., Semmrich, C. *et al.* (2010). *Nature*, **467**, 73–77.

[63] Köhler, S., Schaller, V., and Bausch, A. R. (2011). *Nature Mater.*, **10**, 462–468.

[64] Minc, N., Burgess, D., and Chang, F. (2011). *Cell*, **144**, 414–26.

[65] Manley, S., Gillette, J. M., Patterson, G. H. *et al.* (2008). *Nature Methods*, **5**, 155–7.

[66] Lautenschlaeger, F. and Piel, M. (2012). *Curr. Opin. Cell Biol.*, **25**, 116–124.

[67] Théry, M. (2010). *J. Cell Sci.*, **123**, 4201–13.

[68] Eyckmans, J., Boudou, T., Yu, X. *et al.* (2011). *Dev. Cell*, **21**, 35–47.

[69] Even-Ram, S. and Yamada, K. M (2005). *Curr. Opin. Cell Biol.*, **17**, 524–32.

[70] Kruse, K., Joanny, J.-F., Jülicher, F. *et al.* (2006). *Phys. Biol.*, **3**, 130–137.

[71] Lämmermann, T., Bader, B. L, Monkley, S. J. *et al.* (2008). *Na-*

ture, **453**, 51–55.

[72] Malawista, S. E. and de Boisfleury Chevance, A. (1997). *Proc. Natl. Acad. Sci. USA*, **94**, 11577–11582.

[73] Faure-André, G., Vargas, P., Yuseff, M.-I. *et al.* (2008). *Science*, **322**, 1705–1710.

[74] Renkawitz, J., Schumann, K., Weber, M. *et al.* (2009). *Nature Cell Biol.*, **11**, 1438–1443.

[75] Hawkins, R. J., Piel, M., Faure-Andre, G. *et al.* (2009). *Phys. Rev. Lett.*, **102**, 058103.

[76] Alberts, B. (2002). *Molecular Biology of the cell* (4th edn). Garland Science, New York.

[77] Serrador, J. M., Nieto, M., and Sánchez-Madrid, F. (1999). *Trends Cell Biol.*, **9**, 228–233.

[78] Gerbal, F., Chaikin, P., Rabin, Y. *et al.* (2000). *Biophys. J.*, **79**, 2259–2275.

[79] Callan-Jones, A. C., Joanny, J.-F., and Prost, J. (2008). *Phys. Rev. Lett.*, **100**, 258106.

[80] Theriot, J. A. and Mitchison, T. J. (1991). *Nature*, **352**, 126–131.

[81] van Helden, S. F. G., Krooshoop, D. J. E. B., Broers, K. C. M. *et al.* (2006). *J. Immunol.*, **177**, 1567–1574.

[82] Bryant, D. M. and Mostov, K. E. (2008). *Nature Rev. Mol. Cell Biol.*, **9**, 887–901.

[83] Slaughter, B. D., Smith, S. E., and Li, R. (2009). *Cold Spring Harb. Perspect. Biol.*, **1**, a003384.

[84] Franca-Koh, J., Kamimura, Y., and Devreotes, P. (2006). *Curr. Opin. Genet. Dev.*, **16**, 333–8.

[85] Martin, S. G. (2009). *Trends Cell Biol.*, **19**, 447–54.

[86] Mori, Y., Jilkine, A., and Edelstein-Keshet, L. (2011). *SIAM J. Appl. Math.*, **71**(4), 1401–1427.

[87] Goryachev, A. B. and Pokhilko, A. V. (2008). *FEBS Lett.*, **582**(10), 1437–43.

[88] Wedlich-Soldner, R., Altschuler, S., Wu, L. *et al.* (2003). *Science*, **299**, 1231–1235.

[89] Marco, E., Wedlich-Soldner, R., Li, R. *et al.* (2007). *Cell*, **129**, 411–422.

[90] Goehring, N. W., Trong, P. K., Bois, J. S. *et al.* (2011). *Science*, **334**, 1137–41.

[91] Howard, J., Grill, S. W., and Bois, J. S. (2011). *Nature Rev. Mol. Cell Biol.*, **12**, 392–398.

[92] Johnson, J. M., Jin, M., and Lew, D. J. (2011). *Curr. Opin. Genet. Dev.*, **21**, 740–6.

[93] Baumgart, T., Hammond, A. T., Sengupta, P. *et al.* (2007). *Proc. Natl. Acad. Sci. USA*, **104**, 3165–70.

[94] Solon, J., Streicher, P., Richter, R. *et al.* (2006). *Proc. Natl. Acad. Sci. USA*, **103**, 12382–12387.

[95] Barral, Y., Mermall, V., Mooseker, M. S. *et al.* (2000). *Mol. Cell*, **5**, 841–51.

[96] Howell, A. S., Savage, N. S., Johnson, S. A. *et al.* (2009). *Cell*, **139**,

731–43.

[97] Paluch, E., van der Gucht, J., and Sykes, C. (2006). *J. Cell Biol.*, **175**, 687–92.

[98] Tinevez, J.-Y., Schulze, U., Salbreux, G. *et al.* (2009). *Proc. Natl. Acad. Sci. USA*, **106**, 18581–6.

[99] Hawkins, R. J., Poincloux, R., Bénichou, O. *et al.* (2011). *Biophys. J.*, **101**, 1041–1045.

[100] Poincloux, R., Collin, O., Lizárraga, F. *et al.* (2011). *Proc. Natl. Acad. Sci. USA*, **108**, 1943–1948.

[101] Bretscher, A., Edwards, K., and Fehon, R. G. (2002). *Nature Rev. Mol. Cell Biol.*, **3**(8), 586–99.

[102] Caudron, M., Bunt, G., Bastiaens, P. *et al.* (2005). *Science*, **309**(5739), 1373–6.

[103] Schuyler, S. C. and Pellman, D. (2001). *J. Cell Sci.*, **114**(Pt 2), 247–55.

[104] Yuseff, M.-I., Reversat, A., Lankar, D. *et al.* (2011). *Immunity*, **35**(3), 361–74.

[105] Salbreux, G., Joanny, J.-F., Prost, J. *et al.* (2007). *Phys. Biol.*, **4**, 268–284.

[106] Iglesias, P. A. and Devreotes, P. N. (2008). *Curr. Opin. Cell Biol.*, **20**, 35–40.

[107] Loose, M., Kruse, K., and Schwille, P. (2011). *Annu. Rev. Biophys.*, **40**, 315–336.

[108] Irazoqui, J. E., Howell, A. S., Theesfeld, C. L. *et al.* (2005). *Mol. Biol. Cell*, **16**, 1296–1304.

[109] Bouzigues, C., Morel, M., Triller, A. *et al.* (2007). *Proc. Natl. Acad. Sci. USA*, **104**, 11251–11256.

[110] Altschuler, S. J., Angenent, S. B., Wang, Y. *et al.* (2008). *Nature*, **454**, 886–889.

[111] Hawkins, R. J., Benichou, O., Piel, M. *et al.* (2009). *Phys. Rev. E*, **80**, 040903–4.

[112] Loverdo, C., Bénichou, O., Moreau, M. *et al.* (2008). *Nature Phys.*, **4**, 134–137.

[113] Mimori-Kiyosue, Y. and Tsukita, S. (2003). *J. Biochem.*, **134**, 321–326.

[114] Karpova, T. S., McNally, J. G., Moltz, S. L. *et al.* (1998). *J. Cell Biol.*, **142**, 1501–1517.

[115] Jaqaman, K., Kuwata, H., Touret, N. *et al.* (2011). *Cell*, **146**(4), 593–606.

[116] Jilkine, A. and Edelstein-Keshet, L. (2011). *PLoS Comput. Biol.*, **7**, e1001121.

Biological Fluid Interfaces and Membranes

13

Aidan T. Brown[1,2] and Pietro Cicuta[1]

(1) Cavendish Laboratory and Nanoscience Centre
University of Cambridge, Cambridge CB3 0HE, UK
(2) School of Physics and Astronomy
University of Edinburgh, Edinburgh EH9 3JZ, UK

Introduction

Surfactant molecules can self-assemble, or assemble onto existing amphiphilic interfaces, adopting a wide range of complex shapes and structures, ordered in 1, 2, or 3 dimensions, over length scales spanning from nanometers to microns. As well as displaying a range of different possible structural phases, there are numerous examples of hierarchical order. For example, surfactants can form bulk phases which are fluid at the molecular scale, yet highly ordered at longer length scales. A key self-assembled structure is the bilayer; this is itself a very rich motif. In multi-component bilayer systems the two leaflets can have asymmetric compositions, and can exhibit patterning due to lateral segregation and mutual repulsion of fluid domains. These features give membrane-based systems the capacity to deform spontaneously into complex shapes, or undergo significant morphological changes in response to applied stresses, fields, chemical or thermal gradients, or other stimuli. In addition they can exhibit flow and self-healing, while maintaining their internal structure, patterning, and compartmentalization. These remarkable structural and dynamical features have permitted the biological evolution of complex living cells, and hold great promise for future biotechnological and biomedical applications.

This chapter describes both monolayers and bilayers; we will call *films* the very asymmetric structures that form at the interface between coexisting bulk phases, in a multiphase system. There are several examples of interfaces between biological fluids; the most relevant to human physiology are probably the liquid/air interfaces in the lungs and on the surface of the eyes. Both of these feature films with remarkable properties of compressibility and self-healing which will be discussed below. In contrast to a film, a *membrane bilayer* can separate two compartments which may have the same, or similar, chemical nature. Membranes, and in particular phospholipid bilayers, are a structural feature present in

[1] The constituent molecules of films and interfaces are surface active, and the reader is referred to Chapter 2: Surfactants in this Handbook for general properties of this class of molecules. The surfactants used in biological systems are a restricted class, limited both by the biosynthesis pathways available and by self-assembly constraints discussed below.

the cells of every biological organism on earth. Starting from the 1920s, they were recognized as structural and functional units regulating crosstransport, and in many cases providing a scaffold for interactions to take place. Membranes allow the fine regulation of concentration levels which are essential to maintain the biochemical reaction networks that underpin life. Indeed, membranes have been discussed as the fundamental structure that enabled cellular life to originate [1].

In such a vast and interdisciplinary topic, we have made choices of content and covered various aspects in differing depths. This chapter aims to summarize the current knowledge on surface films and membranes, giving context for their role in biology, but particularly overviewing the fundamental physical ideas that underpin fundamental studies of *in-vitro* model systems.[1] The theoretical background to describe membranes from a modern continuum and statistical mechanics point of view developed starting from the 1970s when Canham and Helfrich introduced a powerful description of the membrane in terms of its configuration-dependent free energy, giving insight into the role that this description can play in understanding aspects of cell biology.

Constituent molecules of biological interfaces

Biological membranes are essential in defining compartments, isolating chemical species, and providing a barrier to control chemical gradients and differences in composition within cells and between different cells. Membranes are made of opposed single layers of phospholipids, making a bilayer, as sketched in Figure 13.1. The balance between various lipid-lipid interaction forces, including their interaction with the surrounding aqueous medium and the entropic drive to disperse, results in making this bilayer structure the thermodynamically stable state, for a certain class of surfactants. Single monolayers of phospholipids are found in cer-

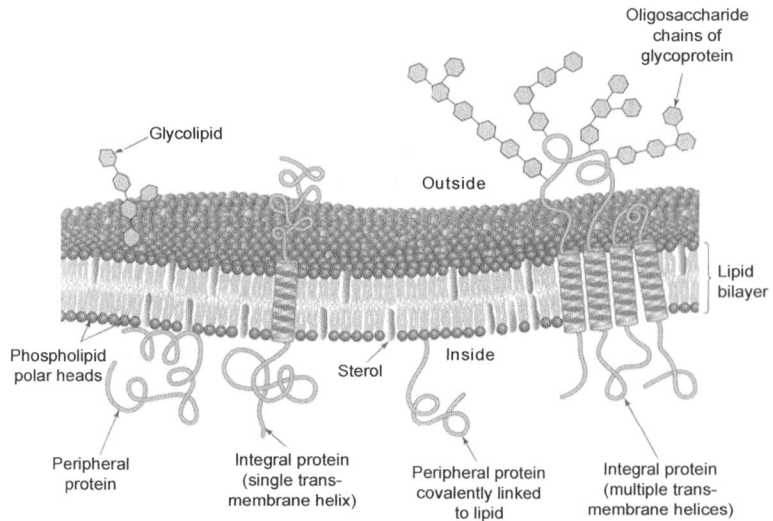

Fig. 13.1 Schematic diagram of the plasma membrane of an animal cell, showing various lipid and protein classes. The hydrophilic headgroups face outward, toward the aqueous phase, while the hydrophobic chains are shielded from water. Redrawn from Nelson and Cox [2].

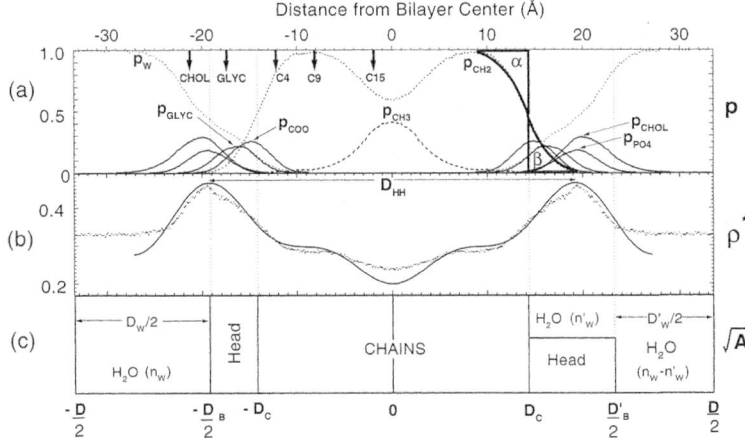

Fig. 13.2 Three representations of structure of DPPC bilayers in the L_α fluid phase. (a) Probability distribution functions p for different component groups; the downward pointing arrows show the peak locations determined by neutron diffraction with 25% water. (b) Electron density profile ρ^* from X-ray studies (solid line) and from simulations (dots). (c) Two alternative volumetric pictures: on the left monolayer is a simple three compartment representation; on the right monolayer a more realistic representation of the interfacial headgroup region. Reproduced with permission from [3], ©2000, Elsevier.

tain organs, most notably on the interface between the aqueous internal medium and air. Single layers are also commonly studied as the simplest experimental model system on which to understand the physical chemistry of films, for example the intermolecular interactions, packing, and the interaction of a film with other molecules adsorbing from solution.

Surfactant molecules are defined as having the tendency to partition in-between two co-existing phases (amphiphilic), most commonly an aqueous phase and a non-polar phase such as air or an organic liquid. Typically these molecules have a polar "head group" which interacts favorably with water, and a non-polar hydrophobic "tail". The most common surfactants have one or two tails of varying length. In general, the more chemical groups are in the tail, the lower the solubility in the aqueous phase. Short-tail molecules are commonly used as detergents, and there is a dynamical exchange between each molecule being adsorbed on the interface and it being in solution.[2]

Amphiphilic molecules by their nature tend to aggregate to minimize unfavorable interactions with the solvent. This typically takes place above a threshold concentration (the critical micelle concentration, CMC). The criteria for understanding of the CMC on general thermodynamic grounds, and to determine the possible shapes of aggregates formed from molecules of different average shape, are very well introduced in [4]. Single-chain lipids tend to form aggregates with a high intrinsic curvature; they are present in cells but cannot form extended bilayers by themselves.

In most cases the molecules that make up the continuous scaffold in biological films and interfaces are the phospholipids (also called glycerophospholipids) and sphingolipids. Sterols are smaller lipid molecules (such as cholesterol) which are also present, often at high concentration, but could not form bilayers without the phospholipid scaffold. Single chain lipids (such as lysolipids and fatty acids) are also present and influence the permeability of membranes to transport of other molecules. From the point of view of self-assembly (thermodynamic stability) and

[2]This is known as Gibbs adsorption equilibrium, and is discussed further in the "Surfactants" chapter of this Handbook. In contrast, longer tailed molecules are essentially irreversibly adsorbed. The molecules used by biological organisms are generally long-tail but need to strike a balance: long enough to guarantee stability and integrity to membranes, over reasonable timescales; short enough to enable their synthesis by the organism (also avoiding waste of resources) and to allow processes like bending, dissociation of membranes and pore formation.

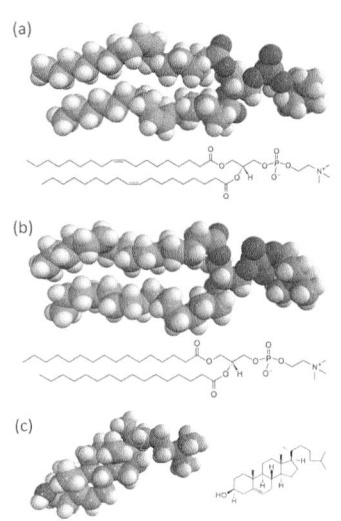

Fig. 13.3 Three common lipids, discussed in various sections of this chapter, shown in two standard representations. (a) 18:1 PC, full name: 1,2-dioleoyl-sn-glycero-3-phosphocholine (DOPC); (b) 16:0 PC, full name: 1,2-dipalmitoyl-sn-glycero-3-phosphocholine (DPPC); (c) cholesterol. DOPC is a saturated phospholipid, while DPPC is unsaturated; this difference is crucial for the miscibility of phospholipid species, and is shown schematically in Figure 13.4. Cholesterol is a sterol with many biological roles.

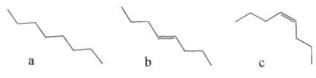

Fig. 13.4 Phospholipids may have tail groups which are either a) saturated, b) trans-unsaturated, or c) cis-unsaturated.

mechanics, the lipid scaffold is already a complex and important system to be understood. Proteins and carbohydrates are also present in biological films and membranes, sometimes accounting for over half of the weight fraction. Membrane proteins (either permanently anchored via some trans-membrane groups or in temporarily association with the membrane) are involved in many key biochemical reactions [5]. Membrane lipids are themselves often also directly involved in these reactions; for example, binding of the protein AKT to the phospholipid PIP$_3$ is a signal that can trigger a range of events downstream, including the onset of cell growth and protein production [6]. The fact that biological cell membranes are bilayers was proved experimentally in the 1920s by Gorter and Grendel, but the generality of this finding remained under discussion into the 1970s, until it was accepted as the "Fluid Mosaic Model" of Singer and Nicolson [7]. Contemporaries are now thinking well beyond the fluid mosaic model; the role of the lipid scaffold is receiving renewed attention, as it becomes clear that the overall lipid composition is tightly regulated, and that the physical properties of the lipid phase can themselves control and regulate protein function. In many membranes, the lipid scaffold itself is neither spatially homogeneous nor static in composition over time.

The structure of single-component lipid bilayers is well understood: many systems have been investigated by neutron diffraction, enabling to locate fairly precisely the distribution within the bilayer thickness of the key chemical groups; there is reasonable agreement also with numerical simulations, see Figure 13.2 and [3]. These profiles provide atomistic information which can in principle be used to calculate the mechanical behavior of the layer under tension, or under a bending deformation.

A detailed introduction to the chemical nature of phospholipids is beyond the scope of this chapter, and readers should consult dedicated monographs [2, 8]. One essential feature of the phospholipids is that they have two acyl chains, each a linear chain of typically between 12 and 24 carbon atoms, see Figure 13.3. The C–C bonds can be saturated or unsaturated; in the latter case, biological organisms typically produce a bond with a *cis* configuration [9], i.e. there is a kink in the linear chain (see Figure 13.4). The introduction of a *trans* double bond to a saturated phospholipid lowers the fluid to gel transition (so-called 'main' transition) temperature of the lipid because of the reduced packing ability due to the loss of freedom of rotation around the double bond. The kink produced by the *cis* bond further reduces the packing ability and the transition temperature. Biologically produced unsaturated phospholipids typically only have *cis* double bonds [9]. Mixtures of saturated and unsaturated lipids have a tendency to phase separate into coexisting domains. The chemical nomenclature of phospholipids (e.g. 1-palmitoyl-2-oleoyl-sn-glycero-3-phosphocholine, also known as POPC, the most abundant species in many cell membranes) accounts for the polar head group, the length of the chains, and the saturation of bonds (see http://www.chem.qmul.ac.uk/iupac/lipid/ and references within).

Outline

We have introduced lipids, and particularly phospholipids, which are the main components of biological interface films and membranes. They are the biological building blocks, providing the interactions and structures that can be self-assembled into higher-order structures (mesophases). The rest of this chapter is structured as follows: Section 13.1 addresses fluid interfaces, introducing fundamental concepts to describe the behavior of two-dimensional systems under deformation, and presenting pulmonary surfactant and tear film as biological examples of films with remarkable dynamical properties. We then go into more depth of the mechanics and dynamics of bilayers, because this is a particularly well developed but at the same time still an active research area within Soft Matter physics, both in terms of theory and experiment. In contrast to monolayers, bilayers resist bending deformation. Section 13.2 introduces the fundamental concept of the Helfrich free energy, from which the continuum mechanics of the membrane and membrane shapes, can be obtained. Membrane fluctuations are discussed including dynamical aspects in Section 13.3, coupling the energy concepts introduced above with low Reynolds' number hydrodynamics. Biological membranes are multi-component mixtures, a fact which is proving to have broad implications; the thermodynamics and phase behavior of lipid mixtures are the topic of Section 13.4. In Section 13.5 we provide a brief introduction to the behavior of lipids within cells. This section is no substitute for cell biology textbooks [5, 10], but provides some specific references and a personal overview of processes within cell biology where membrane physics is known to play a significant role; it is hoped it can establish a common ground for physicists and biologists to discuss membrane behavior. Finally we conclude by pointing to some open questions in this broad and exciting field.

References will be given throughout, but it is worth highlighting a set of recent reviews, focusing on specific aspects of this field. A broad overview of interfaces in soft matter is described in [11]. Fundamental aspects of physics including advanced theoretical treatment of membrane shape and mechanics are reviewed in [12, 13]. A monograph with a similar scope to the current chapter is [14], while a source of further information on the biochemistry and molecular structure of biomembranes and membrane proteins is [8]. Particularly important areas of membrane *in-vitro* research addresses the action of drugs [15], drug delivery [16] and specifically anesthetics [17]. The importance and complexity of lipid membranes in biology, well beyond their function as chemical barriers, are presented in [18]. Our approach to membrane biological physics is in the spirit of [19], with a focus on slightly different examples.

13.1 Fluid interfaces and films

The first record of scientific study of surface monolayer films is Benjamin Franklin's report to the Royal Society in 1774 describing the effects on

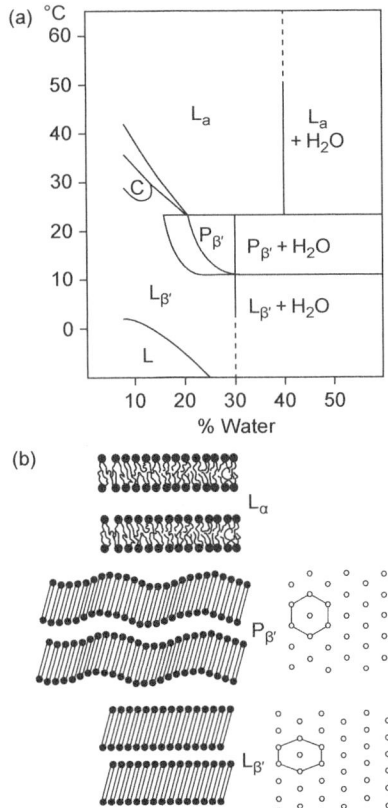

Fig. 13.5 Lipid bilayers exist in several phases, determined by the degree and type of in-plane ordering. Bulk lipid samples show similar phase behavior, but they additionally show complex phase behavior in the out of plane ordering. (a) Phase diagram of DMPC in water from [22], with (b) diagrammatic representations of the various lipid phases. The three phases $L_{\beta'}$, $P_{\beta'}$ and L_α correspond to bulk lamellar phases with the intra-bilayer structure as shown in (b). $L_{\beta'} + H_2O$, $P_{\beta'} + H_2O$ and $L_\alpha + H_2O$ are isolated planar membranes or vesicles, with the corresponding intra-bilayer structures. The shaded regions are other bulk phases, such as cubic and hexagonal. In the isolated and bulk lamellar phases, the concentration of water is largely irrelevant to the internal structure of the bilayer, which is determined by intermolecular interactions within the bilayers; here the relevant parameter is temperature, as can be seen from the horizontal phases boundaries. Figure redrawn from [22].

the surface waves of spreading a teaspoon of oil on a pond [20]. Various other historical scientific figures, often best known today for discoveries in other fields, have since then been interested in oil films. Lord Rayleigh's paper [in *Philosophical Magazine* **48**, 337, 1899] establishes that surface films are one molecule thick. Interface films are also part of everyday products, including detergents and foods, and indeed in 1891 Agnes Pockels became a pioneer of interface science by observing oil films in her home kitchen, resulting in the modern technique of surface concentration control by sweeping barriers [as reported in *Nature* **43**, 437, 1891]. Cooking has been carried out for thousands of years, clearly predating physiology as a science, and many foods are multiphase systems (like mayonnaise and whipped cream) where surface mono- and bilayers hold the key to texture and mechanics.

Modern monolayer studies are mainly performed on well-defined flat surfaces, but the real systems one is modeling are the interfacial monolayers that enable the formation of large surface area systems (e.g. the lung alveoli), or determine long-time stability in heterogeneous liquids, like liquid-liquid emulsions or gas-liquid dispersions (such as foams) by preventing coagulation [21]. Considering that the interfacial area in such systems is large enough that a significant mass fraction of material is confined to surface films, it is no surprise that the flow properties of the whole system can be effectively determined by the two-dimensional film viscosity and elasticity.

The scientific curiosity in studying surface films stems partly from the comparisons that are possible between the usual everyday bulk materials and the properties of the corresponding two-dimensional systems. In many cases, one can study the same molecules or colloids as fluids or suspensions both in three and in two dimensions. Phase transitions (gas, liquid, solid phases can be identified), mechanical properties (e.g. compressibility) and flow behavior, can all be accessed and measured in two as well as in three dimensions. Although the experiments are performed with very different equipment, the theoretical ideas and physical explanations have much in common. There are however some important exceptions where the reduced dimensionality leads to singular behavior, such as the existence of intermediate phases between gas, liquid, and solid in two-dimensional systems of hard disks [23] or rods [24], and in the dynamical aspects of two-dimensional systems [11]. An underlying theme throughout this chapter is the transition in two dimensions from a fluid to a solid in systems that do not crystallize, or crystallize partially similar to bulk liquid crystals, see Figure 13.5.

The basic statistical physics of interfaces, as introduced by Safran [25] and the chapter 2 of this Handbook, are relevant to understand the basic behavior of surface films. Surfactants are subject to the sources of interaction common in many colloidal systems, such as the effect of excluded volume, charge interactions, counterion entropy, hydrophobic effects, etc. To give a measure of the amount of relevant theoretical progress that occurred after Gaines' 1966 seminal monograph on surface films [26], one only needs to remember that the understanding of

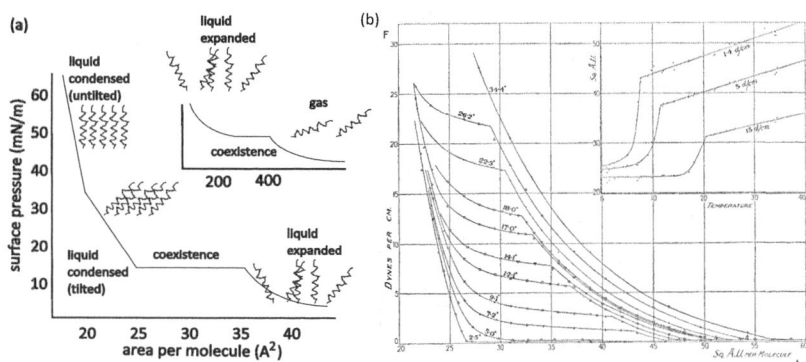

Fig. 13.6 (a) Schematic diagram of the main phase transitions that take place in a lipid monolayer. In the 'gas' phase, the lipid molecules are fairly flat on the interface, occupying a large area per molecule, and exerting therefore a very low osmotic surface pressure. (b) The phases of surface lipid monolayers have been known and studied with the Langmuir trough technique since the early days of surface science [27]. Visible here, at the larger surface areas (labeled as A) and extending to more compact systems and higher pressure (labeled F for force) the higher the temperature, is the 'liquid expanded' (or LE) phase described by equation (13.11). At the highest temperature, this layer is always in this disordered phase.

critical phenomena through renormalization group theory, providing a background for scaling analysis in complex fluids, and indirectly for so much of soft matter physics, occurred later in the 1970s. In terms of the dynamics in these systems, our understanding is much more limited; molecules in surface films can pack very densely, with excluded volume playing a very strong role in limiting lipid motility. A general theory of dynamical arrest (dynamics near the glass transition) would be relevant here but is still elusive.

General concepts

This section summarizes the background and nomenclature used for surface films, much of which is carried through to membranes, thus covering the whole chapter. The first of these is the surface tension γ: a thermodynamic quantity defined as:

$$\gamma \equiv \frac{\partial F}{\partial A}, \tag{13.1}$$

where $F = F(A, T)$ is the Helmholtz free energy of the surface. This definition of tension is clear in the context of bulk phases coexisting in equilibrium; out of equilibrium, or for decorated surfaces and bilayer films, there are as yet unresolved issues, discussed in Section 13.4. Consider a surface (or interface) of area A on which a known amount of molecules has been deposited to form a monolayer. For this system the measurable surface (or interface) tension γ is usually reduced compared to the clean tension γ_0, due to the presence of the monolayer. This reduction is identified as a surface pressure Π (it is an osmotic pressure) exerted by the layer: $\Pi \equiv \gamma_0 - \gamma$.

A measurement of Π as a function of A at constant T is referred to as an *isotherm* and it is a very basic experimental characterization of the monolayer phase behavior, see Figure 13.6. Together with x-ray diffrac-

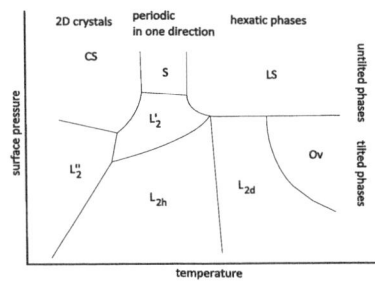

Fig. 13.7 Many phases can be identified in monolayers, based on the degree of order. The figure illustrates qualitatively the classic example of the liquid-crystalline monolayer phases of Behenic acid (C_{22}) monolayers, adapted from [29].

tion, the increasing molecular order of molecules on compression has been mapped, see Figure 13.7. Under compression, some phospholipid layers can reduce the surface tension of water, from its value of around 72×10^{-3} N/m to almost zero. Shorter chain molecules will reach a maximum pressure ('collapse') beyond which material is lost into the subphase, or the layer ceases to be a flat monolayer. The surface concentration Γ is inversely proportional to A, since phospholipids typically give rise to *insoluble* films, i.e. a monolayer stable in time with respect to desorption. This can be achieved for many systems, in which the surface concentration can remain *out* of equilibrium with the concentration of molecules dissolved in the subphase because the energy barrier for desorption is very high. The origin of this barrier can be the di-block nature of the molecule (for example if it has a polar and a non-polar side), or purely geometrical in the case of colloidal sized macromolecules [28].

The macroscopic dynamical response of monolayers is usually classified in terms of constitutive coefficients such as elastic moduli [11]. A different problem is the fundamental physical explanation of these coefficients, for example their relation to molecular structure. Elements of surface rheology (flow behavior) are summarized in the reviews by Joly [30] and Miller *et al.* [31] that cover the field starting from the earliest contributions by Boussinesq. Many aspects are common between the rheology of three-dimensional systems and two-dimensional films, but there are two very significant differences:

- In three dimensions the gas phase is compressible but the liquid and solid phases are almost incompressible, whereas in two dimensions all three phases (which can still be distinguished by the structural correlations and by the flow properties) are compressible due to the greater freedom of molecular rearrangement on a surface, so that three-dimensional techniques have to be adapted with caution to two dimensions. As a simple example, one can obtain the shear viscosity in a three-dimensional liquid by measuring the drag on a sphere, but when moving a disk in the plane of a monolayer attention has to be given to the possibility of causing compression as well as shear.
- The flow of surface monolayers is always coupled to some degree with flow of the liquid subphase, and for each experimental technique this has to be taken into account to recover the physical parameters of the two-dimensional layer itself.

In-plane deformation. One can imagine three kinds of deformation to occur in the surface plane, as shown in Figure 13.8: dilation (or compression), shear, and uniaxial extension. The last two occur with no area change. In practice, pure dilation experiments can be realized by changing the volume of a pendant drop covered by a monolayer; shear experiments can be done in linear canal or concentric ring geometries; pure extension (which would be analogous to 3D experiments to measure Young's modulus) is very difficult to achieve on a surface film, since the film usually has to be confined by a frame on all sides. Note that dilation and shear are the basic independent modes of deformation. The mechanical response to compressions is proportional to the dilatational elastic modulus ε, defined as:

$$\varepsilon \equiv -A \frac{\partial \Pi}{\partial A}. \quad (13.2)$$

It is the inverse of a compressibility and it can be measured if Π is known as a function of A. The resistance to dilatational flow is characterized by the dilatational viscosity η_d defined as

$$\eta_d \equiv A \frac{\Delta \Pi}{\partial A / \partial t}, \quad (13.3)$$

and there are methods to measure η_d specifically.

As in bulk rheology, the shear elastic modulus G is defined as the ratio between the increment of shear stress and the increment of strain, and the shear elastic viscosity η_s as the ratio between the shear force and the rate of shear. Shear elasticity only appears when a solid structure is formed on the interface between fluid phases.

Out-of-plane deformation. It is clear that if a surface which is initially flat is deformed while holding the boundaries fixed, then the surface area increases. This deformation will require an energy proportional to the increase itself and to the surface tension γ. It is worth recalling here that the surface tension is equal to the surface energy per unit area only for the case of a 'clean' interface. If the deformations are small, then the surface can be described by the deviation $h = h(x, y)$ from the plane $z = 0$, as sketched in Figure 13.9.

Phospholipid *monolayers* have a very small bending modulus, but in *bilayers* the energy required for bending becomes significant; in Section 13.4 it is shown how to calculate the average mean-square amplitude of these fluctuations, in the general case of an interface film or membrane that resists both stretching and bending.

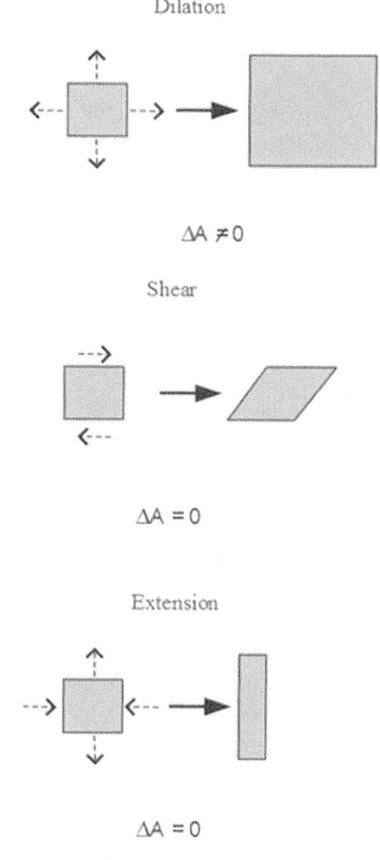

Fig. 13.8 Simple geometries for in-plane surface deformations. Note that extension, which is a simple mechanical test in 3D yielding the Young's modulus of the material, is experimentally challenging for liquid surface films. Dilation can be realized by inflating a spherical bubble, coated with the film to be studied; special Langmuir troughs with square or circular perimeters have also been designed. Excellent approximations to shear deformation are possible, see main text.

Linear rheology. This is the framework that holds when the response of the system is proportional to the stress which is applied. If the response is linear, it follows that superposition is valid for successive events. Assuming a monolayer of area A_0 and small time-dependent area changes, $A(t)/A_0 = 1 + \Delta A/A_0 \exp(-i\omega t)$ where ΔA is the amplitude and ω is the frequency, under linear response theory the pressure (this is the stress) can be written as:

$$\Pi(t) = \varepsilon^*(\omega) \frac{\Delta A}{A_0} \exp(i\omega t), \quad (13.4)$$

where $\varepsilon^*(\omega)$ is the complex dilatational modulus, defined as:

$$\varepsilon^*(\omega) = i\omega \int_0^\infty \varepsilon(t) \exp(-i\omega t) dt. \quad (13.5)$$

This is a frequency-dependent quantity, which can be written as:

$$\varepsilon^*(\omega) = \varepsilon'(\omega) + i\varepsilon''(\omega) = \varepsilon'(\omega) + i\omega\, \eta_d(\omega), \quad (13.6)$$

where the real part is identified with the elastic component of the modulus and the imaginary part with the dissipative component.

In the same way as introduced above for dilation, in an oscillatory measurement one can measure the complex shear modulus:

$$G^*(\omega) = G'(\omega) + i\omega\,\eta_s(\omega) = G'(\omega) + i\,G''(\omega). \tag{13.7}$$

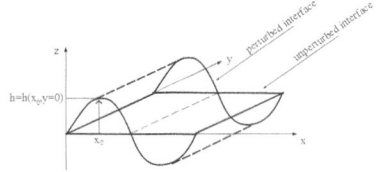

Fig. 13.9 Diagram of an out-of-plane deformation, not to scale. Typical thermal fluctuations roughen the interface with an amplitude of a few Å, and over a wide range of lengthscales from the molecular scale up to the system size. Gravity provides the restoring force for large wavelength fluctuations, while surface tension forces are more important at shorter wavelengths. The crossover between these two regimes occurs at the capillary lengthscale $\ell_c = \sqrt{\gamma/(\Delta\rho g)}$, where γ is the surface tension, $\Delta\rho$ is the density difference between the upper and lower phases, and g is the gravitational acceleration.

Many dilute monolayer systems behave as purely elastic under dilation ($\eta_d = 0$) and have a negligible shear modulus (like in an ideal fluid). In this case ε can be simply derived from the uniaxial compression isotherm and this equilibrium modulus will be called *static*, ε_{st}, because of the very low frequencies that correspond to usual compression speeds. Techniques based on probing the wave or fluctuation dynamics, e.g. surface light scattering, measure the sum $G^*(\omega) + \varepsilon^*(\omega)$. However it is usually the case that $G^*(\omega)$ is negligible for the type of systems where these techniques function best (films which are in the gas phase, or in any case at low concentration), and therefore results are in most cases discussed as if $\varepsilon^*(\omega)$ alone is measured [32]. The interfacial stress rheometer (ISR) described in [33] isolates specifically the shear modulus $G^*(\omega)$.

Statistical mechanics of surface films

Statistical mechanics is the theoretical framework used to connect the observable equilibrium macroscopic thermodynamic variables (in particular here the easily accessible surface pressure) to the system's symmetries and microscopic properties. For example, one would like to understand the following points, starting from measurable macroscopic quantities:

- the conformation of molecules at the interface (flat, upright, coiled, rod-like, entangled, aggregated, etc.);
- the inter-molecular interactions, for example between charged lipids, proteins, or colloid particles;
- the physical state of the system (gas, liquid, glass, etc.).

Experiments typically measure either equilibrium properties like the osmotic pressure and its dependence on the concentration, temperature and subphase conditions, or the dynamical behavior (properties of flow), by observing for example the relaxation following thermal fluctuations or an external deformation.

It is useful to recall how surface-active molecules lower the surface tension, and this section follows the discussion in [25]. If N surfactant molecules of radius a_0 are put on a surface of area A, then if the molecules do not interact (or in the very dilute limit), the Helmholtz total free energy of the surface can be written as:

$$F_{surf} = A\left[\gamma_0 + \frac{k_B T}{a_0^2}\phi\,(\log\phi - 1) + \frac{u_0}{a_0^2}\phi\right], \tag{13.8}$$

where $\phi = Na_0^2/A$ is the covered area fraction. The first term in equation (13.8) is the energy of the clean surface, the second term is the ideal gas entropy, and in the third term u_0 is the energy gain per molecule at the surface. If the surfactant were *soluble*, the surface concentration would be determined by an equilibrium between the surface and bulk chemical potentials, which would, in general, depend on a balance between u_0 and the entropy in the bulk, as described by the Gibbs absorption equation. However for the case of *insoluble* molecules, somewhat counter-intuitively, the surface anchoring energy u_0 plays no part in reducing the surface tension. This can be seen by applying the definition of surface tension $\gamma = \partial F_{surf}/\partial A$ to equation (13.8) (at constant N), thereby obtaining:

$$\gamma = \gamma_0 - \frac{k_B T}{a_0^2}\phi. \quad (13.9)$$

Equation (13.9) is the 2D equivalent of van't Hoff's law, which says that the osmotic pressure Π of a dilute system is the same as the pressure of an ideal gas. If excluded volume effects are considered, equation (13.8) is modified to:

$$F_{surf} = A\left[\gamma_0 + \frac{k_B T}{a_0^2}[\phi\log\phi + (1-\phi)\log(1-\phi)] + \frac{u_0}{a_0^2}\phi\right], (13.10)$$

and the surface tension is:

$$\gamma = \gamma_0 + \frac{k_B T}{a_0^2}\log(1-\phi). \quad (13.11)$$

The remarkably simple form of equation (13.11) is essentially the same as used to fit isotherm data in the literature. While strictly this treatment is describing the 'gas' phase at lowest concentration, this form also fits very well the 'liquid expanded' phase of lipid monolayers, see Figure 13.6. The reason for this is that the 'Liquid Expanded' phase is essentially another 'gas' phase, in which the lipid molecules are oriented perpendicular to the interface, and a phospholipid in this condition (e.g. DPPC) has a typical area per molecule[3] of 75Å^2.

A connection with the physical conditions in bilayers may be useful. Taking DPPC (the major component of lung surfactant) as an example, its area per molecule in free bilayer films is measured by x-rays to be around 64Å^2. At this area per molecule, the compression isotherm is very temperature dependent, and at body temperature, the LE/LC coexistence has almost disappeared: the phase is LE, the pressure is around 20×10^{-3} N/m, and with a little further compression down to 50Å^2 the layer is close packed entering a solid phase. The comparison of area per molecule is meaningful, in that we understand from this that in the self-assembled bilayer state there remains some flexibility for the molecules to flow, and adjust their tilt and hence area per molecule and bilayer thickness.

---

[3] It is instructive to consider the effect of molecular size on the pressure unit scale $k_B T/a_0^2$. For

- small surfactant molecules: $a_0 \simeq 5$Å $\leftrightarrows (k_B T/a_0^2) \simeq 20$mN/m;
- phospholipids in LE: $a_0 \simeq 9$Å $\leftrightarrows (k_B T/a_0^2) \simeq 5$mN/m;
- compact proteins: $a_0 \simeq 50$Å $\leftrightarrows (k_B T/a_0^2) \simeq 0.2$mN/m;
- colloidal spheres (or disks) visible in a microscope: $a_0 \simeq 500$nm $\leftrightarrows (k_B T/a_0^2) \simeq 2 \times 10^{-5}$mN/m;

and it should be kept in mind that 0.2mN/m is roughly the sensitivity of instruments commonly used to measure surface tension, so that it is not currently possible to measure the surface pressure of a 2D gas of colloidal particles.

Viscosity of particle suspensions

For solid inclusions in the surface film (or in the membrane bilayer), the hydrodynamics needs to take account of various boundaries: inclusion/membrane; inclusion/surrounding liquid; membrane/surrounding liquid. This last contribution is very important: the flow set up in the membrane is coupled to bulk flow. The situation has been studied in various relevant limits. If the two-dimensional membrane shear viscosity is $\eta_m^{(2d)}$ (typical values in the L_o phase are $10^{-8} - 10^{-6}$ Ns/m [34]) and the bulk viscosity is $\eta^{(3d)}$ (typically 10^{-3} Ns/m^2), the ratio $\eta_m^{(2d)}/\eta^{(3d)}$ defines a lengthscale ℓ_m ($10^{-5} - 10^{-3}$ m in the $\eta_m^{(2d)}$ range above). Saffman and Delbrück [35] analyzed the case of small inclusions of radius $a < \ell_m$, for which although the main part of the dissipation takes place in the membrane; the coupling to the bulk phase remains important. The diffusion coefficient has a particularly weak dependence on the inclusion radius:

$$D(a) = \frac{k_B T}{4\pi \eta_m^{(2d)}} \left[\ln\left(\frac{\ell_m}{a} - \gamma + s\right) \right], \quad (13.12)$$

where γ is Euler's constant 0.5772... and s depends on the boundary conditions: $s = 0$ for solid domains and $s = 1/2$ for liquid inclusions. The Saffman and Delbrück result is valid for small inclusions (and high viscosity membranes) [35]. It has been tested on the diffusion of micrometer-sized domains in a background of liquid-ordered phase [34]. This hydrodynamical result will eventually fail on inclusions approaching the single lipid molecular scale, as was verified experimentally with small peptides [36].

For larger inclusions ($a \gtrsim \ell_m$), the bulk dissipation dominates, and indeed in the large a limit the diffusion coefficient is independent of the membrane viscosity [37]:

$$D(a) = \frac{k_B T}{16\pi \eta^{(3d)}} \frac{1}{a}. \quad (13.13)$$

Readers should follow [38] for a treatment of the cross-over region $a \sim \ell_m$. From the point of view of hydrodynamics, the case of an inclusion moving within a viscous monolayer, surrounded by less viscous phases, is the same as that of a bilayer system.

By taking into account the drag due to flow of a solvent of viscosity η_s around a single sphere, Einstein calculated the excess viscosity of a dilute suspension of spheres in the bulk, $\eta = -\eta_s(1 + 2.5\Phi)$, where η is the viscosity of the suspension and Φ is the volume fraction of spheres. As the concentration is increased, hydrodynamic interactions between particles become important and they contribute higher-order corrections. For many concentrated systems the viscosity is well approximated empirically by the Krieger-Dougherty equation [39]:

$$\eta_0 = \eta_s \left(1 - \frac{\Phi}{\Phi_{max}}\right)^2, \quad (13.14)$$

where Φ_{max} is the close packing fraction.

For particles confined to the two-dimensional surface of a bulk fluid, it is expected that the same scaling forms should hold.

Examples of biological films

Interface films assemble between liquids and gases. Skin, and more generally epithelial tissues, act as interfaces, but they are stiff enough that the science of liquid interface films described here may not be relevant. In contrast, other interfaces have to remain soft, or even liquid, and are continually dynamically assembling and regenerating over time: In the human body, very important examples are the interface films in the lungs (pulmonary surfactant film) and on the surface of the eyes (tear film).

The alveoli in lungs have a very large surface area, and one function of lung surfactant is to allow mechanical stability of these structures. Lung surfactant is however a mixture, with some remarkable properties. In general as discussed previously when the surface density of surfactants is increased, there is a series of transitions toward liquid and solid phases. At some point of compression, the in-plane pressure reaches a threshold where it becomes favorable for the layer to "collapse", i.e. to lose its two-dimensional character. For most surfactants the collapse process corresponds to a loss of material into the subphase, and this process is irreversible on short timescales since, on expansion, the adsorption is limited by diffusion in the bulk. Some layers however have enough intermolecular coherence that they collapse in a different way, by "buckling" and forming folds out of the plane; this process is immediately reversible on expansion. Lung surfactant (LS) has this property, see Figure 13.10. LS is a mixture of lipids and proteins that coats the alveoli. On exhalation, the monolayer is compressed, and eventually collapses by forming folds and wrinkles. Attaining a low surface tension at collapse is crucial for lessening the work of breathing, and the protein fraction is essential for proper functioning of the lung, providing the essential ability to fold/unfold and thereby reincorporate collapsed materials upon inhalation [40].

Another example of lipid fluid interface with even more remarkable rheological properties, and achieved without an input from proteins, is tear film, a stratified system sketched in Figure 13.11. An in-depth review of the fluid dynamics aspects is given in [41]. Over fractions of a second, this film is capable of re-spreading over the surface of the cornea, making an optically smooth interface with air. The biochemical composition of this structure is presented in detail in [42]: a lipid layer forms the outer interface between an aqueous layer of the tear film and air, playing a large role in maintaining tear film stability. The conventional view is that below the aqueous layer lies a mucus layer in contact with the cell surface, although there may not be a sharp distinction between aqueous and mucus layers. "Meibomian lipids" are the primary component of the lipid layer, and their physical properties appear to

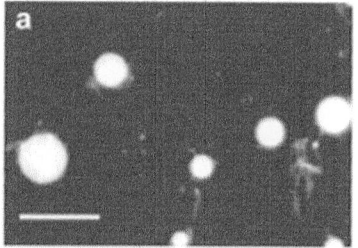
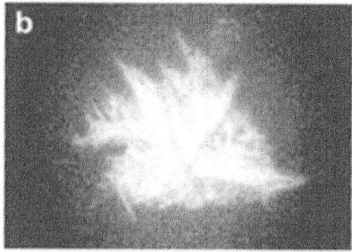
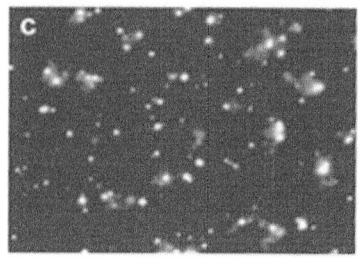

Fig. 13.10 Fluorescence micrographs of surface monolayer mixtures mimicking lung surfactant film show different behaviors at collapse. (a, b) show monolayer collapse from single component films of different fluidity; (c) a binary mixture shows a different squeeze out of one component. In (a) POPG (low viscosity) forms bilayer disks and tubes during collapse. In (b) the rigid palmitic acid film on water forms dendritic growth of a crystalline phase. In (c) the more fluid component, POPG, is squeezed out from a binary mixture of DPPG and POPG leaving the film enriched in the more rigid component, DPPG. The scale bar is $50\,\mu$m. This mechanism is thought to enable the film to quickly extend to a large coverage on expansion. Reproduced with permission from The Biophysical Society [40].

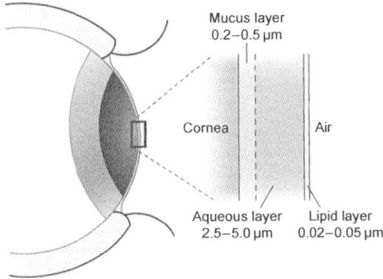

Fig. 13.11 Diagram of the three-layer structure of the fluid film on the surface of the cornea. This remarkable structure is able to reform every time the eye blinks. Redrawn from [41].

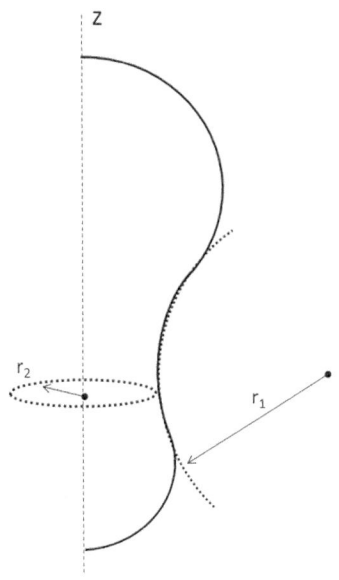

Fig. 13.12 Diagram showing the extremum curvatures, $c_1 = -1/r_1$, $c_2 = 1/r_2$ for a membrane which is axisymmetric around the z-axis. The extremum curvatures are for general surfaces defined as the curvature of the circles of maximum and minimum radius which are tangent to the surface at a particular point. Curvature is defined as positive if it is concave toward the interior of the membrane, negative otherwise.

be particularly crucial to the functionality of the tear film [43]. The lipid layer decreases the surface tension of the air/tear-film interface to around 45×10^{-3} N/m, and reduces evaporation, stabilizing the tear film against rupture (referred to as tear-film breakup in the eye literature) [41]. Film rupture occurs physiologically around once a minute in healthy eyes, and a blink (either complete or partial) of the eyelid, occurring over around 0.1 s, restores the film integrity. The spreading process of this thin layer is complex, and a key role is played by the lipid film, which is compressed into a liquid condensed (LC) phase on closing the lids, and re-spreads onto clean aqueous layer under its own pressure. A recent study [43] has suggested that on expansion the lipids remain locally in their LC gel-like state, with the film percolating over the available area by expanding while maintaining a connected network; this is consistent also with the relatively high surface viscosity measured for the lipid film.

13.2 Membrane mechanics

The basic physics required to calculate the material response functions and the amplitude of thermal fluctuations for these systems is calculated in this section and in Section 13.3, where membrane dynamics is considered. These results are the backbone enabling data on the material properties of bilayers to be recovered from experiments that probe membrane mechanics. Only a narrow range of bilayer material properties allows biological processes as we know them to take place, as discussed in a later section.

The Helfrich free energy

As with many other condensed matter systems, it is possible, and useful, to treat lipid bilayers in a continuum limit, where the identity and location of individual molecules ceases to be relevant [25]. In this limit, a particular lipid membrane can be described by a set of moduli and friction coefficients, coupling to various types of motion. The values of these parameters are determined by the molecular interactions in the bilayer (see for example [44]), but here they are treated as phenomenological constants. In the continuum limit, the energy density of a lipid bilayer has two contributions: the intrinsic curvature energy of the surface, and various fields on the surface, such as the lipid surface density and, for multi-component membranes, the lipid composition. The simplest curvature energy density up to quadratic order in the curvatures, which is invariant with respect to changes of coordinate, and symmetric between the two extremum curvatures c_1 and c_2 (see Figure 13.12) can be written as [45]:

$$f = 2\kappa (H - C_0)^2 + \kappa_G K_G, \tag{13.15}$$

where $H = (c_1 + c_2)/2$ is the mean curvature, $K_G = c_1 c_2$ is the Gaussian curvature, and C_0 is a spontaneous curvature.

On top of this intrinsic curvature energy, additional fields couple to each other and to the curvature. The simplest such field is the lipid surface density in the inner (ρ_-) and outer (ρ_+) monolayers, where the densities are defined as the number density of lipids at the dividing surface between the two monolayers. To quadratic order, these density terms couple to themselves and to the curvature to give [46]:

$$f = 2\left(\kappa + \frac{KD^2}{4}\right)(H - C_0)^2 + \frac{K\rho^2}{2} + \kappa_G K_G, \quad (13.16)$$

where

$$\rho = \frac{\rho_+ - \rho_-}{2\rho_0}, \quad (13.17)$$

$$\bar{\rho} = \frac{\rho_+ + \rho_- - 2\rho_0}{2\rho_0}, \quad (13.18)$$

and ρ_0 is the mean density of lipids per monolayer, D is the distance between the surfaces around which the two monolayers pivot, approximately corresponding to the distance between the headgroups, and K is the area expansion modulus of the bilayer. This coupling is illustrated in Figure 13.13.

The free energy of a membrane can be described by an integral of the local energy density over the membrane surface S:

$$F_{local} = \int_S 2\left(\kappa + \frac{KD^2}{4}\right)(H - C_0)^2 + \frac{K\rho^2}{2} dS, \quad (13.19)$$

where the Gaussian curvature term has disappeared since, according to the Gauss Bonnet theorem, an integral of the Gaussian curvature over a surface depends only on the topology of the surface, which will not matter for local minimizations of curvature energy, or for thermal fluctuations.

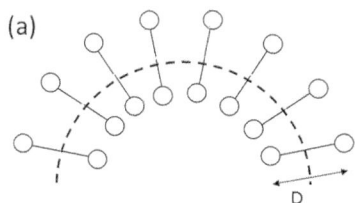
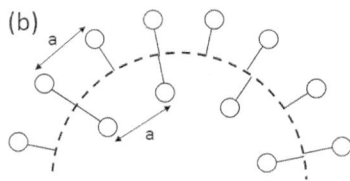

Fig. 13.13 How lipid density influences bilayer bending. Bending of (a) a membrane at constant ρ^+ and ρ^- (the inner and outer monolayer density projected onto the bilayer center marked by the dashed line), (b) an incompressible monolayer: the headgroups, or some other pivot point a distance $D/2$ from the midplane, maintain a fixed spacing a.

Vesicle shapes

The surface integral of equation (13.19) is not, on its own, sufficient to work out the equilibrium shape of a closed membrane system. Certain quantities, typically the number of particles in a given state, are strictly conserved over experimental timescales. These strict conservation laws are not directly useful in a continuum model, because individual molecules are not identified in the continuum limit. However, these constraints can be translated into global potentials coupling to, for example, the total area or volume of the bilayer, or the difference between the areas of the inner and outer monolayer. Though total lipid number may be strictly conserved, fluctuations of the total mean area are still permitted because of the finite value of the area expansion modulus of the membrane. However, if the modulus of this global potential is found to be sufficiently high, the quantity of interest itself, such as the area or volume of the membrane, can be approximated as strictly conserved. Furthermore, the Lagrangian multiplier technique can be used to transform this conservation law into a term in the local energy density. Because of the high bulk modulus of water, and the high area compressibility modulus K of lipid bilayers, the constraints on solute and total lipid molecule number can be translated into strict constraints on volume and area, respectively, in this fashion, and the total energy becomes:

$$F = \int_S \sigma + 2\left(\kappa + \frac{KD^2}{4}\right)(H - C_0)^2 + \frac{K\rho^2}{2} dS - \int_V P dV, \quad (13.20)$$

where the second integral is over the volume V, bounded by the surface S, and σ and P are Lagrangian multipliers which maintain the given area and volume, respectively. The Lagrange multiplier approach is an approximation toward the more rigorous method of imposing hard constraints on the volume and surface area by Dirac delta functions in the partition function [48]. This approximation is exact for determining the equilibrium shape, because the determination of the equilibrium shape only depends on the first variation of the Hamiltonian. However, the approximation is not necessarily valid for all thermal fluctuations, as discussed in [48].

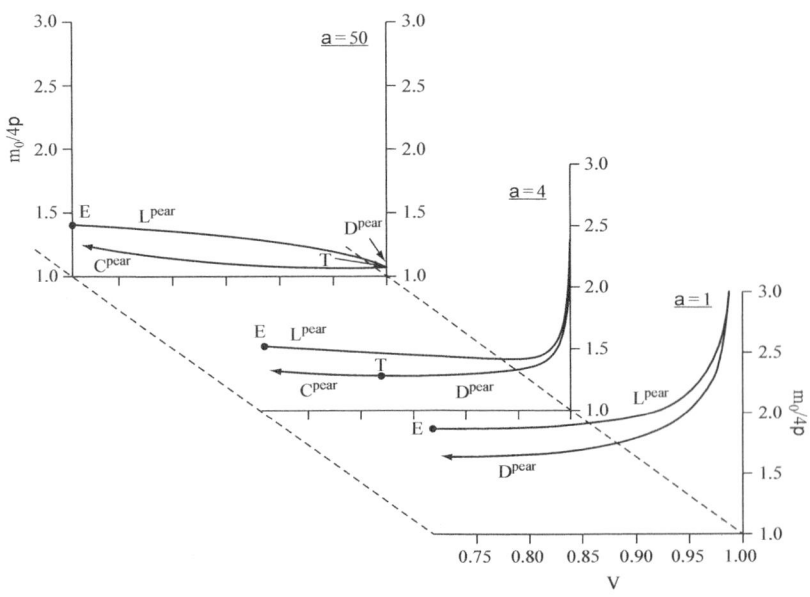

Fig. 13.14 The phases in the vesicle shape phase diagram depend on the area difference elasticity α of the ADE model, as shown in the figure [47]. m_0 corresponds to the equilibrium area difference ΔA_0, and v is the reduced volume. C_{pear} and D_{pear} are continuous and discontinuous phase boundaries between pear shaped (above) and prolate (below) vesicles, respectively. Continuous and discontinuous transitions meet at the tricritical points labeled T. For $\alpha = 1$, this transition is discontinuous over the range of the figure, whereas for $\alpha = 50$, the transition is continuous throughout. L_{pear} is the limiting boundary of the pear shape, corresponding to two spheres connected by a vanishingly small neck. These phase boundaries correspond to the prolate/pear and the pear/bud boundaries in Figure 13.15, respectively. At the points marked E, the calculation of the phase boundary was halted, because more complicated shapes become relevant there, including the non-axisymmetric starfish shapes shown in Figure 13.15. Figure reprinted with permission from [47]; Copyright (1994) by the American Physical Society.

The third and more controversial constraint is the constraint on the difference between the number of molecules in the two monolayers. It is not clear whether this constraint on monolayer number can be converted immediately into a strict constraint on the area difference between the two monolayers, since the modulus of this quantity is difficult to measure. Historically, three different approaches have been used for this problem. The earliest models, called 'bilayer couple' (BC) [49], assumed that the constraint on monolayer number could indeed be converted immediately into a hard constraint on area difference, and hence expressed through Lagrangian multipliers as a local energy density. The difficulty with this approach is that, in contrast to the area of the vesicle, the fractional variation in area difference can be very large even for a small degree of stretching, because the area difference is typically very small to start with (for instance, in a planar bilayer, the area difference is zero). The Area Difference Elasticity (ADE) model [50] takes this into account by adding a new term to equation (13.20), proportional to the squared variation of the area difference between the two monolayers:

$$F_{ADE} = F_0 + \frac{\kappa \alpha \pi}{8AD^2} (\Delta A - \Delta A_0)^2 , \qquad (13.21)$$

where α is the dimensionless area difference elasticity, A is the membrane area, ΔA is the difference between the area of the two monolayers, taken at the lipid-water interface, $D/2$ from the bilayer center, and ΔA_0 is the equilibrium difference between the area of the two monolayers. ΔA_0 is proportional to the difference in the number of molecules between the two membranes. For artificial vesicles, this number difference will depend on the method of production. For example, one would expect large vesicles formed from the fusion of smaller vesicles, as in [53], to display a large excess of molecules in the outer monolayer because of the relatively higher area difference in the smaller vesicles.

Within the ADE model, a particular value for α is often chosen on some experimental basis, but the whole phase diagram can also be mapped with α as a variable parameter, see Figure 13.14. The BC model is a special case of the ADE model with $\alpha \to \infty$. The third well-studied model, the spontaneous curvature (SC) model, does not treat the constraint on area difference directly, but simply includes a finite and constant spontaneous curvature. The most general, physically relevant model, would be the area difference elasticity model combined with a finite constant curvature, as described in equation (13.21). However, when a finite C_0 is included in the ADE model with finite α, it is found to be equivalent to an ADE model with $C_0 = 0$ and a renormalized equilibrium area difference ΔA_0 [48]. When $\alpha = 0$ however, the spontaneous curvature cannot be included in this way. Hence, the three models can all be included in either a single ADE model, with variables α and ΔA_0, or a special case for $\alpha = 0$ and C_0 finite. Furthermore, the bilayer nature of the membrane implies $\alpha > 0$, so this last special case is not physically relevant. Experimentally, α has been found to be of order unity [47], so that the two terms in equation (13.21) will have a similar influence on the phase diagram.

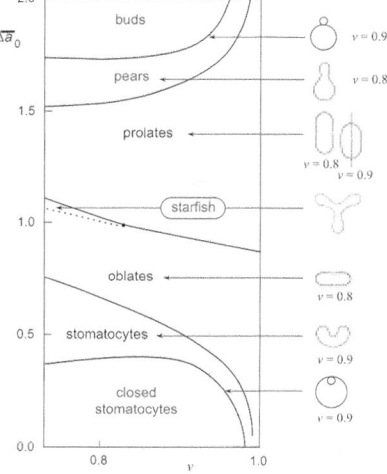

Fig. 13.15 Theoretical vesicle shape phase diagrams predict a wide range of equilibrium shapes. A section of the phase diagram of vesicle shape in the ADE model in [51]. $\Delta \bar{a}_0$ is a combination of the equilibrium area difference ΔA_0 and the area difference elasticity, α. ν is the reduced volume, which is the volume divided by the volume of a sphere of the same area. The symmetry axis on each of the axisymmetric shapes is vertical, whereas the starfish shape is illustrated by a cross section through the equator. The starfish region of the phase diagram represents a class of similar shapes, having various numbers of arms, which all have similar energy, and which have been observed experimentally in [52]. Reproduced from [51] with permission from EDP Sciences.

In the ADE model, the shapes of minimum energy are then calculated for given values of the four parameters: area A, volume V, equilibrium area difference ΔA_0, and area difference elasticity α. This four-dimensional phase diagram can be collapsed into three dimensions by forming dimensionless parameters with the area A. Cuts through this full 3D phase diagram at constant α are shown in Figure 13.14 and a typical 2D section of this phase diagram is shown in Figure 13.15. Varying α gives slightly different phase boundaries (see [47] for some typical examples). However, the stationary shapes are very similar. In fact, in [47] it has been shown that there is a direct correlation between the stationary shapes in each phase diagrams for different values of α. Once the phase diagram has been found for one value of α, it can be mapped to all values of α: the stationary shapes remain the same. The stability of a particular stationary shape, the location of the phase boundaries, and the order of phase transitions do all vary with α.

These theoretical shape phase diagrams can be directly compared with experiment. Figure 13.16 shows a comparison between the predictions of the BC model and equilibrium vesicle shapes obtained by temperature control. Of particular note is the discoid shape (bottom right; note that the symmetry axis runs from left to right). This is the shape of a healthy human red blood cell (RBC), and its appearance in this catalog of equilibrium shapes suggests that its shape might be maintained without active control by the protein cytoskeleton. In fact, it has been suggested that the shape is maintained by the spontaneous curvature produced by experimentally observed phospholipid asymmetry [55]. Throughout Figure 13.16, there is a striking resemblance between the shapes determined from theory, and those found experimentally. However, there is not as yet a good correspondence between the experimental

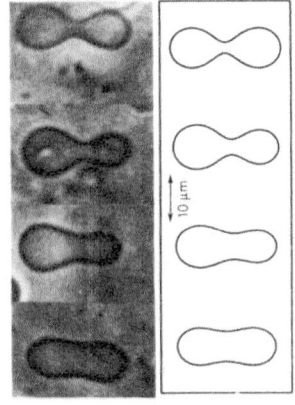
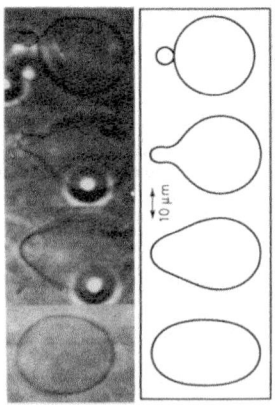
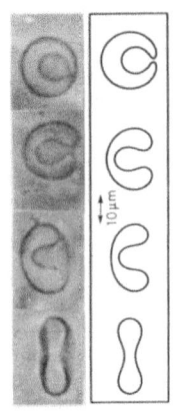

Fig. 13.16 Many of the predicted equilibrium shapes have been observed experimentally. This figure illustrates a comparison between experiment and the BC model, from [54]. The experimental shapes are obtained by temperature control of three DMPC vesicles in a specialized cell which prevents thermal convection, allowing observation of freely floating vesicles. The symmetry axis in each vesicle runs from left to right. The left column shows transitions between pear shaped and prolate vesicles. The middle column shows a transition from prolate (bottom) through pear shaped to budded (top) vesicles. The right column shows the transition from oblate (bottom) to stomatocyte vesicles. All these transitions are shown in Figure 13.15. Reproduced from [54] with permission of EDP Sciences.

conditions under which a particular shape occurs, and the point on theoretical phase diagrams at which that same shape occurs. In particular, vesicle shape is often controlled by varying temperature, but there is not yet sufficiently accurate information about the relevant physical temperature coefficients of the lipid material parameters, to link a particular experimental shape with a particular point on the temperature trajectory. The basic difficulty facing the mapping of an experimental phase diagram, is the fact that C_0, ΔA_0, and ΔA are not easily measureable or easily controllable.

Nevertheless, there have been recent promising experimental developments. The spontaneous curvature can be controlled most easily by asymmetric chemical conditions on either side of the bilayer. In [51] a technique was developed to rapidly change the external pH through a light-sensitive reaction. The authors find that when exposed to increased pH, initially prolate vesicles move to become more pear-like, see Figure 13.15. They suggest that this is due to an increase in spontaneous curvature of the external monolayer, since they also measure the reduced volume of the vesicles, and find that this remains constant. If the spontaneous curvature could be measured simultaneously, and really were entirely decoupled from reduced volume, this method would constitute a clean and effective way of moving through the phase diagram. The equilibrium area difference has been measured, but not controlled, by extruding thin tethers from vesicles held in micropipettes [56], and found to have a mean value of 0, with a wide spread. However difficult it is to map the complete phase diagram experimentally, certain qualitative differences between the experiments and the predictions of the various theories can be assessed. In particular, the order of the shape phase transitions, or in other words, whether they are continuous or discontinuous, varies between models, and with the value of α [47]. For example, in [54], only continuous budding transitions were observed experimentally, whereas in [57], discontinuous budding transitions were observed for a similar vesicle system. Of the theoretical models, the SC model predicts only discontinuous budding transitions and the BC model predicts only continuous budding transitions, whereas the ADE model allows for both orders of transition to coexist for a particular value of α. It is worth noting that the order of shape transitions can be difficult to probe experimentally, since the dynamics are slow even for first-order transitions. Furthermore, theoretical shape phase diagrams are typically produced under the assumption of zero thermal fluctuations, and thermal fluctuations can significantly modify the shape phase diagram by taking up excess area into suboptical ripples, as will be seen in the following section.[4]

A related subject is the formation of narrow tethers from lipid membranes [59], which can be produced by applying a point force to a giant vesicle or cell, with, for example, optical tweezers [60]. Cells also naturally produce membrane tethers by pulling with membrane motors and pushing with polymerized molecular fibres such as tubulin and actin [61]. Tethers can be used to determine membrane bending rigidity, as the ra-

[4] In addition, when budding is involved, the high curvatures in the neck region may go beyond the low curvature regime where the Helfrich Hamiltonian is applicable. Budding, vesiculation (the detachment of the bud from its parent vesicle to form a new daughter vesicle), and fusion are complex, dynamic processes. Fusion and vesiculation necessarily go beyond the continuum model because they involve transfer of lipid molecules between the two monolayers. Fusion is commonly understood to involve the formation of a fusion pore, a channel between the fusing vesicles, on the scale of a few lipid molecules [58], which then swells to become the neck between two connected vesicles. Fusion and vesicle formation in the context of cell biology are discussed further in Section 13.5.

dius r of the tether is given simply by $r = \sqrt{\sigma/2\kappa}$ [61]. This radius can be measured by detecting the fluorescence intensity, whereas the force f on the tether, given by $f = 2\pi\sqrt{2\sigma\kappa}$, can be measured using optical tweezers [61]. Membrane tethers are of approximately constant radius throughout most of their length, but close to their base, they show complex ripples, which have been analyzed in detail in [59].

Thermal fluctuations

As common in many soft matter systems, mechanical characterization is possible via the observation of thermal displacement. This method allows non-contact measurements and is particularly useful in studying low-dimensionality and living systems, both of which are easily adversely affected by probe contact. This section illustrates how to calculate the spectrum of fluctuations of a membrane, which is directly related to the bending modulus; this underpins a very popular set of experiments known as the analysis of membrane flickering.

The bending rigidity κ of phospholipid bilayers is typically only a few tens of $k_B T$, so that membrane shape fluctuations around equilibrium are easily visible using optical microscopy. To fully understand these fluctuations involves the non-trivial problem of calculating the partition function:

$$Z = \int D\left[\vec{X}\right] \exp\left(-\frac{F(\vec{X})}{k_B T}\right), \qquad (13.22)$$

where $D\left[\vec{X}\right]$ represents a functional integral over each particular membrane configuration $\vec{X}$, weighted so that the configurations $\vec{X}$ are represented with the correct probability.[5] Even at first order in $k_B T/\kappa$, Z is difficult to calculate for fluctuations around general shapes, because there is no analytical way to calculate the curvature eigenmodes of non-trivial shapes such as the discocytic RBC. However, on certain simple surfaces of constant curvature: sphere, planes, and cylinders, the Hamiltonian can be diagonalized in a natural way [62].

Consider first an isolated rectangular membrane, with sides of length L_x and L_y. We wish to calculate the expected fluctuations of the membrane. Such a membrane has no contained volume, and the spontaneous curvature must be zero for the plane to be an equilibrium surface. In addition, it has been shown that small static fluctuations around a plane or sphere are decoupled from lipid density fluctuations [63], so we use the simplified Hamiltonian:

$$F = \int_S dS \left[2\kappa H^2 + \sigma\right], \qquad (13.23)$$

where the integral runs over the whole fluctuating surface of the membrane. The equilibrium surface S_0 is a flat membrane of sides L_x and L_y, and we consider small fluctuations h around this equilibrium surface. The surface can be parameterized in the Monge gauge, where the

---

[5] Deciding on the correct weighting for each configuration is a non-trivial problem, which we shall defer until later in the discussion. For now, we will deal with the well-established low-temperature approximation, where Z is only expanded to first order in $k_B T/\kappa$.

normal distance h from the flat surface is a function of coordinates x and y in the plane. The Hamiltonian is given by:

$$F = \int dxdy \left[\frac{\kappa}{2} \left[\nabla \cdot \left(\frac{\nabla h}{\sqrt{1+(\nabla h)^2}} \right) \right]^2 + \sigma \right] \sqrt{1+(\nabla h)^2} \,, \quad (13.24)$$

where the integral over the whole area of the fluctuating surface has been transformed into an integral over the basal rectangle $L_x \times L_y$. Taking only the terms up to quadratic order in h, and neglecting constant terms gives the variation in the Hamiltonian:

$$\Delta F \approx \frac{1}{2} \int dxdy \left[\kappa (\nabla^2 h)^2 + \sigma (\nabla h)^2 \right] . \quad (13.25)$$

There are no terms linear in h since the plane is an equilibrium shape. The eigenvectors of this approximate Hamiltonian are the eigenfunctions of the Laplacian ∇^2, which are the Fourier components $h_{\vec{q}}$ for the allowed wavevector $\vec{q} = 2\pi \left(\frac{m}{L_x}, \frac{n}{L_y} \right)$, with m and n positive or negative integers. Performing the Fourier integrals[6] gives:

$$\Delta F \approx \frac{L_x L_y}{2} \sum_q \left(\sigma q^2 + \kappa q^4 \right) h_{\vec{q}}^2, \quad (13.26)$$

where $q = |\vec{q}|$. Applying the equipartition theorem gives the mean-squared amplitude of each mode:

$$\langle |h_{\vec{q}}|^2 \rangle = \frac{k_B T}{L_x L_y} \frac{1}{\sigma q^2 + \kappa q^4} . \quad (13.27)$$

[6] It is convenient to calculate these energies defining Fourier transforms for continuum variables:

$$h_{\vec{q}} = 1/(L_x L_y) \int dxdy\, h(x,y) \exp(-i\vec{q}\vec{x})$$

and

$$h(x,y) = (L_x L_y)/(2\pi)^2 \int d\vec{q}\, h_{\vec{q}} \exp(i\vec{q}\vec{x}).$$

The planar membrane is often used as a model for more complex shapes. For example, equatorial fluctuations of a sphere are experimentally accessible, and they can in fact be treated exactly. However, in the literature [64], the equatorial fluctuations are often treated in a planar approximation. On a plane, the equator corresponds to a line in the plane, say $y = 0$. The length of this line is set equal to the vesicle circumference $L_x = 2\pi a$, where a is the sphere radius. Counter-intuitively, the spectrum of observed fluctuations along the line is not the same as the spectrum of fluctuations in the plane, since any fluctuation along a line is in fact composed of several sinusoidal fluctuations at various orientations within the plane. The spectrum along a line is calculated by summing over the components of $\vec{q}$ in the orthogonal direction to the line. This summation gives [64]:

$$\langle |h(q_x, y=0)|^2 \rangle = \frac{k_B T}{L_x} \frac{1}{2\sigma} \left[\frac{1}{q_x} - \frac{1}{\sqrt{(\sigma/\kappa) + q_x^2}} \right] . \quad (13.28)$$

Equation (13.28) has been used by various researchers to study vesicle bilayers and soft cell membranes, in particular the plasma membrane of red blood cells, see Figure 13.17.[7] The general fluctuations spectrum of

[7] Note that there is no dependence on L_y in equation (13.28). It is surprising that although the amplitude of each 2D mode [equation (13.27)] depends on the total area of the membrane, the fluctuations seen along a line through that plane depend only on the length of the line and not on the size of the membrane in the orthogonal direction. Physically, this is because more modes can be fitted onto a wider strip of membrane, but the energy cost of exciting each mode increases proportionally. This is why it does not matter that a square membrane is typically used to approximate a spherical membrane, even though the area of the two membranes is not the same. This result also suggests that for non-spherical axisymmetric shapes, such as red blood cells, the fluctuations on the equator may depend only on the circumference of the equator, and not on the orthogonal profile of the membrane.

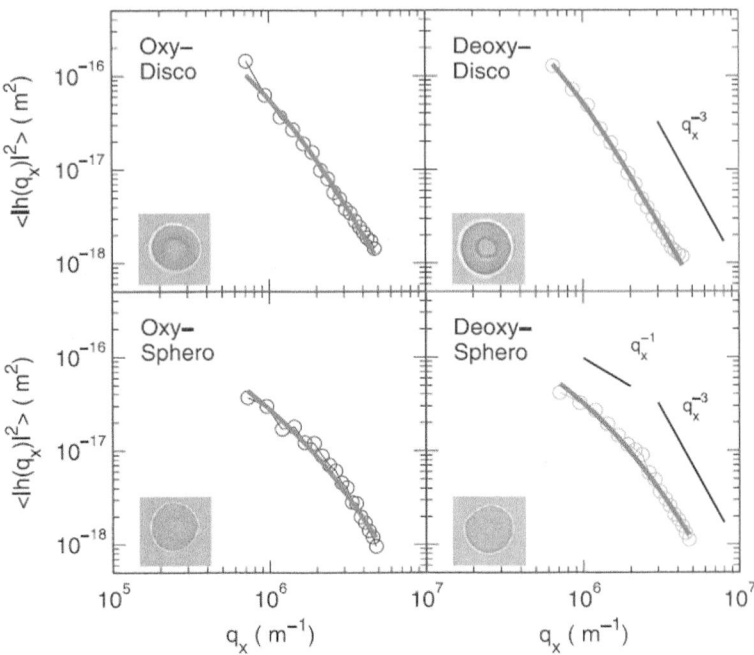

Fig. 13.17 Measuring the mean-square amplitude of contour fluctuation modes versus the equatorial wave vector q_x, and analyzing the data with the theory described here [equation (13.28)] allows the bending modulus and membrane tension to be recovered. Here the experiments are performed on red blood cells under diverse conditions (oxygenated and deoxygenated states, obtained by addition of $Na_2S_2O_4$) obtaining κ between 20 and 30×10^{-20} J and σ between 5 and 10×10^{-7} N/m, depending on conditions. Cells have different shapes, as marked, and this feature correlates to their mechanical properties. Figure reproduced from [65] with permission of The Biophysical Society.

axisymmetric vesicles remains an open question.

From the fluctuation spectrum, one can calculate various properties of the surface, such as the correlation between normal vectors at two points on the surface, and the excess area produced by the fluctuations. The correlation distance is the two dimensional analog of the persistence length of a polymer. A typical bilayer with a bending rigidity of $\kappa = 20k_B T$ and a molecular cut-off at $a = 1$ nm has correlation length $\sim 10^{26}$ m. As pointed out in [66], lipid bilayers are extremely stiff.

The excess area Δ is defined as:

$$\Delta = \frac{A - A_0}{A_0}, \qquad (13.29)$$

where A is the total surface area of the membrane and A_0 is the surface area of the basal plane. In the Monge gauge, on a square membrane of side L, the excess area is given approximately by:

$$\Delta = \frac{k_B T}{4\pi} \int_{q_{min}}^{q_{max}} \frac{q}{\sigma + \kappa q^2} dq \qquad (13.30)$$

where $q_{min} = \pi/L$, $q_{max} = \pi/a$, and a is the molecular length scale. Solving equation (13.30) gives:

$$\Delta = \frac{k_B T}{8\pi \kappa} \log \left[\frac{\tilde{\sigma} + L^2/a^2}{\tilde{\sigma} + 1} \right] \qquad (13.31)$$

where $\tilde{\sigma} = \sigma \pi^2 L^2 / \kappa$ is the dimensionless tension. In a vesicle which has been allowed to equilibrate and to control its own volume by the permeation of water through the membrane, the tension should go to zero. Evaluating equation (13.31) in this tensionless state gives approximately:

$$\Delta \approx \frac{k_B T}{4\pi \kappa} \log \frac{L}{a}. \qquad (13.32)$$

The thermal fluctuations occur on all lengthscales, many of them below the normal optical resolution, so that the thermal fluctuations are capable of hiding large amounts of excess area [67]. This explains why, at low pressures, in micropipette suction experiments, additional vesicle area appears to be produced from apparently tense vesicles, to an extent much larger than would be permitted by the high bulk modulus K of the membrane [67].

Fluctuations around a sphere

The thermal fluctuations can also be treated analytically in a spherical geometry. This can be important because topologically spheres are more realistic models than planes for many biological membranes. Also, the sphere is the simplest minimal shape of the Helfrich Hamiltonian, and the simplest to obtain and study experimentally in artificial vesicles. The analysis of thermal fluctuations is analogous (the eigenmodes on the sphere are the spherical harmonics) to the analysis for a planar membrane. The thermal fluctuation spectrum is derived in [68], and the fluctuations can be projected onto the equator, again by summing over modes in the perpendicular direction. The fluctuations projected onto the equator (which are in most cases the experimentally observable motions) have mean-squared amplitude:

$$\frac{\langle |h_m|^2 \rangle}{a^2} = k_B T \sum_l \frac{Y_l^m(\pi/2, 0)^2}{(l-1)(l+2)\sigma a^2 + (l-1)l(l+1)(l+2)\kappa}. \qquad (13.33)$$

For high degree l, the fluctuation amplitudes given in equation (13.33) approach the equivalent amplitudes in the expression for a planar membrane, equation (13.28), justifying the approximation of the equator of the spherical membrane by a line in a plane. However, for the long wavelength fluctuations, the two equations do not match, and attempting to fit the low wavelength fluctuations of the sphere to those of a planar membrane will result in an apparent surface tension term of order κ/a^2, and an incorrect estimate of κ. While by approximating non-spherical axisymmetric vesicles as spherical (see above) it can be possible to extract κ from the fluctuations, the same approximation to extract the tension is likely to be worse.[8]

[8]The error in this case would likely be of order κ/r_c^2, where r_c is a typical radius of curvature of the non-spherical membrane. For example, for an RBC, a typical curvature would be the radius of curvature perpendicular to the equator, given by $r_c \sim 1\,\mu m$ [69]. With a typical RBC bending modulus, $\kappa \sim 10^{-19}$ J [65], the apparent tension could be expected to be of order 10^{-7} Nm^{-1}.

Beyond first-order perturbation theory

The above discussion provides an overview of the first-order fluctuation theory of lipid bilayer membranes. This theory is well developed, and has been used widely to interpret various experiments on membrane

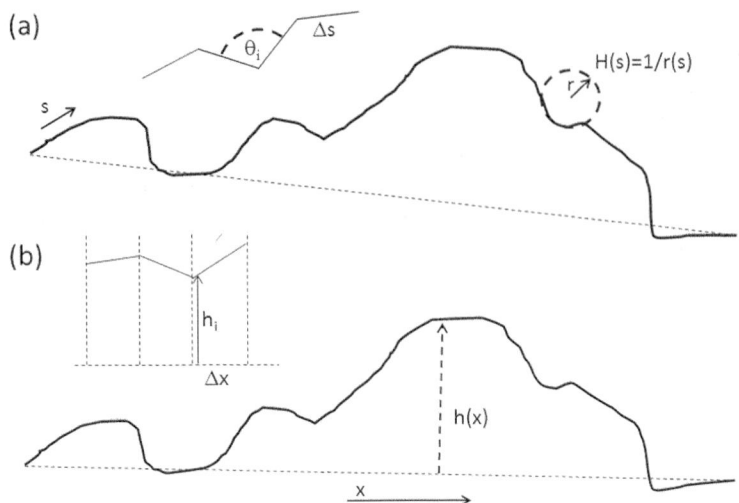

Fig. 13.18 Two ways to parameterize a 1D membrane, which can lead to different predictions for the renormalized bending modulus κ'. Parametrization of a 1D membrane in (a) curvature H, as a function of arc length s, or (b) height h as a function of displacement x along the basal line. In the insets are the equivalent discrete coordinate systems. In (a), the bond angle θ_i, and in (b), the bond height h_i, vary as a function of the bond index i.

thermal fluctuations. However calculating the fluctuation spectra more accurately (to account for mode-coupling) it is found that a continued expansion in the small parameter $k_B T/\kappa$ is non-trivial, see Figure 13.18. It even remains unclear if the fluctuations of the membrane themselves stiffen or soften the membrane. The second-order effects are typically expressed as a renormalization of the physical parameters κ and σ which are expected when experimental results are fitted with first-order predictions. In general, the experimental results at second order: E_2, are treated as though they were experimental results at first order: E_1, but with renormalized physical parameters, here indicated by $'$:

$$E_2(\kappa, \sigma, \kappa_G...) = E_1(\kappa', \sigma', \kappa'_G...). \tag{13.34}$$

For example, equation (13.27) would become to 2nd order:

$$\langle |h_{\vec{q}}|^2 \rangle = \frac{k_B T}{L_x L_y} \frac{1}{\sigma' q^2 + \kappa' q^4}. \tag{13.35}$$

The same convention is applied to membranes out of thermal equilibrium, where an effective temperature is used in an effort to retain the benefits of the equipartition function in a system where it does not strictly apply [70]. These conventions may be dangerous, in that they ignore the potential for novel terms, for example an additional q^6 term in equation (13.27), which cannot be attributed to one of the physical parameters of the membrane. Furthermore, several different experimental geometries exist, from which the physical parameters of the membrane can be extracted, and the renormalized physical parameters extracted from these measurements are not necessarily identical. This is particularly evident for membrane tension, which, when measured by micropipette experiments, may be expected to differ from the tension measured through thermal fluctuations, both of these experimental tensions differing from the Lagrangian multiplier σ. In [71], five types of surface tension are defined, all of which are relevant in different experiments.

The results of the second-order perturbation theory remain controversial, both for κ and σ, so here we will merely cite the relevant references, and discuss the persistent difficulties which arise in this field. There have been numerous calculations of the renormalized bending modulus κ' [62, 72–75], by which is meant the bending modulus appearing in equation (13.35). All authors agree on the functional form of the renormalized bending modulus:

$$\kappa' = \kappa - \left(\frac{k_B T}{4\pi}\right) I \log\left(\frac{L}{a}\right). \qquad (13.36)$$

This expression shows close similarity to the expressions for the excess area in equation (13.32), which is to be expected, since some components of the renormalization come from projecting the fluctuations of a membrane of area A, onto a plane of area A_0. However, there is disagreement over the value of the constant I. Some disparity between the earliest calculations in [72–74] was found to be the result of mathematical error, owing to the complexity of the calculation in 2D [75]. However, there was also a more fundamental difference in method between [72] and the later references: in [72] it was assumed that the correct statistical measure in the partition function is the mean curvature of each mode, whereas in [73–75] a normal displacement measure was employed. The use of the curvature measure leads to a stiffening of the membrane, with $I = -1$ [62], whereas the normal displacement measure leads instead to softening, with $I = 3$ [75].

A softening effect was seen in simulations [76], consistent with the use of the normal displacement measure, whereas [77], using density matrix renormalization to calculate thermal fluctuations with both normal displacement, and curvature measure, showed that the former does indeed produce softening, whereas the latter produces stiffening, as predicted by theory.

The choice of measure may cause similar problems in the calculation of the renormalized tension. However, the argument has followed a different course. It has recently been demonstrated that the quadratic approximation (13.25) of the Helfrich Hamiltonian is not rotationally invariant [71]. Calculations made using this non-rotationally invariant Hamiltonian (for example [78, 79]) lead to a different set of relations between the renormalized and bare tension, compared to the results obtained using a rotationally symmetric, curvature-based Hamiltonian [71, 80]. For example, the former method predicts a difference between the tension measured with a micropipette and the tension measured from the thermal fluctuation spectrum, while the latter finds the two quantities equal. However, this rotational invariance is not a necessary feature of a Helfrich Hamiltonian based on normal displacement, but rather a result of the incorrect truncation of the Hamiltonian to quadratic order [71]. It remains to be seen whether the debate develops, as for the bending modulus, into a discussion over the correct choice of measure.

13.3 Dynamics

The dynamics of membrane motion has been much studied, but is more difficult than the study of equilibrium deformations and small equilibrium fluctuations. Issues include the larger number of dynamic parameters, the coupling between curvature and membrane fluctuations, and the hydrodynamic interactions between distant parts of the membrane. There are many ways in which a fluctuation of membrane shape can relax, and each relaxation pathway has its own viscosity. Dissipation occurs with any of the stretching, splay, twist, and bend deformations, as well as intermonolayer sliding, combining to produce more complex multimodal relaxation. However, these viscous terms are not all of comparable value and not all are visible in a given situation (e.g. membrane contour fluctuations). Here, we will focus first on the equations of motion of single eigenmodes on a plane membrane without density fluctuations, before looking at the coupling between curvature and density fluctuations, and finally examining how the geometry of the membrane and the applied deformation affects the membrane response.

Incompressible planar membranes

On the cellular scale and below, membrane motion can be well approximated as lying in the low Reynolds number regime of fluid dynamics. The only dynamic forces come from viscosity coupling to gradients of velocity, within the lipid bilayer and in the bulk fluid. To obtain the equation of motion of the membrane, these viscous forces must be balanced with the internal elastic forces of the membrane, and external forces from Brownian motion, or from artificial driving, with, for example, optical traps or bulk shear flow. Here, we will describe the equation of motion for small motions of a planar membrane. For an ideal plane membrane with no density fluctuations, the dynamic eigenmodes are again the simple Fourier modes. Since the dynamic and elastic modes are identical, each mode has its own equation of motion of the general form:

$$\zeta_n \frac{\partial U_n}{\partial t} = -\frac{\partial \Delta F}{\partial U_n} + f_n, \qquad (13.37)$$

where U_n is the amplitude of mode n, ζ_n is the frictional term damping motion in U_n, ΔF is the curvature Hamiltonian of the membrane, discussed in the previous section, and f_n is the nth component of an externally applied force, with no coupling between the equations of motion for different modes. In a harmonic potential $\Delta F = 1/2 k_n U_n^2$, and equation (13.37) becomes linear in the variable U_n.

The evolution of each of the eigenvalues U_n is formally equivalent to the well-known problem of a particle moving in a harmonic trap. Assuming that f_n corresponds to the force produced by uncorrelated Brownian motion, the time-correlation of each mode is given by:

$$\langle U_n(t) U_n(t+\tau) \rangle = \langle U_n(t)^2 \rangle \exp\left(-\frac{\tau}{\tau_n}\right), \qquad (13.38)$$

where $\tau_n = \zeta_n/k_n$, and $\langle\ldots\rangle$ represents a time (t) average or equivalently an ensemble average. For the planar membrane, the eigenmodes can be labeled by the wavenumber q, and the frictional term in equation (13.37) is then given by $\zeta_q = 4\eta q L^{-1}$ [81]. The increase of ζ_q with increasing wavenumber is due to the higher gradients of velocity which are present at higher mode numbers. To calculate ζ_q requires solving for the motion of the bulk fluid as well as of the membrane. A harmonic membrane deformation decays exponentially into the bulk with a decay length given by q^{-1}. The gradient of the elastic potential corresponding to equation (13.26) is $\partial \Delta F/\partial U_q = L^2 \left(\sigma q^2 + \kappa q^4\right) U_q$, so that the timescale of each mode is given by:

$$\tau_q = \frac{4\eta q}{\sigma q^2 + \kappa q^4}. \tag{13.39}$$

As usual, this planar approximation is also valid for high wavenumber fluctuations of a spherical or quasi-spherical membrane.

Equivalently, taking the Fourier transform of the equation of motion gives the so-called power spectrum $P(\omega) = \langle|U(\omega)|^2\rangle$. In frequency space, equation (13.37) reads:

$$i\omega U_n(\omega)\zeta_n = -k_n U_n(\omega) + f_n(\omega), \tag{13.40}$$

so that:

$$P(\omega) = \frac{\langle|f_n(\omega)|^2\rangle}{k_n^2 + \omega^2 \zeta_n^2}. \tag{13.41}$$

The $\langle|f_n(\omega)|^2\rangle$ term is in fact a constant across the frequency space, if it is provided by a simple Brownian motion uncorrelated in time. The value of this constant can be determined by relating the power spectrum to the static displacement:

$$\langle|U_n|^2\rangle = \int_{-\infty}^{\infty} P(\omega)d\omega, \tag{13.42}$$

to give:

$$P(\omega) = \frac{\tau_n \langle|U_n|^2\rangle}{\pi \left(1 + \omega^2 \tau_n^2\right)}, \tag{13.43}$$

which, for the mode of wavenumber q, in the zero tension limit is:

$$P(\omega) = \frac{4\eta k_B T}{\pi L^5 q \left[\kappa^2 q^6 + 16\omega^2 \eta^2\right]}. \tag{13.44}$$

As for the static case, isolated Fourier modes on plane membranes or spherical vesicles are often not readily accessible. The relaxation behavior of fluctuations on the equator of a sphere, for example, is given by a sum over the relaxation behavior of modes in a perpendicular direction to give a multi-exponential decay [65]:

$$\langle h_{q_x}(t) h_{q_x}(t+\tau)\rangle_{y=0} = \sum_{q_y} \langle|h_q(t)|^2\rangle \exp\left(-\frac{\tau}{\tau_q}\right). \tag{13.45}$$

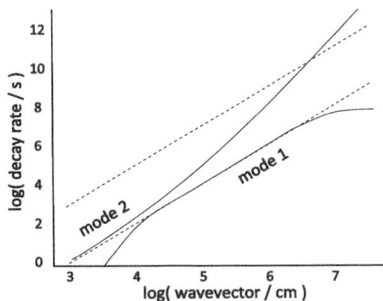

Fig. 13.19 Coupling between density and curvature fluctuations gives rise to a bimodal relaxation spectrum. In [63], the two decay constants of this bimodal relaxation, γ_1 and γ_2 are calculated. They are reproduced here qualitatively as a function of wavenumber q. The modeled membrane parameters are $\kappa = 25$ k$_B$T, $K = 0.07$ Nm^{-1}, $\eta = 10^{-3}$ Pas, $\eta_m = 10^{-10}$ Nsm^{-1}, $b = 10^8$ Nsm^{-3}, $D = 1$ nm. At high wavevector q, the decay constant of the "fast" mode 2 scales as $\tilde{\kappa}q^3/(4\eta)$. At high q, neutron spin echo studies of multilamellar phospholipid membranes have observed mode 1 [83]: This experimental method probes density fluctuations in the plane of the bilayer. The asymptotic regimes of the two decay timescales, at low, intermediate and high wavenumber, are listed in Table 13.1. Adapted from [63].

In practice, the higher q modes in this summation rapidly decay away so that the long-term correlation behaves like the longest wavelength $q = q_x$ mode, which on a sphere is a purely azimuthal $Y_l^l \sim \exp{(il\phi)}$ mode. A second typical example is the power spectrum of fluctuations of a single point $P_P(\omega)$. For an isolated planar membrane, this power spectrum is given by integrating equation (13.44) in both directions in the plane. In the high frequency limit, $P_P \propto \omega^{-5/3}$. This $\omega^{-5/3}$ dependency has been verified over frequencies ranging from 1 Hz to 1 kHz using weak laser diffraction from the edge of a fluctuating red blood cells [82].

Coupling to lipid density and membrane viscosity

So far, we have considered the bilayer as an ideal membrane of infinitesimal thickness and no compressibility. However, the situation is complicated by the bilayer nature of the membrane and by the ability of lipids to flow in the plane of the membrane. This flow is driven by fluctuations of the lipid density ρ, which couple to the curvature as in the Hamiltonian in equation (13.16), and are dissipated by an in-plane viscosity η (η_m later), as well as by drag between the two monolayers with friction constant b.

The relaxation behavior in this complex situation has been calculated in detail for planar membranes in [63]. The main results of this calculation are that there are two relaxation pathways for each q-mode. For *long wavelength* fluctuations, the two modes are separated into a purely displacement mode, which relaxes with inverse timescale γ_1 and a purely density mode which relaxes with a faster inverse timescale γ_2. Because the density modes decay faster than the displacement modes, the displacement modes decay at the relaxed density, meaning that the density has time to flow and to equilibrate to any long wavelength fluctuation of the displacement before that displacement fluctuation can decay. This means that in the long wavelength limit, the displacement modes behave as if the membrane were ideal, with inverse timescale $\gamma_1 = \kappa q^3/(4\eta)$. Conversely, for *very short wavelength* fluctuations, the displacement modes relax faster than the lipid density can equilibrate. Again, the displacement and density modes are uncoupled, but now the γ_2 mode is the pure displacement mode, with $\gamma_2 = \tilde{\kappa}q^3/(4\eta)$. In this condition the effective bending modulus $\tilde{\kappa}$ is higher than κ, because the bilayers do not have time to flow past each other before the displacement relaxes. Comparing with equation (13.16), we find $\tilde{\kappa} = \kappa + Kd^2$. At *intermediate wavelengths*, the density and displacement modes couple, and one expects strong bimodal behavior within this regime. The various dynamic regimes are listed in Table 13.1, and the complete dispersion relation is shown in Figure 13.19.

The bending modulus at non-relaxed density, $\tilde{\kappa}$ is closely related to the area difference elasticity α described in Section 13.2. Suppose that we have a spherical membrane of radius a, and membrane thickness D, with zero equilibrium area difference ΔA_0. The area difference between the inner and outer leaflets ΔA would be given by $\Delta A = 4\pi aD$, to lowest

Table 13.1 Summary of dynamic regimes extracted from [63]. γ_1 and γ_2 are the decay constants of two modes combining bending and slipping between the monolayers. The crossover wavevectors are: $q_1 = \eta K/(b\tilde{\kappa})$ and $q_2 = \sqrt{2b/\mu}$. There is no change in the behavior of γ_2 at q_2.

Mode number range	γ_1	γ_2
$q < q_1$	$\kappa q^3/(4\eta)$	$Kq^2/(4b)$
$q_1 < q < q_2$	$K\kappa q^2/(4b\tilde{\kappa})$	$\tilde{\kappa} q^3/(4\eta)$
$q_2 < q$	$K\kappa/(2\mu\tilde{\kappa})$	$\tilde{\kappa} q^3/(4\eta)$

order in D/a. In the ADE Hamiltonian, equation (13.21), the curvature energy F of this membrane would be given by: $F = 2\pi\kappa(1 + \pi\alpha)$. This same sphere also has an entirely non-relaxed density distribution, in that the spacing of lipids at the midplane is the same in inner and outer monolayers, but the headgroups are closer together in the inner and further apart in the outer, monolayer, as in Figure 13.13(a). This means that the energy would be given by the non-relaxed density bending modulus $\tilde{\kappa}$, so that $F = 2\pi\tilde{\kappa}$. Therefore, $\tilde{\kappa} = \kappa(1 + \pi\alpha)$. Hence, the bending modulus observed for short wavelength (high frequency) membrane fluctuations is closely related to the constant α, which determines the shape phase diagram.

Non-planar membranes

In this section, we discuss how the membrane dynamics is modified by the shape of the membrane for three cases: spheres, cylinders, and closely spaced parallel planes.

The equations of motion for spherical membranes have been calculated analytically in [68]. Again, the eigenmodes of the equations of motion are the same spherical harmonics, so there is no coupling between modes. The effect of the spherical geometry is felt most strongly in the modes of comparable wavelength to the size of the sphere. For shorter wavelength fluctuations, the dynamics approach those of fluctuations of the same wavelength on planar membranes.

For cylinders, a new phenomenon emerges, which is a long wavelength curvature instability leading to a series of spherical vesicles connected by narrow tethers. For sufficiently high tension, greater than $\sigma_c = 3\kappa/2R_0^2$, where R_0 is the undisturbed radius of the membrane [84], longitudinal modes of wavelength greater than $\lambda_c = 2\pi R_0$ become thermodynamically unstable [85], and grow until the membrane has the form of a string of beads. This is called the pearling instability. It has shown [85] that for long wavelength λ and for high tension, the modes initially grow with inverse timescale:

$$\gamma = \frac{\sigma}{\eta R_0} \left(1 - k^2\right) k^2, \qquad (13.46)$$

where $k = 2\pi\lambda/R_0$ is a dimensionless wavenumber. Hence, the fastest growing mode has $k = 1/\sqrt{2}$. It is the wavelength of this mode which determines the wavelength of the final string of beads, regardless of the fact that the most energetically unstable modes have much longer wavelengths. This has been confirmed experimentally in [86], where optical tweezers are used to temporarily move the membrane into the high tension regime. This is an example of a phase transition where the eventual state is controlled by the dynamics, rather than by the thermodynamics of the situation.

In the case of two parallel planar membranes, placed in close proximity, a new, peristaltic mode emerges involving hydrodynamic interactions between fluctuations of the two membranes [81]. Modes of longer wavelength than the inter-membrane spacing will show a decay constant $\gamma_1 \propto q^2$, rather than q^3 as for an isolated membrane. As the wavelength shortens, the decay constant will return to the isolated membrane behavior. This is because the penetration depth of surface fluctuations is proportional to their wavelength, so fluctuations at longer wavelength reach further into the fluid toward the opposing membrane, and cause hydrodynamic coupling between the two membranes. This peristaltic mode leads to a point-like power spectrum $P_P \propto \omega^{-4/3}$, rather than $\propto \omega^{-5/3}$ for a free membrane [81]. However, at sufficiently high frequency, the free membrane result should again be recovered, even for closely spaced planes, since at high frequencies the short wavelength, non-peristaltic modes dominate. This $\omega^{-4/3}$ decay has been seen in the height fluctuation of the flat, central region of RBCs, which can be approximated as two parallel planes separated by approximately 1 μm [81], whereas fluctuations on RBC equators do show the free membrane response [82], presumably because there is no nearby opposing membrane with which the fluctuations can couple.

Non-thermal forces

So far, we have only discussed the dynamic response of membranes to thermal forces induced by Brownian motion. A number of experiments have also been performed where the membrane fluctuations are driven by artificial means, for example, flow in microfluidic channels [87], or manipulation with optical tweezers [88–91]. These active experiments have the advantage that they can look, in principle, at a single mode, and can do so in a time-resolved way, enabling the study of the phase, as well as the amplitude, of the response. These features enable higher resolution to be obtained in the quantity to be measured, than through inherently noisy thermal fluctuations. Another fundamental and important advantage of active techniques is that they enable any non-thermal source of noise in the membrane to be isolated, by comparing the amplitude of spontaneous fluctuation with the real material susceptibility measured in the forced deformation [88, 92].

Even within the linear response regime, the geometry of the deformation has a strong effect on the qualitative form of the response. In [93], the response of a membrane driven by oscillations of a small microscopic bead attached to the membrane is calculated by summing over the stiffnesses of the normal modes of the membrane, and calculating the contribution each mode makes to the displacement of the membrane. The complex response modulus K^* of the membrane with respect to this point-like-force is shown to scale as:

$$K^*(\omega) \sim \frac{\kappa}{a^2 \int_0^1 dp\, (p^3 - i\delta)^{-1}}, \qquad (13.47)$$

where a is the radius of a microscopic bead attached to the membrane and $\delta = 27\pi^2 a^3 \omega \eta/(4\kappa)$. p is a dummy variable, a rescaled version of the wavenumber q. In the limit of infinitesimal a, representing a point-like force, and high-frequency ω, $K^*(\omega)$ scales as $\omega^{2/3}\left(1/2 + i\sqrt{3}/2\right)$. Interestingly, in this high-frequency limit, the response is expected not to be completely out of phase with the driving force, but instead suffers a $\pi/3$ retardation. This occurs because however high ω is, the elastic terms, scaling as κq^4, continue to dominate over the viscous terms, preventing the response from having an entirely viscous character. When a is finite, modes having a wavelength shorter than a are damped out, and the low q, viscous terms dominate, the retardation approaches $\pi/2$, and the complex modulus scales as ω^1, just as for a bead in an optical trap. When the same bead experiences thermal fluctuations, the high frequency power spectrum scales as ω^{-2}, rather than $\omega^{-5/3}$ as in [82], because the high q modes which are present in the membrane as a whole, are not transmitted to the bead's motion [89, 93]. To our knowledge, no experiments have so far measured the phase of response to point-like deformations of a membrane or vesicle. In practice, it is very hard to obtain point-like forces with optical traps since their focal volume is diffraction limited.

Meanwhile, active response experiments have been performed by driving sinusoidal deformations directly with rings of optical traps around the equator of a GUV. There, the complex modulus K^* of the membrane response was predicted to scale as $\omega/\log\omega$ at high frequency [88]. In these experiments, the amplitude and phase were well described by the linear theory, but the experiments did not extend to high enough frequency to verify the $\omega/\log\omega$ scaling. Again, the finite size of the optical traps is expected to modify the high-frequency scaling back to ω^1 in practice.

Various experiments have measured dynamics beyond the linear regime. In [90], optical traps are used to stretch flexible GUVs by several μm, and the timescale of their relaxation is measured. The timescale of relaxation is found to be longer than would be predicted from the thermal fluctuations, indicating non-linear coupling between the harmonic modes. In [94], large scale membrane deformation is driven by the lo-

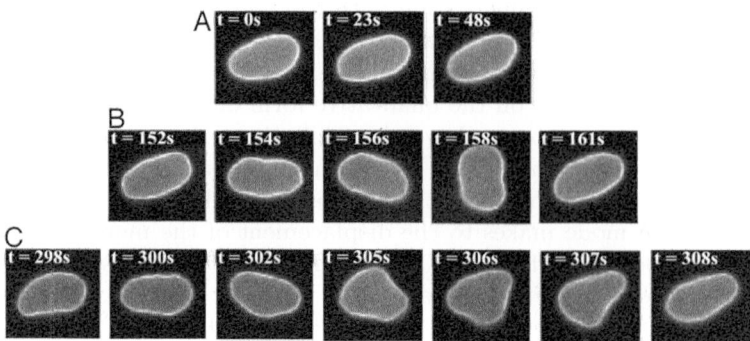

Fig. 13.20 Experimental observation of vesicle behavior controlled by flow in a four roll mill. A) Tank treading; B) Tumbling; C) Trembling. A single vesicle is made to switch between these three regimes by altering the amount of vorticity and shear in flow through a four roll mill. These images are from a single time sequence. Figure reproduced from [87] with permission.

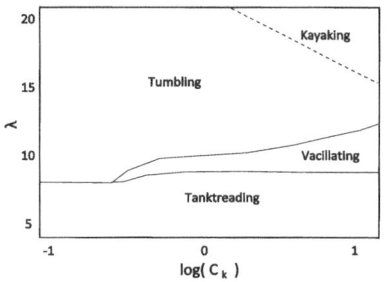

Fig. 13.21 A phase diagram of the behavior of vesicles in shear flow. Experimental observations of the three phases, tank treading, tumbling and vacillating (or trembling), are shown in Figure 13.20. C_k is a dimensionless quantity given by the shear rate multiplied by constant parameters of the vesicle. λ is the ratio of internal to external viscosity. For comparison, λ lies in the range of 10–50 for an RBC in an aqueous solution with a similar viscosity to water [65, 95–97]. The shape of the phase diagram varies with the third significant parameter, the excess area Δ. $\Delta = 0.43$ here. The dashed line is a qualitative boundary, based on the results of [98]. The solid lines are derived from theory [99] and agree with finite element simulations [98].

cal application of an acid to a particular location on a vesicle using a micropipette. The acid changes the spontaneous curvature of the membrane, producing a budding deformation. The relaxation of this bud is found to be strongly dependent on the inter-monolayer friction, which is otherwise difficult to measure.

Membranes under flow

The study of vesicles and RBCs under flow is a highly developed field. There is a complex phase diagram of vesicle behaviors, determined both by parameters of the flow and of the membrane. The most typical behaviors are: tank treading (TT), where the vesicle maintains a constant angle to the flow; tumbling (TU), where the vesicle turns over on the axis perpendicular to the flow direction and the shear direction; and trembling (TR), where the vesicle oscillates with a limited amplitude around a fixed orientation. These behaviors are displayed in Figure 13.20. The phase diagram can be explored by modifying the physical parameters of the vesicle, typically the ratio between the internal and external viscosity, or the excess area. However, these parameters cannot be controlled for a single vesicle; instead, the phase diagram can be explored by modifying the flow behavior, controlling the ratio of rotational to shear flow in a microfluidic four roll mill [87].

One peculiar property of the flow phase diagram is its low dimensionality. That is to say, there are a large number of vesicle and flow parameters which can be measured, but theoretically [100] and experimentally [87], the phase diagram collapses into just two dimensions, based on groups of these parameters [87], see Figure 13.21. This is an effect which is predicted by quasi-spherical approximations of the flow, but it is peculiar that this feature continues even to high excess area. Recent simulations have actually found that additional dimensions are in fact required at high excess area [98]. This contrast with experimental results may be due to the effect of thermal fluctuations, which were not included in the simulations [98]. The flow behavior of individual vesicles has been linked to the overall rheological behavior of dilute vesicle solutions [101], with transitions between the different regimes (TT, TR, TU) in the individual vesicle behavior corresponding to phase transitions in the bulk viscosity. This could have particular implications for the fluid mechanics of blood circulation.

Diffusion in membranes

There is no comprehensive model for lipid flow within cellular membranes. Below the main-chain phase transition temperature, there is local order, and the motion of a lipid molecule is confined by its neighbors. Above the transition, in the fluid state, even the mechanisms of basic dynamic processes, such as lateral diffusion of lipids, are poorly understood. At short time and small length scales, atomic-scale molecular dynamics simulations support a collective mechanism for lipid diffusion [102], rather than the traditional working hypothesis of a rattle-in-cage process. These simulations show the lipid and its nearest neighbors moving in unison (fast and directionally), forming loosely defined clusters, with correlations over tens of nanometers. The picture is supported by very recent quasi-elastic neutron scattering experiments [103]. At longer times, much slower motion is observed: Random walks more typical of individually diffusing molecules, from which diffusion coefficients can be obtained [104, 105]. In this regime, a fluid phase can be described by its viscosity. The temperature dependence of the diffusion coefficient is typically Arrhenius-like, from which an activation energy is obtained. However in view of the fact that recent evidence points to the absence of a "simple" barrier-hopping mechanism, the interpretation of this activation energy remains unclear at the fundamental level. Furthermore, as a function of temperature, more than one regime has been reported within the same fluid phase [38].

The understanding of how membrane inclusions diffuse is better developed, since for most relevant processes the membrane can be treated as a thin Newtonian layer. The discussion in Section 13.3 addressed the case of single inclusions, in flat and un-bounded films or membranes. Interactions between multiple inclusions or obstacles, membrane curvature, and finite size are all physical effects that can affect the diffusion coefficients. Most of these effects can be accounted for dimensionally via a lengthscale (e.g. the radius of curvature; membrane size; separation distance), which can be compared to ℓ_m – physical effects that have lengthscales as small as a or ℓ_m can be important for the flow properties [107]. The generalization to viscoelastic membranes, surrounded by viscous phases, has been considered theoretically [93].

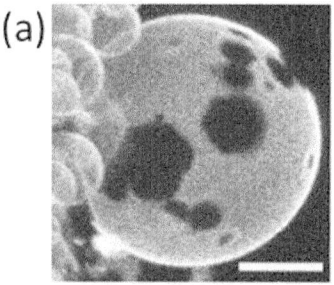

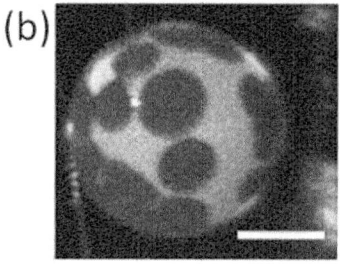

Fig. 13.22 Comparison of domains in (a) DPPC:DPPS and (b) DPPC:DPPE. In (a) the dark, polygonal domains are thought to be mostly DPPC in the hexagonally ordered L'_β phase, whereas in (b) the dark, circular domains are thought to be DPPE rich L_β, which shows no long-range ordering. In both cases, the bright phase is a fluid phase: the solid domains appear dark because a lipid dye is excluded from the solid domains in favor of the more disordered fluid domain. Scale bars are 10 μm. Figure reproduced by permission of IOP Publishing from [106], ©IOP Publishing. All rights reserved.

13.4 Multicomponent membranes

The net interaction between lipids in a bilayer is the result of many contributions, including short-range van der Waals forces, electrostatics, and "hydrophobic" forces (i.e. the complex result of accounting for interactions and configurations of water molecules near the surface). In multi-component mixtures, these contributions can result in a preference

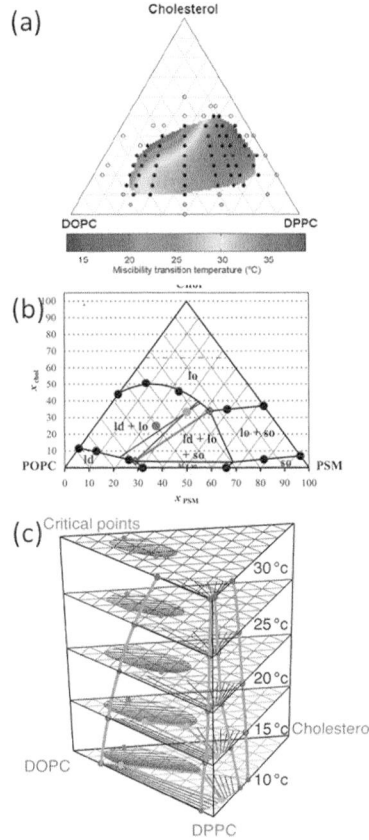

Fig. 13.23 The experimental ternary phase diagrams in (a) and (b) are among the first to have been measured in detail, using fluorescence microscopy (later validated by NMR). The different ternary systems (DOPC:DPPC:cholesterol and POPC:PSM:cholesterol) share the same qualitative features, as discussed in the text. (c) Recent calculation of the phase diagram [108] shows the expected tie lines and temperature dependence. Figures (a,b) reproduced with permission of The Biophysical Society [109, 110]; Figure (c) reprinted with permission from [108]; Copyright (2011) by the American Physical Society.

for demixing, particularly between saturated and unsaturated species. Balancing against entropic contributions accounting for the number of independent configurations, this leads to the common ("regular solution") behavior, with a single mixed phase at high temperature and phase separation below a critical temperature. As well as composition and temperature, pressure is an important thermodynamic parameter in these systems.

In **binary mixtures**, phase separation can take place between a saturated and an unsaturated species, leading to a region of phase coexistence. Likewise, binary mixtures of a saturated phospholipid and a sterol can also show phase separation into two phases with differing compositions. In all these cases, the phase rich in saturated lipid has a "solid" character. As was discussed for monolayers (Figure 13.7), there are a vast number of such solid phases, differing in the liquid-crystalline arrangements, with varying degrees of local symmetry. The presence or absence of long-range order has visible consequences in the shape of domains in multi-component vesicles. L'_β domains in -PC vesicles have a polygonal, usually hexagonal shape, reflecting the underlying hexagonal symmetry; whereas L_β domains in -PE vesicles are circular because there is no long range order, see Figure 13.22.

In **ternary mixtures** in which a sterol, such as cholesterol, is added into the saturated/unsaturated mixture, the system can phase separate into two separate, but both fluid coexisting phases. This case is important because of its biological relevance. Of these two liquid phases, one is referred to as liquid ordered L_o, and the other as liquid disordered L_α. The L_o phase is more ordered in the sense that the components are packed more tightly, and it is thicker than the disordered phase. Most lipid dyes and other impurities will tend to partition into the disordered L_α phase (however, specific lipid dyes have also been developed to label the L_o phase) [111].

Phase separation in ternary saturated/unsaturated + cholesterol mixtures has been characterized for various systems, and linked with possible signaling and regulatory roles in the cell membrane [109, 110]. The phase diagrams depend on the concentration of all three species, and on temperature, see Figure 13.23. For example, at a 1:1 concentration of DPPC:DOPC, and 30% molar cholesterol, the mixture is uniform at temperatures above $\sim 30^\circ C$, and phase separates on cooling through this temperature. At a fixed temperature, phase separation can also be induced by a composition change, for example in the balance of saturated:unsaturated phospholipids, or in the concentration of sterol. The general features of these diagrams are well understood, and recent work has proposed a particularly simple Gibbs free energy approach to calculating the phase boundaries [108].

Biological cell membranes are mixtures of typically thousands of lipids, but rather than displaying a vast number of phases and phase transitions (as would be allowed by the Gibbs phase rule) it seems that these complex mixtures have the same thermodynamic behavior as a ternary system [109]. This is consistent with the simple picture that demixing only occurs between saturated and unsaturated phospholipids, and that different types of saturated lipids (and different types of unsaturated lipids) remain mixed.

Much work has gone into understanding in detail the phase behavior [109] in multicomponent vesicles. This interest stems from both the theoretical challenges inherent in this type of two dimensional system, and from the fact that fluid-fluid phase separation, and the phenomenology that comes with having a thermodynamic system close to a critical point [112], is likely to play a role in biological membranes. Lipid-protein interactions are poorly understood, but likely to regulate protein function. At the very least, the bilayer viscosity determines the motility of membrane bound species [36], and the bilayer deformation induced by protein inclusions (see Figure 13.24) will lead to interactions between these inclusions [113, 114], the details of which depend on the bilayer mechanics, which is a function of composition. More sophisticated regulation processes, linked to the multicomponent lipid nature of the bilayer, are also plausible: close to the point of demixing, the free energy cost of forming domains rich in a particular species, or in a sterol, is small, and indeed spontaneous thermal fluctuations give rise to transient and localized concentration fluctuations.

Physical systems in the vicinity of critical points have general behavior that depends only on the symmetry of the order parameter and the dimensionality of the system, and not on the detailed molecular interactions. In general, near a second-order phase transition (such as the demixing that takes place in the ternary lipid mixtures) the susceptibility diverges, and fluctuations become large and long lived; this is referred to as 'critical behavior'. There are specific laws (common to wide classes of systems) to describe the critical behavior of thermodynamic parameters as a function of the distance to the critical point, in particular here the temperature difference.

The video analysis of the shape fluctuations of domain shape in model membranes (or vesicles obtained from cell blebbing) can be used to measure the line tension of the linear interface between phases, and thereby to characterize the critical behavior, see Figure 13.25 and [115]. In these conditions, both temperature and composition are the control parameters that tune the characteristic size and lifetime of fluctuations. Such a mechanism could underlie the formation of "lipid rafts" in biological membranes [116, 117]. The current definition of a lipid raft is a small aggregate of a particular lipid or group of lipids on a cell membrane surface [118]. They are thought to affect cell signalling and transport by inducing the co-localization of specific membrane proteins. Evidence for lipid rafts has been seen in eukaryotic [119] and bacterial [120] cells, and in virus assembly [121], but this remains an area of active work (see also Section 13.5).

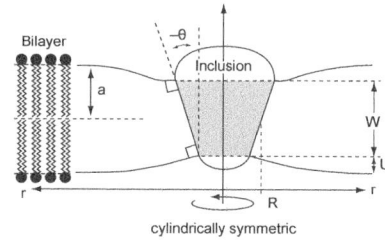

Fig. 13.24 Schematic diagram showing a possible distortion field induced by a membrane protein. Here Wiggins & Phillips [113] addressed the role of ion channels in mechano-sensing. Distortions such as those illustrated here can also lead to bending of the bilayer, or to aggregation of the inclusions, as discussed in the text. Figure redrawn from [113]

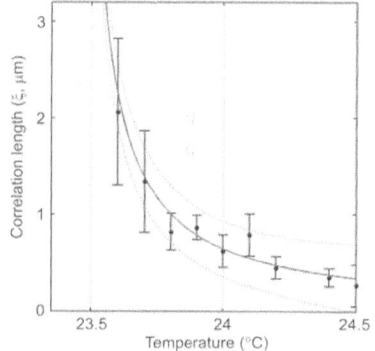

Fig. 13.25 The correlation length is the characteristic size of transient concentration fluctuations; for ternary mixtures being cooled toward the critical point, ξ diverges as a power of the temperature difference to the critical temperature. Micron sized domains are clearly visible in fluorescence microscopy, and are relatively long-lived. This critical behavior can be extrapolated away from the critical point, predicting that domains several nanometers in size (able to interact with proteins, as discussed in the text) should exist at physiological temperature. Figure reprinted with permission from [115]. Copyright (2008) American Chemical Society.

Shapes and dynamics of multicomponent membranes

The Canham-Helfrich free energy, introduced in Section 13.2, is a free energy functional, i.e. it is a function that depends on the shape of the membrane, which is itself a function of the position. The equilibrium shape is that which minimizes the functional, subject to boundary conditions and any physical constraints, and in general this is only accessible via numerical methods. However under some conditions of high symmetry, it is possible to derive simple differential equations (which can be solved numerically) that describe the membrane shapes. This enabled a calculation of the point of budding-off of a coexisting domain, for permeable membranes, as a function of the excess area, the mean and Gaussian bending moduli and the line tension [111, 122]. Similar calculations describe the shape and force necessary to pull a tether out of a vesicle [59]. The recent work of various groups, summarized in [123], provides the relevant equations for the calculation of shapes for axisymmetric membranes, including the case of coexisting multiple domains. The shape satisfies:

$$\ddot{\psi}\cos\psi = -\frac{1}{2}\dot{\psi}^2\sin\psi - \frac{\cos^2\psi}{r}\dot{\psi} + \frac{\sin\psi}{2}\left(\frac{\sin\psi}{r} - C_0\right)^2$$
$$+\frac{\cos^2\psi}{r^2}\sin\psi + \frac{\sigma}{\kappa}\sin\psi + \frac{p}{2\kappa}r + \frac{1}{2\pi\kappa}\frac{f}{r}, \quad (13.48)$$

where r is the radius in cylindrical coordinates, ψ is the tangent angle [see Figure 13.26(a) for notation], C_0 is the spontaneous curvature, σ and κ are the tension and bending moduli. respectively, p is the pressure across the membrane, and f is any applied force. The differential equation looks formidable, but it is solvable for various problems of interest. For example, for the case of a vesicle in which two domains coexist, it is well known that the equilibrium shape is not spherical if there is a line tension present at the boundary between the two domains. Membrane parameters, including the line tension and the elusive Gaussian modulus, could be extracted from experimentally obtained shapes, see Figure 13.26(c) and [124]. As another example, the relations between tether radius R_0 and applied force f_0:

$$R_0 = \sqrt{\frac{\kappa}{2\sigma}} \quad \text{and} \quad f_0 = 2\pi\sqrt{2\kappa\sigma}, \quad (13.49)$$

known from [59], can also be obtained readily from equation (13.48).

Tethers are relevant in biology, since small tubular structures exist within cells. Membrane nanotubes have also been shown to exist between cells, facilitating intracellular transport of vesicles or ionic species, and even viruses [125]. There is also a potential for engineering complex chemical reaction networks [126].

The presence of more than two domains is also an interesting case: if there is sufficient excess area (or no volume constraint), then the domains with different bending moduli are curved to a different extent. This membrane deformation induces an interaction between domains,

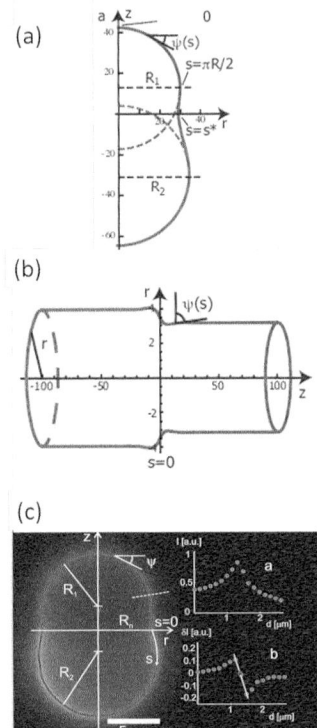

Fig. 13.26 For sufficiently symmetric shapes it is possible to parameterize the Helfrich free energy functional and calculate equilibrium solutions. (a,b) illustrate the notation used in the text, from [123]; (c) shows that the description adequately fits a phase separated giant vesicle. This analysis allows the line tension and both mean and Gaussian bending moduli to be measured. Figures (a,b) are reprinted with permission from [123]; Copyright (2011) Springer-Verlag. Figure (c) reprinted with permission from [124]; Copyright (2008) by the American Physical Society.

which can stabilize small domains against coalescence and potentially can lead to a self-organized sorting into arrays of different domain size, as observed in a model system [127]. The diffusive dynamics of micron-sized domains within the membrane plane has been studied in [34].

Coupling of membrane curvature with partitioning on different leaflets is very weak; because of the small lipid area, even lipids with very different spontaneous curvature have a weak tendency to segregate between leaflets for biologically relevant radii of membrane curvature. The evidence for this is very strong [128–130]. This implies that the lipid composition asymmetry commonly observed in biology, especially when the radius of curvature is large as in the plasma membrane of cells, is maintained by means other than membrane shape [130]. By a similar argument, lipid sorting within the plane is only very weakly coupled to curvature, and significant sorting is only observed for either very tight radii of curvature, or close to a critical point for demixing, where the free energy for sorting becomes negligible [128]. Tight enough radii of curvature can occur in membrane tethers and in the bicontinuous cubic phase. Larger objects, such as membrane proteins, can respond to much weaker membrane curvature, and partition accordingly [129].

13.5 Membranes in cell biology

Nature has developed membranes in the form of lipid bilayers; these can incorporate specialized and highly functional biomolecules. The

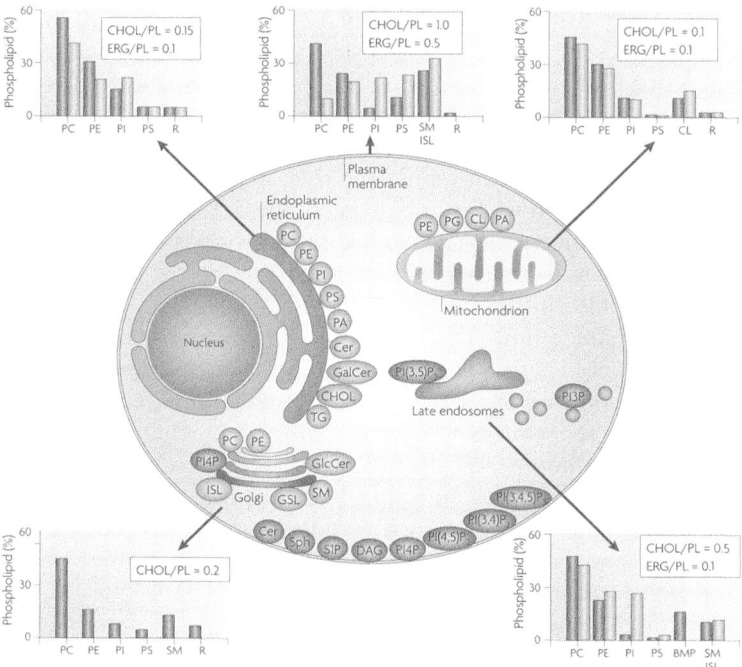

Fig. 13.27 Diagram of a cell, showing various organelles discussed in the text, with the molar fractions of the main lipid types. It is remarkable how cells maintain lipid homeostasis not just globally, but within each compartment of a very dynamic environment characterized by constant flow of lipid vesicles between different membranes. Reproduced with permission from [131], ©2008, Rights Managed by Nature Publishing Group.

most fundamental function of cell membranes is to compartmentalize cell space, and cells from each other, regulating transport. This is achieved thanks to the low permeability of lipid membranes to ions and solutes, allowing concentration gradients of ionic species (and hence electrostatic voltages) to be maintained across the membrane. A simplistic view is to consider membrane proteins as merely using the bilayer as a scaffold, and providing functionality on their own. Instead, as the structures and roles of membrane proteins are becoming better understood, it is becoming clear that lipid self-assembly plays a key role in controlling their function [132]; consequently in many cases the behavior of the proteins and lipids has to be addressed together. The richness of lipid species in a cell and the degree of control on the local composition are remarkable, see Figure 13.27. In this section we aim to highlight behavior that depends on this rich organization. We first present lipid-based processes in the cell (many of which depend of course on proteins), and then introduce membrane proteins explicitly on page 577.

Biological membranes vary significantly in topological complexity from small, highly curved single bilayer structures (e.g. synaptic vesicles), to extended highly convoluted organelles (e.g. cubic membranes in the endoplasmic reticulum [133] and the Golgi apparatus), extended single bilayers (e.g. the plasma membrane) and 'bulk' liquid-crystalline phases (e.g. the myelin sheath). The energy related to curvature and the intrinsic curvature of bilayers, are important in defining the shape of the biological membrane [134]. Many vital cell processes involve dynamic interconversions between these different morphologies, for example by membrane fusion, fission, or budding, on timescales spanning milliseconds to days or longer (10^{-3} - 10^6 s). Lipid asymmetry across the bilayer, lateral organization into domains, and curvature are all known to play crucial roles in maintaining these structures and their associated functionalities [131].

As well as the properties that determine structure and stability, the physical parameters linked to transport and dissipation within the membrane also play an important regulating role; in particular the membrane viscosity determines how fast objects confined to the membrane can move. Typical "objects" are the membrane proteins and assemblies of these proteins. Their function is to regulate processes both within the membrane, and also in the bulk fluid inside and outside the cell [5]. For example, there are pumps that regulate the concentration of calcium, potassium, pH, etc. across the membrane. Other proteins promote the formation and budding off of small membrane vesicles, which are essential for the directed transport of molecules to particular areas of the cells [135]. A very important class of membrane proteins are those that act as receptors, binding to specific chemicals, and triggering a particular response; the sequence of such responses within the cell is called a signalling pathway (or transduction cascade). The molecules that relay signals from receptors on the cell surface to target molecules in the cell cytoplasm or nucleus inside the cell are called second messengers (the first messengers being the signal molecules that arrive on the cell). While many of these processes are very specific and the biochemical details are different for each set of coupled chemical reactions, signaling pathways are themselves an area where general physical principles are important.

Lipid metabolism and turnover

Animals obtain lipids both from their diet and by synthesis. Four species of phospholipid dominate the plasma membrane of mammalian cells: phosphatidyl choline, sphingomyelin, phosphatidyl serine, and phosphatidyl ethanolamine. Their synthesis occurs in the cytosol adjacent to the endoplasmic reticulum (ER) membrane, see Figure 13.27. Here there is a high concentration of proteins that act in phospholipid synthesis (GPAT and LPAAT acyl transferases, phosphatase, and choline phosphotransferase) and allocation (flippase and floppase). Eventually a vesicle will bud off from the ER, containing phospholipids destined for the cytoplasmic cellular membrane on its outer leaflet and phospholipids destined for the exoplasmic cellular membrane on its inner leaflet [10].

Sterols have very low solubility, so in order for them to be circulated from the liver (where they are mainly synthesized) to other tissues, they form complexes with lipoproteins, which are soluble and can be carried in circulation. For example, a "low-density lipoprotein (LDL) particle" is formed by over 1000 cholesterol molecules binding to a single apolipoprotein B-100 molecule (a large protein with mass of 514 kDa); LDL has a hydrophobic core consisting of the polyunsaturated fatty acid linoleate, and this core is surrounded by a shell of phospholipids. LDL particles are approximately 22 nm in diameter, and are mainly secreted by the liver. Other cells in the body can regulate their cholesterol intake by expressing receptors to LDL, which then trigger binding and endocytosis. Sterols are a sub-group of steroids. In humans and other animals, steroids can come from the diet but can also be synthesized within cells. Biosynthesis of steroids follows a common set of chemical metabolic reactions, known as the "mevalonate pathway", resulting in lanosterol, which is then transformed into all the other steroids.

Organelles are very dynamic structures, constantly undergoing fission and fusion events, and therefore they require a constant and well-regulated supply of phospholipids; some organelles, e.g. mitochondria, are themselves also involved in lipid synthesis by transferring intermediate species. Various lipids are specifically transformed during signaling, for example by the enzymes phospholipase C (PLC), a class of enzymes that cleave phospholipids just before the phosphate group; the products of this reaction become components in downstream signaling cascades [5]. These processes imply a colossal traffic of membranes within cells, and specific lipid sorting.

Organelles

Prokaryotes (bacteria and archaea) are single-cell microorganisms, although from most biological points of view, from evolutionary to eco-

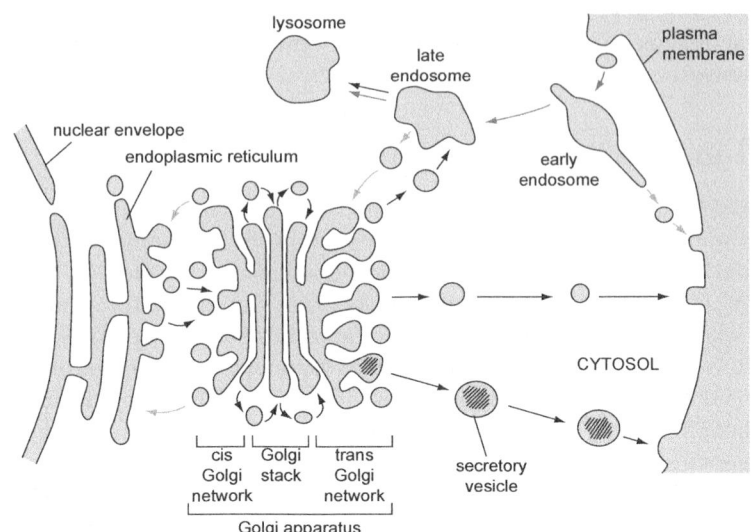

Fig. 13.28 The Golgi apparatus is an organelle present in most cell types, in which a complex arrangement of membrane stacks performs the remarkable function of accepting proteins from the endoplasmic reticulum, modifying them, and packaging them in three different vesicle types, for transport to different areas of the cell. This organelle disappears at cell division and spontaneously self-assembles in the daughter cells. Redrawn from [5].

logical, it is important to identify a colony of these genetically identical cells as the biological entity. Prokaryotic cells are therefore identified by an outer plasma membrane, which in some species is coupled to a cell wall. In a prokaryotic cell there is no nucleus and no complex lipid structures within the cell. In contrast, eukaryotic cells (animals, plants and fungi) have a nucleus and abundant lipid structures are visible by optical and electron microscopy; these are called organelles [5, 134]. The nuclear envelope is a double lipid bilayer, which allows and regulates transport of mRNA out of the nucleus and signaling molecules (which can regulate transcription) into the nucleus. The outer bilayer is continuous with the ER membranes, while the interior is a closed surface. The membrane protein structure which allows diffusive transport of small molecules, and regulates transport of larger cargo, is the nuclear pore complex [136].

One of the largest, and universal, organelles in eukaryotic cells is the Golgi apparatus, sketched in Figure 13.28. This is a hierarchical structure, composed of many folded membranes. It is clear that here, as well as in other cell biological processes, membrane vesicles are constantly being created, moved, fused into other membranes. These vesicles carry proteins, mainly synthesized in the ER, and within the Golgi apparatus these proteins are sorted for transport to appropriate locations in the cell [5].

Organelles are positioned within the cell (in particular at cell division) by the cytoskeleton and molecular motors (kinesin and dynein). It is also possible that the microtubule network, coupled to membranes by proteins, has a role in generating the membrane structures of the endoplasmic reticulum and the Golgi apparatus; certainly it has been shown that membrane tubular structures can be formed in a minimal *in-vitro* system consisting of a lipid reservoir, kinesin-coated beads, microtubules, and ATP [137].

Another organelle present in all nucleated eukaryotic cells is the mitochondrion. At least one of these organelles is essential for energy generation within a cell; human liver cells have typically over 1000. Each mitochondrion is roughly one micron in length; it is enclosed by a membrane and has a large interior area of tightly folded membrane. The fundamental role of mitochondria is to degrade nutrient molecules by oxidation, producing ATP, the chemical energy source ("fuel") for most other driven processes within a cell. The large membrane area and insulating properties of the membrane bilayer are essential in ATP synthesis; ATP is generated utilizing the electrochemical proton motive force (Δp) providing the energy to drive protons against their concentration gradient across the inner mitochondrial membrane (out of the mitochondrial cytoplasm). This results in a net accumulation of H^+ outside the membrane, which then flows back into the mitochondria through the ATP-generating ATP-synthase. ATP-synthase molecules are housed in the membrane. Δp is a combination of both the mitochondrial membrane potential (ϕ_m, a charge gradient) and the mitochondrial pH gradient (ΔpH_m, an H^+ concentration gradient); physiological values are $\phi_m = 150\,\mathrm{mV}$ and $\Delta pH_m = 0.5$.

Chloroplasts should not be forgotten since they are crucially important for life on our planet, and are present in organisms that carry out photosynthesis [5].

Inter-leaflet asymmetry and in-plane inhomogeneity

Biological membranes are made not of one single phospholipid but of many hundreds of chemically different ones, see Figure 13.27 for some examples; the composition of inner and outer leaflets in the bilayer is very often asymmetric, from the plasma membrane of the simplest cells (erythrocytes), to the vesicles extruded from the ER. Lipids can diffuse readily in the plane, but there is a large energy barrier for a lipid to pass from one leaflet to the other (flip-flop), because it is energetically unfavorable for the polar head group to pass through the hydrophobic core of the bilayer. This barrier implies long relaxation times, which seems to allow the possibility of maintaining asymmetry once it has been formed. However this may not hold in biological systems, where the bilayer holds a variety of inclusions that can probably act as defects to promote flip-flop [138]. In any case, the asymmetry of lipid composition cannot be attributed simply to the curvature, as discussed in Section 13.3, implying that it is the result of a non-equilibrium state, assembled and maintained by the cell at some energy cost. Indeed the appearance of phosphatidylserine on the exoplasmic face is an early sign of cell death by apoptosis.[9]

The multiphase system of the cell membrane has so many components that a complete phase diagram is impossibly complicated to measure or calculate. However there is a very important simplification. As described earlier in Section 13.1, phospholipids fall into two classes: saturated and

[9] There are protein machines (flippases and floppases) that promote (and regulate) the transfer of phospholipids synthesized by the cell (and therefore present mostly on the interior cytoplasmic leaflet) to the exoplasmic leaflet. Flippases use ATP to move the aminophospholipids and phosphatidylethanolamine, from the outer leaflet to the inner leaflet of the plasma membrane, against a concentration gradient. Floppases use ATP to transport phosphatidylcholine, sphingolipid and cholesterol against concentration gradients in the opposite direction [139]. Despite this understanding at the molecular level, a full understanding of how asymmetry is maintained at steady state in cell membranes is still lacking.

unsaturated. Saturated lipids pack better and have a higher viscosity.

In many biological membranes there is roughly a ratio 1:1 of saturated:unsaturated phospholipids. There is also a high concentration of cholesterol, up to 50% molar; this is a much smaller lipid that is "dissolved" in the hydrophobic layer. Such systems have a remarkably simple phase diagram, separating into two coexisting fluid phases at a temperature typically below 30 to 25°C, see phase diagrams in Figure 13.23. Lipid composition is critical to many biochemical processes, and lipid homeostasis is important to enable cell functions in general. In particular, there is growing evidence that in living cells the lipid composition is regulated to maintain a certain distance to the critical point [140, 141], a fact that is being noted in the biological literature [118] in connection with the concept of lipid rafts. Lipid rafts represent the well known fact that biological membranes present domains enriched with particular lipids and that this heterogeneity couples to partitioning or adhesion of specific proteins to those regions. For example, some membrane receptors are known to cluster in lipid domains, affecting signalling [142]. It is also clear that protein components of the cytoskeleton, in particular the cortical cytoskeleton, can couple to the lipid composition fluctuations [143]. The physiological proximity of the membrane composition and temperature to the critical point, allows composition fluctuations to occur spontaneously or with very low energy cost; elucidating the biological consequences of this, and looking for general principles of membrane protein regulation by lipid composition, remain active areas of research.

Vesicular traffic in the cell

It is straightforward from equation (13.19) to calculate the energy required to form a closed spherical vesicle, bending from a planar configuration. With constant curvature the free energy cost is $8\pi\kappa$, independent of vesicle radius. With bending moduli typically $\sim 20\, k_B T$, this results in $\sim 500\, k_B T$ of energy, i.e. many orders of magnitude more than what is available from thermal energy. This energy needs to be found in order for example for a cell membrane to "bud" a vesicle for trafficking. The continuous creation of vesicles within a cell is possible because of the presence of membrane proteins that promote curvature, and essentially are catalysts for bending. Vesicles are used to carry macromolecular cargo into, out of, and within cells.

Endocytosis is a very common mechanism by which a cell incorporates material from outside the cell into the cytosol. It is a fairly well understood process, involving a range of proteins. Of great interest in terms of membrane physics is the role of clathrin. This is a protein that induces curvature, and also spontaneously diffuses toward curved regions, forming "clathrin coated pits". These pits allow the invagination of a vesicle into the cytoplasm. At that point a key role is played by dynamin, a mechano-enzyme that polymerizes, forming helices at the neck of the endocytic buds [144]. Torsion of the helix under hydrolysis of GTP constricts the neck, which breaks, releasing a free vesicle. The clathrin then disassembles from the vesicle. Details of this process are still being debated, for example whether there is a correlation between the locations at which pits form (called "hotspots") and the structure of the underlying cortical cytoskeleton; two classes of accessory proteins have been identified in clathrin-mediated endocytosis (CME): nucleation factors and nucleation organizers. Recent work proposed that hotspots are specialized cortical actin patches that organize the nucleation proteins from within the cell [145].

A textbook example of vesicle traffic out of a cell is the synaptic junction. This is the narrow gap between the end of the axon in a neuron cell, and a target cell, for example, a muscle cell. Vesicles loaded with neurotransmitter molecules are formed and maintained near the tip of the axon. On arrival of an electrical signal (the action potential), the vesicles fuse to the axon's membrane, releasing the neurotransmitter (e.g. acetylcholine) into the junction. The muscle cell has receptors to the neurotransmitter. These are ion channels that open on receiving acetylcholine, temporarily depolarizing the muscle cell, and triggering a contraction [19].

Vesicles continually bud off from one membrane and fuse with another, carrying membrane components and soluble molecules referred to as cargo. This membrane traffic flows along highly organized, directional routes, maintained by the cell cytoskeleton network. The biosynthetic-secretory pathway leads outward from the ER toward the Golgi apparatus and the cell surface, with a side route leading to lysosomes, while the endocytic pathway leads inward from the plasma membrane [5]. The flow of membrane between compartments is balanced, with retrieval pathways balancing the flow in the opposite direction, bringing membrane and selected proteins back to the compartment of origin. The balance achieved by the vesicle traffic is remarkable: transport vesicles must bud taking only appropriate proteins, and must fuse only with the appropriate target membrane [5].

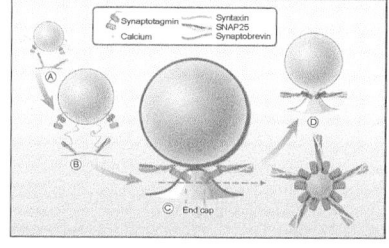

Fig. 13.29 Membrane buckling by C2 domains as a trigger for fusion. As the vesicle approaches the membrane in this model, the vesicular SNARE component binds to its SNARE counterparts on the target membrane, resulting in the formation of a complex that pulls the two membranes into close apposition (steps A and B). The C2 domains of synaptotagmin bind to the SNARE complex, potentially helping to complete their zippering into a continuous helix. The C2 domains also insert into the target membrane in a Ca^{2+}-dependent manner, resulting in membrane buckling and an unstable membrane region optimally localized for fusion (jagged membrane in step C). As the fusion pore opens, the C2 domains would still be localized to the neck, where they might promote the early stages of fusion pore opening (step D). Figure redrawn from [135].

Membrane proteins

About 50% by mass of a biological membrane is composed of proteins (as little as 25% in the insulating myelin sheath and as much as 75% in the membranes of mitochondria and chloroplasts). Some membrane proteins are transmembrane, extending across the bilayer. Others reside in the cytosol and are either anchored into a leaflet by one or more fatty acid chains, or covalently to a lipid in the leaflet. The latter ones are known as integral membrane proteins, to distinguish them from peripheral proteins that bind to the bilayer non-covalently. Many membrane proteins are common between prokaryotes and eukaryotes. The transmembrane section of proteins is often one or more α-helix segments, which are relatively hydrophobic. The β–barrel is also a common motif, as in the porins, which are discussed below [5].

A large number of transmembrane proteins regulate the flux of small molecules, and some actively pump ions and protons across the bilayer, maintaining gradients in the ionic strength of specific ions and gradients in pH. The porins are typical of passive membrane channel proteins. They allow molecules 5000 Daltons or less in molecular weight to freely diffuse from one side of the membrane to the other. A well-studied active membrane pump is Bacteriorhodopsin, which is a light-activated membrane proton-pump, found in a bacterial species [5]. A key pump in the plasma membrane of all animal cells is the sodium-potassium pump, also known as the Na,K-ATPase. It functions in the active transport of sodium and potassium ions across the cell membrane against their concentration gradients; for each ATP the pump breaks down, two potassium ions are transported into the cell and three sodium ions out of the cell. For non-excitable cells, this results in a constant steady state, with typical potential differences in the range 70 to 80 mV [5]. These ionic gradients across membranes are a very important mechanism for energy storage: it was mentioned above how they enable ATP synthesis in the mitochondrion, and the role in triggering muscle contractions.

Various membrane proteins are known to connect to the actin cytoskeleton, and can be involved in cell motility [146]. A vast variety of such proteins exists in humans, and many are very specific to particular cell types. For example, the band 3 transmembrane protein is found only in the plasma membrane of erythrocytes (red blood cells), where it interconnects through ankyrin to the spectrin cortical cytoskeleton [5].

Cells are able to sense forces, and the mechanical properties of the environment, with consequences on morphogenesis and stem cell differentiation [147]. The coupling of lipid composition to cytoskeletal arrangement deserves some comment, as it represents one possible mechanism to rationalize how cells (via the cytoskeleton) can sense forces, turning these into chemical signals. For example, the importance of a specific phospholipid, PIP_2 [148], in cytoskeletal attachment was demonstrated directly by elegant experiments using laser tweezers [149]. The same lipid has been reported to form patches in the plasma membrane, linked to the assembly of clathrin lattices [150]. It has even been proposed that a whole set of proteins (termed the GMC family) in which the N-termini contain either a myristate or two palmitates chains (saturated acyl chains) are localized in small cholesterol-enriched domains or rafts in unfixed cells, all bind PIP_2, and thus localize PIP_2 [148, 151]. Hence, lipid composition is affected by, and acts on, the structure of the underlying cytoskeleton via a wide range of mechanisms.

The role of clathrin in promoting vesiculation is discussed above; clathrin assembly is promoted by attractive interactions mediated by membrane curvature. More generally, there will be interactions between any protein species that deforms the membrane, either by inducing curvature or by modifying locally the natural thickness of the bilayer. Symmetric deformations will lead to attraction [114]. A more subtle effect which has been proposed theoretically is that there will be an additional, weak attraction between any pair of inclusions, because of the loss of fluctuation degrees of freedom in the space between those inclusions, via a process that is analogous to the Casimir effect [152].

Examples are known of proteins that bind to membranes of high curvature: dynamin binds and polymerizes on tight membrane tubes and necks, facilitating tubes formation and vesicle fission [144]. The opposite process of vesicle fusion into membranes, which involves neck formation, is facilitated by SNARE proteins, see Figure 13.29 [135]. The BAR proteins are also known to polymerize on membranes, and induce curvature [153].

It is not clear if the coupling of protein function to lipid composition represents a general mechanism, but specific examples are abundant. PIP_3 is a minority component, accounting with PIP_2 (mentioned above) for around 1% of membrane phospholipids. Yet they are involved in a variety of processes [154]. The localization of PIP_3 has been studied in depth, in relation to the question of cell polarization and eukaryotic chemotaxis [155]. In response to a weak chemotactic gradient, a phase separation process is triggered, localizing PIP_3 on the plasma membrane side exposed to the highest chemoattractant concentration, and the PIP_3-degrading enzyme PTEN and its product PIP_2 to a complementary pattern [156]. This is an amplifying response, which may then have cytoskeletal consequences as discussed above.

An alternative classification of membrane proteins is in terms of integral or amphitropic proteins; the amphitropic proteins can exist in two states (water soluble and lipid bound) and can switch between these by tuning their affinity for the membrane [157]. These proteins are particularly sensitive to membrane composition [158].

A number of membrane proteins respond to the global mechanical properties of the membrane in which they are embedded. The mechanics (e.g. state of tension and bending modulus) can in turn be determined by composition, thus providing another mechanism coupling protein function to composition. An important example of this is represented by the mechano-sensitivity of ion channels [159, 160].

Mechanics of the cell plasma membrane

Section 13.2 presented a general continuum theory for the mechanics of thin, fluid membrane. In the case of uniform lipid bilayer membranes, the key quantities, such as the bending modulus κ, can be calculated from the knowledge of the forces between lipids across the membrane. However it is obvious that biological membranes are far more complex. The main difference is not due to the presence of embedded proteins, or in-plane variation, since such a structure would still behave as a thin sheet; κ would then be the effective bending modulus of this more complex assembly. The main mechanical difference is that a cell membrane is embedded within a viscoelastic medium. In particular, the plasma membrane of a cell is covered on the outside surface by carbohydrate groups, from glycoproteins and glycolipids, and on the inside surface by a layer of cortical cytoskeleton which dominates the mechanical response. In cells with a complete thick cytoskeleton, the mechanical properties of the cytoskeleton entirely outweigh the contribution from the thin plasma membrane.

The structurally simplest eukaryotic cell is the erythrocyte (red blood cell), which in mammals does not have a nucleus nor a system of microtubules. The only cytoskeletal structure is a very thin and sparse cortical cytoskeleton [161]. This enables the cell to be soft and deformable, key characteristics that allow it to flow in thin capillaries. The mechanics of such a cell is only in part consistent with the behavior of a thin material sheet. On the one hand, the fluctuations of cell shape at the micron scale are similar to those of model giant vesicles, see Figure 13.17. The simple bilayer model can be extended to account explicitly for the connection to the cytoskeleton via periodic anchoring sites [162], predicting a dampening of larger wavelength modes. In the planar approximation, this modifies equation (13.26) to:

$$F^{(2)} = \frac{L_x L_y}{2} \sum_q \left(\kappa q^4 + \sigma q^2 + \gamma\right) h_{\vec{q}}^2. \tag{13.50}$$

This additional γ term constrains each point on the membrane to lie on average within a distance $d \sim \sqrt{k_B T} (\kappa \gamma)^{-1/2}$ of the background plane [161]. This is consistent with the RBC appearing to have a large shear modulus $\mu \sim 4 \times 10^{-6}$ Jm^{-2} under static deformation, when pulled by tweezers [163, 164] but a vanishing shear modulus when the thermal fluctuations are observed, since γ has a negligible effect for small wavelength fluctuations [161].

On the other hand, the RBC displays complex behavior which is qualitatively different from that of lipid bilayers: As soon as the cell is strained beyond the spontaneous fluctuation amplitudes, for example by optical tweezers, the dynamical response becomes non-linear [165]. There is also an open question as to whether the fluctuations of this cell show the presence of non-thermal active processes, as appears in some experiments [82, 166, 167] but not in others [89, 168]. Active motion has been clearly observed in other more complex cell types [92].

An important issue, difficult to account for in mechanical models, is that the molecular bonds in the cytoskeleton are fragile, and can dynamically break and reform as the cell is deformed. This is probably true also in the cortical cytoskeleton. One approach to this problem is the so-called soft glassy rheology (SGR) model [169], which was developed to understand the dynamics of colloidal crystals and glasses, in which the configuration are trapped. The SGR model consists of a series of shallow potential wells of different depths, which in the context of cells could correspond to configurations of the cytoskeletal network, where bonds are continuously breaking and reforming. The presence of an external stress would favor breaking and restructuring of the network [170]. The model also leads to anomalous exponents in fluctuation power spectra, so the model has been invoked to explain the anomalous power laws of some RBC fluctuations [165].

Outlook

In such a wide topic, there are clearly various quite distinct research themes. Particularly an outlook reflects the opinions, knowledge, and interests of the authors. At the interface between the physical and life sciences there are often two opposite driving forces: the motivation to understand the biology may dominate, possibly leading to exploring reductionist model systems that highlight and isolate some process or functional aspect; on the contrary, a technological motivation may prevail, within which one takes inspiration from the biological process, aiming to reproduce some aspects synthetically in order to create functional materials. Both these approaches thrive in the area of films and membranes.

As one of the simplest platforms to study systems in two-dimensions (or quasi-two dimensions, if out of plane tilt or dynamics are relevant), surface films remain an active area of research in the fundamentals of soft matter physics. In this aspect, the frontier trends are the behavior of dense systems, as they cross a point of dynamical arrest, and the control of inter-molecular interactions in order to self-assemble desired patterns and structure. Biological or biologically-inspired molecules (both lipids and proteins) are very commonly used.

Surface films are also themselves present in biology (two examples, tear films and lung surfactant films have been discussed here), and various aspects of their dynamics, phase behavior, and deformation under strain (buckling) are still being explored. Very complex interfaces also exist, involving complex viscoelastic subphases such as mucus layers; the flow and deformation behavior of these 'surface+bulk' structures is still poorly understood.

Structured bulk phases, and vesicles in suspension, are easily formed by phospholipids. This has great potential, still barely tapped, as a basis for artificial functional materials and systems. One can envision materials with tunable pore size for protein crystallization, or for chemical sensing. Fine control of membrane topology and the assembly of hierarchical structures, such as connected networks of many vesicles, would allow a new network-connectivity control parameter to fine tune chemical reactions spatially.

In biology, while we approach the centenary of the discovery of cell membranes, their role and function are very much in the spotlight. The importance of lipid-protein interactions has been suggested by a few researchers, but this is a radically new concept in molecular biology and represents an area of which very little is known. At a larger scale, of the entire cell, an active research community is investigating cell mechanics, focusing on the cytoskeleton, and much progress has been made in understanding the role of actin filaments, crosslinking proteins, and molecular motors. In cells, the cortical actin cytoskeleton is anchored to the membrane. It is likely that various aspects of current knowledge, as well as possibly mechanisms still to be uncovered, will need to be brought together to gain a full understanding of force transduction and force signalling between cells, two processes that have important implications both for the subtle processes of cell differentiation and the runaway division of cancerous cells.

Acknowledgments

We would like to thank our closest co-workers over the last few years in this area, Y.-Z. Yoon, S.L. Veatch and S.L. Keller, and P.D. Olmsted, O. Ces, A. Roux, G. Cristofolini, S. Mahajan and U.F. Keyser for carefully reading this manuscript and providing many useful comments. Funding for this work is from ITN-Transpol and EPSRC-Capitals.

References

[1] Schrum, J.P., Zhu, T.F, and Szostak, J.W. (2010). *Cold Spring Harb. Perspect. Biol.*, **2**, a002212.
[2] Nelson, D.L. and Cox, M.M. (2004). *Lehninger Principles of Biochemistry* (4th edn). W.H. Freeman, New York.
[3] Nagle, J. F. and Tristram-Nagle, S. (2000). *Biochim. Biophys. Acta*, **1469**, 159 – 195.
[4] Israelachvili, J. (2011). *Intermolecular & Surface Forces, 3rd ed.* Academic Press, San Diego.
[5] Alberts, B., Bray, D., Lewis, J. et al. (1994). *Molecular Biology of the Cell*. Garland Publishing, New York.
[6] Franke, T.F., Kaplan, D.R., Cantley, L.C. et al. (1997). *Science*, **275**, 665–668.
[7] Singer, S. J. and Nicolson, G. L. (1972). *Science*, **175**, 720.
[8] Luckey, M. (2008). *Membrane Structural Biology: Biochemical and Biophysical Foundations*. Cambridge University Press, New York.
[9] Lehninger, A. (1975). *Biochemistry*. Worth Publishers, New York.
[10] Lodish, H., Berk, A., Kaiser, C.A. et al. (2007). *Molecular Cell Biology*. W.H. Freeman, New York.
[11] Sagis, L. M. C. (2011). *Rev. Mod. Phys.*, **83**, 1367–1403.
[12] Seifert, U. (1997). *Adv. Phys.*, **46**, 13.
[13] Safran, S. A. (1999). *Adv. Phys.*, **48**, 395.
[14] Heimburg, T. (2007). *Thermal Biophysics of Membranes*. Wiley-VCH Verlag, Weinheim (Germany).
[15] Seddon, A. M., Casey, D., Law, R. V. et al. (2009). *Chem. Soc. Rev.*, **38**, 2509.
[16] Allen, T. M. (1996). *Curr. Opin. Coll. & Interface Sci.*, **1**, 645.
[17] Franks, N. P. and Lieb, W. R. (1994). *Nature*, **367**, 607.
[18] Mouritsen, O. G. (2004). *Life - as a Matter of Fat: the Emerging Science of Lipidomics*. Springer, Berlin.
[19] Phillips, R., Kondev, J., and Theriot, J. (2008). *Physical Biology of the Cell*. Garland, New York.
[20] Tanford, C. (1989). *Ben Franklin stilled the waves*. Duke University Press, Durham, NC.
[21] Stone, H. A., Koehler, S. A., Hilgenfeldt, S. et al. (2003). *J. Phys.: Cond. Matter*, **15**, 283.

[22] Janiak, M. J., Small, D. M., and Shipley, G. G. (1979). *J. Biol. Chem.*, **254**, 6068.
[23] Binder, K., Sengupta, S., and Nielaba, P. (2002). *J. Phys.: Cond. Matter*, **14**, 2323.
[24] Bates, M. A. and Frenkel, D. (2000). *J. Chem. Phys.*, **112**, 10034.
[25] Safran, S. (1994). *Frontiers in Physics 90: Statistical Thermodynamics of Surfaces, Interfaces, and Membranes*. Addison-Wesley, Reading, MA.
[26] Gaines, G. L. (1960). *Insoluble Monolayers at Liquid-Gas Interfaces*. Wiley, New York.
[27] Adam, N.K. and Jessop, G. (1926). *Proc. R. Soc. Lond. A*, **112**, 362–375.
[28] Pieranski, P. (1980). *Phys. Rev. Lett.*, **45**, 569.
[29] Kaganer, V. M., Möhwald, H., and Dutta, P. (1999). *Rev. Mod. Phys.*, **71**, 779.
[30] Joly, M. (1969). In *Surface and Colloid Science. Vol. 5* (ed. E. Matijević). Wiley, New York.
[31] Miller, R., Wüstneck, R., Krägel, J. *et al.* (1996). *Colloids and Surfaces A*, **111**, 75.
[32] Cicuta, P. and Hopkinson, I. (2004). *Coll. Surf. A: Physicochem. Eng. Aspects*, **233**, 97.
[33] Brooks, C. F., Fuller, G. G., Curtis, C. W. *et al.* (1999). *Langmuir*, **15**, 2450.
[34] Cicuta, P., Keller, S. L., and Veatch, S. L. (2007). *J. Phys. Chem. B*, **111**, 3328.
[35] Saffman, P.G. and Delbrück, M. (1975). *Proc. Natl. Acad. Sci.*, **72**, 3111.
[36] Gambin, Y., Lopez-Esparza, R., Reffay, M. *et al.* (2006). *Proc. Natl. Acad. Sci.*, **103**, 2098.
[37] Hughes, B.D., Pailthorpe, B.A., and White, L.R. (1981). *J. Fluid Mech.*, **110**, 349.
[38] Petrov, E.P. and Schwille, P. (2008). *Biophys J.*, **94**, 41–43.
[39] Krieger, I.M. and Dougherty, T.J. (1959). *Trans. Soc. Rheol.*, **3**, 137–152.
[40] Lee, K. Y. C. (2008). *Ann. Rev. Phys. Chem.*, **59**, 771–91.
[41] Braun, R. J. (2012). *Ann. Rev. Fluid Mech.*, **44**, 267–297.
[42] Butovich, I.A. (2011). *Prog. Lipid Res.*, **50**, 278–301.
[43] Leiske, D.L., Raju, S.R., Ketelson, H.A. *et al.* (2010). *Exp. Eye Res.*, **90**, 598–604.
[44] Rawicz, W., Olbrich, K. C., McIntosh, T. *et al.* (2000). *Biophys. J.*, **79**, 328.
[45] Helfrich, W. (1973). *Z. Naturforsch. C*, **28**, 693.
[46] Seifert, U. (1993). *Phys. Rev. Lett.*, **70**, 1335.
[47] Miao, L., Seifert, U., Wortis, M. *et al.* (1994). *Phys. Rev. E*, **49**, 5389.
[48] Seifert, U. (1995). *Z. Physik B: Cond. Mat.*, **97**, 299.
[49] Svetina, S. and Žekš, B. (1989). *Eur. Biophys. J.*, **17**, 101–111.
[50] Seifert, U., Miao, L., Dobereiner, H. G. *et al.* (1992). In *Springer*

Proceedings in Physics, Vol. 66 (ed. R. Lipowsky, D. Richter, and K. Kremer), p. 93. Springer, Berlin.

[51] Petrov, P. G., Lee, J. B., and Döbereiner, H.-G. (1999). *Europhys. Lett.*, **48**, 435.

[52] Wintz, W., Döbereiner, H-G., and Seifert, U. (1996). *Europhys. Lett.*, **33**, 403.

[53] Montes, L.-R., Alonso, A., Goñi, F. M. *et al.* (2007). *Biophys. J.*, **93**, 3548.

[54] Berndl, K., Kas, J., Lipowsky, R. *et al.* (1990). *Europhys. Lett.*, **13**, 659.

[55] Seigneuret, M. and Devaux, P. F. (1984). *Proc. Natl. Acad. Sci.*, **81**, 3751.

[56] Evans, E. and Yeung, A. (1994). *Chem. Phys. Lipids*, **73**, 39.

[57] Käs, J. and Sackmann, E. (1991). *Biophys. J.*, **60**, 825.

[58] Jahn, R. and Grubmüller, H. (2002). *Curr. Op. Cell Biol.*, **14**, 488–495.

[59] Powers, T. R., Huber, G., and Goldstein, R. E. (2002). *Phys. Rev. E*, **65**, 041901.

[60] Jesorka, A. and Orwar, O. (2008). *Ann. Rev. Anal. Chem.*, **1**, 801–832.

[61] Derenyi, I., Koster, G., Van Duijn, M. M. *et al.* (1992). In *Controlled Nanoscale Motion* (ed. H. Linke and A. Mansson), pp. 141–159. Springer-Verlag, Berlin.

[62] Pinnow, H. A. and Helfrich, W. (2000). *Eur. Phys. J. E*, **3**, 149.

[63] Seifert, U. and Langer, S. A. (1993). *Europhys. Lett.*, **23**, 71.

[64] Pécréaux, J., Dobereiner, H. G., Prost, J. *et al.* (2007). *Eur. Phys. J. E*, **13**, 277.

[65] Yoon, Y-Z., Hong, H., Brown, A. *et al.* (2009). *Biophys. J.*, **97**, 1606.

[66] De Gennes, P. G. and Taupin, C. (1982). *J. Phys. Chem.*, **86**, 2294–2304.

[67] Fournier, J.-B., Ajdari, A., and Peliti, L. (2001). *Phys. Rev. Lett.*, **86**, 4970.

[68] Milner, S. T. and Safran, S. A. (1987). *Phys. Rev. A*, **36**, 4371.

[69] Evans, E. A. and Fung, Y. C. (1972). *Microvasc. Res.*, **4**, 335.

[70] Girard, P., Prost, J., and Bassereau, P. (2005). *Phys. Rev. Lett.*, **94**, 088102.

[71] Farago, O. (2011). *Phys. Rev. E*, **84**, 051914.

[72] Helfrich, W. (1985). *J. Phys. (France)*, **46**, 1263.

[73] Peliti, L. and Leibler, S. (1985). *Phys. Rev. Lett.*, **54**, 1690.

[74] Forster, D. (1986). *Phys. Lett.*, **114A**, 115.

[75] Kleinert, H. (1986). *Phys. Lett.*, **114A**, 263.

[76] Gompper, G. and Kroll, D. M. (1996). *J. Phys. I (France)*, **6**, 1305.

[77] Nishiyama, Y. (2002). *Phys. Rev. E*, **66**, 061907.

[78] Imparato, A. (2006). *J. Chem. Phys.*, **124**, 154714.

[79] Fournier, J-B. and Barbetta, C. (2008). *Phys. Rev. Lett.*, **100**, 078103.

[80] Schmid, F. (2011). *Eur. Phys. Lett.*, **95**, 28008.
[81] Brochard, F. and Lennon, J. F. (1975). *J. Phys. France*, **36**, 1035.
[82] Betz, T., Lenz, M., Joanny, J.-F. *et al.* (2009). *Proc. Natl. Acad. Sci. USA*, **106**, 15312.
[83] Pfeiffer, W., König, S., Legrand, J. F. *et al.* (1993). *Europhys. Lett.*, **23**, 457.
[84] Granek, R. and Olami, Z. (1995). *J. Phys. II France*, **5**, 1349–1370.
[85] Nelson, P., Powers, T., and Seifert, U. (1995). *Phys. Rev. Lett.*, **74**, 3384.
[86] Bar-Ziv, R. and Moses, E. (1994). *Phys. Rev. Lett.*, **73**, 1392.
[87] Deschamps, J., Kantsler, V., Segre, E. *et al.* (2009). *Proc. Natl. Acad. Sci. USA*, **106**, 11444.
[88] Brown, A., Kotar, J., and Cicuta, P. (2011). *Phys. Rev. E*, **84**, 021930.
[89] Yoon, Y-Z., Kotar, J., Brown, A. T. *et al.* (2011). *Soft Matter*, **7**, 2042.
[90] Zhou, H., Gabilondo, B. B., Losert, W. *et al.* (2011). *Phys. Rev. E*, **83**, 011905.
[91] Poole, C. and Losert, W. (2007). In *Methods in Molecular Biology 400: Methods in Membrane Lipids.* (ed. A. M. Dopico), p. 389. Humana Press, Totowa, NJ.
[92] Mizuno, D., Bacabac, R., Tardin, C. *et al.* (2009). *Phys. Rev. Lett.*, **102**, 168102.
[93] Levine, A. J. and MacKintosh, F. C. (2002). *Phys. Rev. E*, **66**, 061606.
[94] Bitbol, A.-F., Fournier, J.-B., Angelova, M. I. *et al.* (2010). *J. Phys. Cond. Matt.*, **23**, 284102.
[95] McClain, B. L., Finkelstein, I. J., and Fayer, M. D. (2004). *Chem. Phys. Lett.*, **392**, 324.
[96] Kelemen, C., Chien, S., and Artmann, G. (2001). *Biophys. J.*, **80**, 2622–2630.
[97] Hochmuth, R., Buxbaum, K., and Evans, E. (1980). *Biophys. J.*, **29**, 177–182.
[98] Biben, T., Farutin, A., and Misbah, C. (2011). *Phys. Rev. E*, **83**, 031921.
[99] Farutin, A., Biben, T., and Misbah, C. (2010). *Phys. Rev. E*, **81**, 061904.
[100] Lebedev, V., Turitsyn, K., and Vergeles, S. (2007). *Phys. Rev. Lett.*, **99**, 218101.
[101] Danker, G. and Misbah, C. (2007). *Phys. Rev. Lett.*, **98**, 088104.
[102] Falck, E., Róg, T., Karttunen, M. *et al.* (2008). *J. Am. Chem. Soc.*, **130**, 44–45.
[103] Busch, S., Smuda, C., Pardo, L. C. *et al.* (2010). *J. Am. Chem. Soc.*, **132**(10), 3232–3233.
[104] Rubenstein, J. L. R., Smith, B. A., and McConnell, H. M. (1979). *Proc. Natl. Acad. Sci.*, **76**, 15.
[105] Murcia, M.J., Garg, S., and Naumann, C.A. (2007). In *Methods in Molecular Biology 400: Methods in Membrane Lipids.* (ed. A. M.

Dopico), pp. 277–294. Humana Press, Totowa, NJ.

[106] Gordon, V. D., Beales, P. A., Zhao, Z. *et al.* (2006). *J. Phys.: Cond. Matter*, **18**, L415.

[107] Gov, N. S. (2006). *Phys. Rev. E*, **73**, 041918.

[108] Wolff, J., Marques, C. M., and Thalmann, F. (2011). *Phys. Rev. Lett.*, **106**, 128104.

[109] Veatch, S. L. and Keller, S. L. (2003). *Biophys. J.*, **85**, 3074.

[110] de Almeida, R.F.M., Federov, A., and Prieto, M. (2003). *Biophys. J.*, **85**, 24062416.

[111] Baumgart, T., Hess, S., and Webb, W.W. (2003). *Nature*, **425**, 821–824.

[112] Honerkamp-Smith, A. R., Cicuta, P., Collins, M. D. *et al.* (2008). *Biophys. J.*, **95**, 236–246.

[113] Wiggins, P. and Phillips, R. (2004). *Proc. Natl. Acad. Sci.*, **101**, 4071–4076.

[114] Fournier, J-B., Dommersnes, P.-G., and Galatola, P. (2003). *C. R. Biologies*, **326**, 467.

[115] Veatch, S. L., Cicuta, P., Sengupta, P. *et al.* (2008). *ACS Chemical Biology*, **3**, 287.

[116] Honerkamp-Smith, A. R., Veatch, S. L., and Keller, S. L. (2009). *Biochim. Biophys. Acta - Biomembranes*, **1788**, 53.

[117] Simons, K. and Ikonen, E. (1997). *Nature*, **387**, 569.

[118] Simons, K. and Gerl, M. J. (2010). *Nature*, **11**, 688.

[119] Friedrichson, T. and Kurzchalia, T. V. (1998). *Nature*, **394**, 802.

[120] Lopez, D. and Kolter, R. (2010). *Genes Dev.*, **24**, 1893.

[121] Polozov, I. V., Bezrukov, L., Gawrisch, K. *et al.* (2008). *Nat. Chem. Biol.*, **4**, 248–55.

[122] Jülicher, F. and Lipowsky, R. (1993). *Phys. Rev. Lett.*, **70**, 2964.

[123] Idema, T. and Storm, C. (2011). *Eur. Phys. J. E*, **34**, 67.

[124] Semrau, S., Idema, T., Holtzer, L. *et al.* (2008). *Phys. Rev. Lett.*, **100**, 088101.

[125] Sowinski, S., Jolly, C., Berninghausen, O. *et al.* (2008). *Nature Cell Biol.*, **10**, 211 – 219.

[126] Karlsson, A., Karlsson, R., Karlsson, M. *et al.* (2001). *Nature*, **409**, 150.

[127] Idema, T., Semrau, S., Storm, C. *et al.* (2010). *Phys. Rev. Lett.*, **104**, 198102.

[128] Sorre, B., Callan-Jones, A., Manneville, J.-B. *et al.* (2009). *Proc. Natl. Acad. Sci.*, **106**(14), 5622–5626.

[129] Tian, A. and Baumgart, T. (2009). *Biophys J.*, **96**, 2676–88.

[130] Kamal, M. M., Mills, D., Grzybek, M. *et al.* (2009). *Proc. Natl. Acad. Sci.*, **106**(52), 22245–22250.

[131] van Meer, G., Voelker, D. R., and Feigenson, G. W. (2008). *Nature Rev. Mol. Cell Biol.*, **9**, 112–124.

[132] Bagatolli, L. A., Ipsen, J. H., Simonsen, A. C. *et al.* (2010). *Prog. Lip. Research*, **49**, 378–89.

[133] Lingwood, D., Schuck, S., Ferguson, C. *et al.* (2009). *J. Biol. Chem.*, **284**, 12041–12048.

[134] Voeltz, G. K. and Prinz, W. A. (2007). *Nature Rev. Mol. Cell Biol.*, **8**, 258–264.

[135] McMahon, H. T., Kozlov, M., and Martens, S. (2010). *Cell*, **140**, 601–605.

[136] Görlich, D. and Kutay, U. (1999). *Ann. Rev. Cell Develop. Biol.*, **15**, 607–660.

[137] Roux, A., Cappello, G., Cartaud, J. *et al.* (2002). *Proc. Natl. Acad. Sci. USA*, **99**, 5394–5399.

[138] Sanyal, S. and Menon, A. K. (2009). *ACS Chem. Biol.*, **4**, 895–909.

[139] Clark, M.R. (2011). *Nature Immunol.*, **12**, 373–375.

[140] Jin, A.J., Edidin, M., Nossal, R. *et al.* (1999). *Biochemistry*, **38**, 13275–8.

[141] Baumgart, T., Hammond, A., Sengupta, P. *et al.* (2007). *Proc. Natl. Acad. Sci.*, **104**, 3165–3170.

[142] Bethani, I., Skånland, S. S., Dikic, I. *et al.* (2010). *EMBO J.*, **18**, 2677–2688.

[143] Machta, B.B., Papanikolaou, S., Sethna, J.P. *et al.* (2011). *Biophys J.*, **100**, 1668–77.

[144] Roux, A., Koster, G., Lenz, M. *et al.* (2010). *Proc. Natl. Acad. Sci.*, **107**, 4141–4146.

[145] Nunez, D., Antonescu, C., Mettlen, M. *et al.* (2011). *Traffic*, **12**, 1868–1878.

[146] Sechi, A.S. and Wehland, J. (2000). *J. Cell Sci.*, **113**, 3685–3695.

[147] Engler, A.J., Sen, S., Sweeney, H.L. *et al.* (2006). *Cell*, **126**(4), 677–689.

[148] McLaughlin, S., Wang, J., Gambhir, A. *et al.* (2002). *Ann. Rev. Biophys. Biomol. Structure*, **31**, 151–75.

[149] Raucher, D., Stauffer, T., Chen, W. *et al.* (2000). *J. Cell Sci.*, **100**, 221–228.

[150] Gillooly, D.J. and Stenmark, H. (2001). *J. Cell Sci.*, **291**, 993–994.

[151] Laux, T., Fukami, K., Thelen, M. *et al.* (2000). *J. Cell Sci.*, **149**, 1455–1472.

[152] Bitbol, A-F., Dommersnes, P. G., and Fournier, J.-B. (2010). *Phys. Rev. E*, **81**, 050903.

[153] Frost, A., Perera, R., Roux, A. *et al.* (2008). *Cell*, **132**(5), 807–817.

[154] Czech, M.P. (2000). *Cell*, **100**, 603606.

[155] Gamba, A., Kolokolov, I. Lebedev, V., and Ortenzi, G. (2009). *J. Stat. Mech.*, **P02019**.

[156] Gamba, A., de Candia, A., Di Talia, S. *et al.* (2005). *Proc. Natl. Acad. Sci.*, **102**, 1692716932.

[157] Kinnunen, P.K.J., Kõiv, A., Lehtonen, J.Y.A. *et al.* (1994). *Chem. Phys. Lipids*, **73**, 181–207.

[158] Johnson, J.E. and Cornell, R.B. (1999). *Mol. Membrane Biol.*, **16**, 217–235.

[159] Ursell, T., Phillips, R., Kondev, J. *et al.* (2008). In *Mechanosensitivity in Cells and Tissues 1: Mechanosensitive Ion Channels* (ed. A. Kamkin and I. Kiseleva), p. 37. Springer-Verlag, Berlin.

[160] Phillips, R., Ursell, T., Wiggins, P. *et al.* (2009). *Nature*, **459**, 379–385.
[161] Auth, T., Safran, S. A., and Gov, N. S. (2007). *Phys. Rev. E*, **76**, 051910.
[162] Gov, N. and Safran, S. (2004). *Phys. Rev. E*, **69**, 011101.
[163] Sleep, J., Wilson, D., Simmons, R. *et al.* (1999). *Biophys. J.*, **77**, 3085.
[164] Hénon, S., Lenormand, G., Richert, A. *et al.* (1999). *Biophys. J.*, **76**, 1145.
[165] Yoon, Y. Z., Kotar, J., Yoon, G. *et al.* (2008). *Phys. Biol.*, **5**, 036007.
[166] Tuvia, S., Almagor, A., Bitler, A. *et al.* (1997). *Proc. Natl. Acad. Sci.*, **94**, 5045–5049.
[167] Tuvia, S., Levin, S., Bitler, A. *et al.* (1999). *Biophys. J.*, **141**, 1151.
[168] Evans, J., Gratzer, W., Mohandas, N. *et al.* (2008). *Biophys. J.*, **94**, 4134.
[169] Sollich, P., Lequeux, F., Hébraud, P. *et al.* (1997). *Phys. Rev. Lett.*, **78**, 2020.
[170] Kroy, K. (2006). *Curr. Op. Coll. Interface Sci.*, **11**, 56–64.

Index

Actin
  filaments, 499
  gel, nonlinear elasticity, 505
  treadmilling, 500, 506
Active transport, 508, 524
Amphiphilic systems, 51, 319, 537

Backflow effect, 129
Binodal line, 26, 348
Bjerrum length, 384
Block copolymers
  strong segregation, 275
  weak segregation, 273
Bragg diffraction, 302
Bravais lattices, 302, 303
Brownian motion
  free diffusion, 3, 7
  on membrane, 560
  definition, 3
  sedimentation, 3

Capillary length, 76, 153, 544
Capillary number, 75
Cell membrane, 502, 571, 579
Cell motility, 515
Cell polarity, 519
Cholesteric liquid crystals, 101
Chromonic liquid crystals, 97
Coacervation, 482
Coexistence, 12, 14, 67, 440
Collagen
  crimp, 486
  hierarchical structure, 481, 484, 486
  triple helix, 477
Collagen and elastin, 475, 489
Colloid
  definition, 1
  gelation, 25
  kinetic arrest, 23
  non-Newtonian media, 41
  equation of state, 5
Constitutive equation, 32, 190, 257, 505, 507
Contact angle, 51, 60
Continuum mechanics, 452
Critical micelle concentration, 53, 64, 537
Crystallographic symmetry, 306, 308, 310
Cycle rank, in polymer network, 338

Darcy law, 459

Debye frequency, 11
Debye layer, 17, 79, 399, 403
Depletion interaction, 14, 15, 86, 110, 518
Dielectric anisotropy, 114
Diffusion
  in block-copolymers, 290, 291
  on membrane, 546, 567
  reptation, 245, 248
  Rouse chains, 421
Dilatation (compression) modulus, 57, 543
DLVO interactions, 79

Elasticity
  bulk and shear moduli, 453, 456
  monolayers, 57, 542, 548
  reference frame, 452
Elastin, 479, 482, 487
  disorder and elasticity, 488
Elastomers
  extensional elasticity, 344, 360
  filled composites, 368
  Gaussian theories, 351
  liquid crystalline, 365
  network crosslinking, 334, 340
  non-Gaussian theories, 353
  swelling by solvent, 346, 362
Electro-optical effect, 96, 99, 131
Electro-osmosis, 462
Electrophoresis, 128
Emulsions
  Bancroft rule, 82
  coalescence, 85
Ensemble
  canonical, 8, 9
  grand canonical, 8
  microcanonical (NVE), 206
  stress, 214, 215, 217
  volume, canonical, 211, 212, 217
Entanglements, 234, 242, 341, 422
Ergodicity, 24, 205
Ewald sphere, 312
Excluded volume, 7, 11, 109, 372, 405, 438, 545

Fabric tensor, 191, 193, 195
Flory-Huggins theory, 272, 429
Foam
  Surface Evolver, 152
  coarsening, 155

definition, 147
drainage equation, 156
dry and wet, 148
particle stabilized, 162
Plateau borders, 149
Force chains, 180, 195, 197
Force dipole, 508
Fractal clusters, 26

Gels
  active, 508
  electro-mechanical coupling, 463
  polar, 507, 519
  polyelectrolyte, 433, 451
  solvent diffusion, 461
  swelling and collapse, 362, 451
Glass transition
  hard-sphere colloids, 19
  Mode coupling theory, 21
Granular matter
  definition, 168, 169
  jamming transition, 182
Granular system
  compactivity, 206, 211
  stress distribution, 180
  stress tensor, 192, 197
Gyroid phases, 278

Helfrich free energy (membrane), 548
Helfrich-Hurault instability, 116
Herschel-Bulkley relation, 158

Intermediate filaments, 501
Isostaticity, 189, 192, 200

Jamming
  colloids, 19, 23, 27
  granular systems, 170
  marginal rigidity, 182, 188

Kibble mechanism, 119
Krafft temperature, 54, 67

Lamellar phases, 54, 67, 71, 105, 274, 277, 282
Lamellipodia, 514
Laplace pressure
  in foams, 151
  in monolayers, 62, 75
Lennard-Jones potential, 13

Lindemann melting criterion, 11
Lipid bilayer, 502, 536
Lipid bilayers, 540
Lipid monolayer
   ordered phases, 542
   rheology, 543
Liquid crystals
   anisotropic viscosity, 129
   Blue phases, 102
   colloids, 41, 125
   elastomer, 365
   nematics, 98
   smectics, 105
Lyotropic liquid crystals, 96

Maxwell relaxation model, 21, 505
Mean field potential, 8
Membrane
   curvature elasticity, 58, 548
   film rupture, 79
   fluctuation spectrum, 554, 557, 561, 568
   phase separation, 544
   phospholipid, 536
Membrane proteins, 577
Micellar aggregation, 63, 65, 68
Microphase separation, 272, 277, 289, 431
Microtubules, 501
Molecular motors, 500, 501
Mooney-Rivlin equation, 344

Nematic director, 95
Nematics
   Frank elasticity, 111
   Frederiks effect, 99, 130
   Landau-de Gennes theory, 108
   order parameter, 107
   surface anchoring, 117
   uniaxial, biaxial, 98

Onsager theory, 109
Osmotic pressure, 7, 9, 419, 502, 541

Packings in 2D, 177
Packings in 3D, 178
Partition function
   canonical, 8, 9
   grand canonical, 8
   structural, 211, 213, 217
Peclet number, 29
Phase diagram
   active medium, 511
   bicontinuous phases, 67, 71, 72, 278
   block-copolymers, 285
   hard sphere colloid, 10
   lipid membrane, 542, 568
   microemulsions, 70
   microphase separation, 274, 277
   phase coexistence, 14, 67, 440
   polyeletrolyte chains, 392, 410
Phospholipids, 538
Pickering emulsion, 41
Polyelectrolyte
   definition, 383
   free energy, 386
   gels swelling, 433
   globular, 388
   persistence length, 415
   phase separation, 429
   reptation dynamics, 423
   Rouse dynamics, 421
Polyeletrolyte
   counterion condensation, 393
   ionization equilibrium, 398
   overlap concentration, 411
   persistence length, 401, 403
   swelling by solvent, 405
Polymers
   blobs, 385, 413
   branched, 250
   constraint release, 248, 258
   entanglements, 234, 237, 242, 341
   reptation, 236, 245, 248
   swelling by solvent, 237, 384, 405
   tube model, 235, 242, 258
Primitive path, 234, 235, 240, 246

Quadrons, 173, 175, 211

Reciprocal lattice, 304, 308, 310
Rheology
   complex moduli G', G'', 33, 245, 292, 504, 505, 544
   foams, 157
   yielding, 32, 158

Scattering
   form factor, 303, 319
   Laue, 302
   neutrons, 300
   non-periodic systems, 319
   small-angle, 299, 312, 319
   theory, 297
Scattering vector, 298
Scattering, form factor, 310
Scleroproteins (fibrous), 475
Shear modulus, 21, 57, 158, 361, 435, 440, 453
Shear thinning, 29, 258
Smectic layer elasticity, 115
Soft glassy rheology, 36
Spinodal line, 26, 274, 283, 348, 430
Strain invariants, 358
Stress stiffening, 505
Structural degrees of freedom, 208
Structure factor, 274, 298, 303
   dynamic, 291
   polyelectrolytes, 420
Surface isotherm, 542, 544
Surface pressure, 52, 541
Surface tension
   dynamic, 55
   Gibbs equation, 53, 537
   microemulsions, 74
Surfactant
   definition, 51, 319
   examples, 55
   HLB, 83
Surfactants
   examples, 538

Topological defects
   disclinations, 118
   in smectics, 123
   point defects, 120
Turing mechanism, 523

Uniaxial extension/compression, 258, 345, 370, 543, 544

Vesicle
   shape fluctuations, 557
Vesicles, 66
   shapes, phase diagram, 549
   shear flow, 566
Virial coefficient, 10, 60, 387, 405, 409
Von Neumann law, 155

Wetting and dewetting, 74, 76

Young's modulus, 201, 456, 484, 490, 543